5 Reasons to Use MyMathLab® for Applied Calculus

1 **Thousands of high-quality exercises.** Algorithmic exercises of all types and difficulty levels are available to meet the needs of students with diverse mathematical backgrounds.

2 **Helps students help themselves.** Homework isn't effective if students don't do it. MyMathLab not only grades homework, but it also does the more subtle task of providing specific feedback and guidance along the way. As an instructor, you can control the amount of guidance students receive.

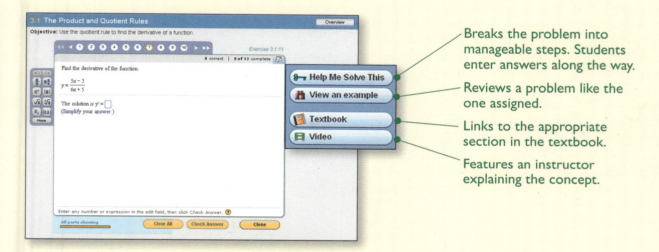

Breaks the problem into manageable steps. Students enter answers along the way.

Reviews a problem like the one assigned.

Links to the appropriate section in the textbook.

Features an instructor explaining the concept.

3 **Addresses gaps in algebra skills.** Our "Getting Ready for Applied Calculus" content addresses gaps in prerequisite skills that can impede student success. Use the built-in diagnostic tests—at the beginning of the course or at the chapter level—to identify precise areas of weakness. MyMathLab will then automatically provide remediation specific to each student.

4 **New ways to visualize concepts.** Where a pen and whiteboard fall short, we have added Interactive Figures to help you incorporate the geometric aspects of the course. Assignable Interactive Figure Exercises help students better understand the course content and use the figures for maximum effect.

5 **Helps motivate students.** New MathTalk videos show the relevance of calculus in business settings. Online exercises that accompany these videos help students apply the ideas introduced there. MathTalk videos can also be used as lecture starters or in a flipped classroom setting.

Since 2001, more than 15 million students at more than 1,950 colleges have used MyMathLab. Users have reported significant increases in pass rates and retention. Why? With immediate feedback, students are compelled to do more work and get help when they need it. See www.mymathlab.com/success_report.html for the latest information on how schools are successfully using MyMathLab.

Learn more at www.mymathlab.com

Calculus
AND ITS APPLICATIONS
ELEVENTH EDITION

Marvin L. Bittinger
Indiana University Purdue University Indianapolis

David J. Ellenbogen
Community College of Vermont

Scott A. Surgent
Arizona State University

PEARSON

Boston   Columbus   Indianapolis   New York   San Francisco
Amsterdam   Cape Town   Dubai   London   Madrid   Milan   Munich   Paris   Montréal   Toronto
Delhi   Mexico City   São Paulo   Sydney   Hong Kong   Seoul   Singapore   Taipei   Tokyo

Editor in Chief: Deirdre Lynch
Executive Editor: Jennifer Crum
Editorial Assistant: Joanne Wendelken
Program Manager: Tatiana Anacki
Project Manager: Rachel S. Reeve
Program Management Team Lead: Marianne Stepanian
Project Management Team Lead: Christina Lepre
Media Producer: Jonathan Wooding
TestGen Content Manager: John R. Flanagan
MathXL Content Manager: Kristina Evans
Marketing Manager: Jeff Weidenaar
Marketing Assistant: Brooke Smith
Senior Author Support/Technology Specialist: Joe Vetere

Rights and Permissions Project Manager: Diahanne Lucas
Procurement Specialist: Carol Melville
Associate Director of Design: Andrea Nix
Program Design Lead: Barbara T. Atkinson
Text Design, Art Editing, Photo Research: Geri Davis/The Davis Group, Inc.
Production Coordination: Jane Hoover/Lifland et al., Bookmakers
Composition: Lumina Datamatics, Inc.
Illustrations: Network Graphics, Inc., and William Melvin
Cover Design: Jenny Willingham/Infiniti Design
Cover Image: Christian Mueller/Shutterstock

Library of Congress Cataloging-in-Publication Data
Bittinger, Marvin L.
 Calculus and its applications / Marvin L. Bittinger, Indiana University,
Purdue University, Indianapolis, David J. Ellenbogen, Community College of Vermont,
Scott A. Surgent, Arizona State University. — 11th edition.
  pages cm
 Includes bibliographical references and index.
  ISBN 0-321-97939-7
 1. Calculus—Textbooks. I. Ellenbogen, David. II. Surgent, Scott Adam. III. Title.
QA303.2.B466 2016
515—dc23   2014015389

3 4 5 6 7 8 9 10—V011—18 17 16 15

www.pearsonhighered.com

ISBN 13: 978-0-321-97939-1
ISBN 10: 0-321-97939-7

To Elaine, Victoria, and Beth

Contents

An alternative, expanded version of this text is available and contains the following topics:

**Calculus and Its Applications: Expanded Version
(ISBN: 978-0-134-12349-3)**

Preface

Calculus and Its Applications strives to be the most student-oriented applied calculus text on the market, and this eleventh edition continues to improve in that regard. The authors believe that appealing to students' intuition and speaking in a direct, down-to-earth manner make this text accessible to any student possessing the prerequisite math skills. By presenting more topics in a conceptual and often visual manner and adding student self-assessment and teaching aids, this revision addresses students' needs better than ever before. Tapping into areas of student interest, the authors provide motivation through an abundant supply of examples and exercises rich in real-world data from business, economics, environmental studies, health care, and the life sciences.

New examples cover applications ranging from monthly cell phone traffic (in exabytes) to the growth of Facebook membership. Found in every chapter, realistic applications draw students into the discipline and help them to generalize the material and apply it to new situations. To further spark student interest, hundreds of meticulously rendered graphs and illustrations appear throughout the text, making it a favorite among students who are visual learners.

Appropriate for a one-term course, this text is an introduction to applied calculus. A course in intermediate algebra is a prerequisite, although "Appendix A: Review of Basic Algebra," together with "Chapter R: Functions, Graphs, and Models," provides a sufficient foundation to unify the diverse backgrounds of most students. For schools offering a two-term course, *Calculus and Its Applications, Expanded Version,* by the same author team, is available; it includes chapters covering trigonometric functions, differential equations, sequences and series, and probability distributions.

New to This Edition

With every revision comes the opportunity to improve the presentation of the content in a variety of ways based on feedback from users of the text and author experience. This revision is focused on providing students with *just-in-time* reminders of skills that should be familiar to them, updating and adding new applications—especially business applications—throughout the text, and fine-tuning the pedagogical design elements that make this book so user-friendly. The exercises have been carefully evaluated to be sure they align with the objectives of the text, and data analysis of MyMathLab exercises has helped guide the gradation and organization of the exercises in both the text and MyMathLab. Throughout the preface, the new elements of the text are described and specific new and modified content outlined. Below is a quick snapshot of what's new in this edition.

- Additional coverage of exponential functions has been added to Section R.5, to support students' use of these functions in applications.
- The former Section 3.6, "An Economics Application: Elasticity of Demand," has been moved and is now Section 2.7, allowing students to be introduced to the application sooner so they can use the concept throughout Chapter 3.
- A discussion of annuities has been added to Section 3.5, and a new section dedicated to amortization (with the use of Excel as an option) has been added as Section 3.6.
- Section 5.6, "Volume," now includes discussion on the use of shells, and Section 6.4, "An Application: The Least-Squares Technique," was expanded to include a derivation of an exponential regression by hand.
- Applications have been updated to reflect more current data and trends whenever possible.

- Exercise sets have been carefully evaluated to ensure appropriate gradation of level, odd/even pairing, and specific connection to the objectives being evaluated. In addition, MyMathLab usage data was analyzed to expose any exercises that needed improvement.
- MyMathLab has been greatly improved to include a vast array of new resources for both instructors and students. In addition to a greater quantity and variety of exercises, MyMathLab now includes more videos, additional levels of assessment, and interactive figures to help students gain the skills and knowledge they need to be successful in this and future courses.

Our Approach

Intuitive Presentation

Although the word *intuitive* has many meanings and interpretations, its use here means "experience based, without proof." Throughout the text, when a concept is discussed, its presentation is designed so that the students' learning process is based on their earlier mathematical experience. This is illustrated by the following situations.

- Before the formal definition of *continuity* is presented, an informal explanation is given, complete with graphs that make use of student intuition about ways in which a function could be discontinuous (see pp. 109–110).
- The definition of *derivative*, in Chapter 1, is presented in the context of a discussion of average rates of change (see p. 129). This presentation is more accessible and realistic than the strictly geometric idea of slope.
- When maximization problems involving volume are introduced (see pp. 244–246), a function is derived that is to be maximized. Instead of forging ahead with the standard calculus solution, the text first asks the student to make a table of function values, graph the function, and then estimate the maximum value. This experience provides students with more insight into the problem. They recognize not only that different dimensions yield different volumes, but also that the dimensions yielding the maximum volume may be conjectured or estimated as a result of the calculations.
- The explanation underlying the definition of the number e is presented in Chapter 3 both graphically and through a discussion of continuously compounded interest (see pp. 330–331).
- Within MyMathLab, students and instructors have access to **interactive figures** that illustrate concepts and allow manipulation by the user so that he or she can better predict and understand the underlying concepts. Also available within MyMathLab are questions that provide focus for student use.

Timely Help for Gaps in Algebra Skills

One of the most critical factors underlying success in this course is a strong foundation in algebra skills. We recognize that students start this course with varying degrees of skills, so we have included multiple opportunities in both the text and MyMathLab to help students target their weak areas and remediate or refresh the needed skills.

In the Text

- **Prerequisite Skills Diagnostic Test (Part A).** This portion of the diagnostic test assesses skills refreshed in Appendix A: Review of Basic Algebra. Answers to the questions reference specific examples within the appendix.
- **Appendix A: Review of Basic Algebra.** This 11-page appendix provides examples on topics such as exponents, equations, and inequalities and applied problems. It ends with an exercise set, for which answers are provided at the back of the book so students can check their understanding.
- **Prerequisite Skills Diagnostic Test (Part B).** This portion of the diagnostic test assesses skills that are reviewed in "Chapter R: Functions, Graphs, and Models," and the answers reference specific sections in that chapter. Some instructors may choose

to cover these topics thoroughly in class, making this assessment less critical. Other instructors may use all or portions of this test to determine whether there is a need to spend time remediating before moving on with Chapter 1.

■ **Chapter R: Functions, Graphs, and Models.** This chapter covers basic concepts related to functions, graphing, and modeling. It is an optional chapter based on students' prerequisite skills.

In MyMathLab

■ **Integrated Review.** You can diagnose weak prerequisite skills through built-in diagnostic quizzes. By coupling these quizzes with Personalized Homework, MyMathLab provides remediation for just those skills a student lacks. Even if you choose not to assign these quizzes with Personalized Homework, students can self-remediate through videos and practice exercises provided at the objective level. MyMathLab provides the just-in-time help that students need, so you can focus on the course content.

■ **New! Basic Skills Videos.** Videos are now available within exercises to help refresh the key algebra skill required for a specific exercise.

Exercises and Applications

There are over 3500 assignable section-level homework exercises in this edition. A large percentage of these exercises are rendered algorithmically in MyMathLab. All exercise sets are enhanced by the inclusion of real-world applications, detailed art pieces, and illustrative graphs. There are a variety of types of exercises, too, so different levels of understanding and varying approaches to problems can be assessed. In addition to applications, the exercise sets include *Thinking and Writing*, *Synthesis*, *Technology Connection*, and *Concept Reinforcement* exercises. The exercises in MyMathLab reflect the depth and variety of those in the printed text.

The authors also provide *Quick Check* exercises, following selected examples, to give students the opportunity to check their understanding of new concepts or skills as soon as they learn them and one skill at a time. Instructors may include these as part of a lecture as a means of gauging skills and gaining immediate student feedback. Answers to the Quick Check exercises are provided following the exercise set at the end of each section.

Relevant and factual applications drawn from a broad spectrum of fields are integrated throughout the text as applied examples and exercises, and are also featured in separate application sections. Applications have been updated and expanded in this edition to include even more real data. In addition, each chapter opener features an application that serves as a preview of what students will learn in the chapter.

The applications in the exercise sets in the text and within MyMathLab are grouped under headings that identify them as reflecting real-life situations: Business and Economics, Life and Physical Sciences, Social Sciences, and General Interest. This organization allows the instructor to gear the assigned exercises to specific students and also allows each student to know whether a particular exercise applies to his or her major.

Furthermore, the Index of Applications at the back of the book provides students and instructors with a comprehensive list of the many different fields considered throughout the text.

Opportunities to Incorporate Technology

This edition continues to emphasize mathematical modeling, utilizing the advantages of technology as appropriate. The use of Excel as a tool for solving problems has been expanded in this edition. Though the use of technology is optional with this text, its use meshes well with the text's more intuitive approach to applied calculus.

Technology Connections

Technology Connections are included throughout the text to illustrate the use of technology, including graphing calculators, Excel spreadsheets, and smartphone apps. Whenever appropriate, art that simulates graphs or tables generated by a graphing calculator is included as well. The goal is to take advantage of technology to which many students have access, wherever it makes sense, given the mathematical situation.

Four types of Technology Connections allow students and instructors to explore key ideas:

- **Lesson/Teaching**. These provide students with an example, followed by exercises to work within the lesson.
- **Checking**. These tell the students how to verify a solution within an example by using a graphing calculator.
- **Exploratory/Investigation**. These provide questions to guide students through an investigation.
- **Technology Connection Exercises**. Most exercise sets contain technology-based exercises identified with either an icon or the heading "Technology Connection." This type of exercise also appears in the Chapter Review Exercises, Chapter Tests, and the supplemental Printable Test Forms.

Extended Technology Applications at the end of every chapter use real applications and real data. They require a step-by-step analysis that encourages group work. More challenging in nature, the exercises in these features often involve the use of regression to create models on a graphing calculator. The data in the Extended Technology Applications has been updated wherever possible to keep the applications fresh for instructors and relevant for students.

Use of Art and Color

One of the hallmarks of this text is the pervasive use of color as a pedagogical tool. Color is used in a methodical and precise manner that enhances the readability of the text for students and instructors.

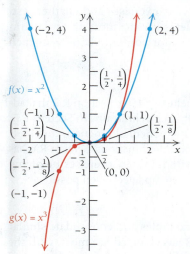

- When two curves are graphed using the same set of axes, one is usually red and the other blue, with the red graph being the curve of major importance. The equation labels are the same color as the curve for clarity (see p. 52).
- When the instructions say "Graph," the dots match the color of the curve. When dots are used for emphasis other than just merely plotting, they are black. Throughout the text, blue is used for secant lines and red for tangent lines (see p. 128).
- Red denotes substitution in equations while blue highlights the corresponding outputs, and the specific use of color is carried out in related figures (see pp. 206–207).
- Beginning with the discussion of integration, an amber color is used to highlight areas in graphs (see p. 400).

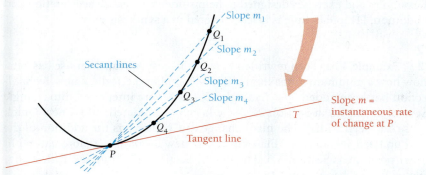

Opportunity for Review and Synthesis

Recognizing that it is often while preparing for exams that concepts gel for students, this text offers abundant opportunities for students to review, analyze, and synthesize recently learned concepts and skills.

- A **Section Summary** precedes every exercise set to assist students in identifying the key topics for each section and to serve as a mini-review.
- A **Chapter Summary** at the end of every chapter includes a section-by-section list of key definitions, concepts, and theorems, with examples for further clarification.
- **Chapter Review Exercises**, which include bracketed references to the section in which the related concept is first introduced, provide comprehensive coverage and appropriate referencing of each chapter's material.
- A **Chapter Test** at the end of every chapter and a **Cumulative Review** at the end of the text give students an authentic exam-like environment for testing their mastery. Answers to the chapter tests and the Cumulative Review are at the back of the text and include section references so students can diagnose their mistakes while preparing for exams.

New and Revised Content

In response to faculty and student feedback, we have made many changes to the text's content for this edition. New examples and exercises have been added throughout each chapter, as well as new problems to each chapter's review exercises and chapter test. Data have been updated wherever achievable, so problems use the most up-to-date information possible. Following is an overview of the major content changes in each chapter.

Chapter R

Several new exercises have been added to this chapter, including 10 in Section R.2 that address how to describe functions verbally and translate them algebraically. In Section R.5, a subsection and several exercises were added that cover exponential functions and their graphs to help bridge the gap in student understanding of those functions. Section R.6 has expanded discussion of exponential functions, along with four new exercises involving exponential models. Throughout, data-driven examples were updated when possible.

Chapter 1

The goal for Chapter 1 was to update the data-based examples and consolidate similar exercises into a more manageable number. Section 1.8 has changed the most, with the addition of examples and exercises designed to help students visualize acceleration and velocity. In addition, l'Hôpital's Rule is briefly covered in a synthesis exercise.

Chapter 2

In Section 2.5, Example 3 has been rewritten to factor in cost, spreadsheet use has been added to show how a minimum or maximum can be found numerically, a new Technology Connection has been added, and Examples 6 and 7 were integrated into a single example. Several examples in Section 2.6 have been consolidated, and a new Quick Check exercise has been added. The main change to Chapter 2 is the addition of the expanded and updated Section 2.7, "Elasticity." This new location is a more natural fit than its former position as Section 3.6. The former Section 2.7, "Implicit Differentiation and Related Rates," has become Section 2.8.

Chapter 3

New material on exponential functions has been added to Section 3.1 based on the expanded content in Section R.5. In Section 3.2, there is more emphasis on the general antiderivative for $1/x$, for all x except $x = 0$, through additional examples and exercises. The Rule of 70 is now included in Section 3.3, which also has a new Technology Connection. Section 3.5 has been expanded to include a discussion on annuities. New Section 3.6 covers the topic of amortization and includes some Excel spreadsheet applications.

Chapter 4

Application examples and exercises were added and updated throughout Chapter 4. In addition, new material on Simpson's Rule was added to Section 4.2, and the topic of recursion was moved to a synthesis exercise in Section 4.6.

Chapter 5

Topics throughout Chapter 5 were expanded, allowing for over 80 new exercises and applications. Improper integration at a vertical asymptote was added to Section 5.3, and finding volume by shells was added to Section 5.6.

Chapter 6

The main changes in Chapter 6 are the addition of exponential regression to Section 6.4 and the addition of average value of a two-variable function to Section 6.6. New application exercises were also added, and data-driven examples and exercises were updated throughout.

Appendix A

New material on the Principle of Square Roots was added as a reference for students.

Supplements

STUDENT SUPPLEMENTS

Student's Solutions Manual
(ISBN: 0-321-99905-3/978-0-321-99905-4)
- Provides detailed solutions to all odd-numbered section exercises, with the exception of the Thinking and Writing exercises

Graphing Calculator Manual for Applied Mathematics (downloadable)
- By Victoria Baker, Nicholls State University
- Contains detailed instruction for using the TI-83/TI-83+/TI-84+C
- Instructions are organized by topic.
- Downloadable from within MyMathLab

Excel Spreadsheet Manual for Applied Mathematics (downloadable)
- By Stela Pudar-Hozo, Indiana University-Northwest
- Contains detailed instruction for using Excel 2013
- Instructions are organized by topic.
- Downloadable from within MyMathLab

Video Lectures with optional captioning (online)
- Complete set of digitized videos for student use anywhere
- Example-level and lecture-level videos available
- Available in MyMathLab

INSTRUCTOR SUPPLEMENTS

Annotated Instructor's Edition
(ISBN: 0-321-99903-7/978-0-321-99903-0)
- Includes numerous Teaching Tips
- Includes all of the answers, usually on the same page as the exercises, for quick reference

Instructor's Solutions Manual (downloadable)
- Provides complete solutions to all text exercises
- Available to qualified instructors through the Pearson Instructor Resource Center, **www.pearsonhighered.com/irc**, and MyMathLab

Printable Test Forms (downloadable)
- Contains four alternative tests per chapter
- Contains four comprehensive final exams
- Includes answer keys
- Available to qualified instructors through the Pearson Instructor Resource Center, **www.pearsonhighered.com/irc**, and MyMathLab

PowerPoint Lecture Presentations
- Classroom presentation software oriented specifically to the text's topic sequence
- Available to qualified instructors through the Pearson Instructor Resource Center, **www.pearsonhighered.com/irc**, and MyMathLab

TECHNOLOGY RESOURCES

MyMathLab® Online Course (access code required)

MyMathLab from Pearson is the world's leading online resource in mathematics, integrating interactive homework, assessment, and media in a flexible, easy-to-use format.

MyMathLab delivers **proven results** in helping individual students succeed.

- MyMathLab has a consistently positive impact on student retention, subsequent success, and overall achievement. MyMathLab can be successfully implemented in any environment—lab-based, hybrid, fully online, or traditional.
- MyMathLab's comprehensive online gradebook automatically tracks students' results on tests, quizzes, homework, and in the study plan. You can use the gradebook to quickly intervene if your students have trouble, or to provide positive feedback on a job well done.

MyMathLab provides **engaging experiences** that personalize, stimulate, and measure learning for each student.

- **Personalized Learning:** MyMathLab's personalized homework and adaptive study plan features allow your students to work on just what they need to learn when it makes the most sense.
  - Chapter skills check quizzes are available (at the chapter level and course-wide) to help students recognize what prerequisite skills they may need to brush up on.
  - The results of the chapter skills check quiz can be tied to personalized homework so that each student is given targeted skills to practice prior to starting course-level work.
- **Exercises:** The homework and practice exercises in MyMathLab are correlated to the exercises in the textbook, and they regenerate algorithmically to give students unlimited opportunity for practice and mastery. The software offers immediate, helpful feedback when students enter incorrect answers.
- **Multimedia Learning Aids:** Exercises include guided solutions, sample problems, animations, videos, and eText access for extra help at point-of-use.
  - Videos include section-level, lecture-style videos as well as example-level videos. Both are helpful resources when instructors or tutors are not available.
  - *MathTalk* videos show how the math the students are learning now may apply to their future careers. These videos could be used to kick off a lecture, in a flipped classroom setting, or as a way to motivate students. Videos can be assigned as homework or as a prerequisite to students getting started on the homework. Premade questions are available for the MathTalk videos.
  - *Interactive figures* illustrate concepts and allow manipulation, so the user can better predict, visualize, and understand the underlying concept. Questions are available within MyMathLab to provide focus for student use.

- **MyMathLab Accessibility:** MyMathLab is compatible with the JAWS screen reader, and enables problems to be read and interacted with via keyboard controls and math notation input. MyMathLab also works with screen enlargers, including ZoomText, MAGic, and SuperNova. And all MyMathLab videos for this course have closed captioning. More information on this functionality is available at **mymathlab.com/accessibility**.

And, MyMathLab comes from an **experienced partner** with educational expertise and an eye on the future. Whether you are just getting started with MyMathLab, or have a question along the way, we're here to help you learn about our technologies and how to incorporate them into your course. Contact us at **www.mymathlab.com**.

MyMathLab® Integrated Review Course (access code required)

These new courses provide students with all the same great MyMathLab features, but make it easier for instructors to get started. Each course includes pre-assigned homework and quizzes to make creating a course even simpler. You can find more details at **www.mymathlab.com**.

MyLabsPlus

MyLabsPlus combines proven results and engaging experiences from MyMathLab® with convenient management tools and a dedicated services team. Designed to support growing math programs, it includes additional features such as:

- **Batch Enrollment:** Your school can create the login name and password for every student and instructor, so everyone can be ready to start class on the first day. Automation of this process is also possible through integration with your school's Student Information System.
- **Login from your campus portal:** You and your students can link directly from your campus portal into your MyLabsPlus courses. A Pearson service team works with your institution to create a single sign-on experience for instructors and students.
- **Advanced Reporting:** MyLabsPlus's advanced reporting allows instructors to review and analyze students' strengths and weaknesses by tracking their performance on tests, assignments, and tutorials. Administrators can review grades and assignments across all courses on your MyLabsPlus campus for a broad overview of program performance.
- **24/7 Support:** Students and instructors receive 24/7 support, 365 days a year, by email or online chat.

MyLabsPlus is available to qualified adopters. For more information, visit our website at **www.mylabsplus.com** or contact your Pearson representative.

MathXL® Online Course (access code required)

MathXL® is the homework and assessment engine that runs MyMathLab. With MathXL, instructors can:

- Create, edit, and assign online homework and tests using algorithmically generated exercises correlated at the objective level to the textbook.
- Create and assign their own online exercises and import TestGen tests for added flexibility.
- Maintain records of all student work tracked in MathXL's online gradebook.

With MathXL, students can:

- Take chapter tests in MathXL and receive personalized study plans and/or personalized homework assignments based on their test results.
- Use the study plan and/or the homework to link directly to tutorial exercises for the objectives they need to study.

- Access supplemental animations and video clips directly from selected exercises.

MathXL is available to qualified adopters. For more information, visit our website at **www.mathxl.com**, or contact your Pearson representative.

TestGen®

TestGen® (**www.pearsoned.com/testgen**) enables instructors to build, edit, print, and administer tests using a computerized bank of questions developed to cover all the objectives of the text. TestGen is algorithmically based, allowing instructors to create multiple but equivalent versions of the same question or test with the click of a button. Instructors can also modify test bank questions or add new questions. The software and testbank are available for instructors to download from Pearson Education's online catalog.

Acknowledgments

As authors, we have taken many steps to ensure the accuracy of this text. Many devoted individuals comprised the team that was responsible for monitoring the revision and production process in a manner that makes this a work of which we can all be proud. We are thankful for our publishing team at Pearson, as well as all of the Pearson representatives who share our book with educators across the country. Many thanks to Michelle Christian, who was instrumental in getting Scott Surgent's first book printed and in bringing him to the attention of the Pearson team. We would like to thank Jane Hoover for her many helpful suggestions, proofreading, and checking of art. Jane's attention to detail and pleasant demeanor made our work as low in stress as humanly possible, given the demands of the production process. Geri Davis deserves credit for both the attractive design of the text and the coordination of the many illustrations, photos, and graphs. She is always a distinct pleasure to work with and sets the standard by which all other art editors are measured. Many thanks also to Jennifer Blue, Lisa Collette, and John Morin for their careful checking of the manuscript and typeset pages. We are grateful to all those who have contributed to the improvements of this text over the years, including those who were instrumental in helping us to shape this 11th edition.

The following individuals provided terrific insights and meaningful suggestions for improving this text. We thank them:

Fernanda Botelho, *University of Memphis*
Hugh Cornell, *University of North Florida*
Marvin Stick, *University of Massachusetts, Lowell*
Timothy D. Sullivan, *Northern Illinois University*
Kimberly Walters, *Mississippi State University*

Prerequisite Skills Diagnostic Test

To the Student and the Instructor

Part A of this diagnostic test covers basic algebra concepts, such as properties of exponents, multiplying and factoring polynomials, equation solving, and applied problems. Part B covers topics, discussed in Chapter R, such as graphs, slope, equations of lines, and functions, most of which come from a course in intermediate or college algebra. This diagnostic test does not cover regression, though it is considered in Chapter R and used throughout the text. This test can be used to assess student needs for this course. Students who miss most of the questions in part A should study Appendix A before moving to Chapter R. Those who miss most of the questions in part B should study Chapter R. Students who miss just a few questions might study the related topics in either Appendix A or Chapter R before continuing with the calculus chapters.

Part A

Answers and locations of worked-out solutions appear on p. A-39.

Express each of the following without an exponent.

1. 4^3 2. $(-2)^5$ 3. $\left(\frac{1}{2}\right)^3$ 4. $(-2x)^1$ 5. e^0

Express each of the following without a negative exponent.

6. x^{-5} 7. $\left(\frac{1}{4}\right)^{-2}$ 8. t^{-1}

Multiply. Express each answer without a negative exponent.

9. $x^5 \cdot x^6$ 10. $x^{-5} \cdot x^6$ 11. $2x^{-3} \cdot 5x^{-4}$

Divide. Express each answer without a negative exponent.

12. $\dfrac{a^3}{a^2}$ 13. $\dfrac{e^3}{e^{-4}}$

Simplify. Express each answer without a negative exponent.

14. $(x^{-2})^3$ 15. $(2x^4 y^{-5} z^3)^{-3}$

Multiply.

16. $3(x - 5)$ 17. $(x - 5)(x + 3)$

18. $(a + b)(a + b)$

19. $(2x - t)^2$ 20. $(3c + d)(3c - d)$

Factor.

21. $2xh + h^2$ 22. $x^2 - 6xy + 9y^2$

23. $x^2 - 5x - 14$ 24. $6x^2 + 7x - 5$

25. $x^3 - 7x^2 - 4x + 28$

Solve.

26. $-\frac{5}{6}x + 10 = \frac{1}{2}x + 2$ 27. $3x(x - 2)(5x + 4) = 0$

28. $4x^3 = x$ 29. $\dfrac{2x}{x - 3} - \dfrac{6}{x} = \dfrac{18}{x^2 - 3x}$

30. $17 - 8x \geq 5x - 4$

31. After a 5% gain in weight, a grizzly bear weighs 693 lb. What was the bear's original weight?

32. Raggs, Ltd., a clothing firm, determines that its total revenue, in dollars, from the sale of x suits is given by $200x + 50$. Determine the number of suits the firm must sell to ensure that its total revenue will be more than $70,050.

Part B

Answers and locations of worked-out solutions appear on p. A-39.

Graph.

1. $y = 2x + 1$ 2. $3x + 5y = 10$

3. $y = x^2 - 1$ 4. $x = y^2$

5. A function f is given by $f(x) = 3x^2 - 2x + 8$. Find each of the following: $f(0), f(-5)$, and $f(7a)$.

6. A function f is given by $f(x) = (x + 3)^2 - 4$. Find all x such that $f(x) = 0$.

7. Graph the function f defined as follows:
$$f(x) = \begin{cases} 4, & \text{for } x \leq 0, \\ 3 - x^2, & \text{for } 0 < x \leq 2, \\ 2x - 6, & \text{for } x > 2. \end{cases}$$

8. Write interval notation for $\{x | -4 < x \leq 5\}$.

9. Find the domain: $f(x) = \dfrac{3}{2x - 5}$.

10. Find the slope and y-intercept of $2x - 4y - 7 = 0$.

11. Find an equation of the line that has slope 3 and contains the point $(-1, -5)$.

12. Find the slope of the line containing the points $(-2, 6)$ and $(-4, 9)$.

Graph.

13. $f(x) = x^2 - 2x - 3$ 14. $f(x) = x^3$ 15. $f(x) = \dfrac{1}{x}$

16. $f(x) = |x|$ 17. $f(x) = -\sqrt{x}$

18. Suppose that $1000 is invested at 5%, compounded annually. How much is the investment worth at the end of 2 yr?

Credits

Chapter R   **p. 1:** The Washington Post/Getty Images. **p. 2:** Berents/Shutterstock. **p. 7:** The Washington Post/Getty Images. **p. 10:** (upper) ZUMA Press, Inc./Alamy; (lower) Bo Bridges/PR Newswire/AP Images. **p. 36:** Felinda/Fotolia. **p. 39:** Kruwt/Fotolia. **p. 41:** Scott Surgent. **p. 46:** Karramba Production/Fotolia. **p. 53:** Graphicus, Serafim Chekalkin. **p. 59:** Kyslynskyy/Fotolia. **p. 70:** Monkey Business Images/Shutterstock. **p. 72:** Brian Jackson/Fotolia. **p. 73:** Blend Images/Shutterstock. **p. 90:** Stephen Coburn/Shutterstock.

Chapter 1   **p. 92:** Scott Boehm/Associated Press. **p. 110:** Bastos/Fotolia. **p. 117:** Kunal Mehta/Shutterstock. **p. 127:** Oliver Hausen/Fotolia. **p. 134:** Graphicus, Serafim Chekalkin. **p. 148:** (upper) Kushnirov Avraham/Fotolia; (lower) Scott Boehm/Associated Press. **p. 152:** Federicocandonifoto/Shutterstock. **p. 155:** Bildagentur Zoonar GmbH/Shutterstock. **p. 159:** Brian Spurlock. **p. 162:** Robnroll/Shutterstock. **p. 174:** Dotshock/Shutterstock. **p. 186:** Bikeriderlondon/Shutterstock. **p. 187:** New York Daily News Archive/Getty Images.

Chapter 2   **p. 188:** Stephen Shepherd/Alamy. **p. 216:** Graphicus, Serafim Chekalkin. **p. 219:** gpointstudio/Shutterstock. **p. 232:** uss Reed/Globe Photos/ZUMAPRESS/Alamy. **p. 243:** Scott Surgent. **p. 247:** (upper) Stephen Shepherd/Alamy; (lower) Screenshots from Microsoft® Excel®. Used by permission of Microsoft Corporation. **p. 253:** Screenshots from Microsoft® Excel®. Used by permission of Microsoft Corporation. **p. 257:** Mi.Ti./Fotolia. **p. 269:** Losevsky Pavel/Alamy. **p. 278:** Minerva Studio/Fotolia. **p. 280:** (left) francesco de marco/Shutterstock; (right) 3D4Medical/Science Source. **p. 292:** Mike Thomas/Fotolia. **p. 293:** (left) njsphotography/Fotolia; (right) mattjeppson/Fotolia.

Chapter 3   **p. 294:** Dragon Images/Shutterstock. **p. 305:** Peter Bernik/Shutterstock. **p. 325:** (left) Glow Images; (right) Dragon Images/Shutterstock. **p. 327:** UPI/Heritage Auctions/Newscom. **p. 332:** (left) Vladimir Mucibabic/Fotolia; (right) EZIO PETERSEN/UPI/Newscom. **p. 333:** Edelweiss/Fotolia. **p. 334:** Chuck Crow/The Plain Dealer/Landov. **p. 335:** Mike Blake/Reuters. **p. 339:** UPI Photo/Debbie Hill/Newscom. **p. 344:** Josemaria Toscano/Fotolia. **p. 348:** Ethan Daniels/Shutterstock. **p. 354:** Tommy E Trenchard/Alamy. **p. 370:** Photos 12/Alamy. **p. 371:** (upper left) PIXAR ANIMATION STUDIOS/WALT DISNEY PICTURES/Alb/Album/Superstock; (bottom right) Mim Friday/Alamy.

Chapter 4   **p. 372:** Transtock/Superstock. **p. 379:** Nenetus/Shutterstock. **p. 380:** Michael Jung/Shutterstock. **p. 383:** PicturenetCorp/Fotolia. **p. 384:** Sira Anamwong/Shutterstock. **p. 399:** Transtock/Superstock. **p. 404:** Karichs/Fotolia. **p. 445:** NASA. **p. 447:** (bottom left) 06photo/Shutterstock; (top right) Nik Wheeler/Alamy. **p. 448:** (top left) Wassiliy/Fotolia; (bottom right) estherpoon/Shutterstock.

Chapter 5   **p. 449:** Anekoho/Fotolia. **p. 455:** Darryl Leniuk/Getty Images. **p. 463:** Sergey Volkov/Fotolia. **p. 466:** Anekoho/Fotolia. **p. 479:** SeanPavonePhoto/Fotolia. **p. 481:** Tetra Images/Alamy. **p. 492:** Jens Kalaene/dpa/picture-alliance/Newscom. **p. 494:** Brian Spurlock. **p. 497:** dvande/Shutterstock. **p. 498:** fotogrammi3/Fotolia. **p. 500:** mirec/Fotolia. **p. 501:** John Elk/Getty Images. **p. 502:** Chris VanLennep Photo/Fotolia. **p. 505:** Popperfoto/Getty Images. **p. 524:** Vladimir Salman/Shutterstock.

Chapter 6   **p. 525:** Tom Grill/Corbis. **p. 528:** aigarsr/Fotolia. **pp. 530–531:** Quick Graph, Kz Labs. **p. 532:** (left) Tom Grill/Corbis; (right) lafoto/Shutterstock. **p. 533:** Corbis. **p. 536:** VIEW Pictures Ltd/Alamy. **p. 541:** Maisie Paterson/Getty Images. **p. 544:** Rick Hanston/Latent Images. **p. 555:** H. Ruckemann UPI Photo Service/Newscom. **p. 560:** ZUMA Press, Inc/Alamy. **p. 565:** picsfive/Fotolia. **p. 579:** Suzanne Tucker/Shutterstock.

Appendix B   **pp. 598–599:** Screenshots from Microsoft® Excel®. Used by permission of Microsoft Corporation.

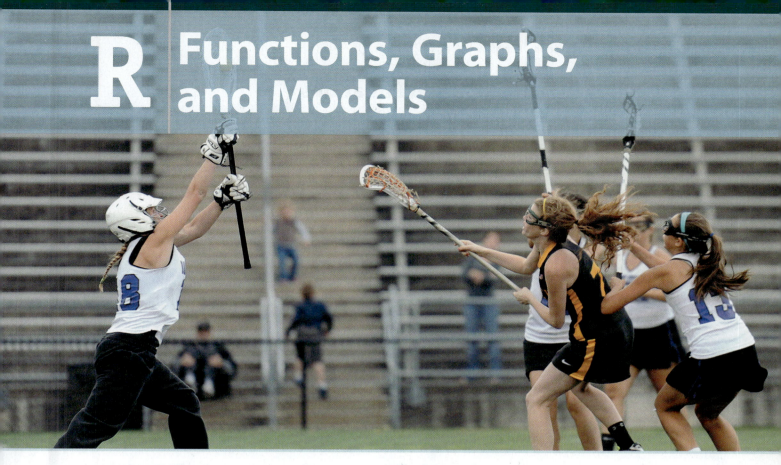

R Functions, Graphs, and Models

What You'll Learn

R.1 Graphs and Equations
R.2 Functions and Models
R.3 Finding Domain and Range
R.4 Slope and Linear Functions
R.5 Nonlinear Functions and Models
R.6 Mathematical Modeling and Curve Fitting

Why It's Important

This chapter introduces functions and covers their notation, graphs, and applications. Also presented are many topics considered often throughout the text: supply and demand; total cost, revenue, and profit; the concept of a mathematical model; and curve fitting.

Skills in using a graphing calculator are introduced in optional Technology Connections. Details on keystrokes are given in the *Graphing Calculator Manual* (GCM).

Part A of the diagnostic test (p. xv), on basic algebra concepts, allows students to determine whether they need to review Appendix A (p. 585) before studying this chapter. Part B, on college algebra topics, assesses the need to study this chapter before the calculus chapters.

Where It's Used

Participation of Females in High School Athletics: What is the predicted number of female high school athletes in 2017? (*This problem appears as Example 5 in Section R.1.*)

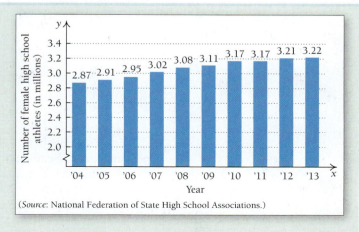

(*Source*: National Federation of State High School Associations.)

1

R.1

- Graph equations.
- Use graphs as mathematical models to make predictions.
- Carry out calculations involving compound interest.

Graphs and Equations

What Is Calculus?

What is calculus? This is a common question at the start of a course like this. Let's consider a simplified answer for now.

The common food can comes in many sizes, as the photo at the left illustrates. The following is a typical problem from an algebra course. Try to solve it. (If you need some algebra review, refer to Appendix A at the end of the book.)

> **Algebra Problem**
>
> A soup can contains a volume of 250 cm^3. If the height of the can is twice the length of the radius of the base, find the dimensions of the can.

From geometry, the volume of a cylinder is $V = \pi r^2 h$, where r is the radius of the circular base in centimeters and h is the height in centimeters. Since $V = 250$ and $h = 2r$, we have $\pi r^2(2r) = 250$, or $2\pi r^3 = 250$. Solving for r, we find that the soup can has a radius of approximately 3.4 cm and a height of 6.8 cm.

The following is a calculus problem that a manufacturer of cans might need to solve:

> **Calculus Problem**
>
> A soup can is to contain a volume of 250 cm^3. If the cost of the material for the two circular ends is $0.0008 per square centimeter and the cost of the material for the side is $0.0015 per square centimeter, what dimensions will minimize the cost of the can?

One way to solve this problem might be to choose several sets of dimensions that give a volume for the can of 250 cm^3, compute the resulting costs, and determine which is the lowest. For any r, we must have $h = 250/\pi r^2$. Using a computer spreadsheet, we could create a table such as the one below. We let $r =$ radius in centimeters and $h =$ height in centimeters.

	A	B	C	
1				
2	Radius	Height	Cost	
3	r	h	C	
4	3.5	6.495277933	0.275868914	
5	3.6	6.139440947	0.273485845	
6	3.7	5.812063892	0.271525071	
7	3.8	5.510190767	0.269961189	
8	3.9	5.231239624	0.268771404	
9	4	4.972947167	0.2679352	
10	4.1	4.733322705	0.267434061	
11	4.2	4.510609676	0.267251237	← Possible lowest cost
12	4.3	4.303253363	0.267371533	
13	4.4	4.109873692	0.267781137	
14	4.5	3.929242206	0.268467467	

From the data in the table, we might conclude that the lowest cost is about $0.26725 per can when the can's radius is 4.2 cm and its height is 4.51 cm. But how can we be certain that no other dimensions will give a lower cost? We need the tools of calculus to answer this. In Chapter 2, we will study such maximum–minimum problems, including a complete solution to this one, which appears as Example 3 in Section 2.5.

Other topics we will consider in calculus are the slope of a curve at a point, rates of change, area under a curve, accumulations of quantities, and some statistical applications.

Graphs

The study of graphs is an essential aspect of calculus. A graph offers the opportunity to visualize relationships. For instance, the graph below shows how life expectancy has changed over time in the United States. One topic that we consider later in calculus is how a change on one axis affects the change on another.

ESTIMATED LIFE EXPECTANCY OF U.S. NEWBORNS BY YEAR OF BIRTH, 1929–2013

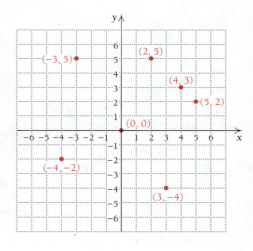

(*Source*: U.S. National Center for Health Statistics.)

Ordered Pairs and Graphs

Each point in a plane corresponds to an ordered pair of numbers. Note in the figure at the right that the point corresponding to the pair $(2, 5)$ is different from the point corresponding to the pair $(5, 2)$. This is why we call a pair like $(2, 5)$ an *ordered pair*. The first number is called the *first coordinate* of the point, and the second number is called the *second coordinate*. Together these are the *coordinates of the point*. The horizontal line is often labeled as the *x-axis*, and the vertical line is often labeled as the *y-axis*. The two axes intersect at the *origin*, $(0, 0)$.

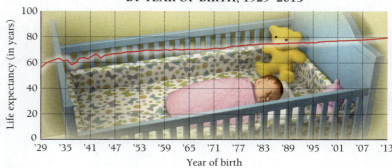

Graphs of Equations

A **solution** of an equation in two variables is an ordered pair of numbers that, when substituted for the variables, forms a true statement. If not directed otherwise, we usually take the variables in *alphabetical* order. For example, $(-1, 2)$ is a solution of the equation $3x^2 + y = 5$, because when we substitute -1 for x and 2 for y, we get a true statement:

$$3x^2 + y = 5$$

$$\begin{array}{c|c} 3(-1)^2 + 2 \ ? \ 5 & \\ 3 + 2 & 5 \\ 5 & 5 \quad \text{TRUE} \end{array}$$

> **DEFINITION**
>
> The **graph** of an equation is a drawing that represents all ordered pairs that are solutions of the equation.

We obtain the graph of an equation by plotting enough ordered pairs (that are solutions) to see a pattern.

EXAMPLE 1 Graph: $y = 2x + 1$.

Solution We first find some ordered pairs that are solutions and arrange them in a table. To find an ordered pair, we can choose any number for x and then determine y. For example, if we choose -2 for x and substitute that value in $y = 2x + 1$, we find that $y = 2(-2) + 1 = -4 + 1 = -3$. Thus, $(-2, -3)$ is a solution. We select both negative numbers and positive numbers, as well as 0, for x.

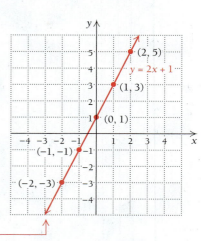

x	y	(x, y)
-2	-3	$(-2, -3)$
-1	-1	$(-1, -1)$
0	1	$(0, 1)$
1	3	$(1, 3)$
2	5	$(2, 5)$

   ↑   ↑   ↑
(1) Choose any x.
(2) Compute y.
(3) Form the pair (x, y).
(4) Plot the points.

Quick Check 1 ✔

Graph: $y = 3 - x$.

After we plot the points, we look for a pattern in the graph. In this case, the points suggest a solid line. We draw the line with a straightedge and label it $y = 2x + 1$.

1 ✔

EXAMPLE 2 Graph: $3x + 5y = 10$.

Solution We could choose x-values, substitute, and solve for y-values, but we first solve for y to ease the calculations.*

$$3x + 5y = 10$$
$$3x + 5y - 3x = 10 - 3x \qquad \text{Subtracting } 3x \text{ from both sides}$$
$$5y = 10 - 3x \qquad \text{Simplifying}$$
$$\tfrac{1}{5} \cdot 5y = \tfrac{1}{5} \cdot (10 - 3x) \qquad \text{Multiplying both sides by } \tfrac{1}{5}, \text{ or dividing both sides by 5}$$
$$y = \tfrac{1}{5} \cdot (10) - \tfrac{1}{5} \cdot (3x) \qquad \text{Using the distributive law}$$
$$= 2 - \tfrac{3}{5}x \qquad \text{Simplifying}$$
$$= -\tfrac{3}{5}x + 2$$

Next we use $y = -\tfrac{3}{5}x + 2$ to find some ordered pairs, choosing multiples of 5 for x to avoid fractions.

---

*Be sure to consult Appendix A, as needed, for a review of algebra.

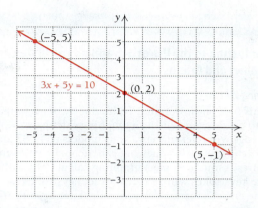

x	y	(x, y)
0	2	$(0, 2)$
5	-1	$(5, -1)$
-5	5	$(-5, 5)$

Quick Check 2 ✔

Graph: $3x - 5y = 10$.

We plot the points, draw the line, and label the graph as shown.   **2** ✔

Examples 1 and 2 show graphs of linear equations. Such graphs are considered in greater detail in Section R.4.

EXAMPLE 3   Graph: $y = x^2 - 1$.

Solution

x	y	(x, y)
-2	3	$(-2, 3)$
-1	0	$(-1, 0)$
0	-1	$(0, -1)$
1	0	$(1, 0)$
2	3	$(2, 3)$

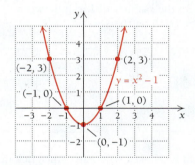

Quick Check 3 ✔

Graph: $y = 2 - x^2$.

This time the pattern of the points is a curve called a *parabola*. We plot enough points to see a pattern and draw the graph.   **3** ✔

EXAMPLE 4   Graph: $x = y^2$.

Solution   In this case, x is expressed in terms of the variable y. Thus, we first choose numbers for y and then compute x.

x	y	(x, y)
4	-2	$(4, -2)$
1	-1	$(1, -1)$
0	0	$(0, 0)$
1	1	$(1, 1)$
4	2	$(4, 2)$

(1) Choose any y.
(2) Compute x.
(3) Form the pair (x, y).
(4) Plot the points.

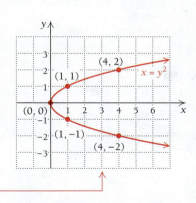

Quick Check 4 ✔

Graph: $x = 1 + y^2$.

We plot these points, keeping in mind that x is still the first coordinate and y the second. We look for a pattern and complete the graph by connecting the points.   **4** ✔

Introduction to the Use of a Graphing Calculator: Windows and Graphs

Viewing Windows

In this first of the optional Technology Connections, we begin to create graphs using a graphing calculator. Most of the coverage will refer to the various models of TI-83 and TI-84 calculators but in a somewhat generic manner, discussing features common to most graphing calculators. Although some keystrokes will be listed, exact keystrokes can be found in the owner's manual for your calculator or in the *Graphing Calculator Manual* (GCM) that accompanies this text.

The **viewing window** is a feature common to all graphing calculators. This is the rectangular screen in which a graph appears. Windows are described by four numbers in the format [**L, R, B, T**], representing the **L**eft and **R**ight endpoints of the *x*-axis and the **B**ottom and **T**op endpoints of the *y*-axis. A WINDOW feature can be used to set these dimensions. Below is a window setting of $[-20, 20, -5, 5]$ with axis scaling denoted as Xscl = 5 and Yscl = 1, which means that there are 5 units between tick marks extending from -20 to 20 on the *x*-axis and 1 unit between tick marks extending from -5 to 5 on the *y*-axis.

 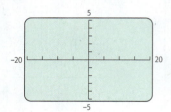

Scales should be chosen with care, since tick marks become blurred and indistinguishable when too many appear. On most graphing calculators, a window setting of $[-10, 10, -10, 10]$, Xscl = 1, Yscl = 1, Xres = 1 is considered **standard**.

Graphs are made up of black rectangular dots called **pixels**. The setting xres allows users to set pixel resolution at 1 through 8 for graphs of equations. At Xres = 1, equations are evaluated and graphed at each pixel on the *x*-axis. At Xres = 8, equations are evaluated and graphed at every eighth pixel on the *x*-axis. The resolution is better for smaller Xres values than for larger ones.

Graphs

Let's use a graphing calculator to graph the equation $y = x^3 - 5x + 1$. The equation can be entered by pressing ⟨ Y= ⟩ and entering x^3−5x+1. We obtain the following graph in the standard viewing window.

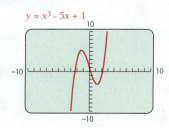

It is often necessary to change viewing windows in order to best reveal the curvature of a graph. For example, each of the following is a graph of $y = 3x^5 - 20x^3$, but with a different viewing window. Which do you think best displays the curvature of the graph?

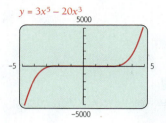

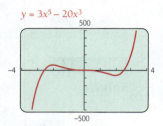 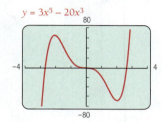

In general, choosing a window that best reveals a graph's characteristics involves some trial and error and, in some cases, some knowledge about the shape of the graph.

To graph an equation like $3x + 5y = 10$, most calculators require that the equation be solved for *y*. Thus, we must rewrite the equation and enter it as

$$y = \frac{-3x + 10}{5}, \quad \text{or} \quad y = -\frac{3}{5}x + 2.$$

(See Example 2.) Its graph is shown below.

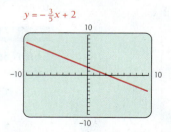

To graph an equation like $x = y^2$, we solve for *y* and get $y = \sqrt{x}$ or $y = -\sqrt{x}$. We then graph both equations, $y_1 = \sqrt{x}$ and $y_2 = -\sqrt{x}$.

EXERCISES

Graph each of the following equations. Select the standard window, $[-10, 10, -10, 10]$, with Xscl = 1 and Yscl = 1.

1. $y = x + 3$ **2.** $y = x - 5$

3. $y = 2x - 1$ **4.** $y = 3x + 1$

5. $y = -\frac{2}{3}x + 4$ **6.** $y = -\frac{4}{5}x + 3$

7. $2x - 3y = 18$ **8.** $5y + 3x = 4$

(continued)

9. $y = x^2$

10. $y = (x + 4)^2$

11. $y = 8 - x^2$

12. $y = 4 - 3x - x^2$

13. $y + 10 = 5x^2 - 3x$

14. $y - 2 = x^3$

15. $y = x^3 - 7x - 2$

16. $y = x^4 - 3x^2 + x$

17. $y = |x|$ (On most calculators, this is entered as $y = \text{abs}(x)$.)

18. $y = |x - 5|$

19. $y = |x| - 5$

20. $y = 9 - |x|$

Mathematical Models

When a real-world situation is described using mathematics, the description is a **mathematical model**. For example, the natural numbers constitute a mathematical model for situations in which counting is essential. Also, the speed at which a body falls due to gravity can be described using a mathematical model.

Mathematical models are abstracted from real-world situations. A mathematical model may give results that allow us to predict what will happen in the real-world situation. If the predictions are too inaccurate or the results of experimentation do not conform to the model, the model must be changed or discarded.

CREATING A MATHEMATICAL MODEL

1. Recognize a real-world problem. ➡ 2. Collect data. ➡ 3. Analyze the data. ➡ 4. Construct a model. ➡ 5. Test and refine the model. ➡ 6. Explain and predict.

Mathematical modeling is often an ongoing process. For example, finding a mathematical model that will provide an accurate prediction of population growth is not a simple task. Any population model that can be devised will need to be reshaped as further information is acquired.

EXAMPLE 5 The graph below shows participation by females in high school athletics from 2004 to 2013.

PARTICIPATION OF FEMALES IN HIGH SCHOOL ATHLETICS

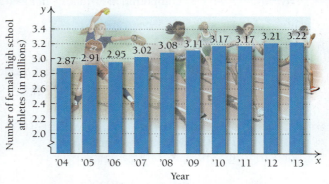

Number of female high school athletes (in millions)

2.87 2.91 2.95 3.02 3.08 3.11 3.17 3.17 3.21 3.22

'04 '05 '06 '07 '08 '09 '10 '11 '12 '13   Year

(*Source*: National Federation of State High School Associations.)

Use the model $N = 0.041t + 2.88$, where t is the number of years after 2004 and N is the number of participants, in millions, to predict the number of female high school athletes in 2017.

Solution Since 2017 is 13 years after 2004, we substitute 13 for t:

$$N = 0.041t + 2.88 = 0.041(13) + 2.88 = 3.41.$$

According to this model, in 2017, approximately 3.41 million females will participate in high school athletics.

Quick Check 5 ✔

Using the model in Example 5, determine the year in which the number of female high school athletes will reach 3.5 million.

5 ✔

As is the case with many kinds of models, the model in Example 5 is not perfect. For example, for $t = 1$, we get $N = 2.921$, a number slightly different from the 2.91 in the original data. But, for purposes of estimating, the model is adequate. The cubic model $N = -0.00043t^3 + 0.003t^2 + 0.044t + 2.87$ also fits the data, at least in the short term: For $t = 1$, we get $N \approx 2.92$, close to the given data value. But for $t = 13$, we have $N = 3.00$, which is quite different from the prediction in Example 5. The difficulty with a cubic model here is that, eventually, its predictions become undependable. For example, the model in Example 5 predicts that there will be 3.54 million female high school athletes in 2020, but the cubic model predicts only 2.58 million. We always have to subject our models to careful scrutiny.

Compound Interest

One important model that is extremely precise involves **compound interest.** Suppose we invest P dollars at interest rate r, expressed as a decimal, and compounded annually. The amount A_1 in the account at the end of the first year, is given by

$$A_1 = P + Pr = P(1 + r).$$ The original amount invested, P, is called the principal.

Going into the second year, we have $P(1 + r)$ dollars, so by the end of the second year, we will have the amount A_2 given by

$$A_2 = A_1 \cdot (1 + r) = [P(1 + r)](1 + r) = P(1 + r)^2.$$

Going into the third year, we have $P(1 + r)^2$ dollars, so by the end of the third year, we will have the amount A_3 given by

$$A_3 = A_2 \cdot (1 + r) = [P(1 + r)^2](1 + r) = P(1 + r)^3.$$

In general, we have the following theorem.

THEOREM 1

If an amount P is invested at interest rate r, expressed as a decimal, and compounded annually, in t years it will grow to the amount A given by

$$A = P(1 + r)^t.$$

EXAMPLE 6   **Business: Compound Interest.**   Suppose $1000 is invested in Fibonacci Investment Fund at 5%, compounded annually. How much is in the account at the end of 2 yr?

Solution   We substitute 1000 for P, 0.05 for r, and 2 for t into the equation $A = P(1 + r)^t$ and get

$$\begin{aligned}
A &= 1000(1 + 0.05)^2 & \text{Substituting} \\
&= 1000(1.05)^2 & \text{Adding terms in parentheses} \\
&= 1000(1.1025) & \text{Squaring} \\
&= \$1102.50. & \text{Multiplying}
\end{aligned}$$

Quick Check 6 ✔

Business. Repeat Example 6 for an interest rate of 6%.

There is $1102.50 in the account after 2 yr.   **6** ✔

For interest that is compounded quarterly (four times per year), we can find a formula like the one above, as illustrated in the following diagram.

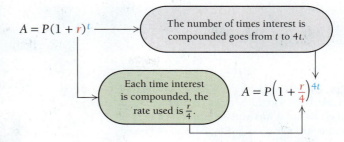

$$A = P(1 + r)^t$$

The number of times interest is compounded goes from t to $4t$.

Each time interest is compounded, the rate used is $\frac{r}{4}$.

$$A = P\left(1 + \frac{r}{4}\right)^{4t}$$

"Compounded quarterly" means that the interest is divided by 4 and compounded four times per year. In general, the following theorem applies.

> **THEOREM 2**
>
> If a principal P is invested at interest rate r, expressed as a decimal, and compounded n times a year, in t years it will grow to an amount A given by
>
> $$A = P\left(1 + \frac{r}{n}\right)^{nt}.$$

EXAMPLE 7 **Business: Compound Interest.** Suppose $1000 is invested in Wellington Investment Fund at 5%, compounded quarterly. How much is in the account at the end of 3 yr?

Solution We use the equation $A = P(1 + r/n)^{nt}$, substituting 1000 for P, 0.05 for r, 4 for n (compounding quarterly), and 3 for t. Then we get

$$A = 1000\left(1 + \frac{0.05}{4}\right)^{4 \cdot 3} \quad \text{Substituting}$$

$$= 1000(1 + 0.0125)^{12}$$

$$= 1000(1.0125)^{12}$$

$$= 1000(1.160754518) \quad \text{Using a calculator to approximate } (1.0125)^{12}$$

$$= 1160.754518$$

$$\approx \$1160.75. \quad \text{The symbol } \approx \text{ means "approximately equal to."}$$

There is $1160.75 in the account after 3 yr. **7 ✔**

Quick Check 7 ✔

Business: Compound Interest. Repeat Example 7 for an interest rate of 6%.

A calculator with a $\boxed{y^x}$ or $\boxed{\wedge}$ key and a ten-digit readout was used to find $(1.0125)^{12}$ in Example 7. The number of places on a calculator may affect the accuracy of the answer. Thus, you may occasionally find that your answers do not agree with those at the back of the book, which were found on a calculator with a ten-digit readout. In general, when using a calculator, do all computations and round only at the end, as in Example 7. Usually, your answer will agree to at least four digits. It is a good idea to consult with your instructor about the accuracy required.

Section Summary

- Most graphs can be created by plotting points and looking for patterns. A graphing calculator can create graphs rapidly.
- Mathematical equations can serve as models of many kinds of applications.

- An example of mathematical model is the formula for compound interest. If P dollars are invested at interest rate r, compounded n times a year for t years, then the amount A at the end of the t years is given by

$$A = P\left(1 + \frac{r}{n}\right)^{nt}.$$

R.1 | Exercise Set

Exercises designated by the symbol ✎ are Thinking and Writing Exercises. They should be answered using one or two English sentences. Because answers to many such exercises will vary, solutions are not given at the back of the book.

Graph. (Unless directed otherwise, assume that "Graph" means "Graph by hand.")

1. $y = x - 1$

2. $y = x + 4$

3. $y = -\frac{1}{4}x$

4. $y = -3x$

5. $y = -\frac{5}{3}x + 3$

6. $y = \frac{2}{3}x - 4$

7. $x + y = 5$

8. $x - y = 4$

9. $6x + 3y = -9$

10. $8y - 2x = 4$

11. $2x + 5y = 10$

12. $5x - 6y = 12$

13. $y = x^2 - 5$

14. $y = x^2 - 3$

15. $x = y^2 + 2$

16. $x = 2 - y^2$

17. $y = |4 - x|$

18. $y = |x|$

19. $y = 7 - x^2$

20. $y = 5 - x^2$

21. $y - 7 = x^3$

22. $y + 1 = x^3$

APPLICATIONS

23. Medicine. Ibuprofen is a medication used to relieve pain. The function

$$A = 0.5t^4 + 3.45t^3 - 96.65t^2 + 347.7t, \ 0 \le t \le 6,$$

can be used to estimate the number of milligrams, A, of ibuprofen in the bloodstream t hours after 400 mg of the medication has been swallowed. (*Source*: Based on data from Dr. P. Carey, Burlington, VT.) How many milligrams of ibuprofen are in the bloodstream 2 hr after 400 mg has been swallowed?

24. Running records. According to at least one study, the world record in any running race can be modeled by a linear equation. In particular, the world record R, in minutes, for the mile run in year x can be modeled by

$$R = -0.006x + 15.714.$$

Use this model to estimate the world records for the mile run in 1954, 2000, and 2015. Round your answers to the nearest hundredth of a minute.

25. Snowboarding in the half-pipe. Shaun White, "The Flying Tomato," won a gold medal in the 2010 Winter Olympics for snowboarding in the half-pipe. He soared an unprecedented 25 ft above the edge of the half-pipe. His speed $v(t)$, in miles per hour, upon reentering the pipe can be approximated by $v(t) = 10.9t$, where t is the number of seconds for which he was airborne. White was airborne for 2.5 sec. (*Source*: "White Rides to Repeat in Halfpipe, Lago Takes Bronze," Associated

Press, 2/18/2010.) How fast was he going when he reentered the half-pipe?

26. Skateboard bomb drop. The distance $s(t)$, in feet, traveled by a body falling freely from rest in t seconds is approximated by

$$s(t) = 16t^2.$$

On April 6, 2006, pro skateboarder Danny Way smashed the world record for the "bomb drop" by free-falling 28 ft from the Fender Stratocaster guitar atop the Hard Rock Hotel & Casino in Las Vegas onto a ramp below. (*Source*: www.skateboardingmagazine.com.) How long did it take until he hit the ramp?

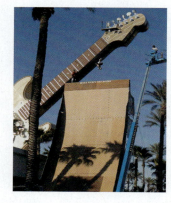

27. Hearing-impaired Americans. The number N, in millions, of hearing-impaired Americans of age x can be approximated by the graph that follows.

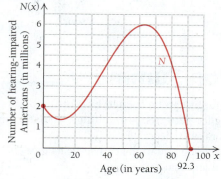

(*Source*: Better Hearing Institute.)

Use the graph to answer the following.

a) Approximate the number of hearing-impaired Americans of ages 20, 40, 50, and 60.

b) For what ages is the number of hearing-impaired Americans approximately 4 million?

c) Estimate the age for which the greatest number of Americans is hearing-impaired.

d) What difficulty do you have in making this determination?

28. Life science: incidence of breast cancer. The following graph approximates the incidence of breast cancer *y*, per 100,000 women, as a function of age *x*, where *x* represents ages 25 to 102.

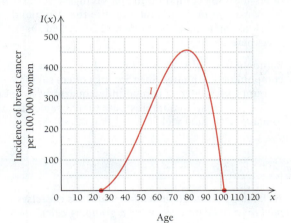

(*Source*: Based on data from the National Cancer Institute.)

a) What is the incidence of breast cancer in 40-yr-old women?

b) For what ages is the incidence of breast cancer about 400 per 100,000 women?

c) Examine the graph and try to determine the age at which the largest incidence of breast cancer occurs.

d) What difficulty do you have making this determination?

29. Compound interest. Southside Investments purchases a $100,000 certificate of deposit from Newton Bank, at 2.8%. How much is the investment worth (rounded to the nearest cent) at the end of 1 yr, if interest is compounded:

a) annually?   **b)** semiannually?
c) quarterly?   **d)** daily (use 365 days
e) hourly?   for 1 yr)?

30. Compound interest. Greenleaf Investments purchases a $300,000 certificate of deposit from Descartes Bank, at 2.2%. How much is the investment worth (rounded to the nearest cent) at the end of 1 yr, if interest is compounded:

a) annually?   **b)** semiannually?
c) quarterly?   **d)** daily (use 365 days
e) hourly?   for 1 yr)?

31. Compound interest. Stateside Brokers deposit $30,000 in Godel Municipal Bond Funds, at 4%. How much is the

investment worth (rounded to the nearest cent) at the end of 3 yr, if interest is compounded:

a) annually?   **b)** semiannually?
c) quarterly?   **d)** daily (use 365 days
e) hourly?   for 1 yr)?

32. Compound interest. The Kims deposit $1000 in Wiles Municipal Bond Funds, at 5%. How much is the investment worth (rounded to the nearest cent) at the end of 4 yr, if interest is compounded:

a) annually?   **b)** semiannually?
c) quarterly?   **d)** daily (use 365 days
e) hourly?   for 1 yr)?

Determining monthly loan payments. *If P dollars are borrowed at an annual interest rate r, the payment M made each month for a total of n months is*

$$M = P \dfrac{\dfrac{r}{12}\left(1 + \dfrac{r}{12}\right)^n}{\left(1 + \dfrac{r}{12}\right)^n - 1}.$$

33. Fermat's Last Bank makes a car loan of $18,000, at 6.4% interest and with a loan period of 3 yr. What is the monthly payment?

34. At Haken Bank, Ken Appel takes out a $100,000 mortgage at an interest rate of 4.8% for a loan period of 30 yr. What is the monthly payment?

Annuities. *If P dollars are invested annually in an annuity (investment fund), after n years, the annuity will be worth*

$$W = P\left[\dfrac{(1 + r)^n - 1}{r}\right],$$

where r is the interest rate, compounded annually.

35. Kate invests $3000 annually in an annuity from Mersenne Fund that earns 6.57% interest. How much is the investment worth after 18 yr? Round to the nearest cent.

36. Paulo establishes an annuity that earns $7\frac{1}{4}$% interest and wants it to be worth $50,000 in 20 yr. How much will he need to invest annually to achieve this goal?

37. Condor population. The condor population in California and Arizona from 2002 to 2012 is approximated in the graph below.

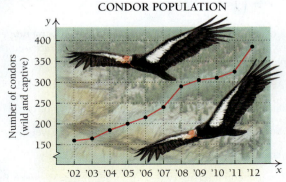

CONDOR POPULATION

(*Source*: Based on data from esasuccess.org.)

a) In what years was the condor population at or above 230?
b) In what year was the condor population at 200?
c) In what year was the condor population the highest?
d) In what year was the condor population the lowest?

38. Speculate as to why the condor population in California and Arizona experienced such a dramatic rise in 2007–2008.

SYNTHESIS

Retirement account. *Sally makes deposits into a retirement account every year from the age of 30 until she retires at age 65.*

39. a) If Sally deposits $1200 per year and the account earns interest at a rate of 4% per year, compounded annually, how much will she have in the account when she retires? (*Hint:* Use the annuity formula given for Exercises 35 and 36.)
   b) How much of that total amount is from Sally's deposits? How much is interest?

40. a) Sally plans to take regular monthly distributions from her retirement account from the time she retires until she is 80 years old, when the account will have a value of $0. How much should she take each month? Assume the interest rate is 4% per year, compounded monthly. (*Hint:* Use the formula given for Exercises 33 and 34 that calculates the monthly payments on a loan.)
   b) What is the total of the payments she will receive? How much of the total will be her own money (see part b of Exercise 39), and how much will be interest?

Annual yield. *The annual interest rate r, when compounded more than once a year, results in a slightly higher yearly interest rate; this is called the* **annual (or effective) yield** *and denoted as Y. For example, $1000 deposited at 5%, compounded monthly for 1 yr (12 months), will have a value of* $A = 1000(1 + \frac{0.05}{12})^{12} = \$1051.16$. *The interest earned is $51.16/$1000, or 0.05116, which is 5.116% of the original deposit. Thus, we say this account has a yield of Y = 0.05116, or 5.116%. The formula for annual yield depends on the annual interest rate r and the compounding frequency n:*

$$Y = \left(1 + \frac{r}{n}\right)^n - 1.$$

For Exercises 41–48, find the annual yield as a percentage, to two decimal places, given the annual interest rate and the compounding frequency.

41. Annual interest rate of 5.3%, compounded monthly

42. Annual interest rate of 4.1%, compounded quarterly

43. Annual interest rate of 3.75%, compounded weekly

44. Annual interest rate of 4%, compounded daily

45. Lena is considering two savings accounts: Western Bank offers 4.5%, compounded annually, on saving accounts, while Commonwealth Savings offers 4.43%, compounded monthly.
   a) Find the annual yield for both accounts.
   b) Which account has the higher annual yield?

46. Chris is considering two savings accounts: Sierra Savings offers 5%, compounded annually, on savings accounts, while Foothill Bank offers 4.88%, compounded weekly.
   a) Find the annual yield for both accounts.
   b) Which account has the higher annual yield?

47. Stockman's Bank will pay 4.2%, compounded annually, on a savings account. A competitor, Mesalands Savings, offers monthly compounding on savings accounts. What is the minimum annual interest rate that Mesalands needs to pay to make its annual yield exceed that of Stockman's?

48. Belltown Bank offers a certificate of deposit at 3.75%, compounded annually. Shea Savings offers savings accounts with interest compounded quarterly. What is the minimum annual interest rate that Shea needs to pay to make its annual yield exceed that of Belltown?

TECHNOLOGY CONNECTION

The Technology Connection heading indicates exercises designed to provide practice using a graphing calculator.

Graph.

49. $y = x - 150$

50. $y = 25 - |x|$

51. $y = x^3 + 2x^2 - 4x - 13$

52. $y = \sqrt{23 - 7x}$

53. $9.6x + 4.2y = -100$

54. $y = -2.3x^2 + 4.8x - 9$

55. $x = 4 + y^2$

56. $x = 8 - y^2$

Answers to Quick Checks

1. $y = 3 - x$

2. $3x - 5y = 10$

3. $y = 2 - x^2$

4. $x = 1 + y^2$

5. 2019   6. There is $1123.60 in the account after 2 yr.

7. There is $1195.62 in the account after 3 yr.

Functions and Models

Identifying Functions

The idea of a *function* is one of the most important concepts in mathematics. Put simply, a function is a special kind of correspondence between two sets. Let's look at the following.

> To each letter on a telephone keypad there corresponds a number.
>
> To each model of television in a store there corresponds its price.
>
> To each real number there corresponds the cube of that number.

In each of these examples, the first set is called the *domain* and the second set is called the *range*. Given a member of the domain, there is *exactly one* member of the range to which it corresponds. This type of correspondence is called a *function*.

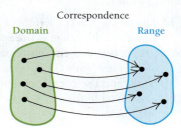

DEFINITION

A **function** is a correspondence between a first set, called the **domain**, and a second set, called the **range**, such that each member of the domain corresponds to *exactly one* member of the range.

EXAMPLE 1   Determine whether or not each correspondence is a function.

a) *Cumulative number of iPhones sold*

Domain	Range
2009	→ 20,400,000
2010	→ 40,000,000
2011	→ 72,000,000
2012	→ 125,400,000

(*Source:* Apple Inc.)

b) *Squaring*

Domain	Range
3	→ 9
4	→ 16
5	
−5	→ 25

c) *Basketball teams*

d) *Basketball teams*

Domain	Range
Knicks	→ New York
Lakers	→ Los Angeles
Clippers	
Hawks	→ Atlanta

Solution

a) The correspondence *is* a function because each member of the domain corresponds (is matched) to only one member of the range.

b) The correspondence *is* a function because each member of the domain corresponds to only one member of the range, even though two members of the domain correspond to 25.

c) The correspondence *is not* a function because one member of the domain, Los Angeles, corresponds to two members of the range, the Lakers and the Clippers.

d) The correspondence *is* a function because each member of the domain corresponds to only one member of the range, even though two members of the domain correspond to Los Angeles. ■

Quick Check 1 ✔

Determine whether or not each correspondence is a function.

a) The domain is the books in a library, the range is the set of positive integers, and the correspondence is the page count of each book.

b) The domain is the letters of the alphabet, the range is the set of people listed in the white pages of the New York City phone book, and the correspondence is the first letter of a person's last name.

EXAMPLE 2   Determine whether or not each correspondence is a function.

Domain	Correspondence	Range
a) A family	Each person's weight	A set of positive numbers
b) The integers $\{\ldots, -3, -2, -1, 0, 1, 2, 3, \ldots\}$	Each number's square	A set of nonnegative integers: $\{0, 1, 4, 9, 16, 25, \ldots\}$
c) The set of all states	Each state's U.S. Senators	The set of all 100 U.S. Senators

Solution

a) The correspondence is a function because each person has *only one* weight.

b) The correspondence is a function because each integer has *only one* square.

c) The correspondence is *not* a function because each state has *two* U.S. Senators.   **1** ✔

Consistent with the definition on p. 13, we regard a function as a set of ordered pairs, such that no two pairs have the same first coordinate paired with different second coordinates. The domain is the set of all first coordinates, and the range is the set of all second coordinates. Function names are usually represented by lowercase letters. Thus, if f represents the function in Example 1(b), we have

$$f = \{(3, 9), (4, 16), (5, 25), (-5, 25)\}$$

and

$$\text{Domain of } f = \{3, 4, 5, -5\}; \qquad \text{Range of } f = \{9, 16, 25\}.$$

Finding Function Values

Many functions can be described by an equation like $y = 2x + 3$ or $y = 4 - x^2$. To graph the function given by $y = 2x + 3$, we find ordered pairs by performing calculations for selected x values.

for $x = 4, y = 2x + 3 = 2 \cdot 4 + 3 = 11$;   The graph includes $(4, 11)$.

for $x = -5, y = 2x + 3 = 2 \cdot (-5) + 3 = -7$;   The graph includes $(-5, -7)$.

for $x = 0, y = 2x + 3 = 2 \cdot 0 + 3 = 3$; and so on.   The graph includes $(0, 3)$.

For $y = 2x + 3$, the **inputs** (members of the domain) are the values of x substituted into the equation. The **outputs** (members of the range) are the resulting values of y. If we call the function f, we can use x to represent an arbitrary input and $f(x)$, read "f of x" or "f at x" or "the value of f at x," to represent the corresponding output. In this notation, the function given by $y = 2x + 3$ is written as $f(x) = 2x + 3$, and the calculations above can be written more concisely as

$$f(4) = 2 \cdot 4 + 3 = 11;$$
$$f(-5) = 2 \cdot (-5) + 3 = -7;$$
$$f(0) = 2 \cdot 0 + 3 = 3; \text{ and so on.}$$

Thus, instead of writing "when $x = 4$, the value of y is 11," we can simply write "$f(4) = 11$," which is most commonly read as "f of 4 is 11."

It helps to think of a function as a machine. Think of $f(4) = 11$ as the result of putting a member of the domain (an input), 4, into the machine. The machine knows the correspondence $f(x) = 2x + 3$, computes $2 \cdot 4 + 3$, and produces a member of the range (the output), 11.

Function: $f(x) = 2x + 3$

Input	Output
4	11
-5	-7
0	3
t	$2t + 3$
$a + h$	$2(a + h) + 3$

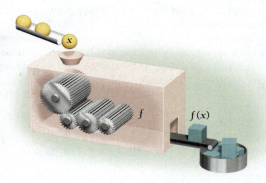

Remember that $f(x)$ *does not mean* "f times x" and should never be read that way.

EXAMPLE 3   The squaring function f is given by

$$f(x) = x^2.$$

Find $f(-3), f(1), f(k), f(\sqrt{k}), f(1 + t)$, and $f(x + h)$.

Solution   We have

$$f(-3) = (-3)^2 = 9;$$
$$f(1) = 1^2 = 1;$$
$$f(k) = k^2;$$
$$f(\sqrt{k}) = (\sqrt{k})^2 = k;$$
$$f(1 + t) = (1 + t)^2 = 1 + 2t + t^2;$$
$$f(x + h) = (x + h)^2 = x^2 + 2xh + h^2.$$

For a review of algebra, see Appendix A on p. 585.

To find $f(x + h)$, remember what the function does: It squares the input. Thus, $f(x + h) = (x + h)^2 = x^2 + 2xh + h^2$. This amounts to replacing x on both sides of $f(x) = x^2$ with $x + h$. **2 ✔**

Quick Check 2 ✔

A function f is given by $f(x) = 3x + 5$. Find $f(4), f(-5), f(0), f(a)$, and $f(a + h)$.

EXAMPLE 4   A function f is given by $f(x) = 3x^2 - 2x + 8$. Find $f(0), f(-5)$, and $f(7a)$.

Solution   One way to find function values when a formula is given is to think of the formula with blanks, or placeholders, as follows:

$$f(\ \ \ ) = 3 \cdot \boxed{}^2 - 2 \cdot \boxed{} + 8.$$

To find an output for a given input, we think: "Whatever goes in the blank on the left goes in the blank(s) on the right."

$$f(0) = 3 \cdot 0^2 - 2 \cdot 0 + 8 = 8$$
$$f(-5) = 3(-5)^2 - 2 \cdot (-5) + 8 = 3 \cdot 25 + 10 + 8 = 75 + 10 + 8 = 93$$
$$f(7a) = 3(7a)^2 - 2(7a) + 8 = 3 \cdot 49a^2 - 14a + 8 = 147a^2 - 14a + 8$$

3 ✔

Quick Check 3 ✔

A function f is given by $f(x) = 3x^2 + 2x - 7$. Find $f(4), f(-5), f(0), f(a)$, and $f(5a)$.

TECHNOLOGY CONNECTION

The TABLE Feature

The TABLE feature is one way to find ordered pairs of inputs and outputs of functions. To see how, consider the function given by $f(x) = x^3 - 5x + 1$. We enter it as $y_1 = x^3 - 5x + 1$. To use the TABLE feature, we access the TABLE SETUP screen and enter the x-value at which the table will start and an increment for the x-value. For this equation, let's set TblStart = 0.3 and ΔTbl = 1. (Other values can be chosen.) This means that the table's x-values will start at 0.3 and increase by 1.

We next set Indpnt and Depend to Auto and then press TABLE. The result is shown below.

X	Y1	
.3	−.473	
1.3	−3.303	
2.3	1.667	
3.3	20.437	
4.3	59.007	
5.3	123.38	
6.3	219.55	
X = .3		

The arrow keys, ⌃ and ⌄, allow us to scroll up and down the table and extend it to other values not initially shown.

X	Y1	
12.3	1800.4	
13.3	2287.1	
14.3	2853.7	
15.3	3506.1	
16.3	4250.2	
17.3	5092.2	
18.3	6038	
X = 18.3		

EXERCISES

Use the function given by $f(x) = x^3 - 5x + 1$ for Exercises 1 and 2.

1. Use the TABLE feature to construct a table starting with $x = 10$ and ΔTbl $= 5$. Find the value of y when x is 10. Then find the value of y when x is 35.

2. Adjust the table settings to Indpnt: Ask. How does the table change? Enter a number of your choice and see what happens. Use this setting to find the value of y when x is 28.

EXAMPLE 5   A function f subtracts the square of an input from the input:
$$f(x) = x - x^2.$$

Find $f(4), f(x + h),$ and $\dfrac{f(x + h) - f(x)}{h}$.

Solution   We have
$$f(4) = 4 - 4^2 = 4 - 16 = -12;$$

and
$$f(x + h) = (x + h) - (x + h)^2$$
$$= x + h - (x^2 + 2xh + h^2) \qquad \text{Squaring the binomial}$$
$$= x + h - x^2 - 2xh - h^2.$$

For the expression $\dfrac{f(x + h) - f(x)}{h}$, we have

$$\frac{f(x + h) - f(x)}{h} = \frac{\overbrace{x + h - x^2 - 2xh - h^2}^{f(x+h),\ \text{found above}} - \overbrace{(x - x^2)}^{f(x)}}{h}$$

$$= \frac{h - 2xh - h^2}{h} \qquad \text{Simplifying}$$

$$= \frac{h(1 - 2x - h)}{h} \qquad \text{Factoring}$$

$$= 1 - 2x - h, \quad \text{for } h \neq 0. \qquad \boxed{4}$$

Quick Check 4 ✔

A function f is given by $f(x) = 2x - x^2$. Find $f(4), f(x + h),$ and $\dfrac{f(x + h) - f(x)}{h}$.

Graphs of Functions

Consider again the function given by $f(x) = x^2$. The input 3 is associated with the output 9. The input–output pair $(3, 9)$ is one point on the *graph* of this function.

> **DEFINITION**
>
> The **graph** of a function f is a drawing that represents all the input–output pairs $(x, f(x))$. In cases where the function is given by an equation, the graph of the function is the graph of the equation $y = f(x)$.

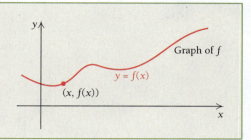

It is customary to locate input values (the domain) on the horizontal axis and output values (the range) on the vertical axis.

EXAMPLE 6   Graph: $f(x) = x^2 - 1$.

Solution

x	$f(x)$	$(x, f(x))$
-2	3	$(-2, 3)$
-1	0	$(-1, 0)$
0	-1	$(0, -1)$
1	0	$(1, 0)$
2	3	$(2, 3)$

(1) Choose any x.
(2) Compute y.
(3) Form the pair (x, y).
(4) Plot the points.

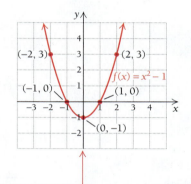

Quick Check 5 ✔

Graph: $f(x) = 2 - x^2$.

We plot the input–output pairs from the table and, in this case, draw a curve to complete the graph.

5 ✔

TECHNOLOGY CONNECTION

Graphs and Function Values

We graphed equations in the Technology Connection in Section R.1. To graph a function, we use the same procedure, after changing the "$f(x) = $" notation to "$y = .$" Thus, to graph $f(x) = 2x^2 + x$, we key in $y_1 = 2x^2 + x$.

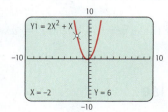

There are several ways to find function values. One is to use the TABLE feature, as previously discussed. Another is

to use the TRACE feature. To do so, graph the function, press TRACE, and either move the cursor or enter any x-value that is in the window. The corresponding y-value appears automatically. Function values can also be found using the VALUE or Y-VARS feature. Consult an owner's manual or the GCM for details.

EXERCISES

1. Graph $f(x) = x^2 + 3x - 4$. Then find $f(-5), f(-4.7), f(11)$, and $f(2/3)$. (*Hint*: To find $f(11)$, be sure that the window dimensions for the x-values include $x = 11$.)

2. Graph $f(x) = 3.7 - x^2$. Then find $f(-5), f(-4.7), f(11)$, and $f(2/3)$.

3. Graph $f(x) = 4 - 1.2x - 3.4x^2$. Then find $f(-5)$, $f(-4.7), f(11)$, and $f(2/3)$.

The Vertical-Line Test

Let's now determine how to look at a graph and decide whether it represents a function. In the graph at the right, note that the input x_1 has *two* outputs. Since a function has exactly *one* output for every input, this graph cannot represent a function. Note that a vertical line can intersect the graph in more than one place.

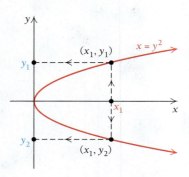

The Vertical-Line Test

A graph represents a function if it is impossible to draw a vertical line that intersects the graph more than once.

EXAMPLE 7   Determine whether each of the following is the graph of a function.

a)

b)

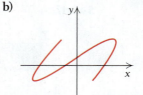

c)

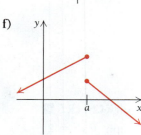

d)

e)

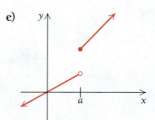

f)

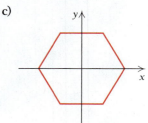

Quick Check 6 ✔

Determine whether each of the following is the graph of a function.

a)

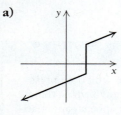

b)

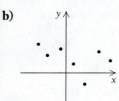

Solution

a) The graph is that of a function. It is impossible to draw a vertical line that intersects the graph more than once.

b) The graph is not that of a function. A vertical line (in fact, many) can intersect the graph more than once.

c) The graph is not that of a function.

d) The graph is that of a function.

e) The graph is that of a function. Note that a vertical line at $x = a$ would intersect the graph once.

f) The graph is not that of a function. Note that a vertical line at $x = a$ would intersect the graph more than once.

6 ✔

Functions Defined Piecewise

Sometimes functions are defined *piecewise*, using different output formulas for different parts of the domain, as in parts (e) and (f) of Example 7. To graph a piecewise-defined function, we usually work from left to right, using the correspondence specified for the *x*-values on each part of the horizontal axis.

EXAMPLE 8 Graph the function defined as follows:

$$f(x) = \begin{cases} 4, & \text{for } x \le 0, \\ 3 - x^2, & \text{for } 0 < x \le 2, \\ 2x - 6, & \text{for } x > 2. \end{cases}$$

This means that for any input x less than or equal to 0, the output is 4.

This means that for any input x greater than 0 and less than or equal to 2, the output is $3 - x^2$.

This means that for any input x greater than 2, the output is $2x - 6$.

Solution Working from left to right, note that for all x-values less than or equal to 0, the graph is the horizontal line $y = 4$. Thus,

$$f(-2) = 4;$$
$$f(-1) = 4;$$

and $\quad f(0) = 4.$

The solid dot indicates that $(0, 4)$ is part of the graph.

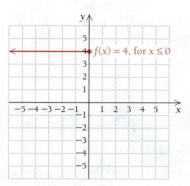

Next, observe that for x-values greater than 0 but not greater than 2, the graph is a portion of the parabola given by $y = 3 - x^2$. Note that for $f(x) = 3 - x^2$,

$$f(0.5) = 3 - 0.5^2 = 2.75;$$
$$f(1) = 2;$$

and $\quad f(2) = -1.$

The open dot at $(0, 3)$ indicates that this point is *not* part of the graph.

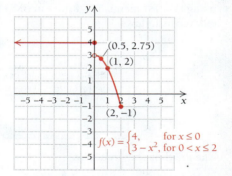

Finally, note that for x-values greater than 2, the graph is the line $y = 2x - 6$.

$$f(2.5) = 2 \cdot 2.5 - 6 = -1;$$
$$f(4) = 2;$$

and $\quad f(5) = 4.$

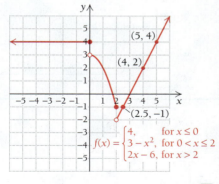

Quick Check 7 ✔

Graph the function defined as follows:

$$f(x) = \begin{cases} 4, & \text{for } x \le 0, \\ 4 - x^2, & \text{for } 0 < x \le 2, \\ 2x - 6, & \text{for } x > 2. \end{cases}$$

7 ✔

EXAMPLE 9 Graph the function defined as follows:

$$g(x) = \begin{cases} 3, & \text{for } x = 1, \\ -x + 2, & \text{for } x \ne 1. \end{cases}$$

Solution The function is defined such that $g(1) = 3$ and for all other x-values (that is, for $x \ne 1$), we have $g(x) = -x + 2$. Thus, to graph this function, we graph the line given by $y = -x + 2$, but with an open dot at the point corresponding to $x = 1$. To complete the graph, we plot the point $(1, 3)$ since $g(1) = 3$.

x	$g(x)$	$(x, g(x))$
-3	$-(-3) + 2$	$(-3, 5)$
0	$-0 + 2$	$(0, 2)$
1	3	$(1, 3)$
2	$-2 + 2$	$(2, 0)$
3	$-3 + 2$	$(3, -1)$

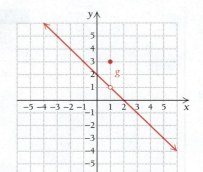

Quick Check 8 ✔

Graph the function defined as follows:

$$f(x) = \begin{cases} 1, & \text{for } x = -2, \\ 2 - x, & \text{for } x \neq -2. \end{cases}$$

8 ✔

TECHNOLOGY CONNECTION

Graphing Functions Defined Piecewise

Graphing functions defined piecewise generally involves the use of inequality symbols, which are often accessed using the TEST menu. The function in Example 8 can be entered as follows:

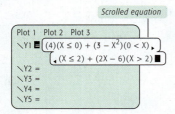

The graph is shown below.

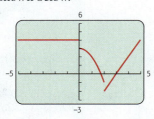

Note that most graphing calculators do not display solid or open dots.

EXERCISES

Graph.

1. $f(x) = \begin{cases} -x - 2, & \text{for } x < -2, \\ 4 - x^2, & \text{for } -2 \leq x < 2, \\ x + 3, & \text{for } x \geq 2 \end{cases}$

2. $f(x) = \begin{cases} x^2 - 2, & \text{for } x \leq 3, \\ 1, & \text{for } x > 3 \end{cases}$

3. $f(x) = \begin{cases} x + 3, & \text{for } x \leq -2, \\ 1, & \text{for } -2 < x \leq 3, \\ x^2 - 10, & \text{for } x > 3 \end{cases}$

Some Final Remarks

We sometimes use the terminology *y is a function of x*. This means that x is an input and y is an output. It also means that x is the **independent variable** because it represents inputs and y is the **dependent variable** because it represents outputs. We may refer to "a function $y = x^2$" without naming it using a letter f.

Section Summary

● *Functions* are a key concept in mathematics.

● The essential trait of a function is that to each number in the *domain* there corresponds one and only one number in the *range*.

R.2 Exercise Set

Note: A review of algebra can be found in Appendix A on p. 585.

Determine whether each correspondence is a function.

1. Domain   Range

2. Domain   Range
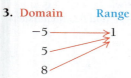

3. Domain   Range

4. Domain   Range

5. Sandwich prices.

DOMAIN	RANGE
Hamburger	→ $0.99
Cheeseburger	→ $1.29
Filet-O-Fish®	→ $3.49
Quarter Pounder® with cheese	→ $3.69
Big Mac®	→ $3.99
Crispy Chicken	→ $4.29
Chicken McNuggets®	
Double Quarter Pounder® with cheese	→ $4.69

(*Source*: www.mcdonalds.com.)

6. Sandwich calorie content.

DOMAIN	RANGE
Hamburger	→ 250
Cheeseburger	→ 300
Quarter Pounder®	→ 410
Double Cheeseburger®	→ 440
Filet-O-Fish®	→ 380
Big Mac®	→ 540
Double Quarter Pounder® with cheese	→ 740

(*Source*: www.mcdonalds.com.)

Determine whether each of the following is a function.

Domain	Correspondence	Range
7. A set of iPods	Each iPod's memory in gigabytes	A set of numbers
8. A set of iPods	Each iPod's serial number	A set of alphanumeric codes
9. A set of iPods	The number of songs on each iPod	A set of numbers
10. A set of iPods	The number of Avril Lavigne songs on each iPod	A set of numbers
11. The set of all real numbers	The square of a number, to which 8 is added	The set of all positive numbers greater than or equal to 8
12. The set of all real numbers	The fourth power of a number	The set of all nonnegative numbers.
13. A set of females	Each person's biological mother	A set of females
14. A set of males	Each person's biological father	A set of males
15. A set of avenues	An intersecting road	A set of cross streets
16. A set of textbooks	An even-numbered page in each book	A set of pages
17. A set of shapes	The area of each shape	A set of area measurements
18. A set of shapes	The perimeter of each shape	A set of length measurements

19. A function f is given by

$$f(x) = 4x - 3.$$

This function takes a number x, multiplies it by 4, and subtracts 3.

a) Complete this table.

x	5.1	5.01	5.001	5
$f(x)$				

b) Find $f(4), f(3), f(-2), f(k), f(1 + t)$, and $f(x + h)$.

20. A function f is given by

$$f(x) = 3x + 2.$$

This function takes a number x, multiplies it by 3, and adds 2.

a) Complete this table.

x	4.1	4.01	4.001	4
$f(x)$				

b) Find $f(5), f(-1), f(k), f(1 + t)$, and $f(x + h)$.

21. A function g is given by

$$g(x) = x^2 - 3.$$

This function takes a number x, squares it, and subtracts 3. Find $g(-1), g(0), g(1), g(5), g(u), g(a + h)$, and $\dfrac{g(a + h) - g(a)}{h}$.

22. A function g is given by

$$g(x) = x^2 + 4.$$

This function takes a number x, squares it, and adds 4. Find $g(-3), g(0), g(-1), g(7), g(v), g(a + h)$, and $\dfrac{g(a + h) - g(a)}{h}$.

23. A function f is given by

$$f(x) = \frac{1}{(x + 3)^2}.$$

This function takes a number x, adds 3, squares the result, and takes the reciprocal of that result.

a) Find $f(4), f(0), f(a), f(t + 4), f(x + h)$, and $\dfrac{f(x + h) - f(x)}{h}$.

b) Note that f could also be given by

$$f(x) = \frac{1}{x^2 + 6x + 9}.$$

Explain what this does to an input number x.

24. A function f is given by

$$f(x) = \frac{1}{(x - 5)^2}.$$

This function takes a number x, subtracts 5 from it, squares the result, and takes the reciprocal of the square.

a) Find $f(3), f(-1), f(k), f(t - 1), f(t - 4)$, and $f(x + h)$.

b) Note that f could also be given by

$$f(x) = \frac{1}{x^2 - 10x + 25}.$$

Explain what this does to an input number x.

25. A function f takes a number x, multiplies it by 4, and adds 2.

a) Write f as an equation.
b) Graph f.

26. A function g takes a number x, multiplies it by -3, and subtracts 4.

a) Write g as an equation.
b) Graph g.

27. A function h takes a number x, squares it, and adds x.

a) Write h as an equation.
b) Graph h.

28. A function k takes a number x, squares it, and subtracts 3 times x.

a) Write k as an equation.
b) Graph k.

Graph each function.

29. $f(x) = 2x - 5$

30. $f(x) = 3x - 1$

31. $g(x) = -4x$

32. $g(x) = -2x$

33. $f(x) = x^2 - 2$

34. $f(x) = x^2 + 4$

35. $f(x) = 6 - x^2$

36. $g(x) = -x^2 + 1$

37. $g(x) = x^3$

38. $g(x) = \frac{1}{2}x^3$

Use the vertical-line test to determine whether each graph is that of a function. (In Exercises 47–50 and 52, the dashed lines are not part of the graphs.)

39.

40.

41.
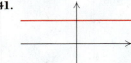

42.

43.

44.
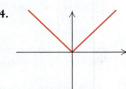

45.

46.
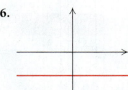

47.
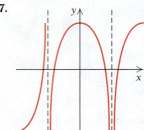

48.
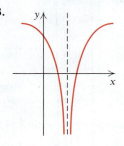

49.

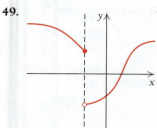

50.

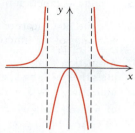

51.

52.

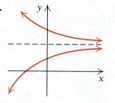

53. a) Graph $x = y^2 - 2$.
 b) Is this a function?

54. a) Graph $x = y^2 - 3$.
 b) Is this a function?

55. For $f(x) = x^2 - 3x$, find $\dfrac{f(x + h) - f(x)}{h}$.

56. For $f(x) = x^2 + 4x$, find $\dfrac{f(x + h) - f(x)}{h}$.

For Exercises 57–60, consider the function f given by

$$f(x) = \begin{cases} -2x + 1, & \text{for } x < 0, \\ 17, & \text{for } x = 0, \\ x^2 - 3, & \text{for } 0 < x < 4, \\ \frac{1}{2}x + 1, & \text{for } x \geq 4. \end{cases}$$

57. Find $f(-1)$ and $f(1)$.

58. Find $f(-3)$ and $f(3)$.

59. Find $f(0)$ and $f(10)$.

60. Find $f(-5)$ and $f(5)$.

Graph.

61. $f(x) = \begin{cases} 1, & \text{for } x < 0, \\ -1, & \text{for } x \geq 0 \end{cases}$

62. $f(x) = \begin{cases} 2, & \text{for } x \leq 3, \\ -2, & \text{for } x > 3 \end{cases}$

63. $f(x) = \begin{cases} 6, & \text{for } x = -2, \\ x^2, & \text{for } x \neq -2 \end{cases}$

64. $f(x) = \begin{cases} 5, & \text{for } x = 1, \\ x^3, & \text{for } x \neq 1 \end{cases}$

65. $g(x) = \begin{cases} -x, & \text{for } x < 0, \\ 4, & \text{for } x = 0, \\ x + 2, & \text{for } x > 0 \end{cases}$

66. $g(x) = \begin{cases} 2x - 3, & \text{for } x < 1, \\ 5, & \text{for } x = 1, \\ x - 2, & \text{for } x > 1 \end{cases}$

67. $g(x) = \begin{cases} \frac{1}{2}x - 1, & \text{for } x < 2, \\ -4, & \text{for } x = 2, \\ x - 3, & \text{for } x > 2 \end{cases}$

68. $g(x) = \begin{cases} x^2, & \text{for } x < 0, \\ -3, & \text{for } x = 0, \\ -2x + 3, & \text{for } x > 0 \end{cases}$

69. $f(x) = \begin{cases} -7, & \text{for } x = 2, \\ x^2 - 3, & \text{for } x \neq 2 \end{cases}$

70. $f(x) = \begin{cases} -6, & \text{for } x = -3, \\ -x^2 + 5, & \text{for } x \neq -3 \end{cases}$

Compound interest. *The amount of money, A(t), in a savings account that pays 3% interest, compounded quarterly for t years, with an initial investment of P dollars, is given by*

$$A(t) = P\left(1 + \frac{0.03}{4}\right)^{4t}.$$

71. If \$500 is invested at 3%, compounded quarterly, how much will the investment be worth after 2 yr?

72. If \$800 is invested at 3%, compounded quarterly, how much will the investment be worth after 3 yr?

Chemotherapy. *In computing the dosage for chemotherapy, the measure of a patient's body surface area is needed. A good approximation of this area s, in square meters (m²), is given by*

$$s = \sqrt{\frac{hw}{3600}},$$

where w is the patient's weight in kilograms (kg) and h is the patient's height in centimeters (cm). (Source: U.S. Oncology.) Use this information for Exercises 73 and 74. Round your answers to the nearest thousandth.

73. Assume that a patient's height is 170 cm. Find the patient's approximate surface area assuming that:
 a) The patient's weight is 70 kg.
 b) The patient's weight is 100 kg.
 c) The patient's weight is 50 kg.

74. Assume that a patient's weight is 70 kg. Approximate the patient's surface area assuming that:
 a) The patient's height is 150 cm.
 b) The patient's height is 180 cm.

75. Scaling stress factors. In psychology a process called *scaling* is used to attach numerical ratings to a group of life experiences. In the following table, various events have been rated on a scale from 1 to 100 according to their stress levels.

EVENT	SCALE OF IMPACT
Death of spouse	100
Divorce	73
Jail term	63
Marriage	50
Lost job	47
Pregnancy	40
Death of close friend	37
Loan over $10,000	31
Child leaving home	29
Change in schools	20
Loan less than $10,000	17
Christmas	12

(*Source:* Thomas H. Holmes, University of Washington School of Medicine.)

a) Does the table represent a function? Why or why not?

b) What are the inputs? What are the outputs?

SYNTHESIS

Solve for y in terms of x, and determine if the resulting equation represents a function.

76. $2x + y - 16 = 4 - 3y + 2x$

77. $2y^2 + 3x = 4x + 5$

78. $(4y^{2/3})^3 = 64x$

79. $(3y^{3/2})^2 = 72x$

80. Explain why the vertical-line test works.

81. Is 4 in the domain of *f* in Exercises 57–60? Explain why or why not.

TECHNOLOGY CONNECTION

In Exercises 82 and 83, use the TABLE *feature to construct a table for the function under the given conditions.*

82. $f(x) = x^3 + 2x^2 - 4x - 13$; TblStart $= -3$; ΔTbl $= 2$

83. $f(x) = \dfrac{3}{x^2 - 4}$; TblStart $= -3$; ΔTbl $= 1$

84. A function *f* is given by

$f(x) = |x - 2| + |x + 1| - 5$.

Find $f(-3), f(-2), f(0)$, and $f(4)$.

85. Graph the function in each of Exercises 82–84.

86. Use the TRACE feature to find several ordered-pair solutions of the function $f(x) = \sqrt{10 - x^2}$.

87. A function *f* takes a number *x*, adds 2, and then multiplies the result by 5, while a function *g* takes a number *x*, multiplies it by 5, and then adds 2.

a) Write *f* and *g* as equations.
b) Graph *f* and *g* on the same axes.
c) Are *f* and *g* the same function?

88. A function *h* takes a number *x*, subtracts 4, and then squares the result, while a function *k* takes a number *x*, squares it, and then subtracts 4.

a) Write *h* and *k* as equations.
b) Graph *h* and *k* on the same axes.
c) Are *h* and *k* the same function?

89. A function *f* takes a number *x*, multiplies it by 3, and then adds 6, while a function *g* takes a number *x*, adds *a* to it, and then multiplies the result by 3. Find *a* if *f* and *g* are the same function.

90. A function *h* takes a number *x*, adds 3, and then squares the result, while a function *k* takes a number *x*, squares it, adds 6 times *x*, and then adds *a* to the result. Find *a* if *h* and *k* are the same function.

Answers to Quick Checks

1. **(a)** The correspondence is a function since each book has one page count associated with it.   **(b)** The correspondence is not a function since more than one person will have a last name that starts with any particular letter. That is, each letter has more than one last name associated with it. **2.** $17, -10, 5, 3a + 5, 3a + 3h + 5$
3. $49, 58, -7, 3a^2 + 2a - 7, 75a^2 + 10a - 7$
4. $-8, 2x + 2h - x^2 - 2xh - h^2, 2 - 2x - h$
5. $f(x) = 2 - x^2$ **6.** **(a)** The graph is not that of a function.   **(b)** The graph is that of a function.

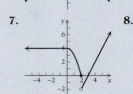

7.

8.

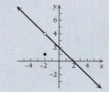

Finding Domain and Range

Set Notation

A **set** is a collection of objects. The set we consider most in calculus is the set of **real numbers**, $\mathbb{R}$. There is a real number for every point on the number line.

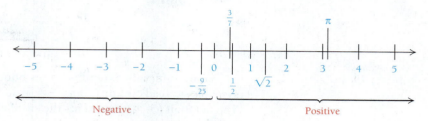

The set consisting of $-\frac{9}{25}$, 0, and $\sqrt{2}$ can be written $\left\{-\frac{9}{25}, 0, \sqrt{2}\right\}$. This method of describing sets is known as the **roster method**. It lists every member of the set. We describe larger sets using **set-builder notation**, which specifies conditions under which an object is in the set. For example, the set of all real numbers less than 4 can be described as follows in set-builder notation:

$$\{x \mid x \text{ is a real number less than } 4\} \text{ or } \{x \mid x < 4\}.$$

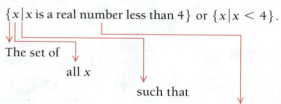

The set of

all x

such that

x is a real number less than 4.

Interval Notation

We can also describe certain sets using **interval notation**. If a and b are real numbers, with $a < b$, we define the interval (a, b) as the set of all numbers between but not including a and b, that is, the set of all x for which $a < x < b$. Thus,

$$(a, b) = \{x \mid a < x < b\}.$$

The points a and b are the **endpoints** of the interval. The parentheses indicate that the endpoints are *not* included in the interval.

The interval $[a, b]$ is defined as the set of all x for which $a \leq x \leq b$. Thus,

$$[a, b] = \{x \mid a \leq x \leq b\}.$$

The brackets indicate that the endpoints *are* included in the interval.*

Be careful not to confuse the *interval* (a, b) with the *ordered pair* (a, b) used to represent a point in the plane, as in Section R.1. The context in which the notation appears makes the meaning clear.

Intervals like $(-2, 3)$, in which neither endpoint is included, are called **open intervals**; intervals like $[-2, 3]$, which include both endpoints, are said to be **closed intervals**. Thus, $[a, b]$ is read "the closed interval a, b" and (a, b) is read "the open interval a, b."

Write interval notation for a set of points.

Find the domain and the range of a function.

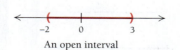

An open interval

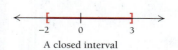

A closed interval

---

*Some books use the representations 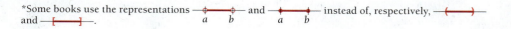 and instead of, respectively, and

Some intervals are **half-open** and include one endpoint but not the other:

$$(a, b] = \{x \mid a < x \leq b\}. \quad \text{The graph excludes } a \text{ and includes } b.$$

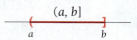

$$[a, b) = \{x \mid a \leq x < b\}. \quad \text{The graph includes } a \text{ and excludes } b.$$

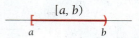

Some intervals extend without bound in one or both directions. We use the symbols ∞, read "infinity," and $-\infty$, read "negative infinity," to name these intervals. The notation $(5, \infty)$ represents the set of all real numbers greater than 5. That is,

$$(5, \infty) = \{x \mid x > 5\}.$$

Similarly, the notation $(-\infty, 5)$ represents the set of all real numbers less than 5. That is,

$$(-\infty, 5) = \{x \mid x < 5\}.$$

The notations $[5, \infty)$ and $(-\infty, 5]$ are used when we want to include an endpoint. The interval $(-\infty, \infty)$ names the set of all real numbers.

$$(-\infty, \infty) = \{x \mid x \text{ is a real number}\}$$

Interval notation is summarized in the following table. Note that the symbols ∞ and $-\infty$ always have a parenthesis next to them; neither of these represents a real number.

Intervals: Notation and Graphs

INTERVAL NOTATION	SET NOTATION	GRAPH
(a, b)	$\{x \mid a < x < b\}$	
$[a, b]$	$\{x \mid a \leq x \leq b\}$	
$[a, b)$	$\{x \mid a \leq x < b\}$	
$(a, b]$	$\{x \mid a < x \leq b\}$	
(a, ∞)	$\{x \mid x > a\}$	
$[a, \infty)$	$\{x \mid x \geq a\}$	
$(-\infty, b)$	$\{x \mid x < b\}$	
$(-\infty, b]$	$\{x \mid x \leq b\}$	
$(-\infty, \infty)$	$\{x \mid x \text{ is a real number}\}$	

EXAMPLE 1   Write interval notation for each set or graph:

a) $\{x | -4 < x < 5\}$

b) $\{x | x \geq -2\}$

c)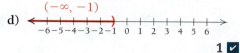

d)

Solution

a) $\{x | -4 < x < 5\} = (-4, 5)$

b) $\{x | x \geq -2\} = [-2, \infty)$

c) $(-2, 4]$

d) $(-\infty, -1)$

1 ✔

Quick Check 1 ✔

Write interval notation for each set, and show it on a graph.

a) $\{x | -2 \leq x \leq 5\}$

b) $\{x | -2 \leq x < 5\}$

c) $\{x | -2 < x \leq 5\}$

d) $\{x | -2 < x < 5\}$

Finding Domain and Range

Recall that when a set of ordered pairs is such that no two different pairs share a common first coordinate, we have a function. The **domain** is the set of all first coordinates, and the **range** is the set of all second coordinates.

EXAMPLE 2   For the function f shown to the right, determine the domain and the range.

Solution   This function consists of just four ordered pairs and can be written as

$$f = \{(-3, 1), (1, -2), (3, 0), (4, 5)\}.$$

We can determine the domain and the range by reading the x- and the y-values directly from the graph.

The domain is the set of all first coordinates, $\{-3, 1, 3, 4\}$. The range is the set of all second coordinates, $\{1, -2, 0, 5\}$.

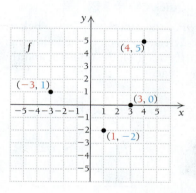

Quick Check 2 ✔

For the function shown below, determine the domain and the range.

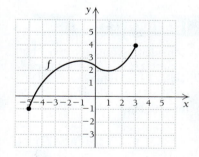

2 ✔

EXAMPLE 3   For the function f shown to the right, determine each of the following.

a) The number in the range that is paired with the input 1. That is, find $f(1)$.

b) The domain of f

c) The number(s) in the domain that is (are) paired with the output 1. That is, find all x-values for which $f(x) = 1$.

d) The range of f

Solution

a) To determine which number is paired with the input 1, we locate 1 on the horizontal axis. Next, we identify the point on the graph of f for which 1 is the first coordinate. From that point, we look to the vertical axis to find the corresponding y-coordinate, 2. The input 1 has the output 2– that is, $f(1) = 2$.

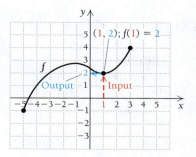

b) The domain of the function is the set of all x-values, or inputs, of the points on the graph. These extend from −5 to 3 and can be viewed as the curve's shadow, or *projection*, onto the x-axis. Thus, the domain is the set $\{x | -5 \le x \le 3\}$, or, in interval notation, $[-5, 3]$.

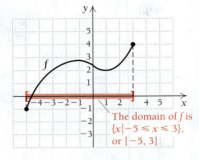

The domain of f is $\{x | -5 \le x \le 3\}$, or $[-5, 3]$

c) To determine which number(s) in the domain is (are) paired with the output 1, we locate 1 on the vertical axis. From there, we look left and right to the graph of f to identify any points for which 1 is the second coordinate. One such point exists: $(-4, 1)$. For this function, we note that $x = -4$ is the only member of the domain paired with the range value 1. For other functions, there might be more than one member of the domain paired with a member of the range.

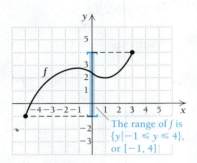

$(-4, 1); f(-4) = 1$

Quick Check 3 ✔

For the function f, shown below, determine each of the following: $f(-1), f(1)$, the domain, and the range.

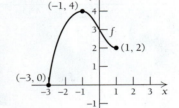

d) The range of the function is the set of all y-values, or outputs, of the points on the graph. These extend from −1 to 4 and can be viewed as the curve's shadow, or projection, onto the y-axis. Thus, the range is the set $\{y | -1 \le y \le 4\}$, or, in interval notation, $[-1, 4]$.

The range of f is $\{y | -1 \le y \le 4\}$, or $[-1, 4]$

3 ✔

TECHNOLOGY CONNECTION 〰

Consider the function given by $f(x) = 1/(x - 3)$. The table below was obtained with ΔTbl set at 0.25. Note that the calculator cannot calculate $f(3)$.

ΔTbl = 0.25

X	Y1
2	−1
2.25	−1.333
2.5	−2
2.75	−4
3	ERR:
3.25	4
3.5	2
3.75	1.3333

X = 2

EXERCISES

1. Make a table for $f(x) = 1/(x^2 - 4)$ from $x = -3$ to $x = 0$ and with ΔTbl set at 0.5.

2. Create a table for the function given in Example 5.

When a function is given by an equation or formula, the domain is understood to be the largest set of real numbers (inputs) for which function values (outputs) can be calculated. That is, the domain is the set of all allowable inputs into the formula. To find the domain, think, "For what input values does the function have an output?"

EXAMPLE 4   Find the domain: $f(x) = |x|$.

Solution   We ask, "What can we substitute?" Is there any number x for which we cannot calculate $|x|$? The answer is no. Thus, the domain of f is the set of all real numbers. ∎

EXAMPLE 5   Find the domain: $f(x) = \dfrac{3}{2x - 5}$.

Solution   We recall that a denominator cannot equal zero and ask, "What can we substitute?" Is there any number x for which we cannot calculate $3/(2x - 5)$? Since $3/(2x - 5)$ cannot be calculated when the denominator, $2x - 5$, is 0, we solve $2x - 5 = 0$ to find those real numbers that must be excluded from the domain of f:

$$2x - 5 = 0 \quad \text{Setting the denominator equal to 0}$$
$$2x = 5 \quad \text{Adding 5 to both sides}$$
$$x = \tfrac{5}{2}. \quad \text{Dividing both sides by 2}$$

Thus, $\frac{5}{2}$ is not in the domain, whereas all other real numbers are. We say that f is *not defined at* $\frac{5}{2}$, or $f\left(\frac{5}{2}\right)$ *does not exist*.

The domain of f is $\left\{x \mid x \text{ is a real number } and \, x \neq \frac{5}{2}\right\}$, or, in interval notation, $\left(-\infty, \frac{5}{2}\right) \cup \left(\frac{5}{2}, \infty\right)$. The symbol $\cup$ indicates the *union* of two sets and means that all elements in both sets are included in the domain. ■

EXAMPLE 6 Find the domain: $f(x) = \sqrt{4 + 3x}$.

Solution We ask, "Is there any number x for which we can't calculate $\sqrt{4 + 3x}$?" Since radicands in even roots cannot be negative, $\sqrt{4 + 3x}$ is not a real number when $4 + 3x$ is negative. The domain is all real numbers for which $4 + 3x \geq 0$. We find the domain by solving the inequality. (See Appendix A for a review of solving inequalities.)

$$4 + 3x \geq 0$$
$$3x \geq -4 \qquad \text{Adding } -4 \text{ to both sides}$$
$$x \geq -\tfrac{4}{3} \qquad \text{Dividing both sides by 3}$$

The domain is $\left\{x \mid -\frac{4}{3} \leq x < \infty\right\}$, or, in interval notation, $\left[-\frac{4}{3}, \infty\right)$. **4** ✔

Quick Check 4 ✔

Find the domain of each function. Express each answer using interval notation.

a) $f(x) = \dfrac{5}{x - 8}$

b) $f(x) = x^3 + |2x|$

c) $f(x) = \sqrt{2x - 8}$

TECHNOLOGY CONNECTION

Determining Domain and Range

Graph each function in the given viewing window. Then determine the domain and the range.

a) $f(x) = 3 - |x|$, $[-10, 10, -10, 10]$

b) $f(x) = x^3 - x$, $[-3, 3, -4, 4]$

c) $f(x) = \dfrac{3}{x - 2}$, $[-14, 14, -14, 14]$

d) $f(x) = x^4 - 2x^2 - 3$, $[-4, 4, -6, 6]$

We have the following.

a)
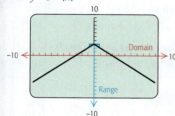
$y = 3 - |x|$

Domain $= \mathbb{R}$ (the real numbers), or $(-\infty, \infty)$
Range $= (-\infty, 3]$

b)
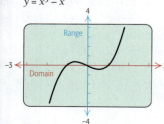
$y = x^3 - x$

Domain $= \mathbb{R}$
Range $= \mathbb{R}$

c)
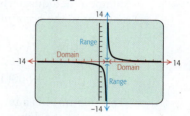
$y = \dfrac{3}{x - 2}$

Xscl = 2, Yscl = 2

The number 2 is excluded as an input.
Domain $= \{x \mid x \text{ is a real number } and \, x \neq 2\}$, or $(-\infty, 2) \cup (2, \infty)$;
Range $= \{y \mid y \text{ is a real number } and \, y \neq 0\}$, or $(-\infty, 0) \cup (0, \infty)$

d)
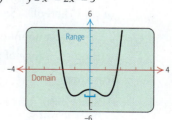
$y = x^4 - 2x^2 - 3$

Domain $= \mathbb{R}$
Range $= [-4, \infty)$

We can confirm our results using the TRACE feature, moving the cursor along the curve or entering any x-value in which we have interest. We can also use the TABLE feature. In Example (d), the domain might not appear to be the set of all real numbers because the graph seems "thin," but careful examination of the formula shows that we can indeed substitute any real number.

(continued)

Determining Domain and Range (*continued*)

EXERCISES

Graph each function in the given viewing window. Then determine the domain and the range.

1. $f(x) = |x| - 4$, $[-10, 10, -10, 10]$

2. $f(x) = 2 + 3x - x^3$, $[-5, 5, -5, 5]$

3. $f(x) = \dfrac{-3}{x + 1}$, $[-20, 20, -20, 20]$

4. $f(x) = x^4 - 2x^2 - 7$, $[-4, 4, -9, 9]$

5. $f(x) = \sqrt{x + 4}$, $[-8, 8, -8, 8]$

6. $f(x) = \sqrt{9 - x^2}$, $[-5, 5, -5, 5]$

7. $f(x) = -\sqrt{9 - x^2}$, $[-5, 5, -5, 5]$

8. $f(x) = x^3 - 5x^2 + x - 4$, $[-10, 10, -20, 10]$

Domains and Ranges in Applications

The domain and the range of a function given by a formula are sometimes affected by the context of an application. Let's look at an example similar to Exercises 71 and 72 in Section R.2.

> **EXAMPLE 7**   **Business: Compound Interest.**   Suppose $500 is invested at 2.5%, compounded quarterly for t years. From Theorem 2 in Section R.1, we know that the amount in the account is given by
>
> $$A(t) = 500\left(1 + \frac{0.025}{4}\right)^{4t}$$
>
> $$= 500(1.00625)^{4t}.$$

The amount A is a function of the number of years t for which the money is invested. Determine the domain of A.

Solution   We can substitute any real number for t into the formula, but a negative number of years is not meaningful. The context of the application excludes negative numbers. Thus, the domain is the set of all nonnegative numbers, $[0, \infty)$.   ∎

> **EXAMPLE 8**   **iPhone Data Plans.**   Recently, Sprint offered a data plan that allows a customer 450 min per month for $80 and charges $0.45 for each additional minute (or part thereof). (*Source*: www.sprint.com, November 2013.)

a) Find the amount of a customer's bill if 500 min are used.

b) Find the range of possible monthly charges if a customer uses up to 500 min.

Solution

a) Since 500 min is 50 min more than 450 min, the customer will be charged $80 + 0.45(50) = \$102.50$.

b) The range is the set of outputs, in this case, the possible monthly charges. The charge amounts rise in increments of $0.45; that is, the amounts can be $80.00, $80.45, $80.90, $81.35, and so on, up through $102.50. Thus, the range is the set $\{80.00, 80.45, 80.90, 81.35, \ldots, 102.50\}$. The ellipses ($\ldots$) indicate that the pattern continues up to and including a charge of $102.50.

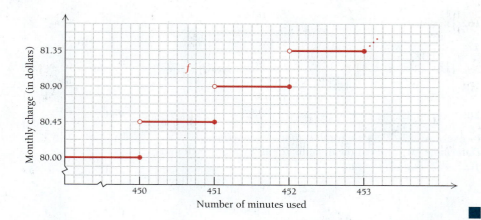

Section Summary

The following is a review of function terminology from Sections R.1–R.3.

Function Concepts

- Formula for f: $f(x) = x^2 - 7$
- For every input of f, there is exactly one output.
- For the input 1, the output is -6.
- $f(1) = -6$
- The point $(1, -6)$ is on the graph.
- Domain = the set of all inputs
  = the set of all real numbers, $\mathbb{R}$
- Range = the set of all outputs
  = $[-7, \infty)$

Graph

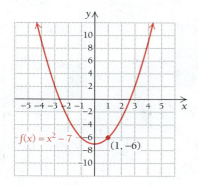

$f(x) = x^2 - 7$ $(1, -6)$

R.3 | Exercise Set

In Exercises 1–10, write interval notation for each graph.

1.
 number line from -5 to 5

2. number line from -5 to 5

3. number line from -1 to 6

4. number line from -1 to 6

5. number line from -10 to -3

6. number line from -10 to -3

7.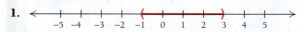
 x .... $x + h$

8. x .... $x + h$

9. q

10. p

Write interval notation for each of the following. Then graph the interval on a number line.

11. The set of all numbers x such that $-2 \leq x \leq 2$

12. The set of all numbers x such that $-5 < x < 5$

13. $\{x \mid 6 < x \leq 20\}$

14. $\{x \mid -4 \leq x < -1\}$

15. $\{x \mid x > -3\}$

16. $\{x | x \le -2\}$

17. $\{x | -2 < x \le 3\}$

18. $\{x | -10 \le x < 4\}$

19. $\{x | x \ge 12.5\}$

20. $\{x | x < 12.5\}$

In Exercises 21–32, each graph is that of a function. Determine (a) $f(1)$; (b) the domain; (c) all x-values such that $f(x) = 2$; and (d) the range.

21.
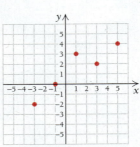

22.
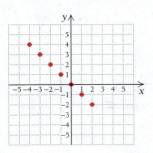

23.
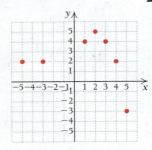

24.
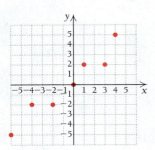

25.
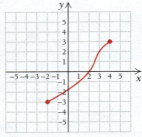

26.
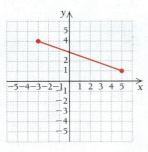

27.
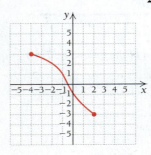

28.
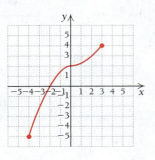

29.
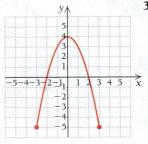

30.
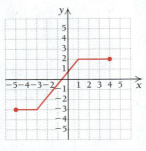

31.
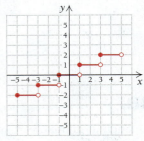

32.
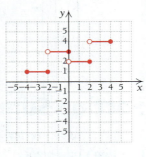

Find the domain of each function given below.

33. $f(x) = \dfrac{6}{2 - x}$

34. $f(x) = \dfrac{2}{x + 3}$

35. $f(x) = \sqrt{2x}$

36. $f(x) = \sqrt{x - 2}$

37. $f(x) = x^2 - 2x + 3$

38. $f(x) = x^2 + 3$

39. $f(x) = \dfrac{x - 2}{6x - 12}$

40. $f(x) = \dfrac{8}{3x - 6}$

41. $f(x) = |x - 4|$

42. $f(x) = |x| - 4$

43. $f(x) = \dfrac{3x - 1}{7 - 2x}$

44. $f(x) = \dfrac{2x - 1}{9 - 2x}$

45. $g(x) = \sqrt{4 + 5x}$

46. $g(x) = \sqrt{2 - 3x}$

47. $g(x) = x^2 - 2x + 1$

48. $g(x) = 4x^3 + 5x^2 - 2x$

49. $g(x) = \dfrac{2x}{x^2 - 25}$ (*Hint*: Factor the denominator.)

50. $g(x) = \dfrac{x - 1}{x^2 - 36}$ (*Hint*: Factor the denominator.)

51. $g(x) = |x| + 1$

52. $g(x) = |x + 7|$

53. $g(x) = \dfrac{2x - 6}{x^2 - 6x + 5}$

54. $g(x) = \dfrac{3x - 10}{x^2 - 4x - 5}$

55. For the function f shown to the right, find all x-values for which $f(x) \leq 0$.

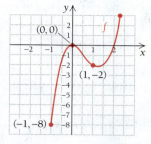

56. For the function g shown to the right, find all x-values for which $g(x) = 1$.

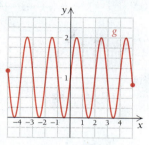

APPLICATIONS

Business and Economics

57. Compound interest. Suppose $5000 is invested at 3.1% interest, compounded semiannually, for t years.

   a) The amount A in the account is a function of time. Find an equation for this function.

   b) Determine the domain of A.

58. Compound interest. Suppose $3000 is borrowed as a college loan, at 5% interest, compounded daily, for t years.

   a) The amount A that is owed is a function of time. Find an equation for this function.

   b) Determine the domain of A.

Life and Physical Sciences

59. Incidence of breast cancer. The following graph (considered in Exercise 28 of Exercise Set R.1 without an equation) approximates the incidence of breast cancer I, per 100,000 women, as a function of age x. The equation for this graph is

$I(x) = -0.0000554x^4 + 0.0067x^3 - 0.0997x^2 - 0.84x - 0.25.$

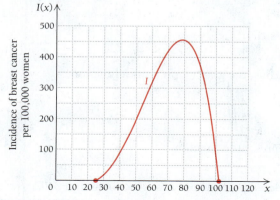

(*Source*: Based on data from the National Cancer Institute.)

a) Use the graph to determine the domain of I.

b) Use the graph to determine the range of I.

c) What 10-yr age interval sees the greatest increase in the incidence of breast cancer? Explain how you determined this.

60. Hearing-impaired Americans. The following graph (considered in Exercise Set R.1) approximates the number N, in millions, of hearing-impaired Americans who are x years old. The equation for this graph is

$N(x) = -0.000065x^3 + 0.0072x^2 - 0.133x + 2.062.$

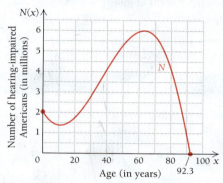

(*Source*: Better Hearing Institute.)

a) Use the graph to determine the domain of N.

b) Use the graph to determine the range of N.

c) If you were marketing a new type of hearing aid, at what age group (expressed as a 10-yr interval) would you target advertisements? Why?

61. Lung cancer. The following graph approximates the incidence of lung and bronchus cancer L, per 100,000 males, as a function of t, the number of years since 1940. The equation for this graph is

$L(t) = -0.00054t^3 + 0.02917t^2 + 1.2329t + 8.$

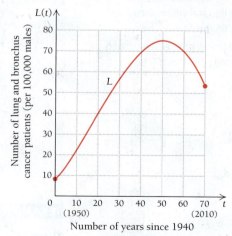

(*Source*: Based on data from the American Cancer Society Surveillance Research, 2005.)

a) Use the graph to estimate the domain of L.

b) Use the graph to estimate the range of L.

62. See Exercise 61.

a) Use the graph to approximate all the *x*-values (years since 1940) for which the cancer rate is 50 per 100,000.

b) Use the graph to approximate all the *x*-values (years since 1940) for which the cancer rate is 70 per 100,000.

c) Use the formula to approximate the lung and bronchus cancer rate in 2010.

SYNTHESIS

 63. Is it possible for the domain and the range of a function to be the same set? Why or why not?

 64. Is there an infinite number of functions for which 3 is not in the domain? Explain why or why not.

TECHNOLOGY CONNECTION

65. Determine the range of each of the functions in Exercises 33, 35, 39, 40, and 47.

66. Determine the range of each of the functions in Exercises 34, 36, 48, 51, and 54.

Answers to Quick Checks

1. (a) $[-2, 5]$, ⟵—[++++++]—⟶
    −2   5

(b) $[-2, 5)$, ⟵—[+++++++)—⟶
    −2   5

(c) $(-2, 5]$, ⟵—(++++++]—⟶
    −2   5

(d) $(-2, 5)$, ⟵—(++++++)—⟶
    −2   5

2. Domain is $\{-3, -1, 1, 2, 3\}$, and range is $\{-2, 1, 2\}$.
3. $f(-1) = 4, f(1) = 2$; domain is $[-3, 1]$, and range is $[0, 4]$ **4. (a)** $(-\infty, 8) \cup (8, \infty)$ **(b)** $(-\infty, \infty)$ **(c)** $[4, \infty)$

R.4

- Graph equations of the form $y = f(x) = c$ and $x = a$.
- Graph linear functions.
- Find an equation of a line when given the slope and one point on the line and when given two points on the line.
- Solve applied problems involving slope and linear functions.

Slope and Linear Functions

Horizontal and Vertical Lines

Let's consider graphs of equations $y = c$ and $x = a$, where c and a are real numbers.

EXAMPLE 1

a) Graph $y = 4$.

b) Decide whether the graph represents a function.

Solution

a) The graph consists of all ordered pairs whose second coordinate is 4. To see how a pair such as $(-2, 4)$ could be a solution of $y = 4$, we can write $y = 4$ as

$$y = 0x + 4.$$

Then $(-2, 4)$ is a solution because

$$0(-2) + 4 = 4$$

is true.

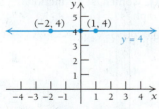

b) The vertical-line test holds. Thus, the graph represents a function.

EXAMPLE 2

a) Graph $x = -3$.

b) Decide whether the graph represents a function.

Solution

a) The graph consists of all ordered pairs whose first coordinate is -3. To see how a pair such as $(-3, 4)$ could be a solution of $x = -3$, we can write the equation as

$$x + 0y = -3.$$

Then $(-3, 4)$ is a solution because

$$(-3) + 0(4) = -3$$

is true.

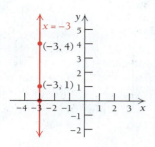

Quick Check 1 ✔

Graph each equation:

a) $x = 4$;

b) $y = -3$.

b) This graph does not represent a function because it fails the vertical-line test. A vertical line at $x = -3$ meets the graph more than once—in fact, infinitely many times. **1** ✔

In general, we have the following.

> **THEOREM 3**
>
> The graph of $y = c$, or $f(x) = c$, a horizontal line, is the graph of a function. Such a function is referred to as a **constant function**. The graph of $x = a$ is a vertical line, and $x = a$ is not a function.

TECHNOLOGY CONNECTION

Visualizing Slope

Exploratory: Squaring A Viewing Window

The standard $[-10, 10, -10, 10]$ viewing window shown below is not scaled identically on both axes. Note that the intervals on the y-axis are about two-thirds the length of those on the x-axis.

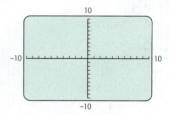

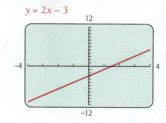

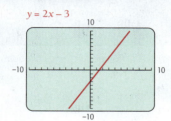

If we change the dimensions of the window to $[-6, 6, -4, 4]$, we get a graph for which the units are visually about the same on both axes.

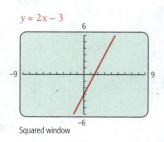

Squared window

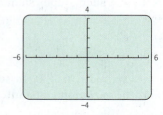

Creating such a window is called **squaring the window**. On many calculators, this is accomplished automatically by selecting the zsquare option of the zoom menu.

Each of the following is a graph of $y = 2x - 3$, but with different viewing windows. When the window is squared, as shown in the final graph, we get the typical representation of a *slope* of 2.

EXERCISES

Use a squared viewing window for each of these exercises.

1. Graph $y = x + 1$, $y = 2x + 1$, $y = 3x + 1$, and $y = 10x + 1$. What do you think the graph of $y = 247x + 1$ looks like?

2. Graph $y = x$, $y = \frac{7}{8}x$, $y = 0.47x$, and $y = \frac{2}{31}x$. What do you think the graph of $y = 0.000018x$ looks like?

3. Graph $y = -x$, $y = -2x$, $y = -5x$, and $y = -10x$. What do you think the graph of $y = -247x$ looks like?

4. Graph $y = -x - 1$, $y = -\frac{3}{4}x - 1$, $y = -0.38x - 1$, and $y = -\frac{5}{32}x - 1$. What do you think the graph of $y = -0.000043x - 1$ looks like?

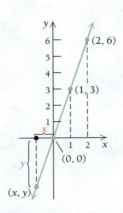

The Equation $y = mx$

Consider the following table of numbers and look for a pattern.

x	1	-1	$-\frac{1}{2}$	2	-2	3	-7	5
y	3	-3	$-\frac{3}{2}$	6	-6	9	-21	15

Note that the ratio of the y-value to the x-value is 3 to 1. That is,

$$\frac{y}{x} = 3, \quad \text{or} \quad y = 3x.$$

Ordered pairs from the table can be used to graph the equation $y = 3x$ (see the figure at the left). Note that this is a function.

> **THEOREM 4**
>
> The graph of the function given by
>
> $$y = mx \qquad \text{or} \qquad f(x) = mx$$
>
> is the straight line through the origin $(0, 0)$ and the point $(1, m)$. The constant m is called the **slope** of the line.

The rows of flowers form lines of equal slope.

Various graphs of $y = mx$ for $m > 0$ are shown on the left below. Such graphs slant up from left to right and rise faster when m is larger.

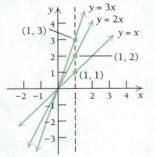

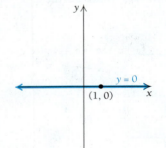

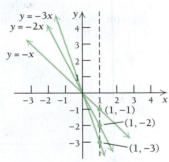

Quick Check 2 ✔

Graph each equation:

a) $y = \frac{1}{2}x$;

b) $y = -\frac{1}{2}x$.

When $m = 0$, we have $y = 0x$, or $y = 0$. In the middle above is a graph of $y = 0$. Note that this is both the x-axis and a horizontal line. Graphs of $y = mx$ for negative values of m are shown on the right above. Note that such graphs slant down from left to right.   **2** ✔

Direct Variation

Many applications involve equations like $y = mx$, with $m > 0$. In such situations, we say that we have **direct variation**, and m is the **variation constant**, or **constant of proportionality**. Generally, we assume $x \geq 0$.

> **DEFINITION**
>
> The variable y **varies directly** with x if there is some positive constant m such that $y = mx$. We also say that y is **directly proportional** to x.

EXAMPLE 3 Life Science: Weight on Earth and the Moon. The weight M, in pounds, of an object on the moon is directly proportional to the weight E of that object on Earth. An astronaut who weighs 180 lb on Earth will weigh 28.8 lb on the moon.

a) Find an equation of variation.

b) An astronaut weighs 120 lb on Earth. How much will the astronaut weigh on the moon?

Solution

a) The equation has the form $M = mE$. To find m, we substitute:

$$M = mE$$
$$28.8 = m \cdot 180$$
$$\frac{28.8}{180} = m$$
$$0.16 = m.$$

Thus, $M = 0.16E$ is the equation of variation.

b) To find the weight on the moon of an astronaut who weighs 120 lb on Earth, we substitute 120 for E in the equation of variation,

$$M = 0.16 \cdot 120 \qquad \text{Substituting 120 for } E$$
$$M = 19.2.$$

Thus, an astronaut who weighs 120 lb on Earth weighs 19.2 lb on the moon. **3** ✔

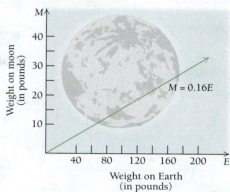

Quick Check 3 ✔

Measurements in inches and centimeters are directly proportional. For example, a length of 12 in. is the same as a length of 30.48 cm.
a) Find an equation of variation.
b) The height of a table is 90 cm. Find this measurement in inches.

The Equation $y = mx + b$

Compare the graphs of the equations

$$y = 3x \quad \text{and} \quad y = 3x - 2$$

(see the following figure). Note that the graph of $y = 3x - 2$ is a shift 2 units down of the graph of $y = 3x$, and that $y = 3x - 2$ has the y-intercept $(0, -2)$. Both graphs represent functions.

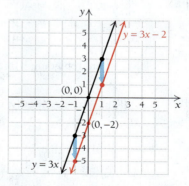

> **DEFINITION**
>
> A **linear function** is any function that can be written in the form
>
> $$y = mx + b \quad \text{or} \quad f(x) = mx + b.$$
>
> Its graph has the same slope, m, as the graph of $y = mx$ and crosses the y-axis at $(0, b)$. The point $(0, b)$ is called the **y-intercept**. (See the figure at the left.)

Two lines that have the same slope but different y-intercepts are **parallel**. The lines $y = mx$ and $y = mx + b$ are examples of parallel lines.

Quick Check 4 ✔

Find the slope and the y-intercept of the graph of $3x - 6y - 7 = 0$.

The Slope–Intercept Equation

Every nonvertical line is uniquely determined by its slope m and its y-intercept $(0, b)$. In other words, the slope describes the "slant" of the line, and the y-intercept locates the point at which the line crosses the y-axis. Thus, we have the following.

> **DEFINITION**
>
> $y = mx + b$ is called the **slope–intercept equation** of a line.

EXAMPLE 4   Find the slope and the y-intercept of the graph of $2x - 4y - 7 = 0$.

Solution   We solve for y:

$$2x - 4y - 7 = 0$$
$$4y = 2x - 7 \qquad \text{Adding 4y to both sides and reversing the equation}$$
$$y = \frac{2}{4}x - \frac{7}{4} \qquad \text{Dividing both sides by 4}$$

Slope: $\frac{1}{2}$ y-intercept: $\left(0, -\frac{7}{4}\right)$ **4** ✔

The Point–Slope Equation

Suppose we know the slope of a line and any point on the line. We can use these to find an equation of the line.

EXAMPLE 5   Find an equation of the line with slope 3 containing the point $(-1, -5)$.

Solution   The slope is given as $m = 3$. From the slope–intercept equation, we have

$$y = 3x + b. \tag{1}$$

To determine b, we substitute -5 for y and -1 for x:

$$-5 = 3(-1) + b$$
$$-5 = -3 + b,$$
so $$-2 = b$$

Then, replacing b in equation (1) with -2, we get $y = 3x - 2$.   ∎

More generally, if a point (x_1, y_1) is on the line given by

$$y = mx + b, \tag{2}$$

it follows that

$$y_1 = mx_1 + b. \tag{3}$$

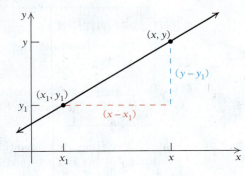

Subtracting the left and right sides of equation (3) from the left and right sides, respectively, of equation (2), we have

$$y - y_1 = (mx + b) - (mx_1 + b)$$
$$= mx + b - mx_1 - b \qquad \text{Simplifying}$$
$$= mx - mx_1 \qquad \text{Combining like terms}$$
$$= m(x - x_1). \qquad \text{Factoring}$$

DEFINITION

$y - y_1 = m(x - x_1)$ is called the **point–slope equation** of a line. The point is (x_1, y_1), and the slope is m.

This definition allows us to write an equation of a line given its slope and the coordinates of *any* point on the line.

EXAMPLE 6   Find an equation of the line with slope $\frac{2}{3}$ containing the point $(-1, -5)$.

Solution   Substituting in

$$y - y_1 = m(x - x_1),$$

we get

$$y - (-5) = \tfrac{2}{3}\big[x - (-1)\big]$$
$$y + 5 = \tfrac{2}{3}(x + 1)$$
$$y + 5 = \tfrac{2}{3}x + \tfrac{2}{3} \qquad \text{Using the distributive law}$$
$$y = \tfrac{2}{3}x + \tfrac{2}{3} - 5 \qquad \text{Subtracting 5 from both sides}$$
$$y = \tfrac{2}{3}x + \tfrac{2}{3} - \tfrac{15}{3}$$
$$y = \tfrac{2}{3}x - \tfrac{13}{3}. \qquad \text{Combining like terms} \qquad \text{5} \checkmark$$

Quick Check 5 ✔

Find the equation of the line with slope $-\frac{2}{3}$ containing the point $(-3, 6)$.

Which lines have the same slope?

Computing Slope

We now determine a method of computing the slope of a line when we know the coordinates of two of its points. Suppose (x_1, y_1) and (x_2, y_2) are the coordinates of two different points, P_1 and P_2, respectively, on a line that is not vertical. Consider a right triangle with legs parallel to the axes, as shown in the following figure.

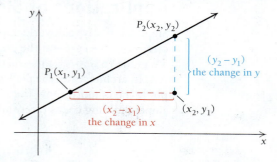

Note that the change in y is $y_2 - y_1$ and the change in x is $x_2 - x_1$. The ratio of these changes is the slope. To see this, consider the point–slope equation,

$$y - y_1 = m(x - x_1).$$

Since (x_2, y_2) is on the line, it must follow that

$$y_2 - y_1 = m(x_2 - x_1). \qquad \text{Substituting}$$

Since the line is not vertical, the two x-coordinates differ; so $x_2 - x_1$ is nonzero, and we can divide the preceding equation by it to get the following theorem.

> **THEOREM 5**
>
> The **slope** of a line containing points (x_1, y_1) and (x_2, y_2) is
> $$m = \frac{y_2 - y_1}{x_2 - x_1} = \frac{\text{change in } y}{\text{change in } x}.$$

EXAMPLE 7 Find the slope of the line containing the points $(-2, 6)$ and $(-4, 9)$.

Solution We have

$$m = \frac{y_2 - y_1}{x_2 - x_1} = \frac{6 - 9}{-2 - (-4)} \qquad \text{Regarding } (-2, 6) \text{ as } P_2 \text{ and } (-4, 9) \text{ as } P_1$$

$$= \frac{-3}{2} = -\frac{3}{2}.$$

It does not matter which point is taken first, as long as we subtract the coordinates in the same order. Thus, we can also find m as follows:

$$m = \frac{9 - 6}{-4 - (-2)} = \frac{3}{-2} = -\frac{3}{2}. \qquad \text{Here, } (-4, 9) \text{ serves as } P_2, \\ \text{and } (-2, 6) \text{ serves as } P_1.$$

6 ✔

Quick Check 6 ✔

Find the slope of the line containing the points $(2, 3)$ and $(1, -4)$.

If a line is horizontal, the change in y for any two points is 0. Thus, a horizontal line has slope 0. If a line is vertical, the change in x for any two points is 0. In this case, the slope is *not defined* because we cannot divide by 0. A vertical line has undefined slope. Thus, "0 slope" and "undefined slope" are two very different concepts.

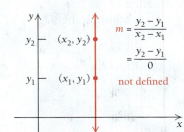

Applications of Slope

Slope has many real-world applications. For example, numbers like 2%, 3%, and 6% are often used to represent the *grade* of a road, a measure of how steep a road on a hill is. A 3% grade ($3\% = \frac{3}{100}$) means that for every horizontal distance of 100 ft, the road rises 3 ft. In architecture, the *pitch* of a roof is a measure of how steeply it is angled—a steep pitch sheds more snow than a shallow pitch. Wheelchair-ramp design also involves slope: Building codes rarely allow the steepness of a wheelchair ramp to exceed $\frac{1}{12}$.

Road grade $= \frac{a}{b}$
(expressed as a percent)

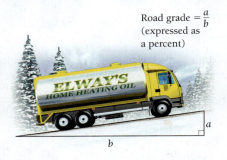

Ski trail difficulty ratings, or *gradients*, are yet another application of slope. The following table presents examples.

Ski Trail Difficulty Ratings in North America

TRAIL RATING	SYMBOL	LEVEL OF DIFFICULTY	DESCRIPTION
Green Circle	🟢	Easiest	A Green Circle trail is the easiest. These trails typically have slope gradients ranging from 6% to 25% (a 100% slope is a 45° angle).
Blue Square	🟦	Intermediate	A Blue Square trail is of intermediate difficulty. These trails have gradients ranging from 25% to 40%.
Black Diamond	◆	Difficult	Black Diamond trails tend to be steep—typically 40% and up.

To estimate a gradient, hold your arm parallel to the ground out from your side—that is a 0% gradient. Hold it at a 45° angle—that is a 100% gradient. A 22.5° angle is a 41% gradient. And, surprisingly, an angle of only 3.5° constitutes a 6% gradient. What do you think is the slope of the steep road near the top of the mountain in the photo?*

Slope can also be considered as an **average rate of change**.

Frenchman Mountain near Las Vegas, Nevada.

EXAMPLE 8 **Life Science: Amount Spent on Cancer Research.** The amount spent on cancer research has increased steadily over the years and is approximated in the following graph. Find the average rate of change of the amount spent on the research.

CANCER RESEARCH

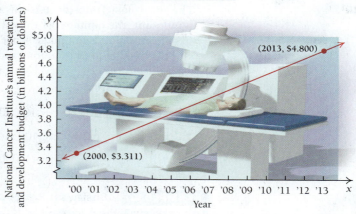

(*Source*: National Cancer Institute.)

Solution First, we determine the coordinates of two points on the graph. Here, they are given as (2000, $3.311) and (2013, $4.800). Then we compute the slope, or rate of change, as follows:

$$\text{Slope} = \text{average rate of change} = \frac{\text{change in } y}{\text{change in } x}$$

$$= \frac{\$4.800 - \$3.311}{2013 - 2000} = \frac{1.489}{13} \approx \$0.1145 \text{ billion/yr.} \quad \blacksquare$$

---

*The road's slope is about 0.35, or a 35% gradient, or an angle of about 20.3°.

Applications of Linear Functions

Many applications are modeled by linear functions.

EXAMPLE 9   Business: Total Cost.   Raggs, Ltd., a clothing firm, has **fixed costs** of $10,000 per year. These costs, such as rent, maintenance, and so on, must be paid no matter how much the company produces. To produce x units of a certain kind of suit, it costs $20 per suit (unit) in addition to the fixed costs. That is, the **variable costs** for producing x of these suits are $20x$ dollars. These costs are due to the amount produced and cover material, wages, fuel, and so on. The **total cost** $C(x)$ of producing x suits in a year is given by

$$C(x) = (\text{Variable costs}) + (\text{Fixed costs}) = 20x + 10{,}000.$$

a) Graph the variable-cost, fixed-cost, and total-cost functions.

b) What is the total cost of producing 100 suits? 400 suits?

Solution

a) The variable-cost and fixed-cost functions appear in the graph on the left below.

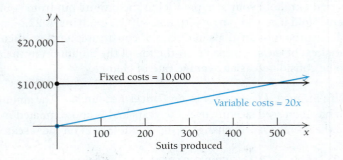

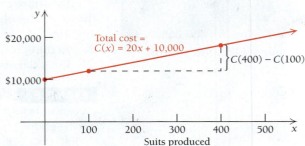

The total-cost function is shown in the graph on the right. From a practical standpoint, the domains of these functions are nonnegative integers 0, 1, 2, 3, and so on, since it does not make sense to make either a negative number or a fractional number of suits. It is common practice to draw such graphs as though the domains were the entire set of nonnegative real numbers.

b) The total cost of producing 100 suits is

$$C(100) = 20 \cdot 100 + 10{,}000 = \$12{,}000.$$

The total cost of producing 400 suits is

$$C(400) = 20 \cdot 400 + 10{,}000 = \$18{,}000.$$

EXAMPLE 10   Business: Profit-and-Loss Analysis.   When a business sells an item, it receives the *price* paid by the consumer (this is normally greater than the *cost* to the business of producing the item).

a) The **total revenue** that a business receives is the product of the number of items sold and the price paid per item. Thus, if Raggs, Ltd., sells x suits at $80 per suit, the total revenue $R(x)$, in dollars, is given by

$$R(x) = \text{Unit price} \cdot \text{Quantity sold} = 80x.$$

If $C(x) = 20x + 10{,}000$ (see Example 9), graph R and C using the same set of axes.

b) The **total profit** that a business makes is the amount left after all costs have been subtracted from the total revenue. Thus, if $P(x)$ represents the total profit when x items are produced and sold, we have

$$P(x) = (\text{Total revenue}) - (\text{Total costs}) = R(x) - C(x).$$

Determine $P(x)$, and draw its graph using the same set of axes used for the graph in part (a).

c) The company will *break even* at that value of x for which $P(x) = 0$ (that is, no profit and no loss). This is the point at which $R(x) = C(x)$. Find the **break-even value** of x.

Solution

a) The graphs of $R(x) = 80x$ and $C(x) = 20x + 10,000$ are shown below. When $C(x)$ is above $R(x)$, a loss will occur. This is shown by the region shaded red. When $R(x)$ is above $C(x)$, a gain will occur. This is shown by the region shaded gray.

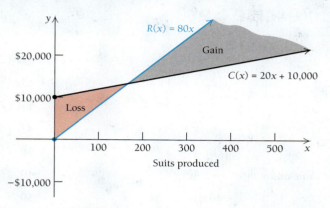

b) To find P, the profit function, we have

$$P(x) = R(x) - C(x) = 80x - (20x + 10,000)$$
$$= 60x - 10,000.$$

The graph of $P(x)$ is shown by the heavy line. The red portion of the line shows a "negative" profit, or loss. The black portion of the heavy line shows a "positive" profit, or gain.

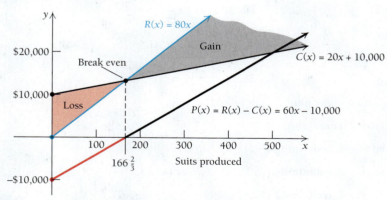

c) To find the break-even value, we solve $R(x) = C(x)$:

$$R(x) = C(x)$$
$$80x = 20x + 10,000$$
$$60x = 10,000$$
$$x = 166\tfrac{2}{3}.$$

How do we interpret the fractional answer, since it is not possible to produce $\tfrac{2}{3}$ of a suit? We simply round up to 167. Estimates of break-even values are usually sufficient since companies want to operate well away from those values in order to maximize profit.

7 ✔

Quick Check 7 ✔

Business. Suppose that in Examples 9 and 10 fixed costs are increased to $20,000. Find:

a) the total-cost, total-revenue, and total-profit functions;

b) the break-even value.

Section Summary

- Graphs of functions that are straight lines (*linear functions*) are characterized by an equation of the type $f(x) = mx + b$, where m is the slope and $(0, b)$ is the *y-intercept*, the point at which the graph crosses the *y*-axis.

- The *point–slope equation* of a line is $y - y_1 = m(x - x_1)$, where (x_1, y_1) is a point on the line and m is the slope.

R.4 Exercise Set

Graph.

1. $x = 5$ **2.** $x = 3$

3. $y = -4$ **4.** $y = -2$

5. $x = -1.5$ **6.** $x = -4.5$

7. $y = 2.25$ **8.** $y = 3.75$

Graph. List the slope and y-intercept.

9. $y = -2x$ **10.** $y = -3x$

11. $f(x) = -0.5x$ **12.** $f(x) = 0.5x$

13. $y = 3x - 4$ **14.** $y = 2x - 5$

15. $g(x) = x - 2.5$ **16.** $g(x) = -x + 3$

17. $y = 7$ **18.** $y = -5$

Find the slope and y-intercept.

19. $y - 4x = 1$ **20.** $y - 3x = 6$

21. $2x + y - 3 = 0$ **22.** $2x - y + 3 = 0$

23. $3x - 3y + 6 = 0$ **24.** $2x + 2y + 8 = 0$

25. $x = 3y + 7$ **26.** $x = -4y + 3$

Find an equation of the line:

27. with $m = 7$, containing $(1, 7)$.

28. with $m = -5$, containing $(-2, -3)$.

29. with $m = -2$, containing $(2, 3)$.

30. with $m = -3$, containing $(5, -2)$.

31. with slope -5, containing $(5, 0)$.

32. with slope 2, containing $(3, 0)$.

33. with y-intercept $(0, -6)$ and slope $\frac{1}{2}$.

34. with y-intercept $(0, 7)$ and slope $\frac{4}{3}$.

35. with slope 0, containing $(4, 8)$.

36. with slope 0, containing $(2, 3)$.

Find the slope of the line containing the given pair of points. If a slope is undefined, state that fact.

37. $(5, -3)$ and $(-2, 1)$ **38.** $(-2, 1)$ and $(6, 3)$

39. $(-3, -5)$ and $(1, -6)$ **40.** $(2, -3)$ and $(-1, -4)$

41. $(3, -7)$ and $(3, -9)$ **42.** $(-4, 2)$ and $(-4, 10)$

43. $\left(-\frac{3}{16}, -\frac{1}{2}\right)$ and $\left(\frac{5}{8}, -\frac{3}{4}\right)$ **44.** $\left(\frac{4}{5}, -3\right)$ and $\left(\frac{1}{2}, \frac{2}{5}\right)$

45. $(2, 3)$ and $(-1, 3)$ **46.** $\left(-6, \frac{1}{2}\right)$ and $\left(-7, \frac{1}{2}\right)$

47. $(x, 4x)$ and $(x + h, 4(x + h))$

48. $(x, 3x)$ and $(x + h, 3(x + h))$

49. $(x, 2x + 3)$ and $(x + h, 2(x + h) + 3)$

50. $(x, 3x - 1)$ and $(x + h, 3(x + h) - 1)$

51–60. *Find an equation of the line containing the pair of points in each of Exercises 37–46.*

61. Find the slope (or grade) of the treadmill.

62. Find the slope of the skateboard ramp.

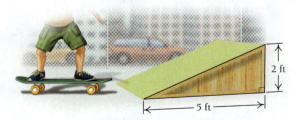

63. Find the slope (or head) of the river. Express the answer as a percentage.

APPLICATIONS

Business and Economics

64. Highway tolls. It has been suggested that since heavier vehicles are responsible for more wear and tear on highways, drivers should pay tolls in direct proportion to the weight of their vehicles. Suppose a Toyota Camry weighing 3350 lb was charged $2.70 for traveling an 80-mile stretch of highway.

a) Find an equation of variation that expresses the amount of the toll T as a function of the vehicle's weight w.

b) What would the toll be if a 3700-lb Jeep Cherokee drove the same stretch of highway?

65. Inkjet cartridges. A registrar's office finds that the number of inkjet cartridges, I, required each year for its copiers and printers varies directly with the number of students enrolled, s.

a) Find an equation of variation that expresses I as a function of s, if the office requires 16 cartridges when 2800 students enroll.

b) How many cartridges would be required if 3100 students enrolled?

66. Profit-and-loss analysis. Boxowitz, Inc., a computer firm, is planning to sell a new graphing calculator. For the first year, the fixed costs for setting up the new production line are $100,000. The variable costs for each calculator are $20. The sales department projects that 150,000 calculators will be sold during the first year at a price of $45 each.

a) Find and graph $C(x)$, the total cost of producing x calculators.

b) Using the same axes as in part (a), find and graph $R(x)$, the total revenue from the sale of x calculators.

c) Using the same axes as in part (a), find and graph $P(x)$, the total profit from the production and sale of x calculators.

d) What profit or loss will the firm realize if the expected sale of 150,000 calculators occurs?

e) How many calculators must the firm sell in order to break even?

67. Profit-and-loss analysis. Red Tide is planning a new line of skis. For the first year, the fixed costs for setting up production are $45,000. The variable costs for producing each pair of skis are estimated at $80, and the selling price will be $450 per pair. It is projected that 3000 pairs will sell the first year.

a) Find and graph $C(x)$, the total cost of producing x pairs of skis.

b) Find and graph $R(x)$, the total revenue from the sale of x pairs of skis. Use the same axes as in part (a).

c) Using the same axes as in part (a), find and graph $P(x)$, the total profit from the production and sale of x pairs of skis.

d) What profit or loss will the company realize if the expected sale of 3000 pairs occurs?

e) How many pairs must the company sell in order to break even?

68. Profit-and-loss analysis. Jamal decides to mow lawns to earn money. The initial cost of his lawnmower is $250. Gasoline and maintenance costs are $4 per lawn.

a) Formulate a function $C(x)$ for the total cost of mowing x lawns.

b) Jamal determines that the total-profit function for the lawn-mowing business is given by $P(x) = 9x - 250$. Find a function for the total revenue from mowing x lawns. How much does Jamal charge per lawn?

c) How many lawns must Jamal mow before he begins making a profit?

69. Straight-line depreciation. Quick Copy buys an office machine for $5200 on January 1 of a given year. The machine is expected to last for 8 yr, at the end of which time its *salvage value* will be $1100. If the company figures the decline in value to be the same each year, then the *straight-line depreciation value*, $V(t)$, after t years, $0 \le t \le 8$, is given by

$$V(t) = C - t\left(\frac{C - S}{N}\right),$$

where C is the original cost of the item, N is the number of years of expected life, and S is the salvage value. $V(t)$ is also called the *book value*.

a) Find the linear function for the straight-line depreciation of the machine.

b) Find the book value of the machine after 0 yr, 1 yr, 2 yr, 3 yr, 4 yr, 7 yr, and 8 yr.

70. Straight-line depreciation. (See Exercise 69.) Hanna's Photography spends $40 per square foot on improvements to a 25,000-ft^2 office space. Under IRS guidelines for straight-line depreciation, these improvements will depreciate completely—that is, have zero salvage value—after 39 yr. Find the depreciated value of the improvements after 10 yr.

71. Straight-line depreciation. The Video Game Wizard buys a new computer system for $60,000 and projects that its book value will be $2000 after 5 yr. Using straight-line depreciation, find the book value after 3 yr.

72. Straight-line depreciation. Tyline Electric uses the function $B(t) = -700t + 3500$ to find the book value, $B(t)$, in dollars, of a photocopier t years after its purchase.

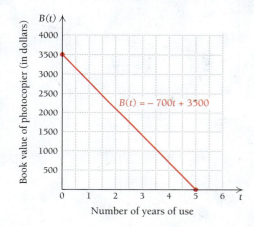

a) What do the numbers −700 and 3500 signify?
b) How long will it take the copier to depreciate completely?
c) What is the domain of *B*? Explain.

General Interest

73. Stair requirements.   A North Carolina state law requires that stairs have minimum treads of 9 in. and maximum risers of 8.25 in. (*Source:* North Carolina Office of the State Fire Marshal.) According to this law, what is the maximum grade of stairs in North Carolina?

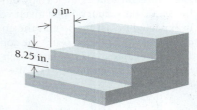

9 in.
8.25 in.

74. Health insurance premiums.   Find the average rate of change in the annual premium for a family's health insurance.

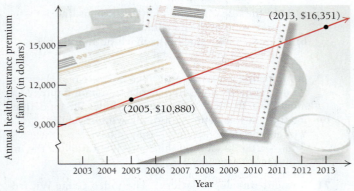

(2013, $16,351)
(2005, $10,880)

(*Source*: The Kaiser Family Foundation; Health Research and Education Trust.)

75. Health insurance premiums.   Find the average rate of change in the annual premium for a single person.

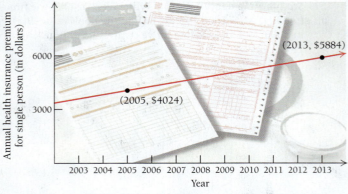

(2013, $5884)
(2005, $4024)

(*Source*: The Kaiser Family Foundation; Health Research and Education Trust.)

76. Two-year college tuitions.   Find the average rate of change of the tuition and fees at public two-year colleges.

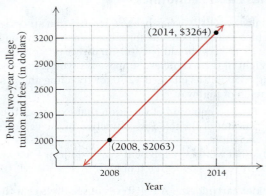

(2014, $3264)
(2008, $2063)

(*Source*: U.S. National Center for Education Statistics, *Digest of Education Statistics*, annual.)

77. Organic food sales.   Find the average rate of change of organic food sales in the United States.

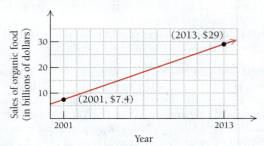

(2013, $29)
(2001, $7.4)

(*Source: Big Green Opportunity Report*, 2013.)

78. Energy conservation.   The R-factor of home insulation is directly proportional to its thickness *T*.
a) Find an equation of variation if $R = 12.51$ when $T = 3$ in.
b) What is the R-factor for insulation that is 6 in. thick?

79. Nerve impulse speed.   Impulses in nerve fibers travel at a speed of 293 ft/sec. The distance *D*, in feet, traveled in *t* sec is given by $D = 293t$. How long would it take an impulse to travel from the brain to the toes of a person who is 6 ft tall?

80. Muscle weight.   The weight *M* of a person's muscles is directly proportional to the person's body weight *W*.

Muscle weight is directly proportional to body weight.

a) It is known that a person weighing 200 lb has 80 lb of muscles. Find an equation of variation expressing M as a function of W.
b) Express the variation constant as a percentage, and interpret the resulting equation.
c) What is the muscle weight of a person weighing 120 lb?

81. Brain weight. The weight B of a person's brain is directly proportional to the person's body weight W.

  a) It is known that a person weighing 120 lb has a brain that weighs 3 lb. Find an equation of variation expressing B as a function of W.
  b) Express the variation constant as a percentage, and interpret the resulting equation.
  c) What is the weight of the brain of a person weighing 160 lb?

82. Stopping distance on glare ice. The stopping distance (at some fixed speed) of regular tires on glare ice is a linear function of the air temperature F,

$$D(F) = 2F + 115,$$

where $D(F)$ is the stopping distance, in feet, when the air temperature is F, in degrees Fahrenheit.

  a) Find $D(0°)$, $D(-20°)$, $D(10°)$, and $D(32°)$.
  b) Explain why the domain should be restricted to the interval $[-57.5°, 32°]$.

83. Reaction time. While driving a car, you see a child suddenly crossing the street. Your brain registers the emergency and sends a signal to your foot to hit the brake. The car travels a reaction distance D, in feet, during this time, where D is a function of the speed r, in miles per hour, that the car is traveling when you see the child. That reaction distance is a linear function given by

$$D(r) = \frac{11r + 5}{10}.$$

  a) Find $D(5)$, $D(10)$, $D(20)$, $D(50)$, and $D(65)$.
  b) Graph $D(r)$.
  c) What is the domain of the function? Explain.

84. Estimating heights. An anthropologist can use certain linear functions to estimate the height of a male or female, given the length of certain bones. The *humerus* is the bone from the elbow to the shoulder. Let $x =$ the length of the humerus, in centimeters. Then the height, in centimeters, of a male with a humerus of length x is given by

$$M(x) = 2.89x + 70.64.$$

The height, in centimeters, of a female with a humerus of length x is given by

$$F(x) = 2.75x + 71.48.$$

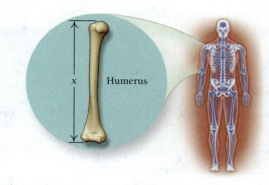

A 26-cm humerus was uncovered in some ruins.

  a) If we assume it was from a male, how tall was he?
  b) If we assume it was from a female, how tall was she?

85. Percentage of young adults using social networking sites. In 2006, the percentage of 18- to 29-year-olds who used social networking sites was 49%. In 2013, that percentage had risen to 89%. (*Source:* pewresearch.org.)

  a) Use the year as the x-coordinate and the percentage as the y-coordinate. Find the equation of the line that contains the data points.
  b) Use the equation in part (a) to estimate the percentage of young adults using social networking sites in 2014.
  c) Use the equation in part (a) to estimate the year in which the percentage of young adults using social networking sites will reach 100%.
  d) Explain why a linear equation cannot be used for years after the year found in part (c).

86. Manatee population. In January 2005, 3143 manatees were counted in an aerial survey of Florida. In January 2011, 4834 manatees were counted. (*Source*: Florida Fish and Wildlife Conservation Commission.)

  a) Using the year as the x-coordinate and the number of manatees as the y-coordinate, find an equation of the line that contains the two data points.
  b) Use the equation in part (a) to estimate the number of manatees counted in January 2010.
  c) The actual number counted in January 2010 was 5067. Does the equation found in part (a) give an accurate representation of the number of manatees counted each year? Why or why not?

87. Urban population. The population of Woodland is P. After growing 2%, the new population is N.

  a) Assuming that N is directly proportional to P, find an equation of variation.
  b) Find N when $P = 200{,}000$.
  c) Find P when $N = 367{,}200$.

88. Median age of women at first marriage. In general, people in our society are marrying at a later age. The median age, $A(t)$, of women at first marriage can be approximated by

$$A(t) = 0.08t + 19.7,$$

where t is the number of years after 1950. Thus, $A(0)$ is the median age of women at first marriage in 1950, $A(50)$ is the median age in 2000, and so on.

a) Find $A(0)$, $A(1)$, $A(10)$, $A(30)$, and $A(50)$.

b) What was the median age of women at first marriage in 2008?

c) Graph $A(t)$.

SYNTHESIS

89. Suppose $(2, 5)$, $(4, 13)$, and $(7, y)$ all lie on the same line. Find y.

90. Describe one situation in which you would use the slope–intercept equation rather than the point–slope equation.

91. Business: daily sales.   Match each sentence below with the most appropriate of the following graphs (I, II, III, or IV).

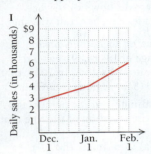

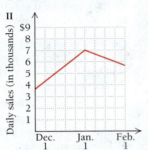

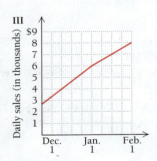

a) After January 1, daily sales continued to rise, but at a slower rate.

b) After January 1, sales decreased faster than they ever grew.

c) The rate of growth in daily sales doubled after January 1.

d) After January 1, daily sales decreased at half the rate that they grew in December.

92. Business: depreciation.   A large crane is being depreciated according to the model $V(t) = 900 - 60t$, where $V(t)$ is in thousands of dollars and t is the number of years since 2005. If the crane is to be depreciated until its value is $0, what is the domain of this model?

93. Graph some of the total-revenue, total-cost, and total-profit functions in this exercise set using the same set of axes. Identify regions of profit and loss.

Answers to Quick Checks

1. (a) 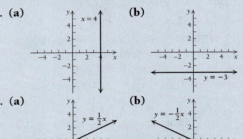 **(b)**

2. (a) 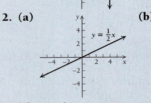 **(b)**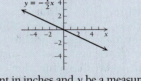

3. (a) Let x be a measurement in inches and y be a measurement in centimeters. Then $y = 2.54x$ or $x = 0.39y$. **(b)** 35.1 in.

4. $m = \frac{1}{2}$, y-intercept: $\left(0, -\frac{7}{6}\right)$ **5.** $y = -\frac{2}{3}x + 4$ **6.** $m = 7$

7. (a) $C(x) = 20x + 20{,}000$; $R(x) = 80x$; $P(x) = R(x) - C(x) = 60x - 20{,}000$; **(b)** 333 suits

---

R.5

Nonlinear Functions and Models

Many functions have graphs that are not lines. In this section, we study some of these **nonlinear functions** that we will use throughout this course.

Quadratic Functions

- Graph nonlinear functions and solve applied problems.
- Manipulate radical expressions and rational exponents.
- Determine the domain of a rational function and graph certain rational functions.
- Find the equilibrium point given a supply function and a demand function.

> **DEFINITION**
>
> A **quadratic function** f is given by
> $$f(x) = ax^2 + bx + c, \quad \text{where } a \neq 0.$$

We have already encountered some quadratic functions—for example, $f(x) = x^2$ and $g(x) = x^2 - 1$. We can create hand-drawn graphs of quadratic functions using the following information.

> The graph of a quadratic function $f(x) = ax^2 + bx + c$ is called a **parabola**.
>
> **a)** It is always a cup-shaped curve, like those in Examples 1 and 2 that follow.
>
> **b)** It opens upward if $a > 0$ and opens downward if $a < 0$.
>
> **c)** It has a turning point, or **vertex**, whose first coordinate is $x = -\dfrac{b}{2a}$.
>
> **d)** The vertical line $x = -b/(2a)$ is the line of symmetry, although it is not part of the graph.

EXAMPLE 1   Graph: $f(x) = x^2 - 2x - 3$.

Solution   Note that for $f(x) = 1x^2 - 2x - 3$, we have $a = 1, b = -2$, and $c = -3$. Since $a > 0$, the graph opens upward. The x-coordinate of the vertex is

$$x = -\frac{b}{2a}$$

$$= -\frac{-2}{2(1)} = 1.$$

Substituting 1 for x, we find the second coordinate of the vertex, $f(1)$:

$$f(1) = 1^2 - 2(1) - 3$$
$$= 1 - 2 - 3$$
$$= -4.$$

The vertex is $(1, -4)$, and the line $x = 1$ is the line of symmetry of the graph. We choose some x-values on each side of the vertex, compute y-values, plot the points, and graph the parabola.

x	$f(x)$	
1	−4	←Vertex
0	−3	←y-intercept
2	−3	
3	0	
4	5	
−1	0	
−2	5	

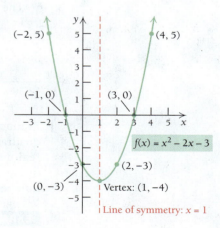

EXAMPLE 2   Graph: $f(x) = -2x^2 + 10x - 7$.

Solution   Note that $a = -2$, and since $a < 0$, the graph will open downward. The x-coordinate of the vertex is

$$x = -\frac{b}{2a}$$

$$= -\frac{10}{2(-2)} = \frac{5}{2}.$$

EXERCISES

Using the procedure of Examples 1 and 2, graph each of the following by hand, using the TABLE feature to create an input–output table for each function. Then press GRAPH to check your sketch.

1. $f(x) = x^2 - 6x + 4$

2. $f(x) = -2x^2 + 4x + 1$

Quick Check 1 ✔

Graph each function:

a) $f(x) = x^2 + 2x - 3$;

b) $f(x) = -2x^2 - 10x - 5$.

Substituting $\frac{5}{2}$ for x in the equation, we find the second coordinate of the vertex:

$$y = f\left(\tfrac{5}{2}\right) = -2\left(\tfrac{5}{2}\right)^2 + 10\left(\tfrac{5}{2}\right) - 7$$
$$= -2\left(\tfrac{25}{4}\right) + 25 - 7$$
$$= \tfrac{11}{2}.$$

The vertex is $\left(\tfrac{5}{2}, \tfrac{11}{2}\right)$, and the line of symmetry is $x = \tfrac{5}{2}$. We choose some x-values on each side of the vertex, compute y-values, plot the points, and graph the parabola:

x	$f(x)$
$\frac{5}{2}$	$\frac{11}{2}$	←Vertex
0	-7
1	1
2	5
3	5
4	1
5	-7

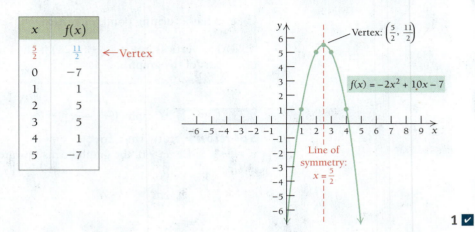

First coordinates of points at which a quadratic function intersects the x-axis (x-intercepts), if they exist, can be found by solving the quadratic equation, $ax^2 + bx + c = 0$. If real-number solutions exist, they can be found using the *quadratic formula*. See Appendix A at the end of the book for additional review of this important result.

> **THEOREM 6**   **The Quadratic Formula**
>
> The solutions of any quadratic equation $ax^2 + bx + c = 0$, $a \neq 0$, are given by
>
> $$x = \frac{-b \pm \sqrt{b^2 - 4ac}}{2a}.$$

To solve a quadratic equation, first try to factor and use the Principle of Zero Products (see Appendix A). When factoring is not possible or seems difficult, use the quadratic formula. It will always give the solutions. When $b^2 - 4ac < 0$, there are no real-number solutions and thus no x-intercepts. There are solutions in an expanded number system called the *complex numbers*. In this text, we will work only with real numbers.

EXAMPLE 3   Solve: $3x^2 - 4x = 2$.

Solution   We first write an equivalent equation of the form $ax^2 + bx + c = 0$, and then determine $a, b,$ and c:

$$3x^2 - 4x - 2 = 0,$$
$$a = 3, \quad b = -4, \quad c = -2.$$

We then use the quadratic formula:

$$x = \frac{-b \pm \sqrt{b^2 - 4ac}}{2a}$$

$$= \frac{-(-4) \pm \sqrt{(-4)^2 - 4(3)(-2)}}{2 \cdot 3} \qquad \text{Substituting}$$

$$= \frac{4 \pm \sqrt{16 + 24}}{6} = \frac{4 \pm \sqrt{40}}{6} \qquad \text{Simplifying}$$

$$= \frac{4 \pm \sqrt{4 \cdot 10}}{6} = \frac{4 \pm 2\sqrt{10}}{6}$$

$$= \frac{2(2 \pm \sqrt{10})}{2 \cdot 3} \qquad \text{Factoring}$$

$$= \frac{2 \pm \sqrt{10}}{3}.$$

The solutions are $(2 + \sqrt{10})/3$ and $(2 - \sqrt{10})/3$, or approximately 1.721 and -0.387, respectively.   **2** ✔

Quick Check 2 ✔

Solve: $3x^2 + 2x = 7$.

Algebraic–Graphical Connection

Let's make an algebraic–graphical connection between the solutions of a quadratic equation and the x-intercepts of a quadratic function.

We just graphed equations of the form $f(x) = ax^2 + bx + c, a \neq 0$. Let's look at the graph of $f(x) = x^2 + 6x + 8$ and its x-intercepts, shown to the right.

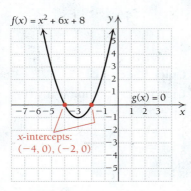

$f(x) = x^2 + 6x + 8$

$g(x) = 0$

x-intercepts: $(-4, 0), (-2, 0)$

The **x-intercepts**, $(-4, 0)$ and $(-2, 0)$, are the points at which the graph crosses the x-axis. These pairs are also the points of intersection of the graphs of $f(x) = x^2 + 6x + 8$ and $g(x) = 0$ (the x-axis). The x-values, -4 and -2, can be found by solving $f(x) = g(x)$:

$$x^2 + 6x + 8 = 0$$
$$(x + 4)(x + 2) = 0 \qquad \text{Factoring; there is no need for the quadratic formula here.}$$
$$x + 4 = 0 \quad \text{or} \quad x + 2 = 0 \qquad \text{Principle of Zero Products}$$
$$x = -4 \quad \text{or} \quad x = -2.$$

The solutions of $x^2 + 6x + 8 = 0$ are -4 and -2, which are the first coordinates of the x-intercepts, $(-4, 0)$ and $(-2, 0)$, of the graph of $f(x) = x^2 + 6x + 8$. A brief review of factoring can be found in Appendix A at the end of the book.

Polynomial Functions

Linear and quadratic functions are part of a general class of *polynomial functions*.

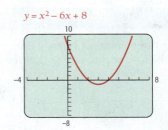

DEFINITION

A **polynomial function** f is given by

$$f(x) = a_n x^n + a_{n-1} x^{n-1} + \cdots + a_2 x^2 + a_1 x^1 + a_0,$$

where n is a nonnegative integer and $a_n, a_{n-1}, \ldots, a_1, a_0$ are real numbers, called the **coefficients**.

The following are examples of polynomial functions:

$$f(x) = -5, \qquad\qquad \text{(A constant function)}$$
$$f(x) = 4x + 3, \qquad\qquad \text{(A linear function)}$$
$$f(x) = -x^2 + 2x + 3, \qquad \text{(A quadratic function)}$$
$$f(x) = 2x^3 - 4x^2 + x + 1. \quad \text{(A cubic, or third-degree, function)}$$

In general, creating graphs of polynomial functions other than linear and quadratic functions is difficult without a calculator. We use calculus to sketch such graphs in Chapter 2. However, some **power functions**, of the form

$$f(x) = ax^n,$$

are relatively easy to graph.

EXAMPLE 4   Using the same set of axes, graph $f(x) = x^2$ and $g(x) = x^3$.

Solution   We set up a table of values, plot the points, and then draw the graphs.

x	x^2	x^3
-2	4	-8
-1	1	-1
$-\frac{1}{2}$	$\frac{1}{4}$	$-\frac{1}{8}$
0	0	0
$\frac{1}{2}$	$\frac{1}{4}$	$\frac{1}{8}$
1	1	1
2	4	8

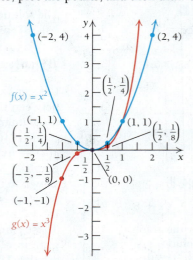

Quick Check 3 ✔

Graph each function using the same set of axes:

$$f(x) = 4 - x^2 \text{ and}$$
$$g(x) = x^3 - 1.$$

3 ✔

TECHNOLOGY CONNECTION

Solving Polynomial Equations

The INTERSECT Feature

Solving $x^3 = 3x + 1$ amounts to finding the x-coordinates of the point(s) of intersection of the graphs of

$$f(x) = x^3 \quad \text{and} \quad g(x) = 3x + 1.$$

We enter the functions as

$$y_1 = x^3 \quad \text{and} \quad y_2 = 3x + 1$$

and then graph. We use a $[-3, 3, -5, 8]$ window to see the curvature and possible points of intersection.

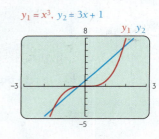

There appear to be at least three points of intersection. Using the INTERSECT feature in the CALC menu, we see that the point of intersection on the left is about $(-1.53, -3.60)$.

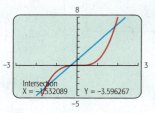

In a similar manner, we approximate the other points of intersection as $(-0.35, -0.04)$ and $(1.88, 6.64)$. The solutions of $x^3 = 3x + 1$ are the x-coordinates of these points, approximately

$$-1.53, \ -0.35, \text{ and } 1.88.$$

The ZERO Feature

The ZERO, or ROOT, feature can be used to solve an equation. The word "zero" in this context refers to an input, or x-value,

(continued)

for which the output of a function is 0. That is, c is a **zero** of the function f if $f(c) = 0$.

To use such a feature requires a 0 on one side of the equation. Thus, to solve $x^3 = 3x + 1$, we obtain $x^3 - 3x - 1 = 0$ by subtracting $3x + 1$ from both sides. Graphing $y = x^3 - 3x - 1$ and using the ZERO feature to find the zeros, we view a screen like the following.

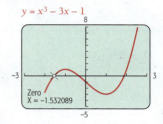

$y = x^3 - 3x - 1$

Zero
X = −1.532089

We see that $x^3 - 3x - 1 = 0$ when $x \approx -1.53$, so -1.53 is an approximate solution of $x^3 = 3x + 1$. In a similar manner, we can approximate the other solutions as -0.35 and 1.88. Note that the points of intersection of the graphs of f and g have the same x-values as the zeros of $x^3 - 3x - 1$.

EXERCISES

Using the INTERSECT feature, solve each equation.

1. $x^2 = 10 - 3x$ **2.** $2x + 24 = x^2$

3. $x^3 = 3x - 2$

4. $x^4 - 2x^2 = 0$

Using the ZERO feature, solve each equation.

5. $0.4x^2 = 280x$
(*Hint:* Use $[-200, 800, -100{,}000, 200{,}000]$.)

6. $\frac{1}{3}x^3 - \frac{1}{2}x^2 = 2x - 1$

7. $x^2 = 0.1x^4 + 0.4$

8. $0 = 2x^4 - 4x^2 + 2$

Find the zeros of each function.

9. $f(x) = 3x^2 - 4x - 2$

10. $f(x) = -x^3 + 6x^2 + 5$

11. $g(x) = x^4 + x^3 - 4x^2 - 2x + 4$

12. $g(x) = -x^4 + x^3 + 11x^2 - 9x - 18$

TECHNOLOGY CONNECTION

App for the iPhone: Graphicus

The advent of the iPhone and other sophisticated mobile devices has made available many inexpensive mathematics applications. One popular and useful app is Graphicus. It has a more visually appealing display than most graphing calculators, but it does have some limitation. For example, it does not do regression, as described in Section R.6.

Graphicus can graph most of the functions we encounter in this text. It can find roots and intersections, and it

excels at many aspects of calculus. Let's look again at the function $f(x) = 2x^3 - x^4$.

To graph and find roots, first open Graphicus. Touch the blank rectangle at the top of the screen and enter the function as y(x)=2x^3-x^4. Press ⊕ in the upper right, and you will see the graph in Fig. 1. Notice the seven icons at the bottom. The one at the far left is for zooming. The fourth icon from the left is for finding roots. Touch it and the roots of the function are highlighted (Fig. 2). Touch the symbol marking each root, and a box appears with its value; see Figs. 3 and 4.

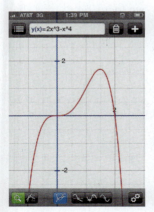

FIGURE 1

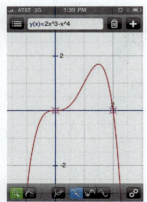

FIGURE 2

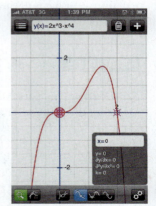

FIGURE 3

FIGURE 4

(*continued*)

App for the iPhone: Graphicus (*continued*)

Piecewise-defined functions cannot be entered. The graph of a function with a missing point, such as $f(x) = \dfrac{x^2 - 9}{x + 3}$, will not show the hole at $x = -3$. Graphicus can graph more than one function on the same set of axes. Graph $g(x) = x^3 - 1$ and $f(x) = 2x^3 - x^4$, and then press the fourth icon to find the points of intersection. Press some of the other icons to further explore Graphicus.

For more information, consult the Graphicus page at the iTunes Store.

EXERCISES

Using Graphicus, repeat Exercises 1–12 in the Technology Connection on page 53.

Rational Functions

> **DEFINITION**
>
> Functions given by the quotient, or ratio, of two polynomials are called **rational functions**.

The following are examples of rational functions:

$$f(x) = \frac{x^2 - 9}{x - 3}, \qquad h(x) = \frac{x - 3}{x^2 - x - 2},$$

$$g(x) = \frac{3x^2 - 4x}{2x + 10}, \qquad k(x) = \frac{x^3 - 2x + 7}{1} = x^3 - 2x + 7.$$

As the function k illustrates, every polynomial function is also a rational function.

The domain of a rational function cannot include any input value that results in division by zero. Thus, for f above, the domain consists of all real numbers except 3. To determine the domain of h, we set the denominator equal to 0 and solve:

$$x^2 - x - 2 = 0$$
$$(x + 1)(x - 2) = 0$$
$$x = -1 \quad or \quad x = 2.$$

Neither -1 nor 2 is in the domain of h. The domain of h is the real numbers except -1 and 2. The numbers -1 and 2 "split," or "separate," the intervals in the domain.

Graphing rational functions can be complicated and is best done using tools of calculus that we will develop in Chapters 1 and 2. For now, we focus on graphs that are fairly basic.

EXAMPLE 5   Graph: $f(x) = \dfrac{x^2 - 9}{x - 3}$.

Solution   This particular function can be simplified before we graph it. We do so by factoring the numerator and removing a factor equal to 1 as follows:

$$f(x) = \frac{x^2 - 9}{x - 3} \qquad \text{\color{red}Noting that 3 is not in the domain of } f$$

$$= \frac{(x - 3)(x + 3)}{x - 3}$$

$$= \frac{x - 3}{x - 3} \cdot \frac{x + 3}{1} \qquad \text{\color{red}Noting that } \frac{x - 3}{x - 3} = 1$$

$$= x + 3, \quad x \neq 3. \qquad \text{\color{red}We must specify } x \neq 3.$$

If $y_1 = (x^2 - 9)/(x - 3)$ and $y_2 = x + 3$ are both graphed, the two graphs appear indistinguishable. To see this, graph both lines and use the arrow keys, ⌃ and ⌄, and the TRACE feature to move the cursor from line to line.

EXERCISES

1. Compare the results of using TRACE and entering the value 3 with y_1 and y_2.

2. Use the TABLE feature with TblStart set at -1 and ΔTbl set at 1. How do y_1 and y_2 differ in the resulting table?

This simplification assumes that x is not 3. By writing $x \neq 3$, we indicate that for any x-value other than 3, the equation $f(x) = x + 3$ is used:

$$f(x) = x + 3, \quad x \neq 3.$$

To find function values, we substitute any value for x other than 3. We make calculations as in the following table and draw the graph. The open dot at $(3, 6)$ indicates that this point is *not* part of the graph.

x	$f(x)$
-3	0
-2	1
-1	2
0	3
1	4
2	5
4	7

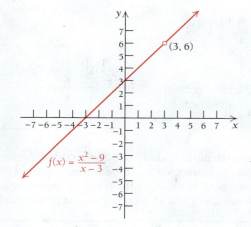

One important class of rational functions is given by $f(x) = k/x$, where k is a constant.

EXAMPLE 6 Graph: $f(x) = 1/x$.

Solution We make a table of values, plot the points, and then draw the graph.

x	$f(x)$
-3	$-\frac{1}{3}$
-2	$-\frac{1}{2}$
-1	-1
$-\frac{1}{2}$	-2
$-\frac{1}{4}$	-4
$\frac{1}{4}$	4
$\frac{1}{2}$	2
1	1
2	$\frac{1}{2}$
3	$\frac{1}{3}$

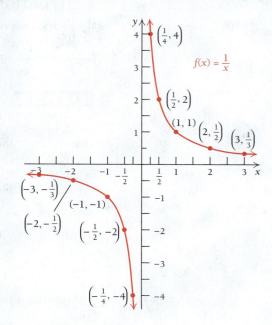

Quick Check 4 ✔

Graph each function:

a) $f(x) = \dfrac{x^2 - 9}{x + 3}$;

b) $f(x) = -\dfrac{1}{x}$.

4 ✔

TECHNOLOGY CONNECTION

Graphs of Rational Functions

Consider two graphs of the function given by

$$f(x) = \frac{2x + 1}{x - 3}.$$

CONNECTED mode: DOT mode:

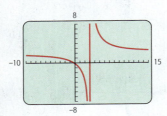   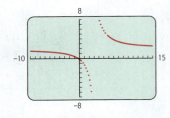

Use of CONNECTED mode leads to an *incorrect* graph in this case. Because, in CONNECTED mode, points are joined with line segments, both branches of the graph are connected, making the vertical line $x = 3$ appear as part of the graph.

On the other hand, in DOT mode, the calculator simply plots dots representing coordinates of points. When graphing rational functions, it is usually best to use DOT mode.

EXERCISES
Graph each of the following using DOT mode.

1. $f(x) = \dfrac{4}{x - 2}$ **2.** $f(x) = \dfrac{x}{x + 2}$

3. $f(x) = \dfrac{x^2 - 1}{x^2 + x - 6}$ **4.** $f(x) = \dfrac{x^2 - 4}{x - 1}$

5. $f(x) = \dfrac{10}{x^2 + 4}$ **6.** $f(x) = \dfrac{8}{x^2 - 4}$

7. $f(x) = \dfrac{2x + 3}{3x^2 + 7x - 6}$ **8.** $f(x) = \dfrac{2x^3}{x^2 + 1}$

In Example 6, note that 0 is not in the domain of f because it would yield a denominator of zero. The function given by $f(x) = 1/x$ is an example of **inverse variation**.

> **DEFINITION**
>
> y **varies inversely** with x if there is some positive number k for which $y = k/x$. We also say that y is **inversely proportional** to x.

EXAMPLE 7   **Business: Stocks and Gold.**   Certain economists theorize that stock prices are inversely proportional to the price of gold. That is, when gold prices go up, stock prices go down, and when gold prices go down, stock prices go up. Assume that the Dow Jones Industrial Average D, an index of the overall prices of stocks, is inversely proportional to the price of gold G, in dollars per ounce. If D was 10,619.70 when G was \$1129.60 per ounce, find the Dow Jones Industrial Average if the price of gold rises to \$1400.

Solution   We assume $D = k/G$, so $10,619.7 = k/1129.6$ and $k = 11,996,013.12$. Thus,

$$D = \frac{11,996,013.12}{G}.$$

Substituting 1400 for G and computing D, we have

$$D = \frac{11,996,013.12}{1400}$$

$$\approx 8568.6.$$

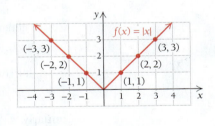

Price of gold (in dollars per ounce)

Quick Check 5 ✔

Business: Stocks and Gold.
Suppose in Example 7 that the price of gold drops to $1000 per ounce. Predict what happens to the Dow Jones Industrial Average. Then compute the value of D.

Warning! Do not put too much "stock" in the equation of this example. It is meant only to give an idea of economic relationships. An equation for predicting the stock market accurately has not been found! 5 ✔

Absolute-Value Functions

The absolute value of a number is its distance from 0 on the number line. We denote the absolute value of a number x as $|x|$. The absolute-value function, given by $f(x) = |x|$, is very important in calculus, and its graph has a distinctive V shape.

EXAMPLE 8 Graph: $f(x) = |x|$.

Solution We make a table of values, plot the points, and then draw the graph.

x	$f(x)$
-3	3
-2	2
-1	1
0	0
1	1
2	2
3	3

TECHNOLOGY CONNECTION

Absolute value is entered as abs and is often accessed from the NUM option of the MATH menu.

EXERCISES
Graph.

1. $f(x) = \left|\frac{1}{2}x\right|$

2. $g(x) = |x^2 - 4|$

We can think of this function as being defined piecewise by considering the definition of absolute value:

$$f(x) = |x| = \begin{cases} x, & \text{if } x \geq 0, \\ -x, & \text{if } x < 0. \end{cases}$$

Square-Root Functions

The following is an example of a square-root function and its graph.

EXAMPLE 9 Graph: $f(x) = -\sqrt{x}$.

Solution The domain of this function is the set of all nonnegative numbers—the interval $[0, \infty)$. You can find approximate values of square roots on your calculator.

We make a table of values, plot the points, and then draw the graph.

x	0	1	2	3	4	5
$f(x) = -\sqrt{x}$	0	−1	−1.4	−1.7	−2	−2.2

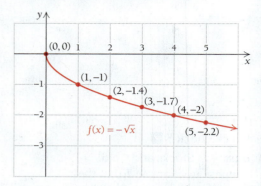

Quick Check 6 ✔

Graph each function:

a) $f(x) = |x + 2|$;

b) $f(x) = \sqrt{x} + 4$.

6 ✔

Power Functions with Rational Exponents

We are motivated to define rational exponents so that the following laws of exponents still hold (also see Appendix A):

> For any nonzero real number a and any integers n and m,
>
> $$a^n \cdot a^m = a^{n+m}; \quad \frac{a^n}{a^m} = a^{n-m}; \quad (a^n)^m = a^{n \cdot m}; \quad a^{-m} = \frac{1}{a^m}.$$

This suggests that $a^{1/2}$ be defined so that $(a^{1/2})^2 = a^{(1/2) \cdot 2} = a^1$. Thus, we define $a^{1/2}$ as $\sqrt{a}$. Similarly, in order to have $(a^{1/3})^3 = a^{(1/3) \cdot 3} = a^1$, we define $a^{1/3}$ as $\sqrt[3]{a}$. In general,

$$a^{1/n} = \sqrt[n]{a}, \quad \text{provided } \sqrt[n]{a} \text{ is defined.}$$

Again, for the laws of exponents to hold, and assuming that $\sqrt[n]{a}$ exists, we have

$$a^{m/n} = (a^m)^{1/n} = \sqrt[n]{a^m} = (\sqrt[n]{a})^m.$$

Similarly, $a^{-m/n}$ is defined by

$$a^{-m/n} = \frac{1}{a^{m/n}} = \frac{1}{\sqrt[n]{a^m}}.$$

EXAMPLE 10   Rewrite each of the following as an equivalent expression using radical notation: **a)** $x^{1/3}$; **b)** $t^{6/7}$; **c)** $x^{-2/3}$; **d)** $r^{-1/4}$.

Solution

a) $x^{1/3} = \sqrt[3]{x}$

b) $t^{6/7} = \sqrt[7]{t^6}$

c) $x^{-2/3} = \dfrac{1}{x^{2/3}} = \dfrac{1}{\sqrt[3]{x^2}}$

d) $r^{-1/4} = \dfrac{1}{r^{1/4}} = \dfrac{1}{\sqrt[4]{r}}, \quad r > 0$ ■

EXAMPLE 11   Rewrite each of the following as an equivalent expression with rational exponents:

a) $\sqrt[4]{x}, \quad$ for $x \geq 0$

b) $\sqrt[3]{r^2}$

c) $\sqrt{x^{10}}, \quad$ for $x \geq 0$

d) $\dfrac{1}{\sqrt[3]{b^5}}, \quad$ for $b \neq 0$

Solution

a) $\sqrt[4]{x} = x^{1/4}$ $x \geq 0$

b) $\sqrt[3]{r^2} = r^{2/3}$

c) $\sqrt{x^{10}} = x^{10/2} = x^5$, $x \geq 0$

d) $\dfrac{1}{\sqrt[3]{b^5}} = \dfrac{1}{b^{5/3}} = b^{-5/3}$, $b \neq 0$ ■

EXAMPLE 12   Simplify: **a)** $8^{5/3}$; **b)** $81^{3/4}$.

Solution

a) $8^{5/3} = \left(8^{1/3}\right)^5 = \left(\sqrt[3]{8}\right)^5 = 2^5 = 32$

b) $81^{3/4} = (81^{1/4})^3 = (\sqrt[4]{81})^3 = 3^3 = 27$ ■

Because even roots (square roots, fourth roots, sixth roots, and so on) of negative numbers are not real numbers, the domain of a radical function may have restrictions.

EXAMPLE 13   Find the domain of the function given by

$$f(x) = \sqrt[4]{2x - 10}.$$

Solution   For $f(x)$ to be a real number, $2x - 10$ cannot be negative. Thus, to find the domain of f, we solve the inequality $2x - 10 \geq 0$:

$$2x - 10 \geq 0$$
$$2x \geq 10 \qquad \text{\color{red}{Adding 10 to both sides}}$$
$$x \geq 5. \qquad \text{\color{red}{Dividing both sides by 2}}$$

The domain of f is $\{x \mid x \geq 5\}$, or, in interval notation, $[5, \infty)$.   **7 ✔**

Power functions of the form $f(x) = ax^k$, with k a fraction, occur in many applications.

Quick Check 7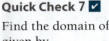

Find the domain of the function given by

$$f(x) = \sqrt{x + 3}.$$

EXAMPLE 14   **Life Science: Home Range.**   The *home range* of an animal is defined as the region to which the animal confines its movements. It has been shown that for carnivorous mammals the area of that region can be approximated by

$$H(w) = 0.11w^{1.36},$$

where w is the mass of the animal, in grams, and $H(w)$ is the area of the home range, in hectares. Graph the function. (*Source*: Based on information in Emlen, J. M., *Ecology: An Evolutionary Approach*, p. 200 (Reading, MA: Addison-Wesley, 1973), and Harestad, A. S., and Bunnel, F. L., "Home Range and Body Weight—A Reevaluation," *Ecology*, Vol. 60, No. 2 (April, 1979), pp. 389–402.)

Solution   We can approximate function values using a power key, usually labeled ⌃ or y^x. Note that $w^{1.36} = w^{136/100} = \sqrt[100]{w^{136}}$.

w	0	700	1400	2100	2800	3500
$H(w)$	0	814.2	2089.9	3627.5	5364.5	7266.5

A lynx in its territorial area.

The graph is shown below. Note that function values increase from left to right. As body weight increases, the area over which the animal moves increases.

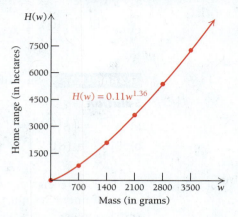

$H(w) = 0.11w^{1.36}$

Home range (in hectares)

Mass (in grams)

Exponential Functions

Sometimes, we consider expressions in which a constant (called the *base*) has a variable exponent. Such expressions are used in *exponential functions*.

> **DEFINITION**
>
> An **exponential function** f is given by
> $$f(x) = a_0 \cdot a^x,$$
> where x is any real number and $a > 0$ and $a \neq 1$. The number a is the **base**.

The following are examples of exponential functions:

$$f(x) = 2^x, \quad f(x) = \left(\frac{1}{2}\right)^x, \quad f(x) = 3(0.4)^x.$$

Note that in contrast to a power function like $y = x^2$, an exponential function has the variable in the exponent, not the base. Let's consider the graphs of exponential functions.

EXAMPLE 15   Graph: $y = f(x) = 2^x$.

Solution   First, we find some function values. Note that 2^x is always positive:

$$x = 0, \qquad y = 2^0 = 1;$$
$$x = \frac{1}{2}, \qquad y = 2^{1/2} = \sqrt{2} \approx 1.4;$$
$$x = 1, \qquad y = 2^1 = 2;$$
$$x = 2, \qquad y = 2^2 = 4;$$
$$x = 3, \qquad y = 2^3 = 8;$$
$$x = -1, \qquad y = 2^{-1} = \frac{1}{2};$$
$$x = -2, \qquad y = 2^{-2}$$
$$= \frac{1}{2^2} = \frac{1}{4}.$$

In this part of the curve, $f(x)$ gets larger without bound.

$f(x) = 2^x$

This part of the curve comes very close to the x-axis, but does not touch or cross it.

y-intercept: $(0, 1)$

x	0	$\frac{1}{2}$	1	2	3	-1	-2
$y = f(x) = 2^x$	1	1.4	2	4	8	$\frac{1}{2}$	$\frac{1}{4}$

Quick Check 8 ✔

For $f(x) = 3^x$, complete this table of function values.

x	$f(x) = 3^x$
0	
1	
2	
3	
-1	
-2	
-3	

Graph $f(x) = 3^x$.

8 ✔

EXAMPLE 16   Graph: $y = g(x) = 3 \cdot \left(\frac{1}{2}\right)^x$.

Solution   We find some function values:

$$x = 0, \quad y = 3 \cdot \left(\frac{1}{2}\right)^0 = 3;$$

$$x = \frac{1}{2}, \quad y = 3 \cdot \left(\frac{1}{2}\right)^{1/2} \approx 2.1;$$

$$x = 1, \quad y = 3 \cdot \left(\frac{1}{2}\right)^1 = \frac{3}{2};$$

$$x = 2, \quad y = 3 \cdot \left(\frac{1}{2}\right)^2 = \frac{3}{4};$$

$$x = -1, \quad y = 3 \cdot \left(\frac{1}{2}\right)^{-1} = 3 \cdot 2 = 6;$$

$$x = -2, \quad y = 3 \cdot \left(\frac{1}{2}\right)^{-2} = 3 \cdot 4 = 12;$$

$$x = -3, \quad y = 3 \cdot \left(\frac{1}{2}\right)^{-3} = 3 \cdot 8 = 24.$$

Quick Check 9 ✔

For $g(x) = 2 \cdot \left(\frac{1}{3}\right)^x$, complete this table of function values.

x	$g(x) = 2 \cdot \left(\frac{1}{3}\right)^x$
0	
1	
2	
3	
−1	
−2	
−3	

Graph $g(x) = 2 \cdot \left(\frac{1}{3}\right)^x$.

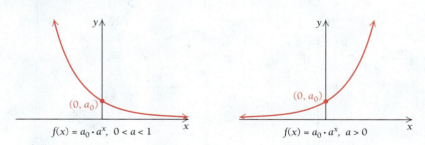
$g(x) = 3 \cdot \left(\frac{1}{2}\right)^x$
y-intercept: (0, 3)

x	0	$\frac{1}{2}$	1	2	−1	−2	−3
$y = f(x) = 3 \cdot \left(\frac{1}{2}\right)^x$	3	2.1	$\frac{3}{2}$	$\frac{3}{4}$	6	12	24

9 ✔

In general, when $0 < a < 1$, as x gets larger, $f(x)$ gets smaller. When $a > 1$, as x gets larger, $f(x)$ also gets larger. The domain of an exponential function is all of the real numbers (this is discussed further in Section 3.1). The range is $y > 0$. Note that when $x = 0$, we have $f(0) = a_0 \cdot a^0 = a_0 \cdot 1 = a_0$. Thus, the y-intercept is $(0, a_0)$, and there are no x-intercepts.

$(0, a_0)$
$f(x) = a_0 \cdot a^x, \ 0 < a < 1$

$(0, a_0)$
$f(x) = a_0 \cdot a^x, \ a > 0$

Supply and Demand Functions

In economics, *demand* is modeled by a decreasing function; that is, as the price x gets larger, the demand $D(x)$ gets smaller. However, *supply* is modeled by an increasing function: As the price x gets larger, so does the supply, $S(x)$.

Demand Functions

Suppose the relationship between the price x of a 5-lb bag of sugar and the quantity q of bags that consumers will demand at that price is given in the table and graph below.

Demand Schedule

Price, x, per 5-lb Bag	Quantity, q, of 5-lb Bags (in millions)
$5	4
4	5
3	7
2	10
1	15

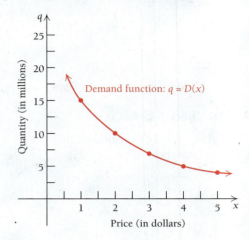

Note that the quantity consumers demand is inversely proportional to the price. As the price goes up, the quantity demanded goes down.

Supply Functions

Furthermore, suppose the relationship between the price x of a 5-lb bag of sugar and the quantity q of bags that sellers are willing to supply, or sell, at that price is given in the table and graph below.

Supply Schedule

Price, x, per 5-lb Bag	Quantity, q, of 5-lb Bags (in millions)
$1	0
2	10
3	16
4	20
5	22

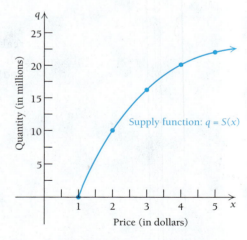

Note that sellers will supply greater quantities at higher prices, and consumers will demand greater quantities at lower prices.

Let's now look at these curves together. As price increases, supply increases and demand decreases; and as price decreases, demand increases but supply decreases. The point of intersection (x_E, q_E) is called the **equilibrium point.** The equilibrium price x_E (in this case, $2 per bag) corresponds to an equilibrium quantity q_E (in this case, 10 million bags). Sellers are willing to sell 10 million bags at $2 per bag, and consumers are willing to buy 10 million bags at that price. The situation is a bit like a buyer and a seller haggling over the price of an item. The equilibrium point, or selling price, is what they finally agree on.

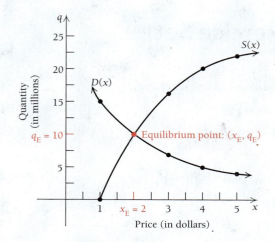

EXAMPLE 17 **Economics: Equilibrium Point.** Find the equilibrium point for the demand and supply functions for the Ultra-Fine coffee maker. Here q represents the number of coffee makers produced, in hundreds, and x is the price of a coffee maker, in dollars.

Demand: $q = 50 - \dfrac{1}{4}x$

Supply: $q = x - 25$

Solution To find the equilibrium point, the quantity demanded must match the quantity produced:

$$50 - \frac{1}{4}x = x - 25 \qquad \text{Setting demand equal to supply}$$

$$50 + 25 = x + \frac{1}{4}x \qquad \text{Adding } 25 + \frac{1}{4}x \text{ to both sides}$$

$$75 = \frac{5}{4}x$$

$$75 \cdot \frac{4}{5} = x \qquad \text{Multiplying both sides by } \frac{4}{5}$$

$$60 = x.$$

Thus, $x_E = 60$. To find q_E, we substitute x_E into either function. We select the supply function:

$$q_E = x_E - 25 = 60 - 25 = 35.$$

Thus, the equilibrium quantity is 3500 units, and the equilibrium point is ($60, 3500).

10 ✔

TECHNOLOGY CONNECTION 📈

EXERCISE

1. Use the INTERSECT feature to find the equilibrium point for the following demand and supply functions.

Demand: $q = 1123.6 - 61.4x$

Supply: $q = 201.8 + 4.6x$

Quick Check 10 ✔

Economics: Equilibrium Point.
Repeat Example 17 for the following functions:

Demand: $q = 70 - \dfrac{1}{5}x$

Supply: $q = x - 20$

Section Summary

- Many types of functions have graphs that are not straight lines; among these are *quadratic functions, polynomial functions, power functions, rational functions, the absolute-value function, square-root functions,* and *exponential functions.*

- Demand is modeled by a decreasing function; that is, as the price x gets larger, the demand $D(x)$ gets smaller. Supply is modeled by an increasing function; that is, as the price x gets larger, so does the supply, $S(x)$. The point of intersection of graphs of demand and supply functions for the same product is called the *equilibrium point.*

R.5   Exercise Set

Graph each pair of equations on one set of axes.

1. $y = \frac{1}{4}x^2$ and $y = -\frac{1}{4}x^2$

2. $y = \frac{1}{2}x^2$ and $y = -\frac{1}{2}x^2$

3. $y = x^2$ and $y = x^2 - 1$

4. $y = x^2$ and $y = x^2 - 3$

5. $y = -3x^2$ and $y = -3x^2 + 2$

6. $y = -2x^2$ and $y = -2x^2 + 1$

7. $y = |x|$ and $y = |x - 3|$

8. $y = |x|$ and $y = |x - 1|$

9. $y = x^3$ and $y = x^3 + 1$

10. $y = x^3$ and $y = x^3 + 2$

11. $y = \sqrt{x}$ and $y = \sqrt{x - 1}$

12. $y = \sqrt{x}$ and $y = \sqrt{x - 2}$

For each of the following, state whether the graph of the function is a parabola. If the graph is a parabola, find its vertex.

13. $f(x) = x^2 + 4x - 7$

14. $f(x) = x^3 - 2x + 3$

15. $g(x) = 2x^4 - 4x^2 - 3$

16. $g(x) = 3x^2 - 6x$

Graph.

17. $y = x^2 - 6x + 5$

18. $y = x^2 - 4x + 3$

19. $y = -x^2 + 2x - 1$

20. $y = -x^2 - x + 6$

21. $f(x) = 3x^2 - 6x + 4$

22. $f(x) = 2x^2 - 6x + 1$

23. $g(x) = -3x^2 - 4x + 5$

24. $g(x) = -2x^2 - 3x + 7$

25. $y = \dfrac{3}{x}$

26. $y = \dfrac{2}{x}$

27. $y = -\dfrac{2}{x}$

28. $y = \dfrac{-3}{x}$

29. $y = \dfrac{1}{x - 1}$

30. $y = \dfrac{1}{x^2}$

31. $y = \sqrt[3]{x}$

32. $y = \dfrac{1}{|x|}$

33. $g(x) = \dfrac{x^2 + 7x + 10}{x + 2}$

34. $f(x) = \dfrac{x^2 + 5x + 6}{x + 3}$

35. $f(x) = \dfrac{x^2 - 1}{x - 1}$

36. $g(x) = \dfrac{x^2 - 25}{x - 5}$

37. $f(x) = (1.25)^x$

38. $g(x) = (1.5)^x$

39. $f(x) = (0.4)^x$

40. $g(x) = (0.25)^x$

41. $f(x) = 3 \cdot (1.1)^x$

42. $g(x) = 2 \cdot (1.45)^x$

43. $f(x) = 4 \cdot (0.8)^x$

44. $g(x) = 2.5 \cdot (0.33)^x$

Solve.

45. $x^2 - 2x = 2$

46. $x^2 - 2x + 1 = 5$

47. $x^2 + 6x = 1$

48. $x^2 + 4x = 3$

49. $4x^2 = 4x + 1$

50. $-4x^2 = 4x - 1$

51. $3y^2 + 8y + 2 = 0$

52. $2p^2 - 5p = 1$

53. $x + 7 + \dfrac{9}{x} = 0$ (*Hint:* Multiply both sides by x.)

54. $1 - \dfrac{1}{w} = \dfrac{1}{w^2}$

Rewrite each of the following as an equivalent expression using radical notation.

55. $x^{1/5}$

56. $t^{1/7}$

57. $y^{2/3}$

58. $t^{2/5}$

59. $t^{-2/5}$

60. $y^{-2/3}$

61. $b^{-1/3}$

62. $b^{-1/5}$

63. $(x^2 - 3)^{-1/2}$

64. $(y^2 + 7)^{-1/4}$

Rewrite each of the following as an equivalent expression with rational exponents.

65. $\sqrt{x^3}$

66. $\sqrt{x^5}$

67. $\sqrt[5]{a^3}$

68. $\sqrt[4]{b^2}$, $b \geq 0$

69. $\sqrt[4]{x^{12}}$, $x \geq 0$

70. $\sqrt[3]{t^6}$

71. $\dfrac{1}{\sqrt{t^5}}$

72. $\dfrac{1}{\sqrt{m^4}}$

73. $\dfrac{1}{\sqrt{x^2 + 7}}$

74. $\sqrt{x^3 + 4}$

Simplify.

75. $9^{3/2}$

76. $16^{5/2}$

77. $64^{2/3}$

78. $8^{2/3}$

Determine the domain of each function.

79. $f(x) = \dfrac{x^2 - 25}{x - 5}$

80. $f(x) = \dfrac{x^2 - 4}{x + 2}$

81. $f(x) = \dfrac{x^3}{x^2 - 5x + 6}$

82. $f(x) = \dfrac{x^4 + 7}{x^2 + 6x + 5}$

83. $f(x) = \sqrt{5x + 4}$

84. $f(x) = \sqrt{2x - 6}$

85. $f(x) = \sqrt[4]{7 - x}$

86. $f(x) = \sqrt[6]{5 - x}$

APPLICATIONS

Business and Economics

Find the equilibrium point for each pair of demand and supply functions.

87. Demand: $q = 1000 - 10x$;   Supply: $q = 250 + 5x$

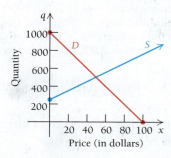

88. Demand: $q = 8800 - 30x$;   Supply: $q = 7000 + 15x$

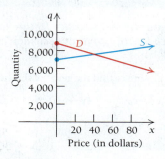

89. Demand: $q = \dfrac{5}{x}$;   Supply: $q = \dfrac{x}{5}$

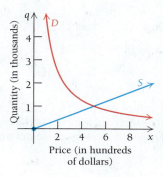

90. Demand: $q = \dfrac{4}{x}$;   Supply: $q = \dfrac{x}{4}$

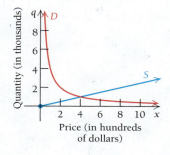

91. Demand: $q = (x - 3)^2$;   Supply: $q = x^2 + 2x + 1$ (assume $x \le 3$)

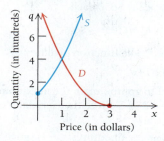

92. Demand: $q = (x - 4)^2$;   Supply: $q = x^2 + 2x + 6$ (assume $x \le 4$)

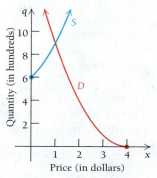

93. Demand: $q = 5 - x$;   Supply: $q = \sqrt{x + 7}$

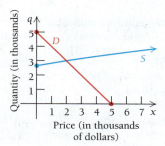

94. Demand: $q = 7 - x$;   Supply: $q = 2\sqrt{x + 1}$

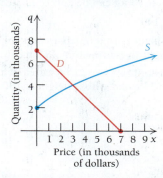

95. Price of admission.   The number of tickets sold for a fund-raiser is inversely proportional to the price of a ticket, p. If 175 tickets can be sold for \$20 each, how many tickets will be sold if the price is \$25 each?

96. Demand. The quantity sold x of a high-definition TV is inversely proportional to the price p. If 85,000 high-definition TVs sold for $1200 each, how many will be sold if the price is $850 each?

Life and Physical Sciences

97. Radar range. The function given by

$$R(x) = 11.74x^{0.25}$$

can be used to approximate the maximum range, $R(x)$, in miles, of ARSR-3 surveillance radar with a peak power of x watts.

a) Determine the maximum radar range when the peak power is 40,000 watts, 50,000 watts, and 60,000 watts.

b) Graph the function.

98. Home range. Refer to Example 14. The home range, in hectares, of an omnivorous mammal of mass w grams is given by

$$H(w) = 0.059w^{0.92}.$$

(*Source*: Harestad, A. S., and Bunnel, F. L., "Home Range and Body Weight—A Reevaluation," *Ecology*, Vol. 60, No. 2 (April, 1979), pp. 389–402.) Complete the table of approximate function values and graph the function.

w	0	1000	2000	3000	4000	5000	6000	7000
$H(w)$	0	34.0						

99. Life science: pollution control. Pollution control has become an important concern worldwide. The function

$$P(t) = 12.85t^{-0.077}$$

describes the average pollution, in micrograms per cubic meter of air, t years after 2000. (*Source*: epa.gov/airtrends/pm.html)

a) Predict the pollution in 2015, 2020, and 2025.
b) Graph the function over the interval [0, 50].

100. Surface area and mass. The surface area of a person whose mass is 75 kg can be approximated by

$$f(h) = 0.144h^{1/2},$$

where $f(h)$ is measured in square meters and h is the person's height in centimeters. (*Source*: U.S. Oncology.)

a) Find the approximate surface area of a person whose mass is 75 kg and whose height is 180 cm.

b) Find the approximate surface area of a person whose mass is 75 kg and whose height is 170 cm.
c) Graph the function f for $0 \le h \le 200$.

SYNTHESIS

101. Zipf's Law. According to Zipf's Law, the number of cities N with a population greater than S is inversely proportional to S. In 2012, there were 285 U.S. cities with a population greater than 100,000. Estimate the number of U.S. cities with a population between 350,000 and 500,000; between 300,000 and 600,000.

102. At most, how many y-intercepts can a function have? Why?

103. What is the difference between a rational function and a polynomial function?

Use the ZERO *feature or the* INTERSECT *feature to approximate the zeros of each function to three decimal places.*

104. $f(x) = x^3 - x$
(Also use algebra to find the zeros of this function.)

105. $f(x) = 2x^3 - x^2 - 14x - 10$

106. $f(x) = \frac{1}{2}(|x - 4| + |x - 7|) - 4$

107. $f(x) = x^4 + 4x^3 - 36x^2 - 160x + 300$

108. $f(x) = \sqrt{7 - x^2} - 1$

109. $f(x) = |x + 1| + |x - 2| - 5$

110. $f(x) = |x + 1| + |x - 2|$

111. $f(x) = |x + 1| + |x - 2| - 3$

112. $f(x) = x^8 + 8x^7 - 28x^6 - 56x^5 + 70x^4 + 56x^3 - 28x^2 - 8x + 1$

113. Find the equilibrium point for the following demand and supply functions, where q is the quantity, in thousands of units, and x is the price per unit, in dollars.

Demand: $q = 15 \cdot (0.95)^x$

Supply: $q = 2.5 \cdot (1.13)^x$

Answers to Quick Checks

1. (a)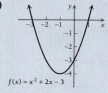

   $f(x) = x^2 + 2x - 3$

   (b)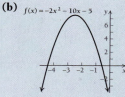

   $f(x) = -2x^2 - 10x - 5$

2. $\dfrac{-1 \pm \sqrt{22}}{3}$,

   or 1.230 and -1.897

3.

   $f(x) = 4 - x^2$

   $g(x) = x^3 - 1$

4. (a)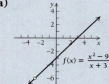

   $f(x) = \dfrac{x^2 - 9}{x + 3}$

   (b)

   $f(x) = -\dfrac{1}{x}$

5. Dow Jones Industrial Average increases; $D = 11{,}996.01$.

6. (a)

   $f(x) = |x + 2|$

   (b)

   $f(x) = \sqrt{x + 4}$

7. $\{x \mid x \geq -3\}$

8.

x	$f(x) = 3^x$
0	1
1	3
2	9
3	27
-1	$\frac{1}{3}$
-2	$\frac{1}{9}$
-3	$\frac{1}{27}$

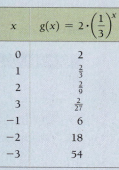

$(0, 1)$

9.

x	$g(x) = 2 \cdot \left(\dfrac{1}{3}\right)^x$
0	2
1	$\frac{2}{3}$
2	$\frac{2}{9}$
3	$\frac{2}{27}$
-1	6
-2	18
-3	54

$(0, 2)$

10. Equilibrium point is ($75, 5500); price is $75, and quantity is 5500.

R.6

Mathematical Modeling and Curve Fitting

Fitting Functions to Data

● Use curve fitting to find a mathematical model for a set of data and use the model to make predictions.

We now have a library of functions that can serve as models for applications, and they are shown below. Cubic and quartic functions are covered in detail in Chapter 2, but we show them for reference. We will not consider rational functions in this section.

Linear function:
$f(x) = mx + b$

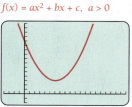

Quadratic function:
$f(x) = ax^2 + bx + c, \ a > 0$

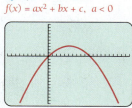

Quadratic function:
$f(x) = ax^2 + bx + c, \ a < 0$

Absolute-value function:
$f(x) = |x|$

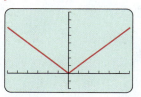

Cubic function:
$f(x) = ax^3 + bx^2 + cx + d, \ a > 0$

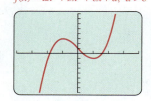

Quartic function:
$f(x) = ax^4 + bx^3 + cx^2 + dx + e, \ a > 0$

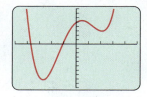

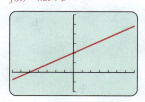

Exponential growth:
$f(x) = a_0 \cdot a^x, \ a > 1$

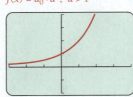

Exponential decay:
$f(x) = a_0 \cdot a^x, \ 0 < a < 1$

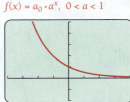

Now let's consider some real-world data. How can we decide which, if any, type of function might fit the data? One simple way is to examine a graph of the data called a **scatterplot**. Then we look for a pattern resembling one of the preceding graphs. If the graph resembles a straight line, the best model might be a linear function. If the graph rises and then falls, or falls and then rises, in a curve resembling a parabola, then the best model might be a quadratic function.

EXAMPLE 1   Choosing Models.   For the following scatterplots and graphs, determine which, if any, of the following functions might be used as a model for the data.

Linear, $f(x) = mx + b$

Quadratic, $f(x) = ax^2 + bx + c, \quad a > 0$

Quadratic, $f(x) = ax^2 + bx + c, \quad a < 0$

Polynomial, neither quadratic nor linear

Exponential, $f(x) = a_0 \cdot a^x, \quad 0 < a < 1 \quad$ or $\quad a > 1$

a)

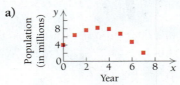

b)

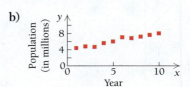

c)

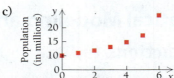

d)

DRIVER FATALITIES BY AGE

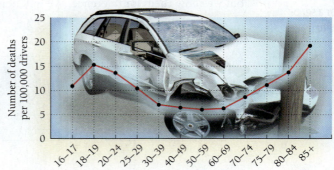

Age of drivers (in years)

(*Source*: Based on data from the American Automobile Association, 2012.)

e)

**ACTUAL AND PROJECTED ANNUAL
SUIT SALES FOR RAGGS LTD.**

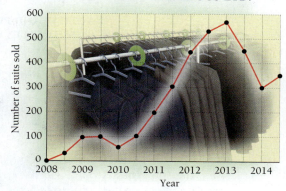

(*Source*: *Business Week* annual executive pay surveys.)

Solution

a) The data rise and then fall in a curved manner fitting a quadratic function,

$$f(x) = ax^2 + bx + c, \quad a < 0.$$

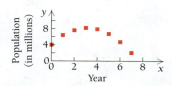

b) The data seem to fit a linear function,

$$f(x) = mx + b.$$

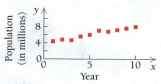

c) The data rise in a manner fitting the right-hand side of a quadratic function,

$$f(x) = ax^2 + bx + c, \quad a > 0.$$

The data may also fit an exponential growth function,

$$f(x) = a_0 \cdot a^x, \quad a > 1.$$

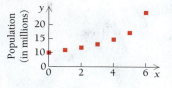

d) The data fall and then rise in a curved manner suggesting a quadratic function,

$$f(x) = ax^2 + bx + c, \quad a > 0.$$

DRIVER FATALITIES BY AGE

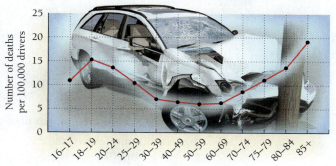

Age of drivers (in years)

(*Source*: Based on data from the American Automobile Association, 2012.)

e) The data rise and fall more than once, so they do not fit a linear or quadratic function but might fit a polynomial function that is neither quadratic nor linear.

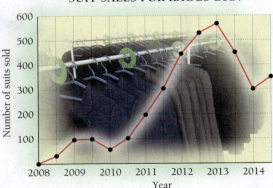

ACTUAL AND PROJECTED ANNUAL SUIT SALES FOR RAGGS LTD.

It is sometimes possible to find a mathematical model by graphing a set of data as a scatterplot, inspecting the graph to see if a known type of function seems to fit, and then using the data points to derive the equation of a specific function.

EXAMPLE 2 **Business: High-Speed Internet Subscribers.** The following table and scatterplot show the number of U.S. households with high-speed Internet service. It appears that the data can be modeled by a linear function.

Years since 2005, x	Number of Households with High-Speed Internet Service (in millions), y	Scatterplot
2005, 0	25.4	
2006, 1	28.9	
2007, 2	35.7	
2008, 3	39.3	
2009, 4	41.8	
2010, 5	44.4	
2011, 6	47.3	
2012, 7	50.3	

(*Source*: National Cable & Telecommunications Association.)

a) Find a linear function that fits the data.

b) Use the model to predict the number of households with high-speed Internet service in 2018.

Solution

a) We can choose any two data points to determine an equation. Let's use $(2, 35.7)$ and $(7, 50.3)$.

We first determine the slope of the line:

$$m = \frac{50.3 - 35.7}{7 - 2} = \frac{14.6}{5} = 2.92.$$

We substitude 2.92 for m and either of the points $(2, 35.7)$ or $(7, 50.3)$ for (x_1, y_1) in the point–slope equation. Using $(7, 50.3)$ we get

$$y - 50.3 = 2.92(x - 7),$$

which simplifies to

$$y = 2.92x + 29.86,$$

where x is the number of years after 2005 and y is in millions. We can then graph the linear equation $y = 2.92x + 29.86$ on the scatterplot to see how it fits the data.

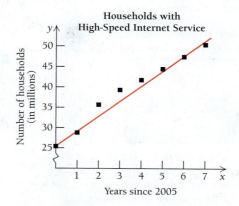

b) We can predict the number of households with high-speed Internet in 2018 by substituting 13 for x (since $2018 - 2005 = 13$) in the model:

$$y = 2.92x + 29.86 \qquad \text{Model}$$
$$= 2.92(13) + 29.86 \qquad \text{Substituting}$$
$$\approx 67.82.$$

We can then predict that there will be about 67.82 million households with high-speed Internet in 2018.

Quick Check 1 ✔

Business. Repeat Example 2 using the data points $(1, 28.9)$ and $(5, 44.4)$. Are the model and the result different from those in Example 2?

1 ✔

Different pairs of points yield different linear models, and *regression* is used to find a line of *best* fit.

TECHNOLOGY CONNECTION

Linear Regression: Fitting a Linear Function to Data

We now consider **linear regression**, the preferred method for fitting a linear function to a set of data. Although the complete basis for this method is discussed in Section 6.4, we consider it here because we can carry out the procedure easily using technology. One advantage of linear regression is that it uses *all* data points rather than just two.

EXAMPLE **Business: High-Speed Internet Subscribers.** Consider the data in Example 2.

a) Find the equation of the regression line for the given data. Then graph both the regression line and the data.

b) Use the model to predict the number of households with high-speed Internet in 2018. Compare your answer to that found in Example 2.

Solution

a) To fit a linear function using regression, we select the EDIT option of the STAT menu. We then enter the data, entering the first coordinate for L1 and the second coordinate for L2.

$$(0, 25.4), (1, 28.9), (2, 35.7), (3, 39.3),$$
$$(4, 41.8), (5, 44.4), (6, 47.3), (7, 50.3)$$

L1	L2	L3
0	25.4	-------
1	28.9	
2	35.7	
3	39.3	
4	41.8	
5	44.4	
6	47.3	
L2(7) = 47.3		

L1	L2	L3 2
1	28.9	-------
2	35.7	
3	39.3	
4	41.8	
5	44.4	
6	47.3	
7	50.3	
L2(8) = 50.3		

(continued)

To view the data points, we turn on PLOT1 by pressing 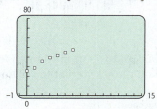 ENTER from y_1 at the Y= screen and then pressing GRAPH. To set the window, we could select the ZoomStat option of ZOOM. Instead, we select a $[-1, 15, 0, 80]$ window.

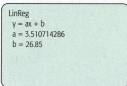

To find the line that best fits the data, we press STAT, select CALC, then LinReg(ax+b), and then press ENTER.

LinReg
y = ax + b
a = 3.510714286
b = 26.85

A convenient way to copy a regression function, in full detail, onto the Y= screen is to press Y=, move the cursor to Y2 (or wherever the function is to appear), press VARS, and select STATISTICS and then EQ and RegEq.

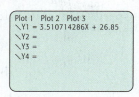

Plot 1 Plot 2 Plot 3
\Y1 = 3.510714286X + 26.85
\Y2 =
\Y3 =
\Y4 =

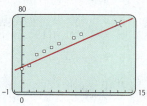

b) To find, or predict, the number of U.S. households with high-speed Internet service in 2018, we press 2ND, CALC, 1, and enter 13.

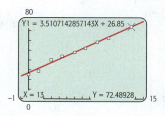

Note that the result, 72.49 million, is more than the 67.96 million found in Example 2. Statisticians might give more credence to the regression value because more data points were used to construct the equation.

EXERCISE

1. **Study time and test scores.** The data in the following table relate study time and test scores.

Study Time (in hours)	Test Grade (in percent)
7	83
8	85
9	88
10	91
11	?

a) Make a scatterplot of the data, fit a regression line to the data, and graph the line with the scatterplot.

b) Use the linear model to predict the test score received when one has studied for 11 hr.

c) Discuss the limitations of a linear model for this application.

EXAMPLE 3   Life Science: Hours of Sleep and Death Rate.   From a study by Dr. Harold J. Morowitz of Yale University, the following data show the relationship between the death rate of men and the average number of hours per day that the men slept.

Average Number of Hours of Sleep, x	Death Rate per 100,000 Males, y
5	1121
6	805
7	626
8	813
9	967

(*Source:* Morowitz, Harold J., "Hiding in the Hammond Report," *Hospital Practice.*)

a) Make a scatterplot of the data, and determine whether the data seem to fit a quadratic function.

b) Find a quadratic function that fits the data.

c) Use the model to find the death rate for males who sleep 2 hr, 8 hr, and 10 hr.

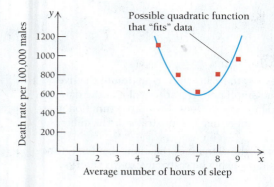

Death rate per 100,000 males

Average number of hours of sleep

Possible quadratic function that "fits" data

Solution

a) The scatterplot is shown to the left. Note that the rate drops and then rises, which suggests that a quadratic function might fit the data.

b) We consider the quadratic model,

$$y = ax^2 + bx + c. \qquad (1)$$

To determine the constants a, b, and c, we choose three of the data points: $(5, 1121)$, $(7, 626)$, and $(9, 967)$. Since these points are to be solutions of equation (1), it follows that

$$1121 = a \cdot 5^2 + b \cdot 5 + c, \quad \text{or} \quad 1121 = 25a + 5b + c,$$
$$626 = a \cdot 7^2 + b \cdot 7 + c, \quad \text{or} \quad 626 = 49a + 7b + c,$$
$$967 = a \cdot 9^2 + b \cdot 9 + c, \quad \text{or} \quad 967 = 81a + 9b + c.$$

We solve this system of three equations in three variables using procedures of algebra and get

$$a = 104.5, \quad b = -1501.5, \quad \text{and} \quad c = 6016.$$

Substituting these values into equation (1), we get the following quadratic function:

$$y = 104.5x^2 - 1501.5x + 6016.$$

c) The death rate for males who sleep 2 hr is

$$y = 104.5(2)^2 - 1501.5(2) + 6016 = 3431 \text{ deaths per 100,000 males.}$$

The death rate for males who sleep 8 hr is

$$y = 104.5(8)^2 - 1501.5(8) + 6016 = 692 \text{ deaths per 100,000 males.}$$

The death rate for males who sleep 10 hr is

$$y = 104.5(10)^2 - 1501.5(10) + 6016 = 1451 \text{ deaths per 100,000 males.}$$

2 ✔

Quick Check 2 ✔

Life Science. See the data on live births in the following Technology Connection. Use the data points $(16, 15.4)$, $(27, 107.2)$, and $(37, 47.2)$ to find a quadratic function that fits the data. Use the model to predict the average number of live births to women age 20.

TECHNOLOGY CONNECTION

Mathematical Modeling Using Regression: Fitting Quadratic and Other Polynomial Functions to Data

Regression extends to quadratic, cubic, and quartic polynomial functions.

EXAMPLE **Life Science: Live Births to Women of Age x.**
The chart below relates the average number of live births to women of a particular age.

Age, x (in years)	Average Number of Live Births per 1000 women
16	15.4
18.5	54.1
22	85.3
27	107.2
32	96.5
37	47.2
42	10.3

(*Source:* CDC, *National Vital Statistics*, Vol. 62, No. 1, June 28, 2013.)

a) Fit a quadratic function to the data using REGRESSION. Then make a scatterplot of the data and graph the quadratic function with the scatterplot.

b) Fit a cubic function to the data using REGRESSION. Then make a scatterplot of the data and graph the cubic function with the scatterplot.

c) Which function seems to fit the data better?

d) Use the function from part (c) to estimate the average number of live births to women of ages 20 and 30.

(*continued*)

Mathematical Modeling Using Regression (*continued*)

Solution   We proceed as follows.

a) To fit a quadratic function using REGRESSION, the procedure is similar to what is outlined in the preceding Technology Connection. We enter the data but select QuadReg instead of LinReg(ax + b). In this case, we use ZoomStat to set the window. Here y_1 uses approximations of a, b, and c from the QuadReg screen.

$$y_1 = -0.54x^2 + 30.56x - 329.5$$

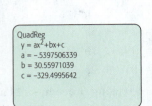 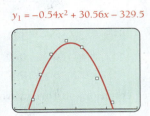

b) To fit a cubic function, we select CubicReg and obtain the following, which we enter as y_2.

$$y_2 = 0.0140646004x^3 \\ - 1.765413134x^2 \\ + 64.29927316x \\ - 620.3338007$$

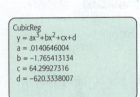 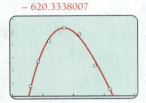

c) The cubic function fits better.

d) We press **2ND** **QUIT** to leave the graph screen. Pressing **VARS** and selecting Y-VARS and then FUNCTION and Y2, we have $y_2(20) \approx 72$ and $y_2(30) \approx 99.5$ as shown.

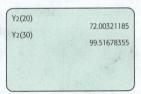

The TRACE feature can also be used. Thus, the average number of live births is 72 per 1000 women age 20 and 99.5 per 1000 women age 30.

EXERCISES

1. **Life science: live births.**

   a) Use REGRESSION to fit a quartic equation to the live-birth data. Graph both the scatterplot of the data and the quartic function. Does the quartic function give a better fit than either the quadratic or the cubic function?

   b) Explain why the domain of the cubic live-birth function should probably be restricted to the interval $[15, 45]$.

2. **Business: median household income by age.**

Age, x	Median Income in 2003
19.5	$27,053
29.5	44,779
39.5	55,044
49.5	60,242
59.5	49,215
65	23,787

   (*Source:* Based on data in the *Statistical Abstract of the United States, 2005.*)

   a) Make a scatterplot of the data and fit a quadratic function to the data using QuadReg. Then graph the quadratic function with the scatterplot.

   b) Fit a cubic function to the data using CubicReg. Then graph the cubic function with the scatterplot.

   c) Fit a quartic function to the data using QuartReg. Then graph the quartic function with the scatterplot.

   d) Which of the quadratic, cubic, or quartic functions seems to best fit the data?

   e) Use the function from part (d) to estimate the median household income of people age 25; of people age 45.

3. **Life science: hours of sleep and death rate.** Solve Example 3 using quadratic regression.

Exponential Models

Data that increase quickly as x increases may fit an exponential model of the form $f(x) = a_0 \cdot a^x$, where $a > 1$. If the data decrease in such a way that they "level off" as x increases, then the exponential model $f(x) = a_0 \cdot a^x$, where $0 < a < 1$, may fit.

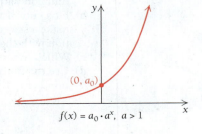 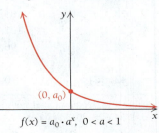

EXAMPLE 4   **Business: Annual Sales of Compact Discs.**   The graph below illustrates the annual sales (in millions of dollars, rounded to the nearest $250 million) of compact discs in the United States between 2007 and 2012.

ANNUAL SALES OF COMPACT DISCS

(*Source*: Based on data from the RIAA.)

a) Find a quadratic model, $f(x) = ax^2 + bx + c$, that fits the data, and use it to estimate annual sales in 2017.

b) Find an exponential model, $f(x) = a_0 \cdot a^x$, that fits the data, and use it to estimate annual sales in 2017.

c) Which model better predicts future annual sales of compact discs?

Solution

a) We find that $f(x) = 223.214x^2 - 2194.643x + 7982.143$ fits this data. Since 2017 is 10 years after 2007, we use $x = 10$. The estimate of annual sales of compact discs for 2017 is $f(10) = 8357.1$, or approximately $8357 million.

b) We find that $f(x) = 7503.176 \cdot (0.791)^x$ also fits the data. Using this model, the annual sales of compact discs for 2017 is $f(10) = 716.47$, or approximately $716.5 million.

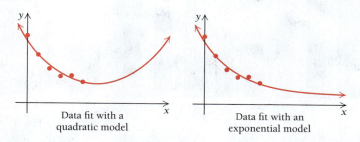

Data fit with a quadratic model   Data fit with an exponential model

c) With the quadratic model, annual sales increase as x (time) increases, so it is probably not a very accurate model for predicting future sales of compact discs. The exponential model yields decreasing sales as x increases. Thus, it is probably a more accurate predictor of future sales of compact discs. ∎

R.6 Exercise Set

Choosing models.   *For the scatterplots and graphs in Exercises 1–9, determine which, if any, of the following functions might be the best model for the data:*

Linear, $f(x) = mx + b$

Quadratic, $f(x) = ax^2 + bx + c, a > 0$

Quadratic, $f(x) = ax^2 + bx + c, a < 0$

Polynomial, neither quadratic nor linear

Exponential, $f(x) = a_0 \cdot a^x, 0 < a < 1$ or $a > 1$

1.

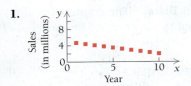

2.

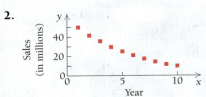

3.

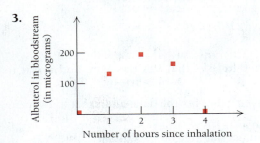

Business and Economics

4.

U.S. TRADE DEFICIT WITH JAPAN

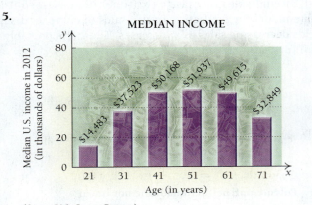

(*Source*: U.S. Census Bureau, Foreign Trade Statistics.)

5.

MEDIAN INCOME

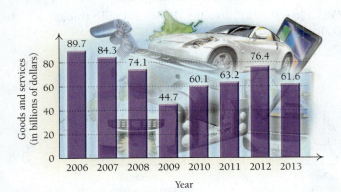

(*Source*: U.S. Census Bureau.)

6.

NBA AVERAGE SALARY

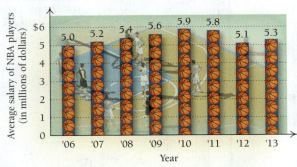

(*Source: The Compendium of Professional Basketball*, by Robert D. Bradley, Xaler Press, 2010; personal communication with author, 2013.)

7.

AVERAGE MONTHLY TEMPERATURE IN DALLAS

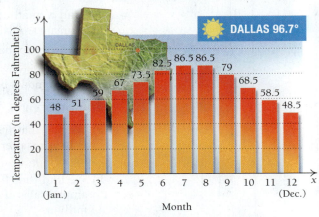

(*Source*: www.weather.com, Dallas, TX, 75235.)

8.

CABLE VIDEO CUSTOMERS

(*Source*: www.ncta.org.)

APPLICATIONS

Business and Economics

9. Cable video customers.

a) For the data shown in Exercise 8, find a linear function that fits the data using the values given for 2006 and 2012. Let x represent the number of years since 2006.

b) Use the linear function to estimate the number of cable video customers in 2018.

10. Average salary in NBA. Use the data from the bar graph in Exercise 6.

a) Find a linear function that fits the data using the average salaries given for the years 2006 and 2013. Use 0 for 2006 and 7 for 2013.

b) Use the linear function to predict average salaries in 2018 and 2025.

c) In what year will the average salary reach 7.0 million?

Life and Physical Sciences

11. Absorption of an asthma medication. Use the data from Exercise 3.

a) Find a quadratic function that fits the data using the data points $(0, 0)$, $(2, 200)$, and $(3, 167)$.

b) Use the function to estimate the amount of albuterol in the bloodstream 4 hr after inhalation.

c) Does it make sense to use this function for $t = 6$? Why or why not?

12. Median income. Use the data given in Exercise 5.

a) Find a quadratic function that fits the data using the data points $(21, 14483)$, $(41, 50168)$, and $(61, 49615)$.

b) Use the function to estimate the median income of Americans who were 55 in 2012.

13. Braking distance.

a) Find a quadratic function that fits the following data.

TRAVEL SPEED (mph)	BRAKING DISTANCE (ft)
20	25
40	105
60	300

(*Source*: New Jersey Department of Law and Public Safety.)

b) Use the function to estimate the braking distance of a car traveling at 50 mph.

c) Does it make sense to use this function when speeds are less than 15 mph? Why or why not?

14. Daytime accidents.

a) Find a quadratic function that fits the following data.

TRAVEL SPEED (in km/h)	NUMBER OF DAYTIME ACCIDENTS (for every 200 million km driven)
60	100
80	130
100	200

b) Use the function to estimate the number of daytime accidents that occur at 50 km/h for every 200 million km driven.

15. High blood pressure in women.

a) Choose two points from the following data and find a linear function that fits the data.

AGE OF FEMALE	PERCENTAGE OF FEMALES WITH HIGH BLOOD PRESSURE
30	1.4
40	8.5
50	19.1
60	31.9
70	53.0

(*Source*: Based on data from Health United States 2005, CDC/NCHS.)

b) Graph the scatterplot and the function on the same set of axes.

c) Use the function to estimate the percentage of 55-yr-old women with high blood pressure.

16. High blood pressure in men.

a) Choose two points from the following data and find a linear function that fits the data.

AGE OF MALE	PERCENTAGE OF MALES WITH HIGH BLOOD PRESSURE
30	7.3
40	12.1
50	20.4
60	24.8
70	34.9

(*Source*: Based on data from Health United States 2005, CDC/NCHS.)

b) Graph the scatterplot and the function on the same set of axes.

c) Use the function to estimate the percent of 55-yr-old men with high blood pressure.

In Exercises 17–20, give the numbers in the exponential function to three decimal places.

17. Population.

a) Find the exponential function that best fits the following data.

YEARS SINCE 1970	POPULATION OF TEXAS (in millions)
0	11.2
10	14.2
20	17.0
30	20.9
43	26.4

(*Source*: www.census.gov.)

b) Graph the scatterplot and the function on the same set of axes.

c) Use the function to estimate the population of Texas in 2020.

18. Population.

a) Find the exponential function that best fits the following data.

YEARS SINCE 1970	POPULATION OF DETROIT (in millions)
0	1.5
10	1.2
20	1.0
30	0.95
40	0.71

(*Source*: www.census.gov.)

b) Graph the scatterplot and the function on the same set of axes.

c) Use the function to estimate the population of Detroit in 2020.

19. Buying power.

a) Find the exponential function that best fits the following data.

YEARS SINCE 1980	EQUIVALENT BUYING POWER OF $100 (in 1980 dollars)
0	$100.00
10	$158.62
20	$208.98
30	$264.63
33	$282.85

(*Source*: http://data.bls.gov/cgi-bin/cpicalc.pl.)

b) Graph the scatterplot and the function on the same set of axes.

c) Use the function to estimate the equivalent buying power of $100 (in 1980 dollars) in 2020.

d) What other models also fit this data? Which model best predicts the equivalent buying power of $100 (in 1980 dollars) for future years? Why?

20. Stock prices.

a) Find the exponential function that best fits the following data.

YEARS SINCE 2010	PRICE OF ONE SHARE OF STARBUCKS STOCK AT BEGINNING OF JANUARY
0	$20.59
1	$30.21
2	$46.61
3	$55.39
4	$76.17

(*Source*: yahoo.finance; Nasdaq.)

b) Graph the scatterplot and the function on the same set of axes.

c) Use the function to estimate the price of one share of Starbucks stock in 2018.

d) What other models also fit this data? Which model best predicts the price of one share of Starbucks stock in future years? Why?

SYNTHESIS

21. Under what conditions might it make better sense to use a linear function rather than a quadratic or cubic function that fits a few data points more closely?

22. For modeling the number of hours of daylight for the dates April 22 to August 22, which would be a better choice: a linear function or a quadratic function? Why?

23. What restrictions should be placed on the domain of the quadratic function found in Exercise 11? Why are such restrictions needed?

24. What restrictions should be placed on the domain of the quadratic function found in Exercise 12? Why are such restrictions needed?

25. Business: cable video customers.

a) Use regression to fit a linear function to the data shown in Exercise 8.

b) Use the function to estimate the number of cable video customers in 2018.

c) Compare your answer for part (b) to that found in Exercise 9. Which is more accurate?

d) Fit an exponential function to the data and use it to estimate the number of cable video customers in 2018.

e) Is a linear or an exponential model more appropriate for this set of data? Why.

26. Business: trade deficit with Japan.

a) Use regression to fit a cubic function to the data in Exercise 4. Let x be the number of years after 2006.

b) Use the function to estimate the trade deficit with Japan in 2017.

c) Why might a linear function be a more logical choice than a cubic function for modeling this set of data?

Answers to Quick Checks

1. $y = 3.875x + 25.025$; 59.9 million high-speed internet subscribers. The model is different, and the prediction is higher.   2. $y = -0.683x^2 + 37.719x - 413.234$; 67.9 live births per 1000 women age 20.

KEY TERMS AND CONCEPTS

EXAMPLES

SECTION R.1

In an **ordered pair** (x, y), the first number is called the **first coordinate** and the second number is called the **second coordinate**. Together, they are the **coordinates** of the point.

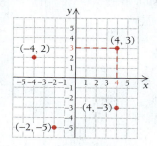

A **solution** of an equation in two variables is an ordered pair of numbers that, when substituted for the variables, forms a true equation. The **graph** of an equation is a drawing that represents all ordered pairs that are solutions of the equation.

This is the graph of the equation $y = x^4 - 2x^2$.

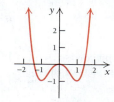

The mathematics used to represent the essential features of an applied problem comprise a **mathematical model**.

Business. The equation

$$A = \$1000\left(1 + \frac{r}{4}\right)^{7 \cdot 4}$$

is used to estimate the amount A, in dollars, to which $1000 will grow in 7 years, at interest rate r, compounded quarterly.

SECTION R.2

A **function** is a correspondence between a first set, called a **domain**, and a second set, called a **range**, for which each member of the domain corresponds to *exactly one* member of the range.

A graph represents a function if it is impossible to draw a vertical line that crosses the graph more than once.

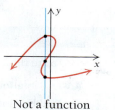

Not a function   A function

Function notation permits us to easily determine what **output**, or member of the range, is paired with a given **input**, or member of the domain.

The function given by $f(x) = x^2 - 5x - 8$ allows us to determine what member of the range is paired with the number 2 of the domain:

$$f(x) = x^2 - 5x - 8$$
$$f(2) = 2^2 - 5(2) - 8$$
$$= 4 - 10 - 8$$
$$= -14.$$

(continued)

KEY TERMS AND CONCEPTS	EXAMPLES

SECTION R.2 (continued)

In calculus, it is important to be able to simplify an expression like

$$\frac{f(x + h) - f(x)}{h}.$$

For $f(x) = x^2 - 5x - 8$,

$$\frac{f(x + h) - f(x)}{h} = \frac{[(x + h)^2 - 5(x + h) - 8] - f(x)}{h}$$

$$= \frac{x^2 + 2xh + h^2 - 5x - 5h - 8 - [x^2 - 5x - 8]}{h}$$

$$= \frac{x^2 + 2xh + h^2 - 5x - 5h - 8 - x^2 + 5x + 8}{h}$$

$$= \frac{2xh + h^2 - 5h}{h}$$

$$= \frac{h(2x + h - 5)}{h}$$

$$= 2x + h - 5, \ h \neq 0.$$

A function that is defined **piecewise** specifies different rules for parts of the domain.

To graph

$$g(x) = \begin{cases} \frac{1}{3}x + 3, & \text{for } x < 3, \\ -x, & \text{for } x \geq 3, \end{cases}$$

we plot $\frac{1}{3}x + 3$ for all inputs x less than 3 and $-x$ for all inputs x greater than or equal to 3.

$g(x) = \frac{1}{3}x + 3$, for $x < 3$

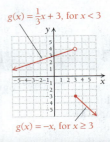

$g(x) = -x$, for $x \geq 3$

SECTION R.3

The domain of any function is a **set**, or collection of objects. In this book, the domain is usually a subset of the real numbers.

There are three ways of describing sets:

Roster Notation: $\{-2, 5, 9, \pi\}$

Set-Builder Notation: $\{x \mid x \text{ is a real number and } x \geq 3\}$

Interval Notation:

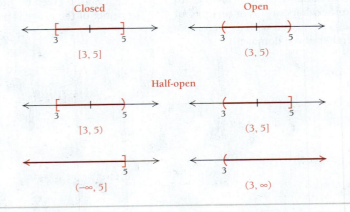

KEY TERMS AND CONCEPTS

EXAMPLES

SECTION R.3 (continued)

For a function given by an equation or formula, the domain is the largest set of real numbers (inputs) for which function values (outputs) can be calculated.

Function

Domain

$$f(x) = \frac{8x}{x^2 - 9} = \frac{8x}{(x - 3)(x + 3)}$$

All real numbers except -3 and 3.

$$g(x) = \sqrt{10 - 5x}$$

All real numbers x for which $10 - 5x \geq 0$ or $10 \geq 5x$ or $2 \geq x$. The domain is $(-\infty, 2]$.

$$h(x) = |x - 3| + 5x^2$$

All real numbers, $\mathbb{R}$.

SECTION R.4

The **slope** m of the line containing the points (x_1, y_1) and (x_2, y_2) is given by $m = \frac{y_2 - y_1}{x_2 - x_1}$ and can be regarded as **average rate of change**. An equation of the form $f(x) = mx + b$ is said to be in **slope–intercept form**. Its graph has **slope** m and **y-intercept** $(0, b)$.

An equation of the form $y - y_1 = m(x - x_1)$ is said to be in **point–slope form**. Its graph is a line with slope m passing through the point (x_1, y_1).

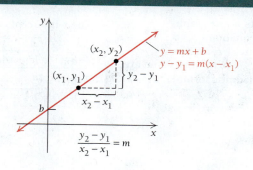

The graph of a **constant function**, given by $f(x) = c$, is a horizontal line.

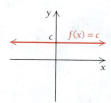

The graph of an equation of the form $x = a$ is a vertical line and is not a function.

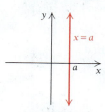

The variable y **varies directly** with x if there exists some positive constant m for which $y = mx$. We say that m is the **variation constant**.

$$y = 3.14x, \qquad A = 1.08t, \qquad I = kR, \qquad y = \frac{1}{3}x$$

Total profit is the difference between **total revenue** and **total cost**.

The value x for which $P(x) = 0$ is the **break-even value**.

For $P(x) = R(x) - C(x)$, $R(x) = 84x$, and $C(x) = 28x + 56{,}000$, the break-even value is found as follows:

$$R(x) = C(x)$$
$$84x = 28x + 56{,}000$$
$$56x = 56{,}000$$
$$x = 10{,}000.$$

(continued)

KEY TERMS AND CONCEPTS	EXAMPLES

SECTION R.5

The graph of a **quadratic function** $f(x) = ax^2 + bx + c$ is a **parabola** with a **vertex** at $\left(-\dfrac{b}{2a}, f\left(-\dfrac{b}{2a}\right)\right)$.

The graph opens **upward** if $a > 0$ and **downward** if $a < 0$. It is an example of a nonlinear function.

The vertical line $x = -\dfrac{b}{2a}$ is the **line of symmetry**.

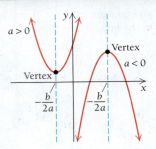

The *x*-intercepts of the graph of a quadratic equation, $f(x) = ax^2 + bx + c$, are the solutions of $ax^2 + bx + c = 0$ and are given by the quadratic formula

$$x = \frac{-b \pm \sqrt{b^2 - 4ac}}{2a}.$$

The *x*-intercepts of $f(x) = x^2 - 6x - 10$ are found by solving $x^2 - 6x - 10 = 0$. They are

$$\frac{-b + \sqrt{b^2 - 4ac}}{2a} = \frac{-(-6) + \sqrt{(-6)^2 - 4(1)(-10)}}{2(1)} = 3 + \sqrt{19}, \text{ and}$$

$$\frac{-b - \sqrt{b^2 - 4ac}}{2a} = \frac{-(-6) - \sqrt{(-6)^2 - 4(1)(-10)}}{2(1)} = 3 - \sqrt{19}.$$

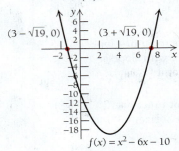

A **polynomial function** is a function like those shown at the right.

$f(x) = -8$, constant;
$g(x) = 3x - 7$, linear;
$h(x) = x^2 - 6x - 10$, quadratic;
$p(x) = 2x^3 - 7x^2 + 0.23x - 10$, cubic.

Power functions are polynomial functions of the form $f(x) = ax^n$, where n is a positive integer.

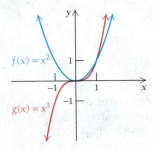

A **rational function** is a function that can be expressed as the quotient of two polynomials.

$$g(x) = \frac{x^2 - 9}{x + 3}, \quad h(x) = \frac{x^3 - 4x^2 + x - 5}{x^2 - 6x - 10},$$

$$f(x) = \frac{x^3 - 4x^2 + x - 5}{1} = x^3 - 4x^2 + x - 5.$$

KEY TERMS AND CONCEPTS	EXAMPLES

SECTION R.5 (continued)

The domain of a rational function is restricted to those input values that do not result in division by zero.

For $f(x) = \dfrac{1}{x - 3}$, the domain is the set of all real numbers except 3, or $(-\infty, 3) \cup (3, \infty)$.

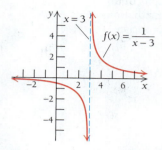

An **absolute-value** function is described using absolute value symbols.

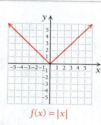
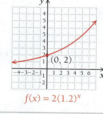

$f(x) = |x|$ $g(x) = |x + 2|$

A **square-root function** is a function described using a square root symbol.

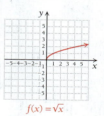
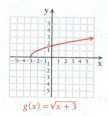

$f(x) = \sqrt{x}$ $g(x) = \sqrt{x + 3}$

An **exponential function** can be written in the form $f(x) = a_0 \cdot a^x$, $a > 0$, $a \neq 1$. The domain of an exponential function is all real numbers.

The range is $y > 0$, and the y-intercept is $(0, a_0)$. There is no x-intercept.

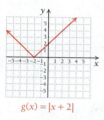

$f(x) = 2(1.2)^x$ $g(x) = 3(0.8)^x$

The point at which demand and supply are the same is called an **equilibrium point**.

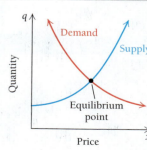

(continued)

KEY TERMS AND CONCEPTS	EXAMPLES

SECTION R.6

When real-world data are available, we can form a **scatterplot** and decide which, if any, of the graphs in this chapter best models the situation. Then we can use an appropriate model to make predictions.

Business.

Average Gasoline Prices, 2003–2013

YEARS SINCE 2003	PRICE PER GALLON
0	1.52
1	1.81
2	2.25
3	2.53
4	2.77
5	3.21
6	2.33
7	2.74
8	3.48
9	3.55
10	3.45

(*Source*: eia.gov/petroleum/gasprices.)

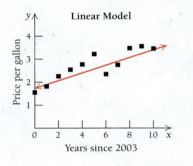

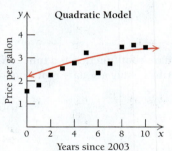

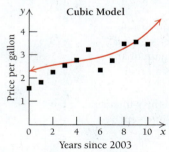

CHAPTER R
Review Exercises

These review exercises are for test preparation. They can also serve as a practice test. Answers are at the back of the book. Red bracketed section references indicate the part(s) of the chapter to restudy if your answer is incorrect.

CONCEPT REINFORCEMENT

For each equation in column A, select the most appropriate graph in column B. [R.1, R.4, R.5]

Column A   **Column B**

1. $y = |x|$ **a)**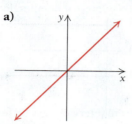

2. $f(x) = x^2 - 1$ **b)**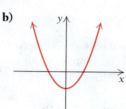

3. $y = -2x - 1$ **c)**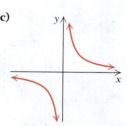

4. $y = x$ **d)**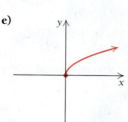

5. $g(x) = \sqrt{x}$ **e)**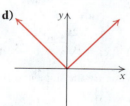

6. $f(x) = 0.5(1.2)^x$ **f)**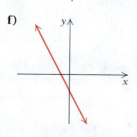

7. $f(x) = \dfrac{1}{x}$ **g)**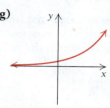

In Exercises 8–14, classify each statement as either true or false.

8. The graph of an equation represents all ordered pairs that are solutions of the equation. [R.1]

9. If $f(-3) = 5$ and $f(3) = 5$, then f cannot be a function. [R.2]

10. The notation $(3, 7)$ can represent a point or an interval. [R.3]

11. An equation of the form $y - y_1 = m(x - x_1)$ has a graph that is a line of slope m passing through (x_1, y_1). [R.4]

12. The graph of an equation of the form $f(x) = ax^2 + bx + c$ has its vertex at $x = b/(2a)$. [R.5]

13. A scatterplot is a random collection of points near a line. [R.6]

14. Unless stated otherwise, the domain of a polynomial function is the set of all real numbers. [R.5]

REVIEW EXERCISES

15. Hearing-impaired Americans. The following graph shows the number of hearing-impaired Americans who are x years old. (*Source*: Better Hearing Institute.) [R.1, R.5]

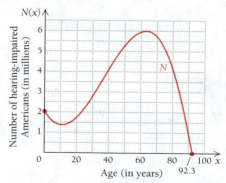

a) What is the number of hearing-impaired Americans aged 40?

b) For what age(s) are about 3,000,000 Americans hearing-impaired?

c) What is a reasonable estimate of the domain of this function? Why?

16. Finance: compound interest. Sam borrows $4000 at 12%, compounded annually. How much does she owe at the end of 2 yr? [R.1]

17. Business: compound interest. Suppose $1100 is invested at 5%, compounded semiannually. How much is in the account at the end of 4 yr? [R.1]

18. Is the following correspondence a function? Why or why not? [R.2]

19. A function is given by $f(x) = -x^2 + x$. Find each of the following. [R.2]

a) $f(3)$ **b)** $f(-5)$

c) $f(a)$ **d)** $f(x + h)$

Graph. [R.5]

20. $f(x) = (x - 2)^2$ **21.** $y = |x + 1|$

22. $f(x) = \dfrac{x^2 - 16}{x + 4}$ **23.** $g(x) = \sqrt{x} + 1$

Use the vertical-line test to determine whether each of the following is the graph of a function. [R.2]

24. 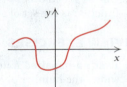   **25.**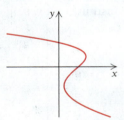

26.    **27.**

28. For the graph of function f shown to the right, determine **(a)** $f(2)$; **(b)** the domain; **(c)** all x-values for which $f(x) = 2$; and **(d)** the range. [R.3]

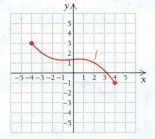

29. Consider the function given by

$$f(x) = \begin{cases} -x^2 + 2, & \text{for } x < 1, \\ 4, & \text{for } 1 \le x < 2, \\ \frac{1}{2}x, & \text{for } x \ge 2. \end{cases}$$

a) Find $f(-1), f(1.5)$, and $f(6)$. [R.2]

b) Graph the function. [R.2]

30. Write interval notation for each graph. [R.3]

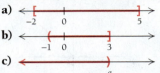

31. Write interval notation for each of the following. Then graph each interval on a number line. [R.3]

a) $\{x | -4 \le x < 5\}$ **b)** $\{x | x > 2\}$

32. For the function graphed below, determine **(a)** $f(-3)$; **(b)** the domain; **(c)** all x-values for which $f(x) = 4$; **(d)** the range. [R.3]

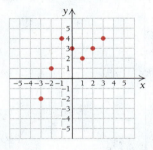

33. Find the domain of f. [R.3, R.5]

a) $f(x) = \dfrac{7}{2x - 10}$

b) $f(x) = \sqrt{x + 6}$

34. What are the slope and the y-intercept of $y = -3x + 2$? [R.4]

35. Find an equation of the line with slope $\frac{1}{4}$, containing the point $(8, -5)$. [R.4]

36. Find the slope of the line containing the points $(2, -5)$ and $(-3, 10)$. [R.4]

Find the average rate of change. [R.4]

37.

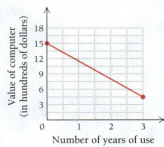

38.

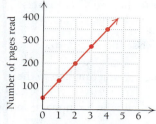

Number of pages read (vertical axis: 100, 200, 300, 400)
Number of days spent reading (horizontal axis: 0 1 2 3 4 5 6)

39. Business: shipping charges. The amount A that Pet-Treats-to-U charges for shipping is directly proportional to the value V of the item(s) being shipped. If the business charges $2.10 to ship a $60 gift basket, find an equation of variation expressing A as a function of V. [R.4]

40. Business: profit-and-loss analysis. The band Soul Purpose has fixed costs of $4000 for producing a new CD. Thereafter, the variable costs are $0.50 per CD, and the CD will sell for $10. [R.4]

a) Find and graph $C(x)$, the total cost of producing x CDs.

b) Find and graph $R(x)$, the total revenue from the sale of x CDs. Use the same axes as in part (a).

c) Find and graph $P(x)$, the total profit from the production and sale of x CDs. Use the same axes as in part (b).

d) How many CDs must the band sell in order to break even?

41. Graph each pair of equations on one set of axes. [R.5]

a) $y = \sqrt{x}$ and $y = \sqrt{x} - 3$

b) $y = x^3$ and $y = (x - 1)^3$

42. Graph each of the following. If the graph is a parabola, identify the vertex. [R.5]

a) $f(x) = x^2 - 6x + 8$ **b)** $g(x) = \sqrt[3]{x} + 2$

c) $y = -\dfrac{1}{x}$ **d)** $y = \dfrac{x^2 + x - 6}{x - 2}$

43. Solve each of the following. [R.5]

a) $5 + x^2 = 4x + 2$

b) $2x^2 = 4x + 3$

44. Rewrite each of the following as an equivalent expression with a rational exponent. [R.5]

a) $\sqrt[5]{x^4}$ **b)** $\sqrt{t^8}$

c) $\dfrac{1}{\sqrt[3]{m^2}}$ **d)** $\dfrac{1}{\sqrt{x^2 - 9}}$

45. Rewrite each of the following as an equivalent expression using radical notation. [R.5]

a) $x^{2/5}$ **b)** $m^{-3/5}$

c) $(x^2 - 5)^{1/2}$ **d)** $\dfrac{1}{t^{-1/3}}$

46. Determine the domain of the function given by $f(x) = \sqrt[8]{2x - 9}$. [R.5]

47. Graph each of the following, and identify the y-intercept. [R.5]

a) $y = \dfrac{1}{2} \cdot (4)^x$ **b)** $y = 3 \cdot \left(\dfrac{1}{4}\right)^x$

48. Economics: equilibrium point. Find the equilibrium point for the given demand and supply functions. [R.5]

Demand: $q = (x - 7)^2$

Supply: $q = x^2 + x + 4$ (assume $x \le 7$)

Quantity (in hundreds) (vertical axis: 4, 8, 12, 16, 20, 24, 28)
Price per item (in dollars) (horizontal axis: 1 2 3 4 5 6 7)

49. Trail maintenance. The amount of time required to maintain a section of the Appalachian Trail varies inversely with the number of volunteers working. If a particular section of trail can be cleared in 4 hr by 9 volunteers, how long would it take 11 volunteers to clear the same section? [R.5]

50. Life science: maximum heart rate. A person exercising should not exceed a maximum heart rate, which depends on his or her gender, age, and resting heart rate. The following table shows data relating resting heart rate and maximum heart rate for a 20-yr-old woman. [R.6]

RESTING HEART RATE, r (in beats per minute)	MAXIMUM HEART RATE, M (in beats per minute)
50	170
60	172
70	174
80	176

(*Source*: American Heart Association.)

a) Using the data points (50, 170) and (80, 176), find a linear function that fits the data.

b) Graph the scatterplot and the function on the same set of axes.

c) Use the function to predict the maximum heart rate of a woman whose resting heart rate is 67.

51. Business: ticket profits. The Spring Valley Drama Troupe is performing a new play. Data relating daily profit P to the number of days after opening night appear below. [R.6]

DAYS, x	0	9	18	27	36	45
PROFIT, P (in dollars)	870	548	−100	−100	510	872

a) Make a scatterplot of the data.

b) Do the data seem to fit a quadratic function?

c) Using the data points $(0, 870)$, $(18, -100)$, and $(45, 872)$, find a quadratic function that fits the data.

d) Use the function from part (c) to estimate the profit made on the 30th day.

e) Make an estimate of the domain of this function. Why must it have restrictions?

SYNTHESIS

52. Economics: demand. The demand function for Clifton Cheddar Cheese is given by

Demand: $q = 800 - x^3$, $0 \le x \le 9.28$,

where x is the price per pound and q is in thousands of pounds. [R.5]

a) Find the number of pounds sold when the price per pound is $6.50.

b) Find the price per pound when 720,000 lb are sold.

TECHNOLOGY CONNECTION

Graph each function and find the zeros, the domain, and the range. [R.5]

53. $f(x) = x^3 - 9x^2 + 27x + 50$

54. $f(x) = \sqrt[3]{|4 - x^2|} + 1$

55. Approximate the point(s) of intersection of the graphs of the two functions in Exercises 53 and 54. [R.5]

56. Life science: maximum heart rate. Use the data in Exercise 50. [R.6]

a) Use regression to fit a linear function to the data.

b) Use the linear function to predict the maximum heart rate of a woman whose resting heart rate is 67.

---

c) Compare your answer to that found in Exercise 50. Are the answers equally reliable? Why or why not?

57. Business: ticket profits. Use the data in Exercise 51. [R.6]

a) Use regression to fit a quadratic function to the data.

b) Use the function to estimate the profit made on the 30th day.

c) What factors might cause the Spring Valley Drama Troupe's profit to drop and then rise?

58. Social sciences: time spent on home computer. The data in the table below relate A, the average number of minutes spent per month on a home computer, to a person's age, x, in years. [R.6]

a) Use regression to fit linear, quadratic, cubic, and quartic functions to the data.

b) Make a scatterplot of the data, and graph each function on the scatterplot.

c) Which function provides the best model for the data? Why?

AGE (in years)	AVERAGE USE (in minutes per month)
6.5	363
14.5	645
21	1377
29.5	1727
39.5	1696
49.5	2052
55	2299

(*Source*: Media Matrix; The PC Meter Company.)

CHAPTER R
Test

1. Business: compound interest. Cecilia invests funds at 6.5% compounded annually. The investment grows to $798.75 in 1 yr. How much was originally invested?

2. A function is given by $f(x) = -x^2 + 5$. Find:

a) $f(-3)$;

b) $f(a + h)$.

3. Find the slope and the y-intercept of the graph of $y = \frac{4}{5}x - \frac{2}{3}$.

4. Find an equation of the line with slope $\frac{1}{4}$, containing the point $(-3, 7)$.

5. Find the slope of the line containing the points $(-9, 2)$ and $(3, -4)$.

Find the average rate of change.

6.

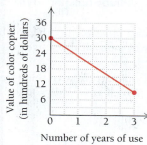

Value of color copier (in hundreds of dollars) vs. Number of years of use

7.

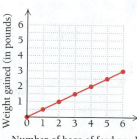

Weight gained (in pounds) vs. Number of bags of feed used

8. Life science: body fluids.   The weight F of fluids in a human is directly proportional to body weight W. It is known that a 180-lb person has 120 lb of fluids. Find an equation of variation expressing F as a function of W.

9. Business: profit-and-loss analysis.   A printing shop has fixed costs of \$8000 for producing a newly designed note card. The variable costs are \$0.08 per card. The revenue from each card will be \$0.50.

a) Find $C(x)$, the total cost of producing x cards.

b) Find $R(x)$, the total revenue from the sale of x cards.

c) Find $P(x)$, the total profit from the production and sale of x cards.

d) How many cards must the company sell in order to break even?

10. Economics: equilibrium point.   Find the equilibrium point for these demand and supply functions:

Demand: $q = (x - 8)^2$, $0 \le x \le 8$,

Supply: $q = x^2 + x + 13$,

given that x is the unit price, in dollars, and q is the quantity demanded or supplied, in thousands.

Use the vertical-line test to determine whether each of the following is the graph of a function.

11.

12.

13. For the following graph of a quadratic function f, determine **(a)** $f(1)$; **(b)** the domain; **(c)** all x-values for which $f(x) = 4$; and **(d)** the range.

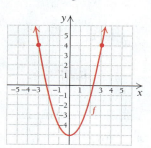

14. Graph: $f(x) = 8/x$.

15. Convert to a rational exponent: $1/\sqrt{t}$.

16. Convert to radical notation: $t^{-3/5}$.

17. Graph: $f(x) = \dfrac{x^2 - 1}{x + 1}$.

Determine the domain of each function.

18. $f(x) = \dfrac{x^2 + 20}{x^2 + 5x - 14}$

19. $f(x) = \dfrac{x}{\sqrt{3x + 6}}$

20. Write interval notation for the following graph.

21. Graph:

$$f(x) = \begin{cases} x^2 + 2, & \text{for } x \ge 0, \\ x^2 - 2, & \text{for } x < 0. \end{cases}$$

22. Graph, and identify the y-intercept:

$$f(x) = \frac{1}{2} \cdot (3)^x.$$

23. Nutrition.   As people age, their daily caloric needs change. The following table shows data for physically active females, relating age, in years, to number of calories needed daily.

AGE	NUMBER OF CALORIES NEEDED DAILY
6	1800
11	2200
16	2400
24	2400
41	2200

(*Source:* Based on data from U.S. Department of Agriculture.)

a) Make a scatterplot of the data.

b) Do the data appear to fit a quadratic function?

c) Using the data points $(6, 1800)$, $(16, 2400)$, and $(41, 2200)$, find a quadratic function that fits the data.

d) Use the function from part (c) to predict the number of calories needed daily by a physically active 30-yr-old woman.

e) What restrictions are reasonable for the domain of the function from part (c)? Why?

SYNTHESIS

24. Simplify: $(64^{4/3})^{-1/2}$.

25. Find the domain and the zeros of the function given by
$$f(x) = (5 - 3x)^{1/4} - 7.$$

26. Write an equation that has exactly three solutions: -3, 1, and 4. Answers may vary.

27. A function's average rate of change over the interval $[1, 5]$ is $-\frac{3}{7}$. If $f(1) = 9$, find $f(5)$.

TECHNOLOGY CONNECTION

28. Graph f and find its zeros, domain, and range:
$$f(x) = \sqrt[3]{|9 - x^2|} - 1.$$

29. **Nutrition.**   Use the data in Exercise 23.

   a) Use REGRESSION to fit a quadratic function to the data.

   b) Use the function from part (a) to predict the number of calories needed daily by a physically active 30-yr-old woman.

   c) Compare your answer from part (b) with that from part (d) of Exercise 23. Which answer do you feel is more accurate? Why?

EXTENDED TECHNOLOGY APPLICATION

Average Price of a Movie Ticket

Extended Technology Applications occur at the ends of all chapters. They consider certain applications in greater depth, make use of calculator skills, and allow for possible group or collaborative learning.

   Have you noticed that the price of a movie ticket seems to increase over time? The table and graph that follow show the average price of a movie ticket for the years 1950 to 2010.

   How much did you pay the last time you went to an evening movie? The average prices in the table may seem low, but they reflect discounts for matinees, children, and senior citizens. Let's use REGRESSION to analyze the data.

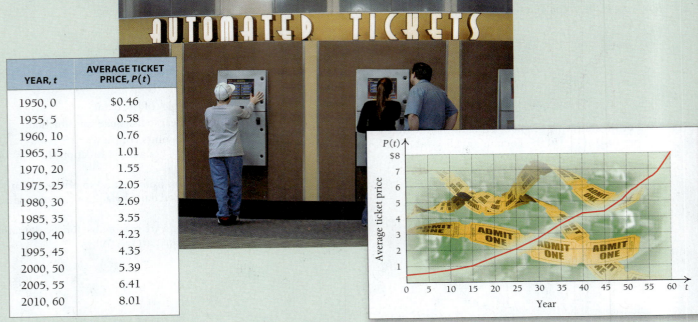

YEAR, t	AVERAGE TICKET PRICE, P(t)
1950, 0	$0.46
1955, 5	0.58
1960, 10	0.76
1965, 15	1.01
1970, 20	1.55
1975, 25	2.05
1980, 30	2.69
1985, 35	3.55
1990, 40	4.23
1995, 45	4.35
2000, 50	5.39
2005, 55	6.41
2010, 60	8.01

(*Source*: Motion Picture Association of America.)

Exercises

1. a) Using REGRESSION, find a linear function that fits the data.
  b) Graph the linear function.
  c) Use the linear function to predict the average price of a movie ticket in 2015 and in 2020. Do these estimates appear reasonable?
  d) Use the function to predict the year when the average price of a ticket will reach $20. Does this estimate seem reasonable?

2. a) Using REGRESSION, find a quadratic function,
$$y = ax^2 + bx + c,$$
  that fits the data.
  b) Graph the quadratic function.
  c) Use the quadratic function to predict the average price of a movie ticket in 2015 and in 2020. Do these estimates appear reasonable?
  d) Use the function to predict the year when the average price of a ticket will reach $20. Does this estimate seem reasonable?

3. a) Using REGRESSION, find a cubic function,
$$y = ax^3 + bx^2 + cx + d,$$
  that fits the data.
  b) Graph the cubic function.
  c) Use the cubic function to predict the average price of a movie ticket in 2015 and in 2020. Do these estimates appear reasonable?
  d) Use the function to predict the year when the average price of a ticket will reach $20. Does this estimate seem reasonable?

4. a) Using REGRESSION, find a quartic function,
$$y = ax^4 + bx^3 + cx^2 + dx + e,$$
  that fits the data.
  b) Graph the quartic function.
  c) Use the quartic function to predict the average price of a movie ticket in 2015 and in 2020. Do these estimates appear reasonable?
  d) Use the function to predict the year when the average price of a ticket will reach $20. Does this estimate seem reasonable?

5. a) Using REGRESSION, find an exponential function,
$$y = a_0 \cdot a^x,$$
  that fits the data.
  b) Graph the exponential function.
  c) Use the exponential function to predict the average price of a movie ticket in 2015 and 2020. Do these estimates appear reasonable?
  d) Use the function to predict the year when the average price of a ticket will reach $20. Does this estimate seem reasonable?

6. You are a research statistician assigned the task of making an accurate prediction of movie ticket prices.
  a) Why might you not use the linear function?
  b) Why might you use the quadratic function rather than the linear function?
  c) Why might you use the exponential function rather than the quadratic function?

There are yet other procedures a statistician uses to choose predicting functions but they are beyond the scope of this text.

1 Differentiation

What You'll Learn

Why It's Important

With this chapter, we begin our study of calculus. The first concepts we consider are *limits* and *continuity*. We apply those concepts to establishing the first of the two main building blocks of calculus: differentiation.

Consider *differentiation*, a process that takes a formula for a function and derives another function, called a *derivative*. A derivative represents an instantaneous rate of change. Throughout the chapter, we develop techniques for finding derivatives and explore their many applications.

Where It's Used

Ticket Prices: The price, *p*, of a ticket to the Super Bowl *t* years after 1967 can be estimated by the model shown in the graph. Find the rate of change of the ticket price with respect to the year, dp/dt. (*This model appears in Exercise 94 in Section 1.5.*)

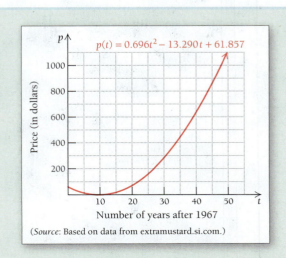

$$p(t) = 0.696t^2 - 13.290t + 61.857$$

Price (in dollars)

Number of years after 1967

(*Source*: Based on data from extramustard.si.com.)

● Find limits of functions, if they exist, using numerical or graphical methods.

Limits: A Numerical and Graphical Approach

In this section, we discuss the concept of a *limit*. The discussion is intuitive—that is, it relies on prior experience and lacks formal proof.

Suppose a football team has the ball on its own 10-yard line. Then, because of a penalty, the referee moves the ball back half the distance to the goal line; the ball is now on the 5-yard line. If the penalty is repeated, the ball will again be moved half the distance to the goal line, to the 2.5-yard line. If the penalty is repeated over and over again, the ball will move steadily *closer* to the goal line but never be placed *on* the goal line. We say that the *limit* of the distance between the ball and the goal line is zero.

Limits

One important aspect of calculus is analysis of how function values (outputs) change as input values change. Basic to this study is the notion of a limit. Suppose a function f is given and we have x-values (the inputs) that are increasingly close to some number a. If the corresponding values of $f(x)$ get closer and closer to some number L, then that number is called the *limit* of $f(x)$ as x approaches a.

For example, let $f(x) = 2x + 3$ and select x-values that get closer and closer to 4. In the table and graph below, we see that as input values approach 4 from the left (that is, are less than 4), output values approach 11, and as input values approach 4 from the right (that is, are greater than 4), output values also approach 11. Thus, we say:

As x approaches 4 from either side, the function $f(x) = 2x + 3$ approaches 11.

Limit Graphically

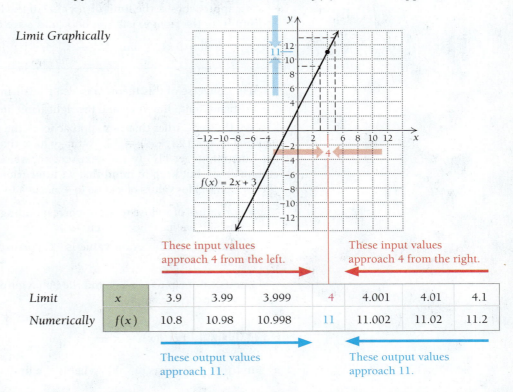

These input values approach 4 from the left.

These input values approach 4 from the right.

Limit	x	3.9	3.99	3.999	4	4.001	4.01	4.1
Numerically	$f(x)$	10.8	10.98	10.998	11	11.002	11.02	11.2

These output values approach 11.

These output values approach 11.

An arrow, →, is often used to stand for the words "approaches from either side." Thus, the statement above can be written:

As $x \to 4$, $2x + 3 = 11$.

The number 11 is said to be the *limit* of $2x + 3$ as x approaches 4 from either side. We can abbreviate this statement as follows:

$$\lim_{x \to 4} (2x + 3) = 11.$$

This is read: "The limit, as x approaches 4, of $2x + 3$ is 11."

DEFINITION

As x approaches a, the **limit** of $f(x)$ is L, written

$$\lim_{x \to a} f(x) = L,$$

if all values of $f(x)$ are close to L for values of x that are sufficiently close, but not necessarily equal, to a. The limit L, if it exists, must be a unique real number.

When we write $\lim_{x \to a} f(x)$, we are indicating that x is approaching a from both sides. If we want to specify the side from which x-values approach a, we write

$\lim_{x \to a^-} f(x)$ to indicate the limit from the left (that is, where $x < a$),

or $\lim_{x \to a^+} f(x)$ to indicate the limit from the right (that is, where $x > a$).

These are called *left-hand* and *right-hand limits*, respectively. For a limit to exist, both the left-hand and right-hand limits must exist and be the same.

THEOREM

As x approaches a, the limit of $f(x)$ is L if the limit from the left exists and the limit from the right exists and both limits are L. That is,

$$\text{if } \lim_{x \to a^+} f(x) = \lim_{x \to a^-} f(x) = L, \text{ then } \lim_{x \to a} f(x) = L.$$

The converse of this theorem is also true: if $\lim_{x \to a} f(x) = L$, then it follows that both the left-hand limit, $\lim_{x \to a^-} f(x)$, and the right-hand limit, $\lim_{x \to a^+} f(x)$, exist and are equal to L.

We showed earlier that as x approaches 4, the value of $2x + 3$ approaches 11, summarized as $\lim_{x \to 4} (2x + 3) = 11$. You may have thought "Why not just substitute 4 into the function to get 11?" In the next section, we see that we can use such shortcuts in certain cases, but keep in mind that we are curious about the behavior of the function $f(x) = 2x + 3$ for values of x close to 4, not at 4 itself. Let's summarize:

- For values of x close to 4, the corresponding values of $f(x)$ are close to 11; this is the *limit* we have been discussing.

- At $x = 4$, the function value is 11, corresponding to the point $(4, 11)$ on the graph of f.

Limits can help us understand the behavior of some functions more clearly. Consider the following example.

EXAMPLE 1   Let $f(x) = \dfrac{x^2 - 1}{x - 1}$.

a) Find $f(1)$, if it exists.   **b)** What is the limit of $f(x)$ as x approaches 1?

Solution

a) $f(1)$ does not exist, since 1 is not in the domain of f.

b) Selecting x-values close to 1 on either side (see the following tables), we see that the function values get close to 2. Thus, the limit of $f(x)$ as x approaches 1 is 2:
$$\lim_{x \to 1} f(x) = 2.$$

TECHNOLOGY CONNECTION

Finding Limits Using the TABLE and TRACE Features

Consider the function given by $f(x) = 3x - 1$. Let's use the TABLE feature to complete the following table. Note that the inputs do not have the same increment from one to the next, and they approach 6 from both the left and the right. We use TblSet and select Indpnt and Ask mode. Then we enter the inputs shown and use the corresponding outputs to complete the table.

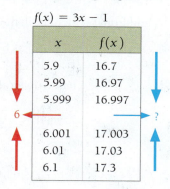

$$f(x) = 3x - 1$$

x	$f(x)$
5.9	16.7
5.99	16.97
5.999	16.997
6	?
6.001	17.003
6.01	17.03
6.1	17.3

Now we set the table in Auto mode and, starting (TblStart) with a number near 6, make tables for some increments (ΔTbl) like 0.1, 0.01, -0.1, -0.01, and so on, to determine $\lim_{x \to 6} f(x)$.

As an alternative, graphical approach, let's use TRACE. We move the cursor from left to right so that the x-coordinate approaches 6 from the left, zooming as needed, to see what happens. In general, TRACE is not an efficient way to find limits, but it will help in visualizing the limit process at this early stage.

Using TABLE and TRACE, we have

$$\lim_{x \to 6^+} f(x) = \boxed{17} \text{ and } \lim_{x \to 6^-} f(x) = \boxed{17}.$$

Thus,

$$\lim_{x \to 6} f(x) = \boxed{17}.$$

EXERCISES

Let $f(x) = 3x - 1$. Use TABLE and TRACE to find each of the following.

1. $\lim_{x \to 2} f(x)$ **2.** $\lim_{x \to -1} f(x)$

Let $g(x) = x^3 - 2x - 2$ for Exercises 3–5.

3. Complete the following table.

x	$g(x)$
7.9	475.24
7.99	492.1
7.999	493.81
8	?
8.001	494.19
8.01	495.9
8.1	513.24

Use TABLE and TRACE to find each of the following.

4. $\lim_{x \to 8} g(x)$ **5.** $\lim_{x \to -1} g(x)$

Quick Check 1 ✔

Let

$$f(x) = \frac{x^2 - 9}{x - 3}.$$

(See Example 5 in Section R.5.)

a) What is $f(3)$?

b) What is the limit of f as x approaches 3?

The graph has a "hole" at the point $(1, 2)$. Thus, even though the function is not defined at $x = 1$, the limit *does* exist as $x \to 1$.

Limit Numerically

$x \to 1^-$ $(x < 1)$	$f(x)$
0.9	1.9
0.99	1.99
0.999	1.999

$x \to 1^+$ $(x > 1)$	$f(x)$
1.1	2.1
1.01	2.01
1.001	2.001

Limit Graphically

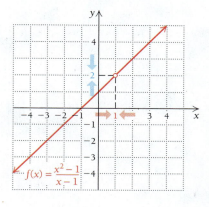

$$f(x) = \frac{x^2 - 1}{x - 1}$$

1 ✔

Limits are also useful when discussing piecewise-defined functions.

EXAMPLE 2   Consider the function H given by

$$H(x) = \begin{cases} 2x + 2, & \text{for } x < 1, \\ 2x - 4, & \text{for } x \geq 1. \end{cases}$$

Graph the function and find the following limits, if they exist, both numerically and graphically.

a) $\lim\limits_{x \to 1} H(x)$

b) $\lim\limits_{x \to -3} H(x)$

Solution   We check the limits from the left and from the right both numerically, with an input–output table, and graphically.

a)   *Limit Numerically*   *Limit Graphically*

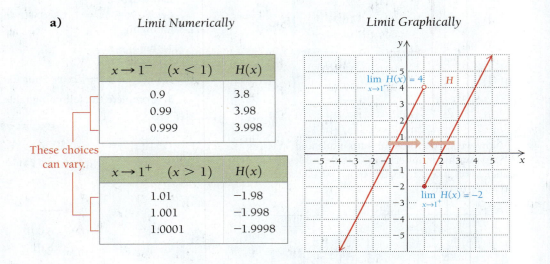

$x \to 1^-$ $(x < 1)$	$H(x)$
0.9	3.8
0.99	3.98
0.999	3.998

These choices can vary.

$x \to 1^+$ $(x > 1)$	$H(x)$
1.01	−1.98
1.001	−1.998
1.0001	−1.9998

As inputs x approach 1 from the left (see the upper table), outputs $H(x)$ approach 4. Thus, the limit from the left is 4. That is,

$$\lim\limits_{x \to 1^-} H(x) = 4.$$

But as inputs x approach 1 from the right (see the lower table), outputs $H(x)$ approach −2. Thus, the limit from the right is −2. That is,

$$\lim\limits_{x \to 1^+} H(x) = -2.$$

Since the limit from the left, 4, is not the same as the limit from the right, −2, we say that

$$\lim\limits_{x \to 1} H(x) \text{ does not exist.}$$

Note that $H(1) = -2$. In this example, the function value exists for $x = 1$, but the limit as $x \to 1$ does not exist.

TECHNOLOGY CONNECTION

Exploratory

Check the results of Examples 1 and 2 using the TABLE feature. See Section R.2 to recall how to graph functions defined piecewise.

Note: Each Technology Connection labeled "Exploratory" is designed to lead you through a discovery process. Because of this, answers are not provided.

Quick Check 2 ✔

Let

$$k(x) = \begin{cases} -x + 4, & \text{for } x \leq 3, \\ 2x + 1, & \text{for } x > 3. \end{cases}$$

Find these limits:

a) $\lim_{x \to 3^-} k(x)$, $\lim_{x \to 3^+} k(x)$,
and $\lim_{x \to 3} k(x)$;

b) $\lim_{x \to 1} k(x)$.

b)

Limit Numerically

$x \to -3^-$ $(x < -3)$	$H(x)$
-3.1	-4.2
-3.01	-4.02
-3.001	-4.002

$x \to -3^+$ $(x > -3)$	$H(x)$
-2.9	-3.8
-2.99	-3.98
-2.999	-3.998

Limit Graphically

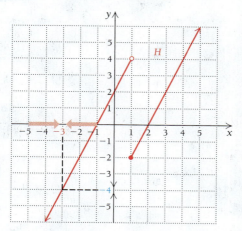

As inputs x approach -3 from the left, outputs $H(x)$ approach -4, so the limit from the left is -4. That is,

$$\lim_{x \to -3^-} H(x) = -4.$$

As inputs x approach -3 from the right, outputs $H(x)$ approach -4, so the limit from the right is -4. That is,

$$\lim_{x \to -3^+} H(x) = -4.$$

Since the limits from the left and the right exist and are the same, we have

$$\lim_{x \to -3} H(x) = -4.$$

2 ✔

The limit of $f(x)$ at a number a *does not depend* on the function value at a or even on whether $f(a)$ exists. That is, the existence of a limit at a has *nothing* to do with the function value $f(a)$.

The "Wall" Method

An alternative approach for Example 2 is to draw a "wall" at $x = 1$, as shown in blue on the graph below. We then trace the graph from left to right with a pencil until we hit the wall and mark the location with an ✕, assuming it can be determined. Then we trace the graph from right to left until we hit the wall and mark that location with an ✕. If the locations are the same, as in the graph to the right below, a limit exists. Thus, for Example 2,

$$\lim_{x \to 1} H(x) \text{ does not exist,} \quad \text{and} \quad \lim_{x \to -3} H(x) = -4.$$

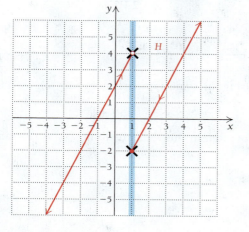

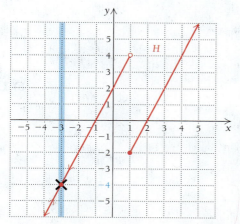

EXAMPLE 3   Consider the function defined as follows:

$$G(x) = \begin{cases} 5, & \text{for } x = 1, \\ x + 1, & \text{for } x \neq 1. \end{cases}$$

Graph the function, and find each of the following limits, if they exist.

a) $\lim\limits_{x \to 1} G(x)$

b) $\lim\limits_{x \to -2} G(x)$

Solution   The graph of G follows.

a) As inputs x approach 1 from the left, outputs $G(x)$ approach 2, so the limit from the left is 2. As inputs x approach 1 from the right, outputs $G(x)$ also approach 2, so the limit from the right is 2. Both limits are the same, so

$$\lim\limits_{x \to 1} G(x) = 2.$$

Note that the limit, 2, is not the same as the function value at 1, which is $G(1) = 5$.

Limit Numerically

$x \to 1^-$ $(x < 1)$	$G(x)$
0.9	1.9
0.99	1.99
0.999	1.999

$x \to 1^+$ $(x > 1)$	$G(x)$
1.1	2.1
1.01	2.01
1.001	2.001

$\lim\limits_{x \to 1} G(x) = 2$

Limit Graphically

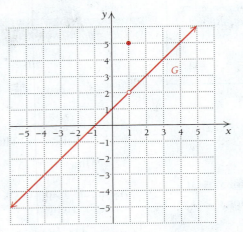

b) Using the same approach as in part (a), we have

$$\lim\limits_{x \to -2} G(x) = -1.$$

Note that in this case, the limit, -1, is the same as the function value, $G(-2)$.

Quick Check 3 ✔

Calculate the following limits based on the graph of f.

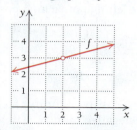

a) $\lim\limits_{x \to 2^-} f(x)$

b) $\lim\limits_{x \to 2^+} f(x)$

c) $\lim\limits_{x \to 2} f(x)$

Limit Numerically

$x \to -2^-$ $(x < -2)$	$G(x)$
-2.1	-1.1
-2.01	-1.01
-2.001	-1.001

$$\lim\limits_{x \to -2} G(x) = -1$$

$x \to -2^+$ $(x > -2)$	$G(x)$
-1.9	-0.9
-1.99	-0.99
-1.999	-0.999

Limit Graphically

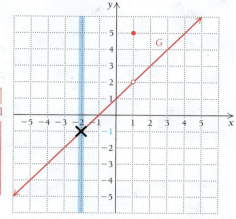

3 ✔

Limits Involving Infinity

Limits also help us understand the role of infinity with respect to some functions. Consider the following example.

EXAMPLE 4 Let $f(x) = \dfrac{1}{x - 2}$.

a) Find $\lim\limits_{x \to 2^-} f(x)$.

b) Find $\lim\limits_{x \to 2^+} f(x)$.

c) Use the information from parts (a) and (b) to form a conclusion about $\lim\limits_{x \to 2} f(x)$.

Solution Note first that $f(2)$ does not exist, since 2 is not in the domain of f.

Limit Numerically

$x \to 2^-$ $(x < 2)$	$f(x)$
1.9	-10
1.99	-100
1.999	$-1,000$
1.9999	$-10,000$

$x \to 2^+$ $(x > 2)$	$f(x)$
2.1	10
2.01	100
2.001	$1,000$
2.0001	$10,000$

Limit Graphically

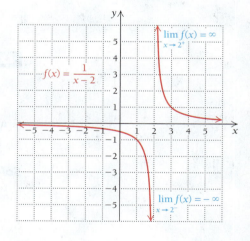

a) The table and graph show that as x approaches 2 from the left, $f(x)$ decreases without bound. We conclude that the left-hand limit is negative infinity; that is, $\lim\limits_{x \to 2^-} f(x) = -\infty$. We symbolize the notion of "infinity" with ∞. The symbol ∞ *does not represent* a real number.

b) The table and graph show that as x approaches 2 from the right, $f(x)$ increases without bound toward positive infinity. The right-hand limit is positive infinity; that is, $\lim\limits_{x \to 2^+} f(x) = \infty$.

c) Since the left-hand and right-hand limits are not equal (and are not finite), $\lim\limits_{x \to 2} f(x)$ does not exist.

Sometimes we need to determine limits as inputs approach positive or negative infinity. In such cases, we are finding *limits at infinity*. Such a limit is expressed as

$$\lim_{x \to \infty} f(x) \qquad \text{or} \qquad \lim_{x \to -\infty} f(x).$$

These limits are approached from one side only: from the left if approaching positive infinity or from the right if approaching negative infinity.

EXAMPLE 5   Let $f(x) = \dfrac{1}{x - 3}$. Find $\lim_{x \to \infty} f(x)$ and $\lim_{x \to -\infty} f(x)$.

Solution   The table shows that as x increases toward positive infinity, $f(x)$ approaches 0. Thus, $\lim_{x \to \infty} f(x) = 0$. As x decreases toward negative infinity, $f(x)$ again approaches 0. Thus, $\lim_{x \to -\infty} f(x) = 0$.

Limit Numerically   *Limit Graphically*

$x \to \infty$	$f(x)$
100	0.0103 . . .
1,000	0.001003 . . .
10,000	0.00010003 . . .

$$\lim_{x \to \infty} f(x) = 0$$

$x \to -\infty$	$f(x)$
−100	−0.0097 . . .
−1,000	−0.000997 . . .
−10,000	−0.00009997 . . .

$$\lim_{x \to -\infty} f(x) = 0$$

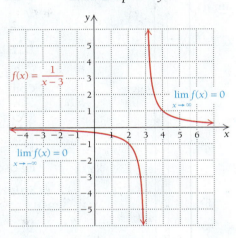

EXAMPLE 6   Consider the function f given by

$$f(x) = \frac{1}{x - 2} + 3.$$

Graph the function, and find each of the following limits, if they exist.

a) $\lim_{x \to 3} f(x)$ **b)** $\lim_{x \to 2} f(x)$

Solution   Note that the graph of f is the same as the graph of $f(x) = \dfrac{1}{x}$ but shifted 2 units to the right and 3 units up.

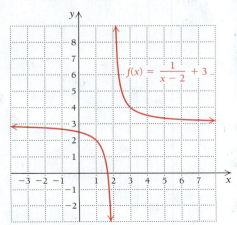

a) As x approaches 3 from the left, $f(x)$ approaches 4, so the limit from the left is 4. As x approaches 3 from the right, $f(x)$ also approaches 4. Since the limit from the left, 4, is the same as the limit from the right, we have

$$\lim_{x \to 3} f(x) = 4.$$

Limit Numerically   *Limit Graphically*

$x \to 3^-$ $(x < 3)$	$f(x)$
2.9	4.1
2.99	$4.\overline{01}$
2.999	$4.\overline{001}$

$$\lim_{x \to 3} f(x) = 4$$

$x \to 3^+$ $(x > 3)$	$f(x)$
3.1	$3.\overline{90}$
3.01	$3.\overline{9900}$
3.001	3.999000

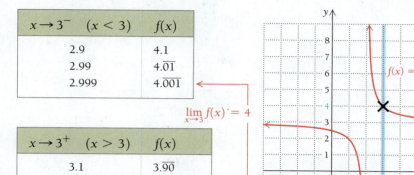

b) As inputs x approach 2 from the left, outputs $f(x)$ become more and more negative, without bound. These numbers do not approach any real number, although it might be said that the limit from the left is negative infinity, $-\infty$. That is,

$$\lim_{x \to 2^-} f(x) = -\infty.$$

As inputs x approach 2 from the right, outputs $f(x)$ become larger and larger, without bound. These numbers do not approach any real number, although it might be said that the limit from the right is infinity, ∞. That is,

$$\lim_{x \to 2^+} f(x) = \infty.$$

Because the left-sided limit differs from the right-sided limit,

$$\lim_{x \to 2} f(x) \text{ does not exist.}$$

Limit Numerically   *Limit Graphically*

$x \to 2^-$ $(x < 2)$	$f(x)$
1.9	-7
1.99	-97
1.999	-997

$$\lim_{x \to 2} f(x) \text{ does not exist.}$$

$x \to 2^+$ $(x > 2)$	$f(x)$
2.1	13
2.01	103
2.001	1003

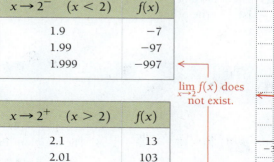

EXAMPLE 7 Consider again the function in Example 6, given by

$$f(x) = \frac{1}{x - 2} + 3.$$

Find $\lim_{x \to \infty} f(x)$.

Solution As x gets larger and larger, $f(x)$ gets closer and closer to 3. Thus,

$$\lim_{x \to \infty} f(x) = 3.$$

Limit Numerically

$x \to \infty$	$f(x)$
10	3.125
100	3.0102 . . .
1000	3.0010 . . .

$$\lim_{x \to \infty} f(x) = 3$$

Limit Graphically

$$f(x) = \frac{1}{x - 2} + 3$$

$$\lim_{x \to \infty} f(x) = 3$$

Quick Check 4 ✔

Let $h(x) = \dfrac{1}{1 - x} + 6$. Find

these limits:

a) $\lim_{x \to 1} h(x)$;   **b)** $\lim_{x \to 2} h(x)$;

c) $\lim_{x \to \infty} h(x)$.

4 ✔

Section Summary

- The *limit* of a function f, as x approaches a, is written $\lim_{x \to a} f(x) = L$. This means that as the values of x approach a, the corresponding values of $f(x)$ approach L. The value L must be a unique real number.
- A *left-hand limit* is written $\lim_{x \to a^-} f(x)$. The values of x are approaching a from the left, that is, $x < a$.
- A *right-hand limit* is written $\lim_{x \to a^+} f(x)$. The values of x are approaching a from the right, that is, $x > a$.
- If the left-hand and right-hand limits (as x approaches a) are *not* equal, the limit does *not* exist. On the other

hand, if the left-hand and right-hand limits are equal and not infinite, the limit does exist.
- A limit $\lim_{x \to a} f(x)$ may exist even though the function value $f(a)$ does not exist. (See Example 1.)
- A limit $\lim_{x \to a} f(x)$ may exist and be different from the function value $f(a)$. (See Example 3a.)
- Graphs and tables are useful tools in determining limits.

1.1   Exercise Set

Complete each of the following statements.

1. As x approaches _____, the value of $-3x$ approaches 6.

2. As x approaches _____, the value of $x - 2$ approaches 5.

3. The notation $\lim_{x \to 4} f(x)$ is read _____.

4. The notation $\lim_{x \to 1} g(x)$ is read _____.

5. The notation $\lim_{x \to 5^-} F(x)$ is read _____.

6. The notation $\lim_{x \to 4^+} G(x)$ is read _____.

7. The notation _____ is read "the limit, as x approaches 2 from the right."

8. The notation _____ is read "the limit, as x approaches 3 from the left."

9. The notation _____ is read "the limit as x approaches 5."

10. The notation _____ is read "the limit as x approaches $\frac{1}{2}$."

For Exercises 11 and 12, consider the function f given by

$$f(x) = \begin{cases} x - 2, & \text{for } x \leq 3, \\ x - 1, & \text{for } x > 3. \end{cases}$$

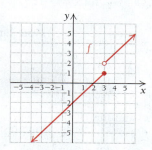

If a limit does not exist, state that fact.

11. Find (a) $\lim_{x \to 3^-} f(x)$; (b) $\lim_{x \to 3^+} f(x)$; (c) $\lim_{x \to 3} f(x)$.

12. Find (a) $\lim_{x \to -1^-} f(x)$; (b) $\lim_{x \to -1^+} f(x)$; (c) $\lim_{x \to -1} f(x)$.

For Exercises 13 and 14, consider the function g given by

$$g(x) = \begin{cases} x + 6, & \text{for } x < -2, \\ -\frac{1}{2}x + 1, & \text{for } x \geq -2. \end{cases}$$

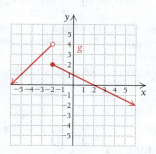

If a limit does not exist, state that fact.

13. Find (a) $\lim_{x \to -2^-} g(x)$; (b) $\lim_{x \to -2^+} g(x)$; (c) $\lim_{x \to -2} g(x)$.

14. Find (a) $\lim_{x \to 4^-} g(x)$; (b) $\lim_{x \to 4^+} g(x)$; (c) $\lim_{x \to 4} g(x)$.

For Exercises 15–22, use the following graph of F to find each limit. When necessary, state that the limit does not exist.

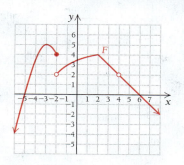

15. $\lim_{x \to 2} F(x)$

16. $\lim_{x \to -3} F(x)$

17. $\lim_{x \to -5} F(x)$

18. $\lim_{x \to -2} F(x)$

19. $\lim_{x \to 6} F(x)$

20. $\lim_{x \to 4} F(x)$

21. $\lim_{x \to -2^-} F(x)$

22. $\lim_{x \to -2^+} F(x)$

For Exercises 23–30, use the following graph of G to find each limit. When necessary, state that the limit does not exist.

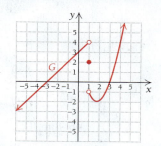

23. $\lim_{x \to 0} G(x)$

24. $\lim_{x \to -2} G(x)$

25. $\lim_{x \to 1^+} G(x)$

26. $\lim_{x \to 1^-} G(x)$

27. $\lim_{x \to 1} G(x)$

28. $\lim_{x \to 3^-} G(x)$

29. $\lim_{x \to 3} G(x)$

30. $\lim_{x \to 3^+} G(x)$

For Exercises 31–40, use the following graph of H to find each limit. When necessary, state that the limit does not exist.

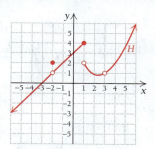

31. $\lim_{x \to -2^-} H(x)$

32. $\lim_{x \to -3} H(x)$

33. $\lim_{x \to -2} H(x)$

34. $\lim_{x \to -2^+} H(x)$

35. $\lim_{x \to 1^+} H(x)$

36. $\lim_{x \to 1^-} H(x)$

37. $\lim_{x \to 1} H(x)$

38. $\lim_{x \to 3^-} H(x)$

39. $\lim_{x \to 3} H(x)$

40. $\lim_{x \to 3^+} H(x)$

For Exercises 41–50, use the following graph of f to find each limit. When necessary, state that the limit does not exist.

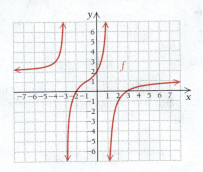

41. $\lim\limits_{x \to 2} f(x)$

42. $\lim\limits_{x \to -1} f(x)$

43. $\lim\limits_{x \to 0} f(x)$

44. $\lim\limits_{x \to -3} f(x)$

45. $\lim\limits_{x \to 1} f(x)$

46. $\lim\limits_{x \to 3^-} f(x)$

47. $\lim\limits_{x \to -2} f(x)$

48. $\lim\limits_{x \to -4} f(x)$

49. $\lim\limits_{x \to -\infty} f(x)$

50. $\lim\limits_{x \to \infty} f(x)$

For Exercises 51–68, graph each function and then find the specified limits. When necessary, state that the limit does not exist.

51. $f(x) = |x|$;   find $\lim\limits_{x \to 0} f(x)$ and $\lim\limits_{x \to -2} f(x)$.

52. $f(x) = x^2$;   find $\lim\limits_{x \to -1} f(x)$ and $\lim\limits_{x \to 0} f(x)$.

53. $g(x) = x^2 - 5$;   find $\lim\limits_{x \to 0} g(x)$ and $\lim\limits_{x \to -1} g(x)$.

54. $g(x) = |x| + 1$;   find $\lim\limits_{x \to -3} g(x)$ and $\lim\limits_{x \to 0} g(x)$.

55. $G(x) = \dfrac{1}{x + 2}$;   find $\lim\limits_{x \to -1} G(x)$ and $\lim\limits_{x \to -2} G(x)$.

56. $F(x) = \dfrac{1}{x - 3}$;   find $\lim\limits_{x \to 3} F(x)$ and $\lim\limits_{x \to 4} F(x)$.

57. $f(x) = \dfrac{1}{x} - 2$;   find $\lim\limits_{x \to \infty} f(x)$ and $\lim\limits_{x \to 0} f(x)$.

58. $f(x) = \dfrac{1}{x} + 3$;   find $\lim\limits_{x \to \infty} f(x)$ and $\lim\limits_{x \to 0} f(x)$.

59. $g(x) = \dfrac{1}{x - 3} + 2$;   find $\lim\limits_{x \to \infty} g(x)$ and $\lim\limits_{x \to 3} g(x)$.

60. $g(x) = \dfrac{1}{x + 2} + 4$;   find $\lim\limits_{x \to \infty} g(x)$ and $\lim\limits_{x \to -2} g(x)$.

61. $F(x) = \begin{cases} 2x + 1, & \text{for } x < 1, \\ x, & \text{for } x \geq 1. \end{cases}$

Find $\lim\limits_{x \to 1^-} F(x)$, $\lim\limits_{x \to 1^+} F(x)$, and $\lim\limits_{x \to 1} F(x)$.

62. $G(x) = \begin{cases} -x + 3, & \text{for } x < 2, \\ x + 1, & \text{for } x \geq 2. \end{cases}$

Find $\lim\limits_{x \to 2^-} G(x)$, $\lim\limits_{x \to 2^+} G(x)$, and $\lim\limits_{x \to 2} G(x)$.

63. $g(x) = \begin{cases} -x + 4, & \text{for } x < 3, \\ x - 3, & \text{for } x > 3. \end{cases}$

Find $\lim\limits_{x \to 3^-} g(x)$, $\lim\limits_{x \to 3^+} g(x)$, and $\lim\limits_{x \to 3} g(x)$.

64. $f(x) = \begin{cases} 3x - 4, & \text{for } x < 1, \\ x - 2, & \text{for } x > 1. \end{cases}$

Find $\lim\limits_{x \to 1^-} f(x)$, $\lim\limits_{x \to 1^+} f(x)$, and $\lim\limits_{x \to 1} f(x)$.

65. $F(x) = \begin{cases} -2x - 3, & \text{for } x < -1, \\ x^3, & \text{for } x > -1. \end{cases}$ Find $\lim\limits_{x \to -1} F(x)$.

66. $G(x) = \begin{cases} x^2, & \text{for } x < -1, \\ x + 2, & \text{for } x > -1. \end{cases}$ Find $\lim\limits_{x \to -1} G(x)$.

67. $H(x) = \begin{cases} x + 1, & \text{for } x < 0, \\ 2, & \text{for } 0 \leq x < 1, \\ 3 - x, & \text{for } x \geq 1. \end{cases}$

Find $\lim\limits_{x \to 0} H(x)$ and $\lim\limits_{x \to 1} H(x)$.

68. $G(x) = \begin{cases} 2 + x, & \text{for } x \leq -1, \\ x^2, & \text{for } -1 < x < 3, \\ 9, & \text{for } x \geq 3. \end{cases}$

Find $\lim\limits_{x \to -1} G(x)$ and $\lim\limits_{x \to 3} G(x)$.

APPLICATIONS

Business and Economics

Taxicab fares.   *In New York City, taxicabs charge passengers $2.50 for entering a cab and then $0.50 for each one-fifth of a mile (or fraction thereof) traveled. (There are additional charges for slow traffic and idle times, but these are not considered in this problem.) If x represents the distance traveled in miles, then C(x) is the cost of the taxi fare, where*

$$C(x) = \$2.50, \quad \text{if} \quad x = 0,$$
$$C(x) = \$3.00, \quad \text{if} \quad 0 < x \leq 0.2,$$
$$C(x) = \$3.50, \quad \text{if} \quad 0.2 < x \leq 0.4,$$
$$C(x) = \$4.00, \quad \text{if} \quad 0.4 < x \leq 0.6,$$

and so on. The graph of C is shown below. (Source: New York City Taxi and Limousine Commission.)

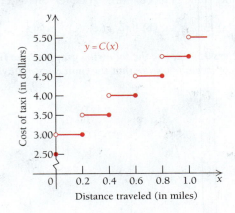

Using the graph of the taxicab fare function, find each of the following limits, if it exists.

69. $\lim\limits_{x\to 0.25^-} C(x)$, $\lim\limits_{x\to 0.25^+} C(x)$, $\lim\limits_{x\to 0.25} C(x)$

70. $\lim\limits_{x\to 0.2^-} C(x)$, $\lim\limits_{x\to 0.2^+} C(x)$, $\lim\limits_{x\to 0.2} C(x)$

71. $\lim\limits_{x\to 0.6^-} C(x)$, $\lim\limits_{x\to 0.6^+} C(x)$, $\lim\limits_{x\to 0.6} C(x)$

The postage function.   *The cost of sending a large envelope via U.S. first-class mail in 2014 was $0.98 for the first ounce and $0.21 for each additional ounce (or fraction thereof). (Source: www.usps.com.) If x represents the weight of a large envelope, in ounces, then p(x) is the cost of mailing it, where*

$$p(x) = \$0.98, \quad if \quad 0 < x \le 1,$$
$$p(x) = \$1.19, \quad if \quad 1 < x \le 2,$$
$$p(x) = \$1.40, \quad if \quad 2 < x \le 3,$$

and so on, up through 13 ounces. The graph of p is shown below.

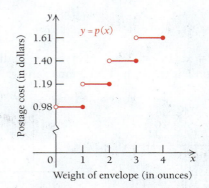

Using the graph of the postage function, find each of the following limits, if it exists.

72. $\lim\limits_{x\to 1^-} p(x)$, $\lim\limits_{x\to 1^+} p(x)$, $\lim\limits_{x\to 1} p(x)$

73. $\lim\limits_{x\to 2^-} p(x)$, $\lim\limits_{x\to 2^+} p(x)$, $\lim\limits_{x\to 2} p(x)$

74. $\lim\limits_{x\to 2.6^-} p(x)$, $\lim\limits_{x\to 2.6^+} p(x)$, $\lim\limits_{x\to 2.6} p(x)$

75. $\lim\limits_{x\to 3} p(x)$

76. $\lim\limits_{x\to 3.4} p(x)$

Tax rate schedule.   *The federal tax rate for single filers is given as a percentage of taxable income in the graph below. (Source: troweprice.com, 2013.)*

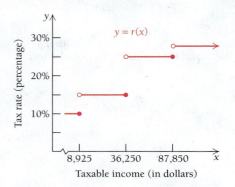

Use the graph for Exercises 77–79.

77. Find $\lim\limits_{x\to 8925^-} r(x)$, $\lim\limits_{x\to 8925^+} r(x)$, and $\lim\limits_{x\to 8925} r(x)$.

78. Find $\lim\limits_{x\to 10,000^-} r(x)$, $\lim\limits_{x\to 10,000^+} r(x)$, and $\lim\limits_{x\to 10,000} r(x)$.

79. Find $\lim\limits_{x\to 50,000} r(x)$ and $\lim\limits_{x\to 87,850} r(x)$.

Tax rate schedule.   *The federal tax rate for heads of household is given in the graph below. (Source: troweprice.com, 2013.)*

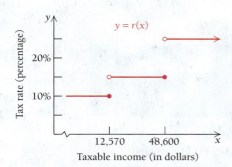

Use the graph for Exercises 80–82.

80. Find $\lim\limits_{x\to 9000^-} r(x)$, $\lim\limits_{x\to 9000^+} r(x)$, and $\lim\limits_{x\to 9000} r(x)$.

81. Find $\lim\limits_{x\to 48,600^-} r(x)$, $\lim\limits_{x\to 48,600^+} r(x)$, and $\lim\limits_{x\to 48,600} r(x)$.

82. Find $\lim\limits_{x\to 60,000} r(x)$ and $\lim\limits_{x\to 12,570} r(x)$.

SYNTHESIS

In Exercises 83–85, fill in each blank so that $\lim\limits_{x\to 2} f(x)$ exists.

83. $f(x) = \begin{cases} \frac{1}{2}x + \underline{}, & for\ x < 2, \\ -x + 6, & for\ x > 2 \end{cases}$

84. $f(x) = \begin{cases} -\frac{1}{2}x + 1, & for\ x < 2, \\ \frac{3}{2}x + \underline{}, & for\ x > 2 \end{cases}$

85. $f(x) = \begin{cases} x^2 - 9, & for\ x < 2, \\ -x^2 + \underline{}, & for\ x > 2 \end{cases}$

86. Graph the function f given by

$$f(x) = \begin{cases} -3, & for\ x = -2, \\ x^2, & for\ x \ne -2. \end{cases}$$

Use GRAPH and TRACE to find each of the following limits. When necessary, state that the limit does not exist.

a) $\lim\limits_{x\to -2^+} f(x)$ **b)** $\lim\limits_{x\to -2^-} f(x)$

c) $\lim\limits_{x\to -2} f(x)$ **d)** $\lim\limits_{x\to 2^+} f(x)$

e) $\lim\limits_{x\to 2^-} f(x)$

f) Does $\lim\limits_{x\to -2} f(x) = f(-2)$?

g) Does $\lim\limits_{x\to 2} f(x) = f(2)$?

In Exercises 87–89, use GRAPH *and* TRACE *to find each limit. When necessary, state that the limit does not exist.*

87. For $f(x) = \begin{cases} x^2 - 2, & \text{for } x < 0, \\ 2 - x^2, & \text{for } x \geq 0, \end{cases}$

find $\lim\limits_{x \to 0} f(x)$ and $\lim\limits_{x \to -2} f(x)$.

88. For $g(x) = \dfrac{20x^2}{x^3 + 2x^2 + 5x}$,

find $\lim\limits_{x \to \infty} g(x)$ and $\lim\limits_{x \to -\infty} g(x)$.

89. For $f(x) = \dfrac{x - 5}{x^2 - 4x - 5}$, find $\lim\limits_{x \to -1} f(x)$ and $\lim\limits_{x \to 5} f(x)$.

> **Answers to Quick Checks**
> **1. (a)** $f(3)$ does not exist; **(b)** 6 **2. (a)** 1, 7, does not exist; **(b)** 3 **3. (a)** 3; **(b)** 3; **(c)** 3 **4. (a)** Limit does not exist; **(b)** 5; **(c)** 6

1.2 Algebraic Limits and Continuity

- Develop and use the Limit Properties to calculate limits.
- Determine whether a function is continuous at a point.

Using numerical and graphical methods for finding limits can be time-consuming. In this section, we develop methods to more quickly evaluate limits for a wide variety of functions. We then use limits to study *continuity*, a concept of great importance in calculus.

Algebraic Limits

Consider the functions given by $f(x) = x$, $g(x) = 3$, and $F(x) = x + 3$, displayed in the following graphs. Note that function F is the sum of functions f and g.

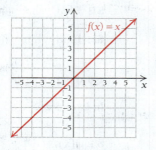 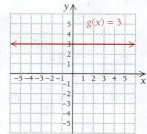 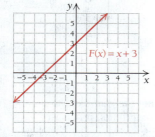

Suppose we are interested in the limits of $f(x)$, $g(x)$, and $F(x)$ as x approaches 2. In Section 1.1, we learned numerical and graphical techniques that can be used to show that

$$\lim_{x \to 2} f(x) = 2, \quad \lim_{x \to 2} g(x) = 3, \quad \text{and} \quad \lim_{x \to 2} F(x) = 5.$$

These techniques work equally well for any value of a. For example, if we choose $a = -1$, we can compute the following limits:

$$\lim_{x \to -1} f(x) = -1, \quad \lim_{x \to -1} g(x) = 3, \quad \text{and} \quad \lim_{x \to -1} F(x) = 2.$$

From these results, the following observations can be made:

1. For any real number a, $\lim\limits_{x \to a} x = a$.
2. For any real number a, $\lim\limits_{x \to a} 3 = 3$.

Recalling that $F(x) = f(x) + g(x)$, we make this reasonable conclusion:

3. For any real number a, $\lim\limits_{x \to a} (x + 3) = a + 3$.

We determined the limits of these functions by observing basic behaviors and making a reasonable generalization. The following list summarizes common limit properties that allow us to calculate limits much more efficiently.

Limit Properties

If $\lim\limits_{x \to a} f(x) = L$ and $\lim\limits_{x \to a} g(x) = M$, and c is any constant, then we have the following limit properties.

L1. The limit of a constant is the constant:

$$\lim\limits_{x \to a} c = c.$$

L2. The limit of a power is the power of that limit, and the limit of a root is the root of that limit:

$$\lim\limits_{x \to a} [f(x)]^m = [\lim\limits_{x \to a} f(x)]^m = L^m, \text{ where } m \text{ is any integer;}$$

$$\lim\limits_{x \to a} \sqrt[n]{f(x)} = \sqrt[n]{\lim\limits_{x \to a} f(x)} = \sqrt[n]{L}, \text{ where } n \geq 2.$$

In the case of the power, we must have $L \neq 0$ if m is negative, and in the case of the root, we must have $L \geq 0$ if n is an even integer.

L3. The limit of a sum or difference is the sum or difference of the limits:

$$\lim\limits_{x \to a} [f(x) \pm g(x)] = \lim\limits_{x \to a} f(x) \pm \lim\limits_{x \to a} g(x) = L \pm M.$$

L4. The limit of a product is the product of the limits:

$$\lim\limits_{x \to a} [f(x) \cdot g(x)] = [\lim\limits_{x \to a} f(x)] \cdot [\lim\limits_{x \to a} g(x)] = L \cdot M.$$

L5. The limit of a quotient is the quotient of the limits:

$$\lim\limits_{x \to a} \frac{f(x)}{g(x)} = \frac{\lim\limits_{x \to a} f(x)}{\lim\limits_{x \to a} g(x)} = \frac{L}{M}, \quad \text{assuming } M \neq 0.$$

L6. The limit of a constant times a function is the constant times the limit of the function:

$$\lim\limits_{x \to a} c \cdot f(x) = c \cdot \lim\limits_{x \to a} f(x) = c \cdot L.$$

Property L6 combines L1 and L4 but is stated separately because it is used so frequently.

EXAMPLE 1 Use the Limit Properties to find $\lim\limits_{x \to 4} (x^2 - 3x + 7)$.

Solution We know that $\lim\limits_{x \to 4} x$ is 4.

By Limit Property L2,

$$\lim\limits_{x \to 4} x^2 = [\lim\limits_{x \to 4} x]^2 = 4^2 = 16,$$

by Limit Property L6,

$$\lim\limits_{x \to 4} (-3x) = -3 \cdot \lim\limits_{x \to 4} x = -3 \cdot 4 = -12,$$

and by Limit Property L1,

$$\lim\limits_{x \to 4} 7 = 7.$$

Using Limit Property L3 to combine these results, we have

$$\lim\limits_{x \to 4} (x^2 - 3x + 7) = 16 - 12 + 7 = 11.$$

■

The result of Example 1 is extended in the following theorem.

THEOREM   **Limits of Rational Functions**

For any rational function F, with a in the domain of F,

$$\lim_{x \to a} F(x) = F(a).$$

Rational functions include all polynomial functions (which include constant functions and linear functions) and ratios composed of such functions (see Section R.5). Thus, the Limit Properties allow us to evaluate limits of rational functions very quickly without tables or graphs, as illustrated by the following examples.

EXAMPLE 2   Find $\lim_{x \to 2} (x^4 - 5x^3 + x^2 - 7)$.

Solution   It follows from the Theorem on Limits of Rational Functions that we can find the limit by substitution:

$$\lim_{x \to 2} (x^4 - 5x^3 + x^2 - 7) = 2^4 - 5 \cdot 2^3 + 2^2 - 7$$
$$= 16 - 40 + 4 - 7$$
$$= -27.$$

EXAMPLE 3   Find $\lim_{x \to 0} \sqrt{x^2 - 3x + 2}$.

Solution   The Theorem on Limits of Rational Functions and Limit Property L2 tell us that we can substitute to find the limit:

$$\lim_{x \to 0} \sqrt{x^2 - 3x + 2} = \sqrt{0^2 - 3 \cdot 0 + 2}$$
$$= \sqrt{2}.$$ 1 ✔

In the following example, a is not in the domain of the function, since it would result in a zero in the denominator. Nevertheless, we can still find the limit.

EXAMPLE 4   Let $r(x) = \dfrac{x^2 - x - 12}{x + 3}$. Find $\lim_{x \to -3} r(x)$.

Solution   We note that $r(-3)$ does not exist, since -3 is not in the domain of r. Since it is impossible to determine this limit by direct evaluation (the assumption for Limit Property L5 is not met), we use a table and a graph. Although $x \neq -3$, x can approach -3 as closely as we wish:

Quick Check 1 ✔

Find these limits and note the Limit Property you use at each step.

a) $\lim_{x \to 1} 2x^3 + 3x^2 - 6$

b) $\lim_{x \to 4} \dfrac{2x^2 + 5x - 1}{3x - 2}$

c) $\lim_{x \to -2} \sqrt{1 + 3x^2}$

Limit Numerically

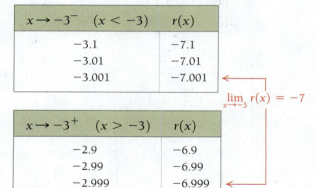

$x \to -3^-$ $(x < -3)$	$r(x)$
-3.1	-7.1
-3.01	-7.01
-3.001	-7.001

$\lim_{x \to -3} r(x) = -7$

$x \to -3^+$ $(x > -3)$	$r(x)$
-2.9	-6.9
-2.99	-6.99
-2.999	-6.999

Limit Graphically

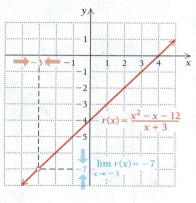

Both the table and the graph suggest that the limit is

$$\lim_{x \to -3} \left(\frac{x^2 - x - 12}{x + 3} \right) = -7.$$

Limit Algebraically. The function is simplified by factoring and noting that the factor $x + 3$ is present in both numerator and denominator. As long as $x \neq -3$, the expression can be simplified:

$$\frac{x^2 - x - 12}{x + 3} = \frac{(x + 3)(x - 4)}{(x + 3)} = x - 4, \qquad \text{for } x \neq -3.$$

We then evaluate the limit using the simplified form:

$$\lim_{x \to -3} (x - 4) = (-3) - 4 = -7.$$

As shown on p. 108, the graph of r is a line with a "hole" at the point $(-3, -7)$. Even though $r(-3)$ does not exist, the limit *does* exist since we are only concerned about the behavior of $r(x)$ as x approaches -3. The decision to simplify was made by noting that $[(-3)^2 - (-3) - 12]/[(-3) + 3] = 0/0$. This *indeterminate form* indicates that the polynomials in the numerator and the denominator share a common factor, in this case, $x + 3$. The $0/0$ form is a hint that a limit may exist. Look for ways to simplify algebraically, or use a table or a graph to determine such a limit. **2** ✔

A common error in determining limits is to assume that all limits can be found by direct evaluation. A student may attempt to find the limit of a function like the one in Example 4, get a zero in the denominator, and then mistakely assume that the limit does not exist. Remember, finding a limit as $x \to a$ focuses on x *close* to a, not *at* a. As Example 4 shows, the function may not be defined at a certain a-value, but its limit as $x \to a$ may still exist.

EXAMPLE 5 Find $\lim\limits_{h \to 0} (3x^2 + 3xh + h^2)$.

Solution We treat x as a constant since we are interested only in the way in which the expression varies when h approaches 0. We use the Limit Properties to find that

$$\begin{aligned} \lim_{h \to 0} (3x^2 + 3xh + h^2) &= 3x^2 + 3x(0) + 0^2 \\ &= 3x^2. \end{aligned}$$ ■

Continuity

The following are examples of graphs of functions that are *continuous* over the whole real number line, that is, over $(-\infty, \infty)$.

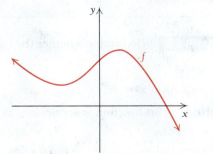
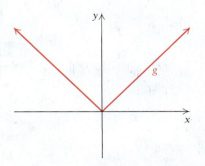

Note that there are no "jumps" or "holes" in the graphs. An intuitive definition of continuity is that a function is **continuous over**, or **on, some interval** if its graph can be traced without lifting pencil from paper. If there is any point in an interval where

---

Quick Check 2

Using a table, a graph, and algebra, find

$$\lim_{x \to 2} \frac{x^2 + 4x - 12}{x^2 - 4}.$$

EXERCISES

Find each limit, if it exists, using the TABLE feature.

1. $\lim\limits_{x \to -2} (x^4 - 5x^3 + x^2 - 7)$

2. $\lim\limits_{x \to 1} \sqrt{x^2 + 3x + 4}$

3. $\lim\limits_{x \to 5} \dfrac{x - 5}{x^2 - 6x + 5}$

4. $\lim\limits_{x \to 3} \dfrac{x - 3}{x^2 - 9}$

a "jump" or a "hole" occurs, then the function is *not continuous* over that interval. The graphs of *F*, *G*, and *H* below show functions that are *not* continuous over $(-\infty, \infty)$.

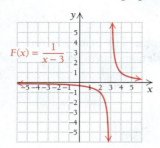

$$F(x) = \frac{1}{x - 3}$$

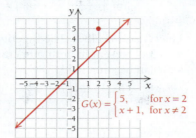

$$G(x) = \begin{cases} 5, & \text{for } x = 2 \\ x + 1, & \text{for } x \neq 2 \end{cases}$$

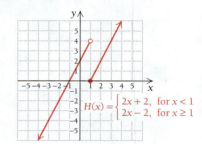

$$H(x) = \begin{cases} 2x + 2, & \text{for } x < 1 \\ 2x - 2, & \text{for } x \geq 1 \end{cases}$$

A continuous curve.

In each case, the graph *cannot* be traced without lifting pencil from paper. However, each case represents a different situation:

- *F* is not continuous over $(-\infty, \infty)$ because 3 is not in the domain. Thus, there is no point to trace at $x = 3$. Note that *F* is continuous over the intervals $(-\infty, 3)$ and $(3, \infty)$.
- *G* is not continuous over $(-\infty, \infty)$ because it is not continuous at $x = 2$. To see this, note that, as *x* approaches 2 from either side, $G(x)$ approaches 3. However, *at* $x = 2$, $G(x)$ *jumps* up to 5. Note that *G* is continuous over $(-\infty, 2)$ and $(2, \infty)$.
- *H* is not continuous over $(-\infty, \infty)$ because it is not continuous at $x = 1$. To see this, trace the graph and note that, as *x* approaches 1 from the left, $H(x)$ approaches 4. However, as *x* approaches 1 from the right, $H(x)$ approaches 0. Note that *H* is continuous over $(-\infty, 1)$ and $(1, \infty)$.

Each of the above graphs has a *point of discontinuity*. The graph of *F* is discontinuous at 3, because $x = 3$ is not in the domain of *F*; *G* is discontinuous at 2, because $\lim\limits_{x \to 2} G(x) \neq G(2)$; and *H* is discontinuous at 1, because $\lim\limits_{x \to 1} H(x)$ does not exist.

DEFINITION

A function *f* is **continuous** at $x = a$ if:

a) $f(a)$ exists,   (The output at *a* exists.)

b) $\lim\limits_{x \to a} f(x)$ exists,   (The limit as $x \to a$ exists.)

and

c) $\lim\limits_{x \to a} f(x) = f(a)$.   (The limit is the same as the output.)

A function is **continuous over an interval *I*** if it is continuous at each point *a* in *I*. If *f* is not continuous at $x = a$, we say that *f* is **discontinuous**, or has a **discontinuity**, at $x = a$.

A visualization of the function in Example 6

EXAMPLE 6   Determine whether the function given by

$$f(x) = 2x + 3$$

is continuous at $x = 4$.

Solution   This function is continuous at $x = 4$ because:

a) $f(4)$ exists, $(f(4) = 11)$

b) $\lim\limits_{x \to 4} f(x)$ exists, $(\lim\limits_{x \to 4} f(x) = 11$ was found on pp. 93–94.)

and

c) $\lim\limits_{x \to 4} f(x) = 11 = f(4)$.

In fact, $f(x) = 2x + 3$ is continuous over the entire real number line. ∎

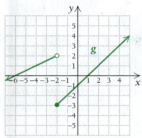

A visualization of the function in Example 7

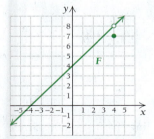

A visualization of the function in Example 8

Quick Check 3 ✔

Let

$$g(x) = \begin{cases} 3x - 5, & \text{for } x < 2, \\ 2x + 1, & \text{for } x \geq 2. \end{cases}$$

Is g continuous at $x = 2$? Why or why not?

EXAMPLE 7 Is the function f given by

$$f(x) = x^2 - 5$$

continuous at $x = 3$? Why or why not?

Solution By the Theorem on Limits of Rational Functions, we have

$$\lim_{x \to 3} f(x) = 3^2 - 5 = 9 - 5 = 4.$$

Since

$$f(3) = 3^2 - 5 = 4,$$

we have

$$\lim_{x \to 3} f(x) = f(3).$$

Since all three parts of the definition of continuity are satisfied, f is continuous at $x = 3$. This function is also continuous over all real x. ∎

EXAMPLE 8 Is the function g, given by

$$g(x) = \begin{cases} \frac{1}{2}x + 3, & \text{for } x < -2, \\ x - 1, & \text{for } x \geq -2, \end{cases}$$

continuous at $x = -2$? Why or why not?

Solution If g is continuous at -2, we must have $\lim_{x \to -2} g(x) = g(-2)$. Thus, we first note that $g(-2) = -2 - 1 = -3$. To find $\lim_{x \to -2} g(x)$, we look at left- and right-hand limits:

$$\lim_{x \to -2^-} g(x) = \frac{1}{2}(-2) + 3 = -1 + 3 = 2; \quad \lim_{x \to -2^+} g(x) = -2 - 1 = -3.$$

Since $\lim_{x \to -2^-} g(x) \neq \lim_{x \to -2^+} g(x)$, we see that $\lim_{x \to -2} g(x)$ does not exist. Thus, g is not continuous at -2. **3** ✔

EXAMPLE 9 Is the function F, given by

$$F(x) = \begin{cases} \dfrac{x^2 - 16}{x - 4}, & \text{for } x \neq 4, \\ 7, & \text{for } x = 4, \end{cases}$$

continuous at $x = 4$? Why or why not?

Solution For F to be continuous at 4, we must have $\lim_{x \to 4} F(x) = F(4)$. Note that $F(4) = 7$. To find $\lim_{x \to 4} F(x)$, note that, for $x \neq 4$,

$$\frac{x^2 - 16}{x - 4} = \frac{(x - 4)(x + 4)}{x - 4} = x + 4.$$

Thus,

$$\lim_{x \to 4} F(x) = 4 + 4 = 8.$$

We see that F is *not* continuous at $x = 4$ since

$$\lim_{x \to 4} F(x) \neq F(4).$$
∎

The Limit Properties and the Theorem on Limits of Rational Functions can be used to show that if f and g are two arbitrary polynomial functions, then f and g are both continuous. Furthermore, $f + g$, $f - g$, $f \cdot g$, and, assuming $g(x) \neq 0$, f/g are also continuous. We can also use the Limit Properties to show that for n, an integer greater than 1, $\sqrt[n]{f(x)}$ is continuous, provided that $f(x) \geq 0$ when n is even.

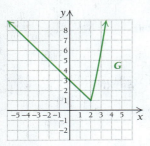

A visualization of the function in Example 10

Quick Check 4 ✔

Let

$$h(x) = \begin{cases} \dfrac{x^2 - 9}{x - 3}, & \text{for } x \neq 3, \\ 7, & \text{for } x = 3. \end{cases}$$

Is h continuous at $x = 3$? Why or why not?

EXAMPLE 10   Is the function G, given by

$$G(x) = \begin{cases} -x + 3, & \text{for } x \leq 2, \\ x^2 - 3, & \text{for } x > 2, \end{cases}$$

continuous for all x? Why or why not?

Solution   For G to be continuous, it must be continuous for all real numbers. Since $y = -x + 3$ is continuous on $(-\infty, 2)$ and $y = x^2 - 3$ is continuous on $(2, \infty)$, we need only to determine whether G is continuous at $x = 2$:

$$G(2) = -2 + 3 = 1;$$
$$\lim_{x \to 2^-} G(x) = -2 + 3 = 1 \quad \text{and} \quad \lim_{x \to 2^+} G(x) = (2)^2 - 3 = 4 - 3 = 1,$$

so $\lim_{x \to 2} G(x) = 1.$

Since $\lim_{x \to 2} G(x) = G(2)$, we have shown that G is continuous at $x = 2$, and we can conclude that G is continuous at all x.

4 ✔

Example 10 shows that a piecewise-defined function may be continuous for all real numbers. In Example 11, we explore a real-world situation involving a piecewise function.

EXAMPLE 11   **Business: Price Breaks.**   Righteous Rocks sells decorative landscape rocks in bulk quantities. For quantities up to and including 500 lb, the company charges $2.50 per pound. For quantities above 500 lb, it charges $2 per pound. The price function can be stated as a piecewise function:

$$p(x) = \begin{cases} 2.50x, & \text{for } 0 < x \leq 500, \\ 2x, & \text{for } x > 500. \end{cases}$$

where p is the price in dollars and x is the quantity in pounds. Is p continuous at $x = 500$? Why or why not?

Solution   The graph of $p(x)$ follows.

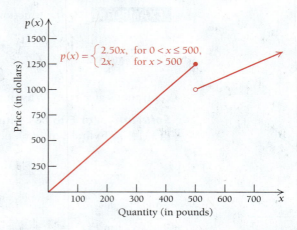

Quick Check 5 ✔

A reservoir is empty at time $t = 0$ minutes. It fills at a rate of 3 gallons of water per minute for 30 minutes. At 30 minutes, the reservoir is no longer being filled and a valve is opened, allowing water to escape at a rate of 4 gallons per minute. The volume v after t minutes is given by the function

$$v(t) = \begin{cases} 3t, & \text{for } 0 \leq t \leq 30, \\ 210 - 4t, & \text{for } t > 30. \end{cases}$$

Find $\lim_{t \to 30^-} v(t)$, $\lim_{t \to 30^+} v(t)$, and $\lim_{t \to 30} v(t)$.

As x approaches 500, we have $\lim_{x \to 500^-} p(x) = 1250$ and $\lim_{x \to 500^+} p(x) = 1000$. Since the left-hand and right-hand limits are not equal, the limit $\lim_{x \to 500} p(x)$ does not exist. Thus, the function is *not* continuous at $x = 500$. This graph literally shows a price "break."

5 ✔

Section Summary

- For a rational function with a in its domain, the limit as x approaches a can be found by evaluating the function at a.
- When evaluating a limit leads to the *indeterminate form* $0/0$, the limit may still exist: algebraic simplification and/or a table and graph are used to find the limit.
- Informally, a function is *continuous* if its graph can be traced without lifting pencil from paper.

- Formally, a function f is continuous at $x = a$ if
  (1) the function value $f(a)$ exists,
  (2) the limit as x approaches a exists, and
  (3) the function value and the limit are equal.
  This is summarized as $\lim_{x \to a} f(x) = f(a)$.
- If any part of the continuity definition fails, then the function is *discontinuous* at $x = a$.

1.2 Exercise Set

Classify each statement as either true or false.

1. $\lim_{x \to 3} 7 = 3$

2. If $\lim_{x \to 2} f(x) = 9$, then $\lim_{x \to 2} \sqrt{f(x)} = 3$.

3. If $\lim_{x \to 1} g(x) = 5$, then $\lim_{x \to 1} [g(x)]^2 = 25$.

4. If $\lim_{x \to 4} F(x) = 7$, then $\lim_{x \to 4} [c \cdot F(x)] = 7c$.

5. If g is discontinuous at $x = 3$, then $g(3)$ must not exist.

6. If f is continuous at $x = 2$, then $f(2)$ must exist.

7. If $\lim_{x \to 4} F(x)$ exists, then F must be continuous at $x = 4$.

8. If $\lim_{x \to 7} G(x)$ equals $G(7)$, then G must be continuous at $x = 7$.

Use the Theorem on Limits of Rational Functions to find each limit. When necessary, state that the limit does not exist.

9. $\lim_{x \to 2} (4x - 5)$

10. $\lim_{x \to 1} (3x + 2)$

11. $\lim_{x \to -1} (x^2 - 4)$

12. $\lim_{x \to -2} (x^2 + 3)$

13. $\lim_{x \to 5} (x^2 - 6x + 9)$

14. $\lim_{x \to 3} (x^2 - 4x + 7)$

15. $\lim_{x \to 2} (2x^4 - 3x^3 + 4x - 1)$

16. $\lim_{x \to -1} (3x^5 + 4x^4 - 3x + 6)$

17. $\lim_{x \to 3} \dfrac{x^2 - 25}{x^2 - 5}$

18. $\lim_{x \to 3} \dfrac{x^2 - 8}{x - 2}$

For Exercises 19–30, the initial substitution of $x = a$ yields the form $0/0$. Look for ways to simplify the function algebraically, or use a table or graph to determine the limit. When necessary, state that the limit does not exist.

19. $\lim_{x \to 3} \dfrac{x^2 - 9}{x - 3}$

20. $\lim_{x \to 5} \dfrac{x^2 - 25}{x - 5}$

21. $\lim_{x \to -2} \dfrac{x^2 - 2x - 8}{x^2 - 4}$

22. $\lim_{x \to 1} \dfrac{x^2 + 5x - 6}{x^2 - 1}$

23. $\lim_{x \to 2} \dfrac{3x^2 + x - 14}{x^2 - 4}$

24. $\lim_{x \to -3} \dfrac{2x^2 - x - 21}{9 - x^2}$

25. $\lim_{x \to 2} \dfrac{x^3 - 8}{2 - x}$

26. $\lim_{x \to 1} \dfrac{x^3 - 1}{x - 1}$

27. $\lim_{x \to 25} \dfrac{\sqrt{x} - 5}{x - 25}$

28. $\lim_{x \to 9} \dfrac{9 - x}{\sqrt{x} - 3}$

29. $\lim_{x \to 2} \dfrac{x^2 + 3x - 10}{x^2 - 4x + 4}$

30. $\lim_{x \to -1} \dfrac{x^2 + 5x + 4}{x^2 + 2x + 1}$

Use the Limit Properties to find the following limits. If a limit does not exist, state that fact.

31. $\lim_{x \to 5} \sqrt{x^2 - 16}$

32. $\lim_{x \to 4} \sqrt{x^2 - 9}$

33. $\lim_{x \to 2} \sqrt{x^2 - 9}$

34. $\lim_{x \to 3} \sqrt{x^2 - 16}$

35. $\lim_{x \to -4^-} \sqrt{x^2 - 16}$

36. $\lim_{x \to 3^+} \sqrt{x^2 - 9}$

Determine whether each of the functions shown in Exercises 37–41 is continuous over the interval $(-6, 6)$.

37.

$y = g(x)$

38.

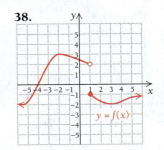

$y = f(x)$

39.

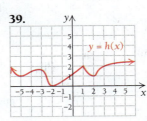

$y = h(x)$

40.

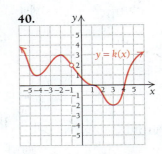

$y = k(x)$

41.

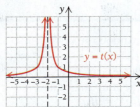

Use the graphs and functions in Exercises 37–41 to answer each of the following. If an expression does not exist, state that fact.

42. a) Find $\lim_{x \to 1^+} g(x)$, $\lim_{x \to 1^-} g(x)$, and $\lim_{x \to 1} g(x)$.
   b) Find $g(1)$.
   c) Is g continuous at $x = 1$? Why or why not?
   d) Find $\lim_{x \to -2} g(x)$.
   e) Find $g(-2)$.
   f) Is g continuous at $x = -2$? Why or why not?

43. a) Find $\lim_{x \to 1^+} f(x)$, $\lim_{x \to 1^-} f(x)$, and $\lim_{x \to 1} f(x)$.
   b) Find $f(1)$.
   c) Is f continuous at $x = 1$? Why or why not?
   d) Find $\lim_{x \to -2} f(x)$.
   e) Find $f(-2)$.
   f) Is f continuous at $x = -2$? Why or why not?

44. a) Find $\lim_{x \to 1} h(x)$.
   b) Find $h(1)$.
   c) Is h continuous at $x = 1$? Why or why not?
   d) Find $\lim_{x \to -2} h(x)$.
   e) Find $h(-2)$.
   f) Is h continuous at $x = -2$? Why or why not?

45. a) Find $\lim_{x \to -1} k(x)$.
   b) Find $k(-1)$.
   c) Is k continuous at $x = -1$? Why or why not?
   d) Find $\lim_{x \to 3} k(x)$.
   e) Find $k(3)$.
   f) Is k continuous at $x = 3$? Why or why not?

46. a) Find $\lim_{x \to 1} t(x)$.
   b) Find $t(1)$.
   c) Is t continuous at $x = 1$? Why or why not?
   d) Find $\lim_{x \to -2} t(x)$.
   e) Find $t(-2)$.
   f) Is t continuous at $x = -2$? Why or why not?

Answer Exercises 47 and 48 using the graphs provided.

47.

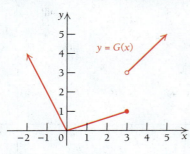

a) Find $\lim_{x \to 3^+} G(x)$.
b) Find $\lim_{x \to 3^-} G(x)$.
c) Find $\lim_{x \to 3} G(x)$.
d) Find $G(3)$.
e) Is G continuous at $x = 3$? Why or why not?
f) Is G continuous at $x = 0$? Why or why not?
g) Is G continuous at $x = 2.9$? Why or why not?

48. Consider the function

$$C(x) = \begin{cases} -1, & \text{for } x < 2, \\ 1, & \text{for } x \geq 2. \end{cases}$$

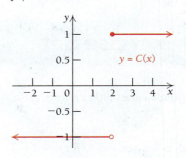

a) Find $\lim_{x \to 2^+} C(x)$.
b) Find $\lim_{x \to 2^-} C(x)$.
c) Find $\lim_{x \to 2} C(x)$.
d) Find $C(2)$.
e) Is C continuous at $x = 2$? Why or why not?
f) Is C continuous at $x = 1.95$? Why or why not?

49. Is the function given by $g(x) = x^2 - 3x$ continuous at $x = 4$? Why or why not?

50. Is the function given by $f(x) = 3x - 2$ continuous at $x = 5$? Why or why not?

51. Is the function given by $G(x) = \dfrac{1}{x}$ continuous at $x = 0$? Why or why not?

52. Is the function given by $F(x) = \sqrt{x}$ continuous at $x = -1$? Why or why not?

53. Is the function given by

$$f(x) = \begin{cases} \frac{1}{2}x + 1, & \text{for } x < 4, \\ -x + 7, & \text{for } x \geq 4, \end{cases}$$

continuous at $x = 4$? Why or why not?

54. Is the function given by

$$g(x) = \begin{cases} \frac{1}{3}x + 4, & \text{for } x \leq 3, \\ 2x - 1, & \text{for } x > 3, \end{cases}$$

continuous at $x = 3$? Why or why not?

55. Is the function given by

$$F(x) = \begin{cases} \frac{1}{3}x + 4, & \text{for } x \leq 3, \\ 2x - 5, & \text{for } x > 3, \end{cases}$$

continuous at $x = 3$? Why or why not?

56. Is the function given by

$$G(x) = \begin{cases} \frac{1}{2}x + 1, & \text{for } x < 4, \\ -x + 5, & \text{for } x > 4, \end{cases}$$

continuous at $x = 4$? Why or why not?

57. Is the function given by

$$g(x) = \begin{cases} \frac{1}{2}x + 1, & \text{for } x < 4, \\ -x + 7, & \text{for } x > 4, \end{cases}$$

continuous at $x = 4$? Why or why not?

58. Is the function given by

$$f(x) = \begin{cases} \frac{1}{3}x + 4, & \text{for } x < 3, \\ 2x - 1, & \text{for } x \geq 3, \end{cases}$$

continuous at $x = 3$? Why or why not?

59. Is the function given by

$$G(x) = \begin{cases} \dfrac{x^2 - 4}{x - 2}, & \text{for } x \neq 2, \\ 5, & \text{for } x = 2, \end{cases}$$

continuous at $x = 2$? Why or why not?

60. Is the function given by

$$F(x) = \begin{cases} \dfrac{x^2 - 1}{x - 1}, & \text{for } x \neq 1, \\ 4, & \text{for } x = 1, \end{cases}$$

continuous at $x = 1$? Why or why not?

61. Is the function given by

$$G(x) = \begin{cases} \dfrac{x^2 - 3x - 4}{x - 4}, & \text{for } x < 4, \\ 2x - 3, & \text{for } x \geq 4, \end{cases}$$

continuous at $x = 4$? Why or why not?

62. Is the function given by

$$f(x) = \begin{cases} \dfrac{x^2 - 4x - 5}{x - 5}, & \text{for } x < 5, \\ x + 1, & \text{for } x \geq 5, \end{cases}$$

continuous at $x = 5$? Why or why not?

63. Is the function given by $g(x) = \dfrac{1}{x^2 - 7x + 10}$ continuous at $x = 5$? Why or why not?

64. Is the function given by $f(x) = \dfrac{1}{x^2 - 6x + 8}$ continuous at $x = 3$? Why or why not?

65. Is the function given by $G(x) = \dfrac{1}{x^2 - 6x + 8}$ continuous at $x = 2$? Why or why not?

66. Is the function given by $F(x) = \dfrac{1}{x^2 - 7x + 10}$ continuous at $x = 4$? Why or why not?

67. Is the function given by $g(x) = \dfrac{1}{x + 5}$ continuous over the interval $(-4, 4)$? Why or why not?

68. Is the function given by $F(x) = -\dfrac{2}{x - 7}$ continuous over the interval $(-5, 5)$? Why or why not?

69. Is the function given by $G(x) = \dfrac{1}{x - 1}$ continuous over the interval $(0, \infty)$? Why or why not?

70. Is the function given by $f(x) = \dfrac{1}{x} + 3$ continuous over the interval $(-7, 7)$? Why or why not?

71. Is the function given by $g(x) = 4x^3 - 6x$ continuous on $\mathbb{R}$?

72. Is the function given by $F(x) = \dfrac{3}{x - 5}$ continuous on $\mathbb{R}$?

APPLICATIONS

Business and Economics

73. The Candy Factory sells candy by the pound, charging $1.50 per pound for quantities up to and including 20 pounds. Above 20 pounds, the Candy Factory charges $1.25 per pound for the entire quantity. If x represents the number of pounds, the price function is

$$p(x) = \begin{cases} 1.50x, & \text{for } x \leq 20, \\ 1.25x, & \text{for } x > 20. \end{cases}$$

Find $\lim\limits_{x \to 20^-} p(x)$, $\lim\limits_{x \to 20^+} p(x)$, and $\lim\limits_{x \to 20} p(x)$.

74. The Copy Shoppe charges $0.08 per copy for quantities up to and including 100 copies. For quantities above 100, the charge is $0.06 per copy. If x represents the number of copies, the price function is

$$p(x) = \begin{cases} 0.08x, & \text{for } x \leq 100, \\ 0.06x, & \text{for } x > 100. \end{cases}$$

Find $\lim\limits_{x \to 100^-} p(x)$, $\lim\limits_{x \to 100^+} p(x)$, and $\lim\limits_{x \to 100} p(x)$.

Life and Physical Sciences

75. A lab technician controls the temperature T inside a kiln. From an initial temperature of 0 degrees Celsius (°C), he allows the temperature to increase by 2°C per minute for the next 60 minutes. After the 60th minute, he allows the temperature to cool by 3°C per minute. If t is the number of minutes, the temperature T is given by

$$T(t) = \begin{cases} 2t, & \text{for } t \leq 60, \\ 300 - 3t, & \text{for } t > 60. \end{cases}$$

Find $\lim\limits_{t \to 60^-} T(t)$, $\lim\limits_{t \to 60^+} T(t)$, and $\lim\limits_{t \to 60} T(t)$.

SYNTHESIS

76. In Exercise 73, let

$$p(x) = \begin{cases} 1.50x, & \text{for } x \le 20, \\ 1.25x + k, & \text{for } x > 20. \end{cases}$$

Find k such that the price function p is continuous at $x = 20$.

77. In Exercise 74, let

$$p(x) = \begin{cases} 0.08x, & \text{for } x \le 100, \\ 0.06x + k, & \text{for } x > 100. \end{cases}$$

Find k such that the price function p is continuous at $x = 100$.

78. Find each limit, if it exists. If a limit does not exist, state that fact.

a) $\lim\limits_{x \to 0} \dfrac{|x|}{x}$

b) $\lim\limits_{x \to -2} \dfrac{x^3 + 8}{x^2 - 4}$

TECHNOLOGY CONNECTION

In Section 1.1, we used the TABLE feature to find limits. Consider

$$\lim_{x \to 0} \frac{\sqrt{1 + x} - 1}{x}.$$

Input–output tables for this function are below. The table on the left uses TblStart $= -1$ and ΔTbl $= 0.5$. By using smaller and smaller step values and beginning closer to 0, we can refine the table and obtain a better estimate of the limit. On the right is an input–output table with TblStart $= -0.03$ and ΔTbl $= 0.01$.

X	Y1
-1	1
-0.5	0.58579
0	ERROR
0.5	0.44949
1	0.41421
1.5	0.38743
2	0.36603
X = -1	

X	Y1
-0.03	0.50381
-0.02	0.50253
-0.01	0.50126
0	ERROR
0.01	0.49876
0.02	0.49752
0.03	0.49631
X = -0.03	

It appears that the limit is 0.5. We can check by graphing

$$y = \frac{\sqrt{1 + x} - 1}{x}$$

and tracing the curve near $x = 0$, zooming in on that portion of the curve.

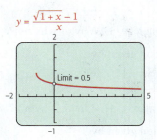

We see that

$$\lim_{x \to 0} \frac{\sqrt{1 + x} - 1}{x} = 0.5.$$

To verify this algebraically, multiply $\dfrac{\sqrt{1 + x} - 1}{x}$ by 1, using $\dfrac{\sqrt{1 + x} + 1}{\sqrt{1 + x} + 1}$. Then simplify the result and find the limit.

In Exercises 79–86, find each limit. Use TABLE *and start with* ΔTbl $= 0.1$. *Then use* 0.01, 0.001, *and* 0.0001. *When you think you know the limit, graph and use* TRACE *to verify your assertion. Then try to verify it algebraically.*

79. $\lim\limits_{x \to 1} \dfrac{\sqrt{x} - 1}{x - 1}$

80. $\lim\limits_{a \to -2} \dfrac{a^2 - 4}{\sqrt{a^2 + 5} - 3}$

81. $\lim\limits_{x \to 0} \dfrac{\sqrt{4 + x} - \sqrt{4 - x}}{x}$

82. $\lim\limits_{x \to 0} \dfrac{\sqrt{3 - x} - \sqrt{3}}{x}$

83. $\lim\limits_{x \to 0} \dfrac{\sqrt{7 + 2x} - \sqrt{7}}{x}$

84. $\lim\limits_{x \to 1} \dfrac{x - \sqrt[4]{x}}{x - 1}$

85. $\lim\limits_{x \to 0} \dfrac{7 - \sqrt{49 - x^2}}{x}$

86. $\lim\limits_{x \to 4} \dfrac{2 - \sqrt{x}}{4 - x}$

Answers to Quick Checks

1. (a) -1; L1, L2, L3, L6; **(b)** $\dfrac{51}{10}$, or 5.1; L1, L2, L3, L5, L6; **(c)** $\sqrt{13}$; L1, L2, L3, L6 **2.** 2 **3.** No, the limit as $x \to 2$ does not exist. **4.** No, the limit as $x \to 3$ is 6, but $h(3) = 7$, so the limit does not equal the function value. **5.** 90, 90, 90

1.3

Average Rates of Change

- Compute an average rate of change.
- Find a simplified difference quotient.

If a car travels 110 mi in 2 hr, its *average rate of change* (*speed*) is 110 mi/2 hr, or 55 mi/hr (55 mph). Suppose you accelerate on a freeway and, glancing at the speedometer, you see that at that instant your *instantaneous rate of change* is 55 mph. These are two quite different concepts. The first you are probably familiar with. The second involves ideas of limits and calculus. Understanding instantaneous rate of change requires a solid understanding of average rate of change.

The following graph shows total production of suits at Raggs, Ltd., during one morning of work.

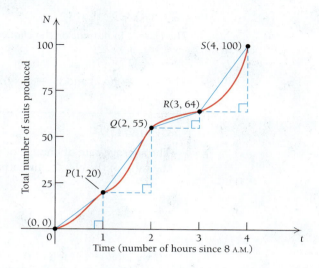

EXAMPLE 1   **Business: Production.**   How many suits were produced at Raggs, Ltd., from 9 A.M. to 10 A.M.?

Solution   At 9 A.M., 20 suits had been produced. At 10 A.M., 55 suits had been produced. In the hour from 9 A.M. to 10 A.M., the number of suits produced was

$$55 \text{ suits} - 20 \text{ suits}, \quad \text{or} \quad 35 \text{ suits}.$$

Note that 35 is the slope of the line segment from P to Q.

EXAMPLE 2   **Business: Average Rate of Change.**   What was the average number of suits produced per hour from 9 A.M. ($t = 1$) to 11 A.M. ($t = 3$)?

Solution   We have

$$\frac{64 \text{ suits} - 20 \text{ suits}}{3 \text{ hr} - 1 \text{ hr}} = \frac{44 \text{ suits}}{2 \text{ hr}}$$

$$= 22 \frac{\text{suits}}{\text{hr}}.$$

Note that 22 is the slope of the line segment from P to R.

Quick Check 1 ✔

State the average rate of change for each situation in a short sentence. Be sure to include units.

a) It rained 4 inches over a period of 8 hours. State your answer in inches per hour.

b) Your car travels 250 miles on 10 gallons of gas. State your answer in miles per gallon.

c) At 2 P.M., the temperature was 82 degrees. At 5 P.M., the temperature was 76 degrees. State your answer in degrees per hour.

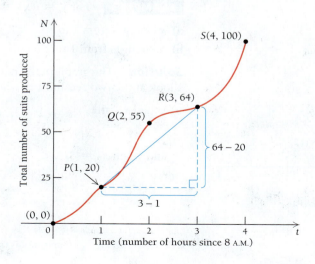

1 ✔

Let's consider a function $y = f(x)$ and two inputs x_1 and x_2. The *change in input*, or the *change in x*, is

$$x_2 - x_1.$$

The *change in output*, or the *change in y*, is

$$y_2 - y_1,$$

where $y_1 = f(x_1)$ and $y_2 = f(x_2)$.

DEFINITION

The **average rate of change of y with respect to x**, as x changes from x_1 to x_2, is the ratio of the change in output to the change in input:

$$\frac{y_2 - y_1}{x_2 - x_1}, \quad \text{where } x_2 \neq x_1.$$

If we look at a graph of the function, we see that

$$\frac{y_2 - y_1}{x_2 - x_1} = \frac{f(x_2) - f(x_1)}{x_2 - x_1},$$

which is both the average rate of change and the slope of the line from $P(x_1, y_1)$ to $Q(x_2, y_2)$.* The line passing through P and Q, denoted $\overleftrightarrow{PQ}$, is called a **secant line**.

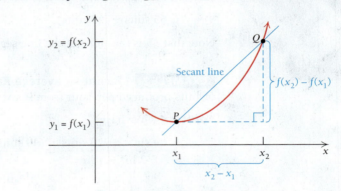

The slope of the secant line represents the average rate of change of f from x_1 to x_2.

EXAMPLE 3   For $y = f(x) = x^2 + 1$, find the average rate of change as:

a) x changes from 1 to 3;

b) x changes from 1 to 2.

Solution   The graph in the margin is not necessary to the computations but gives us a look at the two secant lines whose slopes are being computed.

a) When $x_1 = 1$,

$$y_1 = f(x_1) = f(1) = 1^2 + 1 = 2;$$

and when $x_2 = 3$,

$$y_2 = f(x_2) = f(3) = 3^2 + 1 = 10.$$

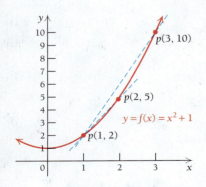

---

*The notation $P(x_1, y_1)$ simply means that point P has coordinates (x_1, y_1).

The average rate of change is

$$\frac{y_2 - y_1}{x_2 - x_1} = \frac{f(x_2) - f(x_1)}{x_2 - x_1}$$

$$= \frac{10 - 2}{3 - 1}$$

$$= \frac{8}{2} = 4.$$

b) When $x_1 = 1$,

$$y_1 = f(x_1) = f(1) = 1^2 + 1 = 2;$$

and when $x_2 = 2$,

$$y_2 = f(x_2) = f(2) = 2^2 + 1 = 5.$$

The average rate of change is

$$\frac{5 - 2}{2 - 1} = \frac{3}{1} = 3.$$

2 ✔

Quick Check 2 ✔

For $f(x) = x^3 + 1$, find the average rate of change between:

a) $x = 1$ and $x = 4$;

b) $x = 1$ and $x = 2$.

For a linear function, the average rate of change is the same for any choice of x_1 and x_2. As we saw in Example 3, a function that is not linear has average rates of change that generally vary with choices of x_1 and x_2.

Difference Quotient as Average Rate of Change

We now develop a notation for average rate of change that does not require subscripts. Instead of x_1, we write x; in place of x_2, we write $x + h$.

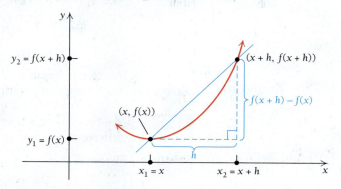

Think of h as the horizontal distance between inputs x_1 and x_2. That is, to get from x_1, or x, to x_2, we move a distance h. Thus, $x_2 = x + h$. Then the average rate of change, also called the *difference quotient*, is given by

$$\frac{y_2 - y_1}{x_2 - x_1} = \frac{f(x_2) - f(x_1)}{x_2 - x_1} = \frac{f(x + h) - f(x)}{(x + h) - x} = \frac{f(x + h) - f(x)}{h}.$$

DEFINITION

The average rate of change of $f(x)$ with respect to x is also called the **difference quotient**. It is given by

$$\frac{f(x + h) - f(x)}{h}, \quad \text{where } h \neq 0.$$

The difference quotient is equal to the slope of the secant line passing through $(x, f(x))$ and $(x + h, f(x + h))$.

EXAMPLE 4 For $f(x) = x^2$, find the difference quotient when:

a) $x = 5$ and $h = 3$. **b)** $x = 5$ and $h = 0.1$. **c)** $x = 5$ and $h = 0.01$.

Solution

a) We substitute $x = 5$ and $h = 3$ into the formula:

$$\frac{f(x + h) - f(x)}{h} = \frac{f(5 + 3) - f(5)}{3} = \frac{f(8) - f(5)}{3}.$$

Since $f(8) = 8^2 = 64$ and $f(5) = 5^2 = 25$, we have

$$\frac{f(8) - f(5)}{3} = \frac{64 - 25}{3} = \frac{39}{3} = 13.$$

The difference quotient is 13. It is the slope of the line passing through $(5, 25)$ and $(8, 64)$.

b) We substitute $x = 5$ and $h = 0.1$ into the formula:

$$\frac{f(x + h) - f(x)}{h} = \frac{f(5 + 0.1) - f(5)}{0.1} = \frac{f(5.1) - f(5)}{0.1}.$$

Since $f(5.1) = (5.1)^2 = 26.01$ and $f(5) = 25$, we have

$$\frac{f(5.1) - f(5)}{0.1} = \frac{26.01 - 25}{0.1} = \frac{1.01}{0.1} = 10.1.$$

c) We substitute $x = 5$ and $h = 0.01$ into the formula:

$$\frac{f(x + h) - f(x)}{h} = \frac{f(5 + 0.01) - f(5)}{0.01} = \frac{f(5.01) - f(5)}{0.01}.$$

Since $f(5.01) = (5.01)^2 = 25.1001$ and $f(5) = 25$, we have

$$\frac{f(5.01) - f(5)}{0.01} = \frac{25.1001 - 25}{0.01} = \frac{0.1001}{0.01} = 10.01.$$ ■

For the function in Example 4, let's find a form of the difference quotient that will allow for more efficient computations.

EXAMPLE 5 For $f(x) = x^2$, find a simplified form of the difference quotient. Then evaluate the difference quotient for $x = 5$ and $h = 0.1$ and for $x = 5$ and $h = 0.01$.

Solution We have

$$f(x) = x^2,$$

so

$$f(x + h) = (x + h)^2 = x^2 + 2xh + h^2. \qquad \text{Multiplying } (x + h)(x + h)$$

Then

$$f(x + h) - f(x) = (x^2 + 2xh + h^2) - x^2 = 2xh + h^2. \qquad \text{The } x^2 \text{ terms add to 0.}$$

Thus,

$$\frac{f(x + h) - f(x)}{h} = \frac{2xh + h^2}{h} = \frac{h(2x + h)}{h} = 2x + h, \quad h \neq 0. \qquad h/h = 1$$

This is the simplified form of this difference quotient. It is important to note that for any difference quotient, we must have $h \neq 0$.

For $x = 5$ and $h = 0.1$, we have

$$\frac{f(x + h) - f(x)}{h} = 2x + h = 2 \cdot 5 + 0.1 = 10 + 0.1 = 10.1.$$

For $x = 5$ and $h = 0.01$, we have

$$\frac{f(x + h) - f(x)}{h} = 2x + h = 2 \cdot 5 + 0.01 = 10.01.$$

Although $2x + h$ is valid only when $h \neq 0$, h can get closer and closer to 0. Perhaps you can sense what value the difference quotient will have in this example when we allow h to approach 0 as a limit. ∎

Compare the results of Example 4(b) and 4(c) and Example 5. In general, computations are easier when a simplified form of a difference quotient is found before any specific calculations are performed.

EXAMPLE 6 For $f(x) = x^3$, find the simplified form of the difference quotient.

Solution For $f(x) = x^3$,

$$f(x + h) = (x + h)^3 = x^3 + 3x^2h + 3xh^2 + h^3, \qquad \text{See Appendix A.}$$
$$f(x + h) - f(x) = (x^3 + 3x^2h + 3xh^2 + h^3) - x^3 = 3x^2h + 3xh^2 + h^3,$$

and

$$\frac{f(x + h) - f(x)}{h} = \frac{3x^2h + 3xh^2 + h^3}{h} \qquad \text{It is understood that } h \neq 0.$$

$$= \frac{h(3x^2 + 3xh + h^2)}{h} \qquad \text{Factoring } h; \; h/h = 1$$

$$= 3x^2 + 3xh + h^2, \quad h \neq 0.$$

Again, this is true *only* for $h \neq 0$. 3 ✔

Quick Check 3 ✔

Use the result of Example 6 to calculate the slope of the secant line (average rate of change) at $x = 2$, for $h = 0.1$ and $h = 0.01$.

The next two examples illustrate the development of the difference quotients for a simple rational function (Example 7) and for the square-root function (Example 8). Both forms are common in calculus.

EXAMPLE 7 For $f(x) = 1/x$, find the simplified form of the difference quotient.

Solution For $f(x) = 1/x$, we have

$$f(x + h) = \frac{1}{x + h}.$$

Then

$$f(x + h) - f(x) = \frac{1}{x + h} - \frac{1}{x}$$

$$= \frac{1}{x + h} \cdot \frac{x}{x} - \frac{1}{x} \cdot \frac{x + h}{x + h} \qquad \text{Multiplying by 1 to get a common denominator}$$

$$= \frac{x - (x + h)}{x(x + h)} \qquad \text{Parentheses are important here.}$$

$$= \frac{x - x - h}{x(x + h)}$$

$$= \frac{-h}{x(x + h)}.$$

Thus,

$$\frac{f(x+h)-f(x)}{h} = \frac{\dfrac{-h}{x(x+h)}}{h}$$

$$= \frac{-h}{x(x+h)} \cdot \frac{1}{h} = \frac{-1}{x(x+h)}, \, h \neq 0.$$

This is true *only* for $h \neq 0$.   ■

EXAMPLE 8   For $f(x) = \sqrt{x}$, find the simplified form of the difference quotient.

Solution   For $f(x) = \sqrt{x}$, we have $f(x+h) = \sqrt{x+h}$, so the difference quotient is

$$\frac{f(x+h)-f(x)}{h} = \frac{\sqrt{x+h}-\sqrt{x}}{h}.$$

Algebraic simplification of this difference quotient leads to

$$\frac{f(x+h)-f(x)}{h} = \frac{1}{\sqrt{x+h}+\sqrt{x}}, \quad h \neq 0.$$

Demonstration of this simplification is left as Exercise 54. This simplified difference quotient appears again in Section 1.5.   ■

In all of the above cases, the simplified difference quotient includes two variables: x and h, where h cannot be 0. Although h cannot be 0, we may let h be as small as we desire. In the next section, we take a further step: allowing h to approach 0 as a limit.

Section Summary

- An *average rate of change* is the slope of a line passing through two points. If the points are (x_1, y_1) and (x_2, y_2), then the average rate of change is $\dfrac{y_2 - y_1}{x_2 - x_1}$.
- If the two points are on the graph of f, an equivalent form of the slope formula is $\dfrac{f(x+h)-f(x)}{h}$, where h is the horizontal difference between the two x-values.

- This is called the *difference quotient*. The line passing through these two points is called a *secant line*.
- The difference quotient gives the *average rate of change* between two points on a graph.
- It is preferable to simplify a difference quotient algebraically before evaluating it for particular values of x and h.

1.3   Exercise Set

For each function in Exercises 1–16, (a) find the simplified form of the difference quotient and then (b) complete the following table.

x	h	$\dfrac{f(x+h)-f(x)}{h}$
5	2	
5	1	
5	0.1	
5	0.01	

1. $f(x) = 5x^2$

2. $f(x) = 4x^2$

3. $f(x) = -5x^2$

4. $f(x) = -4x^2$

5. $f(x) = x^2 - x$

6. $f(x) = x^2 + x$

7. $f(x) = \dfrac{9}{x}$

8. $f(x) = \dfrac{2}{x}$

9. $f(x) = 2x + 3$

10. $f(x) = -2x + 5$

11. $f(x) = 12x^3$

12. $f(x) = 1 - x^3$

13. $f(x) = x^2 - 4x$

14. $f(x) = x^2 - 3x$

15. $f(x) = x^2 - 3x + 5$

16. $f(x) = x^2 + 4x - 3$

APPLICATIONS

Business and Economics

For Exercises 17–24, use each graph to estimate the average rate of change of the percentage of new employees in that type of employment from 2003 to 2007, from 2007 to 2012, and from 2003 to 2012. (Source: Bureau of Labor Statistics.)

17. Total employment.

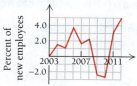

18. Construction.

19. Professional services.

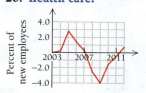

20. Health care.

21. Education.

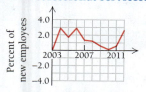

22. Government.

23. Mining and logging.

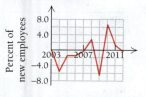

24. Manufacturing.

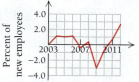

25. Use the following graph to find the average rate of change in U.S. energy consumption from 1982 to 1992, from 1992 to 2002, and from 2002 to 2012.

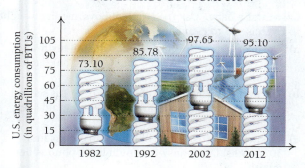

U.S. ENERGY CONSUMPTION

(*Source*: U.S. Energy Information Administration.)

26. Use the following graph to find the average rate of change of the U.S. trade deficit with Japan from 2007 to 2009, from 2009 to 2011, and from 2011 to 2013.

U.S. TRADE DEFICIT WITH JAPAN

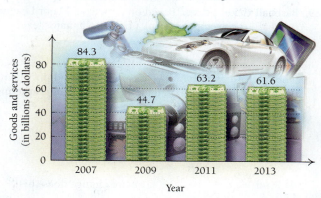

(*Source*: U.S. Census Bureau, *Statistical Abstract of the United States, 2012.*)

27. Utility. Utility is a type of function that occurs in economics. When a consumer receives x units of a product, a certain amount of pleasure, or utility U, is gained. The following represents a typical utility function.

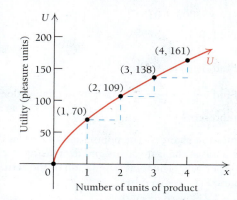

a) Find the average rate of change of U as x changes from 0 to 1; from 1 to 2; from 2 to 3; from 3 to 4.

b) Why do you think the average rates of change are decreasing as x increases?

28. Advertising results. The following graph shows a typical response to advertising. When amount a is spent on advertising, $N(a)$ units of a product are sold.

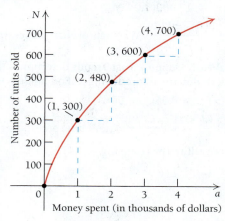

a) Find the average rate of change of N as a changes from 0 to 1; from 1 to 2; from 2 to 3; from 3 to 4.

b) Why do you think the average rates of change are decreasing as x increases?

29. Baseball ticket prices. Based on data from Major League Baseball, the average price of a ticket to a major league game can be approximated by

$$p(x) = 0.06x^3 - 0.5x^2 + 1.64x + 24.76$$

where x is the number of years after 2008 and $p(x)$ is in dollars. (*Source*: Based on data from www.teammarketing .com.)

a) Find $p(4)$.

b) Find $p(6)$.

c) Find $p(6) - p(4)$.

d) Find $\dfrac{p(6) - p(4)}{6 - 4}$. What rate of change does this represent?

30. Compound interest. The amount of money, $A(t)$, in a savings account that pays 6% interest, compounded quarterly for t years, when an initial investment of $2000 is made, is given by

$$A(t) = 2000(1.015)^{4t}.$$

a) Find $A(3)$.

b) Find $A(5)$.

c) Find $A(5) - A(3)$.

d) Find $\dfrac{A(5) - A(3)}{5 - 3}$. What rate of change does this represent?

31. Population change. The population of Payton County was 5400 at the last census and decreasing at the rate of 2.5% per year. The total population of the county after t years, $P(t)$, is given by

$$P(t) = 5400(0.975)^t.$$

Find $\dfrac{P(8) - P(5)}{8 - 5}$. What rate of change does this represent?

32. Population change. The undergraduate population at Harbor College was 17,000 and increasing at the rate of 4.2% per year. The undergraduate population after t years, $P(t)$, is given by

$$P(t) = 17,000(1.042)^t.$$

Find $\dfrac{P(6) - P(2)}{6 - 2}$. What rate of change does this represent?

33. Total cost. Suppose Fast Trends determines that the cost, in dollars, of producing x iPod holders is given by

$$C(x) = -0.05x^2 + 50x.$$

Find $\dfrac{C(305) - C(300)}{305 - 300}$, and interpret the significance of this result to the company.

34. Total revenue. Suppose Fast Trends determines that the revenue, in dollars, from the sale of x iPod holders is given by

$$R(x) = -0.001x^2 + 150x.$$

Find $\dfrac{R(305) - R(300)}{305 - 300}$, and interpret the significance of this result to the company.

Life and Physical Sciences

35. Growth of a baby. The median weights of babies at age t months are graphed below.

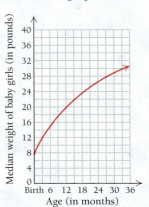 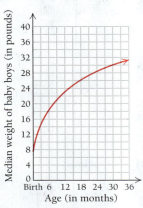

(*Source*: Developed by the National Center for Health Statistics in collaboration with the National Center for Chronic Disease Prevention and Health Promotion, 2000; checked in 2014.)

Use the graph of girls' median weight to estimate:

a) The average growth rate of a girl during her first 12 months. (Your answer should be in pounds per month.)

b) The average growth rate of a girl during her second 12 months.

c) The average growth rate of a girl during her first 24 months.

d) Based on your answers in parts (a)–(c) and the graph, estimate the growth rate of a typical 12-month-old girl, and explain how you arrived at this figure.

e) In what three-month period does a baby girl grow fastest?

36. Growth of a baby. Use the graph of boys' median weight in Exercise 35 to estimate:

a) The average growth rate of a boy during his first 15 months. (Your answer should be in pounds per month.)

b) The average growth rate of a boy during his second 15 months. (Your answer should be in pounds per month.)

c) The average growth rate of a boy during his first 30 months. (Your answer should be in pounds per month.)

d) Based on your answers in parts (a)–(c) and the graph, estimate the growth rate of a typical boy at exactly 15 months, and explain how you arrived at this figure.

37. Home range. It has been shown that the home range, in hectares, of a carnivorous mammal weighing w grams can be approximated by

$$H(w) = 0.11w^{1.36}.$$

(*Source: Based on information in Emlen, J. M., Ecology: An Evolutionary Approach*, p. 200, Reading, MA: Addison-Wesley, 1973; and Harestad, A. S., and Bunnel, F. L., "Home Range and Body Weight—A Reevaluation," *Ecology*, Vol. 60, No. 2, pp. 389–402.)

a) Find the average rate at which a carnivorous mammal's home range increases as the animal's weight grows from 500 g to 700 g.

b) Find $\dfrac{H(300) - H(200)}{300 - 200}$. What does this rate represent?

38. Radar range. The function given by $R(x) = 11.74x^{1/4}$ can be used to approximate the maximum range $R(x)$, in miles, of an ARSR-3 surveillance radar with a peak power of x watts (W). (*Source: Introduction to RADAR Techniques*, Federal Aviation Administration, 1988.)

a) Find the rate at which the maximum radar range changes as peak power increases from 40,000 W to 60,000 W.

b) Find $\dfrac{R(60{,}000) - R(50{,}000)}{60{,}000 - 50{,}000}$. What does this rate represent?

39. Memory. The total number of words, $M(t)$, that a person can memorize in t minutes is shown in the following graph.

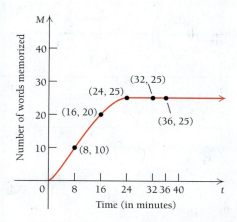

a) Find the average rate of change of M as t changes from 0 to 8; from 8 to 16; from 16 to 24; from 24 to 32; from 32 to 36.

b) Why do the average rates of change become 0 when t exceeds 24 min?

40. Gas mileage. At the beginning of a trip, the odometer on a car reads 30,680, and the car has a full tank of gas. At the end of the trip, the odometer reads 31,077. It takes 13.5 gal of gas to refill the tank.

a) What is the average rate at which the car was traveling, in miles per gallon?

b) What is the average rate of gas consumption in gallons per mile?

41. Average velocity. In t seconds, an object dropped from a certain height will fall $s(t)$ feet, where

$$s(t) = 16t^2.$$

a) Find $s(5) - s(3)$.

b) What is the average rate of change of distance with respect to time during the period from 3 to 5 sec? This is known as **average velocity**, or **speed**.

42. Average velocity. Suppose that in t hours, a truck travels $s(t)$ miles, where

$$s(t) = 10t^2.$$

a) Find $s(5) - s(2)$. What does this represent?

b) Find the average rate of change of distance with respect to time as t changes from $t_1 = 2$ to $t_2 = 5$. This is also *average velocity*.

Social Sciences

43. Population growth. The two curves below describe the numbers of people in two countries at time t, in years.

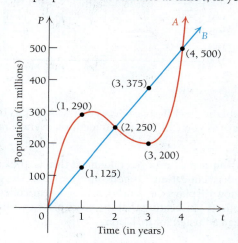

a) Find the average rate of change of each population with respect to time t as t changes from 0 to 4. This is often called the **average growth rate**.

b) If the calculations in part (a) were the only ones made, would we detect the fact that the populations were growing differently? Why or why not?

c) Find the average rates of change of each population as t changes from 0 to 1; from 1 to 2; from 2 to 3; from 3 to 4.

d) For which population does the statement "the population grew consistently at a rate of 125 million per year" convey accurate information? Why?

SYNTHESIS

44. Business: comparing rates of change. The following two graphs show the number of federally insured banks from 1987 to 2013 and the Nasdaq Composite Stock Index over a 6-month period in 2013.

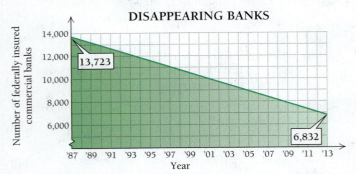

DISAPPEARING BANKS

(*Source*: Federal Deposit Insurance Corp., 12/26/13.)

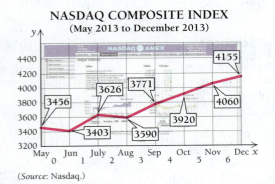

NASDAQ COMPOSITE INDEX
(May 2013 to December 2013)

(*Source*: Nasdaq.)

Explain the difference between these graphs in as many ways as you can. Be sure to mention average rates of change.

45. Rising cost of college. Like most things, the cost of a college education has gone up over time. The graph below displays the yearly costs of 4-year colleges in 2010 dollars—indicating that the costs prior to 2010 have been adjusted for inflation.

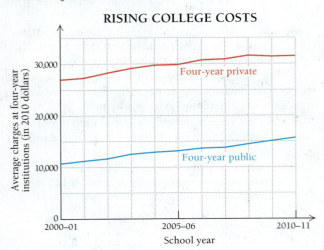

RISING COLLEGE COSTS

(*Source*: National Center for Educational Statistics, *Annual Digest of Educational, Statistics*, 2012.)

a) In what school year(s) did the cost of a private 4-year college appear to increase the most?

b) In what school year(s) did the cost of a public 4-year college appear to increase the most?

c) Assuming an annual inflation rate of 3%, calculate the cost of a year at a public and at a private 4-year college in 2000. Express the costs in 2000 dollars.

Find the simplified difference quotient for each function listed.

46. $f(x) = mx + b$

47. $f(x) = ax^2 + bx + c$

48. $f(x) = ax^3 + bx^2$

49. $f(x) = x^4$

50. $f(x) = x^5$

51. $f(x) = ax^5 + bx^4$

52. $f(x) = \dfrac{1}{x^2}$

53. $f(x) = \dfrac{1}{1 - x}$

54. Below are the steps in the simplification of the difference quotient for $f(x) = \sqrt{x}$ (see Example 8). Provide a brief justification for each step.

$$\frac{f(x + h) - f(x)}{h} = \frac{\sqrt{x + h} - \sqrt{x}}{h}$$

a) $$= \frac{\sqrt{x + h} - \sqrt{x}}{h} \cdot \left(\frac{\sqrt{x + h} + \sqrt{x}}{\sqrt{x + h} + \sqrt{x}} \right)$$

---

b) $$= \frac{x + h + \sqrt{x}\sqrt{x + h} - \sqrt{x}\sqrt{x + h} - x}{h(\sqrt{x + h} + \sqrt{x})}$$

---

c) $$= \frac{x + h - x}{h(\sqrt{x + h} + \sqrt{x})}$$

---

d) $$= \frac{h}{h(\sqrt{x + h} + \sqrt{x})}$$

---

e) $$= \frac{1}{\sqrt{x + h} + \sqrt{x}}$$

For Exercises 55 and 56, find the simplified difference quotient.

55. $f(x) = \sqrt{2x + 1}$

56. $f(x) = \dfrac{1}{\sqrt{x}}$

Answers to Quick Checks

1. (a) It rained $1/2$ inch per hour.   **(b)** Your car gets 25 miles per gallon.   **(c)** The temperature dropped 2 degrees per hour.   **2. (a)** 21;   **(b)** 7   **3.** 12.61, 12.0601

Differentiation Using Limits of Difference Quotients

1.4

- Find derivatives and values of derivatives.
- Find equations of tangent lines.

The *slope of the secant line* passing through two points $(x, f(x))$ and $(x + h, f(x + h))$ on the graph of a function $y = f(x)$ represents the *average rate of change* of $f(x)$ over the interval $[x, x + h]$. This rate is given by the *difference quotient*

$$\frac{f(x + h) - f(x)}{h}, \qquad h \neq 0.$$

In Section 1.3, we simplified difference quotients for several functions. These expressions can be evaluated to calculate the slope of a secant line for given x- and h-values. Recall that h represents the horizontal distance between x and $x + h$. Although h cannot equal zero, we can consider the case in which h approaches 0.

Tangent Lines

A line that touches a circle at exactly one point is called a *tangent line*. The word "tangent" derives from the Latin *tangentem*, meaning "touch," whereas the word "secant" derives from the Latin *secantem*, meaning "cut." In the figure below, the secant line cuts through the circle, while the tangent line touches, but does not cut through, the circle.

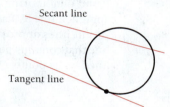

These notions extend to any smooth curve: a tangent line touches a curve at a single point, just as for the circle in the previous figure. In the figure below, the line L touches the curve at P, the *point of tangency*. Although L passes through the curve elsewhere, it is still considered a tangent line to the curve at point P.

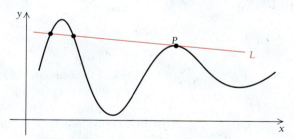

In the next figure, line M crosses the curve only at point P, but is not tangent to the curve; it does not touch the curve in the desired manner.

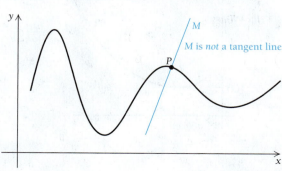

The four-lane highway is tangent to the access loops.

Exploratory

Graph $y_1 = 3x^5 - 20x^3$ with the viewing window $[-3, 3, -80, 80]$, with $\text{Xscl} = 1$ and $\text{Yscl} = 10$. Then graph $y_2 = -7x - 10$, $y_3 = -30x + 13$, and $y_4 = -45x + 28$. Which line appears to be tangent to the graph of y_1 at $(1, -17)$? If necessary, zoom in near $(1, -17)$ to confirm your answer.

In the following figure, all of the lines except for L_1 and L_2 are tangent lines.

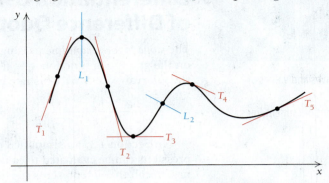

If a curve is smooth (has no corners or breaks), then each point on the curve has a *unique* tangent line; that is, exactly one tangent line can be drawn at any given point.

Differentiation Using Limits

We now define *tangent line* so that it makes sense for *any* curve. To do this, we use the notion of limit.

To obtain the line tangent to the curve at point P, consider secant lines through P and neighboring points Q_1, Q_2, and so on. As the Q's approach P, the secant lines approach line T. Each secant line has a slope. The slopes m_1, m_2, m_3, and so on, of the secant lines approach the slope m of line T. We *define* line T as the **tangent line**, the line that contains point P and has slope m, where m is the limit of the slopes of the secant lines as the points Q approach P.

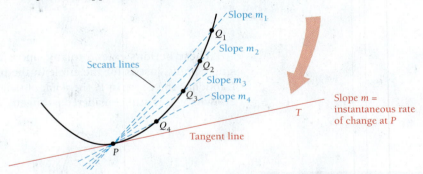

Think of the sequence of secant lines as an animation: as the points Q move closer to P, the resulting secant lines "lie down" on the tangent line.

How might we calculate the limit m? Suppose P has coordinates $(x, f(x))$. Then the first coordinate of Q is x plus some number h, or $x + h$. The coordinates of Q are $(x + h, f(x + h))$, as shown in the figure to the left below.

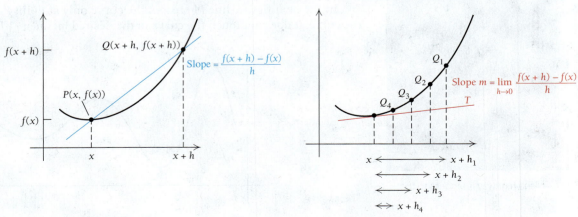

From Section 1.3, we know that the slope of the secant line $\overleftrightarrow{PQ}$ is given by

$$\frac{f(x + h) - f(x)}{h}.$$

Now, as shown in the figure at the bottom right on p. 128, as the Q's approach P, the values of $x + h$ approach x. That is, h approaches 0. Thus, we have the following.

The slope of the tangent line at $(x, f(x))$ is $m = \lim\limits_{h \to 0} \dfrac{f(x + h) - f(x)}{h}$.

This limit is also the **instantaneous rate of change** of $f(x)$ at x.

The formal definition of the *derivative of a function f* can now be given. We will designate the derivative at x as $f'(x)$, rather than m. The notation $f'(x)$ is read "the derivative of f at x," "f prime at x," or "f prime of x."

"Nothing in this world is so powerful as an idea whose time has come."

Victor Hugo

DEFINITION

For a function $y = f(x)$, its **derivative** at x is the function f' defined by

$$f'(x) = \lim\limits_{h \to 0} \frac{f(x + h) - f(x)}{h},$$

provided that the limit exists. If $f'(x)$ exists, then we say that f is **differentiable** at x. We sometimes call f' the **derived function**.

Let's now calculate some formulas for derivatives. That is, given a formula for a function f, we will attempt to find a formula for f'. The formula for f' can then be used to find the slope of a tangent line to $y = f(x)$ at a given x-value.

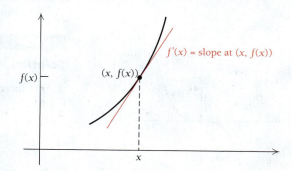

There are three steps in calculating a derivative.

1. Write the difference quotient, $\dfrac{(f(x + h) - f(x))}{h}$.
2. Simplify the difference quotient.
3. Find the limit as h approaches 0.

EXAMPLE 1   For $f(x) = x^2$, find $f'(x)$. Then find $f'(-1)$ and $f'(2)$.

Solution   We carry out the three steps.

1. $\dfrac{f(x + h) - f(x)}{h} = \dfrac{(x + h)^2 - x^2}{h}$

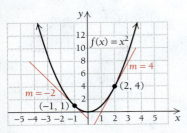

A visualization of Example 1

TECHNOLOGY CONNECTION

Exploratory

To see exactly how the definition of derivative works and to check the results of Example 1, let
$$y_1 = ((-1 + x)^2 - (-1)^2)/x$$
and note that we are using x in place of h. Use the TABLE feature and enter smaller and smaller values for x. Then repeat the procedure for $y_2 = ((2 + x)^2 - 2^2)/x$. How do these results confirm those in Example 1?

2. $\dfrac{f(x + h) - f(x)}{h} = \dfrac{x^2 + 2xh + h^2 - x^2}{h}$ Evaluating $f(x + h)$ and $f(x)$

$\qquad\qquad\qquad = \dfrac{2xh + h^2}{h} = \dfrac{h(2x + h)}{h}$ $\Big\}$ Simplifying

$\qquad\qquad\qquad = 2x + h, \; h \neq 0$

3. $\displaystyle\lim_{h \to 0} \dfrac{f(x + h) - f(x)}{h} = \lim_{h \to 0} (2x + h) = 2x$

Therefore, the derivative of $f(x) = x^2$ is given by

$\qquad f'(x) = 2x.$

Using the fact that $f'(x) = 2x$, it follows that

$\qquad f'(-1) = 2 \cdot (-1) = -2, \quad$ and $\quad f'(2) = 2 \cdot 2 = 4.$

This tells us that at $x = -1$, the curve has a tangent line with slope

$\qquad f'(-1) = -2,$

and at $x = 2$, the tangent line has slope

$\qquad f'(2) = 4.$

We can also say:

- The tangent line to the curve at the point $(-1, 1)$ has slope -2.
- The tangent line to the curve at the point $(2, 4)$ has slope 4.
- The instantaneous rate of change at $x = -1$ is -2.
- The instantaneous rate of change at $x = 2$ is 4.
- f is differentiable at $x = -1$ and $x = 2$. In fact, f is differentiable for all real numbers. ■

EXAMPLE 2 For $f(x) = x^3$, find $f'(x)$. Then find $f'(-1)$ and $f'(1.5)$.

Solution

1. We have

$\qquad \dfrac{f(x + h) - f(x)}{h} = \dfrac{(x + h)^3 - x^3}{h}.$

2. In Example 6 of Section 1.3 (on p. 121), we showed that this simplifies to

$\qquad \dfrac{f(x + h) - f(x)}{h} = 3x^2 + 3xh + h^2, \qquad h \neq 0.$

3. We then have

$\qquad f'(x) = \displaystyle\lim_{h \to 0} \dfrac{f(x + h) - f(x)}{h} = \lim_{h \to 0} (3x^2 + 3xh + h^2) = 3x^2.$

Thus,

$\qquad f'(-1) = 3(-1)^2 = 3 \quad$ and $\quad f'(1.5) = 3(1.5)^2 = 6.75.$ 1 ✔

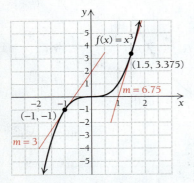

A visualization of Example 2

Quick Check 1 ✔

Use the results from Examples 1 and 2 to find the derivative of $f(x) = x^3 + x^2$, and then calculate $f'(-2)$ and $f'(4)$. Interpret these results.

EXAMPLE 3 For $f(x) = 3x - 4$, find $f'(x)$ and $f'(2)$.

Solution We follow the three steps given above.

1. $\dfrac{f(x + h) - f(x)}{h} = \dfrac{3(x + h) - 4 - (3x - 4)}{h}$ The parentheses are important.

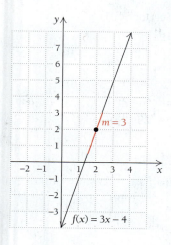

$f(x) = 3x - 4$

2. $\dfrac{f(x + h) - f(x)}{h} = \dfrac{3x + 3h - 4 - 3x + 4}{h}$ Using the distributive law

$\qquad\qquad\qquad = \dfrac{3h}{h} = 3, \quad h \neq 0;$ Simplifying

3. $\displaystyle\lim_{h \to 0} \dfrac{f(x + h) - f(x)}{h} = \lim_{h \to 0} 3 = 3$, since 3 is a constant.

Thus, if $f(x) = 3x - 4$, then $f'(x) = 3$ and $f'(2) = 3$. ∎

The result of Example 3 suggests that, for the graph of any linear function, the slope of a tangent line is the slope of the line itself. That is, the derivative of

$$f(x) = mx + b$$

is $f'(x) = m.$

This formula can be verified in a manner similar to that used in Example 3.

EXAMPLE 4   For $f(x) = \dfrac{1}{x}$:

a) Find $f'(x)$.   **b)** Find $f'(2)$.

c) Find an equation of the tangent line to the curve at $x = 2$.

Solution

a) 1. We have

$$\dfrac{f(x + h) - f(x)}{h} = \dfrac{[1/(x + h)] - (1/x)}{h}.$$

2. On p. 122, we showed that this simplifies to

$$\dfrac{f(x + h) - f(x)}{h} = \dfrac{-1}{x(x + h)}, \quad h \neq 0.$$

3. We want to find

$$\lim_{h \to 0} \dfrac{f(x + h) - f(x)}{h} = \lim_{h \to 0} \dfrac{-1}{x(x + h)}.$$

As $h \to 0$, we have $x + h \to x$. Thus,

$$f'(x) = \lim_{h \to 0} \dfrac{-1}{x(x + h)} = \dfrac{-1}{x \cdot x} = -\dfrac{1}{x^2}.$$

b) Since $f'(x) = -1/x^2$, we have

$$f'(2) = -\dfrac{1}{2^2} = -\dfrac{1}{4}.$$ This is the slope of the tangent line at $x = 2$.

c) We can find an equation of the tangent line at $x = 2$ if we know the line's slope and a point on the line. In part (b), we found that the slope at $x = 2$ is $-\frac{1}{4}$. To find a point on the line, we compute $f(2)$:

$$f(2) = \dfrac{1}{2}.$$ CAUTION! Be careful to use f when computing y-values and f' when computing slope.

We have

Point: $(2, \frac{1}{2})$,   This is $(x_1, f(x_1))$.

Slope: $-\frac{1}{4}$.   This is $f'(x_1)$.

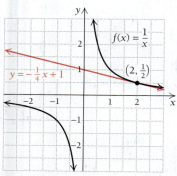

A visualization of Example 4(c)

Quick Check 2 ✔

Repeat Example 4(a) for

$f(x) = -\dfrac{2}{x}$. What are the

similarities between this solution

and that obtained for $f(x) = \dfrac{1}{x}$?

We substitute into the point–slope equation (see Section R.4):

$$y - y_1 = m(x - x_1)$$
$$y - \tfrac{1}{2} = -\tfrac{1}{4}(x - 2)$$
$$y = -\tfrac{1}{4}x + \tfrac{1}{2} + \tfrac{1}{2} \qquad \left.\begin{array}{l} \end{array}\right\} \text{ Rewriting in}$$
$$= -\tfrac{1}{4}x + 1. \qquad\qquad\quad \text{slope–intercept form}$$

The equation of the tangent line to the curve at $x = 2$ is

$$y = -\tfrac{1}{4}x + 1.$$

2 ✔

In Example 4, note that since $f(0)$ does not exist for $f(x) = 1/x$, we cannot evaluate the difference quotient

$$\frac{f(0 + h) - f(0)}{h}.$$

Thus, $f'(0)$ does not exist. We say that "f is not differentiable at 0."

> When a function is not defined at a point, it is not differentiable at that point. In general, if a function is discontinuous at a point, it is not differentiable at that point.

Sometimes a function f is continuous at a point, but its derivative f' does not exist at this point. The function $f(x) = |x|$ is an example.

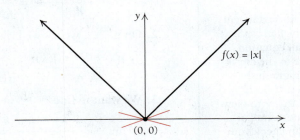

Suppose we try to draw a tangent line at $(0, 0)$. A function like this with a corner seems to have many tangent lines at $(0, 0)$, and thus many slopes. The derivative at such a point would not be unique. Let's try to calculate the derivative at 0.

Since

$$f(x) = |x| = \begin{cases} x, & \text{for } x \geq 0, \\ -x, & \text{for } x < 0, \end{cases}$$

it follows from our earlier work with lines that

$$f'(x) = \begin{cases} 1, & \text{for } x > 0, \\ -1, & \text{for } x < 0. \end{cases}$$

Then, since

$$\lim_{x \to 0^+} f'(x) \neq \lim_{x \to 0^-} f'(x),$$

it follows that $f'(0)$ does not exist.

In general, a function is not differentiable at any corners on its graph.

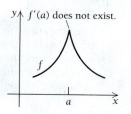

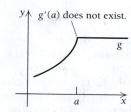

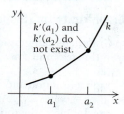

A function is also not differentiable at a point if it has a vertical tangent at that point. For example, the function shown to the right has a vertical tangent at point a. Since the slope of a vertical line is undefined, there is no derivative at such a point.

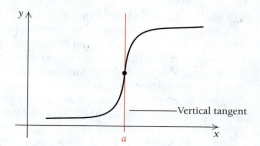

The function given by $f(x) = |x|$ illustrates the fact that although a function may be continuous at each point in an interval I, it may not be differentiable at each point in I. That is, continuity at a point does not imply differentiability at that point. However, differentiability at a point *does* guarantee continuity at that point. That is, if $f'(a)$ exists, then f is continuous at a. The function $f(x) = x^2$ is an example of a function that is differentiable over the interval $(-\infty, \infty)$ and is therefore continuous everywhere. Thus, when a function is differentiable over an interval, it is not just continuous, but is also *smooth* in the sense that there are no corners in its graph.

EXAMPLE 5 Below is the graph of a function $y = t(x)$. List the points in the graph at which the function t is not differentiable.

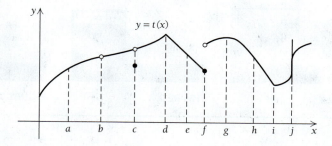

Quick Check 3 ✔

Where is $f(x) = |x + 6|$ not differentiable? Why?

Solution A function is not differentiable at a point if there is a discontinuity, a corner, or a vertical tangent at that point. Therefore, the function $y = t(x)$ is not differentiable at $x = b$, $x = c$, and $x = f$ since the function is discontinuous at these points; it is not differentiable at $x = d$ and $x = i$ since there are corners at these points; and it is not differentiable at $x = j$ as there is a vertical tangent line at this point (the slope is undefined).

3 ✔

Using Graphicus to Find Slopes of Tangent Lines

The app Graphicus can graph most of the functions we encounter in this text and draw tangent lines at points on a curve.

To graph a function and examine its tangent lines, let's consider $f(x) = x^4 - 2x^2$. After opening Graphicus, touch the blank rectangle at the top of the screen and enter the function as y(x)=x^4-2x^2. Press $\boxed{+}$ at the upper right. You will see the graph, as shown in Fig. 1.

Notice the seven icons at the bottom of the screen. The first one is for zooming. Touch it to zoom in and zoom out. Touch the screen with two fingers and move vertically or horizontally to adjust the windows and scaling. The second icon is for visualizing tangent lines. Touch it and then touch the graph at a point, and you will see a highlighted point on the graph and a tangent line to the curve at that point (Fig. 2). Move your finger left to right along the graph and various other points on the curve and tangent lines at those points will be highlighted (Figs. 3 and 4). Think about the slopes of the tangent lines as you move from left to right. When are they positive, negative, or 0?

EXERCISES

Use Graphicus to graph each function. You may need to use the zoom function to alter the window. Display tangent lines, noting where they are positive, negative, and 0. Then try to determine where the slope changes from positive to 0 to negative.

1. $f(x) = 2x^3 - x^4$ 　　　　　　**2.** $f(x) = x(200 - x)$

3. $f(x) = x^3 - 6x^2$

4. $f(x) = -4.32 + 1.44x + 3x^2 - x^3$

5. $g(x) = x\sqrt{4 - x^2}$

6. $g(x) = \dfrac{4x}{x^2 + 1}$ 　　　　　　**7.** $f(x) = \dfrac{x^2 - 3x}{x - 1}$

8. $f(x) = |x + 2| - 3$. What happens to the tangent line at the point $(-2, -3)$?

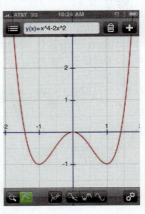

FIGURE 1

FIGURE 2

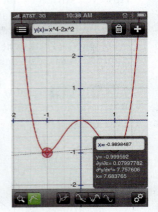

FIGURE 3

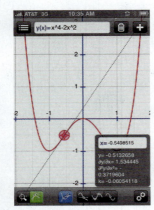

FIGURE 4

Section Summary

- A *tangent line* is a line that touches a smooth curve at a single point, the *point of tangency*. See the first figure on p. 128 for examples of tangent lines and lines that are not tangent lines.
- The *derivative* of a function $f(x)$ is defined by

$$f'(x) = \lim_{h \to 0} \frac{f(x + h) - f(x)}{h}.$$

- The *slope* of the tangent line to the graph of $y = f(x)$ at $x = a$ is the value of the derivative at $x = a$; that is, the slope of the tangent line at $x = a$ is $f'(a)$.
- Slopes of tangent lines are interpreted as *instantaneous rates of change*.
- The equation of a tangent line at $x = a$ is found by simplifying $y - f(a) = f'(a)(x - a)$.

- If a function is differentiable at $x = a$, then it is *continuous* at $x = a$. That is, differentiability at a point implies continuity at that point.
- However, continuity at $x = a$ does *not* imply differentiability at $x = a$. A good example is the absolute-value function, $f(x) = |x|$, or any function whose graph has

a corner. Continuity alone is not sufficient to guarantee differentiability.

- A function is not differentiable at $x = a$ if:
  (1) there is a discontinuity at $x = a$,
  (2) there is a corner at $x = a$, or
  (3) there is a vertical tangent at $x = a$.

1.4 Exercise Set

In Exercises 1–16:

a) *Graph the function.*

b) *Draw tangent lines to the graph at points whose x-coordinates are* $-2, 0,$ *and* $1.$

c) *Find* $f'(x)$ *by determining* $\lim\limits_{h \to 0} \dfrac{f(x + h) - f(x)}{h}.$

d) *Find* $f'(-2), f'(0),$ *and* $f'(1).$ *These slopes should match those of the lines you drew in part* (b).

1. $f(x) = \frac{1}{2}x^2$

2. $f(x) = \frac{3}{2}x^2$

3. $f(x) = -2x^2$

4. $f(x) = -3x^2$

5. $f(x) = -x^3$

6. $f(x) = x^3$

7. $f(x) = 2x + 3$

8. $f(x) = -2x + 5$

9. $f(x) = \frac{3}{4}x - 2$

10. $f(x) = \frac{1}{2}x - 3$

11. $f(x) = x^2 + x$

12. $f(x) = x^2 - x$

13. $f(x) = -5x^2 - 2x + 7$

14. $f(x) = -2x^2 + 3x - 2$

15. $f(x) = \dfrac{1}{x}$

16. $f(x) = \dfrac{2}{x}$

17. Find an equation of the tangent line to the graph of $f(x) = x^3$ at **(a)** $(-2, -8)$; **(b)** $(0, 0)$; **(c)** $(4, 64)$. See Example 2.

18. Find an equation of the tangent line to the graph of $f(x) = x^2$ at **(a)** $(3, 9)$; **(b)** $(-1, 1)$; **(c)** $(10, 100)$. See Example 1.

19. Find an equation of the tangent line to the graph of $f(x) = 2/x$ at **(a)** $(1, 2)$; **(b)** $(-1, -2)$; **(c)** $(100, 0.02)$. See Exercise 16.

20. Find an equation of the tangent line to the graph of $f(x) = -1/x$ at **(a)** $(-1, 1)$; **(b)** $(2, -\frac{1}{2})$; **(c)** $(-5, \frac{1}{5})$.

21. Find an equation of the tangent line to the graph of $f(x) = x^2 - 2x$ at **(a)** $(-2, 8)$; **(b)** $(1, -1)$; **(c)** $(4, 8)$.

22. Find an equation of the tangent line to the graph of $f(x) = 4 - x^2$ at **(a)** $(-1, 3)$; **(b)** $(0, 4)$; **(c)** $(5, -21)$.

23. Find $f'(x)$ for $f(x) = mx + b$.

24. Find $f'(x)$ for $f(x) = ax^2 + bx$.

For Exercises 25–28, list the points in the graph at which each function is not differentiable.

25.

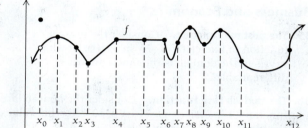

26.

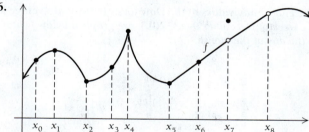

27.

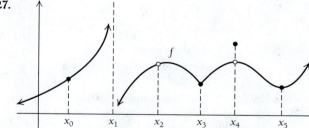

28.

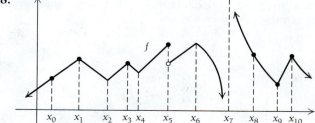

29. Draw a graph that is continuous, but not differentiable, at $x = 3$.

30. Draw a graph that is continuous, with no corners, but not differentiable, at $x = 1$.

31. Draw a graph that has a horizontal tangent line at $x = 5$.

32. Draw a graph that is differentiable and has horizontal tangent lines at $x = 0$, $x = 2$, and $x = 4$.

33. Draw a graph that has horizontal tangent lines at $x = 2$ and $x = 5$ and is continuous, but not differentiable, at $x = 3$.

34. Draw a graph that is continuous for all x, with no corners, but not differentiable at $x = -1$ and $x = 2$.

APPLICATIONS

Business and Economics

35. The postage function. Consider the postage function in Exercises 72–76 on p. 105. At what values in the domain is the function not differentiable?

36. The taxicab fare function. Consider the taxicab fare function in Exercises 69–71 on p. 105. At what values is the function not differentiable?

The end-of-day values of the Dow Jones Industrial Average for the week of December 9–13, 2013, are graphed below, where x is the day of the month.

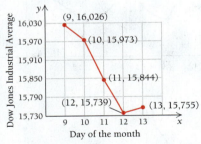

(*Source:* Google Finance.)

37. On what day did the Dow Jones Industrial Average show the greatest rate of increase? On what day did it show the greatest rate of decrease? Give the rates.

38. Is the function in Exercise 37 differentiable at the given x-values? Why or why not?

SYNTHESIS

39. Which of the lines in the following graph appear to be tangent lines? Why or why not?

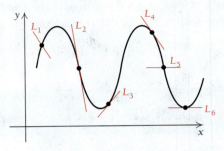

40. On the following graph, use a colored pencil to draw each secant line from point P to the points Q. Then use a different colored pencil to draw a tangent line to the curve at P. Describe what happens.

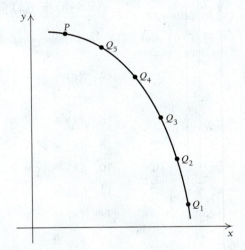

For Exercises 41–48, find $f'(x)$ for the given function.

41. $f(x) = x^4$ (See Exercise 49 in Section 1.3.)

42. $f(x) = \dfrac{1}{1 - x}$ (See Exercise 53 in Section 1.3.)

43. $f(x) = x^5$ (See Exercise 50 in Section 1.3.)

44. $f(x) = \dfrac{1}{x^2}$ (See Exercise 52 in Section 1.3.)

45. $f(x) = \sqrt{x}$ (See Example 8 in Section 1.3.)

46. $f(x) = \sqrt{2x + 1}$ (See Exercise 55 in Section 1.3.)

47. $f(x) = \dfrac{1}{\sqrt{x}}$ (See Exercise 56 in Section 1.3.)

48. $f(x) = ax^2 + bx + c$ (See Exercise 47 in Section 1.3.)

49. Consider the function f given by

$$f(x) = \frac{x^2 - 9}{x + 3}.$$

a) For what x-value(s) is this function not differentiable?

b) Find $f'(4)$.

50. Consider the function g given by

$$g(x) = \frac{x^2 + x}{2x}.$$

a) For what x-value(s) is this function not differentiable?

b) What is $g'(3)$? Describe the simplest way to determine this.

51. Consider the function k given by

$$k(x) = |x - 3| + 2.$$

a) For what x-value(s) is the function not differentiable?

b) Evaluate $k'(0)$, $k'(1)$, $k'(4)$, and $k'(10)$.

52. Consider the function k given by

$$k(x) = 2|x + 5|.$$

a) For what x-value(s) is this function not differentiable?

b) Evaluate $k'(-10)$, $k'(-7)$, $k'(-2)$, and $k'(0)$.

53. Let $f(x) = \dfrac{x^2 + 4x + 3}{x + 1}$. A student recognizes that this function can be simplified as

$$f(x) = \frac{x^2 + 4x + 3}{x + 1} = \frac{(x + 1)(x + 3)}{x + 1} = x + 3.$$

Since $y = x + 3$ is a line with slope 1, the student makes the following conclusions: $f'(-2) = 1$, $f'(-1) = 1$, $f'(0) = 1$, $f'(1) = 1$. Where did the student make an error?

54. Let $g(x) = \sqrt[3]{x}$. A student graphs this function, and the graph appears to be continuous for all real numbers x. The student concludes that g is differentiable for all x, which is false. Identify the error, and explain why the conclusion is false. What is the correct conclusion regarding the differentiability of g?

55. Let F be a function given by

$$F(x) = \begin{cases} x^2 + 1, & \text{for } x \le 2, \\ 2x + 1, & \text{for } x > 2. \end{cases}$$

a) Verify that F is continuous at $x = 2$.

b) Is F differentiable at $x = 2$? Why or why not?

56. Let G be a function given by

$$G(x) = \begin{cases} x^3, & \text{for } x \le 1, \\ 3x - 2, & \text{for } x > 1. \end{cases}$$

a) Verify that G is continuous at $x = 1$.

b) Is G differentiable at $x = 1$? Why or why not?

57. Let H be a function given by

$$H(x) = \begin{cases} 2x^2 - x, & \text{for } x \le 3, \\ mx + b, & \text{for } x > 3. \end{cases}$$

Determine the values of m and b that make H differentiable at $x = 3$.

TECHNOLOGY CONNECTION

58–63. Use a calculator to check your answers to Exercises 17–22.

64. Business: growth of an investment. A company determines that the value of an investment after t years is V, in millions of dollars, where V is given by

$$V(t) = 5t^3 - 30t^2 + 45t + 5\sqrt{t}.$$

Note: Calculators often use only the variables y and x, so you may need to change the variables.

a) Graph V over the interval $[0, 5]$.

b) Find the equation of the secant line passing through the points $(1, V(1))$ and $(5, V(5))$. Then graph this secant line using the same axes as in part (a).

c) Find the average rate of change of the investment between year 1 and year 5.

d) Repeat parts (b) and (c) for the following pairs of points: $(1, V(1))$ and $(4, V(4))$; $(1, V(1))$ and $(3, V(3))$; $(1, V(1))$ and $(1.5, V(1.5))$.

e) What *appears* to be the slope of the tangent line to the graph at the point $(1, V(1))$?

f) Approximate the rate at which the value of the investment is changing at $t = 1$ yr.

65. Use a calculator to determine where $f'(x)$ does not exist, if $f(x) = \sqrt[3]{x - 5}$.

Answers to Quick Checks

1. $f'(x) = 3x^2 + 2x$; $f'(-2) = 8$, $f'(4) = 56$

2. $f'(x) = \dfrac{2}{x^2}$ **3.** At $x = -6$, the graph has a corner.

1.5

- Differentiate using the Power Rule or the Sum–Difference Rule.
- Differentiate a constant or a constant times a function.
- Determine points at which a tangent line has a specified slope.

Historical Note: The German mathematician and philosopher Gottfried Wilhelm von Leibniz (1646–1716) and the English mathematician, philosopher, and physicist Sir Isaac Newton (1642–1727) are both credited with the invention of calculus, though each worked independently. Newton used the dot notation $\dot{y}$ for dy/dt, where y is a function of time; this notation is still used, though it is not as common as Leibniz notation.

Differentiation Techniques: The Power and Sum–Difference Rules

Leibniz Notation

When y is a function of x, one way to express "the derivative of y with respect to x" is the notation

$$\frac{dy}{dx}.$$

This notation was invented by the German mathematician Gottfried von Leibniz. Using this notation, if $y = f(x)$, we can write the derivative of y with respect to x as

$$\frac{dy}{dx} = f'(x).$$

We also use *prime notation*, such as y' or $f'(x)$, to represent a derivative when there is no confusion as to which variables are involved. The dy/dx notation has the same meaning.

We may also write the value of a derivative at a number, such as $f'(2)$, as

$$\frac{dy}{dx}\bigg|_{x=2}.$$

The vertical line is interpreted as "evaluated at," so the above expression is read as "the derivative of y with respect to x evaluated at $x = 2$."

We can also express $f'(x)$ as

$$\frac{d}{dx}f(x).$$

This is identical in meaning to dy/dx and is another way to denote the derivative of the function. When placed next to an expression, d/dx indicates the expression's derivative. Thus, using earlier notation, we can write

$$\frac{d}{dx}x^2 = 2x, \qquad \frac{d}{dx}x^3 = 3x^2, \qquad \frac{d}{dx}\left(\frac{1}{x}\right) = -\frac{1}{x^2}, \qquad \text{and so on.}$$

We will use all of these derivative forms often, and with practice, their use will become natural. In Chapter 2, we will attach additional meanings to dy and dx.

The Power Rule

In Section 1.4, we found the derivatives of certain power functions. Look at the following table and see if you can identify a pattern:

FUNCTION	DERIVATIVE	
x^2	$2x^1$	Example 1 in Section 1.4
x^3	$3x^2$	Example 2 in Section 1.4
x^4	$4x^3$	Exercise 41 in Section 1.4
$\dfrac{1}{x} = x^{-1}$	$-1 \cdot x^{-2} = \dfrac{-1}{x^2}$	Example 4 in Section 1.4
$\dfrac{1}{x^2} = x^{-2}$	$-2 \cdot x^{-3} = \dfrac{-2}{x^3}$	Exercise 44 in Section 1.4
$\sqrt{x} = x^{1/2}$	$\dfrac{1}{2}x^{-1/2} = \dfrac{1}{2\sqrt{x}}$	Exercise 45 in Section 1.4

The pattern can be described as follows: "To find the derivative of a power function, bring the exponent to the front of the variable as a coefficient and reduce the exponent by 1."

Write the exponent as the coefficient.

$$\frac{d}{dx}x^k = k \cdot x^{k-1}$$

Subtract 1 from the exponent.

This rule is summarized as the following theorem.

> **THEOREM 1   The Power Rule**
>
> For any real number k, if $y = x^k$, then
>
> $$\frac{d}{dx}x^k = k \cdot x^{k-1}.$$

EXAMPLE 1   Differentiate each of the following:

a) $y = x^5$;   **b)** $y = x$;   **c)** $y = x^{-4}$.

Solution

a) $\dfrac{d}{dx}x^5 = 5 \cdot x^{5-1} = 5x^4$ Using the Power Rule

b) $\dfrac{d}{dx}x = 1 \cdot x^{1-1} = 1 \cdot x^0 = 1$

c) $\dfrac{d}{dx}x^{-4} = -4 \cdot x^{-4-1} = -4x^{-5}$,   or $-4 \cdot \dfrac{1}{x^5}$,   or $-\dfrac{4}{x^5}$ **1 ✔**

Quick Check 1 ✔

Differentiate:
a) $y = x^{15}$;
b) $y = x^{-7}$.

The Power Rule also allows us to differentiate expressions with rational exponents.

EXAMPLE 2   Differentiate:

a) $y = \sqrt[5]{x}$;   **b)** $y = x^{0.7}$.

Solution

a) $\dfrac{d}{dx}\sqrt[5]{x} = \dfrac{d}{dx}x^{1/5} = \dfrac{1}{5} \cdot x^{(1/5)-1}$

$\qquad = \dfrac{1}{5}x^{-4/5}$,   or $\dfrac{1}{5} \cdot \dfrac{1}{x^{4/5}}$,   or $\dfrac{1}{5} \cdot \dfrac{1}{\sqrt[5]{x^4}}$,   or $\dfrac{1}{5\sqrt[5]{x^4}}$

Quick Check 2 ✔

Differentiate:
a) $y = \sqrt[4]{x}$;
b) $y = x^{-1.25}$.

b) $\dfrac{d}{dx}x^{0.7} = 0.7x^{(0.7)-1} = 0.7x^{-0.3}$,   or $\dfrac{7}{10x^{0.3}}$ **2 ✔**

Numerical Differentiation and Tangent Lines

Consider $f(x) = x\sqrt{4 - x^2}$, graphed below.

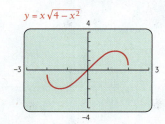

To find the value of dy/dx at a point, we select dy/dx from the CALC menu and key in an x-value or use the arrow keys to move the cursor to a desired point. We then press **ENTER** to obtain the value of the derivative at the given x-value.

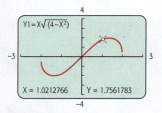
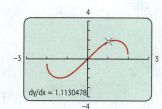

We can also use the TANGENT feature from the DRAW menu to draw the tangent line at any point where the derivative was found. Both the line and its equation will appear.

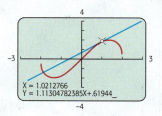

EXERCISES

For each function, use dy/dx to find the derivative, and then draw the tangent line at the given point. When selecting the viewing window, include the specified x-values.

1. $f(x) = x(200 - x)$;
$x = 24, x = 138, x = 150, x = 190$

2. $f(x) = x^3 - 6x^2$;
$x = -2, x = 0, x = 2, x = 4, x = 6.3$

3. $f(x) = -4.32 + 1.44x + 3x^2 - x^3$;
$x = -0.5, x = 0.5, x = 2.1$

The Power Rule can be used to confirm results from Section 1.3.

- $$\frac{d}{dx}\left(\frac{1}{x}\right) = \frac{d}{dx}(x^{-1}) = -1x^{-2} = -\frac{1}{x^2}$$

- $$\frac{d}{dx}\sqrt{x} = \frac{d}{dx}(x^{1/2}) = \frac{1}{2}x^{-1/2} = \frac{1}{2x^{1/2}} = \frac{1}{2\sqrt{x}}$$

The Derivative of a Constant Function

Consider the constant function given by $F(x) = c$. Note that the slope of the tangent line at each point on its graph is 0.

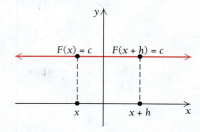

This suggests the following theorem.

THEOREM 2

The derivative of a constant function is 0. That is, $\dfrac{d}{dx}c = 0$.

Proof: Let F be the function given by $F(x) = c$. Then

$$\frac{d}{dx}c = F'(x) = \lim_{h \to 0} \frac{F(x+h) - F(x)}{h} \qquad \text{Using the definition of derivative}$$

$$= \lim_{h \to 0} \frac{c - c}{h}$$

$$= \lim_{h \to 0} \frac{0}{h}$$

$$= 0. \qquad \text{Using Limit Property L1}$$

The difference quotient is always zero. Thus, as h approaches 0, the limit of the difference quotient is 0, so $F'(x) = 0$. ■

The Derivative of a Constant Times a Function

Now let's differentiate functions such as

$$f(x) = 5x^2 \quad \text{and} \quad g(x) = -7x^4.$$

We already know how to differentiate x^2 and x^4. Let's look for a pattern in the results of Section 1.4 and its exercise set.

FUNCTION	DERIVATIVE
$5x^2$	$10x$
$3x^{-1}$	$-3x^{-2}$
$\frac{3}{2}x^2$	$3x$
$1 \cdot x^3$	$3x^2$

Perhaps you have discovered the following theorem.

> **THEOREM 3**
>
> The derivative of a constant times a function is the constant times the derivative of the function. Using derivative notation, we can write this as
>
> $$\frac{d}{dx}[c \cdot f(x)] = c \cdot \frac{d}{dx}f(x).$$

Proof: Let $F(x) = cf(x)$. Then

$$\frac{d}{dx}cf(x) = F'(x)$$

$$= \lim_{h \to 0} \frac{F(x+h) - F(x)}{h} \qquad \text{Using the definition of derivative}$$

$$= \lim_{h \to 0} \frac{cf(x+h) - cf(x)}{h} \qquad \text{Substituting}$$

$$= c \cdot \lim_{h \to 0} \left[\frac{f(x+h) - f(x)}{h} \right] \qquad \text{Factoring and using Limit Properties}$$

$$= c \cdot f'(x). \qquad \text{Using the definition of derivative} \quad ■$$

Combining this rule with the Power Rule allows us to find many derivatives.

EXAMPLE 3   Find each of the following derivatives:

a) $\dfrac{d}{dx} 7x^4$;   b) $\dfrac{d}{dx}(-9x)$;   c) $\dfrac{d}{dx}\left(\dfrac{1}{5x^2}\right)$.

Solution

a) $\dfrac{d}{dx}7x^4 = 7\dfrac{d}{dx}x^4 = 7 \cdot 4 \cdot x^{4-1} = 28x^3$ With practice, this may be done in one step.

b) $\dfrac{d}{dx}(-9x) = -9\dfrac{d}{dx}x = -9 \cdot 1 = -9$

c) $\dfrac{d}{dx}\left(\dfrac{1}{5x^2}\right) = \dfrac{d}{dx}\left(\dfrac{1}{5}x^{-2}\right) = \dfrac{1}{5} \cdot \dfrac{d}{dx}x^{-2}$

$= \dfrac{1}{5}(-2)x^{-2-1}$

$= -\dfrac{2}{5}x^{-3}$,   or $-\dfrac{2}{5x^3}$ **3** ✔

Quick Check 3 ✔

Differentiate each of the following:

a) $y = 10x^9$;
b) $y = \pi x^3$;
c) $y = \dfrac{2}{3x^4}$.

Recall that a function's derivative at x is also its instantaneous rate of change at x.

EXAMPLE 4   **Life Science: Volume of a Tumor.**   The volume V of a spherical tumor can be approximated by

$$V(r) = \tfrac{4}{3}\pi r^3,$$

where r is the radius of the tumor, in centimeters.

a) Find the rate of change of the volume with respect to the radius.
b) Find the rate of change of the volume at $r = 1.2$ cm.

Solution

a) $\dfrac{dV}{dr} = V'(r) = 3 \cdot \dfrac{4}{3} \cdot \pi r^2 = 4\pi r^2$

(This expression turns out to be equal to the tumor's surface area.)

b) $V'(1.2) = 4\pi(1.2)^2 = 5.76\pi \approx 18\,\dfrac{\text{cm}^3}{\text{cm}} = 18\text{ cm}^2$

When the radius is 1.2 cm, the volume is changing at the rate of 18 cm³ for every change of 1 cm in the radius.   ■

The Derivative of a Sum or a Difference

In Exercise 11 of Exercise Set 1.4, we found that the derivative of

$$f(x) = x^2 + x$$

is $f'(x) = 2x + 1$.

Note that the derivative of x^2 is $2x$, the derivative of x is 1, and the sum of these derivatives is $f'(x)$. This illustrates the following theorem.

Exploratory

Let $y_1 = x(100 - x)$, and $y_2 = x\sqrt{100 - x^2}$. Using the Y-VARS option from the VARS menu, enter Y3 as Y1+Y2. Find the derivative of each of the three functions at $x = 8$ using the numerical differentiation feature.

Compare your answers. How do you think you can find the derivative of a sum?

THEOREM 4   The Sum–Difference Rule

Sum.   The derivative of a sum is the sum of the derivatives:

$$\dfrac{d}{dx}[f(x) + g(x)] = \dfrac{d}{dx}f(x) + \dfrac{d}{dx}g(x).$$

Difference.   The derivative of a difference is the difference of the derivatives:

$$\dfrac{d}{dx}[f(x) - g(x)] = \dfrac{d}{dx}f(x) - \dfrac{d}{dx}g(x).$$

Proof: The proof of the Sum Rule relies on the fact that the limit of a sum is the sum of the limits. Let $F(x) = f(x) + g(x)$. Then

$$\lim_{h \to 0} \frac{F(x + h) - F(x)}{h} = \lim_{h \to 0} \frac{[f(x + h) + g(x + h)] - [f(x) + g(x)]}{h}$$

$$= \lim_{h \to 0} \left[\frac{f(x + h) - f(x)}{h} + \frac{g(x + h) - g(x)}{h} \right] = f'(x) + g'(x).$$

To prove the Difference Rule, we note that

$$\frac{d}{dx}(f(x) - g(x)) = \frac{d}{dx}(f(x) + (-1)g(x)) = \frac{d}{dx}f(x) + \frac{d}{dx}(-1)g(x)$$

$$= f'(x) + (-1)\frac{d}{dx}g(x) = f'(x) - g'(x). \quad \blacksquare$$

Any function that is a sum or difference of several terms can be differentiated term by term.

EXAMPLE 5   Find each of the following derivatives:

a) $\dfrac{d}{dx}(5x^3 - 7)$;

b) $\dfrac{d}{dx}\left(24x - \sqrt{x} + \dfrac{5}{x}\right)$.

Solution

a) $\dfrac{d}{dx}(5x^3 - 7) = \dfrac{d}{dx}(5x^3) - \dfrac{d}{dx}(7)$

$$= 5\frac{d}{dx}x^3 - 0$$

$$= 5 \cdot 3x^2$$

$$= 15x^2$$

b) $\dfrac{d}{dx}\left(24x - \sqrt{x} + \dfrac{5}{x}\right) = \dfrac{d}{dx}(24x) - \dfrac{d}{dx}(\sqrt{x}) + \dfrac{d}{dx}\left(\dfrac{5}{x}\right)$

$$= 24 \cdot \frac{d}{dx}x - \frac{d}{dx}x^{1/2} + 5 \cdot \frac{d}{dx}x^{-1}$$

$$= 24 \cdot 1 - \frac{1}{2}x^{(1/2)-1} + 5(-1)x^{-1-1}$$

$$= 24 - \frac{1}{2}x^{-1/2} - 5x^{-2}$$

$$= 24 - \frac{1}{2\sqrt{x}} - \frac{5}{x^2}$$

Quick Check 4 ✔

Differentiate:

$$y = 3x^5 + 2\sqrt[3]{x} + \frac{1}{3x^2} + \sqrt{5}.$$

4 ✔

A word of caution! The derivative of

$$f(x) + c,$$

a function *plus* a constant, is just the derivative of $f(x)$:

$$f'(x).$$

The derivative of

$$c \cdot f(x),$$

a function *times* a constant, is the constant times the derivative of $f(x)$:

$$c \cdot f'(x).$$

That is, the constant is retained for a product, but not for a sum or a difference.

Slopes of Tangent Lines

It is important to be able to determine points at which the tangent line to a curve has a certain slope, that is, points at which the derivative has a certain value.

EXAMPLE 6   Find any points on the graph of $f(x) = -x^3 + 6x^2$ at which (a) the tangent line is horizontal; (b) the tangent line has slope 9.

Solution

a) The derivative is used to find the slope of a tangent line. Since a horizontal tangent line has slope 0, we seek all x for which $f'(x) = 0$:

$$f'(x) = 0 \qquad \text{Setting the derivative equal to 0}$$

$$\frac{d}{dx}(-x^3 + 6x^2) = 0$$

$$-3x^2 + 12x = 0. \qquad \text{Differentiating}$$

We factor and solve:

$$-3x(x - 4) = 0$$

$$-3x = 0 \quad or \quad x - 4 = 0$$

$$x = 0 \quad or \quad x = 4. \qquad \text{Using the Principle of Zero Products}$$

We are to find points *on the graph*, so we must determine the second coordinates from the original equation, $f(x) = -x^3 + 6x^2$.

$$f(0) = -0^3 + 6 \cdot 0^2 = 0.$$

$$f(4) = -4^3 + 6 \cdot 4^2 = -64 + 96 = 32.$$

Thus, the points we are seeking are $(0, 0)$ and $(4, 32)$, as shown on the graph.

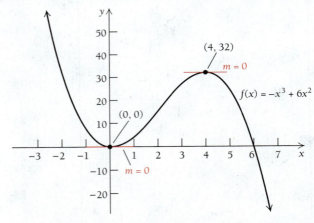

b) We want to find values of x for which $f'(x) = 9$. That is, we want to find x such that

$$-3x^2 + 12x = 9. \qquad \text{As in part (a), note that } \frac{d}{dx}(-x^3 + 6x^2) = -3x^2 + 12x.$$

To solve, we add -9 on both sides and get

$$-3x^2 + 12x - 9 = 0.$$

We then multiply both sides of the equation by $-\frac{1}{3}$, giving

$$x^2 - 4x + 3 = 0,$$

which is factored as follows:

$$(x - 3)(x - 1) = 0.$$

We have two solutions: $x = 1$ or $x = 3$. We need the actual coordinates: when $x = 1$, we have $f(1) = -(1)^3 + 6(1)^2 = 5$. Therefore, at the point $(1, 5)$ on the graph of $f(x)$, the tangent line has a slope of 9. In a similar way, we can state that the tangent line at the point $(3, 27)$ has a slope of 9 as well. All of this is illustrated in the following graph.

Quick Check 5 ✔

For the function in Example 6, find the points for which $f'(x) = -15$.

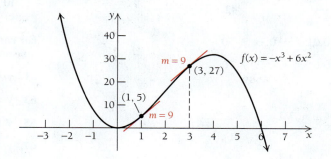

5 ✔

Section Summary

- Common forms of notation for the derivative of a function are y', $f'(x)$, $\dfrac{dy}{dx}$, and $\dfrac{d}{dx} f(x)$.

- The *Power Rule* for differentiation is $\dfrac{d}{dx}[x^k] = kx^{k-1}$, for all real numbers k.

- The derivative of a constant is zero: $\dfrac{d}{dx} c = 0$.

- The derivative of a constant times a function is the constant times the derivative of the function:
$$\frac{d}{dx}[c \cdot f(x)] = c \cdot \frac{d}{dx} f(x).$$

- The derivative of a sum (or difference) is the sum (or difference) of the derivatives of the terms:
$$\frac{d}{dx}[f(x) \pm g(x)] = \frac{d}{dx} f(x) \pm \frac{d}{dx} g(x).$$

1.5 | Exercise Set

Find $\dfrac{dy}{dx}$.

1. $y = x^8$

2. $y = x^7$

3. $y = -0.5x$

4. $y = -3x$

5. $y = 7$

6. $y = 12$

7. $y = 3x^{10}$

8. $y = 2x^{15}$

9. $y = x^{-8}$

10. $y = x^{-6}$

11. $y = 3x^{-5}$

12. $y = 4x^{-2}$

13. $y = x^4 - 7x$

14. $y = x^3 + 3x^2$

15. $y = 4\sqrt{x}$

16. $y = 8\sqrt{x}$

17. $y = x^{0.7}$

18. $y = x^{0.9}$

19. $y = -4.8x^{1/3}$

20. $y = \frac{1}{2}x^{4/5}$

21. $y = \dfrac{6}{x^4}$

22. $y = \dfrac{7}{x^3}$

23. $y = \dfrac{3x}{4}$

24. $y = \dfrac{4x}{5}$

Find each derivative.

25. $\dfrac{d}{dx}\left(\sqrt[4]{x} - \dfrac{3}{x}\right)$

26. $\dfrac{d}{dx}\left(\sqrt[3]{x} + \dfrac{4}{\sqrt{x}}\right)$

27. $\dfrac{d}{dx}\left(-2\sqrt[3]{x^5}\right)$

28. $\dfrac{d}{dx}\left(-\sqrt[4]{x^3}\right)$

29. $\dfrac{d}{dx}(5x^2 - 7x + 3)$

30. $\dfrac{d}{dx}(6x^2 - 5x + 9)$

Find $f'(x)$.

31. $f(x) = 0.3x^{1.2}$

32. $f(x) = 0.6x^{1.5}$

33. $f(x) = \dfrac{3x}{4}$

34. $f(x) = \dfrac{2x}{3}$

35. $f(x) = \dfrac{2}{5x^6}$

36. $f(x) = \dfrac{4}{7x^3}$

37. $f(x) = \dfrac{4}{x} - x^{3/5}$ **38.** $f(x) = \dfrac{5}{x} - x^{2/3}$

39. $f(x) = 7x - 14$ **40.** $f(x) = 4x - 7$

41. $f(x) = \dfrac{x^{3/2}}{3}$ **42.** $f(x) = \dfrac{x^{4/3}}{4}$

43. $f(x) = -0.01x^2 + 0.4x + 50$

44. $f(x) = -0.01x^2 - 0.5x + 70$

Find y'.

45. $y = x^{-3/4} - 3x^{2/3} + x^{5/4} + \dfrac{2}{x^4}$

46. $y = 3x^{-2/3} + x^{3/4} + x^{6/5} + \dfrac{8}{x^3}$

47. $y = \dfrac{x}{7} + \dfrac{7}{x}$ **48.** $y = \dfrac{2}{x} - \dfrac{x}{2}$

49. If $f(x) = \sqrt{x}$, find $f'(4)$.

50. If $f(x) = x^2 + 4x - 5$, find $f'(10)$.

51. If $y = x + \dfrac{2}{x^3}$, find $\dfrac{dy}{dx}\Big|_{x=1}$

52. If $y = \dfrac{4}{x^2}$, find $\dfrac{dy}{dx}\Big|_{x=-2}$

53. If $y = \sqrt[3]{x} + \sqrt{x}$, find $\dfrac{dy}{dx}\Big|_{x=64}$

54. If $y = x^3 + 2x - 5$, find $\dfrac{dy}{dx}\Big|_{x=-2}$

55. If $y = \dfrac{2}{5x^3}$, find $\dfrac{dy}{dx}\Big|_{x=4}$

56. If $y = \dfrac{1}{3x^4}$, find $\dfrac{dy}{dx}\Big|_{x=-1}$

57. Find an equation of the tangent line to the graph of
$f(x) = x^2 - \sqrt{x}$
a) at $(1, 0)$;
b) at $(4, 14)$;
c) at $(9, 78)$.

58. Find an equation (in $y = mx + b$ form) of the tangent line to the graph of $f(x) = x^3 - 2x + 1$
a) at $(2, 5)$;
b) at $(-1, 2)$;
c) at $(0, 1)$.

59. Find the equation of the tangent line to the graph of
$g(x) = \sqrt[3]{x^2}$
a) at $(-1, 1)$;
b) at $(1, 1)$;
c) at $(8, 4)$.

60. Find an equation of the tangent line to the graph of
$f(x) = \dfrac{1}{x^2}$
a) at $(1, 1)$;
b) at $(3, \frac{1}{9})$;
c) at $(-2, \frac{1}{4})$.

For each function, find the points on the graph at which the tangent line is horizontal. If none exist, state that fact.

61. $y = -x^2 + 4$ **62.** $y = x^2 - 3$

63. $y = x^3 - 2$ **64.** $y = -x^3 + 1$

65. $y = 5x^2 - 3x + 8$ **66.** $y = 3x^2 - 5x + 4$

67. $y = -0.01x^2 + 0.4x + 50$

68. $y = -0.01x^2 - 0.5x + 70$

69. $y = -2x + 5$ **70.** $y = 2x + 4$

71. $y = -3$ **72.** $y = 4$

73. $y = -\tfrac{1}{3}x^3 + 6x^2 - 11x - 50$

74. $y = -x^3 + x^2 + 5x - 1$

75. $y = x^3 - 6x + 1$ **76.** $y = \tfrac{1}{3}x^3 - 3x + 2$

77. $f(x) = \tfrac{1}{3}x^3 - 3x^2 + 9x - 9$

78. $f(x) = \tfrac{1}{3}x^3 + \tfrac{1}{2}x^2 - 2$

For each function, find the points on the graph at which the tangent line has slope 1.

79. $y = 6x - x^2$ **80.** $y = 20x - x^2$

81. $y = -0.01x^2 + 2x$ **82.** $y = -0.025x^2 + 4x$

83. $y = \tfrac{1}{3}x^3 - x^2 - 4x + 1$ **84.** $y = \tfrac{1}{3}x^3 + 2x^2 + 2x$

APPLICATIONS

Life Sciences

85. Healing wound. The circumference C, in centimeters, of a healing wound is approximated by
$$C(r) = 6.28r,$$
where r is the wound's radius, in centimeters.
a) Find the rate of change of the circumference with respect to the radius.
b) Find $C'(4)$.
c) Explain the meaning of your answer to part (b).

86. Healing wound. The circular area A, in square centimeters, of a healing wound is approximated by
$$A(r) = 3.14r^2,$$
where r is the wound's radius, in centimeters.
a) Find the rate of change of the area with respect to the radius.
b) Find $A'(3)$.
c) Explain the meaning of your answer to part (b).

87. Growth of a baby. The median weight of a boy whose age is between 0 and 36 months is approximated by

$$w(t) = 8.15 + 1.82t - 0.0596t^2 + 0.000758t^3,$$

where t is in months and w is in pounds.

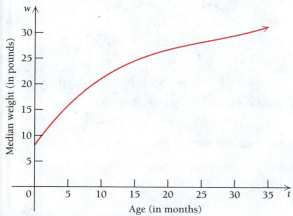

(*Source*: Centers for Disease Control. Developed by the National Center for Health Statistics in collaboration with the National Center for Chronic Disease Prevention and Health Promotion, 2000, Rechecked 2014)

Use this approximation to find the following:

a) The rate of change of weight with respect to time
b) The weight of a boy at age 10 months
c) The rate of change of a boy's weight with respect to time at age 10 months

88. Temperature during an illness. The temperature T of a person during an illness is given by

$$T(t) = -0.1t^2 + 1.2t + 98.6,$$

where T is the temperature, in degrees Fahrenheit, at time t, in days.

a) Find the rate of change of the temperature with respect to time.
b) Find the temperature at $t = 1.5$ days.
c) Find the rate of change at $t = 1.5$ days.

89. Heart rate. The equation

$$R(v) = \frac{6000}{v}$$

can be used to determine the heart rate, R, of a person whose heart pumps 6000 milliliters (mL) of blood per minute and v milliliters of blood per beat. (*Source*: Mathematics Teacher, Vol. 99, No. 4, November 2005.)

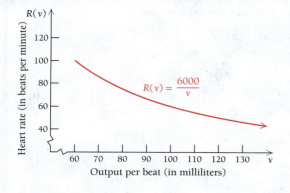

a) Find the rate of change of heart rate with respect to v, the output per beat.
b) Find the heart rate at $v = 80$ mL per beat.
c) Find the rate of change at $v = 80$ mL per beat.

90. Blood flow resistance. The equation

$$S(r) = \frac{1}{r^4}$$

can be used to determine the resistance to blood flow, S, of a blood vessel that has radius r, in millimeters (mm). (*Source*: Mathematics Teacher, Vol. 99, No. 4, November 2005.)

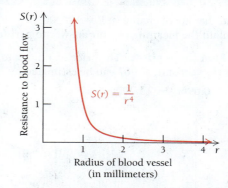

a) Find the rate of change of resistance with respect to r, the radius of the blood vessel.
b) Find the resistance at $r = 1.2$ mm.
c) Find the rate of change of S with respect to r when $r = 0.8$ mm.

Social Sciences

91. Population growth rate. In t years, the population of Kingsville grows from 100,000 to a size P given by

$$P(t) = 100,000 + 2000t^2.$$

a) Find the growth rate, dP/dt.
b) Find the population after 10 yr.
c) Find the growth rate at $t = 10$.
d) Explain the meaning of your answer to part (c).

92. Median age of women at first marriage. The median age of women at first marriage is approximated by

$$A(t) = 0.08t + 19.7,$$

where $A(t)$ is the median age of women marrying for the first time at t years after 1950.

a) Find the rate of change of the median age A with respect to time t.
b) Explain the meaning of your answer to part (a).

General Interest

93. View to the horizon. The view V, or distance in miles, that one can see to the horizon from a height h, in feet, is given by

$$V = 1.22\sqrt{h}.$$

a) Find the rate of change of V with respect to h.
b) How far can one see to the horizon from an airplane window at a height of 40,000 ft?

c) Find the rate of change at $h = 40,000$.
d) Explain the meaning of your answer to part (c).

94. **Super Bowl ticket prices.** The price of a ticket to the Super Bowl t years after 1967 can be estimated by

$$p(t) = 0.696t^2 - 13.290t + 61.857.$$

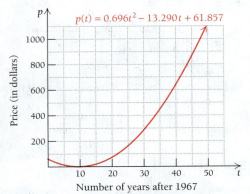

$p(t) = 0.696t^2 - 13.290t + 61.857$

Number of years after 1967

(*Source*: Based on data from extramustard.si.com.)

a) Use the function to predict the price of a Super Bowl ticket in 2014.
b) Find the rate of change of the ticket price with respect to the year, dp/dt.
c) At what rate were ticket prices changing in 2014?

SYNTHESIS

For Exercises 95 and 96, find the interval(s) for which $f'(x)$ is positive.

95. $f(x) = x^2 - 4x + 1$

96. $f(x) = \frac{1}{3}x^3 - x^2 - 3x + 5$

97. Find the points on the graph of
$$y = x^4 - \frac{4}{3}x^2 - 4$$
at which the tangent line is horizontal.

98. Find the points on the graph of
$$y = 2x^6 - x^4 - 2$$
at which the tangent line is horizontal.

99. Use the derivative to help explain why $f(x) = x^5 + x^3$ increases for all x in $(-\infty, \infty)$.

100. Use the derivative to help explain why $g(x) = -2x - x^3$ decreases for all x in $(-\infty, \infty)$.

101. Use the derivative to help explain why $k(x) = \dfrac{1}{x^2}$ decreases for all x in $(0, \infty)$.

102. Use the derivative to help explain why $f(x) = x^3 + ax$ increases for all x in $(-\infty, \infty)$ when $a \geq 0$ but not when $a < 0$.

Find dy/dx. Each function can be differentiated using the rules developed in this section, but some algebra may be required beforehand.

103. $y = (x + 3)(x - 2)$ 104. $y = (x - 1)(x + 1)$

105. $y = \dfrac{x^5 - x^3}{x^2}$ 106. $y = \dfrac{x^5 + x}{x^2}$

107. $y = \sqrt{7x}$ 108. $y = \sqrt[3]{8x}$

109. $y = \left(\sqrt{x} - \dfrac{1}{\sqrt{x}}\right)^2$ 110. $y = (x + 1)^3$

111. When might Leibniz notation be more convenient than function notation?

TECHNOLOGY CONNECTION

Graph each of the following. Then estimate the x-values at which tangent lines are horizontal.

112. $f(x) = x^4 - 3x^2 + 1$

113. $f(x) = \dfrac{5x^2 + 8x - 3}{3x^2 + 2}$

For each of the following, graph f and f' and then determine $f'(1)$. For Exercises 116 and 117, use nDeriv on the TI-83.

114. $f(x) = 20x^3 - 3x^5$

115. $f(x) = x^4 - x^3$

116. $f(x) = \dfrac{4x}{x^2 + 1}$

117. $f(x) = \dfrac{5x^2 + 8x - 3}{3x^2 + 2}$

Differentiation Techniques: The Product and Quotient Rules

- Differentiate using the Product and Quotient Rules.
- Use the Quotient Rule to differentiate the average cost, revenue, and profit functions.

The Product Rule

A function can be written as the product of two other functions. For example, $F(x) = (x^2 + 3x)\sqrt{x}$ can be viewed as the product of $f(x) = x^2 + 3x$ and $g(x) = \sqrt{x}$. However, the derivative of a product of two functions is *not* the product of their derivatives.

> **THEOREM 5   The Product Rule**
>
> Let $F(x) = f(x) \cdot g(x)$. Then
>
> $$F'(x) = \frac{d}{dx}[f(x) \cdot g(x)] = f(x) \cdot \left[\frac{d}{dx}g(x)\right] + g(x) \cdot \left[\frac{d}{dx}f(x)\right].$$
>
> The derivative of a product is the first factor times the derivative of the second factor, plus the second factor times the derivative of the first factor.

The proof of the Product Rule is outlined in Exercise 75 at the end of this section.

To find the derivative of the product $f(x) \cdot g(x)$, we perform the following steps:

1. Write the first factor, $f(x)$.
2. Multiply it by the derivative of the second factor, $g'(x)$.
3. Write the second factor, $g(x)$.
4. Multiply it by the derivative of the first factor, $f'(x)$.
5. Add the result of steps (1) and (2) to the result of steps (3) and (4).

Usually we try to write the results in simplified form. In Examples 1 and 2, we do not simplify in order to better emphasize the steps being performed.

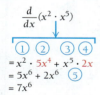

$\dfrac{d}{dx}(x^2 \cdot x^5)$

① ② ③ ④

$= x^2 \cdot 5x^4 + x^5 \cdot 2x$ ⑤

$= 5x^6 + 2x^6$

$= 7x^6$

EXAMPLE 1   Find $\dfrac{d}{dx}[(x^4 - 2x^3 - 7)(3x^2 - 5x)]$. Do not simplify.

Solution   We let $f(x) = x^4 - 2x^3 - 7$ and $g(x) = 3x^2 - 5x$. We differentiate each of these, obtaining $f'(x) = 4x^3 - 6x^2$ and $g'(x) = 6x - 5$. By the Product Rule, the derivative of the given function is then

$$\underbrace{\frac{d}{dx}[\overbrace{(x^4 - 2x^3 - 7)}^{f(x)}\overbrace{(3x^2 - 5x)}^{g(x)}]} = \overbrace{(x^4 - 2x^3 - 7)}^{f(x)} \cdot \overbrace{(6x - 5)}^{g'(x)} + \overbrace{(3x^2 - 5x)}^{g(x)} \cdot \overbrace{(4x^3 - 6x^2)}^{f'(x)}$$

In this example, we could have first multiplied the polynomials and then differentiated. Both methods give the same solution after simplification. ∎

It makes no difference which factor of the given function is regarded as $f(x)$ or $g(x)$. We often regard the first function as $f(x)$ and the second function as $g(x)$, but if we switch the roles, the process still gives the same answer.

Quick Check 1 ✔

Use the Product Rule to differentiate each of the following functions. Do not simplify.

a) $y = (2x^5 + x - 1)(3x - 2)$

b) $y = (\sqrt{x} + 1)(\sqrt[5]{x} - x)$

EXAMPLE 2   For $F(x) = (x^2 + 4x - 11)(7x^3 - \sqrt{x})$, find $F'(x)$. Do not simplify.

Solution   We rewrite this as

$$F(x) = (x^2 + 4x - 11)(7x^3 - x^{1/2}).$$

Then, using the Product Rule, we have

$$F'(x) = (x^2 + 4x - 11)(21x^2 - \tfrac{1}{2}x^{-1/2}) + (7x^3 - x^{1/2})(2x + 4).$$ **1** ✔

The Quotient Rule

The derivative of a quotient is *not* the quotient of the derivatives. To see why, consider x^5 and x^2 and assume $x \neq 0$. The quotient x^5/x^2 is x^3, and the derivative of this quotient is $3x^2$. The derivatives of the numerator and denominator are $5x^4$ and $2x$, respectively, and the quotient of these derivatives, $5x^4/(2x)$, is $(5/2)x^3$, which is not $3x^2$.

The rule for differentiating quotients is as follows.

THEOREM 6   The Quotient Rule

If $Q(x) = \dfrac{N(x)}{D(x)}$,   then $Q'(x) = \dfrac{D(x) \cdot N'(x) - N(x) \cdot D'(x)}{[D(x)]^2}$.

The derivative of a quotient is the denominator times the derivative of the numerator, minus the numerator times the derivative of the denominator, all divided by the square of the denominator.

A proof of this result is outlined in Exercise 96 of Section 1.7 (on p. 166).

The Quotient Rule is illustrated below.

$$\frac{d}{dx}\left[\frac{N(x)}{D(x)}\right] = \frac{D(x) \cdot N'(x) - N(x) \cdot D'(x)}{[D(x)]^2}$$

There are six steps:

1. Write the denominator.
2. Multiply the denominator by the derivative of the numerator.
3. Write a minus sign.
4. Write the numerator.
5. Multiply it by the derivative of the denominator.
6. Divide the expression resulting from steps 1–5 by the square of the denominator.

EXAMPLE 3 For $Q(x) = x^5/x^2$, find $Q'(x)$.

Solution We have already seen that $x^5/x^2 = x^3$ and $\dfrac{d}{dx}x^3 = 3x^2$, but we wish to practice using the Quotient Rule. We have $D(x) = x^2$ and $N(x) = x^5$:

$$Q'(x) = \frac{\overbrace{x^2 \cdot 5x^4}^{D(x) \cdot N'(x)} - \overbrace{x^5 \cdot 2x}^{N(x) \cdot D'(x)}}{\underbrace{(x^2)^2}_{[D(x)]^2}}$$

$$= \frac{5x^6 - 2x^6}{x^4} = \frac{3x^6}{x^4} = 3x^2, \, x \ne 0. \qquad \text{This checks with the result above.}$$

EXAMPLE 4 Differentiate: $f(x) = \dfrac{1 + x^2}{x^3 + 1}$.

Solution We let $N(x) = 1 + x^2$ and $D(x) = x^3 + 1$. We have

$$f'(x) = \frac{(x^3 + 1)(2x) - (1 + x^2)(3x^2)}{(x^3 + 1)^2} \qquad \text{Using the Quotient Rule}$$

$$= \frac{2x^4 + 2x - 3x^2 - 3x^4}{(x^3 + 1)^2} \qquad \text{Using the distributive law}$$

$$= \frac{-x^4 - 3x^2 + 2x}{(x^3 + 1)^2}. \qquad \text{Simplifying}$$

EXAMPLE 5 Differentiate: $f(x) = \dfrac{x^2 - 3x}{x - 1}$.

Solution We have

$$f'(x) = \frac{(x - 1)(2x - 3) - (x^2 - 3x) \cdot 1}{(x - 1)^2} \qquad \text{Using the Quotient Rule}$$

$$= \frac{2x^2 - 5x + 3 - x^2 + 3x}{(x - 1)^2} \qquad \text{Using the distributive law}$$

$$= \frac{x^2 - 2x + 3}{(x - 1)^2}. \qquad \text{Simplifying}$$

2 ✔

Quick Check 2 ✔

a) Differentiate: $f(x) = \dfrac{1 - 3x}{x^2 + 2}$.

Simplify your result.

b) Show that
$$\frac{d}{dx}\left[\frac{ax + 1}{bx + 1}\right] = \frac{a - b}{(bx + 1)^2}.$$

Application of the Quotient Rule

The total cost, total revenue, and total profit functions, discussed in Section R.4, pertain to the accumulated cost, revenue, and profit when x items are produced. Because of economies of scale and other factors, it is common for the cost, revenue (price), and profit for, say, the 10th item to differ from those for the 1000th item. For this reason, businesses often focus on *average* cost, revenue, and profit.

TECHNOLOGY CONNECTION

Checking Derivatives Graphically

To check Example 5, we first enter the function:

$$y_1 = \frac{x^2 - 3x}{x - 1}.$$

Then we enter the possible derivative:

$$y_2 = \frac{x^2 - 2x + 3}{(x - 1)^2}.$$

As a third function, we enter

$$y_3 = \text{nDeriv}(y_1, x, x).$$

Next, we deselect y_1 and graph y_2 and y_3. We use different graph styles and the **Sequential** mode to see each graph as it appears on the screen.

$$y_2 = \frac{x^2 - 2x + 3}{(x - 1)^2}, \quad y_3 = \text{nDeriv}(y_1, x, x)$$

Since the graphs appear to coincide, it appears that $y_2 = y_3$ and we have a check. This is considered a partial

check, however, because the graphs might not coincide with a different viewing window.

We can also use a table to check that $y_2 = y_3$.

X	Y2	Y3
5.97	1.081	1.081
5.98	1.0806	1.0806
5.99	1.0803	1.0803
6	1.08	1.08
6.01	1.0797	1.0797
6.02	1.0794	1.0794
6.03	1.079	1.079

X = 5.97

EXERCISES

1. For the function

$$f(x) = \frac{x^2 - 4x}{x + 2},$$

use graphs and tables to determine which of the following seems to be the correct derivative.

a) $f'(x) = \dfrac{-x^2 - 4x - 8}{(x + 2)^2}$

b) $f'(x) = \dfrac{x^2 - 4x + 8}{(x + 2)^2}$

c) $f'(x) = \dfrac{x^2 + 4x - 8}{(x + 2)^2}$

2–5. Graphically check the results of Examples 1–4.

DEFINITION

If $C(x)$ is the cost of producing x items, then the **average cost** of producing x items is $\dfrac{C(x)}{x}$.

If $R(x)$ is the revenue from the sale of x items, then the **average revenue** from selling x items is $\dfrac{R(x)}{x}$.

If $P(x)$ is the profit from the sale of x items, then the **average profit** from selling x items is $\dfrac{P(x)}{x}$.

EXAMPLE 6   **Business.**   Paulsen's Greenhouse finds that the cost, in dollars, of growing x hundred geraniums is modeled by

$$C(x) = 200 + 100\sqrt[4]{x}.$$

If revenue from the sale of x hundred geraniums is modeled by

$$R(x) = 120 + 90\sqrt{x},$$

find each of the following.

a) The average cost, average revenue, and average profit when x hundred geraniums are grown and sold

b) The rate at which average profit is changing when 300 geraniums have been grown and sold

Solution

a) We let A_C, A_R, and A_P represent average cost, average revenue, and average profit, respectively. Then

$$A_C(x) = \frac{C(x)}{x} = \frac{200 + 100\sqrt[4]{x}}{x};$$

$$A_R(x) = \frac{R(x)}{x} = \frac{120 + 90\sqrt{x}}{x};$$

$$A_P(x) = \frac{P(x)}{x} = \frac{R(x) - C(x)}{x} = \frac{-80 + 90\sqrt{x} - 100\sqrt[4]{x}}{x}.$$

b) To find the rate at which average profit is changing when 300 geraniums are being grown, we calculate $A_{P'}(3)$ (remember that x is in hundreds):

$$A_{P'}(x) = \frac{d}{dx}\left[\frac{-80 + 90x^{1/2} - 100x^{1/4}}{x}\right]$$

$$= \frac{x\left(\frac{1}{2}\cdot 90x^{1/2-1} - \frac{1}{4}\cdot 100x^{1/4-1}\right) - \left(-80 + 90x^{1/2} - 100x^{1/4}\right)\cdot 1}{x^2}$$

$$= \frac{45x^{1/2} - 25x^{1/4} + 80 - 90x^{1/2} + 100x^{1/4}}{x^2} = \frac{75x^{1/4} - 45x^{1/2} + 80}{x^2};$$

$$A_{P'}(3) = \frac{75\sqrt[4]{3} - 45\sqrt{3} + 80}{3^2} \approx 11.20.$$

When 300 geraniums have been grown and sold, average profit is increasing by $11.20 per hundred plants, or about 11.2 cents per plant. ∎

TECHNOLOGY CONNECTION

Using Y-VARS

One way to save keystrokes on most calculators is to use the Y-VARS option on the VARS menu.

    To check Example 6(a), we let $y_1 = 200 + 100x^{0.25}$ and $y_2 = 120 + 90x^{0.5}$. To express the profit function as y_3, we press $\boxed{\text{Y=}}$ and move the cursor to enter Y3. Next we press $\boxed{\text{VARS}}$ and select Y-VARS and then FUNCTION. From the FUNCTION menu we select Y2, which then appears on the $\boxed{\text{Y=}}$ screen. After pressing $\boxed{-}$, we repeat the procedure to get Y1 on the $\boxed{\text{Y=}}$ screen.

```
Plot 1   Plot 2   Plot 3
\Y1 = 200 + 100X^0.25
\Y2 = 120 + 90X^0.5
\Y3 = Y2 – Y1
\Y4 =
\Y5 =
\Y6 =
```

EXERCISES

1. Use the Y-VARS option to enter $y_4 = y_1/x$, $y_5 = y_2/x$, and $y_6 = y_3/x$, and explain what each of the functions represents.

2. Use nDeriv from the MATH menu or dy/dx from the CALC menu to check Example 6(b).

Section Summary

- The *Product Rule* is

$$\frac{d}{dx}[f(x)\cdot g(x)] = f(x)\cdot\frac{d}{dx}[g(x)] + g(x)\cdot\frac{d}{dx}[f(x)].$$

- The *Quotient Rule* is

$$\frac{d}{dx}\left[\frac{f(x)}{g(x)}\right] = \frac{g(x)\cdot\frac{d}{dx}[f(x)] - f(x)\cdot\frac{d}{dx}[g(x)]}{[g(x)]^2}.$$

- Be careful to note the order in which the factors are written when using the Quotient Rule. Because the Quotient Rule involves subtraction, the order in which the operations are performed is important.

1.6 | Exercise Set

Differentiate two ways: first, by using the Product Rule; then, by multiplying the expressions before differentiating. Compare your results as a check. Use a graphing calculator to check your results.

1. $y = x^9 \cdot x^4$ **2.** $y = x^5 \cdot x^6$

3. $f(x) = (2x + 5)(3x - 4)$

4. $g(x) = (3x - 2)(4x + 1)$

5. $F(x) = 3x^4(x^2 - 4x)$

6. $G(x) = 4x^2(x^3 + 5x)$

7. $y = (3\sqrt{x} + 2)x^2$

8. $y = (4\sqrt{x} + 3)x^3$

9. $f(x) = (2x + 5)(3x^2 - 4x + 1)$

10. $g(x) = (4x - 3)(2x^2 + 3x + 5)$

11. $F(t) = (\sqrt{t} + 2)(3t - 4\sqrt{t} + 7)$

12. $G(t) = (2t + 3\sqrt{t} + 5)(\sqrt{t} + 4)$

Differentiate two ways: first, by using the Quotient Rule; then, by dividing the expressions before differentiating. Compare your results as a check. Use a graphing calculator to check your results.

13. $y = \dfrac{x^6}{x^4}$ **14.** $y = \dfrac{x^7}{x^3}$

15. $g(x) = \dfrac{3x^7 - x^3}{x}$ **16.** $f(x) = \dfrac{2x^5 + x^2}{x}$

17. $G(x) = \dfrac{8x^3 - 1}{2x - 1}$ **18.** $F(x) = \dfrac{x^3 + 27}{x + 3}$

19. $y = \dfrac{t^2 - 16}{t + 4}$ **20.** $y = \dfrac{t^2 - 25}{t - 5}$

Differentiate each function.

21. $g(x) = (5x^2 + 4x - 3)(2x^2 - 3x + 1)$

22. $f(x) = (3x^2 - 2x + 5)(4x^2 + 3x - 1)$

23. $y = \dfrac{5x^2 - 1}{2x^3 + 3}$ **24.** $y = \dfrac{3x^4 + 2x}{x^3 - 1}$

25. $F(x) = (-3x^2 + 4x)(7\sqrt{x} + 1)$

26. $G(x) = (8x + \sqrt{x})(5x^2 + 3)$

27. $g(t) = \dfrac{t}{3 - t} + 5t^3$ **28.** $f(t) = \dfrac{t}{5 + 2t} - 2t^4$

29. $G(x) = (5x - 4)^2$

30. $F(x) = (x + 3)^2$
     [Hint: $(x + 3)^2 = (x + 3)(x + 3)$.]

31. $y = (x^3 - 4x)^2$

32. $y = (3x^2 - 4x + 5)^2$

33. $f(x) = 6x^{-4}(6x^3 + 10x^2 - 8x + 3)$

34. $g(x) = 5x^{-3}(x^4 - 5x^3 + 10x - 2)$

35. $F(t) = \left(t + \dfrac{2}{t}\right)(t^2 - 3)$

36. $G(t) = (3t^5 - t^2)\left(t - \dfrac{5}{t}\right)$

37. $y = \dfrac{x^3 - 1}{x^2 + 1} + 4x^3$ **38.** $y = \dfrac{x^2 + 1}{x^3 - 1} - 5x^2$

39. $y = \dfrac{\sqrt[3]{x} - 7}{\sqrt{x} + 3}$ **40.** $y = \dfrac{\sqrt{x} + 4}{\sqrt[3]{x} - 5}$

41. $f(x) = \dfrac{x^{-1}}{x + x^{-1}}$ **42.** $f(x) = \dfrac{x}{x^{-1} + 1}$

43. $F(t) = \dfrac{1}{t - 4}$ **44.** $G(t) = \dfrac{1}{t + 2}$

45. $f(x) = \dfrac{3x^2 - 5x}{x^2 - 1}$ **46.** $f(x) = \dfrac{3x^2 + 2x}{x^2 + 1}$

47. $g(t) = \dfrac{-t^2 + 3t + 5}{t^2 - 2t + 4}$ **48.** $f(t) = \dfrac{3t^2 + 2t - 1}{-t^2 + 4t + 1}$

49. Find an equation of the tangent line to the graph of $y = 8/(x^2 + 4)$ at **(a)** $(0, 2)$; **(b)** $(-2, 1)$.

50. Find an equation of the tangent line to the graph of $y = \sqrt{x}/(x + 1)$ at **(a)** $x = 1$; **(b)** $x = \frac{1}{4}$.

51. Find an equation of the tangent line to the graph of $y = x^2 + 3/(x - 1)$ at **(a)** $x = 2$; **(b)** $x = 3$.

52. Find an equation of the tangent line to the graph of $y = 4x/(1 + x^2)$ at **(a)** $(0, 0)$; **(b)** $(-1, -2)$.

APPLICATIONS

Business and Economics

53. Average cost. Preston's Leatherworks finds that the cost, in dollars, of producing x belts is given by $C(x) = 750 + 34x - 0.068x^2$. Find the rate at which average cost is changing when 175 belts have been produced.

54. Average cost. Tongue-Tied Sauces, Inc., finds that the cost, in dollars, of producing x bottles of barbecue sauce is given by $C(x) = 375 + 0.75x^{3/4}$. Find the rate at which average cost is changing when 81 bottles of barbecue sauce have been produced.

55. Average revenue. Preston's Leatherworks find that the revenue, in dollars, from the sale of x belts is given by $R(x) = 45x^{9/10}$. Find the rate at which average revenue is changing when 175 belts have been produced and sold.

56. Average revenue. Tongue-Tied Sauces, Inc., finds that the revenue, in dollars, from the sale of x bottles of barbecue sauce is given by $R(x) = 7.5x^{0.7}$. Find the rate at which average revenue is changing when 81 bottles of barbecue sauce have been produced and sold.

57. Average profit. Use the information in Exercises 53 and 55 to determine the rate at which Preston's Leatherworks' average profit per belt is changing when 175 belts have been produced and sold.

58. Average profit. Use the information in Exercises 54 and 56 to determine the rate at which Tongue-Tied Sauces' average profit per bottle of barbecue sauce is changing when 81 bottles have been produced and sold.

59. Average profit. Sparkle Pottery has determined that the cost, in dollars, of producing x vases is given by

$$C(x) = 4300 + 2.1x^{0.6}.$$

If the revenue from the sale of x vases is given by $R(x) = 65x^{0.9}$, find the rate at which average profit per vase is changing when 50 vases have been made and sold.

60. Average profit. Cruzin' Boards has found that the cost, in dollars, of producing x skateboards is given by

$$C(x) = 900 + 18x^{0.7}.$$

If the revenue from the sale of x skateboards is given by $R(x) = 75x^{0.8}$, find the rate at which average profit per skateboard is changing when 20 skateboards have been built and sold.

61. Gross domestic product. The U.S. gross domestic product (in billions of dollars) can be approximated by

$$P(t) = 567 + t(36t^{0.6} - 104),$$

where t is the number of the years since 1960. (*Source:* U.S. Bureau of Economic Analysis.)

a) Find $P'(t)$.

b) Find $P'(45)$.

c) In words, explain what $P'(45)$ represents.

Social Sciences

62. Population growth. The population P, in thousands, of the town of Coyote Wells is given by

$$P(t) = \frac{500t}{2t^2 + 9},$$

where t is the time, in years.

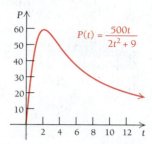

a) Find the growth rate.

b) Find the population after 12 yr.

c) Find the growth rate at $t = 12$ yr.

Life and Physical Sciences

63. Temperature during an illness. Gloria's temperature T during a recent illness is given by

$$T(t) = \frac{4t}{t^2 + 1} + 98.6,$$

where T is the temperature, in degrees Fahrenheit, at time t, in hours.

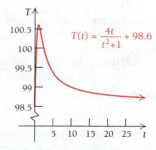

a) Find the rate of change of the temperature with respect to time.

b) Find the temperature at $t = 2$.

c) Find the rate of change of the temperature at $t = 2$.

SYNTHESIS

Differentiate each function.

64. $f(x) = \dfrac{7 - \dfrac{3}{2x}}{\dfrac{4}{x^2} + 5}$ (*Hint:* Simplify before differentiating.)

65. $y(t) = 5t(t - 1)(2t + 3)$

66. $f(x) = x(3x^3 + 6x - 2)(3x^4 + 7)$

67. $g(x) = (x^3 - 8) \cdot \dfrac{x^2 + 1}{x^2 - 1}$

68. $f(t) = (t^5 + 3) \cdot \dfrac{t^3 - 1}{t^3 + 1}$

69. $f(x) = \dfrac{(x - 1)(x^2 + x + 1)}{x^4 - 3x^3 - 5}$

70. Let $f(x) = \dfrac{x}{x + 1}$ and $g(x) = \dfrac{-1}{x + 1}$.

  a) Compute $f'(x)$.

  b) Compute $g'(x)$.

  c) What can you conclude about f and g on the basis of your results from parts (a) and (b)?

71. Let $f(x) = \dfrac{x^2}{x^2 - 1}$ and $g(x) = \dfrac{1}{x^2 - 1}$.

  a) Compute $f'(x)$.

  b) Compute $g'(x)$.

  c) What can you conclude about the graphs of f and g on the basis of your results from parts (a) and (b)?

72. Write a rule for finding the derivative of $f(x) \cdot g(x) \cdot h(x)$. Describe the rule in as few words as possible.

73. Is the derivative of the reciprocal of $f(x)$ the reciprocal of the derivative of $f(x)$? Why or why not?

74. Sensitivity. The reaction R of the body to a dose Q of medication is often represented by the general function

$$R(Q) = Q^2\left(\frac{k}{2} - \frac{Q}{3}\right),$$

where k is a constant and R is in millimeters of mercury (mmHg) if the reaction is a change in blood pressure or in degrees Fahrenheit (°F) if the reaction is a change in temperature. The rate of change dR/dQ is defined to be the body's *sensitivity* to the medication.

  a) Find a formula for the sensitivity.

  b) Explain the meaning of your answer to part (a).

75. A proof of the Product Rule appears below. Provide a justification for each step.

  a) $\dfrac{d}{dx}[f(x) \cdot g(x)] = \lim\limits_{h \to 0} \dfrac{f(x + h)g(x + h) - f(x)g(x)}{h}$

  b) $= \lim\limits_{h \to 0} \dfrac{f(x + h)g(x + h) - f(x + h)g(x) + f(x + h)g(x) - f(x)g(x)}{h}$

  c) $= \lim\limits_{h \to 0} \dfrac{f(x + h)g(x + h) - f(x + h)g(x)}{h} + \lim\limits_{h \to 0} \dfrac{f(x + h)g(x) - f(x)g(x)}{h}$

  d) $= \lim\limits_{h \to 0}\left[f(x + h) \cdot \dfrac{g(x + h) - g(x)}{h}\right] + \lim\limits_{h \to 0}\left[g(x) \cdot \dfrac{f(x + h) - f(x)}{h}\right]$

  e) $= f(x) \cdot \lim\limits_{h \to 0} \dfrac{g(x + h) - g(x)}{h} + g(x) \cdot \lim\limits_{h \to 0} \dfrac{f(x + h) - f(x)}{h}$

  f) $= f(x) \cdot g'(x) + g(x) \cdot f'(x)$

  g) $= f(x) \cdot \left[\dfrac{d}{dx}g(x)\right] + g(x) \cdot \left[\dfrac{d}{dx}f(x)\right]$

TECHNOLOGY CONNECTION

76. Business. Refer to Exercises 54, 56, and 58. At what rate is Tongue-Tied Sauces' profit changing at the break-even point? At what rate is the average profit per bottle of barbecue sauce changing at that point?

77. Business. Refer to Exercises 53, 55, and 57. At what rate is Preston's Leatherworks' profit changing at the break-even point? At what rate is the average profit per belt changing at that point?

For the function in each of Exercises 78–83, graph f and f'. Then estimate points at which the tangent line to f is horizontal. If no such point exists, state that fact.

78. $f(x) = x^2(x - 2)(x + 2)$

79. $f(x) = \left(x + \dfrac{2}{x}\right)(x^2 - 3)$

80. $f(x) = \dfrac{x^3 - 1}{x^2 + 1}$

81. $f(x) = \dfrac{0.01x^2}{x^4 + 0.0256}$

82. $f(x) = \dfrac{0.3x}{0.04 + x^2}$

83. $f(x) = \dfrac{4x}{x^2 + 1}$

84. Use a graph to decide which of the following seems to be the correct derivative of the function in Exercise 83.

$$y_1 = \frac{2}{x}$$

$$y_2 = \frac{4 - 4x}{x^2 + 1}$$

$$y_3 = \frac{4 - 4x^2}{(x^2 + 1)^2}$$

$$y_4 = \frac{4x^2 - 4}{(x^2 + 1)^2}$$

Answers to Quick Checks

1. (a) $y' = (2x^5 + x - 1)(3) + (3x - 2)(10x^4 + 1);$

(b) $y' = \left(\sqrt{x} + 1 \right)\left(\frac{1}{5\sqrt[5]{x^4}} - 1 \right) + (\sqrt[5]{x} - x)\left(\frac{1}{2\sqrt{x}} \right)$

2. (a) $y' = \dfrac{3x^2 - 2x - 6}{(x^2 + 2)^2};$

(b) $\dfrac{(bx + 1)(a) - (ax + 1)(b)}{(bx + 1)^2} = \dfrac{abx + a - abx - b}{(bx + 1)^2}$

$$= \frac{a - b}{(bx + 1)^2}$$

1.7

- Find the composition of two functions.
- Differentiate using the Extended Power Rule or the Chain Rule.

The Chain Rule

The Extended Power Rule

Consider the function given by $y = (1 + x^2)^3$. How do we determine its derivative? We might guess the following:

$$\frac{d}{dx}[(1 + x^2)^3] \stackrel{?}{=} 3(1 + x^2)^2. \qquad \text{Remember, this is a } \textit{guess.}$$

To check this, we first expand the function $y = (1 + x^2)^3$:

$$\begin{aligned} y &= (1 + x^2)^3 \\ &= (1 + x^2) \cdot (1 + x^2) \cdot (1 + x^2) \\ &= (1 + 2x^2 + x^4) \cdot (1 + x^2) \qquad \text{Multiplying the first two factors} \\ &= 1 + 3x^2 + 3x^4 + x^6. \qquad \text{Multiplying by the third factor} \end{aligned}$$

Taking the derivative of this function, we have

$$y' = 6x + 12x^3 + 6x^5.$$

Now we can factor out $6x$:

$$y' = 6x(1 + 2x^2 + x^4).$$

We rewrite $6x$ as $3 \cdot 2x$ and factor the trinomial:

$$y' = 3(1 + x^2)^2 \cdot 2x. \qquad \text{This } \textit{is} \text{ the derivative.}$$

Thus, it seems our original *guess* was close: it lacked only the extra factor, $2x$, which is the derivative of the expression inside the parentheses. This result suggests a general pattern for differentiating functions of this form, in which an expression is raised to a power k.

THEOREM 7   **The Extended Power Rule**

Suppose that $g(x)$ is a differentiable function of x. Then, for any real number k,

$$\frac{d}{dx}[g(x)]^k = k[g(x)]^{k-1} \cdot \frac{d}{dx} g(x).$$

A proof for the case where k is a positive integer is outlined in Exercise 95.

The Extended Power Rule allows us to differentiate functions such as $y = (1 + x^2)^{89}$ without expanding $1 + x^2$ to the 89th power (very time-consuming) and functions such as $y = (1 + x^2)^{1/3}$, for which expanding is impossible.

Let's differentiate $(1 + x^3)^5$. There are three steps to carry out.

1. Mentally block out the "inside" function, $1 + x^3$. $(1 + x^3)^5$
2. Differentiate the "outside" function, $(1 + x^3)^5$. $5(1 + x^3)^4$
3. Multiply by the derivative of the "inside" function. $5(1 + x^3)^4 \cdot 3x^2$

$$= 15x^2(1 + x^3)^4 \quad \text{Simplified}$$

Step (3) is commonly overlooked. Do not forget it!

EXAMPLE 1   Differentiate: **(a)** $f(x) = (1 + x^3)^{1/2}$;
(b) $y = (1 - x^2)^3 + (5 + 4x)^2$.

Solution   **a)** We have

$$\frac{d}{dx}(1 + x^3)^{1/2} = \frac{1}{2}(1 + x^3)^{1/2-1} \cdot 3x^2$$

$$= \frac{3x^2}{2}(1 + x^3)^{-1/2}$$

$$= \frac{3x^2}{2\sqrt{1 + x^3}}.$$

b) Here we combine the Sum–Difference Rule and the Extended Power Rule:

$$\frac{dy}{dx} = 3(1 - x^2)^2(-2x) + 2(5 + 4x)^1 \cdot 4 \quad \text{Differentiating each term using the Extended Power Rule}$$

$$= -6x(1 - x^2)^2 + 8(5 + 4x)$$

$$= -6x(1 - 2x^2 + x^4) + 40 + 32x$$

$$= -6x + 12x^3 - 6x^5 + 40 + 32x \quad \left.\right\} \text{Simplifying}$$

$$= -6x^5 + 12x^3 + 26x + 40.$$ **1** ✔

EXAMPLE 2   Differentiate: **(a)** $f(x) = (3x - 5)^4(7 - x)^{10}$; **(b)** $f(x) = \sqrt[4]{\dfrac{x + 3}{x - 2}}$.

Solution

a) Here we combine the Product Rule and the Extended Power Rule:

$$f'(x) = (3x - 5)^4 \cdot 10(7 - x)^9(-1) + (7 - x)^{10}4(3x - 5)^3(3)$$

$$= -10(3x - 5)^4(7 - x)^9 + (7 - x)^{10}12(3x - 5)^3$$

$$= 2(3x - 5)^3(7 - x)^9[-5(3x - 5) + 6(7 - x)] \quad \begin{array}{l}\text{Factoring out} \\ 2(3x - 5)^3(7 - x)^9\end{array}$$

$$= 2(3x - 5)^3(7 - x)^9(-15x + 25 + 42 - 6x)$$

$$= 2(3x - 5)^3(7 - x)^9(67 - 21x).$$

b) Here we use the Quotient Rule to differentiate the inside function:

$$\frac{d}{dx}\sqrt[4]{\frac{x + 3}{x - 2}} = \frac{d}{dx}\left(\frac{x + 3}{x - 2}\right)^{1/4} = \frac{1}{4}\left(\frac{x + 3}{x - 2}\right)^{1/4-1}\left[\frac{(x - 2)1 - 1(x + 3)}{(x - 2)^2}\right]$$

$$= \frac{1}{4}\left(\frac{x + 3}{x - 2}\right)^{-3/4}\left[\frac{x - 2 - x - 3}{(x - 2)^2}\right]$$

$$= \frac{1}{4}\left(\frac{x + 3}{x - 2}\right)^{-3/4}\left[\frac{-5}{(x - 2)^2}\right], \quad \text{or} \quad \frac{-5}{4(x + 3)^{3/4}(x - 2)^{5/4}}.$$ **2** ✔

Composition of Functions and the Chain Rule

Before discussing the Chain Rule, let's consider *composition of functions*.

One author of this text exercises three times a week at a local YMCA. When he recently bought a pair of running shoes, the box indicated equivalent men's shoe sizes in five countries.

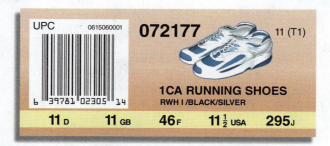

Author Marv Bittinger and his size-$11\frac{1}{2}$ running shoes

This suggests that there are functions that convert men's shoe sizes used in one country to those used in another country. There is, indeed, a function g that gives a correspondence between men's shoe sizes in the United States and those in France:

$$g(x) = \frac{4x + 92}{3},$$

where x is the U.S. size and $g(x)$ is the French size. Thus, U.S. size $11\frac{1}{2}$ corresponds to French size

$$g\left(11\tfrac{1}{2}\right) = \frac{4 \cdot 11\frac{1}{2} + 92}{3}, \quad \text{or } 46.$$

There is also a function f that gives a correspondence between shoe sizes in France and those in Japan. The function is given by

$$f(x) = \frac{15x - 100}{2},$$

where x is the French size and $f(x)$ is the corresponding Japanese size. Thus, French size 46 corresponds to Japanese size

$$f(46) = \frac{15 \cdot 46 - 100}{2}, \quad \text{or } 295.$$

It seems reasonable to conclude that U.S. shoe size $11\frac{1}{2}$ corresponds to Japanese size 295 and that some function h describes this correspondence. Can we find a formula for h?

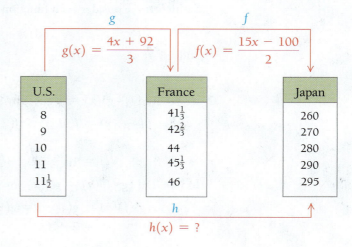

A U.S. size x corresponds to a French size $g(x) = (4x + 92)/3$. If we replace x in $f(x)$ with $(4x + 92)/3$, we can find the corresponding shoe size in Japan:

$$f(g(x)) = \frac{15[4x + 92]/3 - 100}{2}$$

$$= \frac{5(4x + 92) - 100}{2} = \frac{20x + 460 - 100}{2}$$

$$= \frac{20x + 360}{2} = 10x + 180.$$

This gives a formula for h: $h(x) = 10x + 180$. As a check, a U.S. size of $11\frac{1}{2}$ corresponds to a Japanese size $h(11\frac{1}{2}) = 10(11\frac{1}{2}) + 180 = 295$. The function h is the *composition* of f and g, symbolized by $f \circ g$ and read as "f composed with g."

DEFINITION

The **composed** function $f \circ g$, the **composition** of f and g, is defined as

$$(f \circ g)(x) = f(g(x)).$$

We can visualize the composition of functions as shown below.

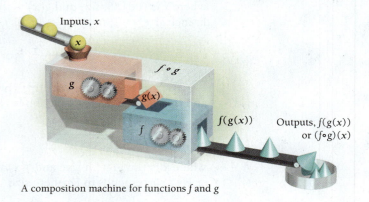

A composition machine for functions f and g

To find $(f \circ g)(x)$, we substitute $g(x)$ for x in $f(x)$. The function $g(x)$ is nested within $f(x)$.

EXAMPLE 3   For $f(x) = x^4$ and $g(x) = 2 + x^3$, find **(a)** $(f \circ g)(x)$ and **(b)** $(g \circ f)(x)$.

Solution   Consider each function separately:

$$f(x) = x^4 \qquad \text{This function raises each input to the 4th power.}$$

and

$$g(x) = 2 + x^3. \qquad \text{This function adds 2 to the cube of each input.}$$

a) We find $(f \circ g)(x) = f(g(x))$ by substituting $g(x)$ for x in $f(x)$:

$$(f \circ g)(x) = f(g(x)) = f(2 + x^3)$$
$$= (2 + x^3)^4 \qquad \text{Using } g(x) \text{ as an input}$$
$$= 16 + 32x^3 + 24x^6 + 8x^9 + x^{12}.$$

b) We find $(g \circ f)(x) = g(f(x))$ by substituting $f(x)$ for x in $g(x)$:

$$(g \circ f)(x) = g(f(x)) = g(x^4)$$
$$= 2 + (x^4)^3 = 2 + x^{12}. \qquad \text{Using } f(x) \text{ as an input}$$

Quick Check 3 ☑

Let $f(x) = x^2 + x$ and $g(x) = 3x + 2$.

Find:

a) $(f \circ g)(x)$;

b) $(g \circ f)(x)$.

EXAMPLE 4   For $f(x) = \sqrt{x}$ and $g(x) = x - 1$, find $(f \circ g)(x)$ and $(g \circ f)(x)$.

Solution

$$(f \circ g)(x) = f(g(x)) = f(x - 1) = \sqrt{x - 1}$$
$$(g \circ f)(x) = g(f(x)) = g(\sqrt{x}) = \sqrt{x} - 1$$

3 ☑

Examples 3 and 4 show that, in general, $(f \circ g)(x) \neq (g \circ f)(x)$.

How do we differentiate a composition of functions? The following theorem tells us.

THEOREM 8   **The Chain Rule**

The derivative of the composition $f \circ g$ is given by

$$\frac{d}{dx}[(f \circ g)(x)] = \frac{d}{dx}[f(g(x))] = f'(g(x)) \cdot g'(x).$$

As noted earlier, the Extended Power Rule is a special case of the Chain Rule. Consider $f(x) = x^k$. For any other function $g(x)$, we have $(f \circ g)(x) = [g(x)]^k$, and the derivative of the composition is

$$\frac{d}{dx}[g(x)]^k = k[g(x)]^{k-1} \cdot g'(x).$$

The Chain Rule can appear in another form. If $y = f(u)$ and $u = g(x)$, then

$$\frac{dy}{dx} = \frac{dy}{du} \cdot \frac{du}{dx}.$$

The proof of Theorem 8 is beyond the scope of this text, but is illustrated in the following example.

EXAMPLE 5   **Business: Rate of Profit.**   GameBoss Video's profit, in dollars, is given by $y = f(u) = 0.1u^2 - 500$, where u is the number of units sold. Its sales are given by $u = g(x) = 125x + 40$, where x is the number of days. Find the rate at which profit is changing on the 5th day.

Solution   We first find the rate of change in profit, y, on the xth day. That is, we want to find dy/dx. Thus, we form the composition of f and g, because $y = (f \circ g)(x)$ relates profit y to number of days x:

$$y = (f \circ g)(x) = f(g(x))$$
$$= 0.1(125x + 40)^2 - 500. \quad \text{Substituting}$$

Differentiating, we have

$$\frac{dy}{dx} = \frac{d}{dx}[(f \circ g)(x)] = \frac{d}{dx}[0.1(125x + 40)^2 - 500]$$

$$= 2 \cdot 0.1(125x + 40) \cdot 125 \qquad \text{Using the Extended Power Rule}$$

$$= 3125x + 1000. \qquad \text{Simplifying}$$

On the 5th day, the company's profit is changing by $3125(5) + 1000 = \$16,625$ per day.

Note that we can also find dy/dx, the rate of change in profit per day, by finding the product of dy/du, the rate of change in profit per unit, and du/dx, the rate of change in units sold per day.

We have

$$\frac{dy}{du} = \frac{d}{du}(0.1u^2 - 500) = 0.2u \, \frac{\text{dollars}}{\text{unit}}, \quad \text{and} \quad \frac{du}{dx} = \frac{d}{dx}(125x + 40) = 125 \, \frac{\text{units}}{\text{day}}.$$

Thus,

$$\frac{dy}{dx} = \frac{dy}{du} \cdot \frac{du}{dx}$$
$$= (0.2u)(125)$$
$$= 25u$$
$$= 25(125x + 40). \qquad \text{Substituting}$$

For $x = 5$, we have $dy/dx = 25(125(5) + 40) = \$16,625$ per day.

Comparing units, we have

$$\frac{dy}{dx} = \frac{dy}{du} \cdot \frac{du}{dx}$$

or

$$\frac{\text{dollars}}{\text{day}} = \frac{\text{dollars}}{\text{unit}} \cdot \frac{\text{units}}{\text{day}}.$$

In a manner of speaking, the units "simplify," leaving dollars/day.   4 ✔

Quick Check 4 ✔

If $y = u^2 + u$ and $u = x^2 + x$, find $\dfrac{dy}{dx}$.

EXAMPLE 6   **Business.**   A new smartwatch is placed on the market. Its quantity sold N is given as a function of time t, in weeks:

$$N(t) = \frac{250,000t^2}{(2t + 1)^2}, \qquad t > 0.$$

Find $N'(t)$. Then find $N'(52)$ and $N'(208)$, and interpret these results.

Solution   To find $N'(t)$, we use the Quotient Rule along with the Extended Power Rule:

$$N'(t) = \frac{d}{dt}\left[\frac{250,000t^2}{(2t + 1)^2}\right] = \frac{(2t + 1)^2 \cdot \frac{d}{dt}[250,000t^2] - 250,000t^2 \cdot \frac{d}{dt}[(2t + 1)^2]}{[(2t + 1)^2]^2}$$

$$= \frac{(2t + 1)^2 \cdot (500,000t) - 250,000t^2 \cdot 2(2t + 1)^1 \cdot 2}{(2t + 1)^4} \qquad \begin{array}{l}\text{Using the}\\ \text{Extended}\\ \text{Power}\\ \text{Rule}\end{array}$$

$$= \frac{(2t + 1)^2(500,000t) - 1,000,000t^2(2t + 1)}{(2t + 1)^4}$$

$$= \frac{500,000t(2t + 1)[(2t + 1) - 2t]}{(2t + 1)^4}. \qquad \begin{array}{l}\text{Factoring}\\ 500,000t(2t + 1)\\ \text{in the numerator}\end{array}$$

Therefore,

$$N'(t) = \frac{500,000t}{(2t + 1)^3}. \qquad \text{Simplifying}$$

We evaluate $N'(t)$ at $t = 52$:

$$N'(52) = \frac{500,000(52)}{(2(52) + 1)^3} \approx 22.5.$$

Thus, after 52 weeks (1 yr), the number of smartwatches sold is increasing by about 22.5 per week.

For $t = 208$ weeks (4 yr), we get

$$N'(208) = \frac{500,000(208)}{(2(208) + 1)^3} \approx 1.4.$$

After 4 yr, the number of smartwatches sold is increasing by about 1.4 per week. What is happening here? Consider the graph of $N(t)$:

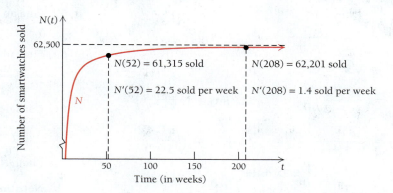

We see that the slopes of the tangent lines, representing the change in the number of smartwatches sold per week level off as t increases. Perhaps the market is becoming saturated with smartwatches; that is, although sales continue to grow, they do so at a slower and slower rate. ∎

Section Summary

- The *Extended Power Rule* tells us that if $y = [f(x)]^k$, then

$$y' = \frac{d}{dx}[f(x)]^k = k[f(x)]^{k-1} \cdot f'(x).$$

- The *composition* of $f(x)$ with $g(x)$ is written $(f \circ g)(x)$ and is defined as $(f \circ g)(x) = f(g(x))$.
- In general, $(f \circ g)(x) \neq (g \circ f)(x)$.

- The *Chain Rule* is used to differentiate a composition of functions. If

$$F(x) = (f \circ g)(x) = f(g(x)),$$

then

$$F'(x) = \frac{d}{dx}[(f \circ g)(x)] = f'(g(x)) \cdot g'(x).$$

1.7   Exercise Set

Differentiate each function.

1. $y = (3 - 2x)^2$ ⎫
2. $y = (2x + 1)^2$ ⎬ Check by expanding and then differentiating.

3. $y = (7 - x)^{55}$

4. $y = (8 - x)^{100}$

5. $y = \sqrt{1 - x}$

6. $y = \sqrt{1 + 8x}$

7. $y = \sqrt{3x^2 - 4}$

8. $y = \sqrt{4x^2 + 1}$

9. $y = (4x^2 + 1)^{-50}$

10. $y = (8x^2 - 6)^{-40}$

11. $y = (x - 4)^8(2x + 3)^6$

12. $y = (x + 5)^7(4x - 1)^{10}$

13. $y = \dfrac{1}{(4x + 5)^2}$

14. $y = \dfrac{1}{(3x + 8)^2}$

15. $y = \dfrac{4x^2}{(7 - 5x)^3}$

16. $y = \dfrac{7x^3}{(4 - 9x)^5}$

17. $f(x) = (3 + x^3)^5 - (1 + x^7)^4$

18. $f(x) = (1 + x^3)^3 - (2 + x^8)^4$

19. $f(x) = x^2 + (200 - x)^2$

20. $f(x) = x^2 + (100 - x)^2$

21. $G(x) = \sqrt[3]{2x - 1} + (4 - x)^2$

22. $g(x) = \sqrt{x} + (x - 3)^3$

23. $f(x) = -5x(2x - 3)^4$

24. $f(x) = -3x(5x + 4)^6$

25. $F(x) = (5x + 2)^4(2x - 3)^8$

26. $g(x) = (3x - 1)^7(2x + 1)^5$

27. $f(x) = x^2\sqrt{4x - 1}$

28. $f(x) = x^3\sqrt{5x + 2}$

29. $F(x) = \sqrt[4]{x^2 - 5x + 2}$

30. $G(x) = \sqrt[3]{x^5 + 6x}$

31. $f(x) = \left(\dfrac{3x - 1}{5x + 2}\right)^4$ **32.** $f(x) = \left(\dfrac{2x}{x^2 + 1}\right)^3$

33. $g(x) = \sqrt{\dfrac{3 + 2x}{5 - x}}$ **34.** $g(x) = \sqrt{\dfrac{4 - x}{3 + x}}$

35. $f(x) = (2x^3 - 3x^2 + 4x + 1)^{100}$

36. $f(x) = (7x^4 + 6x^3 - x)^{204}$

37. $h(x) = \left(\dfrac{1 - 3x}{2 - 7x}\right)^{-5}$ **38.** $g(x) = \left(\dfrac{2x + 3}{5x - 1}\right)^{-4}$

39. $f(x) = \sqrt{\dfrac{x^2 + x}{x^2 - x}}$ **40.** $f(x) = \sqrt[3]{\dfrac{4 - x^3}{x - x^2}}$

41. $f(x) = \dfrac{(5x - 4)^7}{(6x + 1)^3}$ **42.** $f(x) = \dfrac{(2x + 3)^4}{(3x - 2)^5}$

43. $f(x) = 12(2x + 1)^{2/3}(3x - 4)^{5/4}$

44. $y = 6\sqrt[3]{x^2 + x}(x^4 - 6x)^3$

Find $\dfrac{dy}{du}, \dfrac{du}{dx}, and \dfrac{dy}{dx}$.

45. $y = \dfrac{15}{u^3}$ and $u = 2x + 1$

46. $y = \sqrt{u}$ and $u = x^2 - 1$

47. $y = u^{50}$ and $u = 4x^3 - 2x^2$

48. $y = \dfrac{u + 1}{u - 1}$ and $u = 1 + \sqrt{x}$

49. $y = (u + 1)(u - 1)$ and $u = x^3 + 1$

50. $y = u(u + 1)$ and $u = x^3 - 2x$

Find $\dfrac{dy}{dx}$ for each pair of functions.

51. $y = 5u^2 + 3u$, where $u = x^3 + 1$

52. $y = u^3 - 7u^2$, where $u = x^2 + 3$

53. $y = \sqrt{7 - 3u}$, where $u = x^2 - 9$

54. $y = \sqrt[3]{2u + 5}$, where $u = x^2 - x$

55. Find $\dfrac{dy}{dt}$ if $y = \dfrac{1}{u^2 + u}$ and $u = 5 + 3t$.

56. Find $\dfrac{dy}{dt}$ if $y = \dfrac{1}{3u^5 - 7}$ and $u = 7t^2 + 1$.

57. Find an equation for the tangent line to the graph of $y = (x^3 - 4x)^{10}$ at the point $(2, 0)$.

58. Find an equation for the tangent line to the graph of $y = \sqrt{x^2 + 3x}$ at the point $(1, 2)$.

59. Find an equation for the tangent line to the graph of $y = x\sqrt{2x + 3}$ at the point $(3, 9)$.

60. Find an equation for the tangent line to the graph of $y = \left(\dfrac{2x + 3}{x - 1}\right)^3$ at the point $(2, 343)$.

61. Consider
$$g(x) = \left(\dfrac{6x + 1}{2x - 5}\right)^2.$$

   a) Find $g'(x)$ using the Extended Power Rule.

   b) Note that $g(x) = \dfrac{36x^2 + 12x + 1}{4x^2 - 20x + 25}$.
Find $g'(x)$ using the Quotient Rule.

   c) Compare your answers to parts (a) and (b). Which approach was easier, and why?

62. Consider
$$f(x) = \dfrac{x^2}{(1 + x)^5}.$$

   a) Find $f'(x)$ using the Quotient Rule and the Extended Power Rule.

   b) Note that $f(x) = x^2(1 + x)^{-5}$. Find $f'(x)$ using the Product Rule and the Extended Power Rule.

   c) Compare your answers to parts (a) and (b).

63. Let $f(u) = u^3$ and $g(x) = u = 2x^4 + 1$.
Find $(f \circ g)'(-1)$.

64. Let $f(u) = \dfrac{u + 1}{u - 1}$ and $g(x) = u = \sqrt{x}$.
Find $(f \circ g)'(4)$.

65. Let $f(u) = \sqrt[3]{u}$ and $g(x) = u = 1 + 3x^2$.
Find $(f \circ g)'(2)$.

66. Let $f(u) = 2u^5$ and $g(x) = u = \dfrac{3 - x}{4 + x}$.
Find $(f \circ g)'(-10)$.

For Exercises 67–70, use the Chain Rule to differentiate each function. You may need to apply the rule more than once.

67. $f(x) = (2x^3 + (4x - 5)^2)^6$

68. $f(x) = (-x^5 + 4x + \sqrt{2x + 1})^3$

69. $f(x) = \sqrt{x^2 + \sqrt{1 - 3x}}$

70. $f(x) = \sqrt[3]{2x + (x^2 + x)^4}$

APPLICATIONS

Business and Economics

71. Total revenue. A total-revenue function is given by
$$R(x) = 1000\sqrt{x^2 - 0.1x},$$
where $R(x)$ is the total revenue, in thousands of dollars, from the sale of x airplanes. Find the rate at which total revenue is changing when 20 airplanes have been sold.

72. Total cost. A total-cost function is given by

$$C(x) = 2000(x^2 + 2)^{1/3} + 700,$$

where $C(x)$ is the total cost, in thousands of dollars, of producing x airplanes. Find the rate at which total cost is changing when 20 airplanes have been produced.

73. Total profit. Use the total-cost and total-revenue functions in Exercises 71 and 72 to find the rate at which total profit is changing when x airplanes have been produced and sold.

74. Total cost. A company determines that its total cost, in thousands of dollars, for producing x chairs is

$$C(x) = \sqrt{5x^2 + 60},$$

and it plans to boost production t months from now according to the function

$$x(t) = 20t + 40.$$

How fast will costs be rising 4 months from now?

75. Consumer credit. The total outstanding consumer credit of the United States (in billions of dollars) can be modeled by the function

$$C(x) = 9.26x^4 - 85.27x^3 + 287.24x^2 - 309.12x + 2651.4,$$

where x is the number of years since 2008.

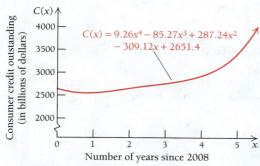

(*Source*: Based on data from federalreserve.gov.)

a) Find dC/dx.
b) Interpret the meaning of dC/dx.
c) Using this model, estimate how quickly outstanding consumer credit was rising in 2014.

76. Utility. Utility is a type of function that occurs in economics. When a consumer receives x units of a product, a certain amount of pleasure, or utility, U, is derived. Suppose the utility related to the number of tickets x sold for a ride at a county fair is

$$U(x) = 80\sqrt{\frac{2x + 1}{3x + 4}}.$$

Find the rate at which utility changes with respect to number of tickets bought.

77. Compound interest. If $1000 is invested at interest rate r, compounded quarterly, in 5 yr it will grow to an amount, A, given by (see Section R.1)

$$A = \$1000\left(1 + \frac{r}{4}\right)^{20}.$$

a) Find the rate of change, dA/dr.
b) Interpret the meaning of dA/dr.

78. Compound interest. If $1000 is invested at interest rate r, compounded annually, in 3 yr it will grow to an amount A given by (see Section R.1)

$$A = \$1000(1 + r)^3.$$

a) Find the rate of change, dA/dr.
b) Interpret the meaning of dA/dr.

79. Business profit. French's Electronics is selling laptop computers. It determines that its total profit, in dollars, is given by

$$P(x) = 0.08x^2 + 80x,$$

where x is the number of units produced and sold. Suppose that x is a function of time, in months, where $x = 5t + 1$.

a) Find the total profit as a function of time t.
b) Find the rate of change of total profit when $t = 48$ months.

80. Consumer demand. Suppose the demand function for a new autobiography is given by

$$D(p) = \frac{80,000}{p},$$

and that price p is a function of time, given by $p = 1.6t + 9$, where t is in days.
a) Find the demand as a function of time t.
b) Find the rate of change of the quantity demanded when $t = 100$ days.

Life and Physical Sciences

81. Chemotherapy. The dosage for Carboplatin chemotherapy drugs depends on several parameters for the particular drug as well as the age, weight, and sex of the patient. For female patients, the formulas giving the dosage for such drugs are

$$D = 0.85A(c + 25) \quad \text{and} \quad c = (140 - y)\frac{w}{72x},$$

where A and x depend on which drug is used, D is the dosage in milligrams (mg), c is called the creatine clearance, y is the patient's age in years, and w is the patient's weight in kilograms (kg). (*Source*: *U.S. Oncology.*)

a) Suppose a patient is a 45-year-old woman and the drug has parameters $A = 5$ and $x = 0.6$. Use this information to write formulas for D and c that give D as a function of c and c as a function of w.
b) Use your formulas from part (a) to compute dD/dc.
c) Use your formulas from part (a) to compute dc/dw.
d) Compute dD/dw.
e) Interpret the meaning of the derivative dD/dw.

SYNTHESIS

If $f(x)$ is a function, then $(f \circ f)(x) = f(f(x))$ is the composition of f with itself. This is called an iterated function, and the composition can be repeated many times. For example, $(f \circ f \circ f)(x) = f(f(f(x)))$. Iterated functions are very useful in many areas, including finance (compound interest is a simple case) and the sciences (in weather forecasting, for example). For each function, use the Chain Rule to find the derivative.

82. If $f(x) = x^2 + 1$, find $\dfrac{d}{dx}[(f \circ f)(x)]$.

83. If $f(x) = x + \sqrt{x}$, find $\dfrac{d}{dx}[(f \circ f)(x)]$.

84. If $f(x) = x^2 + 1$, find $\dfrac{d}{dx}[(f \circ f \circ f)(x)]$

85. If $f(x) = \sqrt{3x}$, find $\dfrac{d}{dx}[(f \circ f \circ f)(x)]$.

Do you see a shortcut?

Differentiate.

86. $y = \sqrt{(2x - 3)^2 + 1}$

87. $y = \sqrt[3]{x^3 + 6x + 1} \cdot x^5$

88. $y = \left(\dfrac{x}{\sqrt{x - 1}}\right)^3$

89. $y = (x\sqrt{1 + x^2})^3$

90. $y = \dfrac{\sqrt{1 - x^2}}{1 - x}$

91. $y = \left(\dfrac{x^2 - x - 1}{x^2 + 1}\right)^3$

92. $g(x) = \sqrt{\dfrac{x^2 - 4x}{2x + 1}}$

93. $f(t) = \sqrt{3t + \sqrt{t}}$

94. $F(x) = [6x(3 - x)^5 + 2]^4$

95. The Extended Power Rule (for positive integer powers) can be verified using the Product Rule. For example, if $y = [f(x)]^2$, then the Product Rule is applied by recognizing that $[f(x)]^2 = [f(x)] \cdot [f(x)]$. Therefore,

$$\dfrac{d}{dx}([f(x)] \cdot [f(x)]) = f(x) \cdot f'(x) + f'(x) \cdot f(x)$$

$$= 2f(x) \cdot f'(x).$$

a) Use the Product Rule to show that $\dfrac{d}{dx}[f(x)]^3 = 3[f(x)]^2 \cdot f'(x)$.

b) Use the Product Rule to show that $\dfrac{d}{dx}[f(x)]^4 = 4[f(x)]^3 \cdot f'(x)$.

96. The following is the beginning of an alternative proof of the Quotient Rule that uses the Product Rule and the Power Rule. Complete the proof, giving reasons for each step.

Proof: Let

$$Q(x) = \dfrac{N(x)}{D(x)}.$$

Then

$$Q(x) = N(x) \cdot [D(x)]^{-1}.$$

Therefore, … .

For the function in each of Exercises 97 and 98, graph f and f' over the given interval. Then estimate points at which the line tangent to f is horizontal.

97. $f(x) = 1.68x\sqrt{9.2 - x^2}; \quad [-3, 3]$

98. $f(x) = \sqrt{6x^3 - 3x^2 - 48x + 45}; \quad [-5, 5]$

Find the derivative of each of the following functions. Then use a calculator to check the results.

99. $f(x) = x\sqrt{4 - x^2}$

100. $f(x) = (\sqrt{2x - 1} + x^3)^5$

Answers to Quick Checks

1. (a) $y' = 3(x^4 + 2x^2 + 1)^2(4x^3 + 4x)$;
(b) The result lacks parentheses around $2x + 4$. It should be written: $4(x^2 + 4x + 1)^3(2x + 4)$.

2. $y' = \dfrac{-36x^5 + 24x^3 + 8x}{(3x^4 + 2)^3}$

3. (a) $f(g(x)) = 9x^2 + 15x + 6$;
(b) $g(f(x)) = 3x^2 + 3x + 2$

4. $\dfrac{dy}{dx} = \dfrac{dy}{du} \cdot \dfrac{du}{dx} = (2u + 1)(2x + 1) =$
$(2(x^2 + x) + 1)(2x + 1) = (2x^2 + 2x + 1)(2x + 1)$

Higher-Order Derivatives

1.8

- Find derivatives of higher order.
- Given a formula for distance, find velocity and acceleration.

Consider the function given by

$$y = f(x) = x^5 - 3x^4 + x.$$

Its derivative f' is given by

$$y' = f'(x) = 5x^4 - 12x^3 + 1.$$

The derivative, f', can also be differentiated. We can think of the derivative of f' as the rate of change of the slope of the tangent lines of f, or the rate at which $f'(x)$ is changing. We use the notation f'' to represent $(f')'$. That is,

$$f''(x) = \frac{d}{dx} f'(x).$$

We call f'' the *second derivative* of f. If $f'(x) = 5x^4 - 12x^3 + 1$, then

$$y'' = f''(x) = 20x^3 - 36x^2.$$

Continuing in this manner, we have

$$f'''(x) = 60x^2 - 72x, \qquad \text{The third derivative of } f$$
$$f''''(x) = 120x - 72, \qquad \text{The fourth derivative of } f$$
$$f'''''(x) = 120. \qquad \text{The fifth derivative of } f$$

When this type of derivative notation gets lengthy, we abbreviate it using a number in parentheses. Thus, $f^{(n)}(x)$ is the nth derivative of f. For the function above,

$$f^{(4)}(x) = 120x - 72,$$
$$f^{(5)}(x) = 120,$$
$$f^{(6)}(x) = 0, \qquad \text{and}$$
$$f^{(n)}(x) = 0, \qquad \text{for any integer } n \geq 6.$$

Leibniz notation for the second derivative of a function given by $y = f(x)$ is

$$\frac{d^2y}{dx^2}, \quad \text{or} \quad \frac{d}{dx}\left(\frac{dy}{dx}\right),$$

read "the second derivative of y with respect to x." The 2's in this notation are *not* exponents. If $y = x^5 - 3x^4 + x$, then

$$\frac{d^2y}{dx^2} = 20x^3 - 36x^2.$$

Leibniz notation for the third derivative is d^3y/dx^3; for the fourth derivative, d^4y/dx^4; and so on:

$$\frac{d^3y}{dx^3} = 60x^2 - 72x,$$
$$\frac{d^4y}{dx^4} = 120x - 72,$$
$$\frac{d^5y}{dx^5} = 120.$$

EXAMPLE 1 For $y = 1/x$, find d^2y/dx^2.

Solution We have $y = x^{-1}$, so

$$\frac{dy}{dx} = -1 \cdot x^{-1-1} = -x^{-2}, \quad \text{or} \quad -\frac{1}{x^2}.$$

Then

$$\frac{d^2y}{dx^2} = (-2)(-1)x^{-2-1} = 2x^{-3}, \quad \text{or} \quad \frac{2}{x^3}.$$

EXAMPLE 2 For $y = (x^2 + 10x)^{20}$, find y' and y''.

Solution To find y', we use the Extended Power Rule:

$$\begin{aligned}
y' &= 20(x^2 + 10x)^{19}(2x + 10) \\
&= 20(x^2 + 10x)^{19} \cdot 2(x + 5) \qquad \text{Factoring} \\
&= 40(x^2 + 10x)^{19}(x + 5).
\end{aligned}$$

To find y'', we use the Product Rule and the Extended Power Rule:

$$\begin{aligned}
y'' &= \frac{d}{dx}[40(x^2 + 10x)^{19} \cdot (x + 5)] \\
&= 40(x^2 + 10x)^{19} \cdot \frac{d}{dx}(x + 5) + (x + 5) \cdot \frac{d}{dx}[40(x^2 + 10x)^{19}] && \text{Using the Product Rule} \\
&= 40(x^2 + 10x)^{19} \cdot (1) + (x + 5) \cdot 19 \cdot 40(x^2 + 10x)^{18}(2x + 10) && \text{Using the Extended Power Rule} \\
&= 40(x^2 + 10x)^{19} + (40 \cdot 19 \cdot 2)(x + 5)^2(x^2 + 10x)^{18} \\
&= 40(x^2 + 10x)^{18}[(x^2 + 10x) + 38(x + 5)^2] && \text{Factoring} \\
&= 40(x^2 + 10x)^{18}[x^2 + 10x + 38(x^2 + 10x + 25)] \\
&= 40(x^2 + 10x)^{18}(39x^2 + 390x + 950).
\end{aligned}$$

Quick Check 1 ✔

a) Find y'':

(i) $y = -6x^4 + 3x^2$;

(ii) $y = \dfrac{2}{x^3}$;

(iii) $y = (3x^2 + 1)^2$.

b) Find $\dfrac{d^4}{dx^4}\left[\dfrac{1}{x}\right]$.

1 ✔

Velocity and Acceleration

We have already seen that a function's derivative represents its instantaneous rate of change. When the function relates distance traveled to time, the instantaneous rate of change is called *speed*, or *velocity*.* The letter v is generally used to stand for velocity.

DEFINITION

The **velocity** of an object that is $s(t)$ units from a starting point at time t is given by

$$\text{Velocity} = v(t) = s'(t) = \lim_{h \to 0} \frac{s(t + h) - s(t)}{h}.$$

Often velocity itself is a function of time. When a jet takes off or a vehicle comes to a sudden stop, the change in velocity is easily felt by passengers. The rate at which velocity changes is called *acceleration*. If Car A requires 8.4 sec to reach 65 mi/hr and Car B requires 8 sec; then B has *faster acceleration* than A. We generally use a to represent acceleration, which we regard as the rate at which velocity is changing.

---

*In this text, the words "speed" and "velocity" are used interchangeably. In physics and engineering, this is not done, since velocity requires direction and speed does not.

> **DEFINITION**
>
> Acceleration $= a(t) = v'(t) = s''(t)$.

EXAMPLE 3 **Free Fall.** When an object is dropped, the distance it falls in t seconds, assuming negligible air resistance, is given by

$$s(t) = 4.905t^2,$$

where $s(t)$ is in meters (m). If a stone is dropped from a cliff, find each of the following, assuming that air resistance is negligible: **(a)** how far the stone has traveled 5 sec after being dropped, **(b)** how fast it is traveling 5 sec after being dropped, and **(c)** its acceleration after it has been falling for 5 sec.

Solution

a) After 5 sec, the stone has traveled

$$s(5) = 4.905(5)^2 = 4.905(25) = 122.625 \text{ m}.$$

b) The speed at which the stone is traveling is given by

$$v(t) = s'(t) = 9.81t.$$

Thus,

$$v(5) = 9.81 \cdot 5 = 49.05 \text{ m/sec}.$$

c) The stone's acceleration after t sec is constant:

$$a(t) = v'(t) = s''(t) = 9.81 \text{ m/sec}^2.$$

Thus, $s''(5) = 9.81 \text{ m/sec}^2.$ 2 ✔

Quick Check 2 ✔

A pebble is dropped from a hot-air balloon. Find how far it has fallen, how fast it is falling, and its acceleration after 3.5 sec. Let $s(t) = 16t^2$, where t is in seconds and s is in feet.

If the graph of $y = s(t)$ represents an object's distance traveled, then the object's velocity at any time is the slope of the tangent line at that point. If the object "speeds up" (accelerates) or "slows down" (decelerates), these changes in velocity will appear as upward and downward bends, respectively, in the graph.

EXAMPLE 4 **Analyzing Velocity and Acceleration Graphically.** Kimberly leaves her home (distance $= 0$) for a 2-hr bicycle ride. The graph of her distance traveled with respect to time t is shown below.

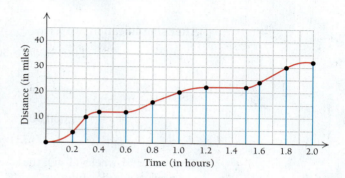

Find the interval(s) in which Kimberly is:

a) traveling at constant velocity;
b) accelerating (speeding up);
c) decelerating (slowing down).

Solution

a) Kimberly is traveling at constant velocity when the slopes of the tangent lines do not change, that is, when the graph is linear. Thus, she is traveling at a constant velocity in the intervals $(0.2, 0.3)$, $(0.4, 0.6)$, $(0.8, 1.0)$, $(1.2, 1.5)$, and $(1.6, 1.8)$. Note that in the intervals $(0.4, 0.6)$ and $(1.2, 1.5)$, the slopes of the tangent lines are zero, indicating that she has stopped.

b) Kimberly is speeding up during the intervals $(0, 0.2)$, $(0.6, 0.8)$, and $(1.5, 1.6)$. This acceleration appears as an upward bend in the graph.

c) Kimberly is slowing down during the intervals $(0.3, 0.4)$, $(1.0, 1.2)$, and $(1.8, 2.0)$. This deceleration (or negative acceleration) appears as a downward bend in the graph. ■

The upward and downward bends in the graph in Example 4 are examples of *concavities*, which are discussed in Chapter 2.

In Example 6 of Section 1.7, $N(t)$ represented the number of smartwatches sold after t weeks. Its first derivative was always positive (always increasing) indicating that sales were always increasing. But sales were leveling off. In the following example, we see how the second derivative can help us understand this observation.

EXAMPLE 5   Business.   In Example 6 of Section 1.7, the function given by

$$N(t) = \frac{250,000t^2}{(2t + 1)^2}, t > 0,$$ represented the number of smartwatches sold after t weeks.

Its derivative is

$$N'(t) = \frac{500,000t}{(2t + 1)^3}.$$

Find $N''(t)$; then use it to calculate $N''(52)$ and $N''(208)$ and interpret these results.

Solution   We use both the Quotient Rule and the Extended Power Rule:

$$\frac{d}{dt}[N'(t)] = \frac{(2t + 1)^3 \cdot \frac{d}{dt}[500,000t] - (500,000t) \cdot \frac{d}{dt}[(2t + 1)^3]}{[(2t + 1)^3]^2}$$ Using the Extended Power Rule

$$= \frac{(2t + 1)^3(500,000) - (500,000t)(3)(2t + 1)^2 \cdot 2}{[(2t + 1)^3]^2}$$

$$= \frac{(2t + 1)^2[(2t + 1)(500,000) - 6(500,000t)]}{(2t + 1)^6}.$$ Factoring out $(2t + 1)^2$

After simplification, we have

$$N''(t) = \frac{-2,000,000t + 500,000}{(2t + 1)^4}.$$

At $t = 52$, we have

$$N''(52) = \frac{-2,000,000(52) + 500,000}{[2(52) + 1]^4} \approx -0.851.$$

Thus, after 52 weeks (1 yr), the rate of the rate of sales is decreasing at -0.851 units per week per week. In other words, although sales are increasing during the 52nd week (remember, $N'(52) > 0$), the rate at which sales are increasing is decreasing. That is, sales are increasing but not as fast as before.

When $t = 208$,

$$N''(208) = \frac{-2,000,000(208) + 500,000}{[2(208) + 1]^4} \approx -0.014.$$

After 208 weeks (4 yr), sales have nearly leveled off. Remember, sales are increasing at the 208th week (recall that $N'(208) > 0$), but since the market is nearly saturated with this product, the rate at which sales are increasing has slowed to near 0.

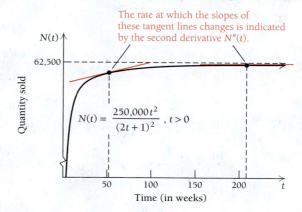

The rate at which the slopes of these tangent lines changes is indicated by the second derivative $N''(t)$.

$$N(t) = \frac{250,000\,t^2}{(2t+1)^2}, \; t > 0$$

Many important real-world applications make use of the second derivative. In Chapter 2, we will examine applications in the fields of economics, health care, and the natural and physical sciences.

TECHNOLOGY CONNECTION

Exploratory

A calculator with a tangent-drawing feature can be used to explore the behavior of the second derivative. Graph $f(x) = x^2$ in the standard window, select DRAW, and then select Tangent. Choose an x-value by keying it in or using the arrow keys to trace along the graph. Press **ENTER**, and a tangent line will be drawn at the selected x-value. In the lower-left corner of the screen, the equation of the tangent line is given. The slope of the line is the coefficient of x. You can create a table of slope values at various x-values, and from this table, see how the second derivative describes the shape of a graph. (This activity can be easily adapted for Graphicus.)

EXERCISES

1. Let $f(x) = x^2$. Use the tangent-drawing feature to complete the following table:

x	-4	-2	0	2	4
Slope at x					

2. The x-values in the table are increasing. What is the corresponding behavior of the slopes (increasing or decreasing)?

3. Is the graph bending up or down? What conclusion can you make about the behavior of the slopes as x increases in value?

4. Make a table of slopes for $f(x) = -x^2$. Analyze their behavior relative to the shape of this function's graph.

5. What general conclusion can you make about the second derivative and the shape of a graph?

Section Summary

- The *second derivative* is the derivative of the first derivative of a function. In symbols, $f''(x) = [f'(x)]'$.
- The second derivative represents the rate of change of the rate of change. In other words, it represents the rate of change of the first derivative.
- A real-life example of a second derivative is *acceleration*. If $s(t)$ represents distance traveled as a function of time for a moving object, then $v(t) = s'(t)$ represents speed (velocity). Any change in speed is the acceleration: $a(t) = v'(t) = s''(t)$.
- The common notation for the nth derivative of a function is $f^{(n)}(x)$, or $\dfrac{d^n}{dx^n} f(x)$.

1.8 Exercise Set

Find d^2y/dx^2.

1. $y = x^4 - 7$

2. $y = x^5 + 9$

3. $y = 2x^4 - 5x$

4. $y = 5x^3 + 4x$

5. $y = 4x^2 - 5x + 7$

6. $y = 4x^2 + 3x - 1$

7. $y = 7x + 2$

8. $y = 6x - 3$

9. $y = \dfrac{1}{x^3}$

10. $y = \dfrac{1}{x^2}$

11. $y = \sqrt{x}$

12. $y = \sqrt[4]{x}$

Find $f''(x)$.

13. $f(x) = x^3 - \dfrac{5}{x}$

14. $f(x) = x^4 + \dfrac{3}{x}$

15. $f(x) = x^{1/5}$

16. $f(x) = x^{1/3}$

17. $f(x) = 2x^{-2}$

18. $f(x) = 4x^{-3}$

19. $f(x) = (x^2 + 3x)^7$

20. $f(x) = (x^3 + 2x)^6$

21. $f(x) = (3x^2 + 2x + 1)^5$

22. $f(x) = (2x^2 - 3x + 1)^{10}$

23. $f(x) = \sqrt[4]{(x^2 + 1)^3}$

24. $f(x) = \sqrt[3]{(x^2 - 1)^2}$

Find y''.

25. $y = x^{3/2} - 5x$

26. $y = x^{2/3} + 4x$

27. $y = (x^3 - x)^{3/4}$

28. $y = (x^4 + x)^{2/3}$

29. $y = 3x^{4/3} - x^{1/2}$

30. $y = 2x^{5/4} + x^{1/2}$

31. $y = \dfrac{2}{x^3} + \dfrac{1}{x^2}$

32. $y = \dfrac{3}{x^4} - \dfrac{1}{x}$

33. $y = (x^3 - 2)(5x + 1)$

34. $y = (x^2 + 3)(4x - 1)$

35. $y = \dfrac{3x + 1}{2x - 3}$

36. $y = \dfrac{2x + 3}{5x - 1}$

37. For $y = x^5$, find d^4y/dx^4.

38. For $y = x^4$, find d^4y/dx^4.

39. For $y = x^6 - x^3 + 2x$, find d^5y/dx^5.

40. For $y = x^7 - 8x^2 + 2$, find d^6y/dx^6.

41. For $f(x) = x^{-3} + 2x^{1/3}$, find $f^{(5)}(x)$.

42. For $f(x) = x^{-2} - x^{1/2}$, find $f^{(4)}(x)$.

43. For $g(x) = x^4 - 3x^3 - 7x^2 - 6x + 9$, find $g^{(6)}(x)$.

44. For $g(x) = 6x^5 + 2x^4 - 4x^3 + 7x^2 - 8x + 3$, find $g^{(7)}(x)$.

APPLICATIONS

Life and Physical Sciences

45. Given
$$s(t) = -10t^2 + 2t + 5,$$
where $s(t)$ is in meters and t is in seconds, find each of the following.
a) $v(t)$
b) $a(t)$
c) The velocity and acceleration when $t = 1$ sec

46. Given
$$s(t) = t^3 + t,$$
where $s(t)$ is in feet and t is in seconds, find each of the following.
a) $v(t)$
b) $a(t)$
c) The velocity and acceleration when $t = 4$ sec

47. Given
$$s(t) = 3t + 10,$$
where $s(t)$ is in miles and t is in hours, find each of the following.
a) $v(t)$
b) $a(t)$
c) The velocity and acceleration when $t = 2$ hr
d) When the distance function is given by a linear function, we have *uniform motion*. What does uniform motion mean in terms of velocity and acceleration?

48. Given
$$s(t) = t^2 - \dfrac{1}{2}t + 3,$$
where $s(t)$ is in meters and t is in seconds, find each of the following.
a) $v(t)$
b) $a(t)$
c) The velocity and acceleration when $t = 1$ sec

49. Free fall. When an object is dropped, the distance it falls in t seconds, assuming negligible air resistance, is given by
$$s(t) = 16t^2,$$
where $s(t)$ is in feet. Suppose a medic's reflex hammer falls from a hovering helicopter. Find **(a)** how far the hammer falls in 3 sec, **(b)** how fast the hammer is traveling 3 sec after being dropped, and **(c)** the hammer's acceleration after it has been falling for 3 sec.

50. Free fall. (See Exercise 49.) Suppose a worker drops a bolt from a scaffold high above a work site. Assuming negligible air resistance, find **(a)** how far the bolt falls in 2 sec, **(b)** how fast the bolt is traveling 2 sec after being dropped, and **(c)** the bolt's acceleration after it has been falling for 2 sec.

51. Free fall. Find the velocity and acceleration of the stone in Example 3 after it has been falling for 2 sec.

52. Free fall. Find the velocity and acceleration of the stone in Example 3 after it has been falling for 3 sec.

53. The following graph describes a bicycle racer's distance from a roadside television camera.

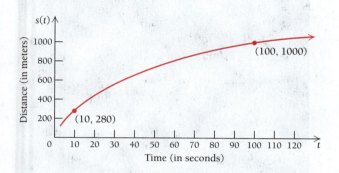

a) When is the bicyclist's velocity the greatest? How can you tell?

b) Is the bicyclist's acceleration positive or negative? How can you tell?

54. The following graph describes an airplane's distance from its last point of rest.

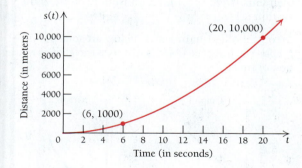

a) Is the plane's velocity greater at $t = 6$ sec or $t = 20$ sec? How can you tell?

b) Is the plane's acceleration positive or negative? How can you tell?

55. Sales. The following graph represents the sales, y, of a new video game after t weeks on the market.

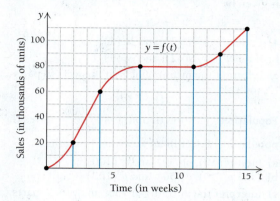

a) In what open interval(s) is $f'(t) = 0$?

b) In what open interval(s) is $f''(t) = 0$?

c) In what open interval(s) is $f''(t) > 0$?

d) In what open interval(s) is $f''(t) < 0$?

e) Describe in words and graphically the difference in meaning between "sales are increasing" and "the rate of sales is increasing."

56. Velocity and acceleration. The following graph describes the distance of Jesse's car from home as a function of time t.

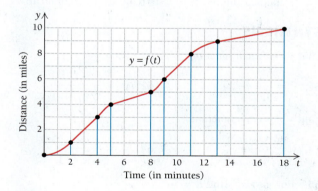

a) In what open interval(s) is Jesse's car accelerating?

b) In what open interval(s) is Jesse's car decelerating?

c) In what open interval(s) is Jesse's car maintaining a constant velocity?

57. Sales. A company determines that monthly sales $S(t)$, in thousands of dollars, after t months of marketing a product are given by

$$S(t) = 2t^3 - 40t^2 + 220t + 160.$$

a) Find $S'(1)$, $S'(2)$, and $S'(4)$.

b) Find $S''(1)$, $S''(2)$, and $S''(4)$.

c) Interpret the meaning of your answers to parts (a) and (b).

58. Sales. Nadia's fashions discovers that the number of items sold t days after launching a new sales promotion is given by

$$N(t) = 2t^3 - 3t^2 + 2t.$$

a) Find $N'(1)$, $N'(2)$, and $N'(4)$.

b) Find $N''(1)$, $N''(2)$, and $N''(4)$.

c) Interpret the meaning of your answers to parts (a) and (b).

59. Population. The function $p(t) = \dfrac{2000t}{4t + 75}$ gives the

population of deer in an area after t months.

a) Find $p'(10)$, $p'(50)$, and $p'(100)$.

b) Find $p''(10)$, $p''(50)$, and $p''(100)$.

c) Interpret the meaning of your answers to parts (a) and (b). What is happening to this population of deer in the long term?

60. Medicine. A medication is injected into the blood-stream, where it is quickly metabolized. The percent concentration p of the medication after t minutes in the bloodstream is modeled by the function $p(t) = \dfrac{2.5t}{t^2 + 1}$.

a) Find $p'(0.5)$, $p'(1)$, $p'(5)$, and $p'(30)$.

b) Find $p''(0.5)$, $p''(1)$, $p''(5)$, and $p''(30)$.

c) Interpret the meaning of your answers to parts (a) and (b). What is happening to the concentration of medication in the bloodstream in the long term?

SYNTHESIS

Find y''' for each function.

61. $y = \dfrac{1}{1 - x}$

62. $y = \dfrac{1}{\sqrt{2x + 1}}$

Find y'' for each function.

63. $y = \dfrac{\sqrt{x} + 1}{\sqrt{x} - 1}$

64. $y = \dfrac{x}{\sqrt{x} - 1}$

65. For $y = x^k$, find d^5y/dx^5.

66. For $y = ax^3 + bx^2 + cx + d$, find d^3y/dx^3.

Find the first through the fourth derivatives. Be sure to simplify each derivative before continuing.

67. $f(x) = \dfrac{x - 1}{x + 2}$

68. $f(x) = \dfrac{x + 3}{x - 2}$

69. Free fall. On Earth, all free-fall distance functions are of the form $s(t) = 4.905t^2$, where t is in seconds and $s(t)$ is in meters. The second derivative always has the same value. What does that value represent?

70. Free fall. On the moon, all free-fall distance functions are of the form $s(t) = 0.81t^2$, where t is in seconds and $s(t)$ is in meters. An object is dropped from a height of 200 meters above the moon. After $t = 2$ sec,

a) How far has the object fallen?

b) How fast is it traveling?

c) What is its acceleration?

d) Explain the meaning of the second derivative of this free-fall function.

71. Hang time. On Earth, an object travels 4.905 m after 1 sec of free fall. Thus, by symmetry, an athlete would require 1 sec to jump 4.905 m high, and another second to come back down. Is it possible for a person to stay in the air for (have a "hang time" of) 2 sec? Can a person have a hang time of 1.5 sec? 1 sec? What do you think is the longest possible hang time achievable by humans jumping from level ground?

72. Free fall. Skateboarder Danny Way free-fell 28 ft from the Fender Stratocaster Guitar atop the Hard Rock Hotel & Casino in Las Vegas onto a ramp below. The distance $s(t)$, in feet, traveled by a body falling freely from rest in t seconds is approximated by $s(t) = 16t^2$. Estimate Way's velocity at the moment he touched down onto the ramp. (*Note:* Use the result from Exercise 26 in Section R.1.)

73. An object rolls 1 m in 1 min. Below are four possible graphs showing the object's distance traveled as a function of time.

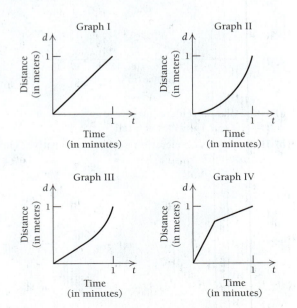

Match each graph with one of the following descriptions:

a) The object rolled unimpeded along a level table.

b) The object rolled along a level concrete walkway, then rolled along a level grassy lawn.

c) The object rolled down an incline.

d) The object rolled along a level table, then down an incline.

74. A bicyclist's distance from her starting point is given by $y = s(t)$. Suppose the graph of s has a corner. Give three situations in which a corner, rather than a smooth curve, can occur. What can you conclude about $s'(t)$ and $s''(t)$ at the corner of the graph of s?

Indeterminate forms and l'Hôpital's Rule. *Let f and g be differentiable over an open interval containing $x = a$. If*

$$\lim_{x \to a}\left(\frac{f(x)}{g(x)}\right) = \frac{0}{0} \quad \text{or} \quad \lim_{x \to a}\left(\frac{f(x)}{g(x)}\right) = \frac{\pm\infty}{\pm\infty}$$

and if $\lim\limits_{x \to a}\left(\dfrac{f'(x)}{g'(x)}\right)$ exists, then

$$\lim_{x \to a}\left(\frac{f(x)}{g(x)}\right) = \lim_{x \to a}\left(\frac{f'(x)}{g'(x)}\right).$$

The forms $0/0$ and $\pm\infty/\pm\infty$ are said to be indeterminate. In such cases, the limit may exist, and l'Hôpital's Rule offers a way to find the limit using differentiation. For example, in Example 1 of Section 1.1, we showed that

$$\lim_{x \to 1}\left(\frac{x^2 - 1}{x - 1}\right) = 2.$$

Since, for $x = 1$, we have $(x^2 - 1)(x - 1) = 0/0$, we differentiate the numerator and denominator separately, and reevaluate the limit:

$$\lim_{x \to 1}\left(\frac{x^2 - 1}{x - 1}\right) = \lim_{x \to 1}\left(\frac{2x}{1}\right) = 2.$$

Use this method to find the following limits. Be sure to check that the initial substitution results in an indeterminate form.

75. $\lim\limits_{x \to 5}\left(\dfrac{x^2 - 25}{2x - 10}\right)$ **76.** $\lim\limits_{x \to -2}\left(\dfrac{x^2 - 4}{x + 2}\right)$

77. $\lim\limits_{x \to 1}\left(\dfrac{x^3 + 2x - 3}{x^2 - 1}\right)$ **78.** $\lim\limits_{x \to 3}\left(\dfrac{x^3 - x - 24}{x^2 - 9}\right)$

79. $\lim\limits_{x \to -3}\left(\dfrac{x^2 - 9}{x + 3}\right)$ **80.** $\lim\limits_{x \to -4}\left(\dfrac{x^2 + x - 12}{x + 4}\right)$

81. $\lim\limits_{x \to 2}\left(\dfrac{x^3 + 5x - 18}{2x^2 - 8}\right)$ **82.** $\lim\limits_{x \to 10}\left(\dfrac{x^2 + x - 110}{x - 10}\right)$

83. $\lim\limits_{x \to \infty}\left(\dfrac{4x^2 + x - 3}{2x^2 + 1}\right)$ **84.** $\lim\limits_{x \to -\infty}\left(\dfrac{3x^3 + x + 11}{6x^3 + x + 2}\right)$

TECHNOLOGY CONNECTION

For the distance function in each of Exercises 85–88, graph s, v, and a over the given interval. Then use the graphs to determine the point(s) at which the velocity switches from increasing to decreasing or from decreasing to increasing.

85. $s(t) = 0.1t^4 - t^2 + 0.4;\quad [-5, 5]$

86. $s(t) = -t^3 + 3t;\quad [-3, 3]$

87. $s(t) = t^4 + t^3 - 4t^2 - 2t + 4;\quad [-3, 3]$

88. $s(t) = t^3 - 3t^2 + 2;\ [-2, 4]$

Answers to Quick Checks

1. (a) (i) $y'' = -72x^2 + 6$, (ii) $y'' = \dfrac{24}{x^5}$,

(iii) $y'' = 108x^2 + 12$; **(b)** $y^{(4)} = \dfrac{24}{x^5}$

2. Distance $= s(3.5) = 196$ ft; velocity $= s'(3.5) = 112$ ft/sec; acceleration $= s''(3.5) = 32$ ft/sec^2

KEY TERMS AND CONCEPTS

EXAMPLES

SECTION 1.1

As x approaches (but is not equal to) a, the **limit** of $f(x)$, if it exists, is L, which is written as

$$\lim_{x \to a} f(x) = L.$$

$$\lim_{x \to 4} (2x + 3) = 11 \qquad\qquad \lim_{x \to 1} \frac{x^2 - 1}{x - 1} = 2$$

Limit Numerically   *Limit Graphically*

	x	$2x + 3$
$x < 4$	3.9	10.8
	3.99	10.98
	3.999	10.998
$x > 4$	4.1	11.2
	4.01	11.02
	4.001	11.002

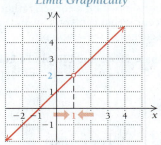

If x approaches a from the left $(x < a)$, and a **left-hand limit** exists, it is written as

$$\lim_{x \to a^-} f(x).$$

If x approaches a from the right $(x > a)$, and a **right-hand limit** exists, it is written as

$$\lim_{x \to a^+} f(x).$$

If the left-hand and right-hand limits are equal as x approaches a, then the limit as x approaches a exists.

If the left-hand and right-hand limits are *not* equal, then the limit as x approaches a does *not* exist.

Consider the function G given by

$$G(x) = \begin{cases} 4 - x, & \text{for } x < 3, \\ \sqrt{x - 2} + 1, & \text{for } x \geq 3. \end{cases}$$

Graph the function and find each limit, if it exists.

a) $\displaystyle\lim_{x \to 1} G(x)$ **b)** $\displaystyle\lim_{x \to 3} G(x)$

   We check the limits from the left and from the right, both numerically and graphically.

a)   *Limit Numerically*   *Limit Graphically*

$x \to 1^-$ $(x < 1)$	$G(x)$
0.9	3.1
0.99	3.01
0.999	3.001

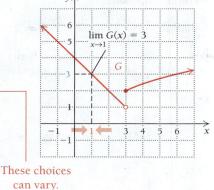

$x \to 1^+$ $(x > 1)$	$G(x)$
1.1	2.9
1.01	2.99
1.001	2.999

These choices can vary.

Both the tables and the graph show that as x gets closer to 1, the outputs $G(x)$ get closer to 3. Thus, $\displaystyle\lim_{x \to 1} G(x) = 3$.

| **KEY TERMS AND CONCEPTS** | **EXAMPLES** |

SECTION 1.1 (continued)

b)

| *Limit Numerically* | | *Limit Graphically* |

$x \to 3^-$ $(x < 3)$	$G(x)$
2.9	1.1
2.99	1.01
2.999	1.001

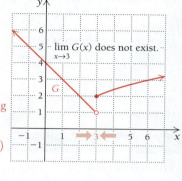

$\lim\limits_{x \to 3} G(x)$ does not exist.

G

$x \to 3^+$ $(x > 3)$	$G(x)$
3.1	2.0488
3.01	2.0050
3.001	2.0005

These are approaching different values, so $\lim\limits_{x \to 3} G(x)$ does not exist.

Both the tables and the graph indicate that $\lim\limits_{x \to 3^-} G(x) \neq \lim\limits_{x \to 3^+} G(x)$. Since the left-hand and right-hand limits differ, $\lim\limits_{x \to 3} G(x)$ does not exist.

SECTION 1.2

For any rational function F (see Section R.5) with a in its domain, we have

$$\lim_{x \to a} F(x) = F(a).$$

For $f(x) = 2x^2 + 3x - 1$ and $a = 2$, we have

$$\lim_{x \to 2} f(x) = f(2) = 2(2)^2 + 3(2) - 1 = 13.$$

For $g(x) = \dfrac{x^2 - 16}{x + 4}$ and $a = 6$, we have

$$\lim_{x \to 6} g(x) = \frac{(6)^2 - 16}{(6) + 4} = \frac{20}{10} = 2.$$

A function f is **continuous** at $x = a$ if the following three conditions are met:

1. $f(a)$ exists. (The output at a exists.)
2. $\lim\limits_{x \to a} f(x)$ exists. (The limit as $x \to a$ exists.)
3. $\lim\limits_{x \to a} f(x) = f(a)$. (The limit is the same as the output.)

If any one of these conditions is not fulfilled, the function is **discontinuous** at $x = a$.

Is the function g given by $g(x) = \dfrac{x^2 - 3x - 4}{x + 1}$ continuous over $[-3, 3]$?

For g to be continuous over $[-3, 3]$, it must be continuous at each point in $[-3, 3]$. Note that

$$g(x) = \frac{x^2 - 3x - 4}{x + 1}$$
$$= \frac{(x + 1)(x - 4)}{x + 1}$$
$$= x - 4, \quad \text{provided } x \neq -1.$$

Since -1 is not in the domain of g, it follows that $g(-1)$ does not exist. Thus, g is not continuous over $[-3, 3]$.

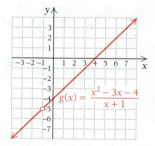

$g(x) = \dfrac{x^2 - 3x - 4}{x + 1}$

SECTION 1.3

The **average rate of change** of y with respect to x between two points (x_1, y_1) and (x_2, y_2) is the slope of the line connecting the points:

$$\frac{y_2 - y_1}{x_2 - x_1}.$$

Business. At 1 P.M., a bookstore had revenue of \$570 for the day, and at 4 P.M., it had revenue of \$900 for the day. Therefore, the average rate of change of revenue with respect to time is

$$\frac{900 - 570}{4 - 1} = \frac{330}{3} = 110,$$

or \$110 dollars per hour for the period of time between 1 P.M. and 4 P.M.

(continued)

KEY TERMS AND CONCEPTS	EXAMPLES

SECTION 1.3 (continued)

Let f be a function. Any line connecting two points on the graph of f is called a **secant line**. Its slope is the average rate of change of f, which is given by the **difference quotient**:

$$\frac{f(x + h) - f(x)}{h},$$

where h is the difference between the two input x-values.

Let $f(x) = 3x^2$. Then

$$f(x + h) = 3(x + h)^2 = 3x^2 + 6xh + 3h^2.$$

The difference quotient for this function simplifies to

$$\frac{f(x + h) - f(x)}{h} = \frac{3x^2 + 6xh + 3h^2 - 3x^2}{h} = 6x + 3h, \quad h \neq 0.$$

Therefore, if $x = 2$ and $h = 0.05$, the slope of the secant line is $6(2) + 3(0.05) = 12.15$.

SECTION 1.4

The **derivative** of a function f is

$$f'(x) = \lim_{h \to 0} \frac{f(x + h) - f(x)}{h}.$$

The derivative, if it exists, gives the slope of the tangent line to f at $x = a$, and that slope is the **instantaneous rate of change** at x. The process of finding a derivative is called **differentiation**.

If $f'(a)$ exists, then f is **differentiable** at $x = a$.

Let $f(x) = 3x^2$. Its simplified difference quotient is $6x + 3h$, $h \neq 0$ (see above). Therefore, the derivative is

$$f'(x) = \lim_{h \to 0} (6x + 3h) = 6x.$$

The slope of the line tangent to f at $x = 2$ is

$$f'(2) = 6(2) = 12.$$

For $f(x) = -x^2 + 5$, find $f'(x)$ and $f'(2)$.
We have

$$\frac{f(x + h) - f(x)}{h} = \frac{-(x + h)^2 + 5 - (-x^2 + 5)}{h}$$

$$= \frac{-(x^2 + 2xh + h^2) + 5 + x^2 - 5}{h}$$

$$= \frac{-2xh - h^2}{h}$$

$$= -2x - h, \quad h \neq 0.$$

Since

$$f'(x) = \lim_{h \to 0} \frac{f(x + h) - f(x)}{h} = \lim_{h \to 0} (-2x - h),$$

we have

$$f'(x) = -2x.$$

It follows that $f'(2) = -2 \cdot 2 = -4$ and f is differentiable at $x = 2$. In fact, f is differentiable for all real numbers x.

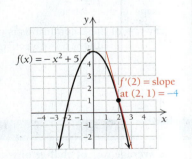

KEY TERMS AND CONCEPTS	**EXAMPLES**

SECTION 1.4 (*continued*)

Continuity (see Section 1.2):

1. If a function f is differentiable at $x = a$, then it is continuous at $x = a$. (Differentiability implies continuity.)

2. Continuity of a function f at $x = a$ does *not* necessarily mean that f is differentiable at $x = a$. Any function whose graph has a corner is continuous but not differentiable at the corner.

3. If a function f is discontinuous at $x = a$, then it is not differentiable at $x = a$.

1. Let $f(x) = 3x^2$. Since we know the derivative is $f'(x) = 6x$ and the derivative at $x = 2$ exists, we can conclude that $f(x)$ is continuous at $x = 2$.

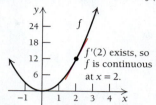

2. The absolute-value function is continuous at $x = 0$ but not differentiable at $x = 0$, since there is a corner at $x = 0$.

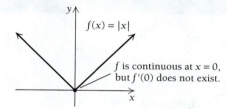

3. The function $g(x) = \dfrac{x^2 - 16}{x + 4}$ is discontinuous at $x = -4$; therefore, g is not differentiable at $x = -4$. (Note that the derivative is defined at other values of x.)

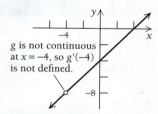

SECTION 1.5

If $y = f(x)$, the derivative in Leibniz notation is written $\dfrac{dy}{dx}$ or $\dfrac{d}{dx}f(x)$. Each has the same meaning as $f'(x)$.

Let $y = x^3$. Then, in Leibniz notation, $\dfrac{dy}{dx} = 3x^2$.

The **Power Rule:**

For any real number k,

$$\frac{d}{dx}x^k = k \cdot x^{k-1}.$$

- $\dfrac{d}{dx}x^7 = 7x^6$

- $\dfrac{d}{dx}\sqrt{x} = \dfrac{d}{dx}x^{1/2} = \dfrac{1}{2}x^{-1/2} = \dfrac{1}{2\sqrt{x}}$

- $\dfrac{d}{dx}\left(\dfrac{1}{x}\right) = \dfrac{d}{dx}x^{-1} = -1 \cdot x^{-2} = -\dfrac{1}{x^2}$

The derivative of a constant is

$$\frac{d}{dx}c = 0.$$

- $\dfrac{d}{dx}34 = 0$

- $\dfrac{d}{dx}\sqrt{2} = 0$

(*continued*)

KEY TERMS AND CONCEPTS	**EXAMPLES**

SECTION 1.5 (continued)

The derivative of a constant times a function is

$$\frac{d}{dx}[c \cdot f(x)] = c \cdot \frac{d}{dx}f(x).$$

- $\frac{d}{dx}3x^8 = 3 \cdot \frac{d}{dx}x^8 = 3 \cdot 8x^7 = 24x^7$
- $\frac{d}{dx}\left(\frac{2}{3x^2}\right) = \frac{2}{3} \cdot \frac{d}{dx}\left(\frac{1}{x^2}\right) = \frac{2}{3} \cdot (-2x^{-3}) = -\frac{4}{3x^3}$

The **Sum–Difference Rule**:

$$\frac{d}{dx}[f(x) \pm g(x)] = \frac{d}{dx}f(x) \pm \frac{d}{dx}g(x).$$

- $\frac{d}{dx}(x^7 + 3x) = 7x^6 + 3$
- $\frac{d}{dx}(5x - x^4) = 5 - 4x^3$

SECTION 1.6

The **Product Rule**:

$$\frac{d}{dx}[f(x) \cdot g(x)]$$
$$= f(x) \cdot g'(x) + g(x) \cdot f'(x).$$

$$\frac{d}{dx}\left[(2x + 3)\sqrt{x}\right] = (2x + 3)\frac{1}{2}x^{-1/2} + \sqrt{x} \cdot 2$$
$$= \frac{2x + 3}{2\sqrt{x}} + 2\sqrt{x}$$

The **Quotient Rule**:

$$\frac{d}{dx}\left[\frac{f(x)}{g(x)}\right] = \frac{g(x) \cdot f'(x) - f(x) \cdot g'(x)}{[g(x)]^2}.$$

$$\frac{d}{dx}\left(\frac{3x - 1}{2x + 5}\right) = \frac{(2x + 5)3 - (3x - 1)2}{(2x + 5)^2} = \frac{17}{(2x + 5)^2}$$

SECTION 1.7

The **Extended Power Rule**:

$$\frac{d}{dx}[g(x)]^k = k[g(x)]^{k-1} \cdot \frac{d}{dx}g(x).$$

$$\frac{d}{dx}(2x^5 + 4x)^7 = 7(2x^5 + 4x)^6(10x^4 + 4)$$
$$= 14(2x^5 + 4x)^6(5x^4 + 2)$$

The **Chain Rule**:

$$\frac{d}{dx}[(f \circ g)(x)] = \frac{d}{dx}[f(g(x))]$$
$$= f'(g(x)) \cdot g'(x).$$

- $\frac{d}{dx}\sqrt{3x^2 + 4} = \frac{d}{dx}(3x^2 + 4)^{1/2} = \frac{1}{2}(3x^2 + 4)^{-1/2}(6x) = \frac{3x}{\sqrt{3x^2 + 4}}$
- $\frac{d}{dx}\left(\frac{1}{8x + 1}\right) = \frac{d}{dx}(8x + 1)^{-1} = -1(8x + 1)^{-2}(8) = -\frac{8}{(8x + 1)^2}$

SECTION 1.8

The **second derivative** is the derivative of the first derivative:

$$\frac{d}{dx}[f'(x)] = [f'(x)]' = f''(x).$$

The second derivative indicates the rate of change of the derivative.

Let $f(x) = 2x^5 + 20$. Then

$$\frac{d}{dx}f(x) = f'(x) = 10x^4 + 20,$$

and therefore,

$$\frac{d^2}{dx^2}f(x) = f''(x) = 40x^3.$$

KEY TERMS AND CONCEPTS	EXAMPLES

SECTION 1.8 (continued)

Higher-order derivatives include second, third, fourth, and so on, derivatives of a function. The *n*th derivative of a function is written as

$$\frac{d^n y}{dx^n} = f^{(n)}(x).$$

For $y = 5x^7 + 20x$, we have the following:

$$\frac{d^3 y}{dx^3} = f^{(3)}(x) = 1050x^4,$$

$$\frac{d^4 y}{dx^4} = f^{(4)}(x) = 4200x^3,$$

$$\frac{d^5 y}{dx^5} = f^{(5)}(x) = 12{,}600x^2, \text{ and so on.}$$

A real-life application of the second derivative is **acceleration**. If $s(t)$ represents distance as a function of time, then velocity is $v(t) = s'(t)$ and acceleration is the change in velocity: $a(t) = v'(t) = s''(t)$.

Physical Sciences. A particle moves according to the distance function $s(t) = 5t^3$, where t is in seconds and $s(t)$ in feet. Therefore, the particle's velocity is given by $v(t) = s'(t) = 15t^2$ and its acceleration by $a(t) = v'(t) = s''(t) = 30t$. At $t = 2$ sec, the particle is $s(2) = 40$ ft from the starting point, traveling at $v(2) = 60$ ft/sec and accelerating at $a(2) = 60$ ft/sec^2 (it's speeding up).

CHAPTER 1
Review Exercises

These review exercises are for test preparation. They can also be used as a practice test. Answers are at the back of the book. The red bracketed section references indicate the section(s) to restudy if your answer is incorrect.

CONCEPT REINFORCEMENT

Classify each statement as either true or false.

1. If $\lim\limits_{x \to 5} f(x)$ exists, then $f(5)$ must exist. [1.1]

2. If $\lim\limits_{x \to 2} f(x) = L$, then $L = f(2)$. [1.1]

3. If f is continuous at $x = 3$, then $\lim\limits_{x \to 3} f(x) = f(3)$. [1.2]

4. A function's average rate of change over the interval $[2, 8]$ is the same as its instantaneous rate of change at $x = 5$. [1.3, 1.4]

5. A function's derivative at a point, if it exists, can be found as the limit of a difference quotient. [1.4]

6. For $f'(5)$ to exist, f must be continuous at 5. [1.4]

7. If f is continuous at 5, then $f'(5)$ must exist. [1.4]

8. The acceleration function is the derivative of the velocity function. [1.8]

Match each function in column A with the most appropriate rule to use for differentiating the function. [1.5, 1.6]

Column A

9. $f(x) = x^7$

10. $g(x) = x + 9$

11. $F(x) = (5x - 3)^4$

12. $G(x) = \dfrac{2x + 1}{3x - 4}$

13. $H(x) = f(x) \cdot g(x)$

14. $f(x) = 2x - 7$

Column B

a) Extended Power Rule

b) Product Rule

c) Sum Rule

d) Difference Rule

e) Power Rule

f) Quotient Rule

For Exercises 15–17, consider

$$\lim_{x \to -7} f(x), \text{ where } f(x) = \frac{x^2 + 4x - 21}{x + 7}.$$

15. **Limit numerically.** [1.1]

   a) Find the limit by completing the following input–output tables.

$x \to -7^-$	$f(x)$
-7.1	
-7.01	
-7.001	

$x \to -7^+$	$f(x)$
-6.9	
-6.99	
-6.999	

   b) Find $\lim_{x \to -7^-} f(x)$, $\lim_{x \to -7^+} f(x)$, and $\lim_{x \to -7} f(x)$, if each exists.

16. **Limit graphically.** Find the limit by graphing the function. [1.1]

17. **Limit algebraically.** Find the limit algebraically. Show your work. [1.2]

Find each limit, if it exists. If a limit does not exist, state that fact. [1.1, 1.2]

18. $\lim_{x \to -2} \dfrac{8}{x}$

19. $\lim_{x \to 1} (4x^3 - x^2 + 7x)$

20. $\lim_{x \to -7} \dfrac{x^2 + 2x - 35}{x + 7}$

21. $\lim_{x \to \infty} \dfrac{1}{x} + 3$

For Exercises 22–30, consider the function g graphed below.

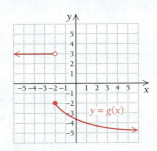

22. Find $\lim_{x \to 1} g(x)$. [1.1]

23. Find $g(1)$. [1.1]

24. Is g continuous at 1? Why or why not? [1.2]

25. Find $\lim_{x \to -2} g(x)$. [1.1]

26. Find $g(-2)$. [1.1]

27. Is g continuous at -2? Why or why not? [1.2]

28. Find the average rate of change between $x = -2$ and $x = 1$. [1.3]

29. Find $g'(-4)$. [1.4]

30. For which value(s) is $g'(x)$ not defined? Why? [1.4]

For Exercises 31–34, consider the function f graphed below.

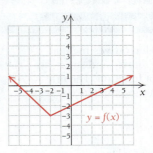

31. For which value(s) is $f'(x)$ not defined? Why? [1.4]

32. Find $\lim_{x \to 0} f(x)$. [1.1]

33. Find $\lim_{x \to -2} f(x)$. [1.1]

34. Is f continuous at $x = -2$? [1.4]

35. For $f(x) = x^3 + x^2 - 2x$, find the average rate of change as x changes from -1 to 2. [1.3]

36. Find a simplified difference quotient for $g(x) = -3x^2 + 2$. [1.3]

37. Find a simplified difference quotient for

   $$f(x) = 2x^5 - 3. \quad [1.3]$$

38. Find an equation of the tangent line to the graph of $y = x^2 + 3x$ at the point $(-1, -2)$. [1.4]

39. Find the point(s) on the graph of $y = -x^2 + 8x - 11$ at which the tangent line is horizontal. [1.5]

40. Find the point(s) on the graph of $y = 5x^2 - 49x + 12$ at which the tangent line has slope 1. [1.5]

Find dy/dx.

41. $y = 9x^5$ [1.5]

42. $y = 8\sqrt[3]{x}$ [1.5]

43. $y = \dfrac{-3}{x^8}$ [1.5]

44. $y = 15x^{2/5}$ [1.5]

45. $y = 0.1x^7 - 3x^4 - x^3 + 6$ [1.5]

Differentiate.

46. $f(x) = \dfrac{5}{12}x^6 + 8x^4 - 2x$ [1.5]

47. $y = (x^3 - 5)(\sqrt{x} + 4x)$ [1.5, 1.6]

48. $y = \dfrac{x^2 + 8}{8 - x}$ [1.6]

49. $g(x) = (5 - x)^2(2x - 1)^5$ [1.6]

50. $f(x) = (x^5 - 3)^7$ [1.7]

51. $f(x) = x^2(4x + 2)^{3/4}$ [1.7]

52. For $y = x^3 - \dfrac{2}{x}$, find $\dfrac{d^4y}{dx^4}$. [1.8]

53. For $y = \dfrac{3}{42}x^7 - 10x^3 + 13x^2 + 28x - 2$, find y''. [1.8]

54. Social science: growth rate. The population of Lawton grows from an initial size of 10,000 to a size P, given by $P = 10,000 + 50t^2$, where t is in years. [1.5]

a) Find the growth rate.

b) Find the number of people in Lawton after 20 yr (at $t = 20$).

c) Find the growth rate at $t = 20$.

For Exercises 55–58, consider the graph of $y = s(t)$, the distance a jogger has run after t minutes. [1.8]

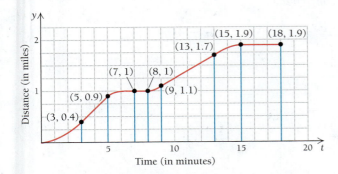

55. In what open interval(s) is the jogger running at a constant speed?

56. In what open interval(s) is she stopped?

57. In what open interval(s) is she accelerating?

58. In what open interval(s) is she decelerating?

59. For $s(t) = t + t^4$, with t in seconds and $s(t)$ in feet, find each of the following. [1.8]

a) $v(t)$

b) $a(t)$

c) The velocity and the acceleration when $t = 2$ sec

60. Business: average revenue, cost, and profit. Given revenue and cost functions $R(x) = 40x$ and $C(x) = 5\sqrt{x} + 100$, find each of the following. Assume $R(x)$ and $C(x)$ are in dollars and x is the number of lamps produced. [1.6]

a) The average cost, the average revenue, and the average profit when x lamps are produced and sold

b) The rate at which average cost is changing when 9 lamps are produced

61. Find $\dfrac{d}{dx}(f \circ g)(x)$ and $\dfrac{d}{dx}(g \circ f)(x)$, given $f(x) = x^2 + 5$ and $g(x) = 1 - 2x$. [1.7]

SYNTHESIS

62. Differentiate $y = \dfrac{x\sqrt{1 + 3x}}{1 + x^3}$. [1.7]

63. Find $\dfrac{d}{dx}(f \circ f \circ f \circ f \circ f)(x)$, where $f(x) = \sqrt[3]{x}$. [1.7]

TECHNOLOGY CONNECTION

Create an input–output table to determine each of the following limits. Start with ΔTbl $= 0.1$ and then go to 0.01, 0.001, and 0.0001. When you think you know the limit, graph the functions, and use TRACE *to verify your assertion.*

64. $\displaystyle\lim_{x \to 1} \dfrac{2 - \sqrt{x + 3}}{x - 1}$ [1.1, 1.5]

65. $\displaystyle\lim_{x \to 11} \dfrac{\sqrt{x - 2} - 3}{x - 11}$ [1.1, 1.5]

66. Graph f and f' over the given interval. Then estimate points at which the tangent line to f is horizontal. [1.5]

$$f(x) = 3.8x^5 - 18.6x^3; [-3, 3]$$

CHAPTER 1
Test

For Exercises 1–3, consider

$$\lim_{x \to 6} f(x), \text{ where } f(x) = \dfrac{x^2 - 36}{x - 6}.$$

1. Numerical limits.

a) Find the limit by completing the following input–output tables.

$x \to 6^-$	$f(x)$
5.9	
5.99	
5.999	

$x \to 6^+$	$f(x)$
6.1	
6.01	
6.001	

b) Find $\displaystyle\lim_{x \to 6^-} f(x)$, $\displaystyle\lim_{x \to 6^+} f(x)$, and $\displaystyle\lim_{x \to 6} f(x)$, if each exists.

2. Graphical limits. Find the limit by graphing the function.

3. Algebraic limits. Find the limit algebraically. Show your work.

Graphical limits.   *For Exercises 4–15, consider the function f graphed below.*

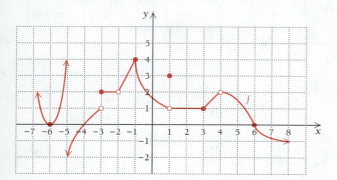

4. $\lim\limits_{x \to -5} f(x)$

5. $\lim\limits_{x \to -4} f(x)$

6. $\lim\limits_{x \to -3} f(x)$

7. $\lim\limits_{x \to -2} f(x)$

8. $\lim\limits_{x \to -1} f(x)$

9. $\lim\limits_{x \to 1} f(x)$

10. $\lim\limits_{x \to 2} f(x)$

11. $\lim\limits_{x \to 4} f(x)$

12. Find $f'(2)$.

13. Find $f'(-6)$.

14. State the value(s) of x at which f is not continuous.

15. State the value(s) of x for which $f'(x)$ is not defined.

Determine whether each function is continuous. If a function is not continuous, state why.

16.

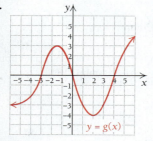

17.

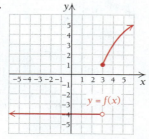

For Exercises 18 and 19, consider the function shown in Exercise 17.

18. a) Find $\lim\limits_{x \to 3} f(x)$.

    b) Find $f(3)$.

    c) Is f continuous at 3?

19. Find $\lim\limits_{x \to 4} f(x)$.

Find each limit, if it exists. If a limit does not exist, state why.

20. $\lim\limits_{x \to 1} (3x^4 - 2x^2 + 5)$

21. $\lim\limits_{x \to 2^+} \dfrac{x - 2}{x(x^2 - 4)}$

22. $\lim\limits_{x \to 0} \dfrac{7}{x}$

23. Find the simplified difference quotient for $f(x) = 2x^2 + 3x - 9$.

24. Find an equation of the line tangent to $y = x + (4/x)$ at the point $(4, 5)$.

25. Find the point(s) on the graph of $y = x^3 - 3x^2$ at which the tangent line is horizontal.

Find dy/dx.

26. $y = x^{23}$

27. $y = 4\sqrt[3]{x} + 5\sqrt{x}$

28. $y = \dfrac{-10}{x}$

29. $y = x^{5/4}$

30. $y = -0.5x^2 + 0.61x + 90$

Differentiate.

31. $y = \dfrac{1}{3}x^3 - x^2 + 2x + 4$

32. $y = (3\sqrt{x} + 1)(x^2 - x)$

33. $f(x) = \dfrac{x}{5 - x}$

34. $f(x) = (x + 3)^4(7 - x)^5$

35. $y = (x^5 - 4x^3 + x)^{-5}$

36. $f(x) = x\sqrt{x^2 + 5}$

37. For $y = x^4 - 3x^2$, find $\dfrac{d^3y}{dx^3}$.

38. Social sciences: memory.   In a certain memory experiment, a person is able to memorize M words after t minutes, where $M = -0.001t^3 + 0.1t^2$.

    a) Find the rate of change of the number of words memorized with respect to time.

    b) How many words are memorized during the first 10 min (at $t = 10$)?

    c) At what rate are words being memorized after 10 min (at $t = 10$)?

39. Business: average revenue, cost, and profit.   Given revenue and cost functions

$$R(x) = 50x \quad \text{and} \quad C(x) = x^{2/3} + 750,$$

where x is the number of Bluetooth speakers produced and $R(x)$ and $C(x)$ are in dollars, find the following:

    a) the average revenue, the average cost, and the average profit when x speakers are produced;

    b) the rate at which average cost is changing when 8 speakers are produced.

For Exercises 40 and 41, let $f(x) = x^2 - x$ and $g(x) = 2x^3$.

40. Find $\dfrac{d}{dx}(f \circ g)(x)$.

41. Find $\dfrac{d}{dx}(g \circ f)(x)$.

42. A ball is placed on an inclined plane and, due to gravity alone, accelerates down the plane. Let $y = s(t)$ represent the ball's distance t seconds after starting to roll. Which graph below best represents s?

a)

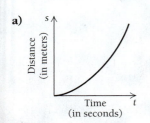

b)

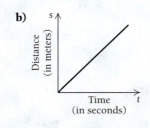

c)

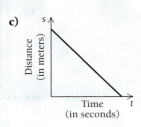

d)

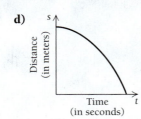

SYNTHESIS

43. Differentiate $y = \sqrt{(1 - 3x)^{2/3}(1 + 3x)^{1/3}}$.

44. Find $\displaystyle\lim_{x \to 3} \frac{x^3 - 27}{x - 3}$.

TECHNOLOGY CONNECTION

45. Graph f and f' over the interval $[0, 5]$. Then estimate points at which the line tangent to f is horizontal.
$$f(x) = 5x^3 - 30x^2 + 45x + 5\sqrt{x}; \quad [0, 5]$$

46. Find the following limit by creating a table of values:
$$\lim_{x \to 0} \frac{\sqrt{5x + 25} - 5}{x}.$$

Start with $\Delta\text{Tbl} = 0.1$ and then go to 0.01 and 0.001. When you think you know the limit, graph
$$y = \frac{\sqrt{5x + 25} - 5}{x},$$

and use TRACE to verify your assertion.

EXTENDED TECHNOLOGY APPLICATION

Path of a Baseball: The Tale of the Tape

Have you ever watched a baseball game and seen a home run ball hit an obstruction after it has cleared the fence? Suppose a ball hits a scoreboard 60 ft above the ground and 400 ft from home plate. An announcer or a message on the scoreboard might proclaim, "According to the tale of the tape, the ball would have traveled 442 ft." How is such a calculation made? The answer is related to the curve formed by the path of a baseball.

The path of a well-hit baseball is *not* the graph of a parabola,
$$f(x) = ax^2 + bx + c.$$

A well-hit baseball follows the path of a "skewed" parabola, as shown at the lower right. One reason that the ball's flight is not parabolic is that it has backspin. This fact, combined with the frictional effect of the ball's stitches with the air, skews the path of the ball in the direction of its landing.

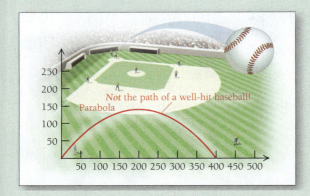

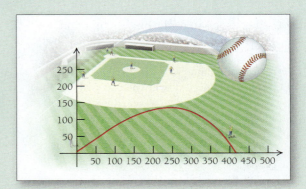

Let's see if we can model the path of a baseball. Consider the following data.

HORIZONTAL DISTANCE, x (In feet)	VERTICAL DISTANCE, y (In feet)
0	4.5
50	43
100	82
200	130
285	142
300	134
360	100
400	60

Assume for the given data that $(0, 4.5)$ is the point at home plate at which the ball is hit, roughly 4.5 ft above the ground. Also, assume the ball hits a billboard 60 ft above the ground and 400 ft from home plate.

Exercises

1. Plot the points and connect them with line segments. This can be done on many calculators by pressing STAT PLOT, turning on PLOT, and selecting the appropriate TYPE.

2. **a)** Use REGRESSION to find a cubic function

   $$y = ax^3 + bx^2 + cx + d$$

   that fits the data.
   b) Graph the function over the interval $[0, 500]$.
   c) Does the function closely model the given data?
   d) Predict the horizontal distance from home plate at which the ball would have hit the ground had it not hit the billboard.
   e) Find the rate of change of the ball's height with respect to its horizontal distance from home plate.
   f) Find the point(s) at which the graph has a horizontal tangent line. Explain the significance of the point(s).

3. **a)** Use REGRESSION to find a quartic function

   $$y = ax^4 + bx^3 + cx^2 + dx + e$$

   that fits the data.
   b) Graph the function over the interval $[0, 500]$.

c) Does the function closely model the given data?
d) Predict the horizontal distance from home plate at which the ball would have hit the ground had it not hit the billboard.
e) Find the rate of change of the ball's height with respect to its horizontal distance from home plate.
f) Find the point(s) at which the graph has a horizontal tangent line. Explain the significance of the point(s).

4. **a)** Although most calculators cannot fit such a function to the data, assume that the equation

   $$y = 0.0015x\sqrt{202{,}500 - x^2}$$

   has been found using a curve-fitting technique. Graph the function over the interval $[0, 500]$.
   b) Predict the horizontal distance from home plate at which the ball would have hit the ground had it not hit the billboard.
   c) Find the rate of change of the ball's height with respect to its horizontal distance from home plate.
   d) Find the point(s) at which the graph has a horizontal tangent line. Explain the significance of the point(s).

5. Look at the answers in Exercises 2(d), 3(d), and 4(b). Compare the merits of the quartic model in Exercise 3 to those of the model in Exercise 4.

Tale of the tape.   Actually, scoreboard operators in the major leagues use different models to predict the distance that a home run ball would have traveled. The models are linear and are related to the trajectory of the ball, that is, how high the ball is hit. See the following graph.

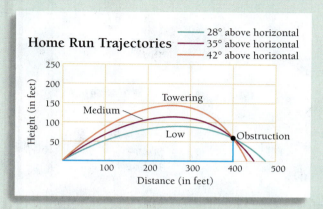

Home Run Trajectories

Suppose a ball hits an obstruction d feet horizontally from home plate at a height of H feet. Then the estimated horizontal distance D that the ball would have traveled, depending on its trajectory type, is

Low trajectory:	$D = 1.1H + d,$
Medium trajectory:	$D = 0.7H + d,$
Towering trajectory:	$D = 0.5H + d.$

Exercises

6. For a ball striking an obstacle at $d = 400$ ft and $H = 60$ ft, estimate how far the ball would have traveled if it were following a low trajectory, a medium trajectory, or a towering trajectory.

7. In 1953, Hall-of-Famer Mickey Mantle hit a towering home run in old Griffith Stadium in Washington, D.C., that hit an obstruction 60 ft high and 460 ft from home plate. Reporters asserted at the time that the ball would have traveled 565 ft. Is this estimate valid?

8. Use the appropriate formula to estimate the distance D for each of the following famous long home runs.

   a) Ted Williams (Boston Red Sox, June 9, 1946): Purportedly the longest home run ball ever hit to right field at Boston's Fenway Park, Williams's ball landed in the stands 502 feet from home plate, 30 feet above the ground. Assume a medium trajectory.

   b) Reggie Jackson (Oakland Athletics, July 13, 1971): Jackson's mighty blast hit an electrical transformer on top of the right-field roof at old Tiger Stadium in the 1971 All-Star Game. The transformer was 380 feet from home plate, 100 feet up. Find the distance the ball would have traveled, assuming a low trajectory and then a medium trajectory. (Jackson's home-run ball left the bat at an estimated 31.5° angle, about halfway between the low and medium range.)

   c) Richie Sexson (Arizona Diamondbacks, April 26, 2004): Sexson hit a drive that caromed off the center-field scoreboard at Bank One Ballpark in Phoenix. The scoreboard is 414 feet from home plate and 75 feet high. Assume a medium trajectory.

The reported distances these balls would have traveled are 527 feet for Williams's home run, 530 feet for Jackson's, and 469 feet for Sexson's. (*Source:* www.hittrackeronline.com.) How close are your estimates?

Many thanks to Robert K. Adair, professor of physics at Yale University, for many of the ideas presented in this application.

2 | Applications of Differentiation

What You'll Learn

2.1 Using First Derivatives to Classify Maximum and Minimum Values and Sketch Graphs

2.2 Using Second Derivatives to Classify Maximum and Minimum Values and Sketch Graphs

2.3 Graph Sketching: Asymptotes and Rational Functions

2.4 Using Derivatives to Find Absolute Maximum and Minimum Values

2.5 Maximum–Minimum Problems; Business, Economics, and General Applications

2.6 Marginals and Differentials

2.7 Elasticity of Demand

2.8 Implicit Differentiation and Related Rates

Why It's Important

In this chapter, we explore many applications of differentiation. We learn to find maximum and minimum values of functions, and that skill allows us to solve many kinds of problems in which we need to find the largest and/or smallest value in a real-world situation. Our differentiation skills will be applied to graphing, and we will use differentials to approximate function values. We will also use differentiation to explore elasticity of demand and related rates.

Where It's Used

Minimizing Cost: Minimizing cost is a common goal in manufacturing. For example, cylindrical food cans come in a variety of sizes. Suppose a soup can is to have a volume of 250 cm^3. The cost of material for the two circular ends is $0.0008/cm^2, and the cost of material for the side is $0.0015/cm^2. What dimensions minimize the cost of material for the soup can? (*This problem appears as Example 3 in Section 2.5.*)

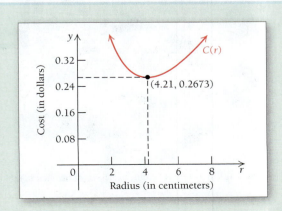

2.1

- Find relative extrema of a continuous function using the First-Derivative Test.
- Sketch graphs of continuous functions.

Using First Derivatives to Classify Maximum and Minimum Values and Sketch Graphs

The graph below shows a typical life cycle of a retail product and is similar to graphs we will consider in this chapter. Note that the number of items sold varies with respect to time. Sales begin at a small level and increase to a point of maximum sales, after which they taper off to a low level, possibly because of new competitive products. The company then rejuvenates the product by making improvements. Think about versions of certain products: televisions can be traditional, flat-screen, or high-definition; music recordings have been produced as phonograph (vinyl) records, audiotapes, compact discs, and MP3 files. Where might each of these products be in a typical product life cycle? Is the curve below appropriate for each product?

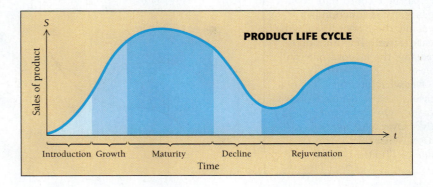

Finding the largest and smallest values of a function—that is, the maximum and minimum values—has extensive applications. The first and second derivatives of a function can provide information for graphing functions and finding maximum and minimum values.

Increasing and Decreasing Functions

If the graph of a function rises from left to right over an interval I, the function is said to be **increasing** on, or over, I. If the graph drops from left to right, the function is said to be **decreasing** on, or over, I.

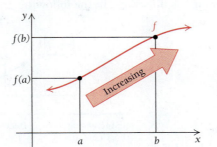

If the input a is less than the input b, then the output for a is less than the output for b.

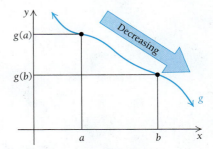

If the input a is less than the input b, then the output for a is greater than the output for b.

Exploratory

Graph the function

$$f(x) = -\frac{1}{3}x^3 + 6x^2 - 11x - 50$$

and its derivative

$$f'(x) = -x^2 + 12x - 11$$

using the window $[-10, 25, -100, 150]$, with Xscl $= 5$ and Yscl $= 25$. Then TRACE from left to right along each graph. Moving the cursor from left to right, note that the x-coordinate always increases. If the function is increasing, the y-coordinate will increase as well. If a function is decreasing, the y-coordinate will decrease.

Over what intervals is f increasing? Over what intervals is f decreasing? Over what intervals is f' positive? Over what intervals is f' negative? What rules might relate the sign of f' to the behavior of f?

We can define these concepts as follows.

DEFINITIONS

A function f is **increasing** over I if, for every a and b in I,

if $a < b$,　then $f(a) < f(b)$.

A function f is **decreasing** over I if, for every a and b in I,

if $a < b$,　then $f(a) > f(b)$.

The above definitions can be restated in terms of secant lines as shown below.

Increasing: $\dfrac{f(b) - f(a)}{b - a} > 0.$　　Decreasing: $\dfrac{f(b) - f(a)}{b - a} < 0.$

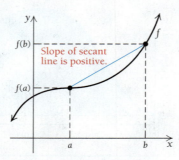
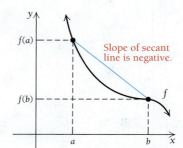

The following theorem shows how we can use the derivative (the slope of a tangent line) to determine whether a function is increasing or decreasing.

THEOREM 1

Let f be differentiable over an open interval I.
If $f'(x) > 0$ for all x in I, then f is increasing over I.
If $f'(x) < 0$ for all x in I, then f is decreasing over I.

Theorem 1 is illustrated in the following graph of $f(x) = \frac{1}{3}x^3 - x + \frac{2}{3}$.

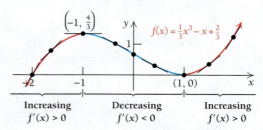

f is increasing over the intervals $(-\infty, -1)$ and $(1, \infty)$; slopes of tangent lines are positive.

f is decreasing over the interval $(-1, 1)$; slopes of tangent lines are negative.

Note in the graph above that $x = -1$ and $x = 1$ are not included in any interval over which the function is increasing or decreasing. These values are examples of *critical values*.

Critical Values

Consider the following graph of a continuous function f.

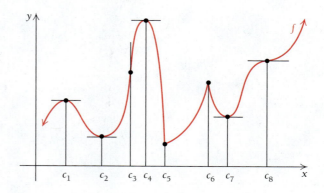

Note the following:

1. $f'(x) = 0$ at $x = c_1, c_2, c_4, c_7,$ and c_8. That is, the tangent line to the graph is horizontal for these values.

2. $f'(x)$ does not exist at $x = c_3, c_5,$ and c_6. The tangent line is vertical at c_3, and there are corners at both c_5 and c_6. (See also the discussion at the end of Section 1.4.)

> **DEFINITION**
>
> A **critical value** of a function f is any number c in the domain of f for which the tangent line at $(c, f(c))$ is horizontal or for which the derivative does not exist. That is, c is a critical value if $f(c)$ exists and
>
> $$f'(c) = 0 \quad \text{or} \quad f'(c) \text{ does not exist.}$$

Thus, in the graph of f above:

- $c_1, c_2, c_4, c_7,$ and c_8 are critical values because $f'(c) = 0$ for each value.
- $c_3, c_5,$ and c_6 are critical values because $f'(c)$ does not exist for each value.

A continuous function can change from increasing to decreasing or from decreasing to increasing *only* at a critical value. In the above graph, $c_1, c_2, c_4, c_5, c_6,$ and c_7 separate intervals over which the function f increases from those in which it decreases or intervals in which the function decreases from those in which it increases. Although c_3 and c_8 are critical values, they do not separate intervals over which the function changes from increasing to decreasing or from decreasing to increasing.

Finding Relative Maximum and Minimum Values

Now consider a graph with "peaks" and "valleys" at $x = c_1, c_2,$ and c_3.

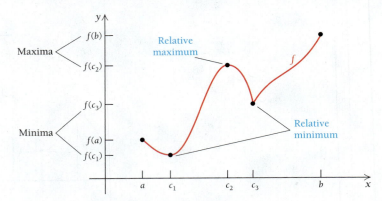

Here, $f(c_2)$ is an example of a **relative maximum** (plural: **maxima**). Each of $f(c_1)$ and $f(c_3)$ is called a **relative minimum** (plural: **minima**). Collectively, maximum and minimum values are called **extrema** (singular: **extremum**).

DEFINITIONS

Let I be the domain of f.

$f(c)$ is a **relative minimum** if there exists within I an open interval I_1 containing c such that $f(c) \leq f(x)$, for all x in I_1;

and

$f(c)$ is a **relative maximum** if there exists within I an open interval I_2 containing c such that $f(c) \geq f(x)$, for all x in I_2.

Look again at the graph above. The x-values at which a continuous function has relative extrema are those values for which the derivative is 0 or for which the derivative does not exist—the critical values.

THEOREM 2

If a function f has a relative extreme value $f(c)$ on an open interval, then c is a critical value, so

$$f'(c) = 0 \quad \text{or} \quad f'(c) \text{ does not exist.}$$

A *relative extreme point*, $(c, f(c))$, is higher or lower than all other points over some open interval containing c. A relative *minimum* point will have a y-value that is *lower* than those of a neighborhood of points both to the left and to the right of it, and, similarly, a relative *maximum* point will have a y-value that is *higher* than those of a neighborhood of points to the left and right of it.

Theorem 2 is useful and important to understand. It says that to find relative extrema, we need only consider inputs for which the derivative is 0 or for which the derivative does not exist. Each critical value is a *candidate* for a value where a relative extremum *might* occur. That is, Theorem 2 does not guarantee that every critical value will yield a relative maximum or minimum. Consider, for example, the graph of

$$f(x) = (x - 1)^3 + 2,$$

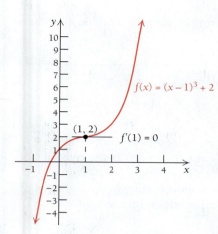

shown at the left. Note that

$$f'(x) = 3(x - 1)^2, \quad \text{and} \quad f'(1) = 3(1 - 1)^2 = 0.$$

Thus, $c = 1$ is a critical value, but f has no relative maximum or minimum at that value.

Theorem 2 does guarantee that if a relative maximum or minimum occurs, then the first coordinate of that extremum is a critical value. How can we tell when the existence of a critical value indicates a relative extremum? The following graph leads us to a test.

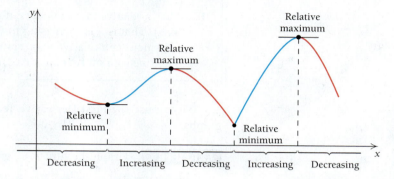

Note that at a critical value where there is a relative minimum, the function is decreasing on the left of the critical value and increasing on the right. At a critical value where there is a relative maximum, the function is increasing on the left of the critical value and decreasing on the right. In both cases, the derivative changes signs on either side of the critical value.

Graph over the interval (a, b)	$f(c)$	Sign of $f'(x)$ for x in (a, c)	Sign of $f'(x)$ for x in (c, b)	Increasing or decreasing
	Relative minimum	$-$	$+$	Decreasing on (a, c); increasing on (c, b)
	Relative maximum	$+$	$-$	Increasing on (a, c); decreasing on (c, b)
	No relative maxima or minima	$-$	$-$	Decreasing on (a, b)
	No relative maxima or minima	$+$	$+$	Increasing on (a, b)

Derivatives tell us when a function is increasing or decreasing. This leads us to the First-Derivative Test.

THEOREM 3 The First-Derivative Test for Relative Extrema

For any continuous function f that has exactly one critical value c in an open interval (a, b):

F1. f has a relative minimum at c if $f'(x) < 0$ on (a, c) and $f'(x) > 0$ on (c, b). That is, f is decreasing to the left of c and increasing to the right of c.

F2. f has a relative maximum at c if $f'(x) > 0$ on (a, c) and $f'(x) < 0$ on (c, b). That is, f is increasing to the left of c and decreasing to the right of c.

F3. f has neither a relative maximum nor a relative minimum at c if $f'(x)$ has the same sign on (a, c) as on (c, b).

We can use the First-Derivative Test to find relative extrema and create accurate graphs.

EXAMPLE 1 Graph the function f given by

$$f(x) = 4x^3 - 9x^2 - 30x + 25,$$

and find any relative extrema.

Solution We plot points and use them to create a rough sketch of the graph, shown as the dashed line in the figure below.

x	$f(x)$
-3	-74
-2	17
-1	42
0	25
1	-10
2	-39
3	-38
4	17

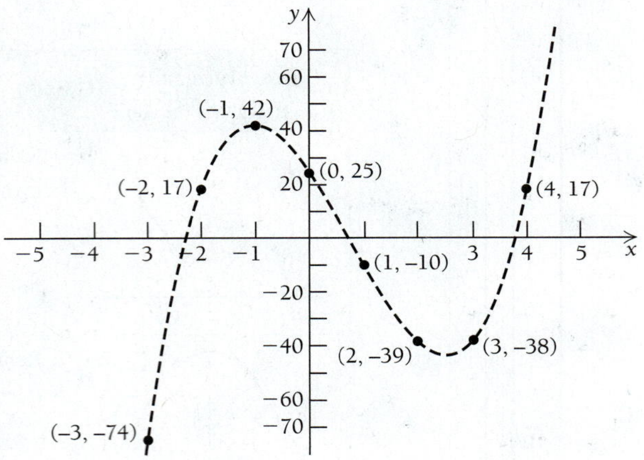

According to this sketch, it *appears* that the graph has a tangent line with slope 0 at $x = -1$ and between $x = 2$ and $x = 3$. To verify this, we use calculus to support our observations. We begin by finding the derivative:

$$f'(x) = 12x^2 - 18x - 30.$$

We next determine where $f'(x)$ does not exist or where $f'(x) = 0$. Since $f'(x)$ is defined for all real numbers, there is no value x for which $f'(x)$ does not exist.

The only possibilities for critical values are those where $f'(x) = 0$. To find such values, we solve $f'(x) = 0$:

$$12x^2 - 18x - 30 = 0$$
$$2x^2 - 3x - 5 = 0 \qquad \text{Dividing both sides by 6}$$
$$(x + 1)(2x - 5) = 0 \qquad \text{Factoring}$$
$$x + 1 = 0 \quad \text{or} \quad 2x - 5 = 0 \qquad \text{Using the Principle of Zero Products}$$
$$x = -1 \quad \text{or} \quad x = \tfrac{5}{2}.$$

The critical values are -1 and $\frac{5}{2}$. Since it is at these values that a relative maximum or minimum might exist, we analyze the sign of the derivative on the intervals $(-\infty, -1)$, $(-1, \frac{5}{2})$, and $(\frac{5}{2}, \infty)$.

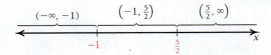

To do so, we choose a test value in each interval and make a substitution. Let's use the values -2, 0, and 4.

$(-\infty, -1)$: Test -2, $\quad f'(-2) = 12(-2)^2 - 18(-2) - 30$
$$= 48 + 36 - 30 = 54 > 0;$$

$(-1, \frac{5}{2})$: $\quad$ Test 0, $\quad f'(0) = 12(0)^2 - 18(0) - 30 = -30 < 0;$

$(\frac{5}{2}, \infty)$: $\quad$ Test 4, $\quad f'(4) = 12(4)^2 - 18(4) - 30$
$$= 192 - 72 - 30 = 90 > 0.$$

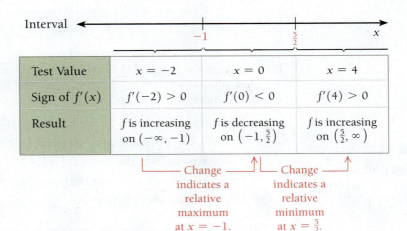

Therefore, by the First-Derivative Test, f has a relative maximum at $x = -1$ given by

$$f(-1) = 4(-1)^3 - 9(-1)^2 - 30(-1) + 25 \qquad \text{Substituting } -1 \text{ for } x \text{ in the } \textit{original} \text{ function}$$

$$= 42 \qquad \text{This is a relative maximum.}$$

and f has a relative minimum at $x = \frac{5}{2}$ given by

$$f\left(\tfrac{5}{2}\right) = 4\left(\tfrac{5}{2}\right)^3 - 9\left(\tfrac{5}{2}\right)^2 - 30\left(\tfrac{5}{2}\right) + 25 = -\tfrac{175}{4}.$$ This is a relative minimum.

Thus, there is a relative maximum at $(-1, 42)$ and a relative minimum at $\left(\tfrac{5}{2}, -\tfrac{175}{4}\right)$, as we suspected from the original sketch of the graph.

The information obtained from the first derivative is very useful in graphing this function. We know that it is continuous and where it is increasing, where it is decreasing, and where it has relative extrema. We complete the graph by calculating additional function values. The graph of the function, shown below in red, is scaled to clearly show its important features.

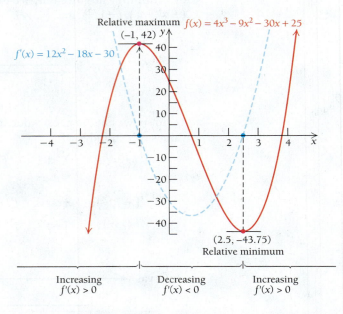

For reference, the graph of the derivative is shown in blue. Note that $f'(x) = 0$ where $f(x)$ has relative extrema. We summarize the behavior of this function by noting where it is increasing or decreasing and by characterizing its critical points:

- f is increasing over the interval $(-\infty, -1)$.
- f has a relative maximum at the point $(-1, 42)$.
- f is decreasing over the interval $\left(-1, \tfrac{5}{2}\right)$.
- f has a relative minimum at the point $\left(\tfrac{5}{2}, -\tfrac{175}{4}\right)$.
- f is increasing over the interval $\left(\tfrac{5}{2}, \infty\right)$.   1 ✔

To use the first derivative for graphing a function f:

1. Find all critical values by determining where $f'(x)$ is 0 and where $f'(x)$ is undefined (but $f(x)$ is defined). Find $f(x)$ for each critical value.
2. Use the critical values to divide the x-axis into intervals and choose a test value in each interval.
3. Find the sign of $f'(x)$ for each test value chosen in step 2, and use this information to determine where $f(x)$ is increasing or decreasing and to classify any extrema as relative maxima or minima.
4. Plot some additional points and sketch the graph.

Quick Check 1 ✔

Graph the function g given by $g(x) = x^3 - 27x - 6$, and find any relative extrema.

> The *derivative f′* is used to find the critical values of *f*. The test values are substituted into the *derivative f′*, and the function values are found using the *original* function *f*.

EXAMPLE 2   Find the relative extrema and sketch the graph of the function *f* given by

$$f(x) = 2x^3 - x^4.$$

Solution   First, we determine the critical values. To do so, we find $f'(x)$:

$$f'(x) = 6x^2 - 4x^3.$$

Next, we find where $f'(x)$ does not exist or where $f'(x) = 0$. Note that $f'(x) = 6x^2 - 4x^3$ is defined for all real numbers x, so the only candidates for critical values are where $f'(x) = 0$:

$$6x^2 - 4x^3 = 0 \qquad \text{Setting } f'(x) \text{ equal to 0}$$
$$2x^2(3 - 2x) = 0 \qquad \text{Factoring}$$
$$2x^2 = 0 \quad \text{or} \quad 3 - 2x = 0$$
$$x^2 = 0 \quad \text{or} \qquad 3 = 2x$$
$$x = 0 \quad \text{or} \qquad x = \tfrac{3}{2}.$$

The critical values are 0 and $\tfrac{3}{2}$. We use these values to divide the x-axis into three intervals: $(-\infty, 0)$, $\left(0, \tfrac{3}{2}\right)$, and $\left(\tfrac{3}{2}, \infty\right)$.

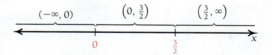

We next find the sign of $f'(x)$ on each interval, using a test value.

$(-\infty, 0)$:   Test -1, $f'(-1) = 6(-1)^2 - 4(-1)^3 = 6 + 4 = 10 > 0$;

$\left(0, \tfrac{3}{2}\right)$:   Test 1, $f'(1) = 6(1)^2 - 4(1)^3 = 6 - 4 = 2 > 0$;

$\left(\tfrac{3}{2}, \infty\right)$:   Test 2, $f'(2) = 6(2)^2 - 4(2)^3 = 24 - 32 = -8 < 0$.

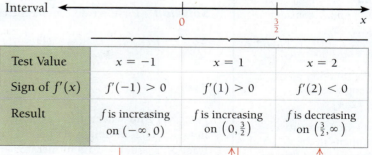

Interval			
Test Value	$x = -1$	$x = 1$	$x = 2$
Sign of $f'(x)$	$f'(-1) > 0$	$f'(1) > 0$	$f'(2) < 0$
Result	f is increasing on $(-\infty, 0)$	f is increasing on $\left(0, \tfrac{3}{2}\right)$	f is decreasing on $\left(\tfrac{3}{2}, \infty\right)$

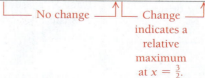

No change — Change indicates a relative maximum at $x = \tfrac{3}{2}$.

Therefore, by the First-Derivative Test, f has no extremum at $x = 0$ (since $f(x)$ is increasing on both sides of 0) and has a relative maximum at $x = \frac{3}{2}$. Thus, $f\left(\frac{3}{2}\right)$, or $\frac{27}{16}$, is a relative maximum.

We use the information obtained to sketch the graph below. Other function values are listed in the table.

x	$f(x)$, approximately
-1	-3
-0.5	-0.31
0	0
0.5	0.19
1	1
1.25	1.46
2	0

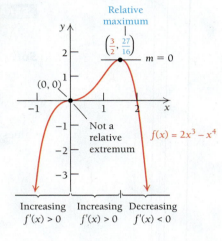

- f is increasing over the interval $(-\infty, 0)$.
- f has a critical point at $(0, 0)$, which is neither a minimum nor a maximum.
- f is increasing over the interval $\left(0, \frac{3}{2}\right)$.
- f has a relative maximum at the point $\left(\frac{3}{2}, \frac{27}{16}\right)$.
- f is decreasing over the interval $\left(\frac{3}{2}, \infty\right)$.

Since f is increasing over the intervals $(-\infty, 0)$ and $\left(0, \frac{3}{2}\right)$, we can say that f is increasing over $\left(-\infty, \frac{3}{2}\right)$ despite the fact that $f'(0) = 0$ within this interval. In this case, we observe that a secant line connecting any two points within this interval has a positive slope.   **2** ✔

Quick Check 2 ✔

Find the relative extrema of the function h given by $h(x) = x^4 - \frac{8}{3}x^3$. Then sketch the graph.

TECHNOLOGY CONNECTION 〰

EXERCISES

In Exercises 1 and 2, consider the function f given by

$$f(x) = 2 - (x - 1)^{2/3}.$$

1. Graph the function using the viewing window $[-4, 6, -2, 4]$.

2. Graph the first derivative. What happens to the graph of the derivative at the critical values?

EXAMPLE 3   Find the relative extrema and sketch the graph of the function f given by

$$f(x) = (x - 2)^{2/3} + 1.$$

Solution   First, we determine the critical values. To do so, we find $f'(x)$:

$$f'(x) = \frac{2}{3}(x - 2)^{-1/3}$$

$$= \frac{2}{3\sqrt[3]{x - 2}}. \qquad \text{Simplifying}$$

Next, we find where $f'(x)$ does not exist or where $f'(x) = 0$. Note that $f'(x)$ does not exist at 2, although $f(x)$ does. Thus, 2 is a critical value. Since the only way for a fraction to be 0 is if its numerator is 0, we see that $f'(x) = 0$ has no solution. Thus, 2 is the only critical value. We use 2 to divide the x-axis into the intervals $(-\infty, 2)$ and $(2, \infty)$.

To determine the sign of $f'(x)$ we choose a test value in each interval. It is not necessary to find an exact value of the derivative; we need only determine the sign. Sometimes we can do this by just examining the formula for the derivative. We choose test values 0 and 3.

$$(-\infty, 2): \quad \text{Test } 0, \quad f'(0) = \frac{2}{3\sqrt[3]{0-2}} < 0;$$

$$(2, \infty): \quad \text{Test } 3, \quad f'(3) = \frac{2}{3\sqrt[3]{3-2}} > 0.$$

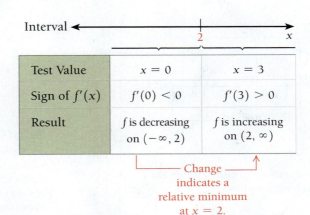

	$x = 0$	$x = 3$
Test Value	$x = 0$	$x = 3$
Sign of $f'(x)$	$f'(0) < 0$	$f'(3) > 0$
Result	f is decreasing on $(-\infty, 2)$	f is increasing on $(2, \infty)$

Change indicates a relative minimum at $x = 2$.

Since we have a change from decreasing to increasing, we conclude from the First-Derivative Test that a relative minimum occurs at $(2, f(2))$, or $(2, 1)$. The graph has *no* tangent line at $(2, 1)$ since $f'(2)$ does not exist.

We use this information and some other function values to sketch the graph.

x	$f(x)$, approximately
-1	3.08
0	2.59
1	2
2	1
3	2
4	2.59

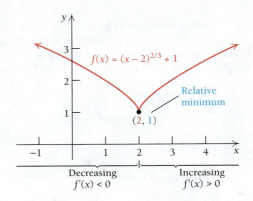

$f(x) = (x-2)^{2/3} + 1$

Relative minimum

$(2, 1)$

Decreasing $f'(x) < 0$ Increasing $f'(x) > 0$

Quick Check 3 ✔

Find the relative extrema of the function g given by $g(x) = 3 - x^{1/3}$. Then sketch the graph.

- f is decreasing over the interval $(-\infty, 2)$.
- f has a relative minimum at the point $(2, 1)$.
- f is increasing over the interval $(2, \infty)$.

3 ✔

Finding Relative Extrema

To explore some methods for approximating relative extrema, let's find the relative extrema of

$$f(x) = -0.4x^3 + 6.2x^2 - 11.3x - 54.8.$$

We first graph the function, using a window that reveals the curvature.

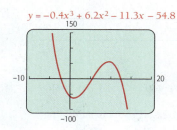

Method 1: TRACE

Using TRACE, we can move the cursor along the curve, noting where relative extrema might occur.

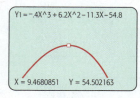

A relative maximum seems to be about $y = 54.5$ at $x = 9.47$. To refine the approximation, we zoom in, press TRACE, and move the cursor, again noting where the y-value is largest. The approximation is about $y = 54.61$ at $x = 9.31$.

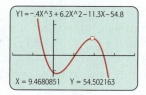

We can repeat to refine the accuracy.

Method 2: TABLE

We can also use the TABLE feature, adjusting starting points and step values to improve accuracy:

$$\text{TblStart} = 9.3 \quad \Delta\text{Tbl} = .01$$

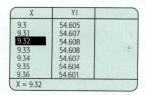

The relative maximum seems to be nearly $y = 54.61$ at an x-value between 9.32 and 9.33. We could next set up a new table showing function values between $f(9.32)$ and $f(9.33)$ to refine the approximation.

Method 3: MAXIMUM, MINIMUM

Using the MAXIMUM option from the CALC menu, we find that a relative maximum of about 54.61 occurs at $x \approx 9.32$.

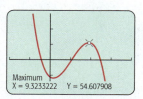

Method 4: fMax or fMin

This feature calculates a relative maximum or minimum value over any closed interval. We see from the initial graph that a relative maximum occurs in $[-10, 20]$. Using the fMax option from the MATH menu, we see that a relative maximum occurs on $[-10, 20]$ when $x \approx 9.32$.

To obtain the maximum value, we evaluate the function at the given x-value, obtaining the following.

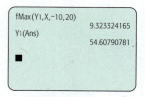

The approximation is about $y = 54.61$ at $x = 9.32$.

Using any of these methods, we find the relative minimum $y \approx -60.30$ at $x \approx 1.01$.

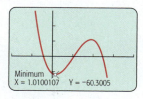

EXERCISE

1. Using one of the methods described above, approximate the relative extrema of the function in Example 1.

Section Summary

- A function f is *increasing* over an interval I if, for all a and b in I such that $a < b$, we have $f(a) < f(b)$. Equivalently, the slope of the secant line connecting a and b is positive:

$$\frac{f(b) - f(a)}{b - a} > 0.$$

- A function f is *decreasing* over an interval I if, for all a and b in I such that $a < b$, we have $f(a) > f(b)$. Equivalently, the slope of the secant line connecting a and b is negative:

$$\frac{f(b) - f(a)}{b - a} < 0.$$

- A function f is *increasing* over an open interval I if, for all x in I, the slope of the tangent line at x is positive; that is, $f'(x) > 0$. Similarly, a function f is *decreasing*

over an open interval I if, for all x in I, the slope of the tangent line is negative; that is, $f'(x) < 0$.

- A *critical value* is a number c in the domain of f such that $f'(c) = 0$ or $f'(c)$ does not exist. The point $(c, f(c))$ is called a *critical point*.
- A relative maximum point is higher than all other points in some interval containing it. Similarly, a relative minimum point is lower than all other points in some interval containing it. The y-value of such a point is called a relative maximum (or minimum) *value* of the function.
- Minimum and maximum points are collectively called *extrema*.
- Critical values are candidates for possible relative extrema. The *First-Derivative Test* is used to classify a critical value as a relative minimum, a relative maximum, or neither.

2.1 | Exercise Set

Find any relative extrema of each function. List each extremum along with the x-value at which it occurs. Then sketch a graph of the function.

1. $f(x) = x^2 + 6x - 3$

2. $f(x) = x^2 + 4x + 5$

3. $f(x) = 2 - 3x - 2x^2$

4. $f(x) = 5 - x - x^2$

5. $F(x) = 0.5x^2 + 2x - 11$

6. $g(x) = 1 + 6x + 3x^2$

7. $g(x) = x^3 + \frac{1}{2}x^2 - 2x + 5$

8. $G(x) = x^3 - x^2 - x + 2$

9. $f(x) = x^3 - 3x^2$

10. $f(x) = x^3 - 3x + 6$

11. $f(x) = x^3 + 3x$

12. $f(x) = 3x^2 + 2x^3$

13. $F(x) = 1 - x^3$

14. $g(x) = 2x^3 - 16$

15. $G(x) = x^3 - 6x^2 + 10$

16. $f(x) = 12 + 9x - 3x^2 - x^3$

17. $g(x) = x^3 - x^4$

18. $f(x) = x^4 - 2x^3$

19. $f(x) = \frac{1}{3}x^3 - 2x^2 + 4x - 1$

20. $F(x) = -\frac{1}{3}x^3 + 3x^2 - 9x + 2$

21. $f(x) = 3x^4 - 15x^2 + 12$

22. $g(x) = 2x^4 - 20x^2 + 18$

23. $G(x) = \sqrt[3]{x + 2}$

24. $F(x) = \sqrt[3]{x - 1}$

25. $f(x) = 1 - x^{2/3}$

26. $f(x) = (x + 3)^{2/3} - 5$

27. $G(x) = \dfrac{-8}{x^2 + 1}$

28. $F(x) = \dfrac{5}{x^2 + 1}$

29. $g(x) = \dfrac{4x}{x^2 + 1}$

30. $g(x) = \dfrac{x^2}{x^2 + 1}$

31. $f(x) = \sqrt[3]{x}$

32. $f(x) = (x + 1)^{1/3}$

33. $g(x) = \sqrt{x^2 + 2x + 5}$

34. $F(x) = \dfrac{1}{\sqrt{x^2 + 1}}$

35–68. Check the results of Exercises 1–34 using a calculator.

For Exercises 69–84, draw a graph to match the description given. Answers will vary.

69. $f(x)$ is increasing over $(-\infty, 2)$ and decreasing over $(2, \infty)$.

70. $g(x)$ is decreasing over $(-\infty, -3)$ and increasing over $(-3, \infty)$.

71. $G(x)$ is decreasing over $(-\infty, 4)$ and $(9, \infty)$ and increasing over $(4, 9)$.

72. $F(x)$ is increasing over $(-\infty, 5)$ and $(12, \infty)$ and decreasing over $(5, 12)$.

73. $g(x)$ has a positive derivative over $(-\infty, -3)$ and a negative derivative over $(-3, \infty)$.

74. $f(x)$ has a negative derivative over $(-\infty, 1)$ and a positive derivative over $(1, \infty)$.

75. $F(x)$ has a negative derivative over $(-\infty, 2)$ and $(5, 9)$ and a positive derivative over $(2, 5)$ and $(9, \infty)$.

76. $G(x)$ has a positive derivative over $(-\infty, -2)$ and $(4, 7)$ and a negative derivative over $(-2, 4)$ and $(7, \infty)$.

77. $f(x)$ has a positive derivative over $(-\infty, 3)$ and $(3, 9)$, a negative derivative over $(9, \infty)$, and a derivative equal to 0 at $x = 3$.

78. $g(x)$ has a negative derivative over $(-\infty, 5)$ and $(5, 8)$, a positive derivative over $(8, \infty)$, and a derivative equal to 0 at $x = 5$.

79. $F(x)$ has a negative derivative over $(-\infty, -1)$ and a positive derivative over $(-1, \infty)$, and $F'(-1)$ does not exist.

80. $G(x)$ has a positive derivative over $(-\infty, 0)$ and $(3, \infty)$ and a negative derivative over $(0, 3)$, but neither $G'(0)$ nor $G'(3)$ exists.

81. $f(x)$ has a negative derivative over $(-\infty, -2)$ and $(1, \infty)$ and a positive derivative over $(-2, 1)$, and $f'(-2) = 0$, but $f'(1)$ does not exist.

82. $g(x)$ has a positive derivative over $(-\infty, -3)$ and $(0, 3)$, a negative derivative over $(-3, 0)$ and $(3, \infty)$, and a derivative equal to 0 at $x = -3$ and $x = 3$, but $g'(0)$ does not exist.

83. $H(x)$ is increasing over $(-\infty, \infty)$, but the derivative does not exist at $x = 1$.

84. $K(x)$ is decreasing over $(-\infty, \infty)$, but the derivative does not exist at $x = 0$ and $x = 2$.

85. Consider this graph.

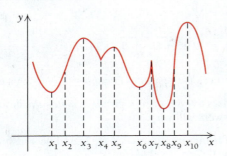

What makes an x-value a critical value? Which x-values above are critical values, and why?

86. Consider this graph.

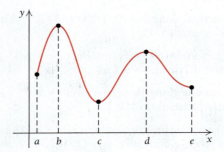

Using the graph and the intervals noted, explain how a function being increasing or decreasing relates to the first derivative.

APPLICATIONS

Business and Economics

87. Employment. According to the U.S. Bureau of Labor Statistics, the number of managerial and professional employees since 2006 is modeled by

$$E(t) = 107.833t^3 - 971.369t^2 + 2657.917t + 50,347.833,$$

where t is the number of years since 2006 and E is thousands of employees. (*Source:* data.bls.gov.) Find the relative extrema of this function, and sketch the graph. Interpret the meaning of the relative extrema.

88. Advertising. Brody Electronics estimates that it will sell N units of a new toy after spending a thousands of dollars on advertising, where

$$N(a) = -a^2 + 300a + 6, \quad 0 \le a \le 300.$$

Find the relative extrema and sketch a graph of the function.

Life and Physical Sciences

89. Solar eclipse. On July 2, 2019, a total solar eclipse will occur over the South Pacific Ocean and parts of South America. (*Source:* eclipse.gsfc.nasa.gov.) The path of the eclipse is modeled by

$$f(t) = 0.00259t^2 - 0.457t + 36.237,$$

where $f(t)$ is the latitude in degrees south of the equator at t minutes after the start of the total eclipse. What is the latitude closest to the equator, in degrees, at which the total eclipse will be visible?

90. Temperature during an illness. The temperature of a person during an intestinal illness is given by

$$T(t) = -0.1t^2 + 1.2t + 98.6, \quad 0 \le t \le 12,$$

where T is the temperature (°F) at t days. Find the relative extrema and sketch a graph of the function.

SYNTHESIS

In Exercises 91–96, the graph of a derivative f' is shown. Use the information in each graph to determine where f is increasing or decreasing and the x-values of any extrema. Then sketch a possible graph of f.

91.

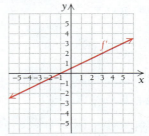

92.

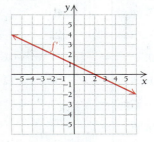

93.

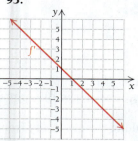

94.

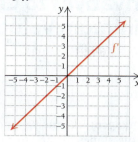

95.

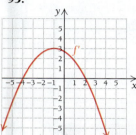

96.

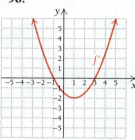

TECHNOLOGY CONNECTION

Graph each function. Then estimate any relative extrema.

97. $f(x) = -x^6 - 4x^5 + 54x^4 + 160x^3 - 641x^2 - 828x + 1200$

98. $f(x) = x^4 + 4x^3 - 36x^2 - 160x + 400$

99. $f(x) = \sqrt[3]{|4 - x^2|} + 1$

100. $f(x) = x\sqrt{9 - x^2}$

Use a calculator's absolute-value feature to graph each function and determine relative extrema and intervals over which the function is increasing or decreasing. State any x-values at which the derivative does not exist.

101. $f(x) = |x - 2|$

102. $f(x) = |2x - 5|$

103. $f(x) = |x^2 - 1|$

104. $f(x) = |x^2 - 3x + 2|$

105. $f(x) = |9 - x^2|$

106. $f(x) = |-x^2 + 4x - 4|$

107. $f(x) = |x^3 - 1|$

108. $f(x) = |x^4 - 2x^2|$

Life science: caloric intake and life expectancy. *The data in the following table give, for various countries for the period 2007–2011, daily caloric intake, projected life expectancy, and under-five mortality. Use the data for Exercises 109 and 110.*

COUNTRY	DAILY CALORIC INTAKE	LIFE EXPECTANCY AT BIRTH (in years)	UNDER-FIVE MORTALITY (number of deaths before age 5 per 1000 births)
Argentina	3030	76	14
Australia	3220	82	5
Bolivia	2100	66	51
Canada	3530	81	6
Dominican Republic	2270	73	25
Germany	3540	80	4
Haiti	1850	62	70
Mexico	3260	77	17
United States	3750	78	8
Venezuela	2650	74	16

(Source: U.N. FAO Statistical Yearbook, 2013.)

109. Life expectancy and daily caloric intake.
   a) Use the regression procedures of Section R.6 to fit a cubic function $y = f(x)$ to the data in the table, where x is daily caloric intake and y is life expectancy. Then fit a quartic function and decide which fits best.
   b) What is the domain of the function?
   c) Does the function have any relative extrema?

110. Under-five mortality and daily caloric intake.
   a) Use the regression procedures of Section R.6 to fit a cubic function $y = f(x)$ to the data in the table, where x is daily caloric intake and y is under-five mortality. Then fit a quartic function and decide which fits best.
   b) What is the domain of the function?
   c) Does the function have any relative extrema?

111. Describe a procedure that can be used to select an appropriate viewing window for the functions given in **(a)** Exercises 1–34 and **(b)** Exercises 97–100.

Answers to Quick Checks

1. Relative maximum at $(-3, 48)$, relative minimum at $(3, -60)$

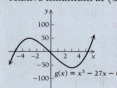

2. Relative minimum at $\left(2, -\frac{16}{3}\right)$

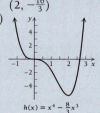

3. There are no extrema.

2.2

Using Second Derivatives to Classify Maximum and Minimum Values and Sketch Graphs

- Classify the relative extrema of a function using the Second-Derivative Test.
- Graph a continuous function in a manner that shows concavity.

The graphs of two functions are shown below. The graph of f bends upward and the graph of g bends downward. Let's relate these observations to the functions' derivatives.

Using a straightedge, draw tangent lines as you move along the graph of f from left to right. What happens to the slopes of the tangent lines? Do the same for the graph of g. Look at how the graphs bend. Do you see a pattern?

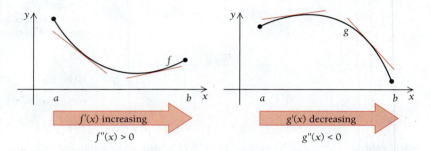

Concavity: Increasing and Decreasing Derivatives

For the graph of f, the slopes of the tangent lines are increasing. That is, f' is increasing over the interval. This can be determined by noting that $f''(x)$ is positive, since the relationship between f' and f'' is like the relationship between f and f'. For the graph of g, the slopes are decreasing. This can be determined by noting that g' is decreasing whenever $g''(x)$ is negative. The bending of each curve is called the graph's *concavity*.

DEFINITION

Suppose that f is a function whose derivative f' exists at every point in an open interval I. Then

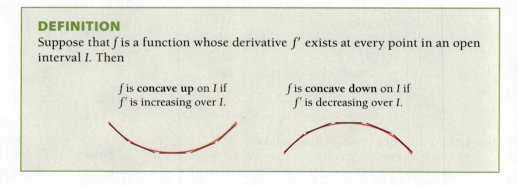

f is **concave up** on I if f' is increasing over I.

f is **concave down** on I if f' is decreasing over I.

The following theorem states how the concavity of a function's graph and the second derivative of the function are related.

THEOREM 4   **A Test for Concavity**

1. If $f''(x) > 0$ on an interval I, then the graph of f is concave up on I.
2. If $f''(x) < 0$ on an interval I, then the graph of f is concave down on I.

Exploratory

Graph

$$f(x) = -\frac{1}{3}x^3 + 6x^2 - 11x - 50$$

and its second derivative,

$$f''(x) = -2x + 12,$$

using the viewing window $[-10, 25, -100, 150]$, with Xscl $= 5$ and Yscl $= 25$.

Over what intervals is the graph of f concave up? Over what intervals is the graph of f concave down? Over what intervals is the graph of f'' positive? Over what intervals is the graph of f'' negative? How do you think the sign of $f''(x)$ relates to the concavity of f?

Now graph the first derivative,

$$f'(x) = -x^2 + 12x - 11,$$

and the second derivative,

$$f''(x) = -2x + 12,$$

using the window $[-10, 25, -200, 50]$, with Xscl $= 5$ and Yscl $= 25$.

Over what intervals is the first derivative f' increasing? Over what intervals is the first derivative f' decreasing? Over what intervals is the graph of f'' positive? Over what intervals is the graph of f'' negative? How do you think the sign of $f''(x)$ relates to the concavity of f?

A function can be decreasing and concave up, decreasing and concave down, increasing and concave up, or increasing and concave down. That is, *concavity* and *increasing/decreasing* are separate concepts. It is the increasing or decreasing aspect of the *derivative* that tells us about the function's concavity.

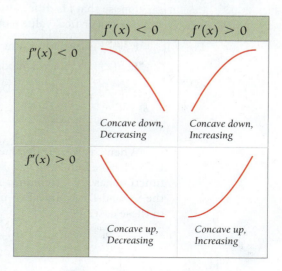

Classifying Relative Extrema Using Second Derivatives

Let's see how we can use second derivatives to determine whether a function has a relative extremum on an open interval.

The following graphs show both types of concavity at a critical value, where $f'(c) = 0$. When the second derivative is positive (graph is concave up) at the critical value, the critical point is a relative minimum point, and when the second derivative is negative (graph is concave down), the critical point is a relative maximum point.

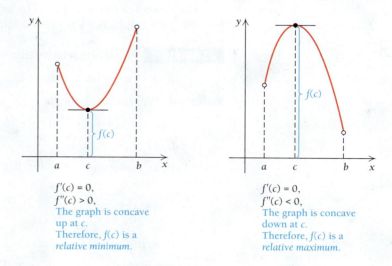

$f'(c) = 0$,
$f''(c) > 0$,
The graph is concave up at c.
Therefore, $f(c)$ is a *relative minimum*.

$f'(c) = 0$,
$f''(c) < 0$,
The graph is concave down at c.
Therefore, $f(c)$ is a *relative maximum*.

This analysis is summarized in Theorem 5:

THEOREM 5 **The Second-Derivative Test for Relative Extrema**

Suppose that f is differentiable for every x in an open interval (a, b) and that there is a critical value c in (a, b) for which $f'(c) = 0$. Then:

1. $f(c)$ is a relative minimum if $f''(c) > 0$.
2. $f(c)$ is a relative maximum if $f''(c) < 0$.

For $f''(c) = 0$, the First-Derivative Test can be used to determine whether $f(x)$ is a relative extremum.

When c is a critical value and $f''(c) = 0$, an extremum may or may not exist at c. Consider the following graphs. In each one, f' and f'' are both 0 at $c = 2$, but the first function has an extremum and the second function does not. An approach other than the Second-Derivative Test must be used to determine whether $f(c)$ is an extremum in these cases.

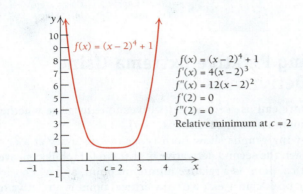

$f(x) = (x - 2)^4 + 1$
$f'(x) = 4(x - 2)^3$
$f''(x) = 12(x - 2)^2$
$f'(2) = 0$
$f''(2) = 0$
Relative minimum at $c = 2$

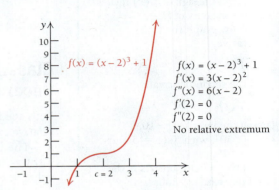

$f(x) = (x - 2)^3 + 1$
$f'(x) = 3(x - 2)^2$
$f''(x) = 6(x - 2)$
$f'(2) = 0$
$f''(2) = 0$
No relative extremum

The second derivative is used to help identify extrema and determine the overall behavior of a graph, as we see in the following examples.

EXAMPLE 1 Find any relative extrema and graph the function f given by

$$f(x) = x^3 + 3x^2 - 9x - 13.$$

Solution We find both the first and second derivatives:

$$f'(x) = 3x^2 + 6x - 9,$$
$$f''(x) = 6x + 6.$$

To find any critical values, we solve $f'(x) = 0$:

$$3x^2 + 6x - 9 = 0$$
$$x^2 + 2x - 3 = 0 \qquad \text{Dividing both sides by 3}$$
$$(x + 3)(x - 1) = 0 \qquad \text{Factoring}$$
$$x + 3 = 0 \quad \text{or} \quad x - 1 = 0 \qquad \text{Using the Principle of Zero Products}$$
$$x = -3 \quad \text{or} \qquad x = 1.$$

We next find second coordinates by substituting the critical values in the original function:

$$f(-3) = (-3)^3 + 3(-3)^2 - 9(-3) - 13 = 14;$$
$$f(1) = (1)^3 + 3(1)^2 - 9(1) - 13 = -18.$$

We now use the Second-Derivative Test with the critical values -3 and 1:

$$f''(-3) = 6(-3) + 6 = -12 < 0; \rightarrow \text{Relative maximum at } x = -3$$
$$f''(1) = 6(1) + 6 = 12 > 0. \longrightarrow \text{Relative minimum at } x = 1$$

Thus, $f(-3) = 14$ is a relative maximum and $f(1) = -18$ is a relative minimum. We plot both $(-3, 14)$ and $(1, -18)$, including short arcs at each point to indicate the concavity. By plotting more points, we can make a sketch, as shown below.

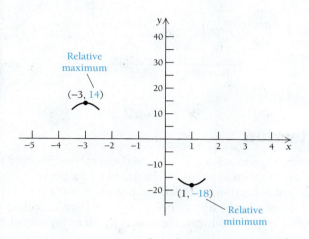

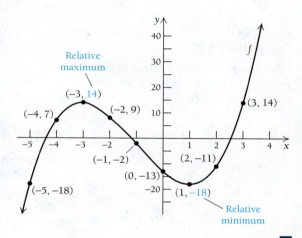

EXAMPLE 2   Find any relative extrema and graph the function f given by

$$f(x) = 3x^5 - 20x^3.$$

Solution   We find both the first and second derivatives:

$$f'(x) = 15x^4 - 60x^2,$$
$$f''(x) = 60x^3 - 120x.$$

Then we solve $f'(x) = 0$ to find any critical values:

$$15x^4 - 60x^2 = 0$$
$$15x^2(x^2 - 4) = 0$$
$$15x^2(x + 2)(x - 2) = 0 \qquad \text{Factoring}$$
$$15x^2 = 0 \quad \text{or} \quad x + 2 = 0 \quad \text{or} \quad x - 2 = 0 \qquad \text{Using the Principle of Zero Products}$$
$$x = 0 \quad \text{or} \qquad x = -2 \quad \text{or} \qquad x = 2.$$

We next find second coordinates by substituting in the original function:

$$f(-2) = 3(-2)^5 - 20(-2)^3 = 64;$$
$$f(0) = 3(0)^5 - 20(0)^3 = 0.$$
$$f(2) = 3(2)^5 - 20(2)^3 = -64;$$

All three of these y-values are candidates for relative extrema.
We now use the Second-Derivative Test with the values $-2, 0,$ and 2:

$$f''(-2) = 60(-2)^3 - 120(-2) = -240 < 0; \rightarrow \text{Relative maximum at } x = -2$$
$$f''(0) = 60(0)^3 - 120(0) = 0. \longrightarrow \text{The Second-Derivative Test fails. Use the First-Derivative Test.}$$
$$f''(2) = 60(2)^3 - 120(2) = 240 > 0; \longrightarrow \text{Relative minimum at } x = 2$$

Thus, $f(-2) = 64$ is a relative maximum, and $f(2) = -64$ is a relative minimum. Since $f'(-1) < 0$ and $f'(1) < 0$, we know that f is decreasing on $(-2, 0)$ and $(0, 2)$. Thus, by the First-Derivative Test, f has no relative extremum at $(0, 0)$. We complete the graph, plotting other points as needed. The extrema are shown in the graph at right.

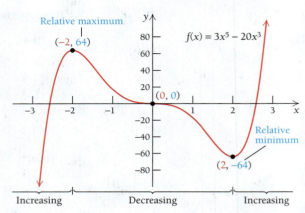

Quick Check 1 ✔

Find any relative extrema and graph the function g given by $g(x) = 10x^3 - 6x^5$.

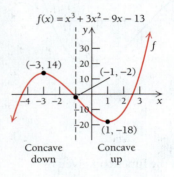

Points of Inflection

Concavity changes from down to up at $(-1, -2)$ in the graph in Example 1 (repeated at the left) and from up to down at $(0, 0)$ in the graph in Example 2 (above). Such points are called **points of inflection**, or **inflection points**. At each such point, the direction of concavity changes. In the figures below, point P is an inflection point. The figures display the sign of $f''(x)$ to indicate the concavity on either side of P.

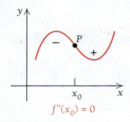

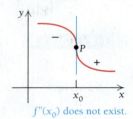

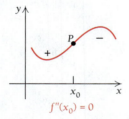

As we move to the right along each curve, the concavity changes at P. Since, as we move through P, the sign of $f''(x)$ changes, either the value of $f''(x_0)$ at P must be 0 or $f''(x_0)$ must not exist.

THEOREM 6   **Finding Points of Inflection**

If f has a point of inflection, it must occur at a value x_0, in the domain of f, where

$$f''(x_0) = 0 \quad \text{or} \quad f''(x_0) \text{ does not exist.}$$

The converse of Theorem 6 is not necessarily true. That is, if $f''(x_0)$ is 0 or does not exist, then there is not necessarily a point of inflection at x_0. There must be a change in the direction of concavity on either side of x_0 for $(x_0, f(x_0))$ to be a point of inflection. For example, for $f(x) = (x - 2)^4 + 1$ (see the graph on p. 206), we have $f''(2) = 0$, but $(2, f(2))$ is not a point of inflection since the graph of f is concave up on both sides of $x = 2$.

To find *candidates* for points of inflection, we look for x-values at which $f''(x) = 0$ or $f''(x)$ does not exist. If $f''(x)$ changes sign on either side of the x-value, we have a point of inflection.

Theorem 6 is analogous to Theorem 2. Whereas Theorem 2 tells us that candidates for extrema occur when $f'(x) = 0$ or $f'(x)$ does not exist, Theorem 6 tells us that candidates for points of inflection occur when $f''(x) = 0$ or $f''(x)$ does not exist.

EXAMPLE 3   Use the second derivative to determine any point(s) of inflection for the function in Example 2.

Solution   The function is given by $f(x) = 3x^5 - 20x^3$, and its second derivative is $f''(x) = 60x^3 - 120x$. We set $f''(x)$ equal to 0 and solve for x:

$$60x^3 - 120x = 0$$
$$60x(x^2 - 2) = 0 \qquad \text{Factoring out } 60x$$
$$60x(x + \sqrt{2})(x - \sqrt{2}) = 0 \qquad \text{Factoring } x^2 - 2 \text{ as a difference of squares}$$
$$x = 0 \quad \text{or} \quad x = -\sqrt{2} \quad \text{or} \quad x = \sqrt{2}. \qquad \text{Using the Principle of Zero Products}$$

Next, we check the sign of $f''(x)$ over the intervals bounded by these three x-values. We are looking for a change in sign from one interval to the next:

Interval

$$-\sqrt{2} \qquad 0 \qquad \sqrt{2}$$

Test Value	$x = -2$	$x = -1$	$x = 1$	$x = 2$
Sign of $f''(x)$	$f''(-2) < 0$	$f''(-1) > 0$	$f''(1) < 0$	$f''(2) > 0$
Result	Concave down on $(-\infty, -\sqrt{2})$	Concave up on $(-\sqrt{2}, 0)$	Concave down on $(0, \sqrt{2})$	Concave up on $(\sqrt{2}, \infty)$

The graph changes from concave down to concave up at $x = -\sqrt{2}$, from concave up to concave down at $x = 0$, and from concave down to concave up at $x = \sqrt{2}$. Therefore, $(-\sqrt{2}, f(-\sqrt{2}))$, $(0, f(0))$, and $(\sqrt{2}, f(\sqrt{2}))$ are points of inflection. Since

$$f(-\sqrt{2}) = 3(-\sqrt{2})^5 - 20(-\sqrt{2})^3 = 28\sqrt{2},$$
$$f(0) = 3(0)^5 - 20(0)^3 = 0,$$
and $$f(\sqrt{2}) = 3(\sqrt{2})^5 - 20(\sqrt{2})^3 = -28\sqrt{2},$$

these points are $(-\sqrt{2}, 28\sqrt{2})$, $(0, 0)$, and $(\sqrt{2}, -28\sqrt{2})$, as shown below.

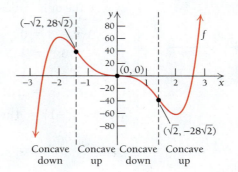

Quick Check 2 ✔

Determine the points of inflection for the function given by $g(x) = 10x^3 - 6x^5$.

2 ✔

Curve Sketching

The first and second derivatives enhance our ability to sketch curves. We use the following strategy:

> ## Strategy for Sketching Graphs*
>
> **a)** *Derivatives and domain.* Find $f'(x)$ and $f''(x)$. Note the domain of f.
>
> **b)** *Critical values of f.* Find the critical values by solving $f'(x) = 0$ and finding where $f'(x)$ does not exist. These numbers yield candidates for relative maxima or minima. Find the function values at these points.
>
> **c)** *Increasing and/or decreasing; relative extrema.* Substitute each critical value, x_0, from step (b) into $f''(x)$. If $f''(x_0) < 0$, then $f(x_0)$ is a relative maximum and f is increasing on the left of x_0 and decreasing on the right. If $f''(x_0) > 0$, then $f(x_0)$ is a relative minimum and f is decreasing on the left of x_0 and increasing on the right.
>
> **d)** *Inflection points.* Determine candidates for inflection points by finding where $f''(x) = 0$ or where $f''(x)$ does not exist. Find the function values at these points.
>
> **e)** *Concavity.* Use the candidates for inflection points from step (d) to define intervals. Substitute test values into $f''(x)$ to determine where the graph is concave up ($f''(x) > 0$) and where it is concave down ($f''(x) < 0$).
>
> **f)** *Sketch the graph.* Sketch the graph using the information from steps (a)–(e), plotting extra points as needed.

The examples that follow apply this step-by-step strategy.

EXAMPLE 4   Find any relative extrema and graph the function f given by

$$f(x) = x^3 - 3x + 2.$$

Solution

a) *Derivatives and domain.* Find $f'(x)$ and $f''(x)$:

$$f'(x) = 3x^2 - 3,$$
$$f''(x) = 6x.$$

Like any polynomial function, f, f', and f'' have as their domain $(-\infty, \infty)$, or $\mathbb{R}$.

b) *Critical values of f.* Find the critical values by determining where $f'(x)$ does not exist and by solving $f'(x) = 0$. We know that $f'(x) = 3x^2 - 3$ exists for all values of x, so the only critical values are where $f'(x)$ is 0:

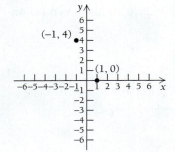

$$3x^2 - 3 = 0 \qquad \text{Setting } f'(x) \text{ equal to } 0$$
$$3x^2 = 3$$
$$x^2 = 1$$
$$x = \pm 1. \qquad \text{Using the Principle of Square Roots}$$

We have $f(-1) = 4$ and $f(1) = 0$, so we plot $(-1, 4)$ and $(1, 0)$.

---

*This strategy is refined further, for rational functions, in Section 2.3.

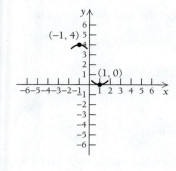

c) *Increasing and/or decreasing; relative extrema.* Substitute the critical values into $f''(x)$:

$$f''(-1) = 6(-1) = -6 < 0,$$

so $f(-1) = 4$ is a relative maximum, with f increasing on $(-\infty, -1)$ and decreasing on $(-1, 1)$.

$$f''(1) = 6 \cdot 1 = 6 > 0,$$

so $f(1) = 0$ is a relative minimum, with f decreasing on $(-1, 1)$ and increasing on $(1, \infty)$.

d) *Inflection points.* Find possible inflection points by finding where $f''(x)$ does not exist and by solving $f''(x) = 0$. We know that $f''(x) = 6x$ exists for all values of x, so we solve $f''(x) = 0$:

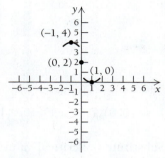

$$6x = 0 \qquad \text{Setting } f''(x) \text{ equal to } 0$$
$$x = 0. \qquad \text{Dividing both sides by 6}$$

We have $f(0) = 2$, which gives us another point, $(0, 2)$, on the graph.

e) *Concavity.* Find the intervals on which f is concave up or concave down. From step (c), we can conclude that f is concave down over the interval $(-\infty, 0)$ and concave up over $(0, \infty)$.

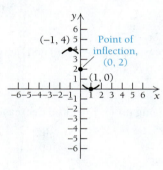

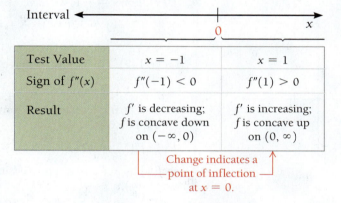

Test Value	$x = -1$	$x = 1$
Sign of $f''(x)$	$f''(-1) < 0$	$f''(1) > 0$
Result	f' is decreasing; f is concave down on $(-\infty, 0)$	f' is increasing; f is concave up on $(0, \infty)$

Change indicates a point of inflection at $x = 0$.

f) *Sketch the graph.* Sketch the graph using the information in steps (a)–(e), plotting some extra values if desired. The graph follows.

TECHNOLOGY CONNECTION

Check the results of Example 4 using a calculator.

x	$f(x)$
-3	-16
-2	0
-1	4
0	2
1	0
2	4
3	20

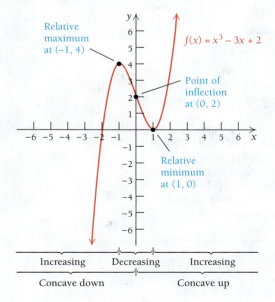

EXAMPLE 5   Find any relative extrema and graph the function f given by

$$f(x) = x^4 - 2x^2.$$

Solution

a) *Derivatives and domain.* Find $f'(x)$ and $f''(x)$:

$$f'(x) = 4x^3 - 4x,$$
$$f''(x) = 12x^2 - 4.$$

The domain of f, f', and f'' is $\mathbb{R}$.

b) *Critical values.* Since the domain of f' is $\mathbb{R}$, the only critical values are where $f'(x) = 0$:

$$4x^3 - 4x = 0 \qquad \color{red}{\text{Setting } f'(x) \text{ equal to } 0}$$
$$4x(x^2 - 1) = 0$$
$$4x = 0 \quad \text{or} \quad x^2 - 1 = 0$$
$$x = 0 \quad \text{or} \qquad x^2 = 1$$
$$x = \pm 1. \qquad \color{red}{\text{Using the Principle of Square Roots}}$$

We have $f(0) = 0$, $f(-1) = -1$, and $f(1) = -1$, so $(0, 0)$, $(-1, -1)$, and $(1, -1)$ are on the graph.

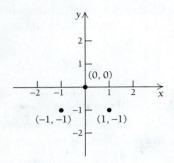

c) *Increasing and/or decreasing; relative extrema.* Substitute the critical values into $f''(x)$. For $x = -1$, we have

$$f''(-1) = 12(-1)^2 - 4 = 8 > 0,$$

so $f(-1) = -1$ is a relative minimum, with f decreasing on $(-\infty, -1)$ and increasing on $(-1, 0)$. For $x = 0$, we have

$$f''(0) = 12 \cdot 0^2 - 4 = -4 < 0,$$

so $f(0) = 0$ is a relative maximum, with f increasing on $(-1, 0)$ and decreasing on $(0, 1)$. For $x = 1$, we have

$$f''(1) = 12 \cdot 1^2 - 4 = 8 > 0,$$

so $f(1) = -1$ is also a relative minimum, with f decreasing on $(0, 1)$ and increasing on $(1, \infty)$.

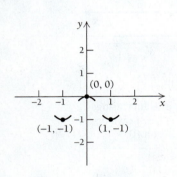

d) *Inflection points.* Find where $f''(x)$ does not exist and where $f''(x) = 0$. Since the domain of f'' is $\mathbb{R}$, we solve $f''(x) = 0$:

$$12x^2 - 4 = 0 \qquad \color{red}{\text{Setting } f''(x) \text{ equal to } 0}$$
$$4(3x^2 - 1) = 0$$
$$3x^2 - 1 = 0 \qquad \color{red}{\text{Dividing both sides by } 4}$$
$$3x^2 = 1$$
$$x^2 = \frac{1}{3}$$
$$x = \pm \frac{1}{\sqrt{3}}. \qquad \color{red}{\text{Using the Principle of Square Roots}}$$

We have

$$f\left(\frac{1}{\sqrt{3}}\right) = \left(\frac{1}{\sqrt{3}}\right)^4 - 2\left(\frac{1}{\sqrt{3}}\right)^2$$
$$= \frac{1}{9} - \frac{2}{3} = -\frac{5}{9}$$

TECHNOLOGY CONNECTION 〜

EXERCISE

1. Consider

$$f(x) = x^3(x - 2)^3.$$

How many relative extrema do you anticipate finding? Where do you think they will be?

Graph f, f', and f'' using $[-1, 3, -2, 6]$ as a viewing window. Estimate any relative extrema and any inflection points of f. Then check your work using the method of Examples 4 and 5.

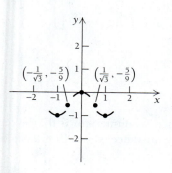

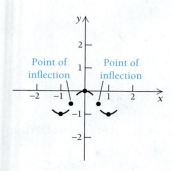

and

$$f\left(-\frac{1}{\sqrt{3}}\right) = -\frac{5}{9}.$$

Thus, $\left(-\dfrac{1}{\sqrt{3}}, -\dfrac{5}{9}\right)$ and $\left(\dfrac{1}{\sqrt{3}}, -\dfrac{5}{9}\right)$ are possible inflection points.

e) *Concavity.* Find the intervals on which f is concave up or concave down. From step (c), we can conclude that f is concave up over the intervals $\left(-\infty, -1/\sqrt{3}\right)$ and $\left(1/\sqrt{3}, \infty\right)$ and concave down over the interval $\left(-1/\sqrt{3}, 1/\sqrt{3}\right)$.

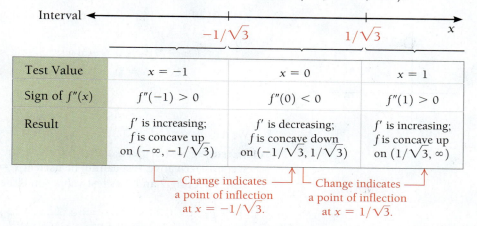

Interval			
	$-1/\sqrt{3}$		$1/\sqrt{3}$
Test Value	$x = -1$	$x = 0$	$x = 1$
Sign of $f''(x)$	$f''(-1) > 0$	$f''(0) < 0$	$f''(1) > 0$
Result	f' is increasing; f is concave up on $(-\infty, -1/\sqrt{3})$	f' is decreasing; f is concave down on $(-1/\sqrt{3}, 1/\sqrt{3})$	f' is increasing; f is concave up on $(1/\sqrt{3}, \infty)$

Change indicates a point of inflection at $x = -1/\sqrt{3}$. Change indicates a point of inflection at $x = 1/\sqrt{3}$.

f) *Sketch the graph.* Sketch the graph using the information in steps (a)–(e). By solving $x^4 - 2x^2 = 0$, we can find the x-intercepts easily. They are $\left(-\sqrt{2}, 0\right)$, $(0, 0)$, and $\left(\sqrt{2}, 0\right)$. Other function values can also be calculated.

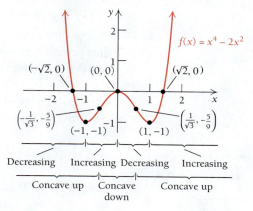

$$f(x) = x^4 - 2x^2$$

Quick Check 3 ✔

Find the relative maxima and minima of the function f given by $f(x) = 1 + 8x^2 - x^4$, and sketch the graph.

3 ✔

EXAMPLE 6   Graph the function f given by

$$f(x) = (2x - 5)^{1/3} + 1.$$

List the coordinates of any extrema and points of inflection. State where the function is increasing or decreasing, as well as where it is concave up or concave down.

Solution

a) *Derivatives and domain.* Find $f'(x)$ and $f''(x)$:

$$f'(x) = \tfrac{1}{3}(2x - 5)^{-2/3} \cdot 2 = \tfrac{2}{3}(2x - 5)^{-2/3}, \text{ or } \frac{2}{3(2x - 5)^{2/3}};$$

$$f''(x) = -\tfrac{4}{9}(2x - 5)^{-5/3} \cdot 2 = -\tfrac{8}{9}(2x - 5)^{-5/3}, \text{ or } \frac{-8}{9(2x - 5)^{5/3}}.$$

The domain of f is $\mathbb{R}$. However, the domain of f' and f'' is $\left(-\infty, \frac{5}{2}\right) \cup \left(\frac{5}{2}, \infty\right)$ since their denominators equal zero when $x = \frac{5}{2}$.

b) *Critical values.* Since

$$f'(x) = \frac{2}{3(2x - 5)^{2/3}}$$

is never 0 (a rational expression equals 0 only when its numerator is 0), the only critical value is when $f'(x)$ does not exist. The only instance where $f'(x)$ does not exist is when its denominator is 0:

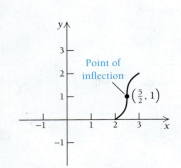

$$3(2x - 5)^{2/3} = 0 \qquad \text{Setting the denominator equal to zero}$$
$$(2x - 5)^{2/3} = 0 \qquad \text{Dividing both sides by 3}$$
$$(2x - 5)^2 = 0 \qquad \text{Cubing both sides}$$
$$2x - 5 = 0 \qquad \text{Using the Principle of Square Roots}$$
$$2x = 5$$
$$x = \tfrac{5}{2}$$

We now have $f\left(\frac{5}{2}\right) = \left(2 \cdot \frac{5}{2} - 5\right)^{1/3} + 1 = 0 + 1 = 1$, so we plot the point $\left(\frac{5}{2}, 1\right)$.

c) *Increasing and/or decreasing; relative extrema.* Since $f''\left(\frac{5}{2}\right)$ does not exist, the Second-Derivative Test cannot be used at $x = \frac{5}{2}$. Instead, we use the First-Derivative Test, selecting 2 and 3 as test values on either side of $\frac{5}{2}$:

$$f'(2) = \frac{2}{3(2 \cdot 2 - 5)^{2/3}} = \frac{2}{3(-1)^{2/3}} = \frac{2}{3 \cdot 1} = \frac{2}{3} > 0,$$

and $f'(3) = \dfrac{2}{3(2 \cdot 3 - 5)^{2/3}} = \dfrac{2}{3 \cdot 1^{2/3}} = \dfrac{2}{3 \cdot 1} = \dfrac{2}{3} > 0.$

Since $f'(x) > 0$ on each side of $x = \frac{5}{2}$, we know that f is increasing on both $\left(-\infty, \frac{5}{2}\right)$ and $\left(\frac{5}{2}, \infty\right)$; thus, $f\left(\frac{5}{2}\right) = 1$ is *not* an extremum.

d) *Inflection points.* Find where $f''(x)$ does not exist and where $f''(x) = 0$. Since $f''(x)$ is never 0 (why?), we only need to find where $f''(x)$ does not exist. Since $f''(x)$ cannot exist for $2x - 5 = 0$, there is a possible inflection point at $\left(\frac{5}{2}, 1\right)$.

e) *Concavity.* We check the concavity on each side of $x = \frac{5}{2}$. We choose $x = 2$ and $x = 3$ as our test values:

$$f''(2) = \frac{-8}{9(2 \cdot 2 - 5)^{5/3}} = \frac{-8}{9(-1)} > 0,$$

so f is concave up on $\left(-\infty, \frac{5}{2}\right)$. And

$$f''(3) = \frac{-8}{9(2 \cdot 3 - 5)^{5/3}} = \frac{-8}{9 \cdot 1} < 0,$$

so f is concave down on $\left(\frac{5}{2}, \infty\right)$.

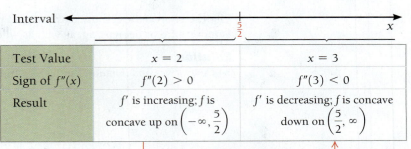

Interval	$\frac{5}{2}$	x
Test Value	$x = 2$	$x = 3$
Sign of $f''(x)$	$f''(2) > 0$	$f''(3) < 0$
Result	f' is increasing; f is concave up on $\left(-\infty, \frac{5}{2}\right)$	f' is decreasing; f is concave down on $\left(\frac{5}{2}, \infty\right)$

Change indicates a point of inflection at $x = 5/2$.

f) *Sketch the graph.* Sketch the graph using the information in steps (a)–(e). By solving $(2x - 5)^{1/3} + 1 = 0$, we can find the *x*-intercept, $(2, 0)$. Other function values can also be calculated.

x	$f(x)$, approximately
0	-0.71
1	-0.44
2	0
$\frac{5}{2}$	1
3	2
4	2.44
5	2.71

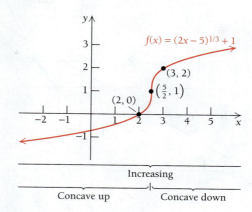

Quick Check 4 ✔

Find any relative extrema and graph the function f given by

$$f(x) = 1 - (x - 2)^{1/3}.$$

4 ✔

The following figures illustrate some information concerning a function f that can be found from its first and second derivatives. Note that the *x*-coordinates of the *x*-intercepts of f' are the critical values of f. Note that the intervals over which f is increasing or decreasing are those intervals for which f' is positive or negative, respectively.

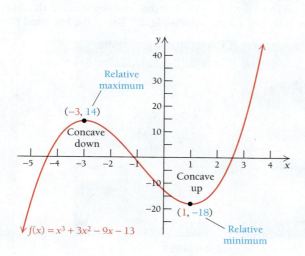

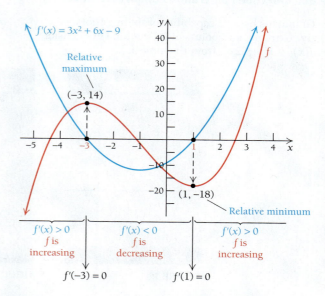

In the figures below, note that the intervals over which f' is increasing or decreasing are, respectively, those intervals over which f'' is positive or negative. And finally, note that when $f''(x) < 0$, the graph of f is concave down, and when $f''(x) > 0$, the graph of f is concave up.

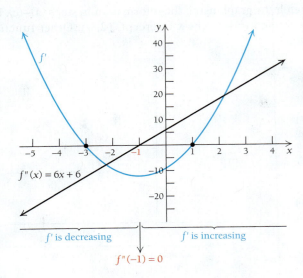

$f''(x) = 6x + 6$

f' is decreasing f' is increasing

$f''(-1) = 0$

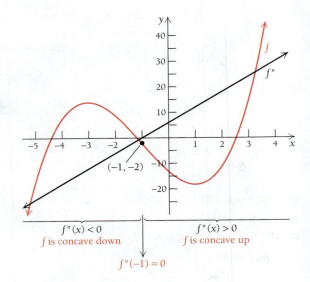

$(-1, -2)$

$f''(x) < 0$
f is concave down

$f''(x) > 0$
f is concave up

$f''(-1) = 0$

TECHNOLOGY CONNECTION

Using Graphicus to Find Roots, Extrema, and Inflection Points and to Graph Derivatives

Graphicus has the capability of finding roots, relative extrema, and points of inflection. Let's consider $f(x) = x^4 - 2x^2$, from Example 5.

After opening Graphicus, touch the blank rectangle at the top of the screen and enter y(x)=x^4-2x^2. Press ⊞ in the upper right. You will see the graph (Fig. 1). Then touch the fourth icon from the left at the bottom, and you will see the roots highlighted on the graph (Fig. 2). Touch the left-hand root

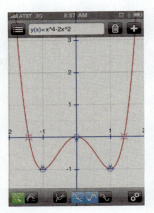

FIGURE 1

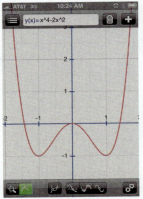

FIGURE 2

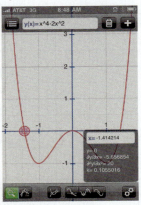

FIGURE 3

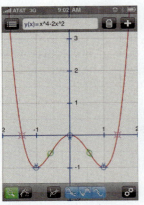

FIGURE 4

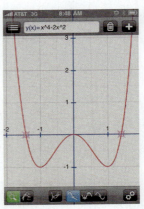

FIGURE 5

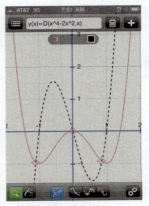

FIGURE 6

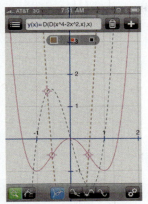

FIGURE 7

(continued)

symbol, and a value for one root, in this case, an approximation, -1.414214, is displayed (Fig. 3).

To find the relative extrema, touch the fifth icon from the left; the relative extrema will be highlighted on the graph in a different color. Touch each point to identify the relative minima at $(-1, -1)$ and $(1, -1)$ and a relative maximum at $(0, 0)$ (Fig. 4).

To find points of inflection, touch the sixth icon from the left, and the points of inflection will be highlighted as shown in Fig. 5. Touch these points and the approximations $(-0.577, -0.556)$ and $(0.577, -0.556)$ are displayed.

Graphicus can graph derivatives. Go back to the original graph of f (Fig. 1). Touch $\boxed{+}$; then press Add derivative, and you will see the original graph and the graph of the first derivative as a dashed line of a different color (Fig. 6). Touch $\boxed{+}$ and Add derivative again, and you will see the graph of the original function, with dashed graphs of the first and second derivatives in different colors (Fig. 7). You can toggle between the derivatives by pressing the colored squares above the graphs.

EXERCISES

Use Graphicus to graph each function and its first and second derivatives. Then find approximations for roots, relative extrema, and points of inflection.

1. $f(x) = 2x^3 - x^4$ **2.** $f(x) = x(200 - x)$

3. $f(x) = x^3 - 6x^2$

4. $f(x) = -4.32 + 1.44x + 3x^2 - x^3$

5. $g(x) = x\sqrt{4 - x^2}$

6. $g(x) = \dfrac{4x}{x^2 + 1}$ **7.** $f(x) = \dfrac{x^2 - 3x}{x - 1}$

8. $f(x) = |x + 2| - 3$ **9.** $f(x) = 3x^5 - 5x^3$

Section Summary

- The second derivative f'' is used to determine the *concavity* of the graph of f.
- If $f''(x) > 0$ for all x in an open interval I, then the graph of f is *concave up* over I.
- If $f''(x) < 0$ for all x in an open interval I, then the graph of f is *concave down* over I.
- If c is a critical value and $f''(c) > 0$, then $f(c)$ is a relative minimum.
- If c is a critical value and $f''(c) < 0$, then $f(c)$ is a relative maximum.

- If c is a critical value and $f''(c) = 0$, the First-Derivative Test must be used to determine if $f(c)$ is a relative extremum.
- If $f''(x_0) = 0$ or $f''(x_0)$ does not exist, and there is a change in concavity on the left and on the right of x_0, then the point $(x_0, f(x_0))$ is called a *point of inflection*.
- Finding relative extrema, intervals over which a function is increasing or decreasing, intervals of upward or downward concavity, and points of inflection form a strategy for accurate curve sketching.

2.2 Exercise Set

For each function, find all relative extrema and classify each as a maximum or minimum. Use the Second-Derivative Test where possible.

1. $f(x) = 4 - x^2$ **2.** $f(x) = 5 - x^2$

3. $f(x) = x^2 + x - 1$ **4.** $f(x) = x^2 - x$

5. $f(x) = -4x^2 + 3x - 1$ **6.** $f(x) = -5x^2 + 8x - 7$

7. $f(x) = x^3 - 12x - 1$

8. $f(x) = 8x^3 - 6x + 1$

Sketch the graph of each function. List the coordinates of any extrema or points of inflection. State where the function is increasing or decreasing and where its graph is concave up or concave down.

9. $f(x) = x^3 - 27x$ **10.** $f(x) = x^3 - 12x$

11. $f(x) = 2x^3 - 3x^2 - 36x + 28$

12. $f(x) = 3x^3 - 36x - 3$ **13.** $f(x) = 80 - 9x^2 - x^3$

14. $f(x) = \frac{8}{3}x^3 - 2x + \frac{1}{3}$ **15.** $f(x) = -x^3 + 3x - 2$

16. $f(x) = -x^3 + 3x^2 - 4$

17. $f(x) = 3x^4 - 16x^3 + 18x^2$
(Round results to three decimal places.)

18. $f(x) = 3x^4 + 4x^3 - 12x^2 + 5$
(Round results to three decimal places.)

19. $f(x) = x^4 - 6x^2$ **20.** $f(x) = 2x^2 - x^4$

21. $f(x) = x^3 - 6x^2 + 9x + 1$

22. $f(x) = x^3 - 2x^2 - 4x + 3$

23. $f(x) = x^4 - 2x^3$ **24.** $f(x) = 3x^4 + 4x^3$

25. $f(x) = x^3 - 6x^2 - 135x$

26. $f(x) = x^3 - 3x^2 - 144x - 140$

27. $f(x) = x^4 - 4x^3 + 10$ **28.** $f(x) = \frac{4}{3}x^3 - 2x^2 + x$

29. $f(x) = x^3 - 6x^2 + 12x - 6$

30. $f(x) = x^3 + 3x + 1$ **31.** $f(x) = 5x^3 - 3x^5$

32. $f(x) = 20x^3 - 3x^5$
(Round results to three decimal places.)

33. $f(x) = x^2(1 - x)^2$
(Round results to three decimal places.)

34. $f(x) = x^2(3 - x)^2$
(Round results to three decimal places.)

35. $f(x) = (x - 1)^{2/3}$ **36.** $f(x) = (x + 1)^{2/3}$

37. $f(x) = (x - 3)^{1/3} - 1$ **38.** $f(x) = (x - 2)^{1/3} + 3$

39. $f(x) = -2(x - 4)^{2/3} + 5$

40. $f(x) = -3(x - 2)^{2/3} + 3$

41. $f(x) = -x\sqrt{1 - x^2}$ **42.** $f(x) = x\sqrt{4 - x^2}$

43. $f(x) = \dfrac{8x}{x^2 + 1}$ **44.** $f(x) = \dfrac{x}{x^2 + 1}$

45. $f(x) = \dfrac{-4}{x^2 + 1}$ **46.** $f(x) = \dfrac{3}{x^2 + 1}$

For Exercises 47–56, sketch a graph that possesses the characteristics listed. Answers may vary.

47. f is decreasing and concave up on $(-\infty, 2)$,
f is decreasing and concave down on $(2, \infty)$.

48. f is increasing and concave up on $(-\infty, 4)$,
f is increasing and concave down on $(4, \infty)$.

49. f is decreasing and concave down on $(-\infty, 3)$,
f is decreasing and concave up on $(3, \infty)$.

50. f is increasing and concave down on $(-\infty, 1)$,
f is increasing and concave up on $(1, \infty)$.

51. f is concave up at $(1, -3)$, concave down at $(8, 7)$, and has an inflection point at $(5, 4)$.

52. f is concave down at $(1, 5)$, concave up at $(7, -2)$, and has an inflection point at $(4, 1)$.

53. $f'(-3) = 0, f''(-3) < 0, f(-3) = 8; f'(9) = 0,$
$f''(9) > 0, f(9) = -6; f''(2) = 0,$ and $f(2) = 1$

54. $f'(-1) = 0, f''(-1) > 0, f(-1) = -5; f'(7) = 0,$
$f''(7) < 0, f(7) = 10; f''(3) = 0,$ and $f(3) = 2$

55. $f'(0) = 0, f''(0) < 0, f(0) = 5; f'(2) = 0, f''(2) > 0,$
$f(2) = 2; f'(4) = 0, f''(4) < 0,$ and $f(4) = 3$

56. $f'(-1) = 0, f''(-1) > 0, f(-1) = -2; f'(1) = 0,$
$f''(1) > 0, f(1) = -2; f'(0) = 0, f''(0) < 0,$
and $f(0) = 0$

APPLICATIONS

Business and Economics

Total revenue, cost, and profit.   Using the same set of axes, graph R, C, and P, the total-revenue, total-cost, and total-profit functions.

57. $R(x) = 50x - 0.5x^2$, $C(x) = 4x + 10$

58. $R(x) = 50x - 0.5x^2$, $C(x) = 10x + 3$

59. **Small business.**   The percentage of the U.S. national income generated by nonfarm proprietors between 1970 and 2000 can be modeled by the function f given by

$$p(x) = \frac{13x^3 - 240x^2 - 2460x + 585,000}{75,000},$$

where x is the number of years since 1970. (*Source:* Based on data from www.bls.gov.) Sketch the graph of this function for $0 \le x \le 40$.

60. **Labor force.**   The percentage of the U.S. civilian labor force aged 45–54 between 1970 and 2000 can be modeled by the function f given by

$$f(x) = 0.025x^2 - 0.71x + 20.44,$$

where x is the number of years after 1970. (*Source:* Based on data from www.bls.gov.) Sketch the graph of this function for $0 \le x \le 30$.

Life and Physical Sciences

61. **Rainfall in Reno.**   The average monthly rainfall in Reno, Nevada, is approximated by

$$R(t) = -0.006t^4 + 0.213t^3 - 1.702t^2 + 0.615t + 27.745,$$

where $R(t)$ is the amount of rainfall in millimeters at t months since December. (*Source:* Based on data from weather.com.)

Find the point(s) of inflection for R. What is the significance of these point(s)?

62. Coughing velocity. A person coughs when a foreign object is in the windpipe. The velocity of the cough depends on the size of the object. Suppose a person has a windpipe with a 20-mm radius. If a foreign object has a radius r, in millimeters, then the velocity V, in millimeters per second, needed to remove the object by a cough is given by

$$V(r) = k(20r^2 - r^3), \quad 0 \le r \le 20,$$

where k is some positive constant. For what size object is the maximum velocity required to remove the object?

SYNTHESIS

 In each of Exercises 63 and 64, determine which graph is the derivative of the other and explain why.

63.

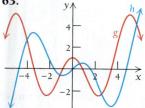

64.

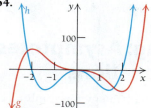

65. Use calculus to prove that the relative minimum or maximum for any function f for which

$$f(x) = ax^2 + bx + c, \quad a \ne 0,$$

occurs at $x = -b/(2a)$.

66. Use calculus to prove that the point of inflection for any function g given by

$$g(x) = ax^3 + bx^2 + cx + d, \quad a \ne 0,$$

occurs at $x = -b/(3a)$.

For Exercises 67–73, assume that f is differentiable over $(-\infty, \infty)$. Classify each of the following statements as either true or false. If a statement is false, explain why.

67. If f has exactly two critical values at $x = a$ and $x = b$, where $a < b$, then there must exist exactly one point of inflection at $x = c$ such that $a < c < b$. In other words, exactly one point of inflection must exist between any two critical points.

68. If f has exactly two critical values at $x = a$ and $x = b$, where $a < b$, then there must exist at least one point of inflection at $x = c$ such that $a < c < b$. In other words, at least one point of inflection must exist between any two critical points.

69. A function f can have no extrema but still have at least one point of inflection.

70. If f has two points of inflection, then there is a critical value located between those points of inflection.

71. The function f can have a point of inflection at a critical value.

72. The function f can have a point of inflection at an extreme value.

73. A function f can have exactly one extreme value but no points of inflection.

 74. Hours of daylight. The number of hours of daylight in Chicago is represented in the graph below. On what dates is the number of hours of daylight changing most rapidly? How can you tell?

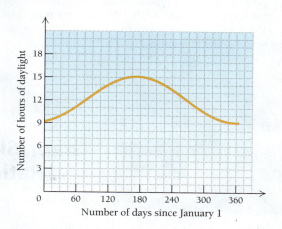

(*Source:* Astronomical Applications Dept., U.S. Naval Observatory.)

TECHNOLOGY CONNECTION

Graph each function. Then estimate any relative extrema. Where appropriate, round to three decimal places.

75. $f(x) = 4x - 6x^{2/3}$

76. $f(x) = 3x^{2/3} - 2x$

77. $f(x) = x^2(1 - x)^3$

78. $f(x) = x^2(x - 2)^3$

79. $f(x) = (x - 1)^{2/3} - (x + 1)^{2/3}$

80. $f(x) = x - \sqrt{x}$

81. Head of household education. The data in the following table relate the percentage of households headed by someone with a bachelor's degree or higher to the number of years since 2000.

NUMBER OF YEARS SINCE 2000	PERCENTAGE OF HOUSEHOLDS
0	66.0
1	75.2
3	78.3
7	84.0
9	88.5
10	89.2
11	89.9

(*Source:* www.census.gov/prod/2013pubs/p20-569.pdf.)

a) Use regression (see Section R.6) to fit linear, cubic, and quartic functions $y = f(x)$ to the data, where x is the number of years since 2000 and y is the percentage of households headed by someone with a bachelor's degree or higher. Which function f best fits the data?

b) What is the domain of f?

c) Does f have any relative extrema? How can you tell?

Answers to Quick Checks

1. Relative minimum: -4 at $x = -1$; relative maximum: 4 at $x = 1$

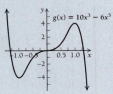

2. $(0, 0)$, $\left(\frac{\sqrt{2}}{2}, 2.475\right)$, and $\left(-\frac{\sqrt{2}}{2}, -2.475\right)$

3. Relative maxima: 17 at $x = -2$ and $x = 2$; relative minimum: 1 at $x = 0$

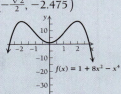

4. No relative extrema

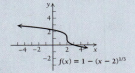

2.3

- Find limits involving infinity.
- Determine the asymptotes of a function's graph.
- Graph rational functions.

Graph Sketching: Asymptotes and Rational Functions

Thus far we have developed a strategy for graphing continuous functions. We now consider discontinuous functions, most of which are rational functions. Our graphing skills must allow for discontinuities and certain lines called *asymptotes*.

Recall the definition of a rational function:

> **DEFINITION**
>
> A **rational function** is a function f that can be described by
>
> $$f(x) = \frac{P(x)}{Q(x)},$$
>
> where $P(x)$ and $Q(x)$ are polynomials, with $Q(x)$ not the zero polynomial. The domain of f consists of all x for which $Q(x) \neq 0$.

We wish to graph rational functions in which the denominator is not a constant. Before doing so, we need to reconsider limits.

Vertical and Horizontal Asymptotes

Consider the rational function

$$f(x) = \frac{x^2 - 1}{x^2 + x - 6} = \frac{(x - 1)(x + 1)}{(x - 2)(x + 3)},$$

whose graph is shown below.

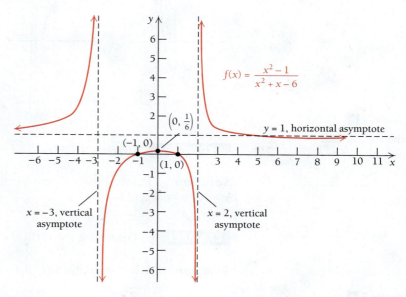

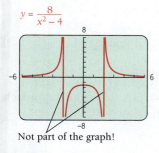
Note that as x gets closer to 2 from the left, the function values approach $-\infty$. As x gets closer to 2 from the right, the function values approach ∞. Thus,

$$\lim_{x \to 2^-} f(x) = -\infty \quad \text{and} \quad \lim_{x \to 2^+} f(x) = \infty.$$

For this graph, we can regard the line $x = 2$ as a "limiting line" called a *vertical asymptote*. The line $x = -3$ is another vertical asymptote.

> **DEFINITION**
>
> The line $x = a$ is a **vertical asymptote** if any of these limit statements is true:
>
> $$\lim_{x \to a^-} f(x) = \infty, \quad \lim_{x \to a^-} f(x) = -\infty, \quad \lim_{x \to a^+} f(x) = \infty, \quad \text{or} \quad \lim_{x \to a^+} f(x) = -\infty.$$

If a rational function f is simplified, meaning that the numerator and denominator have no common factor other than -1 or 1, then if a makes the denominator 0, the line $x = a$ is a vertical asymptote. For example,

$$f(x) = \frac{x^2 - 9}{x - 3} = \frac{(x - 3)(x + 3)}{x - 3}$$

does not have a vertical asymptote at $x = 3$ because $(x - 3)$ is a common factor of the numerator and the denominator. In contrast,

$$g(x) = \frac{x^2 - 4}{x^2 + x - 12} = \frac{(x + 2)(x - 2)}{(x - 3)(x + 4)}$$

has $x = 3$ and $x = -4$ as vertical asymptotes since neither $(x - 3)$ nor $(x + 4)$ is a factor of the numerator.

The figures below show four ways in which a vertical asymptote can occur. The dashed lines represent the asymptotes. They are shown for visual assistance only; they are not part of the graphs of the functions.

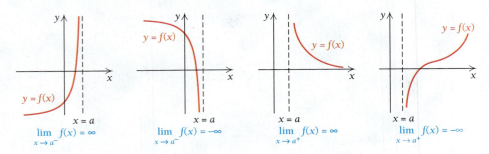

Quick Check 1 ✔

Determine the vertical asymptotes of

$$f(x) = \frac{1}{x(x^2 - 16)}.$$

EXAMPLE 1 Determine the vertical asymptotes of $f(x) = \dfrac{3x - 2}{x(x - 5)(x + 3)}$.

Solution The expression is simplified. Thus, the vertical asymptotes are the lines $x = 0$, $x = 5$, and $x = -3$.   **1** ✔

EXAMPLE 2 Determine the vertical asymptotes of $f(x) = \dfrac{x^2 - 2x}{x^3 - x}$.

Solution We first simplify the expression:

$$f(x) = \frac{x^2 - 2x}{x^3 - x} = \frac{x(x - 2)}{x(x - 1)(x + 1)}$$

$$= \frac{x - 2}{(x - 1)(x + 1)}, \quad x \neq 0.$$

Quick Check 2 ✔

Determine the vertical asymptotes of

$$g(x) = \frac{x - 2}{x^3 - x^2 - 2x}.$$

We now see that the vertical asymptotes are the lines $x = -1$ and $x = 1$.   **2** ✔

Look again at the graph on p. 221. Note that function values approach 1 as x approaches $-\infty$, meaning that $f(x) \to 1$ as $x \to -\infty$. Also, function values approach 1 as x approaches ∞, meaning that $f(x) \to 1$ as $x \to \infty$. Thus,

$$\lim_{x \to -\infty} f(x) = 1 \quad \text{and} \quad \lim_{x \to \infty} f(x) = 1.$$

The line $y = 1$ is called a *horizontal asymptote*.

DEFINITION

The line $y = b$ is a **horizontal asymptote** if either or both of the following limit statements is true:

$$\lim_{x \to -\infty} f(x) = b \quad \text{or} \quad \lim_{x \to \infty} f(x) = b.$$

Horizontal asymptotes occur when the degree of the numerator is less than or equal to the degree of the denominator. (The degree of a polynomial in one variable is the highest power of that variable.) The following figures show three ways in which horizontal asymptotes can occur.

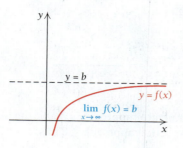
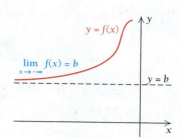
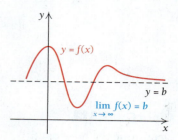

A horizontal asymptote is the limit of a rational function as inputs approach $-\infty$ or ∞.

EXAMPLE 3 Determine the horizontal asymptote of $f(x) = \dfrac{3x - 4}{x}$.

Solution To find the horizontal asymptote, we consider

$$\lim_{x \to \infty} f(x) = \lim_{x \to \infty} \frac{3x - 4}{x}.$$

As the inputs approach ∞, the outputs approach 3. Thus,

$$\lim_{x \to \infty} \frac{3x - 4}{x} = 3.$$

Another way to find this limit is to use the fact that

as $x \to \infty$, we have $\dfrac{1}{x} \to 0$, and more generally, $\dfrac{b}{ax^n} \to 0$,

for any positive integer n and any constants a and b, $a \neq 0$. We multiply by 1, using $(1/x) \div (1/x)$:

$$\lim_{x \to \infty} \frac{3x - 4}{x} = \lim_{x \to \infty} \frac{3x - 4}{x} \cdot \frac{(1/x)}{(1/x)}$$

$$= \lim_{x \to \infty} \frac{\dfrac{3x}{x} - \dfrac{4}{x}}{\dfrac{x}{x}}$$

$$= \lim_{x \to \infty} \frac{3 - \dfrac{4}{x}}{1} \qquad \text{Simplifying, we assume } x \neq 0.$$

$$= \lim_{x \to \infty} \left(3 - \frac{4}{x}\right)$$

$$= 3 - 0 = 3.$$

In a similar manner, it can be shown that

$$\lim_{x \to -\infty} f(x) = 3.$$

The horizontal asymptote is the line $y = 3$.

EXAMPLE 4 Determine the horizontal asymptote of $f(x) = \dfrac{3x^2 + 2x - 4}{2x^2 - x + 1}$.

Solution As in Example 3, the degree of the numerator is the same as the degree of the denominator. Let's adapt the algebraic approach used in that example.

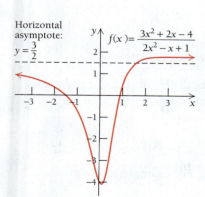

A visualization of Example 3

Horizontal asymptote: $y = \dfrac{3}{2}$

A visualization of Example 4

To do so, we multiply by $(1/x^2) \div (1/x^2)$:

$$f(x) = \frac{3x^2 + 2x - 4}{2x^2 - x + 1} \cdot \frac{\dfrac{1}{x^2}}{\dfrac{1}{x^2}} = \frac{3 + \dfrac{2}{x} - \dfrac{4}{x^2}}{2 - \dfrac{1}{x} + \dfrac{1}{x^2}}.$$

Quick Check 3 ✔

Determine the horizontal asymptote of

$$f(x) = \frac{(2x - 1)(x + 1)}{(3x + 2)(5x + 6)}.$$

As $|x|$ approaches ∞, the numerator approaches 3 and the denominator approaches 2. Therefore, the values of the function approach $\frac{3}{2}$. Thus,

$$\lim_{x \to -\infty} f(x) = \frac{3}{2} \quad \text{and} \quad \lim_{x \to \infty} f(x) = \frac{3}{2}.$$

The line $y = \frac{3}{2}$ is a horizontal asymptote.   **3** ✔

Examples 3 and 4 lead to the following result.

> When the degree of the numerator is the same as the degree of the denominator, the line $y = a/b$ is a horizontal asymptote, where a is the leading coefficient of the numerator and b is the leading coefficient of the denominator.

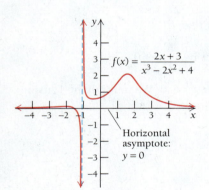

A visualization of Example 5

EXAMPLE 5   Determine the horizontal asymptote of $f(x) = \dfrac{2x + 3}{x^3 - 2x^2 + 4}$.

Solution   Since the degree of the numerator is less than the degree of the denominator, there is a horizontal asymptote. To identify that asymptote, we divide both the numerator and denominator by the highest power of x in the denominator, as in Examples 3 and 4, and find the limits as $|x| \to \infty$:

$$f(x) = \frac{2x + 3}{x^3 - 2x^2 + 4} \cdot \frac{\dfrac{1}{x^3}}{\dfrac{1}{x^3}} = \frac{\dfrac{2}{x^2} + \dfrac{3}{x^3}}{1 - \dfrac{2}{x} + \dfrac{4}{x^3}}.$$

As $|x|$ approaches ∞, the expressions with a denominator that is a power of x approach 0. Thus, the numerator of $f(x)$ approaches 0 as its denominator approaches 1; hence, the entire expression approaches 0. That is, for $x \to -\infty$ or $x \to \infty$,

$$f(x) \approx \frac{0 + 0}{1 - 0 + 0},$$

so $\lim\limits_{x \to -\infty} f(x) = 0$ and $\lim\limits_{x \to \infty} f(x) = 0,$

and the x-axis, the line $y = 0$, is a horizontal asymptote.   ■

> When the degree of the numerator is less than the degree of the denominator, the x-axis, or the line $y = 0$, is a horizontal asymptote.

Slant Asymptotes

Some asymptotes are neither vertical nor horizontal. For example, in the graph of

$$f(x) = \frac{x^2 - 4}{x - 1},$$

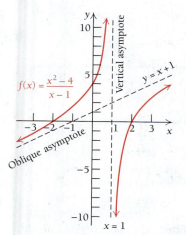

$$f(x) = \frac{x^2 - 4}{x - 1}$$

shown at left, as $|x|$ approaches ∞, the curve approaches the line $y = x + 1$. The line $y = x + 1$ is called a *slant asymptote*, or *oblique asymptote*. In Example 6, we will see how the line $y = x + 1$ is determined.

> **DEFINITION**
>
> A linear asymptote that is neither vertical nor horizontal is called a **slant asymptote**, or an **oblique asymptote**. For any rational function of the form $f(x) = P(x)/Q(x)$, a slant asymptote occurs when the degree of $P(x)$ is exactly 1 more than the degree of $Q(x)$.

We find a slant asymptote by division.

EXAMPLE 6 Find the slant asymptote of $f(x) = \dfrac{x^2 - 4}{x - 1}$.

Solution When we divide the numerator by the denominator, we obtain a quotient of $x + 1$ and a remainder of -3:

$$\begin{array}{r} x + 1 \\ x - 1 \overline{\smash{\big)}\, x^2 \phantom{{}+xx} - 4} \\ \underline{x^2 - x} \phantom{{}-4} \\ x - 4 \\ \underline{x - 1} \\ -3 \end{array}$$

$$f(x) = \frac{x^2 - 4}{x - 1} = (x + 1) + \frac{-3}{x - 1}.$$

Quick Check 4 ✔

Find the slant asymptote of

$$g(x) = \frac{2x^2 + x - 1}{x - 3}.$$

Note that as $|x|$ approaches ∞, $-3/(x - 1)$ approaches 0. Thus, $y = x + 1$ is the slant asymptote.

4 ✔

Intercepts

If they exist, **x-intercepts** occur at values of x for which $y = f(x) = 0$ and are points at which the graph crosses (or touches) the x-axis. If it exists, the **y-intercept** occurs at the value of y for which $x = 0$ and is the point at which the graph crosses (or touches) the y-axis.

EXAMPLE 7 Find the intercepts of the function given by

$$f(x) = \frac{x^3 - x^2 - 6x}{x^2 - 3x + 2}.$$

Solution We factor the numerator and the denominator:

$$f(x) = \frac{x(x + 2)(x - 3)}{(x - 1)(x - 2)}.$$

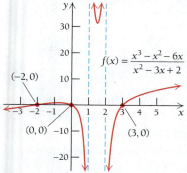

A visualization of Example 7

Quick Check 5 ✔

Find the intercepts of the function given by

$$h(x) = \frac{x^3 - x}{x^2 - 4}.$$

To find the x-intercepts, we solve $f(x) = 0$. Such values occur when the numerator is 0 and the denominator is not. Thus, we solve

$$x(x + 2)(x - 3) = 0.$$

The x-values that make the numerator 0 are 0, -2, and 3. Since none of these make the denominator 0, they yield the x-intercepts $(0, 0)$, $(-2, 0)$, and $(3, 0)$.

To find the y-intercept, we let $x = 0$:

$$f(0) = \frac{0^3 - 0^2 - 6(0)}{0^2 - 3(0) + 2} = 0.$$

In this case, the y-intercept is also an x-intercept, $(0, 0)$.

5 ✔

Sketching Graphs

We can now refine our strategy for graphing.

Strategy for Sketching Graphs

a) *Intercepts.* Find any x- and y-intercepts.

b) *Asymptotes.* Find any vertical, horizontal, or slant asymptotes.

c) *Derivatives and domain.* Find $f'(x)$ and $f''(x)$. Find the domain of f.

d) *Critical values of f.* Find any inputs for which $f'(x)$ is not defined or for which $f'(x) = 0$.

e) *Increasing and/or decreasing; relative extrema.* Substitute each critical value, x_0, from step (d) into $f''(x)$. If $f''(x_0) < 0$, then x_0 yields a relative maximum and f is increasing on the left of x_0 and decreasing on the right. If $f''(x_0) > 0$, then x_0 yields a relative minimum and f is decreasing on the left of x_0 and increasing on the right. On intervals where no critical value exists, use f' and test values to find where f is increasing or decreasing.

f) *Inflection points.* Determine candidates for inflection points by finding x-values for which $f''(x)$ does not exist or for which $f''(x) = 0$. Find the function values at these points. If a function value $f(x)$ does not exist, then the function does not have an inflection point at x.

g) *Concavity.* Use the values from step (f) as endpoints of intervals. Determine the concavity over each interval by checking to see where f' is increasing— that is, where $f''(x) > 0$—and where f' is decreasing—that is, where $f''(x) < 0$. Do this by substituting a test value from each interval into $f''(x)$. Use the results of step (d).

h) *Sketch the graph.* Use the information from steps (a)–(g) to sketch the graph, plotting extra points as needed.

EXAMPLE 8   Sketch the graph of $f(x) = \dfrac{8}{x^2 - 4}$.

Solution

a) *Intercepts.* The x-intercepts occur at values for which the numerator is 0 but the denominator is not. Here, the numerator is a constant, so there are no x-intercepts. To find the y-intercept, we compute $f(0)$:

$$f(0) = \frac{8}{0^2 - 4} = \frac{8}{-4} = -2.$$

This gives us one point on the graph, $(0, -2)$.

b) *Asymptotes.*

Vertical: The denominator, $x^2 - 4 = (x + 2)(x - 2)$, is 0 for x-values of -2 and 2. Thus, the graph has the lines $x = -2$ and $x = 2$ as vertical asymptotes. We draw them as dashed lines (they are guidelines, *not* part of the actual graph).

Horizontal: The degree of the numerator is less than the degree of the denominator, so the x-axis, $y = 0$, is the horizontal asymptote.

Slant: There is no slant asymptote since the degree of the numerator is not 1 more than the degree of the denominator.

c) *Derivatives and domain.* We find $f'(x)$ and $f''(x)$ using the Quotient Rule:

$$f'(x) = \frac{-16x}{(x^2 - 4)^2} \quad \text{and} \quad f''(x) = \frac{16(3x^2 + 4)}{(x^2 - 4)^3}.$$

The domain of f, f', and f'' is $(-\infty, -2) \cup (-2, 2) \cup (2, \infty)$, as determined in step (b).

d) *Critical values of f.* We look for values of x for which $f'(x) = 0$ or for which $f'(x)$ does not exist. From step (c), we see that $f'(x) = 0$ for values of x for which $-16x = 0$, but the denominator is not 0. The only such number is 0 itself. The derivative $f'(x)$ does not exist at -2 and 2, but neither value is in the domain of f. Thus, the only critical value is 0.

e) *Increasing and/or decreasing; relative extrema.* We use the undefined values and the critical values to determine the intervals over which f is increasing and the intervals over which f is decreasing. The values to consider are $-2, 0$, and 2.

Since $f''(0) = \dfrac{16(3 \cdot 0^2 + 4)}{(0^2 - 4)^3} = \dfrac{64}{-64} < 0,$

we know that a relative maximum exists at $(0, f(0))$, or $(0, -2)$. Thus, f is increasing over the interval $(-2, 0)$ and decreasing over $(0, 2)$.

Since $f''(x)$ does not exist for the x-values -2 and 2, we use $f'(x)$ and test values to see if f is increasing or decreasing on $(-\infty, -2)$ and $(2, \infty)$.

Test -3: $f'(-3) = \dfrac{-16(-3)}{[(-3)^2 - 4]^2} = \dfrac{48}{25} > 0$, so f is increasing on $(-\infty, -2)$.

Test 3: $f'(3) = \dfrac{-16(3)}{[(3)^2 - 4]^2} = \dfrac{-48}{25} < 0$, so f is decreasing on $(2, \infty)$.

f) *Inflection points.* We determine candidates for inflection points by finding where $f''(x)$ does not exist and where $f''(x) = 0$. The only values for which $f''(x)$ does not exist are where $x^2 - 4 = 0$, or -2 and 2. Neither value is in the domain of f, so we focus solely on where $f''(x) = 0$, or

$16(3x^2 + 4) = 0.$

Since $16(3x^2 + 4) > 0$ for all real numbers x, there are no points of inflection.

g) *Concavity.* Since no values were found in step (f), the only place where concavity could change is on either side of the vertical asymptotes, $x = -2$ and $x = 2$. To check, we see where $f''(x)$ is positive or negative. The numbers -2 and 2 divide the x-axis into three intervals. We choose a test value in each interval and substitute into $f''(x)$.

Test -3: $f''(-3) = \dfrac{16[3(-3)^2 + 4]}{[(-3)^2 - 4]^3} > 0.$

Test 0: $f''(0) = \dfrac{16[3(0)^2 + 4]}{[(0)^2 - 4]^3} < 0.$ We already knew this from step (e).

Test 3: $f''(3) = \dfrac{16[3(3)^2 + 4]}{[(3)^2 - 4]^3} > 0.$

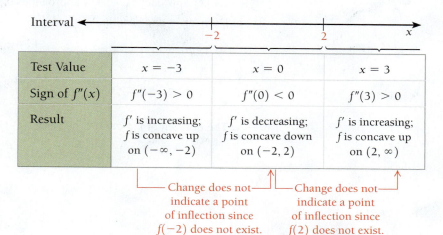

Test Value	$x = -3$	$x = 0$	$x = 3$
Sign of $f''(x)$	$f''(-3) > 0$	$f''(0) < 0$	$f''(3) > 0$
Result	f' is increasing; f is concave up on $(-\infty, -2)$	f' is decreasing; f is concave down on $(-2, 2)$	f' is increasing; f is concave up on $(2, \infty)$

Change does not indicate a point of inflection since $f(-2)$ does not exist. Change does not indicate a point of inflection since $f(2)$ does not exist.

The function is concave up over the intervals $(-\infty, -2)$ and $(2, \infty)$. The function is concave down over $(-2, 2)$.

h) *Sketch the graph.* We sketch the graph using the information in the following table, plotting extra points as needed. The graph is shown below.

x	$f(x)$
-4	$\frac{2}{3}$
-3	1.6
-1	$-2\frac{2}{3}$
0	-2
1	$-2\frac{2}{3}$
3	1.6
4	$\frac{2}{3}$

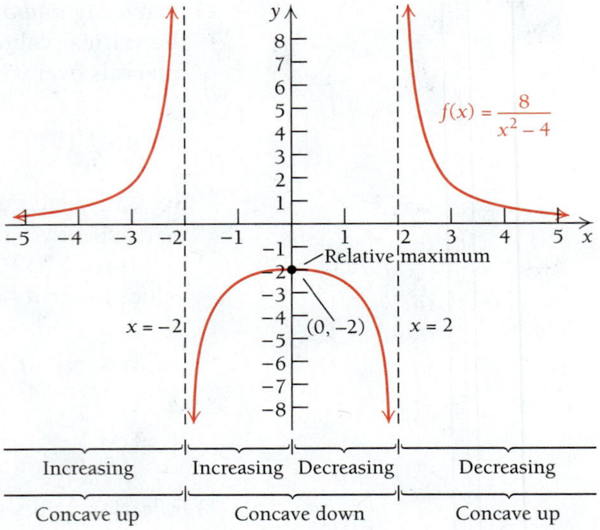

EXAMPLE 9 Sketch the graph of the function given by $f(x) = \dfrac{x^2 + 4}{x}$.

Solution

a) *Intercepts.* The equation $f(x) = 0$ has no solution, so there is no x-intercept. Since 0 is not in the domain, there is no y-intercept either.

b) *Asymptotes.*

Vertical: Since replacing x with 0 makes the denominator 0, the line $x = 0$ (the y-axis) is a vertical asymptote.

Horizontal: The degree of the numerator is greater than the degree of the denominator, so there is no horizontal asymptote.

Slant: The degree of the numerator is 1 greater than the degree of the denominator, so there is a slant asymptote. Note that

$$\frac{x^2 + 4}{x} = \frac{x^2}{x} + \frac{4}{x} = x + \frac{4}{x}.$$

As $|x|$ approaches ∞, $4/x$ approaches 0, so the line $y = x$ is a slant asymptote.

c) *Derivatives and domain.* We find $f'(x)$ and $f''(x)$:

$$f'(x) = 1 - 4x^{-2} = 1 - \frac{4}{x^2};$$

$$f''(x) = 8x^{-3} = \frac{8}{x^3}.$$

The domain of f, f', and f'' is $(-\infty, 0) \cup (0, \infty)$, or all real numbers except 0.

d) *Critical values of f.* We see from step (c) that $f'(x)$ is undefined at $x = 0$, but 0 is not in the domain of f. Thus, to find critical values, we solve $f'(x) = 0$, looking for solutions other than 0:

$$1 - \frac{4}{x^2} = 0 \qquad \text{Setting } f'(x) \text{ equal to 0}$$

$$1 = \frac{4}{x^2}$$

$$x^2 = 4 \qquad \text{Multiplying both sides by } x^2$$

$$x = \pm 2.$$

Thus, -2 and 2 are critical values.

e) *Increasing and/or decreasing; relative extrema.* We use the values found in step (d) to determine where f is increasing or decreasing. The points to consider are -2, 0, and 2.
  Since

$$f''(-2) = \frac{8}{(-2)^3} = -1 < 0,$$

we know that a relative maximum exists at $(-2, f(-2))$, or $(-2, -4)$. Thus, f is increasing over $(-\infty, -2)$ and decreasing over $(-2, 0)$.
  Since

$$f''(2) = \frac{8}{(2)^3} = 1 > 0,$$

we know that a relative minimum exists at $(2, f(2))$, or $(2, 4)$. Thus, f is decreasing over $(0, 2)$ and increasing over $(2, \infty)$.

f) *Inflection points.* We determine candidates for inflection points by finding where $f''(x)$ does not exist or where $f''(x) = 0$. The only value for which $f''(x)$ does not exist is 0, but 0 is not in the domain of f. Thus, the only place an inflection point could occur is where $f''(x) = 0$:

$$\frac{8}{x^3} = 0.$$

But this equation has no solution. Thus, there are no points of inflection.

g) *Concavity.* Since no values were found in step (f), the only place where concavity could change would be on either side of the vertical asymptote, $x = 0$. In step (e), we used the Second-Derivative Test to determine relative extrema. From that work, we know that f is concave down over $(-\infty, 0)$ and concave up over $(0, \infty)$.

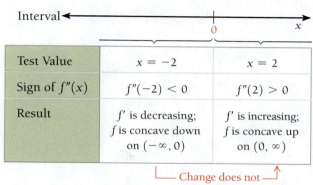

Interval		
Test Value	$x = -2$	$x = 2$
Sign of $f''(x)$	$f''(-2) < 0$	$f''(2) > 0$
Result	f' is decreasing; f is concave down on $(-\infty, 0)$	f' is increasing; f is concave up on $(0, \infty)$

Change does not indicate a point of inflection since $f(0)$ does not exist.

h) *Sketch the graph.* We sketch the graph, shown below, using the preceding information and additional computed values of f.

x	$f(x)$
-4	-5
-2	-4
-1	-5
1	5
2	4
4	5

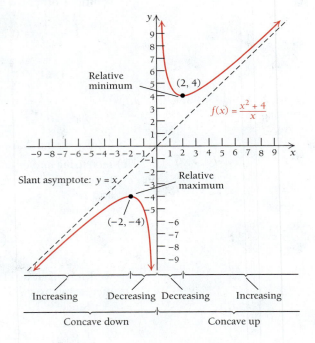

Quick Check 6 ✔

Sketch the graph of the function given by

$$f(x) = \frac{x^2 - 9}{x - 1}.$$

Section Summary

- A line $x = a$ is a *vertical asymptote* if $\lim_{x \to a^-} f(x) = \pm \infty$ or $\lim_{x \to a^+} f(x) = \pm \infty$.
- A line $y = b$ is a *horizontal asymptote* if $\lim_{x \to \infty} f(x) = b$ or $\lim_{x \to -\infty} f(x) = b$.
- A *slant asymptote* occurs when the degree of the numerator is 1 greater than the degree of the denominator.

- Division of polynomials can be used to determine the equation of the slant asymptote.
- Vertical, horizontal, and slant asymptotes can be used as guides for accurate curve sketching. Asymptotes are not a part of a graph but are visual guides only.

2.3 Exercise Set

Determine the vertical asymptote(s) of each function. If none exists, state that fact.

1. $f(x) = \dfrac{x + 4}{x - 2}$

2. $f(x) = \dfrac{2x - 3}{x - 5}$

3. $f(x) = \dfrac{5x}{x^2 - 25}$

4. $f(x) = \dfrac{3x}{x^2 - 9}$

5. $f(x) = \dfrac{x + 3}{x^3 - x}$

6. $f(x) = \dfrac{x + 2}{x^3 - 6x^2 + 8x}$

7. $f(x) = \dfrac{x + 2}{x^2 + 6x + 8}$

8. $f(x) = \dfrac{x + 6}{x^2 + 7x + 6}$

9. $f(x) = \dfrac{7}{x^2 + 49}$

10. $f(x) = \dfrac{6}{x^2 + 36}$

Determine the horizontal asymptote of each function. If none exists, state that fact.

11. $f(x) = \dfrac{6x}{8x + 3}$

12. $f(x) = \dfrac{3x^2}{6x^2 + x}$

13. $f(x) = \dfrac{4x}{x^2 - 3x}$

14. $f(x) = \dfrac{2x}{3x^3 - x^2}$

15. $f(x) = 4 + \dfrac{2}{x}$

16. $f(x) = 5 - \dfrac{3}{x}$

17. $f(x) = \dfrac{6x^3 + 4x}{3x^2 - x}$ **18.** $f(x) = \dfrac{8x^4 - 5x^2}{2x^3 + x^2}$

19. $f(x) = \dfrac{4x^3 - 3x + 2}{x^3 + 2x - 4}$ **20.** $f(x) = \dfrac{6x^4 + 4x^2 - 7}{2x^5 - x + 3}$

21. $f(x) = \dfrac{2x^3 - 4x + 1}{4x^3 + 2x - 3}$ **22.** $f(x) = \dfrac{5x^4 - 2x^3 + x}{x^5 - x^3 + 8}$

Sketch the graph of each function. Indicate where each function is increasing or decreasing, where any relative extrema occur, where asymptotes occur, where the graph is concave up or concave down, where any points of inflection occur, and where any intercepts occur.

23. $f(x) = \dfrac{4}{x}$ **24.** $f(x) = -\dfrac{5}{x}$

25. $f(x) = \dfrac{-2}{x - 5}$ **26.** $f(x) = \dfrac{1}{x - 5}$

27. $f(x) = \dfrac{1}{x - 3}$ **28.** $f(x) = \dfrac{1}{x + 2}$

29. $f(x) = \dfrac{-2}{x + 5}$ **30.** $f(x) = \dfrac{-3}{x - 3}$

31. $f(x) = \dfrac{2x + 1}{x}$ **32.** $f(x) = \dfrac{3x - 1}{x}$

33. $f(x) = x + \dfrac{2}{x}$ **34.** $f(x) = x + \dfrac{9}{x}$

35. $f(x) = \dfrac{-1}{x^2}$ **36.** $f(x) = \dfrac{2}{x^2}$

37. $f(x) = \dfrac{x}{x - 3}$ **38.** $f(x) = \dfrac{x}{x + 2}$

39. $f(x) = \dfrac{1}{x^2 + 3}$ **40.** $f(x) = \dfrac{-1}{x^2 + 2}$

41. $f(x) = \dfrac{x + 3}{x^2 - 9}$ (*Hint:* Simplify.)

42. $f(x) = \dfrac{x - 1}{x^2 - 1}$

43. $f(x) = \dfrac{x - 1}{x + 2}$ **44.** $f(x) = \dfrac{x - 2}{x + 1}$

45. $f(x) = \dfrac{x^2 - 9}{x + 1}$ **46.** $f(x) = \dfrac{x^2 - 4}{x + 3}$

47. $f(x) = \dfrac{x - 3}{x^2 + 2x - 15}$ **48.** $f(x) = \dfrac{x + 1}{x^2 - 2x - 3}$

49. $f(x) = \dfrac{2x^2}{x^2 - 16}$ **50.** $f(x) = \dfrac{x^2 + x - 2}{2x^2 - 2}$

51. $f(x) = \dfrac{10}{x^2 + 4}$ **52.** $f(x) = \dfrac{1}{x^2 - 1}$

53. $f(x) = \dfrac{x^2 + 1}{x}$ **54.** $f(x) = \dfrac{x^3}{x^2 - 1}$

55. $f(x) = \dfrac{x^2 - 16}{x + 4}$ **56.** $f(x) = \dfrac{x^2 - 9}{x - 3}$

APPLICATIONS

Business and Economics

57. Average cost. The total cost, in thousands of dollars, for Acme, Inc., to produce x pairs of rocket skates is given by

$$C(x) = 3x^2 + 80.$$

  a) The *average cost* is given by $A(x) = C(x)/x$. Find $A(x)$.
  b) Graph the average cost.
  c) Find the slant asymptote for the graph of $y = A(x)$, and interpret its significance.

58. Depreciation. Suppose that the value V of the inventory at Fido's Pet Supply, in hundreds of dollars, decreases (depreciates) after t months, where

$$V(t) = 50 - \dfrac{25t^2}{(t + 2)^2}.$$

  a) Find $V(0)$, $V(5)$, $V(10)$, and $V(70)$.
  b) Find the maximum value of the inventory over the interval $[0, \infty)$.
  c) Draw a graph of V.
  d) Does there seem to be a value below which $V(t)$ will never fall? Explain.

59. Total cost and revenue. The total cost and total revenue, in dollars, from producing x couches are given by

$$C(x) = 5000 + 600x \quad \text{and} \quad R(x) = -\tfrac{1}{2}x^2 + 1000x.$$

  a) Find the total-profit function, $P(x)$.
  b) The *average profit* is given by $A(x) = P(x)/x$. Find $A(x)$.
  c) Find the slant asymptote for the graph of $y = A(x)$.
  d) Graph the average profit.

60. Cost of pollution control. Cities and companies find that the cost of pollution control increases along with the percentage of pollutants being removed. Suppose the cost C, in dollars, of removing $p\%$ of the pollutants from a chemical spill is given by

$$C(p) = \dfrac{48{,}000}{100 - p}.$$

  a) Find $C(0)$, $C(20)$, $C(80)$, and $C(90)$.
  b) Find the domain of C.
  c) Draw a graph of C.
  d) Can the company or city afford to remove 100% of the pollutants due to this spill? Explain.

61. Purchasing power. The purchasing power of the U.S. dollar t years after 1990 can be modeled by the function

$$P(x) = \frac{1}{1 + 0.0362x}.$$

(*Source:* Based on data from the Consumer Price Index.)

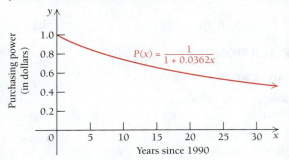

Years since 1990

a) Find $P(5)$, $P(10)$, and $P(25)$.
b) When was the purchasing power $0.50?
c) Find $\lim\limits_{x\to\infty} P(x)$.

Life and Physical Sciences

62. Medication in the bloodstream. After an injection, the amount of a medication A, in cubic centimeters (cc), in the bloodstream decreases with time t, in hours. Suppose that under certain conditions A is given by

$$A(t) = \frac{A_0}{t^2 + 1},$$

where A_0 is the initial amount of the medication. Assume that an initial amount of 100 cc is injected.

a) Find $A(0)$, $A(1)$, $A(2)$, $A(7)$, and $A(10)$.
b) Find the maximum amount of medication in the bloodstream over the interval $[0, \infty)$.
c) Graph the function.
d) According to this function, does the medication ever completely leave the bloodstream? Explain your answer.

General Interest

63. Baseball: earned-run average. A pitcher's *earned-run average* (the average number of runs given up in every 9 innings, or 1 complete game) is given by

$$E = 9 \cdot \frac{r}{n},$$

where r is the number of earned runs allowed in n innings. If we set the number of earned runs allowed at 4 and let n vary, we get a function given by

$$E(n) = 9 \cdot \frac{4}{n}.$$

While pitching for the St. Louis Cardinals in 1968, Bob Gibson had an earned-run average of 1.12, a record low.

a) Complete the following table, rounding to two decimal places:

Innings Pitched, n	9	6	3	1	$\frac{2}{3}$	$\frac{1}{3}$
Earned-Run Average, E						

b) The number of innings pitched n is equivalent to the number of outs that a pitcher is able to get while pitching, divided by 3. For example, a pitcher who gets just 1 out is credited with pitching $\frac{1}{3}$ of an inning. Find $\lim\limits_{n\to 0} E(n)$. Under what circumstances might this limit be plausible?

SYNTHESIS

64. Explain why a vertical asymptote is only a guide and is not part of the graph of a function.

65. Using graphs and limits, explain how three types of asymptotes are used when graphing rational functions.

Find each limit, if it exists.

66. $\lim\limits_{x\to -\infty} \dfrac{-3x^2 + 5}{2 - x}$

67. $\lim\limits_{x\to 0} \dfrac{|x|}{x}$

68. $\lim\limits_{x\to -2} \dfrac{x^3 + 8}{x^2 - 4}$

69. $\lim\limits_{x\to -\infty} \dfrac{-6x^3 + 7x}{2x^2 - 3x - 10}$

70. $\lim\limits_{x\to 1} \dfrac{x^3 - 1}{x^2 - 1}$

TECHNOLOGY CONNECTION

Graph each function using a graphing utility.

71. $f(x) = x^2 + \dfrac{1}{x^2}$

72. $f(x) = \dfrac{x}{\sqrt{x^2 + 1}}$

73. $f(x) = \dfrac{x^3 + 4x^2 + x - 6}{x^2 - x - 2}$

74. $f(x) = \dfrac{x^3 + 2x^2 - 15x}{x^2 - 5x - 14}$

75. $f(x) = \dfrac{x^3 + 2x^2 - 3x}{x^2 - 25}$

76. $f(x) = \left| \dfrac{1}{x} - 2 \right|$

77. Graph the function given by

$$f(x) = \frac{x^2 - 3}{2x - 4}.$$

a) Find any x-intercepts.
b) Find the y-intercept if it exists.
c) Find any asymptotes.

78. Graph the function given by

$$f(x) = \frac{\sqrt{x^2 + 3x + 2}}{x - 3}.$$

a) Estimate $\lim_{x \to \infty} f(x)$ and $\lim_{x \to -\infty} f(x)$ using the graph and input–output tables as needed to refine your estimates.
b) What appears to be the domain of the function? Explain.
c) Find $\lim_{x \to -2^-} f(x)$ and $\lim_{x \to -1^+} f(x)$.

79. Not all asymptotes are linear. Use long division to find an equation for the nonlinear asymptote that is approached by the graph of

$$f(x) = \frac{x^5 + x - 9}{x^3 + 6x}.$$

Then graph the function and its asymptote.

80. Refer to the graph on p. 221. The function is given by

$$f(x) = \frac{x^2 - 1}{x^2 + x - 6}.$$

a) Inspect the graph and estimate the coordinates of any extrema.
b) Find f' and use it to determine the critical values. (*Hint:* you will need the quadratic formula.) Round the x-values to the nearest hundredth.
c) Graph f in the window $[0, 0.2, 0.16, 0.17]$. Use TRACE or MAXIMUM to confirm your results from part (b).
d) Graph f in the window $[9.8, 10, 0.9519, 0.95195]$. Use TRACE or MINIMUM to confirm your results from part (b).

e) How close were your estimates of part (a)? Would you have been able to identify the relative minimum point without calculus?

In Exercises 81–86, determine a rational function that meets the given conditions, and sketch its graph.

81. The function f has a vertical asymptote at $x = 2$, a horizontal asymptote at $y = -2$, and $f(0) = 0$.

82. The function f has a vertical asymptote at $x = 0$, a horizontal asymptote at $y = 3$, and $f(1) = 2$.

83. The function g has vertical asymptotes at $x = -1$ and $x = 1$, a horizontal asymptote at $y = 1$, and $g(0) = 2$.

84. The function g has vertical asymptotes at $x = -2$ and $x = 0$, a horizontal asymptote at $y = -3$, and $g(1) = 4$.

85. The function h has vertical asymptotes at $x = -3$ and $x = 2$, a horizontal asymptote at $y = 0$, and $h(1) = 2$.

86. The function h has vertical asymptotes at $x = -\frac{1}{2}$ and $x = \frac{1}{2}$, a horizontal asymptote at $y = 0$, and $h(0) = -3$.

Answers to Quick Checks

1. The lines $x = 0$, $x = 4$, and $x = -4$ are vertical asymptotes.   **2.** The lines $x = -1$ and $x = 0$ are vertical asymptotes.   **3.** The line $y = \frac{2}{15}$ is a horizontal asymptote.   **4.** The line $y = 2x + 7$ is a slant asymptote.
5. x-intercepts: $(1, 0)$, $(-1, 0)$, $(0, 0)$; y-intercept: $(0, 0)$
6.

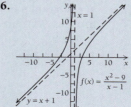

Using Derivatives to Find Absolute Maximum and Minimum Values

2.4

- Find absolute extrema using Maximum–Minimum Principle 1.
- Find absolute extrema using Maximum–Minimum Principle 2.

If an extremum is the highest or lowest value over a function's entire domain, it is called an *absolute extremum*. For example, the parabola given by $f(x) = x^2 + 1$ has a relative minimum at $(0, 1)$. This is also the lowest point for the *entire* graph of f, so it is also the absolute minimum. In applications, we are often interested in absolute extrema.

Absolute Maximum and Minimum Values

A relative minimum may or may not be an absolute minimum, meaning the smallest value of the function over its entire domain. Similarly, a relative maximum may or may not be an absolute maximum, meaning the greatest value of a function over its entire domain.

The function in the following graph has relative minima at c_1 and c_3.

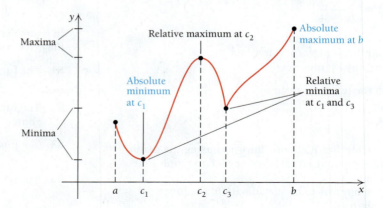

The relative minimum at c_1 is also the absolute minimum. On the other hand, the relative maximum at c_2 is *not* the absolute maximum. The absolute maximum occurs at b.

DEFINITION

Suppose that f is a function with domain I.

$f(c)$ is an **absolute minimum** if $f(c) \leq f(x)$ for all x in I.

$f(c)$ is an **absolute maximum** if $f(c) \geq f(x)$ for all x in I.

Finding Absolute Maximum and Minimum Values over Closed Intervals

Let's consider continuous functions for which the domain is a closed interval. Look at the graphs of f and g below, and try to determine where the absolute maxima and minima (extrema) occur for each function.

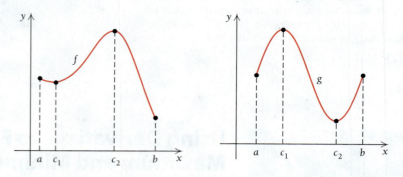

Note that each function does indeed have an absolute maximum value and an absolute minimum value. This leads us to the following theorem.

THEOREM 7   **The Extreme-Value Theorem**

A continuous function f defined over a closed interval $[a, b]$ must have an absolute maximum value and an absolute minimum value over $[a, b]$.

Look again at f and g on the previous page and consider the critical values and the endpoints. The graph of f starts at $f(a)$, falls to $f(c_1)$, rises from $f(c_1)$ to $f(c_2)$, and falls to $f(b)$. The graph of g starts at $g(a)$, rises to $g(c_1)$, falls from $g(c_1)$ to $g(c_2)$, and rises to $g(b)$. It seems reasonable that whatever the maximum and minimum values are, they occur at a critical value or endpoint.

THEOREM 8   Maximum–Minimum Principle 1

Suppose f is a continuous function defined over a closed interval $[a, b]$. To find the absolute maximum and minimum values over $[a, b]$:

a) First find $f'(x)$.

b) Then determine all critical values in $[a, b]$. That is, find all c in $[a, b]$ for which

$$f'(c) = 0 \quad \text{or} \quad f'(c) \text{ does not exist.}$$

c) List the values from step (b) and the endpoints of the interval:

$$a, c_1, c_2, \ldots, c_n, b.$$

d) Evaluate $f(x)$ for each value in step (c):

$$f(a), f(c_1), f(c_2), \ldots, f(c_n), f(b).$$

The largest of these is the **absolute maximum of f over $[a, b]$**. The smallest of these is the **absolute minimum of f over $[a, b]$**.

EXAMPLE 1   Find the absolute maximum and minimum values of

$$f(x) = x^3 - 3x + 2$$

over the interval $\left[-2, \frac{3}{2}\right]$.

Solution   Keep in mind that we are considering only the interval $\left[-2, \frac{3}{2}\right]$.

a) Find $f'(x)$: $f'(x) = 3x^2 - 3$.

b) Find the critical values. The derivative exists for all real numbers. Thus, we solve $f'(x) = 0$:

$$3x^2 - 3 = 0$$
$$3x^2 = 3$$
$$x^2 = 1$$
$$x = \pm 1.$$

c) List the critical values and the endpoints: $-2, -1, 1,$ and $\frac{3}{2}$.

d) Evaluate $f(x)$ at each value in step (c):

$$f(-2) = (-2)^3 - 3(-2) + 2 = -8 + 6 + 2 = 0;$$
$$f(-1) = (-1)^3 - 3(-1) + 2 = -1 + 3 + 2 = 4;$$
$$f(1) = (1)^3 - 3(1) + 2 = 1 - 3 + 2 = 0;$$
$$f\left(\tfrac{3}{2}\right) = \left(\tfrac{3}{2}\right)^3 - 3\left(\tfrac{3}{2}\right) + 2 = \tfrac{27}{8} - \tfrac{9}{2} + 2 = \tfrac{7}{8}$$

The largest of these values, 4, is the maximum. It occurs at $x = -1$. The smallest of these values is 0. It occurs twice: at $x = -2$ and $x = 1$. Thus, over the interval $\left[-2, \frac{3}{2}\right]$, the

$$\text{absolute maximum} = 4 \text{ at } x = -1$$

and the

$$\text{absolute minimum} = 0 \text{ at } x = -2 \text{ and } x = 1.$$

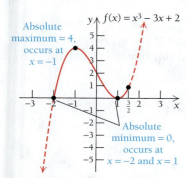

Absolute maximum = 4, occurs at $x = -1$

$f(x) = x^3 - 3x + 2$

Absolute minimum = 0, occurs at $x = -2$ and $x = 1$

A visualization of Example 1

As we see in Example 1, an absolute maximum or minimum value can occur at more than one point.

TECHNOLOGY CONNECTION

Finding Absolute Extrema

To find the absolute extrema in Example 1, we can use any of the methods described in the Technology Connection on p. 200. In this case, we adapt Methods 3 and 4.

Method 3

Method 3 is selected because there are relative extrema in the interval $\left[-2, \frac{3}{2}\right]$. This method gives approximations of these relative extrema.

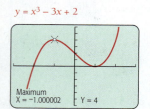
$y = x^3 - 3x + 2$
Maximum
X = -1.000002   Y = 4

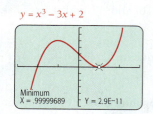
$y = x^3 - 3x + 2$
Minimum
X = .99999689   Y = 2.9E-11

Next, we check function values at these x-values and at the end-points, using Maximum–Minimum Principle 1 to determine the absolute maximum and minimum values over $\left[-2, \frac{3}{2}\right]$.

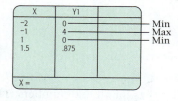

X	Y1	
-2	0	Min
-1	4	Max
1	0	Min
1.5	.875	

X =

Method 4

Example 2 considers the same function as in Example 1, but over a different interval. Because there are no relative extrema, we can use fMax and fMin from the MATH menu. The minimum and maximum values occur at the endpoints, as the following graphs show.

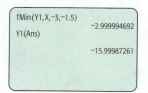
fMin(Y1,X,-3,-1.5)
                    -2.999994692
Y1(Ans)
                    -15.99987261

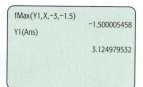
fMax(Y1,X,-3,-1.5)
                    -1.500005458
Y1(Ans)
                    3.124979532

EXERCISE

1. Use a graph to estimate the absolute extrema of $f(x) = x^3 - x^2 - x + 2$, first over the interval $[-2, 1]$ and then over $[-1, 2]$. Then check your work using the methods of Examples 1 and 2.

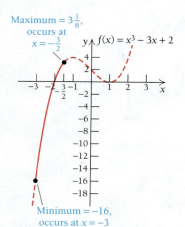
Maximum = $3\frac{1}{8}$, occurs at $x = -\frac{3}{2}$

$y \uparrow f(x) = x^3 - 3x + 2$

Minimum = -16, occurs at $x = -3$

A visualization of Example 2

EXAMPLE 2 Find the absolute maximum and minimum values of

$$f(x) = x^3 - 3x + 2$$

over $\left[-3, -\frac{3}{2}\right]$.

Solution As in Example 1, the derivative is 0 at -1 and 1. But neither -1 nor 1 is in the interval $\left[-3, -\frac{3}{2}\right]$, so there are no critical values in this interval. Thus, the maximum and minimum values occur at the endpoints:

$$f(-3) = (-3)^3 - 3(-3) + 2$$
$$= -27 + 9 + 2 = -16;$$

$$f\left(-\frac{3}{2}\right) = \left(-\frac{3}{2}\right)^3 - 3\left(-\frac{3}{2}\right) + 2$$

$$= -\frac{27}{8} + \frac{9}{2} + 2 = \frac{25}{8} = 3\frac{1}{8}.$$

Thus, the absolute maximum over $\left[-3, -\frac{3}{2}\right]$, is $3\frac{1}{8}$, which occurs at $x = -\frac{3}{2}$, and the absolute minimum is -16, which occurs at $x = -3$.   1 ✓

Quick Check 1 ✓

Find the absolute maximum and minimum values of the function given in Example 2 over the interval $[0, 3]$.

Finding Absolute Maximum and Minimum Values over Other Intervals

When there is only one critical value c in I, we may not need to check endpoint values to determine whether the function has an absolute maximum or minimum value at that point.

> **THEOREM 9   Maximum–Minimum Principle 2**
>
> Suppose f is a function such that $f'(x)$ exists for every x in an interval I and there is *exactly one* (critical) value c in I, for which $f'(c) = 0$. Then
>
> $$f(c) \text{ is the absolute maximum value over } I \text{ if } f''(c) < 0$$
>
> or
>
> $$f(c) \text{ is the absolute minimum value over } I \text{ if } f''(c) > 0.$$

Theorem 9 holds no matter what the interval I is—whether open, closed, or infinite in length. If $f''(c) = 0$, either we must use Maximum–Minimum Principle 1 or we must know more about the behavior of the function over the given interval.

EXAMPLE 3   Find the absolute extrema of $f(x) = 4x - x^2$.

Solution   When no interval is specified, we consider the entire domain of the function. In this case, the domain is the set of all real numbers.

a) Find $f'(x)$:

$$f'(x) = 4 - 2x.$$

b) Find the critical values. The derivative exists for all real numbers. Thus, we need only solve $f'(x) = 0$:

$$4 - 2x = 0$$
$$-2x = -4$$
$$x = 2.$$

c) Since there is only one critical value, we apply Maximum–Minimum Principle 2 using the second derivative:

$$f''(x) = -2.$$

The second derivative is constant. Thus, $f''(2) = -2$, and since this is negative, we have the absolute maximum:

$$f(2) = 4 \cdot 2 - 2^2,$$
$$= 8 - 4 = 4.$$

The function has no minimum, as the graph, shown below, indicates.

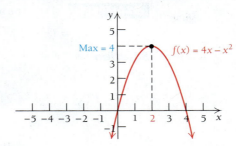

EXAMPLE 4   Find the absolute extrema of $f(x) = 4x - x^2$ over the interval $[1, 4]$.

Solution   From Example 3, we know that the absolute maximum of f on $(-\infty, \infty)$ is $f(2)$, or 4. Since 2 is in $[1, 4]$, we know that the absolute maximum of f over $[1, 4]$ will occur at 2. To find the absolute minimum, we need to check the endpoints:

$$f(1) = 4 \cdot 1 - 1^2 = 3$$

and

$$f(4) = 4 \cdot 4 - 4^2 = 0.$$

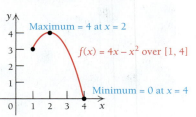

Maximum = 4 at $x = 2$

$f(x) = 4x - x^2$ over $[1, 4]$

Minimum = 0 at $x = 4$

We also see from the graph that the minimum is 0. It occurs at $x = 4$. Thus, the

absolute maximum = 4 at $x = 2$,

and the

absolute minimum = 0 at $x = 4$.   2 ✔

Quick Check 2 ✔

Find the absolute extrema of $f(x) = x^2 - 10x$ over each interval:

a) $[0, 6]$;   **b)** $[4, 10]$.

A Strategy for Finding Absolute Maximum and Minimum Values

The following general strategy can be used when finding absolute maximum and minimum values of continuous functions.

> ### A Strategy for Finding Absolute Maximum and Minimum Values
>
> To find the absolute extrema of a continuous function over an interval:
>
> **a)** Find $f'(x)$.
> **b)** Find the critical values.
> **c)** If the interval is closed and there is more than one critical value, use Maximum–Minimum Principle 1.
> **d)** If the interval is closed and there is exactly one critical value, use either Maximum–Minimum Principle 1 or Maximum–Minimum Principle 2. If it is easy to find $f''(x)$, use Maximum–Minimum Principle 2.
> **e)** If the interval is not closed, such as $(-\infty, \infty)$, $(0, \infty)$, or (a, b), and the function has only one critical value, use Maximum–Minimum Principle 2. In such a case, if the function has a maximum, it will have no minimum; and if it has a minimum, it will have no maximum.

Finding absolute maximum and minimum values when more than one critical value occurs in an interval that is not closed, such as any of those listed in step (e) above, requires a detailed graph or techniques beyond the scope of this book.

EXAMPLE 5   Find the absolute maximum and minimum values of

$$f(x) = (x - 2)^3 + 1.$$

Solution

a) Find $f'(x)$.

$$f'(x) = 3(x - 2)^2.$$

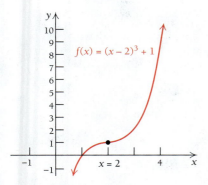

$f(x) = (x - 2)^3 + 1$

$x = 2$

A visualization of Example 5

Quick Check 3

Let $f(x) = x^n$, where n is a positive odd integer. Explain why functions of this form never have an absolute minimum or maximum.

TECHNOLOGY CONNECTION

Finding Absolute Extrema

Let's do Example 6 using MAXIMUM and MINIMUM from the CALC menu. The shape of the graph leads us to see that there is no absolute maximum, but there is an absolute minimum.

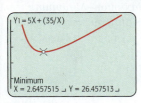

$Y1 = 5X + (35/X)$

Minimum
X = 2.6457515   Y = 26.457513

[0, 10, 0, 50]

Since

$$\sqrt{7} \approx 2.6458,$$

the results in the window confirm the solution found in Example 6.

EXERCISE

1. Use a graph to estimate the absolute extrema of $f(x) = 10x + 1/x$ over $(0, \infty)$. Then check your work using the method of Example 6.

---

b) Find the critical values. The derivative exists for all real numbers. Thus, we solve $f'(x) = 0$:

$$3(x - 2)^2 = 0$$
$$(x - 2)^2 = 0$$
$$x - 2 = 0$$
$$x = 2.$$

c) Since there is only one critical value and there are no endpoints, we can try to apply Maximum–Minimum Principle 2 using the second derivative:

$$f''(x) = 6(x - 2).$$

We have

$$f''(2) = 6(2 - 2) = 0,$$

so Maximum–Minimum Principle 2 does not apply. We cannot use Maximum–Minimum Principle 1 because there are no endpoints. But note that $f'(x) = 3(x - 2)^2$ is never negative. Thus, $f(x)$ is increasing everywhere except at $x = 2$, so there is no maximum and no minimum. For $x < 2$, say $x = 1$, we have $f''(1) = -6 < 0$. For $x > 2$, say $x = 3$, we have $f''(3) = 6 > 0$. Thus, at $x = 2$, the function has a *point of inflection*.   **3** ✔

EXAMPLE 6   Find the absolute maximum and minimum values of

$$f(x) = 5x + \frac{35}{x}$$

over the interval $(0, \infty)$.

Solution

a) Find $f'(x)$. Since $f(x) = 5x + 35x^{-1}$, we have

$$f'(x) = 5 - 35x^{-2}$$
$$= 5 - \frac{35}{x^2}.$$

b) Find the critical values. Since $f'(x)$ exists for all values of x in $(0, \infty)$, the only critical values are those for which $f'(x) = 0$:

$$5 - \frac{35}{x^2} = 0 \qquad \text{Setting } f'(x) \text{ equal to } 0$$
$$5 = \frac{35}{x^2}$$
$$5x^2 = 35 \qquad \text{Multiplying both sides by } x^2, \text{ since } x \neq 0$$
$$x^2 = 7$$
$$x = \sqrt{7}. \qquad \text{Taking the positive root only, since the negative root is not in } (0, \infty)$$

c) Since $(0, \infty)$ is not closed and there is only one critical value, we can apply Maximum–Minimum Principle 2 using the second derivative,

$$f''(x) = 70x^{-3} = \frac{70}{x^3}.$$

Since

$$f''(\sqrt{7}) = \frac{70}{(\sqrt{7})^3} > 0,$$

an absolute minimum occurs at $x = \sqrt{7}$:

$$\text{Absolute minimum} = f(\sqrt{7})$$

$$= 5 \cdot \sqrt{7} + \frac{35}{\sqrt{7}}$$

$$= 5\sqrt{7} + \frac{35}{\sqrt{7}} \cdot \frac{\sqrt{7}}{\sqrt{7}}$$

$$= 5\sqrt{7} + \frac{35\sqrt{7}}{7}$$

$$= 5\sqrt{7} + 5\sqrt{7}$$

$$= 10\sqrt{7} \approx 26.458 \quad \text{at } x = \sqrt{7}.$$

The function has no maximum value, which can happen since the interval $(0, \infty)$ is *not* closed.

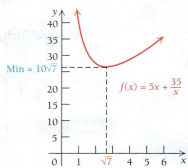

Quick Check 4 ✔

Find the absolute maximum and minimum values of

$$g(x) = \frac{2x^2 + 18}{x}$$

over the interval $(0, \infty)$.

4 ✔

Section Summary

- An *absolute minimum* of f is a value $f(c)$ such that $f(c) \leq f(x)$ for all x in the domain of f.
- An *absolute maximum* of f is a value $f(c)$ such that $f(c) \geq f(x)$ for all x in the domain of f.
- If the domain of f is a closed interval and f is continuous over that domain, then the *Extreme–Value Theorem* guarantees the existence of both an absolute minimum and an absolute maximum.

- Endpoints of a closed interval may be absolute extrema, but not relative extrema.
- If there is exactly one critical value c in the domain of f such that $f'(c) = 0$, then *Maximum–Minimum Principle 2* may be used to find the absolute extrema. Otherwise, *Maximum–Minimum Principle 1* has to be used.

2.4   Exercise Set

1. **Fuel economy.** According to the U.S. Department of Energy, a vehicle's fuel economy, in miles per gallon (mpg), decreases rapidly for speeds over 60 mph.

   a) Estimate the speed at which the absolute maximum gasoline mileage is obtained.

   b) Estimate the speed at which the absolute minimum gasoline mileage is obtained.

   c) What is the mileage obtained at 70 mph?

2. **Fuel economy.** Using the graph in Exercise 1, estimate the absolute maximum and the absolute minimum fuel economy over the interval $[30, 70]$.

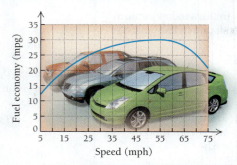

(*Sources*: U.S. Dept. of Energy; a study by West, B.H., McGill, R.N., Hodgson, J.W., Sluder, S.S., and Smith, D.E., Oak Ridge National Laboratory, 1999; www.mpgforspeed.com, 2014.)

Find the absolute maximum and minimum values of each function over the indicated interval, and indicate the x-values at which they occur.

3. $f(x) = 5 + x - x^2$; $[0, 2]$

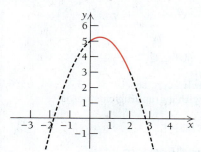

4. $f(x) = 4 + x - x^2$; $[0, 2]$

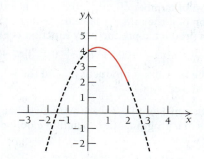

5. $f(x) = x^3 - \frac{1}{2}x^2 - 2x + 5$; $[-2, 1]$

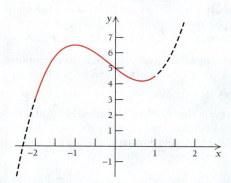

6. $f(x) = x^3 - x^2 - x + 2$; $[-1, 2]$

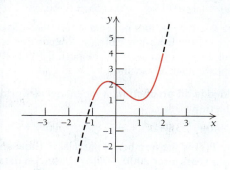

7. $f(x) = x^3 + \frac{1}{2}x^2 - 2x + 4$; $[-2, 0]$

8. $f(x) = x^3 - x^2 - x + 3$; $[-1, 0]$

9. $f(x) = 2x + 4$; $[-1, 1]$

10. $f(x) = 5x - 7$; $[-2, 3]$

11. $f(x) = 7 - 4x$; $[-2, 5]$

12. $f(x) = -2 - 3x$; $[-10, 10]$

13. $f(x) = -5$; $[-1, 1]$

14. $f(x) = x^2 - 6x - 3$; $[-1, 5]$

15. $g(x) = 24$; $[4, 13]$

16. $f(x) = 3 - 2x - 5x^2$; $[-3, 3]$

17. $f(x) = x^2 - 4x + 5$; $[-1, 3]$

18. $f(x) = x^3 - 3x^2$; $[0, 5]$

19. $f(x) = 1 + 6x - 3x^2$; $[0, 4]$

20. $f(x) = x^3 - 3x + 6$; $[-1, 3]$

21. $f(x) = x^3 - 3x$; $[-5, 1]$

22. $f(x) = 3x^2 - 2x^3$; $[-5, 1]$

23. $f(x) = 1 - x^3$; $[-8, 8]$

24. $f(x) = 2x^3$; $[-10, 10]$

25. $f(x) = x^3 - 6x^2 + 10$; $[0, 4]$

26. $f(x) = 12 + 9x - 3x^2 - x^3$; $[-3, 1]$

27. $f(x) = x^3 - x^4$; $[-1, 1]$

28. $f(x) = x^4 - 2x^3$; $[-2, 2]$

29. $f(x) = x^4 - 2x^2 + 5$; $[-2, 2]$

30. $f(x) = x^4 - 8x^2 + 3$; $[-3, 3]$

31. $f(x) = 1 - x^{2/3}$; $[-8, 8]$

32. $f(x) = (x + 3)^{2/3} - 5$; $[-4, 5]$

33. $f(x) = x + \dfrac{4}{x}$; $[-8, -1]$

34. $f(x) = x + \dfrac{1}{x}$; $[1, 20]$

35. $f(x) = \dfrac{x^2}{x^2 + 1}$; $[-2, 2]$

36. $f(x) = \dfrac{4x}{x^2 + 1}$; $[-3, 3]$

37. $f(x) = (x + 1)^{1/3}$; $[-2, 26]$

38. $f(x) = \sqrt[3]{x}$; $[8, 64]$

39–48. Check Exercises 3, 5, 9, 13, 19, 23, 33, 35, 37, and 38 with a graphing calculator.

Find the absolute extrema of each function, if they exist, over the indicated interval. Also indicate the x-value at which each extremum occurs. When no interval is specified, use the real numbers, $(-\infty, \infty)$.

49. $f(x) = 30x - x^2$

50. $f(x) = 12x - x^2$

51. $f(x) = 2x^2 - 40x + 270$

52. $f(x) = 2x^2 - 20x + 340$

53. $f(x) = 16x - \frac{4}{3}x^3$; $(0, \infty)$

54. $f(x) = x - \frac{4}{3}x^3$; $(0, \infty)$

55. $f(x) = x(60 - x)$

56. $f(x) = x(25 - x)$

57. $f(x) = \frac{1}{3}x^3 - 5x$; $[-3, 3]$

58. $f(x) = \frac{1}{3}x^3 - 3x$; $[-2, 2]$

59. $f(x) = -0.001x^2 + 4.8x - 60$

60. $f(x) = -0.01x^2 + 1.4x - 30$

61. $f(x) = -x^3 + x^2 + 5x - 1$; $(0, \infty)$

62. $f(x) = -\frac{1}{3}x^3 + 6x^2 - 11x - 50$; $(0, 3)$

63. $f(x) = 15x^2 - \frac{1}{2}x^3$; $[0, 30]$

64. $f(x) = 4x^2 - \frac{1}{2}x^3$; $[0, 8]$

65. $f(x) = 2x + \dfrac{72}{x}$; $(0, \infty)$

66. $f(x) = x + \dfrac{3600}{x}$; $(0, \infty)$

67. $f(x) = x^2 + \dfrac{432}{x}$; $(0, \infty)$

68. $f(x) = x^2 + \dfrac{250}{x}$; $(0, \infty)$

69. $f(x) = 2x^4 + x$; $[-1, 1]$

70. $f(x) = 2x^4 - x$; $[-1, 1]$

71. $f(x) = \sqrt[3]{x}$; $[0, 8]$

72. $f(x) = \sqrt{x}$; $[0, 4]$

73. $f(x) = (x - 1)^3$

74. $f(x) = (x + 1)^3$

75. $f(x) = 2x - 3$; $[-1, 1]$

76. $f(x) = 9 - 5x$; $[-10, 10]$

77. $f(x) = 2x - 3$; $[-1, 5)$

78. $f(x) = x^{2/3}$; $[-1, 1]$

79. $f(x) = 9 - 5x$; $[-2, 3)$

80. $f(x) = \frac{1}{3}x^3 - x + \frac{2}{3}$

81. $g(x) = x^{2/3}$

82. $f(x) = \frac{1}{3}x^3 - 2x^2 + x$; $[0, 4]$

83. $f(x) = \frac{1}{3}x^3 - \frac{1}{2}x^2 - 2x + 1$

84. $g(x) = \frac{1}{3}x^3 + 2x^2 + x$; $[-4, 0]$

85. $t(x) = x^4 - 2x^2$

86. $f(x) = 2x^4 - 4x^2 + 2$

87–96. Check Exercises 49, 51, 53, 57, 61, 65, 67, 69, 73, and 85 with a graphing calculator.

APPLICATIONS

Business and Economics

97. Monthly productivity. An employee's monthly productivity M, in number of units produced, is found to be a function of t, the number of years of employment. For a certain product, a productivity function is given by

$$M(t) = -2t^2 + 100t + 180, \quad 0 \le t \le 40.$$

Find the maximum productivity and the year in which it is achieved.

98. Advertising. Sound Software estimates that it will sell N units of a program after spending a dollars on advertising, where

$$N(a) = -a^2 + 300a + 6, \quad 0 \le a \le 300,$$

and a is in thousands of dollars. Find the maximum number of units that can be sold and the amount that must be spent on advertising in order to achieve that maximum.

99. Unemployment. The percentage of unemployed workers in service occupations can be modeled by

$$p(x) = -0.039x^3 + 0.594x^2 - 1.967x + 7.555,$$

where x is the number of years since 2003. (*Source:* Based on data from www.bls.gov.) According to this model, in what year during the period 2003–2013 was this percentage a maximum?

100. Labor. The percentage of women aged 21–54 in the U.S. civilian labor force can be modeled by

$$f(x) = -0.0135x^2 + 0.265x + 74.6,$$

where x is the number of years since 1992. (*Source:* based on data from www.bls.gov.) According to this model, in what year during the period 1992–2012 was this percentage a maximum?

101. Worldwide oil production. One model of worldwide oil production is

$$P(t) = 2.69t^4 - 63.941t^3 + 459.895t^2 - 688.692t + 24{,}150.217,$$

where $P(t)$ is the number of barrels, in thousands, produced t years after 2000. (*Source:* Based on data from the U.S. Energy Information Administration.) According to this model, in what year did worldwide oil production achieve an absolute minimum? What was that minimum?

102. Maximizing profit. Corner Stone Electronics determines that its weekly profit, in dollars, from the production and sale of x amplifiers is

$$P(x) = \frac{1500}{x^2 - 6x + 10}.$$

Find the number of amplifiers, x, for which the total weekly profit is a maximum.

Maximizing profit. *The total-cost and total-revenue functions for producing x items are*

$$C(x) = 5000 + 600x \quad \text{and} \quad R(x) = -\frac{1}{2}x^2 + 1000x,$$

where $0 \le x \le 600$. Use these functions for Exercises 103 and 104.

103. a) Find the total-profit function $P(x)$.
   b) Find the number of items, x, for which total profit is a maximum.

104. a) *Average profit* is given by $A(x) = P(x)/x$. Find $A(x)$.
   b) Find the number of items, x, for which average profit is a maximum.

Life and Physical Sciences

105. Blood pressure. For a dosage of x cubic centimeters (cc) of a certain drug, the resulting blood pressure B is approximated by

$$B(x) = 305x^2 - 1830x^3, \quad 0 \le x \le 0.16.$$

Find the maximum blood pressure and the dosage at which it occurs.

SYNTHESIS

106. How is the second derivative useful in finding the absolute extrema of a function?

For Exercises 107–110, find the absolute maximum and minimum values of each function, and sketch the graph.

107. $f(x) = \begin{cases} 2x + 1 & \text{for } -3 \le x \le 1, \\ 4 - x^2, & \text{for } 1 < x \le 2 \end{cases}$

108. $g(x) = \begin{cases} x^2, & \text{for } -2 \le x \le 0, \\ 5x, & \text{for } 0 < x \le 2 \end{cases}$

109. $h(x) = \begin{cases} 1 - x^2, & \text{for } -4 \le x < 0, \\ 1 - x, & \text{for } 0 \le x < 1, \\ x - 1, & \text{for } 1 \le x \le 2 \end{cases}$

110. $F(x) = \begin{cases} x^2 + 4, & \text{for } -2 \le x < 0, \\ 4 - x, & \text{for } 0 \le x < 3, \\ \sqrt{x - 2}, & \text{for } 3 \le x \le 67 \end{cases}$

111. Consider the piecewise-defined function f defined by:

$$f(x) = \begin{cases} x^2 + 2, & \text{for } -2 \le x \le 0, \\ 2, & \text{for } 0 < x < 4, \\ x - 2, & \text{for } 4 \le x \le 6. \end{cases}$$

   a) Sketch its graph.
   b) Identify the absolute maximum.
   c) What is the absolute minimum?

112. Physical science: dry lake elevation. Dry lakes are common in the Western deserts of the United States. These beds of ancient lakes are notable for having perfectly flat terrain. Rogers Dry Lake in California has been used as a landing site for space shuttle missions in recent years. The graph shows the elevation E, in feet, as a function of the distance x, in miles, from a point near one side ($x = 0$) of Rogers Dry Lake to a point near the other side. (*Source:* www.mytopo.com.)

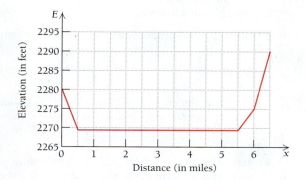

   a) What is the maximum elevation?
   b) What is the minimum elevation?

Find the absolute maximum and minimum values of the function, if they exist, over the indicated interval.

113. $g(x) = x\sqrt{x + 3}; \quad [-3, 3]$

114. $h(x) = x\sqrt{1 - x}; \quad [0, 1]$

115. Business: total cost. Certain costs in business can be separated into two components: those that increase with volume and those that decrease with volume. For example, customer service becomes more expensive as its quality increases, but part of the increased cost is offset by fewer customer complaints. Katie's Clocks determines that its cost of service, $C(x)$, in thousands of dollars, is modeled by

$$C(x) = (2x + 4) + \left(\frac{2}{x - 6}\right), \quad x > 6,$$

where x represents the number of "quality units." Find the number of "quality units" that the firm should use in order to minimize its total cost of service.

116. Let
$$y = (x - a)^2 + (x - b)^2.$$
For what value of x is y a minimum?

117. Business: worldwide oil production. Refer to Exercise 101. In what year(s) during the period 2000–2008 was worldwide oil production increasing most rapidly, and at what rate was it increasing?

118. How is the first derivative useful in finding the absolute extrema of a function?

119. Business: U.S. oil production. One model of oil production in the United States is given by
$$P(t) = 0.0000000219t^4 - 0.0000167t^3 + 0.00155t^2 + 0.002t + 0.22, \quad 0 \leq t \leq 110,$$
where $P(t)$ is the number of barrels of oil, in billions, produced in a year, t years after 1910. (*Source: Beyond Oil*, by Kenneth S. Deffeyes, p. 41, Hill and Wang, New York, 2005.)

a) According to this model, what is the absolute maximum amount of oil produced in the United States and in what year did that production occur?

b) According to this model, at what rate was United States oil production declining in 2010 and in 2015?

Graph each function over the given interval. Visually estimate where any absolute extrema occur. Then use the TABLE *feature to refine each estimate.*

120. $f(x) = x^{2/3}(x - 5); \quad [1, 4]$

121. $f(x) = \frac{3}{4}(x^2 - 1)^{2/3}; \quad [\frac{1}{2}, \infty)$

122. $f(x) = x\left(\frac{x}{2} - 5\right)^4; \quad \mathbb{R}$

123. Life and physical sciences: contractions during pregnancy. The following table gives the pressure of a pregnant woman's contractions as a function of time.

TIME, T (in minutes)	PRESSURE (in millimeters of mercury)
0	10
1	8
2	9.5
3	15
4	12
5	14
6	14.5

Use a calculator that has the REGRESSION option.

a) Fit a linear equation to the data. Predict the pressure of the contractions after 7 min.

b) Fit a quartic polynomial to the data. Predict the pressure of the contractions after 7 min. Find the smallest contraction over the interval $[0, 10]$.

Answers to Quick Checks

1. Absolute maximum is 20 at $x = 3$; absolute minimum is 0 at $x = 1$.   **2.** (a) Absolute maximum is 0 at $x = 0$; absolute minimum is -25 at $x = 5$. (b) Absolute maximum is 0 at $x = 10$; absolute minimum is -25 at $x = 5$.   **3.** The derivative is $f'(x) = nx^{n-1}$. If n is odd, $n - 1$ is even. Thus, nx^{n-1} is always positive or zero, never negative.   **4.** No absolute maximum; absolute minimum is 12 at $x = 3$.

2.5 Maximum–Minimum Problems; Business, Economics, and General Applications

• Solve maximum–minimum problems using calculus.

An important use of calculus is solving maximum–minimum problems, that is, finding the absolute maximum or minimum value of some variable.

EXAMPLE 1 Maximizing Area. A hobby store has 20 ft of fencing to fence off a rectangular area for an electric train in one corner of its display room. The two sides up against the wall require no fence. What dimensions of the rectangle will maximize the area? What is the maximum area?

Solution   Let's first make a drawing and express the area using one variable. If we let $x =$ the length, in feet, of one side and $y =$ the length, in feet, of the other, then, since the sum of the lengths must be 20 ft, we have

$$x + y = 20 \quad \text{and} \quad y = 20 - x.$$

Thus, the area is given by

$$
\begin{aligned}
A &= xy \\
&= x(20 - x) \\
&= 20x - x^2.
\end{aligned}
$$

We are trying to find the maximum value of

$$A(x) = 20x - x^2$$

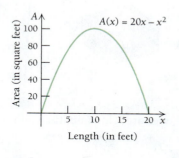

over the interval $(0, 20)$. We consider the interval $(0, 20)$ because x is a length and cannot be negative or 0. Since there is only 20 ft of fencing, x cannot be greater than 20. Also, x cannot be 20 because then the length of y would be 0.

a) We first find $A'(x)$: $A'(x) = 20 - 2x.$

b) This derivative exists for all values of x in $(0, 20)$. Thus, the only critical values are where

$$
\begin{aligned}
A'(x) = 20 - 2x &= 0 \\
-2x &= -20 \\
x &= 10.
\end{aligned}
$$

Since there is only one critical value, we can use the second derivative to determine whether we have a maximum. Note that

$$A''(x) = -2,$$

Since $A''(10)$ is negative, $A(10)$ is a maximum. Now

$$
\begin{aligned}
A(10) &= 10(20 - 10) \\
&= 10 \cdot 10 \\
&= 100.
\end{aligned}
$$

Thus, the maximum area of 100 ft^2 is obtained using 10 ft for the length of one side and $20 - 10$, or 10 ft for the other.   **1** ✔

Quick Check 1 ✔

Repeat Example 1 starting first with 50 ft of fencing, and then with 100 ft of fencing. Do you detect a pattern? If you had n feet of fencing, what would be the dimensions of the maximum area (in terms of n)?

Here is a general strategy for solving maximum–minimum problems.

A Strategy for Solving Maximum–Minimum Problems

1. Read the problem carefully. If relevant, make a drawing.
2. Make a list of appropriate variables and constants, noting what varies, what stays fixed, and what units are used. Label the measurements on your drawing, if one exists.
3. Translate the problem to an equation involving a quantity Q to be maximized or minimized. Try to represent Q in terms of the variables of step 2.
4. Try to express Q as a function of one variable. Use the procedures developed in Sections 2.1–2.4 to determine the maximum or minimum values and the points at which they occur.

EXAMPLE 2   Maximizing Volume. From a thin piece of cardboard 8 in. by 8 in., square corners are cut out so that the sides can be folded up to make a box. What dimensions will yield a box of maximum volume? What is the maximum volume?

Solution   We make a drawing in which x is the length, in inches, of each square to be cut. It is important to note that since the original square is 8 in. by 8 in., after the smaller squares are removed, the lengths of the sides of the box will be $(8 - 2x)$ in. by $(8 - 2x)$ in.

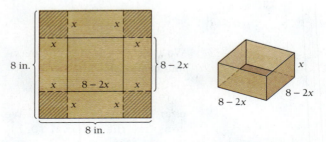

After the four small squares are removed and the sides are folded up, the volume V of the resulting box is

$$V = l \cdot w \cdot h = (8 - 2x) \cdot (8 - 2x) \cdot x,$$

or $V(x) = 4x^3 - 32x^2 + 64x.$

Since $8 - 2x > 0$, this means that $x < 4$. Thus, we need to maximize

$$V(x) = 4x^3 - 32x^2 + 64x \quad \text{over the interval } (0, 4).$$

To do so, we first find $V'(x)$:

$$V'(x) = 12x^2 - 64x + 64.$$

Since $V'(x)$ exists for all x in the interval $(0, 4)$, we can set it equal to zero to find the critical values:

$$12x^2 - 64x + 64 = 0$$
$$4(3x^2 - 16x + 16) = 0$$
$$4(3x - 4)(x - 4) = 0$$
$$3x - 4 = 0 \quad \text{or} \quad x - 4 = 0$$
$$3x = 4 \quad \text{or} \quad x = 4$$
$$x = \tfrac{4}{3} \quad \text{or} \quad x = 4.$$

The only critical value in $(0, 4)$ is $\frac{4}{3}$. Thus, we can use the second derivative,

$$V''(x) = 24x - 64,$$

to determine whether we have a maximum. Since

$$V''(\tfrac{4}{3}) = 24 \cdot \tfrac{4}{3} - 64 = 32 - 64 < 0,$$

we know that $V(\frac{4}{3})$ is a maximum.

Thus, to maximize the box's volume, small squares with edges measuring $\frac{4}{3}$ in., or $1\frac{1}{3}$ in., should be cut from each corner of the original 8 in. by 8 in. piece of cardboard. When the sides are folded up, the resulting box will have sides of length

$$8 - 2x = 8 - 2 \cdot \frac{4}{3} = \frac{24}{3} - \frac{8}{3} = \frac{16}{3} = 5\frac{1}{3} \text{ in.}$$

and a height of $1\frac{1}{3}$ in. The maximum volume is

$$V\left(\frac{4}{3}\right) = 4\left(\frac{4}{3}\right)^3 - 32\left(\frac{4}{3}\right)^2 + 64\left(\frac{4}{3}\right) = \frac{1024}{27} = 37\frac{25}{27} \text{ in}^3.$$

Quick Check 2 ✔

Repeat Example 2 starting with a sheet of cardboard measuring 8.5 in. by 11 in. (the size of a typical sheet of paper). Will this box hold 1 liter (L) of liquid? (*Hint:* 1 L $= 1000$ cm^3 and 1 in$^3 = 16.38$ cm^3.)

2 ✔

In manufacturing, minimizing the amount of material used is always preferred, both from a cost standpoint and in terms of efficiency.

EXAMPLE 3 **Minimizing Cost of Material.** Mendoza Manufacturers specializes in food-storage containers and makes a cylindrical soup can with a volume of 250 cm³. The cost of material for the two circular ends is $0.0008/cm², and the cost of material for the side is $0.0015/cm². What dimensions minimize the cost of material for the soup can? What is the minimum cost?

Solution Recall from Section R.1 that we estimated the dimensions that give the lowest cost using a spreadsheet.

	A	B	C
1			
2	Radius	Height	Cost
3	r	h	C
4	4.1	4.733322705	0.267434061
5	4.2	4.510609676	0.267251237	← Possible minimum cost
6	4.3	4.303253363	0.267371533

The spreadsheet shows that a radius of about 4.2 cm gives a minimal cost of $0.267251237 per can.

We can use calculus to find exact values for the radius and height that will minimize the cost per can. We let h = height of the can and r = radius, both measured in centimeters. The formula for volume of a cylinder is

$$V = \pi r^2 h.$$

Since we know the volume is 250 cm³, this formula allows us to relate h and r, expressing one in terms of the other. We solve for h in terms of r:

$$\pi r^2 h = 250$$

$$h = \frac{250}{\pi r^2}.$$

The can has two circular ends, each with an area of πr^2. Thus, the cost of material for the two circular ends is $2(0.0008)\pi r^2 = 0.0016\pi r^2$.

The side of the can, when disassembled, is a rectangle of area $2\pi rh$, as shown at left. The cost of material for the side is

$$2(0.0015)\pi rh = 0.003\pi rh$$

$$= 0.003\pi r\left(\frac{250}{\pi r^2}\right) \quad \text{Substituting}$$

$$= \frac{0.75}{r}. \quad \text{Simplifying}$$

Thus, the total cost of material for one can is given by

$$C(r) = 0.0016\pi r^2 + \frac{0.75}{r}, \text{ where } r > 0.$$

Differentiating, we have

$$C'(r) = 0.0032\pi r - \frac{0.75}{r^2}.$$

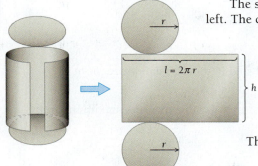

Each of the circular ends of the disassembled can has an area of πr^2, and the side has an area of $2\pi rh$.

We set the derivative equal to zero and solve to find the critical value(s):

$$0.0032\pi r - \frac{0.75}{r^2} = 0$$

$$0.0032\pi r = \frac{0.75}{r^2}$$

$$0.0032\pi r^3 = 0.75 \qquad \text{Multiplying both sides by } r^2$$

$$r^3 = \frac{0.75}{0.0032\pi}$$

$$r = \sqrt[3]{\frac{0.75}{0.0032\pi}}$$

$$r \approx 4.21 \qquad \text{Using a calculator}$$

Note that the critical value $r \approx 4.21$ is the only critical value in the interval $(0, \infty)$. To determine whether this critical value results in a minimum or maximum value of C, we use the Second-Derivative Test:

$$C''(r) = 0.0032\pi + \frac{1.5}{r^3}$$

$$C''(4.21) = 0.0032\pi + \frac{1.5}{(4.21)^3} > 0.$$

The second derivative is positive, so the critical value indicates a minimum point. Thus, the cost of material for one can is minimized when $r = 4.21$ cm. The can will have a height of $h = 250/\pi(4.21)^2 \approx 4.49$ cm, and the cost of material will be $0.2673 per can.

We rounded our answer above. If we do not, we find that a radius of $4.209725752\ldots$ cm will give a minimal cost of $0.26723831\ldots$ per can. A manufacturer who produces millions of cans may be interested in carrying the precision of the calculations to six or seven decimal places.   **3 ✔**

Cost (in dollars), Radius (in centimeters)

$(4.21, 0.2673)$, $C(r)$

Quick Check 3 ✔

Repeat Example 3, assuming that the can has an open top. That is, it consists of one circular end and the side.

EXAMPLE 4   Business: Maximizing Revenue.   Cruzing Tunes determines that in order to sell x units of a new car stereo, the price per unit, in dollars, must be

$$p(x) = 1000 - x.$$

It also determines that the total cost of producing x units is given by

$$C(x) = 3000 + 20x.$$

a) Find the total revenue, $R(x)$.

b) Find the total profit, $P(x)$.

c) How many units must be made and sold in order to maximize profit?

d) What is the maximum profit?

e) What price per unit must be charged in order to make this maximum profit?

Solution

a) $R(x) = $ Total revenue

$= (\text{Number of units}) \cdot (\text{Price per unit})$

$= \qquad x \qquad \cdot \qquad p$

$= \qquad x \qquad \cdot \ (1000 - x) \qquad \text{Substituting}$

$= 1000x - x^2.$

b) $P(x) = $ Total revenue $-$ Total cost

$= R(x) - C(x)$

$= (1000x - x^2) - (3000 + 20x)$

$= -x^2 + 980x - 3000$

c) To find the maximum value of $P(x)$, we first find $P'(x)$:

$$P'(x) = -2x + 980.$$

Note that $x \geq 0$. The critical value(s) are found by solving $P'(x) = 0$:

$$P'(x) = -2x + 980 = 0$$
$$-2x = -980$$
$$x = 490.$$

There is only one critical value. We can therefore try to use the second derivative to determine whether we have an absolute maximum. Note that

$$P''(x) = -2, \text{ a constant.}$$

Since, $P''(490)$ is negative, profit is maximized when 490 units are produced and sold.

d) The maximum profit is given by

$$P(490) = -(490)^2 + 980 \cdot 490 - 3000$$
$$= \$237{,}100.$$

Thus, Cruzing Tunes make a maximum profit of \$237,100 by producing and selling 490 stereos.

e) The price per unit needed to maximize profit is

$$p = 1000 - 490 = \$510.$$ 4 ✔

Quick Check 4 ✔

Repeat Example 4 with the price function

$$p(x) = 1750 - 2x$$

and the cost function

$$C(x) = 2250 + 15x.$$

Round your answers when necessary.

Let's take a general look at the profit, cost, and revenue.

Figure 1 shows an example of total-cost and total-revenue functions. We can estimate what the maximum profit is by looking for the widest gap between $R(x)$ and $C(x)$, when $R(x) > C(x)$. Points B_0 and B_2 are break-even points.

Figure 2 shows the related total-profit function. Note that when production is too low ($< x_0$), there is a loss, perhaps due to high fixed or initial costs and low revenue. When production is too high ($> x_2$), there is also a loss, perhaps due to the increased cost of overtime pay or expansion.

The business operates at a profit everywhere between x_0 and x_2. Note that maximum profit occurs at a critical value x_1. Assuming that $P'(x)$ exists for all x in some interval, usually $[0, \infty)$, the critical value x_1 occurs at some number x such that

$$P'(x) = 0 \quad \text{and} \quad P''(x) < 0.$$

Since $P(x) = R(x) - C(x)$, it follows that

$$P'(x) = R'(x) - C'(x) \quad \text{and} \quad P''(x) = R''(x) - C''(x).$$

Thus, maximum profit occurs at some number x such that

$$P'(x) = R'(x) - C'(x) = 0 \quad \text{and} \quad P''(x) = R''(x) - C''(x) < 0,$$

or $\quad R'(x) = C'(x) \quad \text{and} \quad R''(x) < C''(x).$

In summary, we have the following theorem.

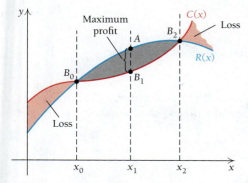

FIGURE 1

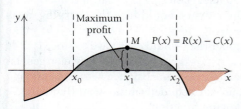

FIGURE 2

THEOREM 10

Maximum profit occurs at those x-values for which

$$R'(x) = C'(x) \quad \text{and} \quad R''(x) < C''(x).^*$$

---

*In Section 2.6, the concepts of *marginal revenue* and *marginal cost* are introduced, allowing $R'(x) = C'(x)$ to be regarded as Marginal revenue = Marginal cost.

You can check that the results in parts (c) and (d) of Example 4 can be easily found using Theorem 10.

EXAMPLE 5   **Business: Determining Ticket Price.**   By keeping records, a theater determines that at a ticket price of $26, it averages 1000 people in attendance. For every drop in price of $1, it gains 50 customers. Each customer spends an average of $4 on concessions. What ticket price should the theater charge in order to maximize total revenue?

Solution   Let x be the number of dollars by which the ticket price of $26 should be decreased. (If x is negative, the price is increased.) We first express the total revenue R as a function of x. Note that the increase in ticket sales is $50x$ when the price drops x dollars:

$$R(x) = (\text{Revenue from tickets}) + (\text{Revenue from concessions})$$
$$= (\text{Number of people}) \cdot (\text{Ticket price}) + (\text{Number of people}) \cdot 4$$
$$= (1000 + 50x)(26 - x) + (1000 + 50x) \cdot 4$$
$$= -50x^2 + 500x + 30,000.$$

To find x such that $R(x)$ is a maximum, we first find $R'(x)$:

$$R'(x) = -100x + 500.$$

This derivative exists for all real numbers x. Thus, the only critical values are where $R'(x) = 0$; so we solve that equation:

$$-100x + 500 = 0$$
$$-100x = -500$$
$$x = 5 \qquad \text{This corresponds to lowering the price by \$5.}$$

Since this is the only critical value, we can use the second derivative,

$$R''(x) = -100,$$

to determine whether we have a maximum. Since $R''(5)$ is negative, $R(5)$ is a maximum. Therefore, in order to maximize revenue, the theater should charge

$$\$26 - \$5, \quad \text{or} \quad \$21 \text{ per ticket.} \qquad \qquad 5 ✔$$

Quick Check 5 ✔

A baseball team charges $30 per ticket and averages 20,000 people in attendance per game. Each person spends an average of $8 on concessions. For every drop of $1 in the ticket price, the attendance rises by 800 people. What ticket price should the team charge to maximize total revenue?

Minimizing Inventory Costs

A retail business outlet needs to be concerned about inventory costs. Suppose, for example, that a home electronics store sells 2500 television sets per year. It *could* operate by ordering all the sets at once. But then the owners would face the carrying costs (insurance, building space, and so on) of storing them all. Thus, they might make several smaller orders, so that the largest number they would ever have to store is 500. However, each time they reorder, there are costs for paperwork, delivery charges, labor, and so on. It seems, therefore, that there must be some balance between carrying costs and reorder costs. Let's see how calculus can help determine what that balance might be. We are trying to minimize the following function:

Total inventory costs = (Yearly carrying costs) + (Yearly reorder costs).

The *lot size* x is the largest number ordered each reordering period. If x units are ordered each period, then during that time somewhere between 0 and x units are in stock. To have a representative expression for the amount in stock at any one time in the period, we can use the average, $x/2$. This represents the average amount held in stock over the course of each time period.

Refer to the graphs shown on the next page. If the lot size is 2500, then during the period between orders, there are somewhere between 0 and 2500 units in stock.

On average, there are 2500/2, or 1250 units in stock. If the lot size is 1250, then during the period between orders, there are somewhere between 0 and 1250 units in stock. On average, there are 1250/2, or 625 units in stock. In general, if the lot size is x, the average inventory is $x/2$.

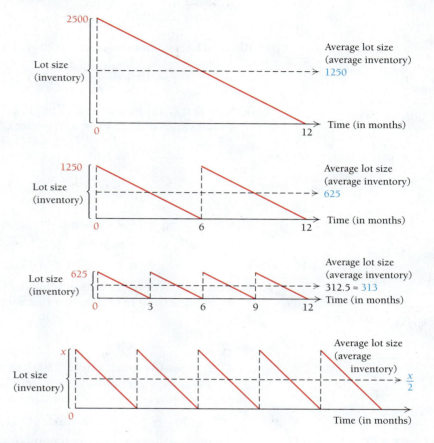

EXAMPLE 6   **Business: Minimizing Inventory Costs.**   Fred's Electronics sells 2500 television sets per year. It costs $20 to store one set for a year. To reorder, there is a fixed cost of $20, plus a fee of $9 per set. How many times per year should the store reorder, and in what lot size, to minimize inventory costs?

Solution   Let $x =$ the lot size. Inventory costs are given by

$$C(x) = (\text{Yearly carrying costs}) + (\text{Yearly reorder costs}).$$

We consider each component of inventory costs separately.

a) *Yearly carrying costs.* The average amount held in stock is $x/2$, and it costs $20 per set for storage. Thus,

$$\text{Yearly carrying costs} = \left(\begin{array}{c}\text{Yearly cost}\\ \text{per item}\end{array}\right) \cdot \left(\begin{array}{c}\text{Average number}\\ \text{of items}\end{array}\right)$$

$$= 20 \cdot \frac{x}{2}.$$

b) *Yearly reorder costs.* We know that x is the lot size, and we let N be the number of reorders each year. Then $Nx = 2500$, and $N = 2500/x$. Thus,

$$\text{Yearly reorder costs} = \left(\begin{array}{c}\text{Cost of each}\\ \text{order}\end{array}\right) \cdot \left(\begin{array}{c}\text{Number of}\\ \text{reorders}\end{array}\right)$$

$$= (20 + 9x)\frac{2500}{x}.$$

c) Thus, we have

$$C(x) = 20 \cdot \frac{x}{2} + (20 + 9x)\frac{2500}{x}$$

$$= 10x + \frac{50,000}{x} + 22,500 = 10x + 50,000x^{-1} + 22,500.$$

d) To find a minimum value of C over $[1, 2500]$, we first find $C'(x)$:

$$C'(x) = 10 - \frac{50,000}{x^2}.$$

e) Note that $C'(x)$ exists for all x in $[1, 2500]$, so the only critical values are those x-values such that $C'(x) = 0$. We solve $C'(x) = 0$:

$$10 - \frac{50,000}{x^2} = 0$$

$$10 = \frac{50,000}{x^2}$$

$$10x^2 = 50,000$$

$$x^2 = 5000$$

$$x = \sqrt{5000} \approx 70.7.$$

Since there is only one critical value in $[1, 2500]$, that is, $x \approx 70.7$, we can use the second derivative to see whether that value yields a maximum or a minimum:

$$C''(x) = \frac{100,000}{x^3}.$$

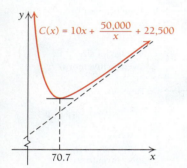

$C''(x)$ is positive for all x in $[1, 2500]$, so we have a minimum at $x \approx 70.7$.

However, it is impossible to reorder 70.7 sets each time, so we consider the two whole numbers closest to 70.7, which are 70 and 71. Since

$$C(70) \approx \$23,914.29 \quad \text{and} \quad C(71) \approx \$23,914.23,$$

it follows that the lot size that will minimize cost is 71, although the difference, $0.06, is not much. (*Note:* Such a procedure will not work for all functions but will work for the type we are considering here.) The number of times an order should be placed is $2500/71$ with a remainder of 15, indicating that 35 orders should be placed. Of those, $35 - 15 = 20$ will be for 71 items and 15 will be for 72 items. **6** ✔

Quick Check 6 ✔

Repeat Example 6 with a storage cost of $30 per set and assuming that the store sells 3000 sets per year.

The lot size that minimizes total inventory costs is often referred to as the *economic ordering quantity*. Three assumptions are made in using the preceding method to determine the economic ordering quantity. First, the demand for the product is the same year round. For television sets, this may be reasonable, but for seasonal items such as clothing or skis, this assumption is unrealistic. Second, the time between placing an order and receiving the product remains consistent year-round. Finally, the costs involved, such as storage, shipping charges, and so on, do not vary. This assumption may not be reasonable in a time of inflation, although variation in these costs can be allowed for by anticipating what they might be and using average costs. Regardless, the model described above is useful, allowing us to analyze a seemingly difficult problem using calculus.

Using Excel Spreadsheets to Numerically Estimate Minimum and Maximum Values

An Excel spreadsheet can numerically estimate absolute extrema. For example, to estimate the values for Example 2, we open a blank spreadsheet and enter text in the cells as shown below:

	A	B	C	D
1	Starting value			
2	Step size:			
3				
4	Height	Length	Width	Volume
5	H=x	L=8–2x	W=8–2x	V=H*L*W

In cell A6, we enter =B1; in cells B6 and C6, we enter =8-2*A6; and in cell D6, we enter =A6*B6*C6. In cell A7, we enter =A6+B2. We copy cells B6, C6, and D6 and paste them into cells B7, C7, and D7, then copy cells A7 through D7 and paste them into cells A8 through D14. This creates nine active rows. More rows can be added if necessary.

In cell B1, we enter a starting x-value, and in cell B2, a step size. For example, if we enter 0 into cell B1 and 0.25 into cell B2, the values in cells A6–A14 will be $x = 0, 0.25, 0.5, 0.75, 1$, and so on. The result is the following table:

	A	B	C	D
1	Starting value	0		
2	Step size:	0.25		
3				
4	Height	Length	Width	Volume
5	H=x	L=8–2x	W=8–2x	V=H*L*W
6	0	8	8	0
7	0.25	7.5	7.5	14.0625
8	0.5	7	7	24.5
9	0.75	6.5	6.5	31.6875
10	1	6	6	36
11	1.25	5.5	5.5	37.8125	← Possible largest volume
12	1.5	5	5	37.5
13	1.75	4.5	4.5	35.4375
14	2	4	4	32

We see that a possible maximum volume of 37.8125 in^2 occurs when $x = 1.25$ in.

To refine our estimate, we enter 1 into cell B1 and 0.1 into cell B2, giving the following table:

	A	B	C	D
1	Starting value	1		
2	Step size:	0.1		
3				
4	Height	Length	Width	Volume
5	H=x	L=8–2x	W=8–2x	V=H*L*W
6	1	6	6	36
7	1.1	5.8	5.8	37.004
8	1.2	5.6	5.6	37.632
9	1.3	5.4	5.4	37.908	← Possible largest volume
10	1.4	5.2	5.2	37.856
11	1.5	5	5	37.5
12	1.6	4.8	4.8	36.864

By reducing the step size, we obtain a new possible maximum volume of 37.908 in^3 when $x = 1.3$ in. Although we could further reduce the step size, we now have an estimate of the maximum volume and the x-value at which it occurs. We use calculus to determine these values exactly.

EXERCISES

1. Use a spreadsheet to verify the results of Example 1.

2. Use a spreadsheet to verify the results of Example 4.

3. Use a spreadsheet to verify the results of Example 5.

4. Use a spreadsheet to verify the results of Example 6.

Section Summary

- In many real-life applications, we need to determine the minimum or maximum value of a function modeling a situation.

- Maximum profit occurs at those x-values for which $R'(x) = C'(x)$ and $R''(x) < C''(x)$.

2.5 | Exercise Set

1. Of all numbers whose sum is 70, find the two that have the maximum product. That is, maximize $Q = xy$, where $x + y = 70$.

2. Of all numbers whose sum is 50, find the two that have the maximum product. That is, maximize $Q = xy$, where $x + y = 50$.

3. Of all numbers whose difference is 6, find the two that have the minimum product.

4. Of all numbers whose difference is 4, find the two that have the minimum product.

5. Maximize $Q = xy^2$, where x and y are positive numbers such that $x + y^2 = 4$.

6. Maximize $Q = xy^2$, where x and y are positive numbers such that $x + y^2 = 1$.

7. Minimize $Q = x^2 + 2y^2$, where $x + y = 3$.

8. Minimize $Q = 2x^2 + 3y^2$, where $x + y = 5$.

9. Maximize $Q = xy$, where x and y are positive numbers such that $x + \frac{4}{3}y^2 = 1$.

10. Maximize $Q = xy$, where x and y are positive numbers such that $\frac{4}{3}x^2 + y = 16$.

11. **Maximizing area.**   A rancher wants to enclose two rectangular areas near a river, one for sheep and one for cattle. There are 240 yd of fencing available. What is the largest total area that can be enclosed?

12. **Maximizing area.**   A lifeguard needs to rope off a rectangular swimming area in front of Long Lake Beach, using 180 yd of rope and floats. What dimensions of the rectangle will maximize the area? What is the maximum area? (Note that the shoreline is one side of the rectangle.)

13. **Maximizing area.**   Hentz Industries plans to enclose three parallel rectangular areas for sorting returned goods. The three areas are within one large rectangular area and 1200 yd of fencing is available. What is the largest total area that can be enclosed?

14. **Maximizing area.**   Grayson Farms plans to enclose three parallel rectangular livestock pens within one large rectangular area using 600 ft of fencing. One side of the enclosure is a pre-existing stone wall.

   a) If the three rectangular pens have their longer sides parallel to the stone wall, find the largest possible total area that can be enclosed.

b) If the three rectangular pens have their shorter sides perpendicular to the stone wall, find the largest possible total area that can be enclosed.

15. Maximizing area. Of all rectangles that have a perimeter of 42 ft, find the dimensions of the one with the largest area. What is its area?

16. Maximizing area. A carpenter is building a rectangular shed with a fixed perimeter of 54 ft. What are the dimensions of the largest shed that can be built? What is its area?

17. Maximizing volume. From a thin piece of cardboard 20 in. by 20 in., square corners are cut out so that the sides can be folded up to make a box. What dimensions will yield a box of maximum volume? What is the maximum volume?

18. Maximizing volume. From a 50-cm-by-50-cm sheet of aluminum, square corners are cut out so that the sides can be folded up to make a box. What dimensions will yield a box of maximum volume? What is the maximum volume?

19. Minimizing surface area. Mendoza Soup Company is constructing an open-top, square-based, rectangular metal tank that will have a volume of 32 ft^3. What dimensions will minimize surface area? What is the minimum surface area?

20. Minimizing surface area. Drum Tight Containers is designing an open-top, square-based, rectangular box that will have a volume of 62.5 in^3. What dimensions will minimize surface area? What is the minimum surface area?

21. Minimizing surface area. Open Air Waste Management is designing a rectangular construction dumpster that will be twice as long as it is wide and must hold 12 yd^3 of debris. Find the dimensions of the dumpster that will minimize its surface area.

22. Minimizing surface area. Ever Green Gardening is designing a rectangular compost container that will be twice as tall as it is wide and must hold 18 ft^3 of composted food scraps. Find the dimensions of the compost container with minimal surface area (include the bottom and top).

APPLICATIONS

Business and Economics

Maximizing profit. *For Exercises 23–28, find the maximum profit and the number of units that must be produced and sold in order to yield the maximum profit. Assume that revenue, R(x), and cost, C(x), are in dollars for Exercises 23–26.*

23. $R(x) = 50x - 0.5x^2, \quad C(x) = 4x + 10$

24. $R(x) = 50x - 0.5x^2, \quad C(x) = 10x + 3$

25. $R(x) = 2x, \quad C(x) = 0.01x^2 + 0.6x + 30$

26. $R(x) = 5x, \quad C(x) = 0.001x^2 + 1.2x + 60$

27. $R(x) = 9x - 2x^2, \quad C(x) = x^3 - 3x^2 + 4x + 1$; assume that $R(x)$ and $C(x)$ are in thousands of dollars, and x is in thousands of units.

28. $R(x) = 100x - x^2, C(x) = \frac{1}{3}x^3 - 6x^2 + 89x + 100$; assume that $R(x)$ and $C(x)$ are in thousands of dollars, and x is in thousands of units.

29. Maximizing profit. Riverside Appliances is marketing a new refrigerator. It determines that in order to sell x refrigerators, the price per refrigerator must be

$$p = 280 - 0.4x.$$

It also determines that the total cost of producing x refrigerators is given by

$$C(x) = 5000 + 0.6x^2.$$

a) Find the total revenue, $R(x)$.
b) Find the total profit, $P(x)$.
c) How many refrigerators must the company produce and sell in order to maximize profit?
d) What is the maximum profit?
e) What price per refrigerator must be charged in order to maximize profit?

30. Maximizing profit. Raggs, Ltd., a clothing firm, determines that in order to sell x suits, the price per suit must be

$$p = 150 - 0.5x.$$

It also determines that the total cost of producing x suits is given by

$$C(x) = 4000 + 0.25x^2.$$

a) Find the total revenue, $R(x)$.
b) Find the total profit, $P(x)$.
c) How many suits must the company produce and sell in order to maximize profit?
d) What is the maximum profit?
e) What price per suit must be charged in order to maximize profit?

31. Maximizing profit. Gritz-Charlston is a 300-unit luxury hotel. All rooms are occupied when the hotel charges $80 per day for a room. For every increase of x dollars in the daily room rate, there are x rooms vacant. Each occupied room costs $22 per day to service and maintain. What should the hotel charge per day in order to maximize profit?

32. Maximizing revenue. Edwards University wants to determine what price to charge for tickets to football games. At a price of $18 per ticket, attendance averages 40,000 people per game. Every decrease of $3 to the ticket price adds 10,000 people to the average attendance. Every person at a game spends an average of $4.50 on concessions. What price per ticket should be charged to maximize revenue? How many people will attend at that price?

33. Maximizing parking tickets. Oak Glen currently employs 8 patrol officers who each write an average of 24 parking tickets per day. For every additional officer placed on patrol, the average number of parking tickets per day written by each officer decreases by 4. How many additional officers should be placed on patrol in order to maximize the number of parking tickets written per day?

34. Maximizing yield. Hood Apple Farm yields an average of 30 bushels of apples per tree when 20 trees are planted on an acre of ground. If 1 more tree is planted per acre, the yield decreases by 1 bushel (bu) per tree as a result of crowding. How many trees should be planted on an acre in order to get the highest yield?

35. Nitrogen prices. During 2001, nitrogen prices fell by 41%. Over the same year, nitrogen demand went up by 12%. (*Source: Chemical Week.*)

a) Assuming a linear change in demand, find the demand function, $q(x)$, by finding the equation of the line that passes through the points $(1, 1)$ and $(0.59, 1.12)$. Here x is the price as a fraction of the January 2001 price, and $q(x)$ is the demand as a fraction of the demand in January.

b) As a percentage of the January 2001 price, what should the price of nitrogen be to maximize revenue?

36. Vanity license plates. According to a pricing model, increasing the fee for vanity license plates by $1 decreases the percentage of a state's population that will request them by 0.04%. (*Source: E. D. Craft, "The demand for vanity (plates): Elasticities, net revenue maximization, and deadweight loss," Contemporary Economic Policy*, Vol. 20, 133–144 (2002).)

a) Recently, the fee for vanity license plates in Maryland was $25, and the percentage of the state's population that had vanity plates was 2.13%. Use this information to construct the demand function, $q(x)$, for the percentage of Maryland's population that will request vanity license plates for a fee of x dollars.

b) Find the fee, x, that will maximize revenue from vanity plates.

37. Maximizing revenue. When the Marchant Theater charges $5 for admission, there is an average attendance of 180 people. For every $0.10 increase in admission, there is a loss of 1 customer from the average number. What admission should be charged in order to maximize revenue?

38. Minimizing costs. A rectangular box with a volume of 320 ft³ is to be constructed with a square base and top. The cost per square foot for the bottom is 15¢, for the top is 10¢, and for the sides is 2.5¢. What dimensions will minimize the cost?

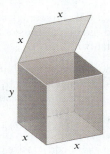

39. Minimizing cost. A rectangular parking area measuring 5000 ft² is to be enclosed on three sides using chain-link fencing that costs $4.50 per foot. The fourth side will be a wooden fence that costs $7 per foot. What dimensions will minimize the total cost to enclose this area, and what is the minimum cost (rounded to the nearest dollar)?

40. Minimizing cost. A rectangular garden measuring 1200 yd² is to be enclosed on two parallel sides by stone wall that costs $35 per yd and the other two sides by wooden fencing that costs $28 per yd. What dimensions will minimize the total cost of enclosing this garden, and what is the minimum cost (rounded to the nearest dollar)?

41. Maximizing area. Bradley Publishing decides that each page in a new book must have an area of 73.125 in², a 0.75-in. margin at the top and at the bottom of each page, and a 0.5-in. margin on each of the sides. What should the outside dimensions of each page be so that the printed area is a maximum?

42. Minimizing inventory costs. A sporting goods store sells 100 pool tables per year. It costs $20 to store one pool table for a year. To reorder, there is a fixed cost of $40 per shipment plus $16 for each pool table. How many times per year should the store order pool tables, and in what lot size, in order to minimize inventory costs?

43. Minimizing inventory costs. A pro shop in a bowling center sells 200 bowling balls per year. It costs $4 to store one bowling ball for a year. To reorder, there is a fixed cost of $1, plus $0.50 for each bowling ball. How many times per year should the shop order bowling balls, and in what lot size, in order to minimize inventory costs?

44. Minimizing inventory costs. A retail outlet for Boxowitz Calculators sells 720 calculators per year. It costs $2 to store one calculator for a year. To reorder, there is a fixed cost of $5, plus $2.50 for each calculator. How many times per year should the store order calculators, and in what lot size, in order to minimize inventory costs?

45. Minimizing inventory costs. Bon Temps Surf and Scuba Shop sells 360 surfboards per year. It costs $8 to store one surfboard for a year. Each reorder costs $10, plus an additional $5 for each surfboard ordered. How many times per year should the store order surfboards, and in what lot size, in order to minimize inventory costs?

46. Minimizing inventory costs. Repeat Exercise 44 using the same data, but assume yearly sales of 256 calculators with the fixed cost of each reorder set at $4.

47. Minimizing inventory costs. Repeat Exercise 45 using the same data, but change the reorder costs from an additional $5 per surfboard to $6 per surfboard.

48. Minimizing surface area. A closed-top cylindrical container is to have a volume of 250 in^2. What dimensions (radius and height) will minimize the surface area?

49. Minimizing surface area. An open-top cylindrical container is to have a volume of 400 cm^2. What dimensions (radius and height) will minimize the surface area?

50. Minimizing cost. Assume that the costs of the materials for making the cylindrical container described in Exercise 48 are $0.005/in^2 for the circular base and top and $0.003/in^2 for the wall. What dimensions will minimize the cost of materials?

51. Minimizing cost. Assume that the costs of the materials for making the cylindrical container described in Exercise 49 are $0.0015/cm^2 for the base and $0.008/cm^2 for the wall. What dimensions will minimize the cost of materials?

General Interest

52. Maximizing volume. The postal service places a limit of 84 in. on the combined length and girth of (distance around) a package to be sent parcel post. What dimensions of a rectangular box with square cross-section will contain the largest volume that can be mailed? (*Hint:* There are two different girths.)

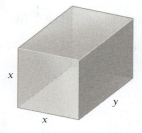

53. Minimizing cost. A rectangular play area of 48 yd^2 is to be fenced off in a person's yard. The next-door neighbor agrees to pay half the cost of the fence on the side of the play area that lies along the property line. What dimensions will minimize the cost of the fence?

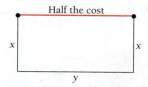

54. Maximizing light. A Norman window is a rectangle with a semicircle on top. Suppose that the perimeter of a particular Norman window is to be 24 ft. What should its dimensions be in order to allow the maximum amount of light to enter through the window?

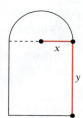

55. Maximizing light. Repeat Exercise 54, but assume that the semicircle is to be stained glass, which transmits only half as much light as clear glass does.

56–110. Use a spreadsheet to numerically verify the result of Exercises 1–55.

SYNTHESIS

111. For what positive number is the sum of its reciprocal and five times its square a minimum?

112. For what positive number is the sum of its reciprocal and four times its square a minimum?

113. Business: maximizing profit. The amount of money that customers deposit in a bank in savings accounts is directly proportional to the interest rate that the bank pays on that money. Suppose a bank is able to loan out all the money deposited in its savings accounts at an interest rate of 18%. What interest rate should it pay on its savings accounts in order to maximize profit?

114. A 24-in. piece of wire is cut in two pieces. One piece is used to form a circle and the other to form a square. How should the wire be cut so that the sum of the areas is a minimum? A maximum?

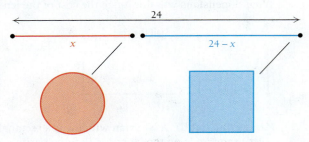

115. Business: minimizing costs. A power line is to be constructed from a power station at point A to an island at point C, which is 1 mi directly out in the water from a point B on the shore. Point B is 4 mi downshore from the power station at A. It costs \$5000 per mile to lay the power line under water and \$3000 per mile to lay the line under ground. At what point S downshore from A should the line come to the shore in order to minimize cost? Note that S could very well be B or A. (*Hint:* The length of CS is $\sqrt{1 + x^2}$.)

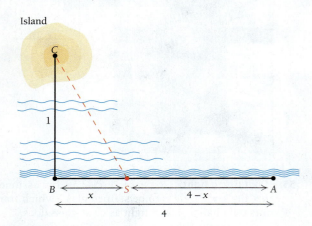

116. Life science: flights of homing pigeons. It is known that homing pigeons tend to avoid flying over water in the daytime, perhaps because downdrafts of air over water make flying difficult. Suppose a homing pigeon is released on an island at point C, which is 3 mi directly out in the water from a point B on shore. Point B is 8 mi downshore from the pigeon's home loft at point A. Assume that a pigeon flying over water uses energy at a rate 1.28 times the rate over land. Toward what point S downshore from A should the pigeon fly in order to minimize the total energy required to get to the home loft at A? Assume that

Total energy =
(Energy rate over water) · (Distance over water)
+(Energy rate over land) · (Distance over land).

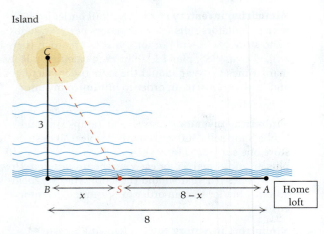

117. A rectangular field is to be divided into two parallel rectangular areas, as shown in the figure below. If the total fencing available is k units, show that the length of each of the three parallel fences will be $k/6$ units.

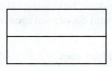

118. Business: minimizing distance. A road is to be built between two cities C_1 and C_2, which are on opposite sides of a river of uniform width r. C_1 is a units from the river, and C_2 is b units from the river, with $a \le b$. A bridge will carry the traffic across the river. Where should the bridge be located in order to minimize the total distance between the cities? Give a general solution using the constants a, b, p, and r as shown in the figure.

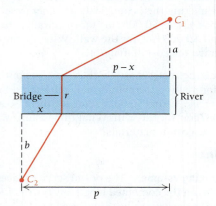

119. Business: minimizing cost. The total cost, in dollars, of producing x units of a certain product is given by

$$C(x) = 8x + 20 + \frac{x^3}{100}.$$

a) Find the average cost, $A(x) = C(x)/x$.
b) Find $C'(x)$ and $A'(x)$.
c) Find the minimum of $A(x)$ and the value x_0 at which it occurs. Find $C'(x_0)$.
d) Compare $A(x_0)$ and $C'(x_0)$.

120. Business: minimizing cost. Consider

$$A(x) = C(x)/x.$$

a) Find $A'(x)$ in terms of $C'(x)$ and $C(x)$.
b) Show that if $A(x)$ has a minimum, then it will occur at that value of x_0 for which

$$C'(x_0) = A(x_0)$$
$$= \frac{C(x_0)}{x_0}.$$

This result shows that if average cost can be minimized, such a minimum will occur when marginal cost equals average cost.

121. Minimize $Q = x^3 + 2y^3$, where x and y are positive numbers, such that $x + y = 1$.

122. Minimize $Q = 3x + y^3$, where $x^2 + y^2 = 2$.

123. Business: minimizing inventory costs—a general solution. A store sells Q units of a product per year. It costs a dollars to store one unit for a year. To reorder, there is a fixed cost of b dollars, plus c dollars for each unit. How many times per year should the store reorder, and in what lot size, in order to minimize inventory costs?

124. Business: minimizing inventory costs. Use the general solution found in Exercise 123 to find how many times per year a store should reorder, and in what lot size, when $Q = 2500$, $a = \$10$, $b = \$20$, and $c = \$9$.

Answers to Quick Checks

1. With 50 ft of fencing, the dimensions are 25 ft by 25 ft (625 ft^2 area); with 100 ft of fencing, they are 50 ft by 50 ft (2500 ft^2 area); in general, n feet of fencing gives $n/2$ ft by $n/2$ ft ($n^2/4$ ft^2 area). **2.** The dimensions are approximately 1.585 in. by 5.33 in. by 7.83 in.; the volume is 66.15 in^3, or 1083.5 cm^3, slightly more than 1 L. **3.** $r \approx 5.3$ cm, $h \approx 2.83$ cm, $c \approx \$0.212$/can **4. (a)** $R(x) = 1750x - 2x^2$; **(b)** $P(x) = -2x^2 + 1735x - 2250$; **(c)** $x = 434$ units; **(d)** maximum profit $= \$374,028$; **(e)** price per unit $= \$882.00$ **5.** $\$23.50$ **6.** $x \approx 63$; the store should place 8 orders for 63 sets and 39 orders for 64 sets.

2.6

Marginals and Differentials

- Find marginal cost, revenue, and profit.
- Find Δy and dy.
- Use differentials for approximations.

In this section, we consider ways of using calculus to make linear approximations. Suppose, for example, that a company is considering an increase in production. Usually the company wants at least an approximation of what the resulting changes in *cost*, *revenue*, and *profit* will be.

Marginal Cost, Revenue, and Profit

Suppose an animal rescue organization that produces its own public service announcements (PSAs) is considering an increase in monthly production from 12 PSAs to 13. To estimate the resulting increase in cost, it would be reasonable to find the rate at which cost is increasing when 12 PSAs are produced and add that to the cost of producing 12 PSAs. That is,

$$C(13) \approx C(12) + C'(12).$$

The number $C'(12)$ is called the *marginal cost at 12*. Remember that $C'(12)$ is the slope of the tangent line at the point $(12, C(12))$. If, for example, this slope is $\frac{3}{4}$, we can regard it as a vertical change of 3 with a horizontal change of 4, or a vertical change of $\frac{3}{4}$ with a horizontal change of 1. It is this latter interpretation that we use for estimating. Graphically, this interpretation can be viewed as shown in the following graph. Note in the figure that $C'(12)$ is slightly more than the difference between $C(13)$ and $C(12)$, or $C(13) - C(12)$. For other curves, $C'(12)$ may be slightly less than $C(13) - C(12)$. Almost always, however, it is simpler to compute $C'(12)$ than it is to compute $C(13) - C(12)$.

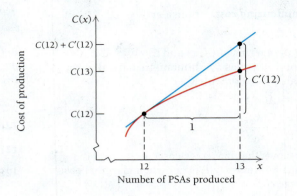

Generalizing, we have the following.

> ### DEFINITIONS
>
> Let $C(x)$, $R(x)$, and $P(x)$ represent, respectively, the total cost, revenue, and profit from the production and sale of x items.
>
> The **marginal cost** at x, given by $C'(x)$, is the approximate cost of the $(x + 1)$st item:
> $$C'(x) \approx C(x + 1) - C(x), \text{ or } C(x + 1) \approx C(x) + C'(x).$$
>
> The **marginal revenue** at x, given by $R'(x)$, is the approximate revenue from the $(x + 1)$st item:
> $$R'(x) \approx R(x + 1) - R(x), \text{ or } R(x + 1) \approx R(x) + R'(x).$$
>
> The **marginal profit** at x, given by $P'(x)$, is the approximate profit from the $(x + 1)$st item:
> $$P'(x) \approx P(x + 1) - P(x), \text{ or } P(x + 1) \approx P(x) + P'(x).$$

The student can confirm that $P'(x) = R'(x) - C'(x)$.

EXAMPLE 1　**Business: Marginal Cost, Revenue, and Profit.**　Given

$$C(x) = 62x^2 + 27{,}500$$
and $$R(x) = x^3 - 12x^2 + 40x + 10,$$

find each of the following.

a) Total profit, $P(x)$

b) Total cost, revenue, and profit from the production and sale of 50 units of the product

c) The marginal cost, revenue, and profit when 50 units are produced and sold

Solution

a)　Total profit $= P(x) = R(x) - C(x)$
$$= x^3 - 12x^2 + 40x + 10 - (62x^2 + 27{,}500)$$
$$= x^3 - 74x^2 + 40x - 27{,}490$$

b) We have

$$C(50) = 62 \cdot 50^2 + 27{,}500 = \$182{,}500 \quad \text{The total cost of producing the first 50 units}$$

$$R(50) = 50^3 - 12 \cdot 50^2 + 40 \cdot 50 + 10 = \$97{,}010 \quad \text{The total revenue from the sale of the first 50 units}$$

$$P(50) = R(50) - C(50)$$
$$= \$97{,}010 - \$182{,}500 \quad \text{We could also use } P(x) \text{ from part (a).}$$
$$= -\$85{,}490 \quad \text{There is a } loss \text{ of \$85,490 when 50 units are produced and sold.}$$

c) We have

$$C'(x) = 124x,$$
$$C'(50) = 124 \cdot 50 = \$6200 \quad \text{Once 50 units have been made, the approximate cost of the 51st unit (marginal cost) is \$6200.}$$

$$R'(x) = 3x^2 - 24x + 40,$$
$$R'(50) = 3 \cdot 50^2 - 24 \cdot 50 + 40 = \$6340 \quad \text{Once 50 units have been sold, the approximate revenue from the 51st unit (marginal revenue) is \$6340.}$$

$$P'(x) = 3x^2 - 148x + 40,$$
$$P'(50) = 3 \cdot 50^2 - 148 \cdot 50 + 40 = \$140 \quad \text{Once 50 units have been produced and sold, the approximate profit from the sale of the 51st item (marginal profit) is \$140.}$$

Quick Check 1 ✔

Given

$$C(x) = 10x + 3$$
and $R(x) = 50x - 0.5x^2,$

find each of the following

a) $P(x)$

b) Total cost, revenue, and profit from the production and sale of 40 units

c) The marginal cost, revenue, and profit when 40 units are produced and sold

1 ✔

To check the accuracy of $R'(50)$ as an estimate of $R(51) - R(50)$, let $y_1 = x^3 - 12x^2 + 40x + 10,$ $y_2 = y_1(x + 1) - y_1(x),$ and $y_3 = \text{nDeriv } (y_1, x, x).$ By using TABLE with Indpnt: Ask, we can display a table in which y_2 (the difference between $y_1(x + 1)$ and $y_1(x)$) can be compared with $y_1'(x).$

X	Y2	Y3
40	3989	3880
48	5933	5800
50	6479	6340

X =

EXERCISE

1. Create a table to check the accuracy of $P'(50)$ as an estimate of $P(51) - P(50).$

Often, in business, formulas for $C(x), R(x),$ and $P(x)$ are not known, but information may exist about the cost, revenue, and profit trends at a particular value $x = a.$ For example, $C(a)$ and $C'(a)$ may be known, allowing a reasonable prediction to be made about $C(a + 1).$ In a similar manner, predictions can be made for $R(a + 1)$ and $P(a + 1).$ In Example 1, formulas *do* exist, so it is possible to see how accurate our predictions were. We check $C(51) - C(50)$ and leave the checks of $R(51) - R(50)$ and $P(51) - P(50)$ to the student (see the Technology Connection in the margin):

$$C(51) - C(50) = 62 \cdot 51^2 + 27{,}500 - (62 \cdot 50^2 + 27{,}500)$$
$$= 6262,$$
whereas $C'(50) = 6200.$

In this case, $C'(50)$ provides an approximation of $C(51) - C(50)$ that is within 1% of the actual value.

Note that marginal cost is different from *average* cost:

$$\text{Average cost per unit for 50 units} = \frac{C(50)}{50} \quad \begin{array}{l}\leftarrow \text{Total cost of 50 units} \\ \leftarrow \text{The number of units, 50}\end{array}$$

$$= \frac{182{,}500}{50} = \$3650,$$

whereas

$$\text{Marginal cost when 50 units are produced} = \$6200$$
$$\approx \text{cost of the 51st unit.}$$

Differentials and Delta Notation

Just as marginal cost $C'(x_0)$ is used to estimate $C(x_0 + 1),$ the value of $f'(x_0),$ can be used to estimate values of $f(x)$ for x-values near $x_0.$ To do so, we need to develop some notation. Recall that the difference quotient,

$$\frac{f(x + h) - f(x)}{h},$$

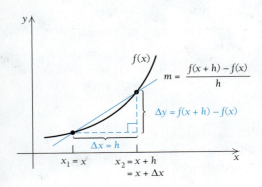

represents the slope of a secant line, as shown in the graph. The number h is regarded as the *change* in x. Another notation for such a change is Δx (read "delta x"), called **delta notation**. The expression Δx is *not* the product of Δ and x, but rather a new type of variable that represents the *change* in the value of x from a *first* value to a *second*. Thus,

$$\Delta x = (x + h) - x = h.$$

If subscripts are used for the first and second values of x, we have

$$\Delta x = x_2 - x_1, \quad \text{or} \quad x_2 = x_1 + \Delta x.$$

Note that the value of Δx can be positive or negative. For example,

$$\text{if } x_1 = 4 \text{ and } \Delta x = 0.7, \text{ then } x_2 = 4.7,$$

and $\quad$ if $x_1 = 4$ and $\Delta x = -0.7$, then $x_2 = 3.3$.

We generally omit the subscripts and use x and $x + \Delta x$. Now suppose we have a function given by $y = f(x)$. A change in x from x to $x + \Delta x$ yields a change in y from $f(x)$ to $f(x + \Delta x)$. The change in y is given by

$$\Delta y = f(x + \Delta x) - f(x).$$

EXAMPLE 2   Find Δy in each case.

a) $y = x^2, x = 4$, and $\Delta x = 0.1$

b) $y = x^3, x = 2$, and $\Delta x = -0.1$

Solution

a) We have

$$\Delta y = (4 + 0.1)^2 - 4^2$$
$$= (4.1)^2 - 4^2 = 16.81 - 16 = 0.81.$$

b) We have

$$\Delta y = [2 + (-0.1)]^3 - 2^3$$
$$= (1.9)^3 - 2^3 = 6.859 - 8 = -1.141.$$

2 ✔

Quick Check 2 ✔

For $y = 2x^4 + x$, $x = 2$, and $\Delta x = -0.05$, find Δy.

Let's now use calculus to predict function values. If delta notation is used, the difference quotient

$$\frac{f(x + h) - f(x)}{h}$$

becomes

$$\frac{f(x + \Delta x) - f(x)}{\Delta x} = \frac{\Delta y}{\Delta x}.$$

We can then express the derivative as

$$\frac{dy}{dx} = \lim_{\Delta x \to 0} \frac{\Delta y}{\Delta x}.$$

For values of Δx close to 0, we have the approximation

$$\frac{dy}{dx} \approx \frac{\Delta y}{\Delta x}, \quad \text{or} \quad f'(x) \approx \frac{\Delta y}{\Delta x}.$$

Multiplying both sides by Δx, we have

$$\Delta y \approx f'(x)\,\Delta x.$$

We can see this in the graph at the left.

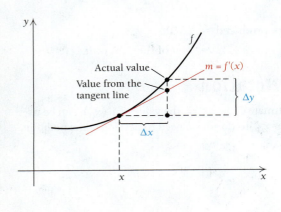

From this graph, it seems reasonable to assume that, for small values of Δx, the y-values on the tangent line can be used to estimate function values on the curve.

For f, a continuous, differentiable function, and small Δx,

$$f'(x) \approx \frac{\Delta y}{\Delta x} \quad \text{and} \quad \Delta y \approx f'(x) \cdot \Delta x.$$

Up to now, we have treated dy/dx as one symbol. We now define dy and dx as separate entities. These symbols are called **differentials**.

DEFINITION

For $y = f(x)$, we define

dx, called the **differential of x**, by $dx = \Delta x$

and dy, called the **differential of y**, by $dy = f'(x)\, dx$.

We can illustrate dx and dy as shown at the right. Note that $dx = \Delta x$, but, in general, $dy \neq \Delta y$, though $dy \approx \Delta y$, for small values of dx.

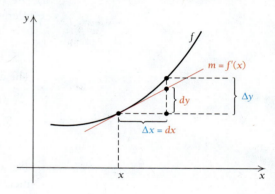

EXAMPLE 3 Consider the function given by $y = f(x) = \sqrt{x}$.

a) Find dy when $x = 9$ and $dx = 1$.

b) Compare dy to Δy.

c) Use the results of parts (a) and (b) to estimate $\sqrt{10}$.

Solution

a) We find dy/dx, then dy:

$$\frac{dy}{dx} = \frac{1}{2\sqrt{x}} \qquad \text{Differentiating } y = x^{1/2} \text{ with respect to } x$$

$$dy = \frac{1}{2\sqrt{x}}\, dx.$$

To find dy when $x = 9$ and $dx = 1$, we substitute:

$$dy = \frac{1}{2\sqrt{9}}(1) = \frac{1}{6} = 0.1666\ldots\,.$$

b) The value Δy is the actual change in y between $x = 9$ and $x = 10$:

$$\Delta y = \sqrt{10} - \sqrt{9}$$

$$= 3.16227767\ldots - 3 \qquad \text{Using a calculator}$$

$$= 0.16227767.$$

We see that $\Delta y = 0.16227767$ and $dy = 0.1666\ldots$ do not differ by much.

For $y = \sqrt{x}$, find dy when $x = 100$ and $dx = -2$. Compare dy to Δy, and estimate the value of $\sqrt{98}$.

c) Note that $\sqrt{10} \approx \sqrt{9} + dy$, so $\sqrt{10} \approx 3 + 0.1666\ldots = 3.1666\ldots$ is a close approximation of $\sqrt{10}$.

3 ✔

We see that the approximation dy and the actual change Δy are reasonably close. It is easier to calculate the approximation since that involves fewer steps, but the trade-off is some loss in accuracy. As long as dx is small, this loss in accuracy is often acceptable, as the following example illustrates.

EXAMPLE 4   **Business: Cost and Tolerance.**   In preparation for laying new tile, Michelle measures the floor of a banquet hall and finds it to be square, measuring 100 ft by 100 ft. Suppose these measurements are accurate to ± 6 in. (the tolerance).

a) Use a differential to estimate the difference in area (dA) due to the tolerance.

b) Compare the result from part (a) with the actual difference in area (ΔA).

c) If each tile covers 1 ft^2 and a box of 12 tiles costs \$24, how many extra boxes of tiles should Michelle buy for covering the floor? What is the extra cost?

Solution
a) The floor is a square, with a presumed measurement of 100 ft per side and a tolerance of ± 6 in. $= \pm 0.5$ ft. The area A in square feet (ft^2) for a square of side length x ft is

$$A(x) = x^2.$$

The derivative is $dA/dx = 2x$, and solving for dA gives the differential of A:

$$dA = 2x\, dx.$$

To find dA, we substitute $x = 100$ and $dx = \pm 0.5$:

$$dA = 2(100)(\pm 0.5) = \pm 100.$$

The value of dA is the approximate difference in area due to the inexactness in measuring. Therefore, if Michelle's measurements are off by half a foot, the total area can differ by approximately $\pm 100\ \text{ft}^2$. A small "error" in measurement can lead to a significant difference in the resulting area.

b) The actual difference in area, ΔA, is calculated. We set $x_1 = 100$ ft, and let x_2 represent the length plus or minus the tolerance.

　　If the true length is at the low end, we have $x_2 = 99.5$ ft. The floor's area is then $99.5^2 = 9900.25\ \text{ft}^2$. The actual difference in area is

$$\begin{aligned} \Delta A &= A(x_2) - A(x_1) \\ &= A(99.5) - A(100) \\ &= 99.5^2 - 100^2 \\ &= 9900.25 - 10{,}000 \\ &= -99.75\ \text{ft}^2. \end{aligned}$$

If the true length is at the high end, we have $x_2 = 100.5$ ft. The floor's area is then $100.5^2 = 10{,}100.25\ \text{ft}^2$. The actual difference in area is

$$\begin{aligned} \Delta A &= A(x_2) - A(x_1) \\ &= A(100.5) - A(100) \\ &= 100.5^2 - 100^2 \\ &= 10{,}100.25 - 10{,}000 \\ &= 100.25\ \text{ft}^2. \end{aligned}$$

Quick Check 4 ✔

The four walls of a room measure 10 ft by 10 ft each, with a tolerance of ± 0.25 ft.

a) Calculate the approximate difference in area, dA, for the four walls.

b) Workers will texture the four walls using "knockdown" spray. Each bottle of knockdown spray costs \$9 and covers 12 ft^2. How many extra bottles should the workers buy to allow for overage in wall area? What will the extra cost be?

We see that ΔA, the actual difference in area, can range from -99.75 ft^2 to 100.25 ft^2. Both values compare well with the approximate value of $dA = \pm 100$ ft^2.

c) The tiles (each measuring 1 ft^2) come 12 to a box. Thus, if the room were exactly 100 ft by 100 ft (an area of 10,000 ft^2), Michelle would need $10{,}000/12 = 833.33\ldots$, or 834 boxes to cover the floor. To take into account the possibility that the room is larger by 100 ft^2, she needs a total of $10{,}100/12 = 841.67\ldots$, or 842 boxes of tiles. Therefore, she should buy $842 - 834$, or 8 extra boxes of tiles, for an extra cost of $(8)(24) = \$192$.   **4** ✔

Historically, differentials were quite valuable when used to make approximations. However, with the advent of computers and graphing calculators, such use has diminished considerably. The use of marginals remains important in the study of business and economics.

Section Summary

- If $C(x)$ represents the cost of producing x items, then *marginal cost* $C'(x)$ is its derivative, and $C'(x) \approx C(x + 1) - C(x)$. Thus, the cost to produce the $(x + 1)$st item can be approximated by $C(x + 1) \approx C(x) + C'(x)$.
- If $R(x)$ represents the revenue from selling x items, then *marginal revenue* $R'(x)$ is its derivative, and $R'(x) \approx R(x + 1) - R(x)$. Thus, the revenue from the $(x + 1)$st item can be approximated by $R(x + 1) \approx R(x) + R'(x)$.
- If $P(x)$ represents profit from selling x items, then *marginal profit* $P'(x)$ is its derivative, and $P'(x) \approx P(x + 1) - P(x)$. Thus, the profit

from the $(x + 1)$st item can be approximated by $P(x + 1) \approx P(x) + P'(x)$.
- In *delta notation*, $\Delta x = (x + h) - x = h$, and $\Delta y = f(x + h) - f(x)$. For small values of Δx, we have $\dfrac{\Delta y}{\Delta x} \approx f'(x)$, which is equivalent to $\Delta y \approx f'(x)\,\Delta x$.
- The *differential* of x is $dx = \Delta x$. Since $\dfrac{dy}{dx} = f'(x)$, we have $dy = f'(x)\,dx$. In general, $dy \approx \Delta y$, and the approximation is often very close for sufficiently small dx.

2.6 Exercise Set

APPLICATIONS

Business and Economics

1. **Marginal revenue, cost, and profit.** Let $R(x)$, $C(x)$, and $P(x)$ be, respectively, the revenue, cost, and profit, in dollars, from the production and sale of x items. If
$$R(x) = 50x - 0.5x^2 \quad \text{and} \quad C(x) = 4x + 10,$$
find each of the following.
   a) $P(x)$
   b) $R(20)$, $C(20)$, and $P(20)$
   c) $R'(x)$, $C'(x)$, and $P'(x)$
   d) $R'(20)$, $C'(20)$, and $P'(20)$

2. **Marginal revenue, cost, and profit.** Let $R(x)$, $C(x)$, and $P(x)$ be, respectively, the revenue, cost, and profit, in dollars, from the production and sale of x items. If
$$R(x) = 5x \quad \text{and} \quad C(x) = 0.001x^2 + 1.2x + 60,$$

find each of the following.
   a) $P(x)$
   b) $R(100)$, $C(100)$, and $P(100)$
   c) $R'(x)$, $C'(x)$, and $P'(x)$
   d) $R'(100)$, $C'(100)$, and $P'(100)$
   e) Describe what each quantity in parts (b) and (d) represents.

3. **Marginal cost.** Suppose the daily cost, in hundreds of dollars, of producing x security systems is
$$C(x) = 0.002x^3 + 0.1x^2 + 42x + 300,$$
and currently 40 security systems are produced daily.
   a) What is the current daily cost?
   b) What would be the additional daily cost of increasing production to 41 security systems daily?
   c) What is the marginal cost when $x = 40$?
   d) Use marginal cost to estimate the daily cost of increasing production to 42 security systems daily.

4. Marginal cost. Suppose the monthly cost, in dollars, of producing x daypacks is

$$C(x) = 0.001x^3 + 0.07x^2 + 19x + 700,$$

and currently 25 daypacks are produced monthly.

a) What is the current monthly cost?
b) What would be the additional cost of increasing production to 26 daypacks monthly?
c) What is the marginal cost when $x = 25$?
d) Use marginal cost to estimate the difference in cost between producing 25 and 27 daypacks per month.
e) Use the answer from part (d) to predict $C(27)$.

5. Marginal revenue. Pierce Manufacturing determines that the daily revenue, in dollars, from the sale of x lawn chairs is

$$R(x) = 0.005x^3 + 0.01x^2 + 0.5x.$$

Currently, Pierce sells 70 lawn chairs daily.

a) What is the current daily revenue?
b) How much would revenue increase if 73 lawn chairs were sold each day?
c) What is the marginal revenue when 70 lawn chairs are sold daily?
d) Use the answer from part (c) to estimate $R(71)$, $R(72)$, and $R(73)$.

6. Marginal profit. For Sunshine Motors, the weekly profit, in dollars, from selling x cars is

$$P(x) = -0.006x^3 - 0.2x^2 + 900x - 1200,$$

and currently 60 cars are sold weekly.

a) What is the current weekly profit?
b) How much profit would be lost if the dealership were able to sell only 59 cars weekly?
c) What is the marginal profit when $x = 60$?
d) Use marginal profit to estimate the weekly profit if sales increase to 61 cars weekly.

7. Marginal revenue. Solano Carriers finds that its monthly revenue, in dollars, from the sale of x carry-on suitcases is

$$R(x) = 0.007x^3 - 0.5x^2 + 150x.$$

Currently Solano is selling 26 carry-on suitcases monthly.

a) What is the current monthly revenue?
b) How much would revenue increase if sales increased from 26 to 28 suitcases?
c) What is the marginal revenue when 26 suitcases are sold?
d) Use the answers from parts (a)–(c) to estimate the revenue resulting from selling 27 suitcases per month.

8. Marginal profit. Crawford Computing finds that its weekly profit, in dollars, from the production and sale of x laptop computers is

$$P(x) = -0.004x^3 - 0.3x^2 + 600x - 800.$$

Currently Crawford builds and sells 9 laptops weekly.

a) What is the current weekly profit?
b) How much profit would be lost if production and sales dropped to 8 laptops weekly?

c) What is the marginal profit when $x = 9$?
d) Use the answers from parts (a)–(c) to estimate the profit resulting from the production and sale of 10 laptops weekly.

9. Sales. Let $N(x)$ be the number of computers sold annually when the price is x dollars per computer. Explain in words what occurs if $N(1000) = 500{,}000$ and $N'(1000) = -100$.

10. Sales. Estimate the number of computers sold in Exercise 9 if the price is raised to $1025.

For Exercises 11–16, assume that $C(x)$ and $R(x)$ are in dollars and x is the number of units produced and sold.

11. For the total-cost function

$$C(x) = 0.01x^2 + 1.6x + 100,$$

find ΔC and $C'(x)$ when $x = 80$ and $\Delta x = 1$.

12. For the total-cost function

$$C(x) = 0.01x^2 + 0.6x + 30,$$

find ΔC and $C'(x)$ when $x = 70$ and $\Delta x = 1$.

13. For the total-revenue function

$$R(x) = 2x,$$

find ΔR and $R'(x)$ when $x = 70$ and $\Delta x = 1$.

14. For the total-revenue function

$$R(x) = 3x,$$

find ΔR and $R'(x)$ when $x = 80$ and $\Delta x = 1$.

15. a) Using $C(x)$ from Exercise 11 and $R(x)$ from Exercise 14, find the total profit, $P(x)$.
b) Find ΔP and $P'(x)$ when $x = 80$ and $\Delta x = 1$.

16. a) Using $C(x)$ from Exercise 12 and $R(x)$ from Exercise 13, find the total profit, $P(x)$.
b) Find ΔP and $P'(x)$ when $x = 70$ and $\Delta x = 1$.

17. Marginal supply. The supply S, of a new rollerball pen is given by

$$S = 0.007p^3 - 0.5p^2 + 150p,$$

where p is the price in dollars.

a) Find the rate of change of quantity with respect to price, dS/dp.
b) How many units will producers want to supply when the price is $25 per unit?
c) Find the rate of change at $p = 25$, and interpret this result.
d) Would you expect dS/dp to be positive or negative? Why?

18. Average cost. The average cost for Turtlehead, Inc., to produce x units of its thermal outerwear is given by the function

$$A(x) = \frac{13x + 100}{x}.$$

Use $A'(x)$ to estimate the change in average cost as production goes from 100 units to 101 units.

19. Marginal productivity. An employee's monthly productivity, M, in number of units produced, is found to be a function of the number of years of service, t. For a certain product, the productivity function is given by

$$M(t) = -2t^2 + 100t + 180.$$

a) Find the productivity of an employee after 5 yr, 10 yr, 25 yr, and 45 yr of service.

b) Find the marginal productivity.

c) Find the marginal productivity at $t = 5$, $t = 10$, $t = 25$, $t = 45$; and interpret the results.

d) Explain how an employee's marginal productivity might be related to experience and to age.

20. Supply. A supply function for a certain product is given by

$$S(p) = 0.08p^3 + 2p^2 + 10p + 11,$$

where $S(p)$ is the number of items produced when the price is p dollars. Use $S'(p)$ to estimate how many more units a producer will supply when the price changes from $18.00 per unit to $18.20 per unit.

21. Gross domestic product. The U.S. gross domestic product, in billions of current dollars, may be modeled by the function

$$P(x) = 567 + x(36x^{0.6} - 104),$$

where x is the number of years since 1960. (*Source:* U.S. Bureau for Economic Analysis.) Use $P'(x)$ to estimate how much the gross domestic product increased from 2014 to 2015.

22. Advertising. Norris Inc. finds that it sells N units of a product after spending x thousands of dollars on advertising, where

$$N(x) = -x^2 + 300x + 6.$$

Use $N'(x)$ to estimate how many more units Norris will sell by increasing its advertising expenditure from $100,000 to $101,000.

Marginal tax rate. *Businesses and individuals are frequently concerned about their marginal tax rate, or the rate at which the next dollar earned is taxed. In progressive taxation, the 80,001st dollar earned is taxed at a higher rate than the 25,001st dollar earned and at a lower rate than the 140,001st dollar earned. Use the following graph, showing the marginal tax rate for 2014 for single filers, to answer Exercises 23–26.*

SINGLE FILERS INCOME BRACKET	RATE
0–9075	10%
9076–36900	15%
36901–89350	25%
89351–186350	28%
186351–405100	33%
405101–406750	35%
406751+	39.6

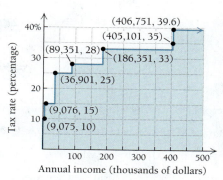

(*Source:* irs.gov, 2014.)

23. Was the taxation in 2014 progressive? Why or why not?

24. Marcy and Tyrone work for the same marketing agency. Because she is not yet a partner, Marcy's year-end income is approximately $95,000; Tyrone's year-end income is approximately $185,000. Suppose one of them is to receive another $5000 in income for the year. Which one would keep more of that $5000 after taxes? Why?

25. Alan earns $85,000 per year and is considering a second job that would earn him another $10,000 annually. At what rate will his tax liability (the amount he must pay in taxes) change if he takes the extra job? Express your answer in tax dollars paid per dollar earned.

26. Iris earns $50,000 per year and is considering extra work that would earn her an extra $3000 annually. At what rate will her tax liability grow if she takes the extra work (see Exercise 25)?

Find Δy and $f'(x)\Delta x$. Round to four and two decimal places, respectively.

27. For $y = f(x) = x^3$, $x = 2$, and $\Delta x = 0.01$

28. For $y = f(x) = x^2$, $x = 2$, and $\Delta x = 0.01$

29. For $y = f(x) = x + x^2$, $x = 3$, and $\Delta x = 0.04$

30. For $y = f(x) = x - x^2$, $x = 3$, and $\Delta x = 0.02$

31. For $y = f(x) = 1/x$, $x = 1$, and $\Delta x = 0.2$

32. For $y = f(x) = 1/x^2$, $x = 1$, and $\Delta x = 0.5$

33. For $y = f(x) = 3x - 1$, $x = 4$, and $\Delta x = 2$

34. For $y = f(x) = 2x - 3$, $x = 8$, and $\Delta x = 0.5$

Use $\Delta y \approx f'(x)\Delta x$ to find a decimal approximation of each radical expression. Round to three decimal places.

35. $\sqrt{26}$ **36.** $\sqrt{8}$

37. $\sqrt{102}$ **38.** $\sqrt{103}$

39. $\sqrt[3]{1005}$ **40.** $\sqrt[3]{28}$

Find dy.

41. $y = \sqrt{3x - 2}$ **42.** $y = \sqrt{x + 1}$

43. $y = (2x^3 + 1)^{3/2}$ **44.** $y = x^3(2x + 5)^2$

45. $y = \dfrac{x^3 + x + 2}{x^2 + 3}$

46. $y = \sqrt[5]{x + 27}$

47. $y = x^4 - 2x^3 + 5x^2 + 3x - 4$

48. $y = (7 - x)^8$

49. In Exercise 48, find dy when $x = 1$ and $dx = 0.01$.

50. In Exercise 47, find dy when $x = 2$ and $dx = 0.1$.

51. For $y = (3x - 10)^5$, find dy when $x = 4$ and $dx = 0.03$.

52. For $y = x^5 - 2x^3 - 7x$, find dy when $x = 3$ and $dx = 0.02$.

53. For $f(x) = x^4 - x^2 + 8$, use a differential to approximate $f(5.1)$.

54. For $f(x) = x^3 - 5x + 9$, use a differential to approximate $f(3.2)$.

SYNTHESIS

Life and Physical Sciences

55. Body surface area. Certain chemotherapy dosages depend on a patient's surface area. According to the Gehan and George model,

$$S = 0.02235 h^{0.42246} w^{0.51456},$$

where h is the patient's height in centimeters, w is his or her weight in kilograms, and S is the approximation to his or her surface area in square meters. (*Source*: www.halls.md.) Joanne is 160 cm tall and weighs 60 kg. Use a differential to estimate how much her surface area changes after her weight decreases by 1 kg.

56. Medical dosage. The function

$$N(t) = \frac{0.8t + 1000}{5t + 4}$$

gives the bodily concentration $N(t)$, in parts per million, of a dosage of medication after t hours. Use differentials to determine whether the concentration changes more from 1.0 hr to 1.1 hr or from 2.8 hr to 2.9 hr.

General Interest

57. Major League ticket prices. The average ticket price of a major league baseball game can be modeled by the function

$$p(x) = 0.06x^3 - 0.5x^2 + 1.64x + 24.76,$$

where x is the number of years after 2008. (*Source*: Major League Baseball.) Use differentials to predict whether ticket price will increase more between 2010 and 2012 or between 2014 and 2016.

58. Suppose a rope surrounds the earth at the equator. The rope is lengthened by 10 ft. By about how much is the rope raised above the earth?

Business and Economics

59. Marginal average cost. In Section 1.6, we defined the average cost of producing x units of a product in terms of the total cost $C(x)$ by $A(x) = C(x)/x$. Find a general expression for *marginal average cost, $A'(x)$*.

60. Cost and tolerance. A firm contracts to paint the exterior of a large water tank in the shape of a half-dome (a hemisphere). The radius of the tank is measured to be 100 ft with a tolerance of ± 6 in. (± 0.5 ft). (The formula for the surface area of a hemisphere is $A = 2\pi r^2$; use 3.14 as an approximation for π.) Each can of paint costs \$30 and covers 300 ft^2.

   a) Calculate dA, the approximate difference in the surface area due to the tolerance.

   b) Assuming the painters cannot bring partial cans of paint to the job, how many extra cans should they bring to cover the extra area they may encounter?

   c) How much extra should the painters plan to spend on paint to account for the possible extra area?

61. Strategic oil supply. The U.S. Strategic Petroleum Reserve (SPR) stores petroleum in large spherical caverns built into salt deposits along the Gulf of Mexico. (*Source*: U.S. Department of Energy.) These caverns can be enlarged by filling the void with water, which dissolves the surrounding salt, and then pumping brine out. Suppose a cavern has a radius of 400 ft, which engineers want to enlarge by 2 ft. Use a differential to estimate how much volume will be added to form the enlarged cavern. (The formula for the volume of a sphere is $V = \frac{4}{3}\pi r^3$; use 3.14 as an approximation for π.)

Marginal revenue. *In each of Exercises 62–66, a demand function, $p = D(x)$, expresses price, in dollars, as a function of the number of items produced and sold. Find the marginal revenue.*

62. $p = 100 - \sqrt{x}$

63. $p = 400 - x$

64. $p = 500 - x$

65. $p = \dfrac{4000}{x} + 3$

66. $p = \dfrac{3000}{x} + 5$

67. Look up "differential" in a book or Web site devoted to math history. In a short paragraph, describe your findings.

68. Explain the uses of the differential.

Elasticity of Demand

2.7

- Find the elasticity of a demand function.
- Find the maximum of a total-revenue function.
- Characterize demand in terms of elasticity.

Retailers and manufacturers often need to know how a small change in price will affect the demand for a product. If a small increase in price produces no change in demand, a price increase may make sense; if a small increase in price creates a large drop in demand, the increase is probably ill advised. To measure the sensitivity of demand to a small percent increase in price, economists calculate the *elasticity of demand*.

Suppose that Klix Video has found that demand for rentals of its DVDs is given by

$$q = D(x) = 120 - 2x,$$

where q is the number of DVDs rented per day at x dollars per rental. Klix Video wants to determine what effect a small increase in the price per rental will have on revenue.

If the price per rental is currently \$2, the demand is $q = D(2) = 120 - 20(2) = 80$ DVDs per day. If the price per rental is raised by a small amount, such as 10%, to \$2.20, the demand will be $q = D(2.2) = 120 - 20(2.2) = 76$ DVDs per day, a 5% decrease. The percent decrease in demand is smaller than the percent increase in price. In this case, the total revenue grows from (\$2 per rental)(80 rentals), or \$160, to (\$2.20 per rental)(76 rentals), or \$167.20.

If the price per rental is currently \$4, the demand is $q = D(4) = 120 - 20(4) = 40$ DVDs per day. If the price per rental is raised by 10% to \$4.40, the demand will be $q = D(4.4) = 120 - 20(4.4) = 32$ DVDs per day, a 20% decrease. Here, the percent decrease in demand is larger thant the percent increase in price. In this case, the total revenue decreases from (\$4 per rental)(40 rentals), or \$160, to (\$4.40 rentals)(32 rentals), or \$140.80.

Let's develop a general formula for elasticity of demand. Suppose q represents a quantity of goods purchased and x is the price per unit of the goods. Recall that q and x are related by the demand function, D, where

$$q = D(x).$$

For any change, Δx, in the price per unit, the percent change in price is

$$\frac{\Delta x}{x} = \frac{\Delta x}{x} \cdot \frac{100}{100} = \frac{\Delta x \cdot 100}{x}\%.$$

A change in price produces a change, Δq, in the quantity demanded, and the percent change in this quantity is

$$\frac{\Delta q}{q} = \frac{\Delta q \cdot 100}{q}\%.$$

The ratio of percent change in quantity sold to percent change in price is

$$\frac{\Delta q/q}{\Delta x/x},$$

which can be expressed as

$$\frac{x}{q} \cdot \frac{\Delta q}{\Delta x}.$$

Since, for differentiable functions,

$$\lim_{\Delta x \to 0} \frac{\Delta q}{\Delta x} = \frac{dq}{dx},$$

we have

$$\lim_{\Delta x \to 0} \frac{x}{q} \cdot \frac{\Delta q}{\Delta x} = \frac{x}{q} \cdot \frac{dq}{dx} = \frac{x}{q} \cdot D'(x) = \frac{x}{D(x)} \cdot D'(x).$$

This result is the basis of the following definition.

DEFINITION

The **elasticity of demand** E is given as a function of price x by

$$E(x) = -\frac{x \cdot D'(x)}{D(x)}.$$

To understand the purpose of the negative sign in the preceding definition, note that the price, x, and the demand, $D(x)$, are both nonnegative. Since $D(x)$ is normally decreasing, $D'(x)$ is usually negative. By inserting a negative sign in the definition, economists make $E(x)$ nonnegative and easier to work with.

EXAMPLE 1 **Economics: Elasticity of Demand for DVD Rentals.** The demand for DVD rentals at Klix Video is given by

$$q = D(x) = 120 - 20x,$$

where q is the number of DVDs rented per day at x dollars per rental. Find each of the following.

a) The elasticity as a function of x

b) The elasticity at $x = 2$ and at $x = 4$. Interpret the meaning of these values of the elasticity.

c) The value of x for which $E(x) = 1$. Interpret the meaning of this price.

d) The total-revenue function, $R(x) = x \cdot D(x)$

e) The price x at which total revenue is a maximum

Solution

a) To find the elasticity, we first find the derivative $D'(x)$:

$$D'(x) = -20.$$

Then we substitute -20 for $D'(x)$ and $120 - 20x$ for $D(x)$ in the expression for elasticity:

$$E(x) = -\frac{x \cdot D'(x)}{D(x)} = -\frac{x \cdot (-20)}{120 - 20x} = \frac{20x}{120 - 20x} = \frac{x}{6 - x}.$$

b) $E(2) = \dfrac{2}{6 - 2} = \dfrac{1}{2}$

At $x = 2$, the elasticity is $\frac{1}{2}$, which is less than 1. Thus, the ratio of percent change in quantity to percent change in price is less than 1. A small percentage increase in price will cause an even smaller percentage decrease in the quantity demanded.

$$E(4) = \frac{4}{6 - 4} = 2$$

At $x = 4$, the elasticity is 2, which is greater than 1. Thus, the ratio of percent change in quantity to percent change in price is greater than 1. A small percentage increase in price will cause a larger percentage decrease in the quantity demanded.

c) We set $E(x) = 1$ and solve for x:

$$\frac{x}{6 - x} = 1$$

$$x = 6 - x \qquad \text{We multiply both sides by } 6 - x, \text{ assuming that } x \neq 6.$$

$$2x = 6$$

$$x = 3.$$

Quick Check 1 ✔

Economics: Demand for DVD Rentals. Internet rentals affect Klix Video in such a way that the demand for rentals of its DVDs changes to

$$q = D(x) = 30 - 5x.$$

a) Find the quantity demanded when the price is $2 per rental, $3 per rental, and $5 per rental.

b) Find the elasticity of demand as a function of x.

c) Find the elasticity at $x = 2$, $x = 3$, and $x = 5$. Interpret the meaning of these values.

d) Find the value of x for which $E(x) = 1$. Interpret the meaning of this price.

e) Find the total-revenue function, $R(x) = x \cdot D(x)$.

f) Find the price x at which total revenue is a maximum.

Thus, when the price is $3 per rental, the ratio of the percent change in quantity to the percent change in price is 1.

d) Recall that total revenue $R(x)$ is given by $x \cdot D(x)$. Then

$$R(x) = x \cdot D(x) = x(120 - 20x) = 120x - 20x^2.$$

e) To find the price x that maximizes total revenue, we find $R'(x)$:

$$R'(x) = 120 - 40x.$$

We see that $R'(x)$ exists for all x in the interval $[0, \infty)$. Thus, we solve:

$$R'(x) = 120 - 40x = 0$$
$$-40x = -120$$
$$x = 3.$$

Since there is only one critical value, we use the second derivative to see if it yields a maximum:

$$R''(x) = -40 < 0.$$

Thus, $R''(3)$ is negative, so $R(3)$ is a maximum. That is, total revenue is a maximum at $3 per rental.

1 ✔

Note in parts (c) and (e) of Example 1 that the value of x for which $E(x) = 1$ is the same as the value of x for which total revenue is a maximum. The following theorem states that this is always the case.

THEOREM 11

Total revenue is increasing at those x-values for which $E(x) < 1$.

Total revenue is decreasing at those x-values for which $E(x) > 1$.

Total revenue is maximized at the value(s) of x for which $E(x) = 1$.

Proof: We know that

$$R(x) = x \cdot D(x),$$

so $$R'(x) = x \cdot D'(x) + D(x) \cdot 1 \qquad \text{Using the Product Rule}$$

$$= D(x)\left[\frac{x \cdot D'(x)}{D(x)} + 1\right] \qquad \text{Factoring; check this by multiplying.}$$

$$= D(x)[-E(x) + 1]$$

$$= D(x)[1 - E(x)].$$

Since we can assume that $D(x) > 0$, it follows that $R'(x)$ is positive for $E(x) < 1$, is negative for $E(x) > 1$, and is 0 when $E(x) = 1$. Thus, total revenue is increasing for $E(x) < 1$, is decreasing for $E(x) > 1$, and is maximized when $E(x) = 1$. ■

In summary, suppose Klix Video in Example 1 raises the price per rental and the total revenue increases. Then we say the demand is *inelastic*. If the total revenue decreases, we say the demand is *elastic*.

In a typical economy, goods with inelastic demands tend to be items people purchase regularly, such as food, clothing, and fuel. These goods tend to be purchased in set amounts, regardless of small fluctuations in price. Goods with elastic demands tend to be larger items purchased occasionally, such as a vehicle or furniture. Purchases of such items can be delayed until the price falls.

Elasticity and Revenue

For a particular value of the price x:

1. The demand is *inelastic* if $E(x) < 1$. An increase in price will bring an increase in revenue. If demand is inelastic, then revenue is increasing.

2. The demand has *unit elasticity* if $E(x) = 1$. The demand has unit elasticity when revenue is at a maximum.

3. The demand is *elastic* if $E(x) > 1$. An increase in price will bring a decrease in revenue. If demand is elastic, then revenue is decreasing.

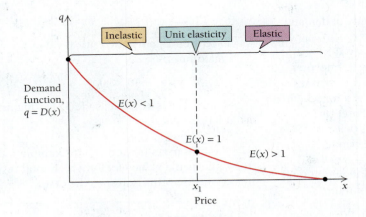

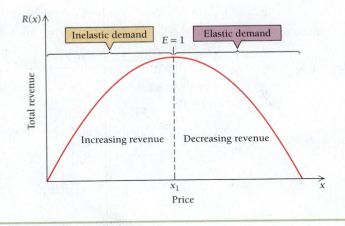

Section Summary

- For a demand function $q = D(x)$, where x is in dollars, the **elasticity of demand** is

$$E(x) = -\frac{xD'(x)}{D(x)}.$$

- Demand is **inelastic** when $E(x) < 1$. If demand is inelastic, then revenue is increasing.

- Demand is **elastic** when $E(x) > 1$. If demand is elastic, then revenue is decreasing.
- Demand has **unit elasticity** when $E(x) = 1$. If demand has unit elasticity at price x, then revenue is maximized.

2.7 | Exercise Set

For the demand function given in each of Exercises 1–10, find the following.

  a) *The elasticity*
  b) *The elasticity at the given price, stating whether the demand is elastic or inelastic*
  c) *The value(s) of x for which total revenue is a maximum (assume that x is in dollars)*

1. $q = D(x) = 400 - x; \quad x = 125$

2. $q = D(x) = 500 - x; \quad x = 38$

3. $q = D(x) = 200 - 4x; \quad x = 46$

4. $q = D(x) = 500 - 2x; \quad x = 57$

5. $q = D(x) = \dfrac{400}{x}; \quad x = 50$

6. $q = D(x) = \dfrac{3000}{x}; \quad x = 60$

7. $q = D(x) = \sqrt{600 - x}; \quad x = 100$

8. $q = D(x) = \sqrt{300 - x}; \quad x = 250$

9. $q = D(x) = \dfrac{100}{(x + 3)^2}; \quad x = 1$

10. $q = D(x) = \dfrac{500}{(2x + 12)^2}; \quad x = 8$

APPLICATIONS

Business and Economics

11. Demand for oil. Suppose you have been hired as an economic consultant concerning the world demand for oil. The demand function is

$$q = D(x) = 50{,}000 + 300x - 3x^2, \quad \text{for } 0 \le x \le 180,$$

where q is measured in millions of barrels of oil per day at a price of x dollars per barrel.

  a) Find the elasticity.
  b) Find the elasticity at a price of $75 per barrel, stating whether demand is elastic or inelastic at that price.
  c) Find the elasticity at a price of $100 per barrel, stating whether demand is elastic or inelastic at that price.
  d) Find the elasticity at a price of $25 per barrel, stating whether demand is elastic or inelastic at that price.
  e) At what price is revenue a maximum?
  f) What quantity of oil will be sold at the price that maximizes revenue? Compare the current world price to your answer.
  g) At a price of $110 per barrel, will a small increase in price cause total revenue to increase or decrease?

12. Demand for chocolate chip cookies. Good Times Bakers determines that the demand function for its chocolate chip cookies is

$$q = D(x) = 967 - 25x,$$

where q is the quantity of cookies sold when the price is x cents per cookie.

  a) Find the elasticity.
  b) At what price is elasticity of demand equal to 1?
  c) At what prices is elasticity of demand elastic?
  d) At what prices is elasticity of demand inelastic?
  e) At what price is the revenue a maximum?
  f) At a price of 20¢ per cookie, will a small increase in price cause total revenue to increase or decrease?

13. Demand for computer games. High Wire Electronics determines the following demand function for a new game:

$$q = D(x) = \sqrt{200 - x^3},$$

where q is the number of games sold per day when the price is x dollars per game.

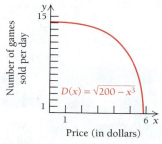

  a) Find the elasticity.
  b) Find the elasticity when $x = 3$.
  c) At $x = 3$, will a small increase in price cause total revenue to increase or decrease?

14. Demand for tomato plants. Sunshine Gardens determines the following demand function during early summer for tomato plants:

$$q = D(x) = \frac{2x + 300}{10x + 11},$$

where q is the number of plants sold per day when the price is x dollars per plant.

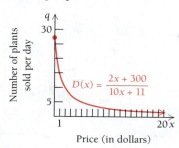

a) Find the elasticity.

b) Find the elasticity when $x = 3$.

c) At \$3 per plant, will a small increase in price cause total revenue to increase or decrease?

SYNTHESIS

15. Business. Tipton Industries determines that the demand function for its sunglass cases is

$$q = D(x) = 180 - 10x,$$

where q is the quantity of sunglass cases sold when the price is x dollars per case.

a) Find the elasticity of demand.

b) Find the elasticity when $x = 8$.

c) At what price is revenue a maximum?

d) If the price is \$8 and Tipton Industries raises the price by 20%, will revenue increase or decrease? Does this contradict the answers to parts (b) and (c)? Why or why not?

16. Economics: constant elasticity curve.

a) Find the elasticity of the demand function

$$q = D(x) = \frac{k}{x^n},$$

where k is a positive constant and n is an integer greater than 0.

b) Is the value of the elasticity dependent on the price per unit?

c) For what value of n does total revenue have a maximum?

17. Explain in your own words the concept of elasticity and its usefulness to economists. Do research or consult an economist to determine when and how this concept was first developed.

18. Explain how the elasticity of demand for a product can be affected by the availability of substitutes for the product.

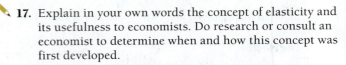

Answers to Quick Check

1. (a) 20, 15, 5;   **(b)** $E(x) = \dfrac{x}{6 - x}$;   **(c)** 0.5, 1, 5

(see Example 1 for the interpretations);

(d) \$3;   **(e)** $R(x) = 30x - 5x^2$;   **(f)** \$3

2.8

Implicit Differentiation and Related Rates*

- Differentiate implicitly.
- Solve related-rate problems.

We often write a function with the output variable (such as y) isolated on one side of the equation. For example, if we write $y = x^3$, we have expressed y as an *explicit* function of x. With an equation like $y^3 + x^2y^5 - x^4 = -11$, it may be difficult to isolate the output variable. In such a case, we have an *implicit* relationship between the variables x and y and can find the derivative of y with respect to x using *implicit differentiation*.

Implicit Differentiation

Consider the equation

$$y^3 = x.$$

This equation *implies* that y is a function of x, for if we solve for y, we get

$$y = \sqrt[3]{x}$$
$$= x^{1/3}.$$

We know from our earlier work that

$$\frac{dy}{dx} = \frac{1}{3}x^{-2/3}. \tag{1}$$

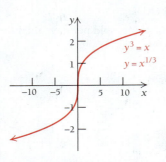

---

*This section can be omitted without loss of continuity.

A method known as **implicit differentiation** allows us to find dy/dx *without* solving for y. To do so, we use the Chain Rule, treating y as a function of x, and differentiate both sides of

$$y^3 = x$$

with respect to x:

$$\frac{d}{dx} y^3 = \frac{d}{dx} x.$$

The derivative on the left side is found using the Extended Power Rule:

$$3y^2 \frac{dy}{dx} = 1.$$ Remembering that the derivative of y with respect to x is written dy/dx

Finally, we solve for dy/dx by dividing both sides by $3y^2$:

$$\frac{dy}{dx} = \frac{1}{3y^2}, \quad \text{or} \quad \frac{1}{3} y^{-2}.$$

We can show that this indeed gives us the same answer as equation (1) by replacing y with $x^{1/3}$:

$$\frac{dy}{dx} = \frac{1}{3} y^{-2} = \frac{1}{3} (x^{1/3})^{-2} = \frac{1}{3} x^{-2/3}.$$

Equations like

$$y^3 + x^2 y^5 - x^4 = -11$$

determine y as a function of x, but are difficult to solve for y. We can nevertheless find a formula for the derivative of y *without* solving for y. To do so usually involves the Extended Power Rule in the form

$$\frac{d}{dx} y^n = ny^{n-1} \cdot \frac{dy}{dx}.$$

TECHNOLOGY CONNECTION

Exploratory

Graphicus can be used to graph equations that relate x and y implicity. Press $\boxed{+}$ and then $\boxed{g(x, y)=0,}$ and enter $y^3 + x^2 y^3 - x^4 = 27$ as y^3+x^2(y^3)-x^4= -11.

EXERCISES

Graph each equation.

1. $y^3 + x^2 y^3 - x^4 = -11$

2. $y^2 x + 2x^3 y^3 = y + 1$

EXAMPLE 1   For $y^3 + x^2 y^5 - x^4 = -11$:

a) Find dy/dx using implicit differentiation.

b) Find the slope of the tangent line to the curve at the point $(2, 1)$.

Solution

a) We differentiate the term $x^2 y^5$ using the Product Rule. Because y is a function of x, it is critical that dy/dx is included as a factor in the result any time a term involving y is differentiated. When an expression involving just x is differentiated, there is no factor dy/dx.

$$\frac{d}{dx} (y^3 + x^2 y^5 - x^4) = \frac{d}{dx} (-11)$$ Differentiating both sides with respect to x

$$\frac{d}{dx} y^3 + \frac{d}{dx} x^2 y^5 - \frac{d}{dx} x^4 = 0$$

$$3y^2 \cdot \frac{dy}{dx} + x^2 \cdot 5y^4 \cdot \frac{dy}{dx} + y^5 \cdot 2x - 4x^3 = 0.$$ Using the Extended Power Rule and the Product Rule

We next isolate those terms with dy/dx as a factor on one side:

$$3y^2 \cdot \frac{dy}{dx} + 5x^2y^4 \cdot \frac{dy}{dx} = 4x^3 - 2xy^5 \qquad \text{Adding } 4x^3 - 2xy^5 \text{ to both sides}$$

$$(3y^2 + 5x^2y^4)\frac{dy}{dx} = 4x^3 - 2xy^5 \qquad \text{Factoring } dy/dx$$

$$\frac{dy}{dx} = \frac{4x^3 - 2xy^5}{3y^2 + 5x^2y^4}. \qquad \begin{array}{l}\text{Solving for } dy/dx \text{ and leaving the} \\ \text{answer in terms of } x \text{ and } y\end{array}$$

b) To find the slope of the tangent line to the curve at $(2, 1)$, we replace x with 2 and y with 1:

$$\frac{dy}{dx} = \frac{4 \cdot 2^3 - 2 \cdot 2 \cdot 1^5}{3 \cdot 1^2 + 5 \cdot 2^2 \cdot 1^4} = \frac{28}{23}.$$

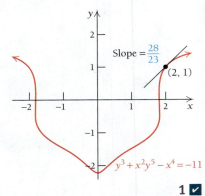

Quick Check 1 ✔

For $y^2x + 2x^3y^3 = y + 1$, find dy/dx using implicit differentiation.

It is common for the expression for dy/dx to contain *both* variables x and y. When this occurs, we calculate a slope by evaluating the derivative at both the x-value and the y-value of the point of tangency.

The steps in Example 1 are typical of those used when differentiating implicitly.

> ### To differentiate implicitly:
>
> **a)** Differentiate both sides of the equation with respect to x (or whatever variable you are differentiating with respect to).
> **b)** Apply the rules for differentiation (the Power, Product, Quotient, and Chain Rules) as necessary. Any time an expression involving y is differentiated, dy/dx will be a factor in the result.
> **c)** Isolate all terms with dy/dx as a factor on one side of the equation.
> **d)** If necessary, factor out dy/dx.
> **e)** If necessary, divide both sides of the equation to isolate dy/dx.

The demand function for a product (see Section R.5) is often given implicitly.

EXAMPLE 2 Differentiate the demand equation $x = \sqrt{200 - p^3}$, implicitly to find dp/dx.

Solution
$$\frac{d}{dx}x = \frac{d}{dx}\sqrt{200 - p^3}$$

$$1 = \frac{1}{2}(200 - p^3)^{-1/2} \cdot (-3p^2) \cdot \frac{dp}{dx} \qquad \begin{array}{l}\text{Using the Extended Power} \\ \text{Rule twice}\end{array}$$

$$1 = \frac{-3p^2}{2\sqrt{200 - p^3}} \cdot \frac{dp}{dx}$$

$$\frac{2\sqrt{200 - p^3}}{-3p^2} = \frac{dp}{dx}$$

Quick Check 2 ✔

Differentiate the equation
$$y = 2\sqrt{1 + p^2}$$
implicitly to find $\dfrac{dp}{dy}$.

2 ✔

Related Rates

Suppose y is a function of x,

$$y = f(x),$$

and x is a function of time, t. Since y depends on x and x depends on t, it follows that y depends on t. The Chain Rule gives the following:

$$\frac{dy}{dt} = \frac{dy}{dx} \cdot \frac{dx}{dt}.$$

Thus, the rate of change of y is *related* to the rate of change of x. When this comes up in problems, it helps to keep in mind that any variable can be thought of as a function of time t. If no expression in terms of t is given, the variable's rate of change with respect to t will be 0.

EXAMPLE 3 **Business: Service Area.** A restaurant supplier services the restaurants in a circular area in such a way that the radius r is increasing at the rate of 2 mi per year at the moment when $r = 5$ mi. At that moment, how fast is the area increasing?

Solution The area A and radius r are related by

$$A = \pi r^2.$$

We take the derivative of both sides with respect to t, and then evaluate, using $dr/dt = 2 \text{ mi/yr}$ and $r = 5$ mi:

$$\frac{dA}{dt} = 2\pi r \cdot \frac{dr}{dt}. \qquad \text{\color{red}The factor } dr/dt \text{ results from the Chain Rule and the}$$
$$\text{\color{red}fact that } r \text{ is assumed to be a function of } t.$$

$$\frac{dA}{dt} = 2\pi (5 \text{ mi}) \left(2 \, \frac{\text{mi}}{\text{yr}} \right) \qquad \text{\color{red}Substituting}$$

$$= 20\pi \, \frac{\text{mi}^2}{\text{yr}} \approx 63 \text{ mi}^2/\text{yr}.$$

The supplier's service area is growing at a rate of about 63 square miles per year when $r = 5$ mi and $dr/dt = 2$ mi/yr.   **3** ✔

Quick Check 3 ✔

A spherical balloon is leaking, losing 20 cm^3 of air per minute. At the moment when the radius of the balloon is 8 cm, how fast is the radius decreasing? (*Hint*: $V = \frac{4}{3}\pi r^3$.)

EXAMPLE 4   Business: Rates of Change of Revenue, Cost, and Profit.   For Luce Landscaping, the total revenue and the total cost for yard maintenance of x homes are given by

$$R(x) = 1000x - x^2, \quad \text{and} \quad C(x) = 3000 + 20x.$$

Suppose that Luce is adding 10 homes per day at the moment when the 400th customer is signed. At that moment, what is the rate of change of (a) total revenue, (b) total cost, and (c) total profit?

Solution

a) We regard $R(x) = 1000x - x^2$ as $R = 1000x - x^2$. Implicitly differentiating with respect to time t, we have

$$\frac{dR}{dt} = 1000 \cdot \frac{dx}{dt} - 2x \cdot \frac{dx}{dt} \qquad \text{Differentiating both sides with respect to time}$$

$$= 1000 \cdot 10 - 2(400)10 \qquad \text{Substituting 10 for } dx/dt \text{ and 400 for } x$$

$$= \$2000 \text{ per day.}$$

b) We regard $C(x) = 3000 + 20x$ as $C = 3000 + 20x$. Implicitly differentiating with respect to time t, we have

$$\frac{dC}{dt} = 20 \cdot \frac{dx}{dt} \qquad \text{Differentiating both sides with respect to time}$$

$$= 20(10)$$

$$= \$200 \text{ per day.}$$

c) Since $P(x) = R(x) - C(x)$, we have

$$\frac{dP}{dt} = \frac{dR}{dt} - \frac{dC}{dt}$$

$$= \$2000 \text{ per day} - \$200 \text{ per day}$$

$$= \$1800 \text{ per day.} \qquad \blacksquare$$

Section Summary

- If variables x and y are related to one another by an equation but neither variable is isolated on one side of the equation, we say that x and y have an implicit relationship. To find dy/dx without solving such an equation for y, we use *implicit differentiation*.
- Whenever we implicitly differentiate y with respect to x, the factor dy/dx will appear as a result of the Chain Rule.

- To determine the slope of a tangent line at a point on the graph of an implicit relationship, we may need to evaluate the derivative by inserting both the x-value and the y-value of the point of tangency.
- If two or more variables are related to one another by an equation, then we may implicity differentiate with respect to time t. The result is a *related rate*.

2.8   Exercise Set

Differentiate implicitly to find dy/dx. Then find the slope of the curve at the given point.

1. $3x^3 - y^2 = 8$; $(2, 4)$

2. $x^3 + 2y^3 = 6$; $(2, -1)$

3. $2x^3 + 4y^2 = -12$; $(-2, -1)$

4. $2x^2 - 3y^3 = 5$; $(-2, 1)$

5. $x^2 + y^2 = 1$; $\left(\dfrac{1}{2}, \dfrac{\sqrt{3}}{2}\right)$

6. $x^2 - y^2 = 1$; $(\sqrt{3}, \sqrt{2})$

7. $2x^3y^2 = -18$; $(-1, 3)$

8. $3x^2y^4 = 12$; $(2, -1)$

9. $x^4 - x^2y^3 = 12$; $(-2, 1)$

10. $x^3 - x^2y^2 = -9$; $(3, -2)$

11. $xy + y^2 - 2x = 0$; $(1, -2)$

12. $xy - x + 2y = 3$; $\left(-5, \dfrac{2}{3}\right)$

13. $4x^3 - y^4 - 3y + 5x + 1 = 0$; $(1, -2)$

14. $x^2y - 2x^3 - y^3 + 1 = 0$; $(2, -3)$

Differentiate implicitly to find dy/dx.

15. $x^2 + 2xy = 3y^2$

16. $2xy + 3 = 0$

17. $x^2 - y^2 = 16$

18. $x^2 + y^2 = 25$

19. $y^3 = x^5$

20. $y^5 = x^3$

21. $x^2y^3 + x^3y^4 = 11$

22. $x^3y^2 + x^5y^3 = -19$

For each demand equation in Exercises 23–30, differentiate implicitly to find dp/dx.

23. $p^2 + p + 2x = 40$

24. $p^3 + p - 3x = 50$

25. $xp^3 = 24$

26. $x^3p^2 = 108$

27. $\dfrac{x^2p + xp + 1}{2x + p} = 1$ (*Hint:* Clear the fraction first.)

28. $\dfrac{xp}{x + p} = 2$ (*Hint:* Clear the fraction first.)

29. $(p + 4)(x + 3) = 48$

30. $1000 - 300p + 25p^2 = x$

31. Nonnegative variable quantities G and H are related by the equation

$$G^2 + H^2 = 25.$$

What is the rate of change dH/dt when $dG/dt = 3$ and $G = 0$? $G = 1$? $G = 3$?

32. Variable quantities A and B are related by the equation

$$A^3 + B^3 = 9.$$

What is the rate of change dA/dt at the moment when $A = 2$ and $dB/dt = 3$?

APPLICATIONS

Business and Economics

Rates of change of total revenue, cost, and profit.   *In Exercises 33–36, find the rates of change of total revenue, cost, and profit with respect to time. Assume that $R(x)$ and $C(x)$ are in dollars.*

33. $R(x) = 50x - 0.5x^2$,

$C(x) = 10x + 3$,

when $x = 10$ and $dx/dt = 5$ units per day

34. $R(x) = 50x - 0.5x^2$,

$C(x) = 4x + 10$,

when $x = 30$ and $dx/dt = 20$ units per day

35. $R(x) = 280x - 0.4x^2$,

$C(x) = 5000 + 0.6x^2$,

when $x = 200$ and $dx/dt = 300$ units per day

36. $R(x) = 2x$,

$C(x) = 0.01x^2 + 0.6x + 30$,

when $x = 20$ and $dx/dt = 8$ units per day

37. **Change of sales.** Suppose that the price p, in dollars, and number of sales, x, of a mechanical pencil are related by

$$5p + 4x + 2px = 60.$$

If p and x are both functions of time, measured in days, find the rate at which x is changing when

$$x = 3, p = 5, \text{and } \frac{dp}{dt} = 1.5.$$

38. **Change of revenue.** For x and p as described in Exercise 37, find the rate at which total revenue $R = xp$ is changing when

$$x = 3, p = 5, \text{and } \frac{dp}{dt} = 1.5.$$

Life and Natural Sciences

39. Rate of change of the Arctic ice cap. In a trend that scientists attribute, at least in part, to global warming, the floating cap of sea ice on the Arctic Ocean has been shrinking since 1980. The ice cap always shrinks in summer and grows in winter. Average minimum size of the ice cap, in square miles, can be approximated by

$$A = \pi r^2.$$

In 2013, the radius of the ice cap was approximately 792 mi and was shrinking at a rate of approximately 4.7 mi/yr. (*Source*: Based on data from nsidc.org.) How fast was the area changing at that time?

40. Rate of change of a healing wound. The area of a healing wound is given by

$$A = \pi r^2.$$

The radius is decreasing at the rate of 1 millimeter per day (-1 mm/day) at the moment when $r = 25$ mm. How fast is the area decreasing at that moment?

41. Body surface area. Certain chemotherapy dosages depend on a patient's surface area. According to the Mosteller model,

$$S = \frac{\sqrt{hw}}{60},$$

where h is the patient's height in centimeters, w is the patient's weight in kilograms, and S is the approximation of the patient's surface area in square meters. (*Source*: www.halls.md.) Assume that Kim's height is a constant 165 cm, but she is losing weight. If she loses 2 kg per month, how fast is her surface area decreasing at the instant she weighs 70 kg?

Poiseuille's Law.   *The flow of blood in a blood vessel is faster toward the center of the vessel and slower toward the outside. The speed of the blood, V, in millimeters per second (mm/sec), is given by*

$$V = \frac{p}{4Lv}(R^2 - r^2),$$

where R is the radius of the blood vessel, r is the distance of the blood from the center of the vessel, and p, L, and v are physical constants related to blood pressure, length of the

blood vessel, and viscosity of the blood, respectively. Assume that dV/dt is measured in millimeters per second squared (mm/sec²). *Use this formula for Exercises 42 and 43.*

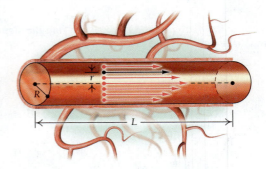

42. Assume that r is a constant as well as p, L, and v.

   a) Find the rate of change dV/dt in terms of R and dR/dt when $L = 80$ mm, $p = 500$, and $v = 0.003$.

   b) A person goes out into the cold to shovel snow. Cold air has the effect of contracting blood vessels far from the heart. Suppose a blood vessel contracts at a rate of

$$\frac{dR}{dt} = -0.0002 \text{ mm/sec}$$

   at a place where the radius of the vessel is $R = 0.075$ mm. Find the rate of change, dV/dt, at that location.

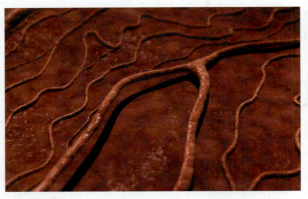

The flow of blood in a blood vessel can be modeled by Poiseuille's Law.

43. Assume that r is a constant as well as p, L, and v.

   a) Find the rate of change dV/dt in terms of R and dR/dt when $L = 70$ mm, $p = 400$, and $v = 0.003$.

   b) When shoveling snow in cold air, a person with a history of heart trouble can develop angina (chest pains) due to contracting blood vessels. To counteract this, he or she may take a nitroglycerin tablet, which dilates the blood vessels. Suppose, after a nitroglycerin tablet is taken, a blood vessel dilates at a rate of

$$\frac{dR}{dt} = 0.00015 \text{ mm/sec}$$

   at a place where the radius of the vessel is $R = 0.1$ mm. Find the rate of change, dV/dt.

General Interest

44. Two cars start from the same point at the same time. One travels north at 25 mph, and the other travels east at 60 mph. How fast is the distance between them increasing at the end of 1 hr? (*Hint:* $D^2 = x^2 + y^2$. To find D after 1 hr, solve $D^2 = 25^2 + 60^2$.)

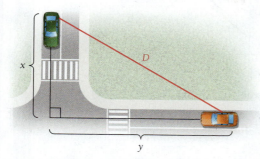

45. A ladder 26 ft long leans against a vertical wall. If the lower end is being moved away from the wall at the rate of 5 ft/sec, how fast is the height of the top changing (this will be a negative rate) when the lower end is 10 ft from the wall?

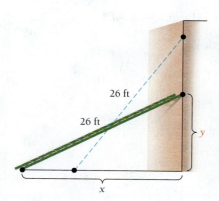

46. An inner city revitalization zone is a rectangle that is twice as long as it is wide. A diagonal through the region is growing at a rate of 90 m per year at a time when the region is 440 m wide. How fast is the area changing at that point in time?

47. The volume of a cantaloupe is approximated by
$$V = \tfrac{4}{3}\pi r^3.$$
The radius is growing at the rate of 0.7 cm/week, at a time when the radius is 7.5 cm. How fast is the volume changing at that moment?

SYNTHESIS

Differentiate implicitly to find dy/dx.

48. $\sqrt{x} + \sqrt{y} = 1$

49. $\dfrac{1}{x^2} + \dfrac{1}{y^2} = 5$

50. $y^3 = \dfrac{x-1}{x+1}$

51. $y^2 = \dfrac{x^2 - 1}{x^2 + 1}$

52. $x^{3/2} + y^{2/3} = 1$

53. $(x - y)^3 + (x + y)^3 = x^5 + y^5$

Differentiate implicitly to find d²y/dx².

54. $xy + x - 2y = 4$

55. $y^2 - xy + x^2 = 5$

56. $x^2 - y^2 = 5$

57. $x^3 - y^3 = 8$

58. Explain the usefulness of implicit differentiation.

59. Look up the word "implicit" in a dictionary. Explain how that definition can be related to the concept of a function that is defined "implicitly."

Graph each of the following equations. Equations must be solved for y before they can be entered into most calculators. Graphicus does not require that equations be solved for y.

60. $x^2 + y^2 = 4$
Note: You will probably need to sketch the graph in two parts: $y = \sqrt{4 - x^2}$ and $y = -\sqrt{4 - x^2}$. Then graph the tangent line to the graph at the point $(-1, \sqrt{3})$.

61. $x^4 = y^2 + x^6$
Then graph the tangent line to the graph at the point $(-0.8, 0.384)$.

62. $y^4 = y^2 - x^2$

63. $x^3 = y^2(2 - x)$

64. $y^2 = x^3$

Answers to Quick Checks

1. $\dfrac{dy}{dx} = \dfrac{y^2 + 6x^2y^3}{1 - 2xy - 6x^3y^2}$ **2.** $\dfrac{dp}{dy} = \dfrac{\sqrt{1 + p^2}}{2p}$

3. Approximately -0.025 cm/min

KEY TERMS AND CONCEPTS

SECTION 2.1

A function is **increasing** over an open interval I if, for all x in I, $f'(x) > 0$.

A function is **decreasing** over an open interval I if, for all x in I, $f'(x) < 0$.

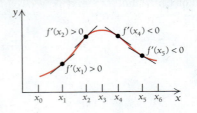

f is increasing over the interval (x_0, x_3) and decreasing over (x_3, x_6).

If f is a continuous function, then a **critical value** is any number c in the domain of f for which $f'(c) = 0$ or $f'(c)$ does not exist.

If $f'(c)$ does not exist, then the graph of f may have a corner or a vertical tangent at $(c, f(c))$.

In both cases, the ordered pair $(c, f(c))$ is called a **critical point**.

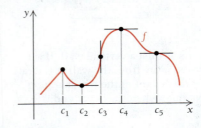

The values c_1, c_2, c_3, c_4, and c_5 are critical values of f.

- $f'(c_1)$ does not exist (corner).
- $f'(c_2) = 0$.
- $f'(c_3)$ does not exist (vertical tangent).
- $f'(c_4) = 0$.
- $f'(c_5) = 0$.

If f is a continuous function, then any **relative extremum** (**maximum** or **minimum**) occurs at a critical value.

The converse is not true: a critical value may not correspond to an extremum.

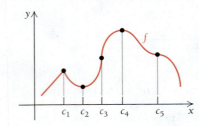

- The critical point $(c_1, f(c_1))$ is a relative maximum.
- The critical point $(c_2, f(c_2))$ is a relative minimum.
- The critical point $(c_3, f(c_3))$ is neither a relative maximum nor a relative minimum.
- The critical point $(c_4, f(c_4))$ is a relative maximum.
- The critical point $(c_5, f(c_5))$ is neither a relative maximum nor a relative minimum.

KEY TERMS AND CONCEPTS	EXAMPLES

SECTION 2.1 (continued)

The **First-Derivative Test** allows us to classify a critical value as a relative maximum, a relative minimum, or neither.

Find the relative extrema of the function given by

$$f(x) = \tfrac{1}{3}x^3 - \tfrac{1}{2}x^2 - 20x + 7.$$

Critical values occur where $f'(x) = 0$ or $f'(x)$ does not exist. The derivative $f'(x) = x^2 - x - 20$ exists for all real numbers, so the only critical values occur when $f'(x) = 0$. Setting $x^2 - x - 20 = 0$ and solving, we have $x = -4$ and $x = 5$ as the critical values.

To apply the First-Derivative Test, we check the sign of $f'(x)$ to the left and right of each critical value, using test values:

Test Value	$x = -5$	$x = 0$	$x = 6$
Sign of $f'(x)$	$f'(-5) > 0$	$f'(0) < 0$	$f'(6) > 0$
Result	f increasing on $(-\infty, -4)$	f decreasing on $(-4, 5)$	f increasing on $(5, \infty)$

Therefore, there is a relative maximum at $x = -4$: $f(-4) = 57\tfrac{2}{3}$. There is a relative minimum at $x = 5$: $f(5) = -63\tfrac{5}{6}$.

SECTION 2.2

The **second derivative**, $f''(x)$, can be used to determine the **concavity** of a graph.

If $f''(x) > 0$ for all x in an open interval I, then the graph of f is **concave up** over I.

If $f''(x) < 0$ for all x in an open interval I, then the graph of f is **concave down** over I.

A **point of inflection** occurs at $(x_0, f(x_0))$ if $f''(x_0) = 0$ and there is a change in concavity on either side of x_0.

The function given by

$$f(x) = \tfrac{1}{3}x^3 - \tfrac{1}{2}x^2 - 20x + 7$$

has the second derivative $f''(x) = 2x - 1$. Setting the second derivative equal to 0, we have $x_0 = \tfrac{1}{2}$. Using test values, we can check the concavity on either side of $x_0 = \tfrac{1}{2}$:

Test Value	$x = 0$	$x = 1$
Sign of $f''(x)$	$f''(0) < 0$	$f''(1) > 0$
Result	f is concave down on $\left(-\infty, \tfrac{1}{2}\right)$	f is concave up on $\left(\tfrac{1}{2}, \infty\right)$

Therefore, the graph of f is concave down over the interval $\left(-\infty, \tfrac{1}{2}\right)$ and concave up over the interval $\left(\tfrac{1}{2}, \infty\right)$. Since there is a change in concavity on either side of $x_0 = \tfrac{1}{2}$, we also conclude that the point $\left(\tfrac{1}{2}, -3\tfrac{1}{12}\right)$ is a point of inflection, where $f\left(\tfrac{1}{2}\right) = -3\tfrac{1}{12}$.

The **Second-Derivative Test** can also be used to classify relative extrema:

If $f'(c) = 0$ and $f''(c) > 0$, then $f(c)$ is a relative minimum.

If $f'(c) = 0$ and $f''(c) < 0$, then $f(c)$ is a relative maximum.

If $f'(c) = 0$ and $f''(c) = 0$, then the First-Derivative Test must be used to classify $f(c)$.

For the function

$$f(x) = \tfrac{1}{3}x^3 - \tfrac{1}{2}x^2 - 20x + 7,$$

evaluating the second derivative, $f''(x) = 2x - 1$, at the critical values yields the following conclusions:

- At $x = -4$, we have $f''(-4) < 0$. Since $f'(-4) = 0$ and the graph is concave down, we conclude that there is a relative maximum at $x = -4$.
- At $x = 5$, we have $f''(5) > 0$. Since $f'(5) = 0$ and the graph is concave up, we conclude that there is a relative minimum at $x = 5$.

(continued)

| KEY TERMS AND CONCEPTS | EXAMPLES |

SECTION 2.3

A line $x = a$ is a **vertical asymptote** if

$$\lim_{x \to a^-} f(x) = \infty,$$

$$\lim_{x \to a^-} f(x) = -\infty,$$

$$\lim_{x \to a^+} f(x) = \infty,$$

or

$$\lim_{x \to a^+} f(x) = -\infty.$$

Consider the function given by

$$f(x) = \frac{x^2 - 1}{x^2 + x - 6}.$$

Factoring, we have

$$f(x) = \frac{(x + 1)(x - 1)}{(x + 3)(x - 2)}.$$

This expression is simplified. Therefore, $x = -3$ and $x = 2$ are vertical asymptotes since

$$\lim_{x \to -3^-} f(x) = \infty \quad \text{and} \quad \lim_{x \to -3^+} f(x) = -\infty$$

and

$$\lim_{x \to 2^-} f(x) = -\infty \quad \text{and} \quad \lim_{x \to 2^+} f(x) = \infty.$$

An asymptote is usually represented as a dashed line; it is not part of the graph itself.

A line $y = b$ is a **horizontal asymptote** if

$$\lim_{x \to -\infty} f(x) = b$$

or

$$\lim_{x \to \infty} f(x) = b.$$

For the function given by

$$f(x) = \frac{x^2 - 1}{x^2 + x - 6},$$

we have

$$\lim_{x \to -\infty} f(x) = 1 \quad \text{and} \quad \lim_{x \to \infty} f(x) = 1.$$

Thus, the line $y = 1$ is a horizontal asymptote.

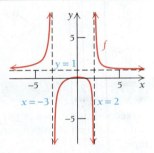

For a rational function of the form $f(x) = P(x)/Q(x)$, a **slant asymptote** occurs if the degree of the numerator is 1 greater than the degree of the denominator.

Let $f(x) = \dfrac{x^2 + 1}{x + 3}$. Division yields

$$f(x) = x - 3 + \frac{10}{x + 3}.$$

As $x \to \infty$ or $x \to -\infty$,

the remainder $\dfrac{10}{x + 3} \to 0.$

Therefore, the slant asymptote is $y = x - 3$.

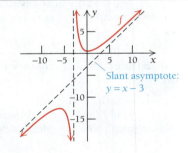

Slant asymptote:
$y = x - 3$

Asymptotes, extrema, x- and y-intercepts, points of inflection, concavity, and intervals over which the function is increasing or decreasing are all used in the strategy for accurate graph sketching.

Consider the function given by
$f(x) = \frac{1}{3}x^3 - \frac{1}{2}x^2 - 20x + 7.$

- f has a relative maximum point at $(-4, 57\frac{2}{3})$ and a relative minimum point at $(5, -63\frac{5}{6})$.
- f has a point of inflection at $(\frac{1}{2}, -3\frac{1}{12})$.
- f is increasing over the intervals $(-\infty, -4)$ and $(5, \infty)$, decreasing over the interval $(-4, 5)$, concave down over the interval $(-\infty, \frac{1}{2})$, and concave up over the interval $(\frac{1}{2}, \infty)$.
- f has a y-intercept at $(0, 7)$.

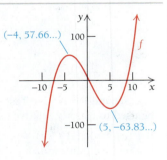

KEY TERMS AND CONCEPTS

EXAMPLES

SECTION 2.3 (*continued*)

Consider the function given by

$$f(x) = \frac{x^2 - 1}{x^2 + x - 6}.$$

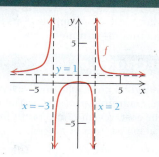

- f has vertical asymptotes $x = -3$ and $x = 2$.
- f has a horizontal asymptote given by $y = 1$.
- f has a y-intercept at $(0, \frac{1}{6})$.
- f has x-intercepts at $(-1, 0)$ and $(1, 0)$.
- f has a relative maximum at $(0.101, 0.168)$ and a relative minimum at $(9.899, 0.952)$.
- f is concave up on $(-\infty, -3)$ and $(2, 14.929)$, and concave down on $(-3, 2)$ and $(14.929, \infty)$.

SECTION 2.4

If f is continuous over $[a, b]$, then the **Extreme-Value Theorem** tells us that f will have both an absolute maximum value and an absolute minimum value over this interval. One or both absolute extrema may occur at an endpoint of this interval.

Maximum–Minimum Principle 1 can be used to determine these absolute extrema: we find all critical values $c_1, c_2, c_3, \ldots, c_n$, in $[a, b]$, then evaluate

$$f(a), f(c_1), f(c_2), f(c_3), \ldots, f(c_n), f(b).$$

The largest of these is the **absolute maximum**, and the smallest is the **absolute minimum**.

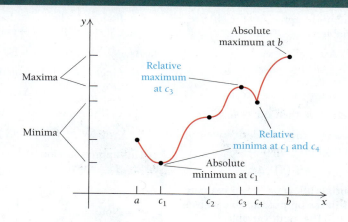

If f is differentiable for all x in an interval I, and there is exactly one critical value c in I such that $f'(c) = 0$, then, according to **Maximum–Minimum Principle 2**, $f(c)$ is an absolute extremum over I.

Let $f(x) = x + \dfrac{2}{x}$, for $x > 0$. The derivative is $f'(x) = 1 - \dfrac{2}{x^2}$. We solve for the critical value:

$$1 - \frac{2}{x^2} = 0$$
$$x^2 = 2$$
$$x = \pm \sqrt{2}.$$

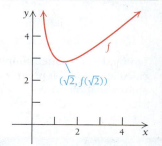

The only critical value over the interval where $x > 0$ is $x = \sqrt{2}$. The second derivative is $f''(x) = \dfrac{4}{x^3}$. We see that

$$f''(\sqrt{2}) = \frac{4}{(\sqrt{2})^3} > 0. \text{ Therefore,}$$

$(\sqrt{2}, f(\sqrt{2}))$ is an absolute minimum.

(*continued*)

KEY TERMS AND CONCEPTS	EXAMPLES

SECTION 2.5

Many real-world applications involve maximum–minimum problems.

See Examples 1–6 in Section 2.5 and the problem-solving strategy on p. 245.

SECTION 2.6

Marginal cost, **marginal revenue**, and **marginal profit** are estimates of the cost, revenue, and profit for the $(x + 1)$st item produced:

- $C'(x) \approx C(x + 1) - C(x)$, so $C(x + 1) \approx C(x) + C'(x)$.
- $R'(x) \approx R(x + 1) - R(x)$, so $R(x + 1) \approx R(x) + R'(x)$.
- $P'(x) \approx P(x + 1) - P(x)$, so $P(x + 1) \approx P(x) + P'(x)$.

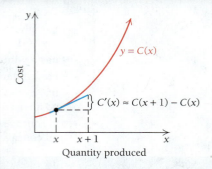

Delta notation represents the change in a variable:

$$\Delta x = x_2 - x_1$$

and

$$\Delta y = f(x_2) - f(x_1).$$

If $x_2 = x_1 + h$, then $\Delta x = h$.

If Δx is small, then the derivative can be used to approximate Δy:

$$\Delta y \approx f'(x) \cdot \Delta x.$$

Differentials allow us to approximate changes in the output variable y given a change in the input variable x:

$$dx = \Delta x$$

and

$$dy = f'(x)\, dx.$$

If Δx is small, then $dy \approx \Delta y$.

In practice, it is often simpler to calculate dy, and it provides a close approximation for Δy.

For $f(x) = x^2$, let $x_1 = 2$ and $x_2 = 2.1$. Then, $\Delta x = 2.1 - 2 = 0.1$. Since $f(x_1) = f(2) = 4$ and $f(x_2) = f(2.1) = 4.41$, we have

$$\Delta y = f(x_2) - f(x_1) = 4.41 - 4 = 0.41.$$

Since $\Delta x = 0.1$ is small, we can approximate Δy by

$$\Delta y \approx f'(x)\, \Delta x.$$

The derivative is $f'(x) = 2x$. Therefore,

$$\Delta y \approx f'(2) \cdot (0.1) = 2(2) \cdot (0.1) = 0.4.$$

This approximation, 0.4, is very close to the actual difference, 0.41.

For $y = \sqrt[3]{x}$, find dy when $x = 27$ and $dx = 2$.

Note that $\dfrac{dy}{dx} = \dfrac{1}{3\sqrt[3]{x^2}}$. Thus, $dy = \dfrac{1}{3\sqrt[3]{x^2}}\, dx$. Evaluating, we have

$$dy = \frac{1}{3\sqrt[3]{(27)^2}}\,(2) = \frac{2}{27}.$$

This result can be used to approximate the value of $\sqrt[3]{29}$, using the fact that $\sqrt[3]{29} \approx \sqrt[3]{27} + dy$:

$$\sqrt[3]{29} \approx \sqrt[3]{27} + dy = 3 + \frac{2}{27} \approx 3.074.$$

Thus, the approximation $\sqrt[3]{29} \approx 3.074$ is very close to the actual value, $\sqrt[3]{29} = 3.07231\ldots$, found with a calculator.

SECTION 2.7

The **elasticity of demand** E is a function of x, the price:

$$E(x) = -\frac{x \cdot D'(x)}{D(x)}.$$

Business. The Leslie Davis Band finds that demand for its CD at performances is given by

$$q = D(x) = 50 - 2x,$$

where x is the price, in dollars, of each CD sold and q is the number of CDs sold at a performance. Find the elasticity when the price is $10 per CD, and interpret the result. Then, find the price at which revenue is maximized.

KEY TERMS AND CONCEPTS	EXAMPLES

SECTION 2.7 (continued)

When $E(x) > 1$, total revenue is decreasing;
when $E(x) < 1$, total revenue is increasing;
and
when $E(x) = 1$, total revenue is maximized.

Elasticity at x is given by

$$E(x) = -\frac{x \cdot D'(x)}{D(x)}$$

$$= -\frac{x(-2)}{50 - 2x}$$

$$= -\frac{-2x}{50 - 2x} = \frac{x}{25 - x}.$$

Thus, $E(10) = \dfrac{10}{25 - 10} = \dfrac{2}{3}.$

Since $E(10)$ is less than 1, the demand for the CD is *inelastic*, and a modest increase in price will increase revenue.

Revenue is maximized when $E(x) = 1$:

$$\frac{x}{25 - x} = 1$$

$$x = 25 - x$$

$$2x = 25$$

$$x = 12.5.$$

At a price of \$12.50 per CD, revenue will be maximized.

SECTION 2.8

If an equation has variables x and y and y is not isolated on one side of the equation, the derivative dy/dx can be found without solving for y using **implicit differentiation**.

Find $\dfrac{dy}{dx}$ if $y^5 = x^3 + 7$.

We differentiate both sides with respect to x, and then solve for $\dfrac{dy}{dx}$.

$$\frac{d}{dx}y^5 = \frac{d}{dx}x^3 + \frac{d}{dx}7$$

$$5y^4 \frac{dy}{dx} = 3x^2$$

$$\frac{dy}{dx} = \frac{3x^2}{5y^4}.$$

A **related rate** occurs when the rate of change of one variable (with respect to time) can be calculated in terms of the rate of change (with respect to time) of another variable of which it is a function.

A large cube of ice is melting, losing 30 cm^3 of its volume, V, per minute. When the length of a side is 20 cm, how fast is that length decreasing?

Since $V = x^3$ and both V and x are changing with time, we differentiate each variable with respect to time:

$$\frac{dV}{dt} = 3x^2 \frac{dx}{dt}.$$

We have $x = 20$ and $\dfrac{dV}{dt} = -30$. Evaluating gives

$$-30 = 3(20)^2 \frac{dx}{dt}$$

or $\dfrac{dx}{dt} = -\dfrac{30}{3(20)^2} = -0.025$ cm/min.

CHAPTER 2
Review Exercises

These review exercises are for test preparation. They can also be used as a practice test. Answers are at the back of the book. The red bracketed section references tell you what part(s) of the chapter to restudy if your answer is incorrect.

CONCEPT REINFORCEMENT

Match each description in column A with the most appropriate graph in column B. [2.1–2.4]

Column A   **Column B**

1. A function with a relative maximum but no absolute extrema

a)

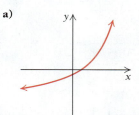

2. A function with both a vertical asymptote and a horizontal asymptote

b)

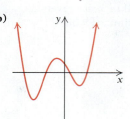

3. A function that is concave up and decreasing

c)

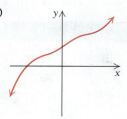

4. A function that is concave up and increasing

d)

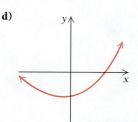

5. A function with three critical values

e)

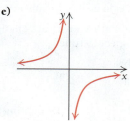

6. A function with one critical value and a second derivative that is always positive

f)

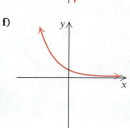

7. A function with a first derivative that is always positive

g)

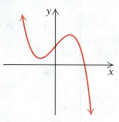

In Exercises 8–16, classify each statement as either true or false.

8. Every continuous function has at least one critical value. [2.1]

9. If a continuous function $y = f(x)$ has extrema, they will always occur where $f'(x) = 0$. [2.1]

10. If $f'(c) = 0$ and $f''(c) > 0$, then $f(c)$ is a relative minimum. [2.2]

11. If $f'(c) = 0$ and $f''(c) = 0$, then $f(c)$ cannot be a relative minimum. [2.2]

12. If the graph of $f(x) = P(x)/Q(x)$ has a horizontal asymptote, then the degree of the polynomial $P(x)$ is less than or equal to that of the polynomial $Q(x)$. [2.3]

13. Absolute extrema of a continuous function f always occur at the endpoints of a closed interval. [2.4]

14. In general, if f is a continuous function, then $\Delta x = dx$ and $\Delta y \geq dy$. [2.6]

15. If the elasticity of demand at a certain price x is 1, then revenue is maximized at this price. [2.7]

16. In a related-rate problem, the variables are commonly differentiated implicitly with respect to time, t. [2.8]

REVIEW EXERCISES

For each function given, find any extrema, along with the x-value at which they occur. Then sketch a graph of the function. [2.1]

17. $f(x) = 4 - 3x - x^2$

18. $f(x) = x^4 - 2x^2 + 3$

19. $f(x) = \dfrac{-8x}{x^2 + 1}$

20. $f(x) = 4 + (x - 1)^3$

21. $f(x) = x^3 + x^2 - x + 3$

22. $f(x) = 3x^{2/3}$

23. $f(x) = 2x^3 - 3x^2 - 12x + 10$

24. $f(x) = x^3 - 3x + 2$

Sketch the graph of each function. List any minimum or maximum values, where they occur, and any points of inflection. State where the function is increasing or decreasing and where it is concave up or concave down. [2.2]

25. $f(x) = \frac{1}{3}x^3 + 3x^2 + 9x + 2$

26. $f(x) = x^2 - 10x + 8$

27. $f(x) = 4x^3 - 6x^2 - 24x + 5$

28. $f(x) = x^4 - 2x^2$

29. $f(x) = 3x^4 + 2x^3 - 3x^2 + 1$ (Round to three decimal places where appropriate.)

30. $f(x) = \frac{1}{5}x^5 + \frac{3}{4}x^4 - \frac{4}{3}x^3 + 8$ (Round to three decimal places where appropriate.)

Sketch the graph of each function. Indicate where each function is increasing or decreasing, the coordinates at which any relative extrema occur, where any asymptotes occur, where the graph is concave up or concave down, and where any intercepts occur. [2.3]

31. $f(x) = \dfrac{2x + 5}{x + 1}$ **32.** $f(x) = \dfrac{x}{x - 2}$

33. $f(x) = \dfrac{5}{x^2 - 16}$ **34.** $f(x) = -\dfrac{x + 1}{x^2 - x - 2}$

35. $f(x) = \dfrac{x^2 - 2x + 2}{x - 1}$ **36.** $f(x) = \dfrac{x^2 + 3}{x}$

For each function, find any absolute extrema over the indicated interval. Indicate the x-value at which each extremum occurs. Where no interval is specified, use the real line. [2.4]

37. $f(x) = x^4 - 2x^2 + 3$; $[0, 3]$

38. $f(x) = 8x^2 - x^3$; $[-1, 8]$

39. $f(x) = x + \dfrac{50}{x}$; $(0, \infty)$

40. $f(x) = x^4 - 2x^2 + 1$

41. Maximize $Q = xy$, where $3x + 4y = 24$. [2.5]

42. Find the minimum value of $Q = x^2 - 2y^2$, where $x - 2y = 1$. [2.5]

43. Business: maximizing profit. If $R(x) = 52x - 0.5x^2$ and $C(x) = 22x - 1$, find the maximum profit and the number of units that must be produced and sold in order to yield this maximum profit. Assume that $R(x)$ and $C(x)$ are in dollars. [2.5]

44. Business: minimizing cost. A rectangular box with a square base and a cover is to have a volume of 2500 ft³. If the cost of material is \$2/ft² for the bottom, \$3/ft² for the top, and \$1/ft² for the sides, what should the dimensions of the box be in order to minimize the cost? [2.5]

45. Business: minimizing inventory cost. Venice Cyclery sells 360 hybrid bicycles per year. It costs \$8 to store

one bicycle for a year. To reorder, there is a fixed cost of \$10, plus \$2 for each bicycle. How many times per year should the shop order bicycles, and in what lot size, in order to minimize inventory costs? [2.5]

46. Business: marginal revenue. Crane Foods determines that its daily revenue, $R(x)$, in dollars, from the sale of x frozen dinners is

$$R(x) = 4x^{3/4}.\ [2.6]$$

a) What is Crane's daily revenue when 81 frozen dinners are sold?

b) What is Crane's marginal revenue when 81 frozen dinners are sold?

c) Use the answers from parts (a) and (b) to estimate $R(82)$.

For Exercises 47 and 48, $y = f(x) = 2x^3 + x$. [2.6]

47. Find Δy and dy, given that $x = 1$ and $\Delta x = -0.05$.

48. a) Find dy.
    b) Find dy when $x = -2$ and $dx = 0.01$.

49. Approximate $\sqrt{83}$ using $\Delta y \approx f'(x)\,\Delta x$. [2.6]

50. Physical science: waste storage. The Waste Isolation Pilot Plant (WIPP) in New Mexico consists of large rooms carved into a salt deposit and is used for long-term storage of radioactive waste. (Source: www.wipp.energy.gov.) A new storage room in the shape of a cube with an edge length of 200 ft is to be carved into the salt. Use a differential to estimate the potential difference in the volume of this room if the edge measurements have a tolerance of ± 2 ft. [2.6]

51. Economics: elasticity of demand. Consider the demand function given by

$$q = D(x) = 400 - 5x.\ [2.7]$$

a) Find the elasticity.

b) Find the elasticity at $x = 10$, and state whether the demand is elastic or inelastic.

c) Find the elasticity at $x = 50$, and state whether the demand is elastic or inelastic.

d) At a price of \$25, will a small increase in price cause total revenue to increase or decrease?

e) Find the price at which total revenue is a maximum.

52. Differentiate the following implicitly to find dy/dx. Then find the slope of the curve at the given point.

$$2x^3 + 2y^3 = -9xy;\quad (-1, -2)\ [2.8]$$

53. A ladder 25 ft long leans against a vertical wall. If the bottom moves away from the wall at the rate of 6 ft/sec, how fast is the height of the top changing when the lower end is 7 ft from the wall? [2.8]

54. Business: total revenue, cost, and profit. Find the rates of change, with respect to time, of total revenue, cost, and profit for

$$R(x) = 120x - 0.5x^2 \quad \text{and} \quad C(x) = 15x + 6,$$

when $x = 100$ and $dx/dt = 30$ units per day. Assume that $R(x)$ and $C(x)$ are in dollars. [2.8]

SYNTHESIS

55. Find the absolute maximum and minimum values, if they exist, over the indicated interval.

$$f(x) = (x - 3)^{2/5}; \quad (-\infty, \infty) \quad [2.4]$$

56. Find the absolute maximum and minimum values of the piecewise-defined function given by

$$f(x) = \begin{cases} 2 - x^2, & \text{for } -2 \le x \le 1, \\ 3x - 2, & \text{for } 1 < x < 2, \\ (x - 4)^2, & \text{for } 2 \le x \le 6. \end{cases} \quad [2.4]$$

57. Differentiate implicitly to find dy/dx:

$$(x - y)^4 + (x + y)^4 = x^6 + y^6. \quad [2.8]$$

58. Find any relative maxima and minima of

$$y = x^4 - 8x^3 - 270x^2. \quad [2.1, 2.2]$$

59. Determine a rational function f whose graph has a vertical asymptote at $x = -2$ and a horizontal asymptote at $y = 3$ and includes the point $(1, 2)$. [2.3]

TECHNOLOGY CONNECTION

Use a calculator to estimate the relative extrema of each function. [2.1, 2.2]

60. $f(x) = 3.8x^5 - 18.6x^3$

61. $f(x) = \sqrt[3]{|9 - x^2|} - 1$

---

62. **Life and physical sciences: incidence of breast cancer.** The following table provides data relating the incidence of breast cancer per 100,000 women of various ages.

a) Use REGRESSION to fit linear, quadratic, cubic, and quartic functions to the data. (The function used in Exercise 28 of Section R.1 was found in this manner.)

AGE	INCIDENCE PER 100,000
0	0
27	10
32	25
37	60
42	125
47	187
52	224
57	270
62	340
67	408
72	437
77	475
82	460
87	420

(*Source:* National Cancer Institute.)

b) Which function best fits the data?

c) Determine the domain of the function on the basis of the function and the problem situation.

d) Determine the maximum value of the function over the domain. At what age is the incidence of breast cancer the greatest? [2.1, 2.2]

CHAPTER 2
Test

Find all relative minimum or maximum values as well as the x-values at which they occur. State where each function is increasing or decreasing. Then sketch a graph of the function.

1. $f(x) = x^2 - 4x - 5$
2. $f(x) = 4 + 3x - x^3$

3. $f(x) = (x - 2)^{2/3} - 4$
4. $f(x) = \dfrac{16}{x^2 + 4}$

Sketch a graph of each function. List any extrema, and indicate any asymptotes or points of inflection.

5. $f(x) = x^3 + x^2 - x + 1$
6. $f(x) = 2x^4 - 4x^2 + 1$

7. $f(x) = (x - 2)^3 + 3$
8. $f(x) = x\sqrt{9 - x^2}$

9. $f(x) = \dfrac{2}{x - 1}$
10. $f(x) = \dfrac{-8}{x^2 - 4}$

11. $f(x) = \dfrac{x^2 - 1}{x}$
12. $f(x) = \dfrac{x - 3}{x + 2}$

Find the absolute maximum and minimum values, if they exist, of each function over the indicated interval. Where no interval is specified, use the real line.

13. $f(x) = x(6 - x)$

14. $f(x) = x^3 + x^2 - x + 1; \quad \left[-2, \frac{1}{2}\right]$

15. $f(x) = -x^2 + 8.6x + 10$

16. $f(x) = -2x + 5; \quad [-1, 1]$

17. $f(x) = -2x + 5$

18. $f(x) = 3x^2 - x - 1$

19. $f(x) = x^2 + \dfrac{128}{x}; \quad (0, \infty)$

20. Maximize $Q = xy^2$, where $x + 2y = 50$.

21. Minimize $Q = x^2 + y^2$, where $x - y = 10$.

22. Business: maximum profit. Find the maximum profit and the number of units, x, of a leather briefcase that must be produced and sold in order to yield the maximum profit. Assume that $R(x)$ and $C(x)$ are the revenue and cost, in dollars, when x units are produced:

$$R(x) = x^2 + 110x + 60,$$
$$C(x) = 1.1x^2 + 10x + 80.$$

23. Business: minimizing cost. From a thin piece of cardboard 60 in. by 60 in., square corners are cut out so that the sides can be folded up to make an open box. What dimensions will yield a box of maximum volume? What is the maximum volume?

24. Business: minimizing inventory costs. Ironside Sports sells 1225 tennis rackets per year. It costs $2 to store one tennis racket for a year. To reorder, there is a fixed cost of $1, plus $0.50 for each tennis racket. How many times per year should Ironside order tennis rackets, and in what lot size, in order to minimize inventory costs?

25. For $y = f(x) = x^2 - 3$, $x = 5$, and $\Delta x = 0.1$, find Δy and $f'(x)\,\Delta x$.

26. Approximate $\sqrt{50}$ using $\Delta y \approx f'(x)\,\Delta x$.

27. For $y = \sqrt{x^2 + 3}$:

a) Find dy.

b) Find dy when $x = 2$ and $dx = 0.01$.

28. Economics: elasticity of demand. Consider the demand function given by

$$q = D(x) = \frac{600}{(x + 4)^2}.$$

a) Find the elasticity.

b) Find the elasticity at $x = 1$, stating whether demand is elastic or inelastic.

c) Find the elasticity at $x = 12$, stating whether demand is elastic or inelastic.

d) At a price of $12, will a small increase in price cause total revenue to increase or decrease?

e) Find the price at which total revenue is a maximum.

29. Differentiate the following implicitly to find dy/dx. Then find the slope of the curve at $(1, 2)$:

$$x^3 + y^3 = 9.$$

30. A spherical balloon has a radius of 15 cm. Use a differential to find the approximate change in the volume of the balloon if the radius is increased or decreased by 0.5 cm. (The volume of a sphere is $V = \frac{4}{3}\pi r^3$. Use 3.14 for π.)

31. A pole 13 ft long leans against a vertical wall. If the lower end moves away from the wall at the rate of 0.4 ft/sec, how fast is the upper end sliding down when the lower end is 12 ft from the wall?

SYNTHESIS

32. Find the absolute maximum and minimum values of the following function, if they exist, over $[0, \infty)$:

$$f(x) = \frac{x^2}{1 + x^3}.$$

33. Business: minimizing average cost. The total cost in dollars of producing x units of a visor is given by

$$C(x) = 100x + 100\sqrt{x} + \frac{\sqrt{x^3}}{100}.$$

How many units should be produced to minimize average cost?

34. Estimate any extrema of the function given by

$$f(x) = 5x^3 - 30x^2 + 45x + 5\sqrt{x}.$$

35. Estimate any extrema of the function given by

$$g(x) = x^5 - x^3.$$

36. Maximize $Q = x^2y^2$, where $x^2 + y = 20$ and x and y are positive.

37. Business: advertising. The business of manufacturing and selling bowling balls is one of frequent changes. Companies introduce new models to the market about every 3 to 4 months. Typically, a new model is created because of advances in technology such as new surface stock or a new way to place weight blocks in a ball. To decide how to best use advertising dollars, companies track sales in relation to amount spent on advertising. Suppose a company has the following data from past sales.

AMOUNT SPENT ON ADVERTISING (in thousands)	NUMBER OF BOWLING BALLS SOLD, N
$ 0	8
50	13,115
100	19,780
150	22,612
200	20,083
250	12,430
300	4

a) Use REGRESSION to fit linear, quadratic, cubic, and quartic functions to the data.

b) Determine the domain of the function in part (a) that best fits the data and the problem situation. How did you arrive at your answer?

c) Determine the maximum value of the function over the domain. How much should the company spend on advertising a new model in order to maximize the number of bowling balls sold?

Maximum Sustainable Harvest

In certain situations, biologists are able to determine what is called a **reproduction curve**. This is a function

$$y = f(P)$$

such that if P is the population at a given point in time, then $f(P)$ is the population a year later. Such a curve is shown below.

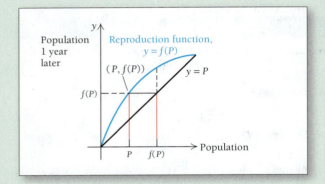

The line $y = P$ is significant because if it coincides with the curve $y = f(P)$, then we know that the population stays the same from year to year. Here the graph of f lies mostly above the line, indicating that the population is increasing.

Too many deer in a forest can deplete the food supply and cause the population to decrease. In such cases, hunters are often allowed to "harvest" some of the deer. Then with a greater food supply, the remaining deer population may prosper and increase.

We know that a population P will grow to a population $f(P)$ in a year. If the population were increasing, then hunters could "harvest" the amount

$$f(P) - P$$

each year without shrinking the initial population P. If the population were remaining the same or decreasing, then such a harvest would deplete the population.

Suppose we want to know the value of P_0 that will allow the harvest to be the largest. If we could determine that P_0, we could let the population grow until it reached that level and then begin harvesting year after year the amount $f(P_0) - P_0$.

Let the harvest function H be given by

$$H(P) = f(P) - P.$$

Then $H'(P) = f'(P) - 1.$

Now, if we assume that $H'(P)$ exists for all values of P and that there is only one critical value, it follows that the *maximum sustainable harvest* occurs at that value P_0 such that

$$H'(P_0) = f'(P_0) - 1 = 0$$

and $H''(P_0) = f''(P_0) < 0.$

Or, equivalently, we have the following.

> ### THEOREM
>
> The **maximum sustainable harvest** occurs at P_0 such that
>
> $$f'(P_0) = 1 \quad \text{and} \quad f''(P_0) < 0,$$
>
> and is given by
>
> $$H(P_0) = f(P_0) - P_0.$$

Exercises

For Exercises 1–3, do the following.

a) *Graph the reproduction curve, the line $y = P$, and the harvest function using the same viewing window.*

b) *Find the population at which the maximum sustainable harvest occurs. Use both a graphical solution and a calculus solution.*

c) *Find the maximum sustainable harvest.*

1. $f(P) = P(10 - P)$, where P is measured in thousands.

2. $f(P) = -0.025P^2 + 4P$, where P is measured in thousands. This is the reproduction curve in the Hudson Bay area for the snowshoe hare.

3. $f(P) = -0.01P^2 + 2P$, where P is measured in thousands. This is the reproduction curve in the Hudson Bay area for the lynx.

For Exercises 4 and 5, do the following.

a) *Graph the reproduction curve, the line $y = P$, and the harvest function using the same viewing window.*

b) *Graphically determine the population at which the maximum sustainable harvest occurs.*

c) *Find the maximum sustainable harvest.*

4. $f(P) = 40\sqrt{P}$, where P is measured in thousands. Assume that this is the reproduction curve for the brown trout population in a large lake.

5. $f(P) = 0.237P\sqrt{2000 - P^2}$, where P is measured in thousands.

6. The table below lists data regarding the reproduction of a certain animal.

a) Use REGRESSION to fit a cubic polynomial to these data.

b) Graph the reproduction curve, the line $y = P$, and the harvest function using the same viewing window.

c) Graphically determine the population at which the maximum sustainable harvest occurs.

POPULATION, P (in thousands)	POPULATION, $f(P)$, 1 YEAR LATER
10	9.7
20	23.1
30	37.4
40	46.2
50	42.6

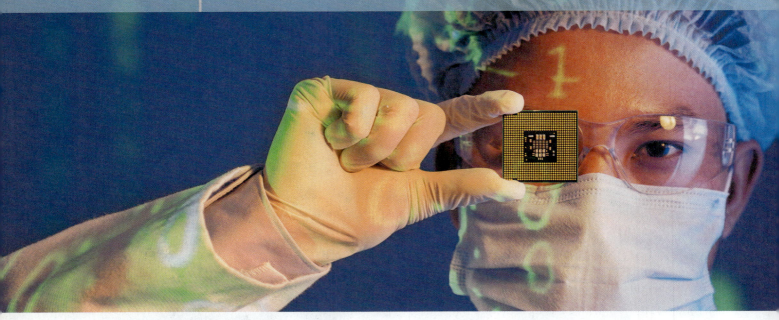

3 | Exponential and Logarithmic Functions

What You'll Learn

3.1 Exponential Functions
3.2 Logarithmic Functions
3.3 Applications: Uninhibited and Limited Growth Models
3.4 Applications: Decay
3.5 The Derivatives of a^x and $\log_a x$; Application: Annuities
3.6 A Business Application: Amortization

Why It's Important

In this chapter, we consider two types of functions that are closely related: *exponential functions* and *logarithmic functions*. After learning to find derivatives of such functions, we will study applications in the areas of population growth and decay, continuously compounded interest, annuities and amortization, spread of disease, and carbon dating.

Where It's Used

Microprocessors: In 1971, Intel Corporation released the Intel 4004 microprocessor, which held 2300 transistors. Since then, the number of transistors on a microprocessor has been growing exponentially at the rate of 0.34, or 34%, per year. Approximately how many transistors will be on a microprocessor in 2020? (*This problem appears as Example 5 in Section 3.3.*)

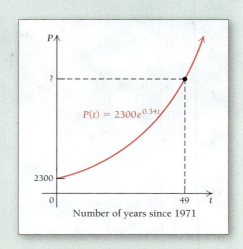

$P(t) = 2300e^{0.34t}$

Number of years since 1971

Exponential Functions

Graphs of Exponential Functions

- Graph exponential functions.
- Differentiate exponential functions.

Consider the following graph. The rapid rise of the graph indicates that it approximates an *exponential function*. In this chapter, we consider such functions and many of their applications.

EXPONENTIAL FUNCTION

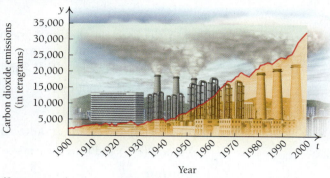

(*Source*: epa.gov.)

Recall from Section R.5 that an **exponential function** is given by $f(x) = a_0 \cdot a^x$, where $a > 0$ and $a \neq 1$. The number a is the **base**, the y-intercept is $(0, a_0)$, there are no x-intercepts, and the function is continuous for all real numbers.

The graphs of $f(x) = a_0 \cdot a^x$, for $a > 1$ and for $0 < a < 1$, are shown below, and common characteristics of each are described.

1. The function given by $f(x) = a_0 \cdot a^x$, with $a > 1$, is a positive, increasing, continuous function. As x gets smaller, $a_0 \cdot a^x$ approaches 0. The graph is concave up, and the x-axis is the horizontal asymptote.

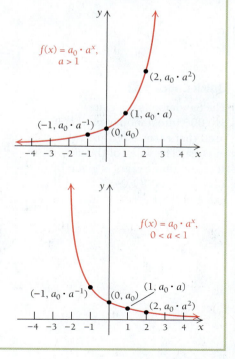

2. The function given by $f(x) = a_0 \cdot a^x$, with $0 < a < 1$, is a positive, decreasing, continuous function. As x gets larger, $a_0 \cdot a^x$ approaches 0. The graph is concave up, and the x-axis is the horizontal asymptote.

For $a = 1$, we have $f(x) = a_0 \cdot a^x = a_0 \cdot 1^x = a_0 \cdot 1 = a_0$; so, in this case, f is a constant function. This is why we are not interested in 1 as the base of an exponential function.

Recall from Theorem 1 in Section R.1 that if an amount P is invested at interest rate r, compounded annually, then in t years, it will grow to an amount A given by the exponential function

$$A(t) = P(1 + r)^t.$$

This formula can be extended to many situations such as population growth and depreciation.

EXAMPLE 1 **Population Growth.** The city of Woodfork had a population of 5200 on January 1, 2010, and was growing by 2.5% per year. The population, P, of Woodfork t years from January 1, 2010, is given by

$$P(t) = 5200(1.025)^t.$$

a) Graph P.

b) Find $P(8)$ and interpret this result.

c) Find $P(-5)$ and interpret this result.

Solution

a) We begin by finding some function values.

t	$P(t)$
$t = 0$	$P(0) = 5200(1.025)^0 = 5200$
$t = 1$	$P(1) = 5200(1.025)^1 = 5330$
$t = 2$	$P(2) = 5200(1.025)^2 \approx 5463$
$t = 3$	$P(3) = 5200(1.025)^3 \approx 5600$
$t = -1$	$P(-1) = 5200(1.025)^{-1} \approx 5073$
$t = -2$	$P(-2) = 5200(1.025)^{-2} \approx 4949$
$t = -3$	$P(-3) = 5200(1.025)^{-3} \approx 4829$

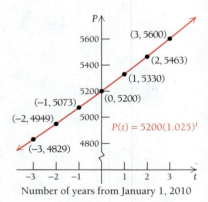

Number of years from January 1, 2010

We plot these points and connect them with a smooth curve, as shown in the graph. The graph is continuous, has a y-intercept at $(0, 5200)$, and is concave up; also, since $a = 1.025 > 0$, the graph is increasing without bound.

b) We have $P(8) = 5200(1.025)^8 \approx 6336$. The population of Woodfork is estimated to be about 6336 on January 1, 2018.

c) We have $P(-5) = 5200(1.025)^{-5} \approx 4596$. The population of Woodfork was about 4596 on January 1, 2005.   **1** ✔

Quick Check 1 ✔

The Sandersons deposit $10,000 into a money market fund that earns 4.8% annual interest. The value of their account after t years is given by $V(t) = 10,000(1.048)^t$. Find the value of their account after 6 yr.

TECHNOLOGY CONNECTION

Exploratory Exercise: Growth

Cut a sheet of $8\frac{1}{2}$-in.-by-11-in. paper into two equal pieces. Then cut these again to obtain four equal pieces. Then cut these to get eight equal pieces, and so on, performing five cutting steps.

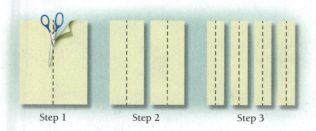

Step 1   Step 2   Step 3

a) Stack the pieces and measure the total thickness.

b) A piece of paper is typically 0.004 in. thick. Check the measurement in part (a) by completing the table.

	t	$0.004 \cdot 2^t$
Start	0	$0.004 \cdot 2^0$, or 0.004
Step 1	1	$0.004 \cdot 2^1$, or 0.008
Step 2	2	$0.004 \cdot 2^2$, or 0.016
Step 3	3	
Step 4	4	
Step 5	5	

c) Graph $f(t) = 0.004(2)^t$.

d) Compute the thickness of the stack (in miles) after 25 steps. (*Hint:* 1 mi = 5280 ft.)

The Number e and the Derivative of e^x

To find the derivative of the exponential function

$$f(x) = a^x,$$

we use the definition of the derivative given by

$$f'(x) = \lim_{h \to 0} \frac{f(x+h) - f(x)}{h}$$ Definition of the derivative

$$= \lim_{h \to 0} \frac{a^{x+h} - a^x}{h}$$ Substituting a^{x+h} for $f(x+h)$ and a^x for $f(x)$

$$= \lim_{h \to 0} \frac{a^x \cdot a^h - a^x \cdot 1}{h}$$ Using the laws for exponents

$$= \lim_{h \to 0} \left(a^x \cdot \frac{a^h - 1}{h} \right)$$ Factoring

$$= a^x \cdot \lim_{h \to 0} \frac{a^h - 1}{h}.$$ Noting that a^x is constant with respect to h, and the limit of a constant times a function is the constant times the limit of that function

Thus, $\quad f'(x) = a^x \cdot \lim_{h \to 0} \frac{a^h - 1}{h}.$

In the case of $g(x) = 2^x$,

$$g'(x) = 2^x \cdot \lim_{h \to 0} \frac{2^h - 1}{h}.$$ Substituting 2 for a

Note that the limit does not depend on the value of x at which we are evaluating the derivative. In order for $g'(x)$ to exist, we must determine whether $\lim_{h \to 0} \dfrac{2^h - 1}{h}$ exists. Let's investigate this question.

We choose a sequence of numbers h approaching 0 and compute $(2^h - 1)/h$, listing the results in a table, as shown at left. It seems reasonable to infer that $(2^h - 1)/h$ has a limit as h approaches 0 and that its approximate value is 0.6931; thus,

$$g'(x) \approx (0.6931)2^x.$$

h	$\dfrac{2^h - 1}{h}$
0.1	0.7177
0.01	0.6956
0.001	0.6934
0.0001	0.6931...
0.00001	0.6931...
-0.001	0.6929
-0.0001	0.6931...
-0.00001	0.6931...

h	$\dfrac{3^h - 1}{h}$
0.1	1.1612
0.01	1.1047
0.001	1.0992
0.0001	1.0987
0.00001	1.0986
−0.001	1.0980
−0.0001	1.0986
−0.00001	1.0986

In other words, the derivative is a constant times 2^x. Similarly, for $t(x) = 3^x$,

$$t'(x) = 3^x \cdot \lim_{h \to 0} \frac{3^h - 1}{h}. \qquad \text{Substituting 3 for } a$$

Again, we can find an approximation for the limit that does not depend on the value of x at which we are evaluating the derivative. Consider the table at left. Again, it seems reasonable to infer that $(3^h - 1)/h$ has a limit as h approaches 0. This time, the limit is approximately 1.0986; thus,

$$t'(x) \approx (1.0986)3^x.$$

In other words, the derivative is a constant times 3^x.

Let's now analyze what we have done. We proved that

$$\text{if } f(x) = a^x, \quad \text{then } f'(x) = a^x \cdot \lim_{h \to 0} \frac{a^h - 1}{h}.$$

Consider $\displaystyle\lim_{h \to 0} \frac{a^h - 1}{h}$.

TECHNOLOGY CONNECTION

EXERCISE
Using TBLSET in Ask mode, produce the tables shown above.

1. For $a = 2$,

$$\lim_{h \to 0} \frac{a^h - 1}{h} = \lim_{h \to 0} \frac{2^h - 1}{h} \approx 0.6931.$$

2. For $a = 3$,

$$\lim_{h \to 0} \frac{a^h - 1}{h} = \lim_{h \to 0} \frac{3^h - 1}{h} \approx 1.0986.$$

It seems reasonable to conclude that, for some choice of a between 2 and 3, we have

$$\lim_{h \to 0} \frac{a^h - 1}{h} = 1.$$

To find that a, it suffices to look for a value such that

$$\frac{a^h - 1}{h} = 1. \qquad (1)$$

Multiplying both sides by h and then adding 1 to both sides, we have

$$a^h = 1 + h.$$

Raising both sides to the power $1/h$, we have

$$a = (1 + h)^{1/h}. \qquad (2)$$

Since equations (1) and (2) are equivalent, it follows that

$$\lim_{h \to 0} \frac{a^h - 1}{h} = \lim_{h \to 0} 1 \quad \text{and} \quad \lim_{h \to 0} a = \lim_{h \to 0} (1 + h)^{1/h}$$

h	$(1 + h)^{1/h}$
0.1	2.5937
0.01	2.7048
0.001	2.7169
0.0001	2.7181
−0.01	2.7320
−0.001	2.7196
−0.0001	2.7184

are also equivalent. Since $\displaystyle\lim_{h \to 0} c = c$,

$$\lim_{h \to 0} \frac{a^h - 1}{h} = 1 \quad \text{and} \quad a = \lim_{h \to 0} (1 + h)^{1/h}$$

are equivalent. Thus, to find the value of a for which $\displaystyle\lim_{h \to 0} \frac{a^h - 1}{h} = 1$, we must have $a = \displaystyle\lim_{h \to 0} (1 + h)^{1/h}$. This last equation gives us the special number we are searching for.* The number is denoted e, in honor of Leonhard Euler (pronounced "Oiler"), the great Swiss mathematician (1707–1783) who did groundbreaking work with it.

DEFINITION

$$e = \lim_{h \to 0} (1 + h)^{1/h} \approx 2.718281828459$$

We call e the *natural base*.

It follows that, for the exponential function $f(x) = e^x$,

$$f'(x) = e^x \cdot \lim_{h \to 0} \frac{e^h - 1}{h}$$

$$= e^x \cdot 1$$

$$= e^x.$$

That is, the derivative of $f(x) = e^x$ is $f'(x) = e^x$.

THEOREM 1

The derivative of the function f given by $f(x) = e^x$ is the function itself:

$$f'(x) = f(x), \quad \text{or} \quad \frac{d}{dx} e^x = e^x.$$

TECHNOLOGY CONNECTION

Exploratory

Graph $f(x) = e^x$ using the window $[-5, 5, -1, 10]$. Trace to a point and note the y-value. Then find dy/dx at that point, using the ⃝CALC key, and compare y and dy/dx.

Repeat this process for three other values of x. What do you observe?

Using Graphicus, graph $y = e^x$. Then use the tangent line feature to move along the curve, noting the x-values, the y-values, and the values of $\partial y/\partial x$, which are the same as dy/dx in this case. How do these values compare?

Theorem 1 says that for $f(x) = e^x$, the derivative at x (the slope of the tangent line) is the same as the function value at x. That is, on the graph of $y = e^x$, at the point $(0, 1)$, the slope is $m = 1$; at the point $(1, e)$, the slope is $m = e$; at the point $(2, e^2)$, the slope is $m = e^2$, and so on. The function $y = e^x$ is the *only* exponential function for which this correlation between the function and its derivative is true.

In Section 3.5, we will develop a formula for the derivative of the more general exponential function given by $y = a^x$.

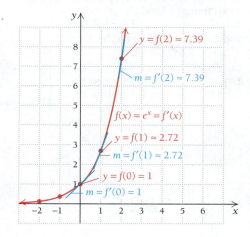

Finding Derivatives of Functions Involving *e*

We can combine Theorem 1 with earlier theorems to differentiate a variety of functions.

EXAMPLE 2   Find dy/dx:

a) $y = 3e^x$;   **b)** $y = x^2 e^x$;   **c)** $y = \dfrac{e^x}{x^3}$.

Solution

a) $\dfrac{d}{dx}(3e^x) = 3\dfrac{d}{dx}e^x$ Recall that $\dfrac{d}{dx}[c \cdot f(x)] = c \cdot f'(x)$.

$\qquad\qquad = 3e^x$ Using Theorem 1

---

*This derivation is based on one presented in Appendix 4 of *e: The Story of a Number,* by Eli Maor (Princeton University Press, 1998).

b) $\dfrac{d}{dx}(x^2 e^x) = x^2 \cdot e^x + e^x \cdot 2x$ \quad Using the Product Rule

$= xe^x(x + 2)$ \quad Factoring

c) $\dfrac{d}{dx}\left(\dfrac{e^x}{x^3}\right) = \dfrac{x^3 \cdot e^x - e^x \cdot 3x^2}{x^6}$ \quad Using the Quotient Rule

$= \dfrac{x^2 e^x(x - 3)}{x^6}$ \quad Factoring

$= \dfrac{e^x(x - 3)}{x^4}$ \quad Simplifying \qquad 2 ✔

Quick Check 2 ✔

Find dy/dx:

a) $y = 6e^x$; \quad **b)** $y = x^3 e^x$;

c) $y = \dfrac{e^x}{x^2}$.

Suppose we have a more complicated function in the exponent, as in

$$h(x) = e^{x^2 - 5x}.$$

This is a composition of functions. For such a function, we have

$$h(x) = g(f(x)) = e^{f(x)}, \quad \text{where} \quad g(x) = e^x \quad \text{and} \quad f(x) = x^2 - 5x.$$

Now $g'(x) = e^x$. Then by the Chain Rule (Section 1.7), we have

$$h'(x) = g'(f(x)) \cdot f'(x)$$
$$= e^{f(x)} \cdot f'(x).$$

Here $f(x) = x^2 - 5x$, so $f'(x) = 2x - 5$. Then

$$h'(x) = g'(f(x)) \cdot f'(x)$$
$$= e^{f(x)} \cdot f'(x)$$
$$= e^{x^2 - 5x}(2x - 5).$$

The next theorem, which we have proven using the Chain Rule, allows us to find derivatives of functions like the one above.

THEOREM 2

The derivative of e to some power is the product of e to that power and the derivative of the power:

$$\frac{d}{dx} e^{f(x)} = e^{f(x)} \cdot f'(x)$$

or

$$\frac{d}{dx} e^u = e^u \cdot \frac{du}{dx}.$$

The following gives us a way to remember this rule.

$$h(x) = e^{x^2 - 5x}$$

Rewrite the original function.

$$h'(x) = e^{x^2 - 5x}(2x - 5) \quad \text{Multiply by the derivative of the exponent.}$$

EXAMPLE 3 Differentiate each of the following with respect to x:

a) $y = e^{8x}$; \qquad **b)** $y = e^{-x^2 + 4x - 7}$; \qquad **c)** $e^{\sqrt{x^2 - 3}}$.

Solution

a) $\dfrac{d}{dx} e^{8x} = e^{8x} \cdot 8$ Using the Chain Rule

$= 8e^{8x}$

b) $\dfrac{d}{dx} e^{-x^2+4x-7} = e^{-x^2+4x-7}(-2x+4)$ Using the Chain Rule

$= -2(x-2)e^{-x^2+4x-7}$

c) $\dfrac{d}{dx} e^{\sqrt{x^2-3}} = \dfrac{d}{dx} e^{(x^2-3)^{1/2}}$

$= e^{(x^2-3)^{1/2}} \cdot \frac{1}{2}(x^2-3)^{-1/2} \cdot 2x$ Using the Chain Rule twice

$= e^{\sqrt{x^2-3}} \cdot x \cdot (x^2-3)^{-1/2}$

$= \dfrac{e^{\sqrt{x^2-3}} \cdot x}{\sqrt{x^2-3}}, \quad \text{or} \quad \dfrac{xe^{\sqrt{x^2-3}}}{\sqrt{x^2-3}}$ **3** ✔

Quick Check 3 ✔

Differentiate:

a) $f(x) = e^{-4x}$;

b) $g(x) = e^{x^3+8x}$;

c) $h(x) = e^{\sqrt{x^2+5}}$.

Graphs of e^x, e^{-x}, and $1 - e^{-kx}$

Knowing how to differentiate $f(x) = e^x$, we now consider the graph of $f(x) = e^x$ with regard to the curve-sketching techniques discussed in Section 2.2.

EXAMPLE 4   Graph: $f(x) = e^x$. Analyze the graph using calculus.

Solution   We simply find some function values using a calculator, plot the points, and sketch the graph as shown below.

x	$f(x)$
-2	0.135
-1	0.368
0	1
1	2.718
2	7.389

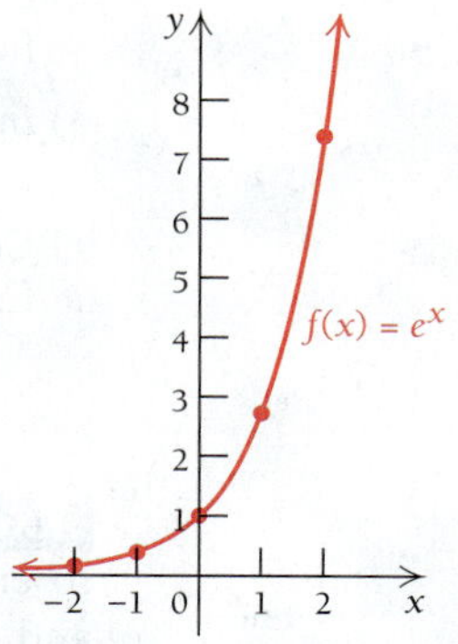

We analyze the graph using calculus as follows.

a) *Derivatives.* Since $f(x) = e^x$, it follows that $f'(x) = e^x$ and $f''(x) = e^x$.

b) *Critical values of f.* Since $f'(x) = e^x > 0$ for all real numbers x, we know that the derivative exists for all real numbers and there is no solution of $f'(x) = 0$. There are no critical values and therefore no maximum or minimum values.

c) *Increasing.* We have $f'(x) = e^x > 0$ for all real numbers x, so the function f is increasing over the entire real line, $(-\infty, \infty)$.

d) *Inflection points.* We have $f''(x) = e^x > 0$ for all real numbers x, so $f''(x) = 0$ has no solution and there are no points of inflection.

e) *Concavity.* Since $f''(x) = e^x > 0$ for all real numbers x, the function f' is increasing and the graph is concave up over the entire real line.   ■

EXAMPLE 5   Graph: $g(x) = e^{-x}$. Analyze the graph using calculus.

Solution   First, we find some function values, plot the points, and sketch the graph.

x	$g(x)$
-2	7.389
-1	2.718
0	1
1	0.368
2	0.135

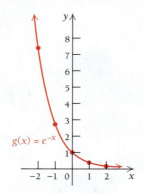

We can then analyze the graph using calculus as follows.

a) *Derivatives*. Since $g(x) = e^{-x}$, we have

$$g'(x) = e^{-x}(-1) = -e^{-x},$$

and

$$g''(x) = -e^{-x}(-1) = e^{-x}.$$

b) *Critical values of g*. Since $e^{-x} = 1/e^x > 0$, we have $g'(x) = -e^{-x} < 0$ for all real numbers x. Thus, the derivative exists for all real numbers, and $g'(x) = 0$ has no solution. There are no critical values and therefore no maximum or minimum values.

c) *Decreasing*. Since the derivative $g'(x) = -e^{-x} < 0$ for all real numbers x, the function g is decreasing over the entire real line.

d) *Inflection points*. We have $g''(x) = e^{-x} > 0$, for all real numbers x, so $g''(x) = 0$ has no solution and there are no points of inflection.

e) *Concavity*. We also know that since $g''(x) = e^{-x} > 0$ for all real numbers x, the function g' is increasing and the graph is concave up over the entire real line.   ■

Functions of the type $f(x) = 1 - e^{-kx}$, with $x \geq 0$, have important applications.

EXAMPLE 6   **Business: Worker Output.**   It is reasonable for a manufacturer to expect the daily output of a new worker to be low at first, increase over time, and then level off. A firm that manufactures 6G smart phones determines that after t working days, the number of phones produced per day by the average worker can be modeled by

$$N(t) = 80 - 70e^{-0.13t}.$$

a) Find $N(0)$, $N(1)$, $N(5)$, $N(10)$, $N(20)$, and $N(30)$.

b) Graph $N(t)$.

c) Find $N'(t)$ and interpret this derivative in terms of rate of change.

d) Find any critical values and intervals over which $N(t)$ is increasing or decreasing.

e) Find $N''(t)$ and any inflection points and intervals over which $N(t)$ is concave up or down.

f) What number of phones seems to determine where worker output levels off?

Solution

a) We make a table of input–output values.

t	0	1	5	10	20	30
$N(t)$	10	18.5	43.5	60.9	74.8	78.6

b) Using these values and/or a graphing calculator, we obtain the graph.

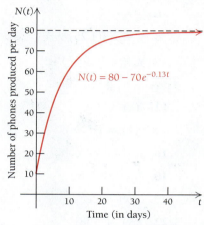

$N(t) = 80 - 70e^{-0.13t}$

c) We have $N'(t) = -70e^{-0.13t}(-0.13) = 9.1e^{-0.13t}$; after t days, the rate of change of number of phones produced per day is given by $9.1e^{-0.13t}$.

d) Since $9.1e^{-0.13t} > 0$ for all real numbers t, we know that $N(t)$ is increasing over the interval $(0, \infty)$. There are no critical values.

e) We have $N''(t) = 9.1(-0.13)e^{-0.13t} = -1.183e^{-0.13t}$. Since $-1.183e^{-0.13t} < 0$ for all real numbers t, there are no points of inflection, and the graph of $N(t)$ is concave down over $(0, \infty)$.

f) Examining the graph and expanding the table of function values, it seems that worker output levels off at no more than 80 phones produced per day.   **4 ✔**

Quick Check 4 ✔

Business. Repeat Example 6 for the output function

$$N(t) = 80 - 60e^{-0.12t}.$$

In general, for $k > 0$, the graph of $h(x) = 1 - e^{-kx}$ is increasing, which we expect since $h'(x) = ke^{-kx}$ is always positive. Note that $h(x)$ approaches 1 as x approaches ∞.

A word of caution! Functions of the type a^x (for example, 2^x, 3^x, and e^x) are different from functions of the type x^a (for example, x^2, x^3, $x^{1/2}$). For a^x, the variable is in the exponent. For x^a, the variable is in the base. The derivative of a^x is not xa^{x-1}.

Section Summary

- An exponential function is given by $f(x) = a_0 \cdot a^x$, where x is any real number, $a > 0$ and $a \neq 1$. The number a is the base, and the y-intercept is $(0, a_0)$.
- If $a > 1$, then $f(x) = a_0 \cdot a^x$ is increasing, concave up, increasing without bound as x increases, and has a horizontal asymptote $y = 0$ as x approaches $-\infty$.
- If $0 < a < 1$, then $f(x) = a_0 \cdot a^x$ is decreasing, concave up, increasing without bound as x approaches $-\infty$, and has a horizontal asymptote $y = 0$ as x approaches infinity.

- The *exponential function* $f(x) = e^x$, where $e \approx 2.71828$, has the derivative $f'(x) = e^x$. That is, the slope of a tangent line to the graph of $y = e^x$ is the same as the function value at x.
- In general, $\dfrac{d}{dx}e^{f(x)} = e^{f(x)} \cdot f'(x)$.
- The graph of $f(x) = e^x$ is an increasing function with no critical values, no maximum or minimum values, and no points of inflection. The graph is concave up, with

$$\lim_{x \to \infty} f(x) = \infty \quad \text{and} \quad \lim_{x \to -\infty} f(x) = 0.$$

3.1 | Exercise Set

Graph.

1. $y = 5^x$

2. $y = 4^x$

3. $y = 2 \cdot 3^x$

4. $y = 3 \cdot 4^x$

5. $y = 5\left(\frac{1}{4}\right)^x$

6. $y = 4\left(\frac{1}{3}\right)^x$

7. $y = 1.3(1.2)^x$

8. $y = 2.4(1.25)^x$

9. $y = 2.6(0.8)^x$

10. $y = 1.13(0.81)^x$

Differentiate.

11. $f(x) = e^x$

12. $f(x) = e^{-x}$

13. $g(x) = e^{2x}$

14. $g(x) = e^{3x}$

15. $f(x) = 6e^x$

16. $f(x) = 4e^x$

17. $F(x) = e^{-7x}$

18. $F(x) = e^{-4x}$

19. $g(x) = 3e^{5x}$

20. $G(x) = 2e^{4x}$

21. $G(x) = -7e^{-x}$

22. $f(x) = -3e^{-x}$

23. $g(x) = \frac{1}{2}e^{-5x}$

24. $f(x) = \frac{1}{3}e^{-4x}$

25. $F(x) = -\frac{2}{3}e^{x^2}$

26. $g(x) = -\frac{4}{5}e^{x^3}$

27. $F(x) = 4 - e^{2x}$

28. $G(x) = 7 + 3e^{5x}$

29. $G(x) = x^3 - 5e^{2x}$

30. $f(x) = x^5 - 2e^{6x}$

31. $g(x) = x^5 e^{2x}$

32. $f(x) = x^7 e^{4x}$

33. $F(x) = \dfrac{e^{2x}}{x^4}$

34. $g(x) = \dfrac{e^{3x}}{x^6}$

35. $f(x) = (x^2 - 2x + 2)e^x$

36. $f(x) = (x^2 + 3x - 9)e^x$

37. $f(x) = \dfrac{e^x}{x^4}$

38. $f(x) = \dfrac{e^x}{x^5}$

39. $f(x) = e^{-x^2+8x}$

40. $f(x) = e^{-x^2+7x}$

41. $f(x) = e^{x^2/2}$

42. $f(x) = e^{-x^2/2}$

43. $y = e^{\sqrt{x-7}}$

44. $y = e^{\sqrt{x-4}}$

45. $y = \sqrt{e^x - 1}$

46. $y = \sqrt{e^x + 1}$

47. $y = e^x + x^3 - xe^x$

48. $y = xe^{-2x} + e^{-x} + x^3$

49. $y = 1 - e^{-3x}$

50. $y = 1 - e^{-x}$

51. $y = 1 - e^{-kx}$

52. $y = 1 - e^{-mx}$

53. $g(x) = (4x^2 + 3x)e^{x^2-7x}$

54. $g(x) = (5x^2 - 8x)e^{x^2-4x}$

Graph each function. Then determine any critical values, inflection points, intervals over which the function is increasing or decreasing, and the concavity.

55. $g(x) = e^{-2x}$

56. $f(x) = e^{2x}$

57. $f(x) = e^{(1/3)x}$

58. $g(x) = e^{(1/2)x}$

59. $f(x) = \frac{1}{2}e^{-x}$

60. $g(x) = \frac{1}{3}e^{-x}$

61. $F(x) = -e^{(1/3)x}$

62. $G(x) = -e^{(1/2)x}$

63. $f(x) = 3 - e^{-x}$, for $x \geq 0$

64. $g(x) = 2(1 - e^{-x})$, for $x \geq 0$

65–74. For each function given in Exercises 55–64, graph the function and its first and second derivatives using a graphing utility.

75. Find the slope of the line tangent to the graph of $f(x) = 2e^{-3x}$ at the point $(0, 2)$.

76. Find the slope of the line tangent to the graph of $f(x) = e^x$ at the point $(0, 1)$.

77. Find an equation of the line tangent to the graph of $f(x) = e^{2x}$ at the point $(0, 1)$.

78. Find an equation of the line tangent to the graph of $G(x) = e^{-x}$ at the point $(0, 1)$.

79. and 80. For each of Exercises 77 and 78, graph the function and the tangent line using a graphing utility.

APPLICATIONS

Business and Economics

81. **U.S. travel exports.** U.S. travel exports (goods and services that international travelers buy while visiting the United States) are increasing exponentially. The value of such exports, t years after 2011, can be approximated by

$$V(t) = 115.32e^{0.094t},$$

where V is in billions of dollars. (*Source:* www.census.gov/foreign-trade/data/index.html.)

a) Estimate the value of U.S. travel exports in 2016 and 2018.

b) Estimate the growth rate for U.S. travel exports in 2016 and 2018.

82. **Organic food.** More Americans are buying organic fruit and vegetables and products made with organic ingredients. The amount $A(t)$, in billions of dollars, spent on organic food and beverages t years after 1995 can be approximated by

$$A(t) = 2.43e^{0.18t}.$$

(*Source: Nutrition Business Journal*, 2004.)

a) Estimate the amount that Americans spent on organic food and beverages in 2009.
b) Estimate the rate at which spending on organic food and beverages was growing in 2006.

83. Marginal cost.   The total cost, in millions of dollars, for Cheevers, Inc., is given by

$$C(t) = 100 - 50e^{-t},$$

where t is the time in years since the start-up date.

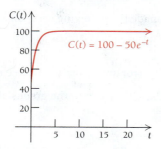

Find each of the following.

a) The marginal cost, $C'(t)$
b) $C'(0)$
c) $C'(4)$ (Round to the nearest thousand.)
d) Find $\lim\limits_{t \to \infty} C(t)$ and $\lim\limits_{t \to \infty} C'(t)$.

84. Marginal cost.   The total cost, in millions of dollars, for Marcotte Industries is given by

$$C(t) = 200 - 40e^{-t},$$

where t is the time in years since the start-up date.

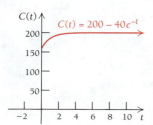

Find each of the following.

a) The marginal cost $C'(t)$
b) $C'(1)$
c) $C'(5)$ (Round to the nearest thousand.)
d) Find $\lim\limits_{t \to \infty} C(t)$ and $\lim\limits_{t \to \infty} C'(t)$.

85. Marginal demand.   At a price of x dollars, the demand, in thousands of units, for a certain turntable is given by the demand function

$$q = 240e^{-0.003x}.$$

a) How many turntables will be bought at a price of $250? Round to the nearest thousand.
b) Graph the demand function for $0 \le x \le 400$.
c) Find the marginal demand, $q'(x)$.
d) Interpret the meaning of the derivative.

86. Marginal supply.   At a price of x dollars, the supply function for the turntable in Exercise 85 is given by

$$q = 75e^{0.004x},$$

where q is in thousands of units.

a) How many turntables will be supplied at a price of $250? Round to the nearest thousand.
b) Graph the supply function for $0 \le x \le 400$.
c) Find the marginal supply, $q'(x)$.
d) Interpret the meaning of the derivative.

For Exercises 87–90, use the Tangent *feature from the* DRAW *menu to find the rate of change in part (b).*

87. Growth of a retirement fund.   Maria deposits $20,000 in an IRA whose value increases by 5.6% every year. The value of the IRA after t years is modeled by

$$V(t) = 20,000(1.056)^t.$$

a) Use the model to estimate the value of Maria's IRA after 7 yr.
b) What is the rate of change in the value of the IRA at the end of 7 yr?
c) When will Maria's IRA have a value of $40,000?

88. Depreciation.   Pelican Fabrics purchases a new video surveillance system. The value of the system is modeled by

$$V(t) = 17,500(0.92)^t,$$

where V is the value of the system, in dollars, t years after its purchase.

a) Use the model to estimate the value of the system 5 yr after it was purchased.
b) What is the rate of change in the value of the system at the end of 5 yr?
c) When will the system be worth half of its original value?

89. Depreciation.   Perriot's Restaurant purchased kitchen equipment on January 1, 2014. The value of the equipment decreases by 15% every year. On January 1, 2016, the value was $14,450.

a) Find an exponential model for the value, V, of the equipment, in dollars, t years after January 1, 2016.
b) What is the rate of change in the value of the equipment on January 1, 2016?
c) What was the original value of the equipment on January 1, 2014?
d) How many years after January 1, 2014 will the value of the equipment have decreased by half?

90. Stock prices. The value (price) of a share of stock in Barrington Gold was $90 on June 15, 2014, and was increasing by 3% every week.

a) Find an exponential model for the value, V, of a share of the stock, in dollars, t weeks after June 15, 2014.

b) What was the rate of change in the value of a share of the stock 6 weeks prior to June 15, 2014?

c) Use the model to estimate the value of a share of the stock 6 weeks prior to June 15, 2014.

d) How many weeks after June 15, 2014, will the stock's share value have doubled?

Life and Physical Sciences

91. Medication concentration. The concentration C, in parts per million, of a medication in the body t hours after ingestion is given by the function

$$C(t) = 10t^2 e^{-t}.$$

a) Find the concentration after 0 hr, 1 hr, 2 hr, 3 hr, and 10 hr.

b) Sketch a graph of the function for $0 \le t \le 10$.

c) Find the rate of change of the concentration, $C'(t)$.

d) Find the maximum value of the concentration and the time at which it occurs.

e) Interpret the meaning of the derivative.

Social Sciences

92. Ebbinghaus learning model. Suppose that you are given the task of learning 100% of a block of knowledge. Human nature is such that we retain only a percentage P of knowledge t weeks after we have learned it. The *Ebbinghaus learning model* asserts that P is given by

$$P(t) = Q + (100 - Q)e^{-kt},$$

where Q is the percentage that we would never forget and k is a constant that depends on the knowledge learned. Suppose that $Q = 40$ and $k = 0.7$.

a) Find the percentage retained after 0 weeks, 1 week, 2 weeks, 6 weeks, and 10 weeks.

b) Find $\lim_{t \to \infty} P(t)$.

c) Sketch a graph of P.

d) Find the rate of change of $P(t)$ with respect to time t.

e) Interpret the meaning of the derivative.

SYNTHESIS

Differentiate.

93. $y = (e^{3x} + 1)^5$

94. $y = \dfrac{e^{3t} - e^{7t}}{e^{4t}}$

95. $y = \dfrac{e^x}{x^2 + 1}$

96. $f(x) = e^{\sqrt{x}} + \sqrt{e^x}$

97. $f(x) = e^{x/2} \cdot \sqrt{x - 1}$

98. $f(x) = \dfrac{xe^{-x}}{1 + x^2}$

99. $f(x) = \dfrac{e^x - e^{-x}}{e^x + e^{-x}}$

100. $f(x) = e^{e^x}$

101. Use the results from Exercises 85 and 86 to determine the *equilibrium point* (the point at which supply equals demand) and the rates at which supply and demand are changing at that point.

Exercises 102 and 103 each give an expression for e. Find the function values that are approximations for e. Round to five decimal places.

102. For $f(t) = (1 + t)^{1/t}$, we have $e = \lim_{t \to 0} f(t)$. Find $f(1)$, $f(0.5)$, $f(0.2)$, $f(0.1)$, and $f(0.001)$.

103. For $g(t) = t^{1/(t-1)}$, we have $e = \lim_{t \to 1} g(t)$. Find $g(0.5)$, $g(0.9)$, $g(0.99)$, $g(0.999)$, and $g(0.9998)$.

104. Find the minimum value of $f(x) = xe^x$ over $[-2, 0]$.

105. A student made the following error on a test:

$$\frac{d}{dx} e^x = xe^{x-1}.$$

Identify the error and explain how to correct it.

106. Describe the differences in the graphs of $f(x) = 3^x$ and $g(x) = x^3$.

TECHNOLOGY CONNECTION

Use a graphing calculator (or Graphicus) to graph each function in Exercises 107 and 108, and find all relative extrema.

107. $f(x) = x^2 e^{-x}$

108. $f(x) = e^{-x^2}$

For each of the functions in Exercises 109–112, graph f, f', and f''.

109. $f(x) = e^x$

110. $f(x) = e^{-x}$

111. $f(x) = 2e^{0.3x}$

112. $f(x) = 1000e^{-0.08x}$

113. Graph

$$f(x) = \left(1 + \frac{1}{x}\right)^x.$$

Use the TABLE feature and very large values of x to confirm that e is approached as a limit.

114. Consider the expression 2^π, where $\pi = 3.1415926....$ Recall that π is an irrational number, that is, a number with an infinite, nonrepeating decimal expansion.

a) Using a calculator, complete the following table, giving answers to eight decimal places.

x	2^x
3	
3.1	
3.14	
3.141	
3.1415	
3.14159	
3.141592	
3.1415926	

b) Based on the results from part (a), estimate the value of 2^π.

c) Using a calculator, find 2^π.

d) Use the results of parts (a)–(c) to explain how a base raised to an irrational number can be found as a numerical limit. Does this work for a base raised to any irrational number? Explain.

Answers to Quick Checks

1. $13,248.53 **2. (a)** $6e^x$; **(b)** $x^2 e^x (x + 3)$;

(c) $\dfrac{e^x (x - 2)}{x^3}$ **3. (a)** $-4e^{-4x}$; **(b)** $e^{x^3 + 8x} (3x^2 + 8)$;

(c) $\dfrac{xe^{\sqrt{x^2 + 5}}}{\sqrt{x^2 + 5}}$ **4. (a)** 20, 26.8, 47.1, 61.9, 74.6, 78.4

(b)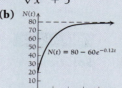

(c) $N'(t) = 7.2e^{-0.12t}$; after t days, the rate of change of number of phones produced per day is given by $7.2e^{-0.12t}$.

(d) $N(t)$ is increasing over $(0, \infty)$. **(e)** $N(t)$ is concave down over $(0, \infty)$. **(f)** 80 phones produced per day

---

3.2

- Convert between exponential and logarithmic equations.
- Solve exponential equations.
- Solve problems involving exponential and logarithmic functions.
- Differentiate functions involving natural logarithms.

Logarithmic Functions

Logarithmic Functions and Their Graphs

Suppose we want to solve

$$10^y = 1000.$$

We are trying to find the power of 10 that will give 1000. Since $10^3 = 1000$, the answer is 3. The number 3 is called "the logarithm, base 10, of 1000."

DEFINITION

A **logarithm** is defined as follows:

$$\log_a x = y \qquad \text{means} \qquad a^y = x, \quad a > 0, a \neq 1.$$

The number $\log_a x$ is the power y to which we raise a to get x. The number a is called the *logarithmic base*. We read $\log_a x$ as "the logarithm, base a, of x."

For logarithms with base 10, $\log_{10} x$ is the power y such that $10^y = x$. Therefore, a logarithm can be thought of as an exponent. We can convert from a logarithmic equation to an exponential equation, and conversely, as follows.

LOGARITHMIC EQUATION	EXPONENTIAL EQUATION
$\log_a M = N$	$a^N = M$
$\log_{10} 100 = 2$	$10^2 = 100$
$\log_5 \frac{1}{25} = -2$	$5^{-2} = \frac{1}{25}$
$\log_{49} 7 = \frac{1}{2}$	$49^{1/2} = 7$

EXAMPLE 1   Solve for x by writing each logarithmic equation as an equivalent exponential equation.

a) $\log_3 9 = x$ **b)** $\log_x 25 = 2$ **c)** $\log_{16} x = \frac{1}{2}$

Solution

a) We write $\log_3 9 = x$ as $3^x = 9$. Therefore, we have $x = 2$, and we say "the logarithm, base 3, of 9, is 2."

b) We write $\log_x 25 = 2$ as $x^2 = 25$. Since the base must be positive, we have $x = 5$.

c) We write $\log_{16} x = \frac{1}{2}$ as $16^{1/2} = x$. Since $16^{1/2} = \sqrt{16}$, we have $x = 4$.   **1** ✔

Quick Check 1 ✔

Solve for x by writing each logarithmic equation as an equivalent exponential equation.

a) $\log_2 16 = x$ **b)** $\log_x 6 = \frac{1}{2}$

c) $\log_7 x = 0$

In order to graph a logarithmic equation, we can graph its equivalent exponential equation.

EXAMPLE 2   Graph: $y = \log_2 x$.

Solution   We first write the equivalent exponential equation:

$$2^y = x.$$

We select values for y and find the corresponding values of 2^y. Then we plot points, remembering that x is still the first coordinate, and connect the points with a smooth curve.

x, or 2^y	y
1	0
2	1
4	2
8	3
$\frac{1}{2}$	−1
$\frac{1}{4}$	−2

① Select y.
② Compute x.

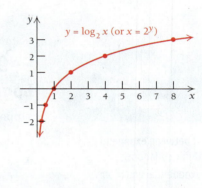

The graphs of $f(x) = 2^x$ and $g(x) = \log_2 x$ are shown below. Note that we can obtain the graph of g by reflecting the graph of f across the line $y = x$. Functions whose graphs can be obtained in this manner are known as *inverses* of each other.

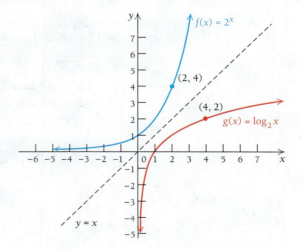

Although we do not develop inverses in detail here, it is important to note that they "undo" each other. For example,

$$f(3) = 2^3 = 8,$$ The input 3 gives the output 8.

and $$g(8) = \log_2 8 = 3.$$ The input 8 gets us back to 3.

Basic Properties of Logarithms

The following are some basic properties of logarithms. The proofs of P1–P3 follow from properties of exponents and are outlined in Exercises 119–121 at the end of this section. Properties P4–P6 follow directly from the definition of logarithm, and a proof of P7 is outlined in Exercise 122.

> **THEOREM 3** **Properties of Logarithms**
>
> For any positive numbers M, N, a, and b, with $a, b \neq 1$, and any real number k:
>
> **P1.** $\log_a (MN) = \log_a M + \log_a N$
>
> **P2.** $\log_a \dfrac{M}{N} = \log_a M - \log_a N$
>
> **P3.** $\log_a (M^k) = k \cdot \log_a M$
>
> **P4.** $\log_a a = 1$
>
> **P5.** $\log_a (a^k) = k$
>
> **P6.** $\log_a 1 = 0$
>
> **P7.** $\log_b M = \dfrac{\log_a M}{\log_a b}$ (The change-of-base formula)

Let's illustrate these properties.

EXAMPLE 3 Given

$$\log_a 2 = 0.301 \quad \text{and} \quad \log_a 3 = 0.477,$$

find each of the following:

a) $\log_a 6$;

b) $\log_a \frac{2}{3}$;

c) $\log_a 81$;

d) $\log_a \frac{1}{3}$;

e) $\log_a \sqrt{a}$;

f) $\log_a (2a)$;

g) $\dfrac{\log_a 3}{\log_a 2}$;

h) $\log_a 5$.

Solution

a) $\log_a 6 = \log_a (2 \cdot 3)$

$\qquad = \log_a 2 + \log_a 3$ Using Property P1

$\qquad = 0.301 + 0.477$

$\qquad = 0.778$

b) $\log_a \frac{2}{3} = \log_a 2 - \log_a 3$ Using Property P2

$\qquad = 0.301 - 0.477$

$\qquad = -0.176$

c) $\log_a 81 = \log_a 3^4$ Recognizing 81 as a power of 3

$\qquad = 4 \log_a 3$ Using Property P3

$\qquad = 4(0.477)$

$\qquad = 1.908$

TECHNOLOGY CONNECTION

Graphing Logarithmic Functions

To graph $y = \log_2 x$, we first graph $y_1 = 2^x$. We next select the DrawInv option from the DRAW menu and then the Y-VARS option from the VARS menu, followed by ①, ①, and **ENTER** to draw the inverse of y_1. Both graphs are drawn together.

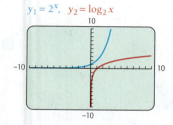

$y_1 = 2^x$, $y_2 = \log_2 x$

EXERCISES

Graph.

1. $y = \log_3 x$

2. $y = \log_5 x$

3. $f(x) = \log_e x$

4. $f(x) = \log_{10} x$

d) $\log_a \frac{1}{3} = \log_a 1 - \log_a 3$ Using Property P2

$\qquad\quad = 0 - 0.477$ Using Property P6

$\qquad\quad = -0.477$

e) $\log_a \sqrt{a} = \log_a (a^{1/2}) = \frac{1}{2}$ Using Property P5

f) $\log_a (2a) = \log_a 2 + \log_a a$ Using Property P1

$\qquad\qquad = 0.301 + 1$ Using Property P4

$\qquad\qquad = 1.301$

g) $\dfrac{\log_a 3}{\log_a 2} = \dfrac{0.477}{0.301} \approx 1.58$

Here we simply divided and used none of the properties.

h) There is no way to find $\log_a 5$ using the properties of logarithms $(\log_a 5 \neq \log_a 2 + \log_a 3)$.   **2** ✔

Quick Check 2 ✔

Given $\log_b 2 = 0.356$ and $\log_b 5 = 0.827$, find each of the following:

a) $\log_b 10$;   **b)** $\log_b \frac{2}{5}$;

c) $\log_b \frac{5}{2}$;   **d)** $\log_b 16$;

e) $\log_b 5b$;   **f)** $\log_b \sqrt{b}$.

Common Logarithms

The number $\log_{10} x$ is the **common logarithm** of x and is abbreviated $\log x$; that is:

> **DEFINITION**
>
> For any positive number x, $\log x = \log_{10} x$.

Thus, when we write "$\log x$" with no base indicated, base 10 is assumed. Note the following comparison of common logarithms and powers of 10.

$1000 = 10^3$	The common	$\log 1000 = 3$
$100 = 10^2$	logarithms at the	$\log 100 = 2$
$10 = 10^1$	right follow from	$\log 10 = 1$
$1 = 10^0$	the powers at the left.	$\log 1 = 0$
$0.1 = 10^{-1}$		$\log 0.1 = -1$
$0.01 = 10^{-2}$		$\log 0.01 = -2$
$0.001 = 10^{-3}$		$\log 0.001 = -3$

Since $\log 100 = 2$ and $\log 1000 = 3$, it seems reasonable that $\log 500$ is somewhere between 2 and 3. Using a calculator with a **LOG** key, we find that $\log 500 \approx 2.6990$.

Before calculators became readily available, common logarithms were listed in tables and used extensively to do certain computations. In fact, computation is the reason logarithms were developed. Since standard notation for numbers is based on 10, it was logical to use base-10, or common, logarithms for computations. Today, computations with common logarithms are mainly of historical interest; the logarithmic functions, base e, are far more important.

Natural Logarithms

The number e, which is approximately 2.71828, was developed in Section 3.1, and has extensive application in many fields. The number $\log_e x$ is the **natural logarithm** of x and is abbreviated $\ln x$.

> **DEFINITION**
>
> For any positive number x, $\ln x = \log_e x$.

The following basic properties of natural logarithms parallel those given earlier for general logarithms.

THEOREM 4   Properties of Natural Logarithms

P1. $\ln (MN) = \ln M + \ln N$ **P5.** $\ln (e^k) = k$

P2. $\ln \dfrac{M}{N} = \ln M - \ln N$ **P6.** $\ln 1 = 0$

P3. $\ln (a^k) = k \cdot \ln a$ **P7.** $\log_b M = \dfrac{\ln M}{\ln b}$ and $\ln M = \dfrac{\log M}{\log e}$

P4. $\ln e = 1$

Let's illustrate the properties given in Theorem 4.

EXAMPLE 4   Given

$$\ln 2 = 0.6931 \quad \text{and} \quad \ln 3 = 1.0986,$$

use the properties of natural logarithms to find each of the following:
a) $\ln 6$;   **b)** $\ln 81$;   **c)** $\ln \frac{1}{3}$;   **d)** $\ln (2e^5)$;   **e)** $\log_2 3$.

Solution
a) $\ln 6 = \ln (2 \cdot 3) = \ln 2 + \ln 3$ Using Property P1

$\qquad\qquad\qquad = 0.6931 + 1.0986$

$\qquad\qquad\qquad = 1.7917$

b) $\ln 81 = \ln (3^4)$

$\qquad\quad = 4 \ln 3$ Using Property P3

$\qquad\quad = 4(1.0986)$

$\qquad\quad = 4.3944$

c) $\ln \frac{1}{3} = \ln 1 - \ln 3$ Using Property P2

$\qquad = 0 - 1.0986$ Using Property P6

$\qquad = -1.0986$

d) $\ln (2e^5) = \ln 2 + \ln (e^5)$ Using Property P1

$\qquad\qquad = 0.6931 + 5$ Using Property P5

$\qquad\qquad = 5.6931$

e) $\log_2 3 = \dfrac{\ln 3}{\ln 2} = \dfrac{1.0986}{0.6931} \approx 1.5851$ Using Property P7   3 ✔

Finding Natural Logarithms Using a Calculator

Using a calculator with an **LN** key allows us to find natural logarithms directly.

EXAMPLE 5   Approximate each of the following to six decimal places:

a) $\ln 5.24$;   **b)** $\ln 0.001278$.

Solution   We use a calculator with an **LN** key.
a) $\ln 5.24 \approx 1.656321$ **b)** $\ln 0.001278 \approx -6.662459$ ■

Exponential Equations

If an equation contains a variable in an exponent, the equation is **exponential**. We can use logarithms to manipulate or solve exponential equations.

EXAMPLE 6 Solve: $e^{-0.04t} = 0.05$.

Solution We have

$$\ln e^{-0.04t} = \ln 0.05 \qquad \text{Taking the natural logarithm on both sides}$$

$$-0.04t = \ln 0.05 \qquad \text{Using Property P5}$$

$$t = \frac{\ln 0.05}{-0.04}$$

$$t \approx \frac{-2.995732}{-0.04} \Bigg\} \quad \text{Using a calculator}$$

$$t \approx 75. \qquad\qquad\qquad\qquad\qquad 4\ ✔$$

Quick Check 4 ✔

Solve each equation:

a) $e^t = 80$;

b) $e^{-0.08t} = 0.25$.

In Example 6, we rounded $\ln 0.05$ to -2.995732 in an intermediate step. When using a calculator, you should find

$$\frac{\ln 0.05}{-0.04}$$

by keying in

$$\boxed{\text{LN}}\ \boxed{.}\ \boxed{0}\ \boxed{5}\ \boxed{)}\ \boxed{÷}\ \boxed{(-)}\ \boxed{.}\ \boxed{0}\ \boxed{4},$$

pressing $\boxed{\text{ENTER}}$, and rounding at the end. Answers at the back of this book have been found in this manner. Remember, the number of places in a table or on a calculator affects the accuracy of the answer. Usually, your answer should agree with that in the Answers section to at least three digits.

TECHNOLOGY CONNECTION

Solving Exponential Equations

Let's solve $e^t = 40$ graphically.

Method 1: The INTERSECT Feature

We change the variable to x and consider the system of equations $y_1 = e^x$ and $y_2 = 40$. We graph the equations in the window $[-1, 8, -10, 70]$ to see the curves and point of intersection.

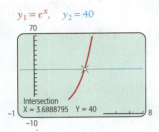

$y_1 = e^x, \quad y_2 = 40$

Intersection
X = 3.6888795   Y = 40

Then we use the INTERSECT option from the CALC menu to find the point of intersection, about $(3.6888795, 40)$. The x-coordinate, 3.6888795, is the approximate solution of $e^t = 40$.

Method 2: The ZERO Feature

We change the variable to x and get a 0 on one side of the equation: $e^x - 40 = 0$. Then we graph $y = e^x - 40$ in the window $[-1, 8, -10, 10]$.

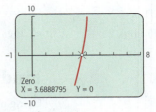

Zero
X = 3.6888795   Y = 0

Using the ZERO option from the CALC menu, we see that the x-intercept is about $(3.6888795, 0)$, so 3.6888795 is the approximate solution.

EXERCISES

Solve graphically using a graphing utility.

1. $e^t = 1000$

2. $e^{-x} = 60$

3. $e^{-0.04t} = 0.05$

4. $e^{0.23x} = 41{,}378$

5. $15e^{0.2x} = 34{,}785.13$

Graphs of Natural Logarithmic Functions

There are two ways in which we might graph $y = f(x) = \ln x$. One is to graph the equivalent equation $x = e^y$ by selecting values for y and calculating the corresponding values of e^y. We then plot points, remembering that x is still the first coordinate.

x, or e^y	y
0.1	−2
0.4	−1
1.0	0
2.7	1
7.4	2
20.1	3

① Select y.
② Compute x.

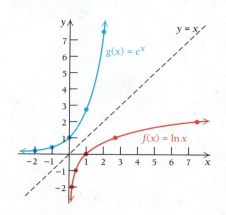

The graph above shows the graph of $g(x) = e^x$ for comparison with that of $f(x) = \ln x$. Note that the functions are inverses of each other. That is, the graph of $y = \ln x$, or $x = e^y$, is a reflection across the line $y = x$ of the graph of $y = e^x$. Any ordered pair (a, b) on the graph of g yields an ordered pair (b, a) on f. Note too that $\lim_{x \to 0^+} \ln x = -\infty$ and the y-axis is a vertical asymptote.

The second method of graphing $y = \ln x$ is to use a calculator to find function values. For example, when $x = 2$, then $y = \ln 2 \approx 0.6931$. This gives the pair $(2, 0.6931)$ shown on the graph.

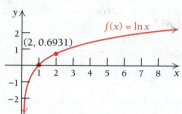

The following properties can be observed from the graph.

> **THEOREM 5**
>
> $\ln x$ exists only for positive numbers x. The domain is $(0, \infty)$.
>
> $\ln x < 0$ for $0 < x < 1$.
>
> $\ln x = 0$ when $x = 1$.
>
> $\ln x > 0$ for $x > 1$.
>
> The function given by $f(x) = \ln x$ is always increasing. The range is the entire real line, $(-\infty, \infty)$, or the set of real numbers, $\mathbb{R}$.

Derivatives of Natural Logarithmic Functions

Let's find the derivative of

$$f(x) = \ln x. \tag{1}$$

We first write its equivalent exponential equation:

$$e^{f(x)} = x. \tag{2}$$

 $\ln x = \log_e x = f(x)$, so $e^{f(x)} = x$, by the definition of logarithm.

Now we differentiate implicitly on both sides of this equation:

$$\frac{d}{dx}e^{f(x)} = \frac{d}{dx}x$$

$$e^{f(x)} \cdot f'(x) = 1 \qquad \text{Using the Chain Rule}$$

$$x \cdot f'(x) = 1 \qquad \text{Substituting } x \text{ for } e^{f(x)} \text{ from equation (2)}$$

$$f'(x) = \frac{1}{x}.$$

Thus, we have the following.

THEOREM 6

For any positive number x,

$$\frac{d}{dx}\ln x = \frac{1}{x}.$$

TECHNOLOGY CONNECTION

Exploratory

Use Graphicus to graph $y = \ln x$. Then touch $+$ and choose **Add derivative**. Using the tangent line feature, move along the curve, noting the x-values, the y-values, and the values of dy/dx, to verify that $\dfrac{dy}{dx} = \dfrac{1}{x}$.

To visualize Theorem 6, look back at the graph of $f(x) = \ln x$ on p. 313. Take a straightedge and place it as if its edge were a tangent line. Start on the left and move the straightedge along the curve toward the right, noting how the tangent lines flatten out. Think about the slopes of these tangent lines. The slopes approach 0 as a limit, though they never actually become 0. This is consistent with the formula

$$\frac{d}{dx}\ln x = \frac{1}{x}, \quad \text{because } \lim_{x \to \infty}\frac{1}{x} = 0.$$

Theorem 6 states that to find the slope of the tangent line at x for the function $f(x) = \ln x$, we need only take the reciprocal of x. This is true only for positive values of x, since $\ln x$ is defined only for positive numbers. Note that

$$\frac{d}{dx}\ln(cx) = \frac{1}{cx} \cdot c = \frac{1}{x}, \quad \text{for } cx > 0$$

and

$$\frac{d}{dx}\ln(-x) = \frac{1}{(-x)} \cdot (-1) = \frac{1}{x}, \quad \text{for } x < 0.$$

Thus, $y = \ln x$ is not the only function for which $dy/dx = 1/x$. In particular, note that for $x < 0$, the function given by $y = \ln x$ is undefined. However, for $x < 0$, the function given by $y = \ln(-x)$ is defined. Since

$$y = \ln|x| \text{ is equivalent to } y = \begin{cases} \ln x, & \text{if } x > 0, \\ \ln(-x), & \text{if } x < 0, \end{cases}$$

it follows that

$$\frac{dy}{dx}\ln|x| = \frac{1}{x}, \quad \text{for all } x \neq 0.$$

This result will be important for our work in Chapter 4. In general, when we write $y = \ln x$, we assume $x > 0$. For example, for the function given by $y = \ln(x^2 + 5x)$, we assume that x is chosen so that $x^2 + 5x > 0$.

Let's find some derivatives.

EXAMPLE 7 Differentiate:

a) $y = 3 \ln x$;　　　　**b)** $y = x^2 \ln x + 5x$;　　　**c)** $y = \dfrac{\ln |2x|}{x^3}$.

Solution

a) $\dfrac{d}{dx}(3 \ln x) = 3 \dfrac{d}{dx} \ln x = \dfrac{3}{x}$

b) $\dfrac{d}{dx}(x^2 \ln x + 5x) = x^2 \cdot \dfrac{1}{x} + 2x \cdot \ln x + 5$　　Using the Product Rule on $x^2 \ln x$

$= x + 2x \cdot \ln x + 5$　　Simplifying

c) $\dfrac{d}{dx}\left(\dfrac{\ln |2x|}{x^3}\right) = \dfrac{x^3 \cdot (1/2x) \cdot 2 - \ln |2x| (3x^2)}{(x^3)^2}$　　Using the Quotient Rule

$= \dfrac{x^2 - 3x^2 \ln |2x|}{x^6}$

$= \dfrac{x^2(1 - 3 \ln |2x|)}{x^6}$　　Factoring

$= \dfrac{1 - 3 \ln |2x|}{x^4}$　　Simplifying　　**5** ✔

Quick Check 5 ✔

Differentiate:

a) $y = 5 \ln x$;

b) $y = x^3 \ln x + 4x$;

c) $y = \dfrac{\ln |4x|}{x^2}$.

Suppose we want to differentiate a function of the form $h(x) = \ln f(x)$, such as

$$h(x) = \ln (x^2 - 8x).$$

This can be regarded as a composition of functions,

$$h(x) = g(f(x)), \quad \text{where} \quad g(x) = \ln x \quad \text{and} \quad f(x) = x^2 - 8x.$$

Now $g'(x) = 1/x$, so by the Chain Rule (Section 1.7), we have

$$h'(x) = g'(f(x)) \cdot f'(x)$$

$$= \dfrac{1}{f(x)} \cdot f'(x).$$

For the above case, $f(x) = x^2 - 8x$, so $f'(x) = 2x - 8$. Then

$$h'(x) = \dfrac{1}{x^2 - 8x} \cdot (2x - 8) = \dfrac{2x - 8}{x^2 - 8x}.$$

The following rule, which results directly from the Chain Rule, allows us to find derivatives of functions like the one above.

THEOREM 7

The derivative of the natural logarithm of a function is the derivative of the function divided by the function:

$$\dfrac{d}{dx} \ln f(x) = \dfrac{1}{f(x)} \cdot f'(x) = \dfrac{f'(x)}{f(x)},$$

or

$$\dfrac{d}{dx} \ln u = \dfrac{1}{u} \cdot \dfrac{du}{dx}.$$

The following gives us a way of remembering this rule.

$$h(x) = \ln \underbrace{(x^2 - 8x)}$$

① Differentiate the "inside" function.

$$h'(x) = \frac{\overbrace{2x - 8}}{\underbrace{x^2 - 8x}}$$

② Divide by the "inside" function.

EXAMPLE 8 Differentiate:

a) $y = \ln (3x)$;   **b)** $y = \ln (x^2 - 5)$;   **c)** $f(x) = \ln \left(\dfrac{x^3 + 4}{x} \right)$.

Solution

a) If $y = \ln (3x)$, then

$$\frac{dy}{dx} = \frac{3}{3x} = \frac{1}{x}.$$

Note that we could have done this using the fact that $\ln (MN) = \ln M + \ln N$:

$$\ln (3x) = \ln 3 + \ln x;$$

then, since $\ln 3$ is a constant, we have

$$\frac{d}{dx} \ln (3x) = \frac{d}{dx} \ln 3 + \frac{d}{dx} \ln x = 0 + \frac{1}{x} = \frac{1}{x}.$$

b) If $y = \ln (x^2 - 5)$, then

$$\frac{dy}{dx} = \frac{2x}{x^2 - 5}.$$

c) If $f(x) = \ln \left(\dfrac{x^3 + 4}{x} \right)$, then, since $\ln \dfrac{M}{N} = \ln M - \ln N$, we have

$$f'(x) = \frac{d}{dx} [\ln (x^3 + 4) - \ln x] \qquad \text{Using P2 avoids use of the Quotient Rule.}$$

$$= \frac{3x^2}{x^3 + 4} - \frac{1}{x}$$

$$= \frac{3x^2}{x^3 + 4} \cdot \frac{x}{x} - \frac{1}{x} \cdot \frac{x^3 + 4}{x^3 + 4} \qquad \text{Finding a common denominator}$$

$$= \frac{(3x^2)x - (x^3 + 4)}{x(x^3 + 4)}$$

$$= \frac{3x^3 - x^3 - 4}{x(x^3 + 4)} = \frac{2x^3 - 4}{x(x^3 + 4)}. \qquad \text{Simplifying}$$

6 ✔

TECHNOLOGY CONNECTION

Exploratory

To check part (a) of Example 8, we let $y_1 = \ln (3x)$, $y_2 = \text{nDeriv}(y_1, x, x)$, and $y_3 = 1/x$. Either GRAPH or TABLE can then be used to show that $y_2 = y_3$. Use this approach to check parts (b) and (c) of Example 8. (Or use Graphicus to check those results.)

Quick Check 6 ✔

Differentiate:

a) $y = \ln 5x$;

b) $y = \ln (3x^2 + 4)$;

c) $y = \ln (\ln 5x)$;

d) $y = \ln \left(\dfrac{x^5 - 2}{x} \right)$.

Applications

EXAMPLE 9 **Social Science: Forgetting.** In a psychological experiment, students were shown a set of nonsense syllables, such as POK, RIZ, DEQ, and so on, and asked to recall them every minute thereafter. The percentage $R(t)$ who retained the syllables after t minutes was found to be given by the logarithmic learning model

$$R(t) = 80 - 27 \ln t, \quad \text{for} \quad t \geq 1.$$

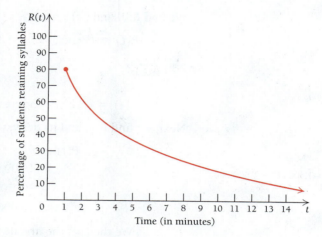

a) What percentage of students retained the syllables after 1 min?

b) Find $R'(2)$, and explain what it represents.

Solution

a) $R(1) = 80 - 27 \cdot \ln 1 = 80 - 27 \cdot 0 = 80\%$

b) $\dfrac{d}{dt}(80 - 27 \ln t) = 0 - 27 \cdot \dfrac{1}{t} = -\dfrac{27}{t}$,

so $R'(2) = -\dfrac{27}{2} = -13.5.$

This result indicates that 2 min after students have been shown the syllables, the percentage of them who remember the syllables is shrinking at the rate of 13.5% per minute. ∎

EXAMPLE 10   **Business: An Advertising Model.**   Sawtelle Inc. begins a radio advertising campaign in New York City to market its new energy drink. The percentage of the "target market" that buys a product is normally a function of the duration of the advertising campaign. The radio station estimates this percentage, as a decimal, by using $f(t) = 1 - e^{-0.04t}$ for this type of product, where t is the number of days of the campaign. The target market is approximately 1,000,000 people, and the price per unit is $1.50. If the campaign costs $1000 per day, how long should it last in order to maximize profit?

Solution   The radio station's model for the percentage of the target market that buys the product seems reasonable if we graph f. The function increases from 0 (0%) toward 1 (100%). The longer the advertising campaign, the larger the percentage of the market that has bought the product.

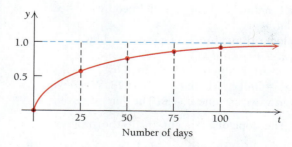

The total-profit function, here expressed in terms of time t, is given by

Profit = Revenue − Cost

$$P(t) = R(t) - C(t).$$

We find $R(t)$ and $C(t)$:

$$R(t) = (\text{Percentage buying}) \cdot (\text{Target market}) \cdot (\text{Price per unit})$$
$$= (1 - e^{-0.04t})(1{,}000{,}000)(1.5) = 1{,}500{,}000 - 1{,}500{,}000e^{-0.04t},$$

and

$$C(t) = (\text{Advertising costs per day}) \cdot (\text{Number of days}) = 1000t.$$

Next, we find $P(t)$ and its derivative:

$$P(t) = R(t) - C(t)$$
$$= 1{,}500{,}000 - 1{,}500{,}000e^{-0.04t} - 1000t,$$
$$P'(t) = -1{,}500{,}000e^{-0.04t}(-0.04) - 1000$$
$$= 60{,}000e^{-0.04t} - 1000.$$

We then set the first derivative equal to 0 and solve:

$$60{,}000e^{-0.04t} - 1000 = 0$$
$$60{,}000e^{-0.04t} = 1000$$
$$e^{-0.04t} = \frac{1000}{60{,}000} = 0.0167$$
$$\ln e^{-0.04t} = \ln 0.0167$$
$$-0.04t = \ln 0.0167$$
$$t = \frac{\ln 0.0167}{-0.04}$$
$$t \approx 102.3.$$

We have only one critical value, so we can use the second derivative to determine whether we have a maximum:

$$P''(t) = 60{,}000e^{-0.04t}(-0.04)$$
$$= -2400e^{-0.04t}.$$

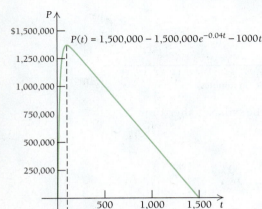

P(t) = 1,500,000 – 1,500,000e^{-0.04t} – 1000t

≈ 102.3

Quick Check 7 ✔

Business: An Advertising Model.
Repeat Example 10 using the function

$$f(t) = 1 - e^{-0.08t}$$

and assuming that the campaign costs $2000 per day.

Since exponential functions are positive, $e^{-0.04t} > 0$ for all numbers t. Thus, since $-2400e^{-0.04t} < 0$ for all t, we have $P''(102.3) < 0$, and we have a maximum.

The advertising campaign should run for 103 days (since we need to round up to a whole day) in order to maximize profit.   **7** ✔

Section Summary

- A logarithmic function $y = \log_a x$ is defined by $a^y = x$, for $a > 0$ and $a \neq 1$.
- The common logarithmic function g is defined by $g(x) = \log_{10} x = \log x$, for $x > 0$.
- The natural logarithm function f is defined by $f(x) = \log_e |x| = \ln |x|$, where $e \approx 2.71828$. When we write $f(x) = \ln x$, we assume $x > 0$. The derivative of f is $f'(x) = \frac{1}{x}$. The slope of a tangent line to the graph of f at x is found by taking the reciprocal of x.

- The graph of $f(x) = \ln x$ is an increasing function with no critical values, no maximum or minimum values, and no points of inflection. The domain is $(0, \infty)$. The range is $(-\infty, \infty)$, or $\mathbb{R}$. The graph is concave down, with

$$\lim_{x \to \infty} f(x) = \infty \quad \text{and} \quad \lim_{x \to 0^+} f(x) = -\infty.$$

- Properties of logarithms are described in Theorems 3 and 4.

3.2 | **Exercise Set**

Write an equivalent exponential equation.

1. $\log_3 81 = 4$ **2.** $\log_2 8 = 3$

3. $\log_{27} 3 = \frac{1}{3}$ **4.** $\log_8 2 = \frac{1}{3}$

5. $\log_a J = K$ **6.** $\log_a K = J$

7. $-\log_b V = w$ **8.** $-\log_{10} h = p$

Solve for x.

9. $\log_7 49 = x$ **10.** $\log_5 125 = x$

11. $\log_x 32 = 5$ **12.** $\log_x 64 = 3$

13. $\log_3 x = 5$ **14.** $\log_6 x = -1$

15. $\log_{11} \sqrt{11} = x$ **16.** $\log_4 1/16 = x$

Write an equivalent logarithmic equation.

17. $e^t = p$ **18.** $e^M = b$

19. $10^3 = 1000$ **20.** $10^2 = 100$

21. $10^{-2} = 0.01$ **22.** $10^{-1} = 0.1$

23. $Q^n = T$ **24.** $M^p = V$

Given $\log_b 3 = 1.099$ *and* $\log_b 5 = 1.609,$ *find each value.*

25. $\log_b \frac{1}{5}$ **26.** $\log_b \frac{5}{3}$

27. $\log_b \sqrt{b^3}$ **28.** $\log_b 15$

29. $\log_b 75$ **30.** $\log_b (5b)$

Given $\ln 4 = 1.3863$ *and* $\ln 5 = 1.6094,$ *find each value.*
Do not use a calculator.

31. $\ln 80$ **32.** $\ln 20$

33. $\ln \frac{1}{5}$ **34.** $\ln \frac{5}{4}$

35. $\ln (4e)$ **36.** $\ln (5e)$

37. $\ln \sqrt{e^8}$ **38.** $\ln \sqrt{e^6}$

39. $\ln \frac{4}{5}$ **40.** $\ln \frac{1}{4}$

41. $\ln \left(\frac{4}{e} \right)$ **42.** $\ln \left(\frac{e}{5} \right)$

Find each logarithm. Round to six decimal places.

43. $\ln 99{,}999$ **44.** $\ln 5894$

45. $\ln 0.0182$ **46.** $\ln 0.00087$

47. $\ln 0.011$ **48.** $\ln 8100$

Solve for t.

49. $e^t = 80$ **50.** $e^t = 10$

51. $e^{3t} = 900$ **52.** $e^{2t} = 1000$

53. $e^{-t} = 0.01$ **54.** $e^{-t} = 0.1$

55. $e^{-0.02t} = 0.06$ **56.** $e^{0.07t} = 2$

Differentiate.

57. $y = -9 \ln x$ **58.** $y = -8 \ln x$

59. $y = 7 \ln |x|$ **60.** $y = 4 \ln |x|$

61. $y = x^6 \ln x - \frac{1}{4} x^4$ **62.** $y = x^4 \ln x - \frac{1}{2} x^2$

63. $f(x) = \ln (9x)$ **64.** $f(x) = \ln (6x)$

65. $f(x) = \ln |5x|$ **66.** $f(x) = \ln |10x|$

67. $g(x) = x^5 \ln (3x)$ **68.** $g(x) = x^2 \ln (7x)$

69. $g(x) = x^4 \ln |6x|$ **70.** $g(x) = x^9 \ln |2x|$

71. $y = \dfrac{\ln x}{x^5}$ **72.** $y = \dfrac{\ln x}{x^4}$

73. $y = \dfrac{\ln |3x|}{x^2}$ **74.** $y = \dfrac{\ln |11x|}{x^8}$

75. $y = \ln \dfrac{x^2}{4}$ $\left(\textit{Hint: } \ln \dfrac{A}{B} = \ln A - \ln B. \right)$

76. $y = \ln \dfrac{x^4}{2}$

77. $y = \ln (3x^2 + 2x - 1)$

78. $y = \ln (7x^2 + 5x + 2)$

79. $f(x) = \ln \left(\dfrac{x^2 - 7}{x} \right)$ **80.** $f(x) = \ln \left(\dfrac{x^2 + 5}{x} \right)$

81. $g(x) = e^x \ln x^2$ **82.** $g(x) = e^{2x} \ln x$

83. $f(x) = \ln (e^x + 1)$

84. $f(x) = \ln (e^x - 2)$

85. $g(x) = (\ln x)^4$ (*Hint:* Use the Extended Power Rule.)

86. $g(x) = (\ln x)^3$ **87.** $f(x) = \ln (\ln (8x))$

88. $f(x) = \ln (\ln (3x))$

89. $g(x) = \ln (5x) \cdot \ln (3x)$

90. $g(x) = \ln (2x) \cdot \ln (7x)$

91. Find the equation of the line tangent to the graph of $y = (x^2 - x) \ln (6x)$ at $x = 2.$

92. Find the equation of the line tangent to the graph of $y = e^{3x} \cdot \ln (4x)$ at $x = 1.$

93. Find the equation of the line tangent to the graph of $y = (\ln x)^2$ at $x = 3.$

94. Find the equation of the line tangent to the graph of $y = \ln(4x^2 - 7)$ at $x = 2$.

APPLICATIONS

Business and Economics

95. Advertising. A model for consumers' response to advertising is given by

$$N(a) = 2000 + 500 \ln a, \quad a \geq 1,$$

where $N(a)$ is the number of units sold and a is the amount spent on advertising, in thousands of dollars.

a) How many units were sold after spending $1000 on advertising?

b) Find $N'(a)$ and $N'(10)$.

c) Find the maximum and minimum values, if they exist.

d) Find $\lim\limits_{a \to \infty} N'(a)$. Discuss whether it makes sense to continue to spend more and more on advertising.

96. Advertising. A model for consumers' response to advertising is given by

$$N(a) = 1000 + 200 \ln a, \quad a \geq 1,$$

where $N(a)$ is the number of units sold and a is the amount spent on advertising, in thousands of dollars.

a) How many units were sold after spending $1000 on advertising?

b) Find $N'(a)$ and $N'(10)$.

c) Find the maximum and minimum values of N, if they exist.

d) Find $N'(a)$. Discuss $\lim\limits_{a \to \infty} N'(a)$. Does it make sense to spend more and more on advertising? Why or why not?

97. An advertising model. Solve Example 10 if the advertising campaign costs $2000 per day.

98. An advertising model. Solve Example 10 if the advertising campaign costs $4000 per day.

99. Marginal revenue. The demand for a new computer game can be modeled by

$$p(x) = 53.5 - 8 \ln x,$$

where $p(x)$ is the price consumers will pay, in dollars, and x is the number of games sold, in thousands. Recall that total revenue is given by $R(x) = x \cdot p(x)$.

a) Find $R(x)$.

b) Find the marginal revenue, $R'(x)$.

c) Is there any price at which revenue will be maximized? Why or why not?

100. Growth of a stock. The value, $V(t)$, in dollars, of a share of Cypress Mills stock t months after it is purchased is modeled by

$$V(t) = 58(1 - e^{-1.1t}) + 20.$$

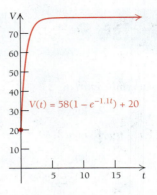

$V(t) = 58(1 - e^{-1.1t}) + 20$

a) Find $V(1)$ and $V(12)$.

b) Find $V'(t)$.

c) After how many months will the value of a share of the stock first reach $75?

d) Find $\lim\limits_{t \to \infty} V(t)$. Discuss the value of a share over a long period of time. Is this trend typical?

101. Marginal profit. The profit, in thousands of dollars, from the sale of x thousand candles can be estimated by

$$P(x) = 2x - 0.3x \ln x.$$

a) Find the marginal profit, $P'(x)$.

b) Find $P'(150)$, and explain what this number represents.

c) How many thousands of candles should be sold to maximize profit?

Life and Physical Sciences

102. Acceptance of a new medicine. The percentage P of doctors who prescribe a certain new medicine is

$$P(t) = 100(1 - e^{-0.2t}),$$

where t is the time, in months.

a) Find $P(1)$ and $P(6)$.

b) Find $P'(t)$.

c) How many months will it take for 90% of doctors to prescribe the new medicine?

d) Find $\lim\limits_{t \to \infty} P(t)$, and discuss its meaning.

Social Sciences

103. Forgetting. Students in a botany class took a final exam. They took equivalent forms of the exam at monthly intervals thereafter. After t months, the average score $S(t)$, as a percentage, was found to be

$$S(t) = 78 - 20 \ln(t + 1), \quad t \geq 0.$$

a) What was the average score when the students initially took the test?

b) What was the average score after 4 months?

c) What was the average score after 24 months?

d) What percentage of their original answers did the students retain after 2 years (24 months)?

e) Find $S'(t)$.

f) Find the maximum value, if one exists.

g) Find $\lim\limits_{t \to \infty} S(t)$, and discuss its meaning.

104. Forgetting. As part of a study, students in a psychology class took a final exam. They took equivalent forms of the exam at monthly intervals thereafter. After t months, the average score $S(t)$, as a percentage, was found to be given by

$$S(t) = 78 - 15 \ln(t + 1), \quad t \geq 0.$$

a) What was the average score when they initially took the test, $t = 0$?
b) What was the average score after 4 months?
c) What was the average score after 24 months?
d) What percentage of their original answers did the students retain after 2 years (24 months)?
e) Find $S'(t)$.
f) Find the maximum and minimum values, if they exist.
g) Find $\lim\limits_{t \to \infty} S(t)$ and discuss its meaning.

105. Walking speed. Bornstein and Bornstein found in a study that the average walking speed v, in feet per second, of a person living in a city of population p, in thousands, is

$$v(p) = 0.37 \ln p + 0.05.$$

(*Source*: M. H. Bornstein and H. G. Bornstein, "The Pace of Life," *Nature*, Vol. 259, pp. 557–559 (1976).)

a) The population of Seattle is 635,000 ($p = 635$). What is the average walking speed of a person living in Seattle?
b) The population of New York is 8,340,000. What is the average walking speed of a person living in New York?
c) Find $v'(p)$.
d) Interpret $v'(p)$ found in part (c).

106. Hullian learning model. A keyboarder learns to type W words per minute after t weeks of practice, where W is given by

$$W(t) = 100(1 - e^{-0.3t}).$$

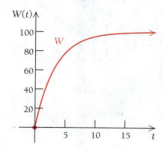

a) Find $W(1)$ and $W(8)$.
b) Find $W'(t)$.
c) After how many weeks will the keyboarder's speed be 95 words per minute?
d) Find $\lim\limits_{t \to \infty} W(t)$, and discuss its meaning.

SYNTHESIS

107. Solve $P = P_0 e^{kt}$ for t.

Differentiate.

108. $f(x) = \ln(x^3 + 1)^5$

109. $g(x) = [\ln(x + 5)]^4$

110. $f(t) = \ln[(t^3 + 3)(t^2 - 1)]$

111. $f(t) = \ln\left|\dfrac{1 - t}{1 + t}\right|$

112. $y = \ln\left|\dfrac{x^5}{(8x + 5)^2}\right|$

113. $f(x) = \log_5 x$

114. $f(x) = \log_7 x$

115. $y = \ln\sqrt{5 + x^2}$

116. $f(t) = \dfrac{\ln t^2}{t^2}$

117. $y = \dfrac{x^{n+1}}{n + 1}\left(\ln x - \dfrac{1}{n + 1}\right)$

118. $f(x) = \ln\dfrac{1 + \sqrt{x}}{1 - \sqrt{x}}$

To prove Properties P1, P2, P3, and P7 of Theorem 3, let $X = \log_a M$ and $Y = \log_a N$, and give reasons for the steps listed in Exercises 119–122.

119. Proof of P1 of Theorem 3.

$M = a^X$ and $N = a^Y$, _____

so $MN = a^X \cdot a^Y = a^{X+Y}$. _____

Thus, $\log_a(MN) = X + Y$ _____

$= \log_a M + \log_a N$. _____

120. Proof of P2 of Theorem 3.

$M = a^X$ and $N = a^Y$, _____

so $\dfrac{M}{N} = \dfrac{a^X}{a^Y} = a^{X-Y}$. _____

Thus, $\log_a \dfrac{M}{N} = X - Y$ _____

$= \log_a M - \log_a N$. _____

121. Proof of P3 of Theorem 3.

$M = a^X$, _____

so $M^k = (a^X)^k$ _____

$= a^{Xk}$. _____

Thus, $\log_a M^k = Xk$ _____

$= k \cdot \log_a M$. _____

122. Proof of P7 of Theorem 3.

Let $\log_b M = R$.

Then $b^R = M$, _____

and $\log_a(b^R) = \log_a M$. _____

Thus, $R \cdot \log_a b = \log_a M$, _____

and $R = \dfrac{\log_a M}{\log_a b}$. _____

It follows that

$\log_b M = \dfrac{\log_a M}{\log_a b}$. _____

123. Find $\lim\limits_{h \to 0} \dfrac{\ln(1 + h)}{h}$.

124. For any $k > 0$, $\ln(kx) = \ln k + \ln x$. Use this fact to show graphically why

$$\frac{d}{dx}\ln(kx) = \frac{d}{dx}\ln x = \frac{1}{x}.$$

125. Use natural logarithms to determine which is larger, 81^{81} or 9^{160}.

126. Explain why $\log_a 0$ is not defined. (*Hint:* Rewrite it as an equivalent exponential expression.)

TECHNOLOGY CONNECTION

Find the minimum value of each function in Exercises 127 and 128. Use a graphing calculator or Graphicus.

127. $f(x) = x \ln x$ **128.** $f(x) = x^2 \ln x$

129. Let $f(x) = \ln|x|$.

   a) Using a graphing utility, sketch the graph of f.
   b) Find the slopes of the tangent lines at $x = -3$, $x = -2$ and $x = -1$.

   c) How do your answers for part (b) compare to the slopes of the tangent lines at $x = 3$, $x = 2$ and $x = 1$?

   d) In your own words, explain how the Chain Rule can be used to show that $\dfrac{d}{dx}\ln(-x) = \dfrac{1}{x}$.

Answers to Quick Checks

1. (a) 4; **(b)** 36; **(c)** 1   **2. (a)** 1.183; **(b)** -0.471; **(c)** 0.471; **(d)** 1.424; **(e)** 1.827; **(f)** $\frac{1}{2}$ **3. (a)** 2.3025; **(b)** 0.9163; **(c)** -0.9163; **(d)** 3.4655; **(e)** 3.6094; **(f)** 0.4307   **4. (a)** $t \approx 4.3820$; **(b)** $t \approx 17.329$

5. (a) $\dfrac{5}{x}$; **(b)** $x^2 + 3x^2 \ln x + 4$; **(c)** $\dfrac{1 - 2\ln|4x|}{x^3}$

6. (a) $\dfrac{1}{x}$; **(b)** $\dfrac{6x}{3x^2 + 4}$; **(c)** $\dfrac{1}{x(\ln 5x)}$; **(d)** $\dfrac{4x^5 + 2}{x(x^5 - 2)}$

7. The advertising campaign should run for about 51 days to maximize profit.

Applications: Uninhibited and Limited Growth Models

3.3

- Find functions that satisfy $dP/dt = kP$.
- Convert between growth rate and doubling time.
- Solve application problems using exponential growth and limited growth models.

Quick Check 1 ✔

Differentiate $f(x) = 5e^{4x}$. Then express $f'(x)$ in terms of $f(x)$.

Exponential Growth

Consider the function given by

$$f(x) = 2e^{3x}.$$

Differentiating, we get

$$f'(x) = 2e^{3x} \cdot 3$$
$$= f(x) \cdot 3.$$

Graphically, this says that the derivative, or slope of the tangent line, is 3 times the function value.   **1** ✔

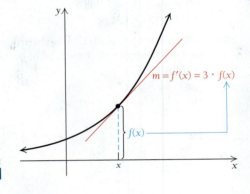

Although we do not prove it here, the exponential function $f(x) = ce^{kx}$ is the only function for which the derivative is a constant times the function itself.

THEOREM 8

A function $y = f(x)$ satisfies

$$\frac{dy}{dx} = ky \qquad \text{or} \qquad f'(x) = k \cdot f(x)$$

if and only if

$$y = ce^{kx} \qquad \text{or} \qquad f(x) = ce^{kx}$$

for some constant c.

EXAMPLE 1 Find the general form of the function that satisfies each equation.

a) $\dfrac{dA}{dt} = 5A$　　　　　**b)** $\dfrac{dP}{dt} = kP$

Solution

a) The desired function is given by $A(t) = ce^{5t}$, where c is an arbitrary constant. As a check, note that

$$A'(t) = ce^{5t} \cdot 5 = 5 \cdot A(t).$$

Quick Check 2 ✔

Find the general form of the function that satisfies the equation

$$\frac{dN}{dt} = kN.$$

b) The desired function is given by $P(t) = ce^{kt}$, where c is an arbitrary constant. As a check, note that

$$\frac{dP}{dt} = ce^{kt} \cdot k = kP.$$

2 ✔

Whereas the solution of an algebraic equation is a number, the solutions of the equations in Example 1 are functions. For example, the solution of $2x + 5 = 11$ is the number 3, and the solution of the equation $dP/dt = kP$ is the function $P(t) = ce^{kt}$. An equation like $dP/dt = kP$, which includes a derivative and which has a function as a solution, is called a *differential equation*.

EXAMPLE 2 Solve the differential equation

$$f'(z) = k \cdot f(z).$$

Solution The solution is $f(z) = ce^{kz}$.　*Check:* $f'(z) = ce^{kz} \cdot k = f(z) \cdot k.$ ■

We will discuss differential equations in more depth in Section 5.7.

Uninhibited Population Growth

The equation

$$\frac{dP}{dt} = kP \quad \text{or} \quad P'(t) = kP(t), \quad \text{with } k > 0,$$

is the basic model of uninhibited (unrestrained) population growth, whether the population is comprised of humans, bacteria in a culture, or dollars invested with interest compounded continuously. In the absence of inhibiting or stimulating factors, a population normally reproduces at a rate proportional to its size, and this is exactly what $dP/dt = kP$ says. The only function that satisfies this differential equation is given by

$$P(t) = ce^{kt},$$

where t is time and k is the rate expressed in decimal notation. Note that

$$P(0) = ce^{k \cdot 0} = ce^0 = c \cdot 1 = c,$$

so c represents the initial population, which we denote as P_0:

$$P(t) = P_0 e^{kt}.$$

The graph of $P(t) = P_0e^{kt}$, for $k > 0$, shows how uninhibited growth produces a "population explosion."

The constant k is the **rate of exponential growth**, or the **exponential growth rate**. This is *not* the instantaneous rate of change of the population size, which varies according to

$$\frac{dP}{dt} = kP,$$

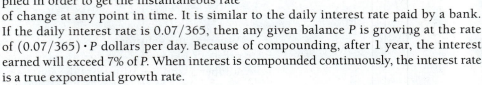

but the constant by which P must be multiplied in order to get the instantaneous rate of change at any point in time. It is similar to the daily interest rate paid by a bank. If the daily interest rate is $0.07/365$, then any given balance P is growing at the rate of $(0.07/365) \cdot P$ dollars per day. Because of compounding, after 1 year, the interest earned will exceed 7% of P. When interest is compounded continuously, the interest rate is a true exponential growth rate.

EXAMPLE 3   **Business: Interest Compounded Continuously.**   Suppose P_0, in dollars, is invested in the Von Neumann Hi-Yield Fund, with interest compounded continuously at 7% per year. That is, the balance P grows at the rate

$$\frac{dP}{dt} = 0.07P.$$

a) Find the function that satisfies the equation. Write it in terms of P_0 and 0.07.

b) Suppose that $100 is invested. What is the balance after 1 yr?

c) In what period of time will an investment of $100 double itself?

Solution

a) $P(t) = P_0e^{0.07t}$ Note that $P(0) = P_0$.

b) $P(1) = 100e^{0.07(1)} = 100e^{0.07}$

$\approx \$107.25$ Using a calculator and rounding

c) We are looking for a number T such that $P(T) = \$200$. The number T is called the **doubling time**. To find T, we solve

$$200 = 100e^{0.07 \cdot T}$$
$$2 = e^{0.07T}$$
$$\ln 2 = \ln e^{0.07T} \qquad \text{Finding the natural logarithm of both sides}$$
$$\ln 2 = 0.07T \qquad \text{Using Property P5 of logarithms: } \ln e^k = k$$
$$\frac{\ln 2}{0.07} = T$$
$$9.9 \approx T.$$

Quick Check 3 ✔

Business: Interest Compounded Continuously. Repeat Example 3 for interest compounded continuously at 4% per year.

Thus, at 7% compounded continuously, $100 will double in approximately 9.9 yr.

3 ✔

To find a general expression relating the exponential growth rate k and the doubling time T, we solve the following:

$$2P_0 = P_0e^{kT}$$
$$2 = e^{kT} \qquad \text{Dividing by } P_0$$
$$\ln 2 = \ln e^{kT} \qquad \text{Finding the natural logarithm of both sides}$$
$$\ln 2 = kT.$$

Note that this relationship between k and T does not depend on P_0. It takes as long for $1 to double as for $1000. We have the following theorem.

> **THEOREM 9**
>
> The *exponential growth rate* k and the *doubling time* T are related by
>
> $$kT = \ln 2 \approx 0.693147, \qquad \text{or} \qquad k = \frac{\ln 2}{T} \approx \frac{0.693147}{T},$$
>
> and $\qquad T = \dfrac{\ln 2}{k} \approx \dfrac{0.693147}{k}.$

Quick Check 4 ✔

Business: Internet Use. Worldwide use of the Internet is increasing at an exponential rate, with traffic doubling every 100 days. What is the exponential growth rate of Internet use?

EXAMPLE 4 **Business: Facebook Membership.** The social-networking Web site Facebook connects people with other members they designate as friends. During its period of heaviest growth, membership in Facebook was doubling every 6 months. What was the exponential growth rate of Facebook membership, as a percentage?

Solution We have

$$k = \frac{\ln 2}{T} \approx \frac{0.693147}{6 \text{ months}} \qquad \text{\small\color{red}If possible, enter the calculation as (ln 2)/6 without approximating the logarithmic value.}$$

$$\approx 0.116 \cdot \frac{1}{\text{month}}.$$

The exponential growth rate of Facebook membership was 11.6% per month. **4** ✔

EXAMPLE 5 **Physical Science: Integrated Circuits.** In 1971, Intel Corporation released the Intel 4004 microprocessor, which held 2300 transistors. Since then, the number of transistors on a microprocessor has been growing exponentially at the rate of 0.34, or 34%, per year. (How was this estimate determined? The answer is in the model we develop in the following Technology Connection.) That is,

$$\frac{dP}{dt} = 0.34P,$$

where t is the number of years since 1971. (*Sources*: www.intel.com; www.motorola.com; www.microsoft.com.)

Quick Check 5 ✔

Life Science: Population Growth in China. In 2006, the population of China was 1.314 billion, and the exponential growth rate was 0.6% per year. Thus,

$$\frac{dP}{dt} = 0.006P,$$

where t is the time, in years, after 2006. Assume uninhibited growth. (*Source: Time Almanac*, 2007.)

a) Find the function that satisfies the equation. Assume $P_0 = 1.314$ and $k = 0.006$.

b) Estimate the population of China at the beginning of 2020.

c) After what period of time will the population be double that in 2006?

a) Find the function that satisfies this equation. Assume $P_0 = 2300$ and $k = 0.34$.

b) Estimate the number of transistors on a microprocessor in 2020.

c) Find the time needed for the number of transistors on a microprocessor to double.

Solution

a) $P(t) = 2300e^{0.34t}$

b) Since 2020 is 49 years after 1971, we have $P(49) = 2300e^{0.34(49)} \approx$ 39.5 billion.

c) $T = \dfrac{\ln 2}{k} = \dfrac{\ln 2}{0.34} \approx 2.04$ yr. The fact that the number of transistors doubles every 2 years is known as Moore's Law, after Intel co-founder Gordon Moore, who foresaw this growth back in 1965. **5** ✔

TECHNOLOGY CONNECTION

Exponential Models Using Regression

Transistors on Microprocessors

The table below shows data regarding the number of transistors on a microprocessors.

YEAR	NUMBER OF TRANSISTORS
1971 (Intel 4004)	2300
1979 (Intel 8088)	29,000
1984 (Motorola 68020)	200,000
1993 (Intel Pentium I)	3,100,000
2000 (Intel Pentium IV)	42,000,000
2010 (Intel Itanium Tukwila)	2,000,000,000
2013 (Microsoft Xbox One)	5,000,000,000

What is the projected number of transistors on a microprocessor in 2020? The graph of this data shows a rapidly growing transistor count that can be modeled with an exponential function. We carry out the regression procedure exactly as we did in Section R.6, but select ExpReg rather than LinReg.

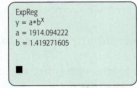

Note that this gives us an exponential model of the type $y = a \cdot b^x$, where y is the number of transistors x years after 1971:

$$y = 1914.094222 \cdot 1.419271605^x.$$

Since $b = e^{\ln b}$, we can write this function as an exponential function, with base e:

$$\begin{aligned} y &= 1914.094222 \cdot 1.419271605^x \\ &= 1914.094222 \left(e^{\ln 1.419271605}\right)^x \\ &= 1914.094222 e^{0.3501437858x} \end{aligned}$$

Thus, in 2020 ($t = 49$), the number of transistors on a microprocessor is estimated to be

$$y = 1914.094222 e^{0.3501437858(49)} = 54.1 \text{ billion.}$$

EXERCISES

Estimate the number of transistors on a microprocessor in each of the following years.

1. 2025 **2.** 2030

3. 2050 **4.** 2100

5. Explain why this model differs slightly from the one used in Example 5. Do both models accurately estimate the number of transistors on a microprocessor over the years 1971–2020?

Projecting College Costs

For Exercises 6 and 7, use the data regarding projected college costs (tuition and room and board) listed in the table below.

SCHOOL YEAR, x YEARS SINCE 2004–2005	COSTS OF ATTENDING A PUBLIC 4-YEAR COLLEGE OR UNIVERSITY (2009–2010 dollars)
2004–2005, 0	12,918
2005–2006, 1	13,188
2006–2007, 2	13,587
2007–2008, 3	13,748
2008–2009, 4	14,400
2009–2010, 5	15,014
2010–2011, 6	15,605

(*Source*: National Center for Education Statistics, *Annual Digest of Education Statistics*: 2014.)

EXERCISES

6. Use REGRESSION to fit an exponential function $y = a \cdot b^x$ to the data. Let the school year 2004–2005 be represented by $x = 0$ and let $y =$ the cost, in dollars.

7. Use the exponential function found in Exercise 6 to estimate college costs in the school years 2016–2017, 2017–2018, and 2020–2021.

It is possible to use two representative data points to determine P_0 and k in $P(t) = P_0 e^{kt}$, as we will see next.

EXAMPLE 6 **Business: Batman Comic Book.** A 1939 comic book with the first appearance of the "Caped Crusader," Batman, sold at auction in Dallas in 2010 for a record $1.075 million. The comic book originally cost 10¢ (or $0.10). Using the two data points (0, $0.10) and (71, $1,075,000), we can model the increasing value of the comic book. The modeling assumption is that the value V of the comic book grows exponentially, as given by

$$\frac{dV}{dt} = kV.$$

(*Source*: Heritage Auction Galleries.)

a) Find the function that satisfies this equation. Assume $V_0 = \$0.10$.

b) Estimate the value of the comic book in 2020.

c) What is the doubling time for the value of the comic book?

d) In what year will the value of the comic book be $30 million, assuming there is no change in the growth rate?

Solution

a) Because of the modeling assumption, we have $V(t) = V_0 e^{kt}$. Since $V_0 = \$0.10$, it follows that

$$V(t) = 0.10 e^{kt}.$$

We have made use of the data point (0, $0.10). Next, we use the data point (71, $1,075,000) to determine k. We solve

$$V(t) = 0.10 e^{kt}, \quad \text{or} \quad 1{,}075{,}000 = 0.10^{k(71)}$$

for k, using natural logarithms:

$$1{,}075{,}000 = 0.10 e^{k(71)} = 0.10 e^{71k}$$

$$\frac{1{,}075{,}000}{0.10} = e^{71k} \qquad \text{\color{red}Dividing to simplify}$$

$$10{,}750{,}000 = e^{71k}$$

$$\ln 10{,}750{,}000 = \ln e^{71k} \qquad \text{\color{red}Finding the logarithm of both sides}$$

$$\ln 10{,}750{,}000 = 71k \qquad \text{\color{red}Using Property P5}$$

$$\frac{\ln(10{,}750{,}000)}{71} = k$$

$$0.228 \approx k. \qquad \text{\color{red}Using a calculator and rounding to the nearest thousandth}$$

The desired function is $V(t) = 0.10 e^{0.228t}$, where V is in dollars and t is the number of years since 1939.

b) To estimate the value of the comic book in 2020, which is $2020 - 1939 = 81$ years after 1939, we substitute 81 for t in the equation:

$$V(t) = 0.10 e^{0.228t}$$

$$V(81) = 0.10 e^{0.228(81)} \approx \$10{,}484{,}567.$$

c) The doubling time T is given by

$$T = \frac{\ln 2}{k} = \frac{\ln 2}{0.228} \approx 3.04 \text{ yr.}$$

Quick Check 6 ✔

Business: Batman Comic Book.
In Example 6, the consigner had bought the comic book in the late 1960s for $100. (*Source*: Heritage Auction Galleries.) Assume that the year of purchase was 1969 and that the value V of the comic book has since grown exponentially, as given by

$$\frac{dV}{dt} = kV,$$

where t is the number of years since 1969.

a) Use the data points $(0, \$100)$ and $(41, \$1,075,000)$ to find the function that satisfies the equation.

b) Estimate the value of the comic book in 2020, and compare your answer to that of Example 6.

c) What is the doubling time for the value of the comic book?

d) In what year will the value of the comic book be $30 million? Compare your answer to that of Example 6.

d) We substitute $30,000,000 for $V(t)$ and solve for t:

$$V(t) = 0.10e^{0.228t}$$
$$30,000,000 = 0.10e^{0.228t}$$
$$\frac{30,000,000}{0.10} = e^{0.228t}$$
$$300,000,000 = e^{0.228t}$$
$$\ln 300,000,000 = \ln e^{0.228t}$$
$$\ln 300,000,000 = 0.228t$$
$$\frac{\ln 300,000,000}{0.228} = t$$
$$86 \approx t. \qquad \text{Using a calculator and rounding to the nearest year}$$

We add 86 to 1939 to get 2025 as the year in which the value of the comic book will reach $30 million.   **6** ✔

Models of Limited Growth

We have seen that the growth model $P(t) = P_0 e^{kt}$ applies to *unlimited*, or *unrestricted*, population growth. However, there are often factors that prevent a population from exceeding some limiting value L—perhaps a limitation on food, living space, or other natural resources. One model of such growth is

$$P(t) = \frac{L}{1 + be^{-kt}}, \quad \text{for} \quad k > 0,$$

which is called the *logistic equation,* or *logistic function.*

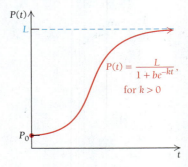

$$P(t) = \frac{L}{1 + be^{-kt}},$$
for $k > 0$

EXAMPLE 7   **Business: Satellite Radio Subscribers.**   Satellite radio companies provide subscribers with clear signals of over 100 radio stations. The providers XM and Sirius at first experienced what seemed like exponential growth, but the slowing of this growth led Sirius to buy out XM in 2008, forming Sirius XM. The combined number of subscribers N, in millions, t years after 2000, can be modeled by the logistic equation

$$N(t) = \frac{24.691}{1 + 34.348e^{-0.585t}}.$$

(*Source*: Sirius XM Radio, Inc.)

a) Find the number of subscribers in 2010 and 2013.

b) Find the rate at which the number of subscribers was growing in 2010.

c) Graph the equation.

d) Explain why an uninhibited growth model is inappropriate but a logistic equation is appropriate to model this growth.

Solution

a) We use a calculator to find the function values:

$$N(10) = 22.468 \text{ million and } N(13) = 24.276 \text{ million.}$$

After 10 yr, there were about 22,468,000 subscribers.
After 13 yr, there were about 24,276,000 subscribers.

b) We find the rate of change using the Quotient Rule:

$$N'(t) = \frac{(1 + 34.348e^{-0.585t}) \cdot 0 - 24.691(34.348e^{-0.585t})(-0.585)}{(1 + 34.348e^{-0.585t})^2}$$

$$= \frac{496e^{-0.585t}}{(1 + 34.348e^{-0.585t})^2}.$$

Next, we use a calculator to evaluate the derivative at $t = 10$:

$$N'(10) = 1.183.$$

In 2010, the number of subscribers was growing at a rate of 1.183 million, or 1,183,000, per year.

c) Using a graphing utility, we obtain the following graph.

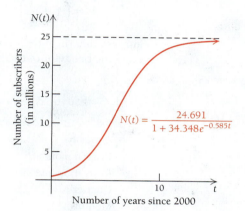

Quick Check 7 ✔

Life Science: Spread of an Epidemic. In a town whose population is 3500, an epidemic of a disease occurs. The number of people N infected t days after the disease first appears is given by

$$N(t) = \frac{3500}{1 + 19.9e^{-0.6t}}.$$

a) How many people are initially infected with the disease $(t = 0)$?

b) Find the number infected after 2 days, 8 days, and 18 days.

c) Graph the equation.

d) Find the rate at which the disease is spreading after 16 days.

e) Will all 3500 residents ever be infected? Why or why not?

d) An uninhibited growth model is inappropriate because as more and more subscribers are added, the population contains fewer who have not subscribed, perhaps because of the cost or a lack of interest, or simply because the population is finite. The logistic equation, graphed in part (c), displays the rapid early rise in the number of subscribers as well as the slower growth in later years. It would appear that the limiting value of subscribers is between 24 and 25 million. **7 ✔**

Another model of limited growth is provided by

$$P(t) = L(1 - e^{-kt}), \text{ for } k > 0,$$

which is graphed below. This function increases over its domain, $[0, \infty)$, but increases most rapidly at the beginning, unlike a logistic function.

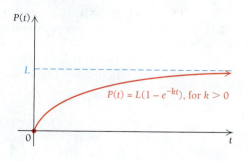

Business Application: An Alternative Derivation of e and $P(t) = P_0 e^{kt}$

The number e can also be found using the *compound-interest formula* (which was developed in Chapter R),

$$A = P\left(1 + \frac{r}{n}\right)^{nt},$$

where A is the amount that an initial investment P will be worth after t years at interest rate r, expressed as a decimal, compounded n times per year.

Suppose that $1 is invested at 100% interest ($r = 100\% = 1$) for 1 yr (though obviously no bank would pay this). The formula becomes

$$A = \left(1 + \frac{1}{n}\right)^n.$$

Suppose that the number of compounding periods, n, increases indefinitely. Let's investigate the behavior of the function. We obtain the following table of values and graph.

COMPOUNDING PERIOD	FREQUENCY, n	$A = \left(1 + \dfrac{1}{n}\right)^n$
Annually	1	$2.00000
Every 6 months	2	$2.25000
Every 4 months	3	$2.37037
Quarterly	4	$2.44141
Monthly	12	$2.61304
Weekly	52	$2.69260
Daily	365	$2.71457
Every hour	8,760	$2.71813
Every minute	525,600	$2.71828

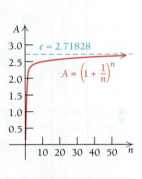

With *continuous* compounding, we have $A = \lim\limits_{n\to\infty}\left(1 + \dfrac{1}{n}\right)^n$, or equivalently,

$$A = \lim_{h\to\infty}\left(1 + \frac{1}{1/h}\right)^{1/h}, \quad \text{or} \quad A = \lim_{h\to 0}(1 + h)^{1/h}.$$

Recall from Section 3.1 that $\lim\limits_{h\to 0}(1 + h)^{1/h} = e$. Thus,

$$\lim_{n\to\infty}\left(1 + \frac{1}{n}\right)^n = e.$$

This result is confirmed by the graph and table above. If $1 were invested at an interest rate of 100% with increasingly frequent compounding periods, the greatest value it could grow to in 1 yr is about $2.7183.

To develop the formula

$$P(t) = P_0 e^{kt},$$

we again start with the compound-interest formula,

$$A = P\left(1 + \frac{r}{n}\right)^{nt},$$

and assume that interest is compounded continuously. Let $P = P_0$ and $r = k$:

$$P(t) = P_0\left(1 + \frac{k}{n}\right)^{nt}.$$

We are interested in what happens as n approaches ∞. To find this limit, we first let

$$\frac{k}{n} = \frac{1}{q}, \quad \text{so that} \quad qk = n.$$

Note that since k is a positive constant, as n gets large, so must q. Thus,

$$P(t) = \lim_{n\to\infty}\left[P_0\left(1 + \frac{k}{n}\right)^{nt}\right] \qquad \text{\color{red}Letting the number of compounding periods become infinite}$$

$$= P_0 \lim_{q\to\infty}\left[\left(1 + \frac{1}{q}\right)^{qkt}\right] \qquad \text{\color{red}Substituting } 1/q \text{ for } k/n \text{ and } qk \text{ for } n \text{ and } q\to\infty \text{ for } n\to\infty$$

$$= P_0\left[\lim_{q\to\infty}\left(1 + \frac{1}{q}\right)^{q}\right]^{kt} \qquad \text{\color{red}Using Limit Property L2 from Section 1.2}$$

$$= P_0[e]^{kt}.$$

Section Summary

- Uninhibited growth can be modeled by a *differential equation* of the form $\dfrac{dP}{dt} = kP$, which has the solution $P(t) = P_0 e^{kt}$.
- The *exponential growth rate* k and the *doubling time* T are related by the equation $T = \dfrac{\ln 2}{k}$, or $k = \dfrac{\ln 2}{T}$.

- Certain kinds of limited growth can be modeled by equations such as $P(t) = \dfrac{L}{1 + be^{-kt}}$ and $P(t) = L(1 - e^{-kt})$, for $k > 0$.

3.3 Exercise Set

1. Find the general form of f if $f'(x) = 4f(x)$.

2. Find the general form of g if $g'(x) = 6g(x)$.

3. Find the general form of the function that satisfies $dA/dt = -9A$.

4. Find the general form of the function that satisfies $dP/dt = -3P(t)$.

5. Find the general form of the function that satisfies $dQ/dt = kQ$.

6. Find the general form of the function that satisfies $dR/dt = kR$.

APPLICATIONS

Business and Economics

7. U.S. patents. The number of applications for patents, N, grew dramatically in recent years, with growth averaging about 5.8% per year. That is,

$$N'(t) = 0.058N(t).$$

(*Source:* Based on data from the U.S. Patent and Trademark office, 2009–2013.)

a) Find the function that satisfies this equation. Assume that $t = 0$ corresponds to 2009, when approximately 483,000 patent applications were received.

b) Estimate the number of patent applications in 2020.

c) Estimate the doubling time for $N(t)$.

8. Franchise expansion. Pete Zah's, Inc., is selling franchises for pizza shops throughout the country. The marketing manager estimates that the number of franchises, N, will increase at the rate of 10% per year, that is,

$$\frac{dN}{dt} = 0.10N.$$

a) Find the function that satisfies this equation. Assume that the number of franchises at $t = 0$ is 50.

b) How many franchises will there be in 20 yr?

c) In what period of time will the initial number of 50 franchises double?

9. Compound interest. If an amount P_0 is invested in the Mandelbrot Bond Fund and interest is compounded continuously at 5.9% per year, the balance P grows at the rate given by

$$\frac{dP}{dt} = 0.059P.$$

a) Find the function that satisfies the equation. Write it in terms of P_0 and 0.059.

b) Suppose $1000 is invested. What is the balance after 1 yr? After 2 yr?

c) When will an investment of $1000 double itself?

10. Compound interest. If an amount P_0 is invested in a savings account and interest is compounded continuously at 4.3% per year, the balance P grows at the rate given by

$$\frac{dP}{dt} = 0.043P.$$

a) Find the function that satisfies the equation. Write it in terms of P_0 and 0.043.

b) Suppose $20,000 is invested. What is the balance after 1 yr? After 2 yr?

c) When will an investment of $20,000 double itself?

11. Bottled water sales. Since 2000, sales of bottled water have increased at the rate of approximately 9.3% per year. That is, the volume of bottled water sold, G, in billions of gallons, t years after 2000 is growing at the rate given by

$$\frac{dG}{dt} = 0.093G.$$

(*Source: The Beverage Marketing Corporation.*)

a) Find the function that satisfies the equation, given that approximately 4.7 billion gallons of bottled water were sold in 2000.

b) Predict the number of gallons of water sold in 2025.

c) What is the doubling time for $G(t)$?

12. Annual net sales. Green Mountain Coffee Roasters produces many varieties of flavored coffees, teas, and K-cups. Since 2008, the net sales S of the company have grown exponentially at the rate of 46.3% per year. This growth can be approximated by

$$\frac{dS}{dt} = 0.463S,$$

where t is the number of years since 2008. (*Source: Green Mountain Coffee Roasters financial statements.*)

a) Find the function that satisfies the equation, given that net sales in 2008 ($t = 0$) were approximately $500,000.

b) Estimate net sales in 2012 and 2016.

c) What is the doubling time for $S(t)$?

13. Annual interest rate. Euler Bank advertises that it compounds interest continuously and that it will double your money in 15 yr. What is its annual interest rate?

14. Annual interest rate. Hardy Bank advertises that it compounds interest continuously and that it will double your money in 12 yr. What is its annual interest rate?

15. Oil demand. The growth rate of the demand for oil in the United States is 10% per year. When will the demand be double that of 2012?

16. Coal demand. The growth rate of the demand for coal in the world is 4% per year. When will the demand be double that of 2006?

Interest compounded continuously. *For Exercises 17–20, complete the following.*

Initial Investment at $t = 0$, P_0	Interest Rate, k	Doubling Time, T (in years)	Amount after 5 yr
17. $75,000	6.2%	____	_____
18. $5,000	____	____	$7,130.90
19. _____	8.4%	____	$11,414.71
20. _____	____	11	$17,539.32

21. Art masterpieces. In 2004, a collector paid $104,168,000 for Pablo Picasso's "Garcon à la Pipe." The same painting sold for $30,000 in 1950. (*Source: BBC News, 5/6/04.*)

Boy with a Pipe (1905), Pablo Picasso.
© 2011 Picasso Estate/ARS

a) Find the exponential growth rate k, to three decimal places, and determine the exponential growth function V, for which $V(t)$ is the painting's value, in dollars, t years after 1950.

b) Predict the value of the painting in 2015.

c) What is the doubling time for the value of the painting?

d) How long after 1950 will the value of the painting be $1 billion?

22. Per capita income. In 2009, U.S. per capita personal income *I* was $48,040. In 2012, it was $52,430. (*Source:* data.worldbank.org.) Assume that the growth of U.S. per capita personal income follows an exponential model.

 a) Letting $t = 0$ be 2009, write the function.
 b) Predict what U.S. per capita income will be in 2020.
 c) In what year will U.S. per capita income be double that of 2009?

23. Federal receipts. In 2011, U.S. federal receipts (money taken in) totaled $2.30 trillion. In 2013, total federal receipts were $2.77 trillion. (*Source*: usgovernmentrevenue.com.) Assume that the growth of total federal receipts, *F*, can be modeled by an exponential function and use 2011 as the base year ($t = 0$).

 a) Find the growth rate *k* to six decimal places, and write the exponential function $F(t)$, for total receipts in trillions of dollars.
 b) Estimate total federal receipts in 2015.
 c) When will total federal receipts be $10 trillion?

24. Consumer price index. The *consumer price index* compares the costs, *c*, of goods and services over various years, where 1983 is used as a base ($t = 0$). The same goods and services that cost $100 in 1983 cost $226 in 2012. (*Source*: Bureau of Labor Statistics.)

 a) Model *c* as an exponential function rounding the growth rate *k* to five decimal places.
 b) Estimate what the goods and services costing $100 in 1983 will cost in 2020.
 c) In what year did the same goods and services cost twice the 1983 price?

Total mobile data traffic. The following graph shows the predicted monthly mobile data traffic for the years 2013–2018. Use these data in Exercises 25 and 26.

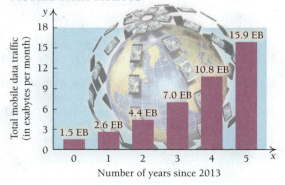

MOBILE DATA TRAFFIC

(*Source*: Cisco VNI Mobile, 2014.)

25. a) Use REGRESSION to fit an exponential function $y = a \cdot b^x$ to the data. Let *y* be the monthly mobile data traffic in exabytes (EB, 1 exabyte = 1 billion gigabytes) and *x* the number of years since 2013. Then convert the function to an exponential function, base *e*, using the fact that $b = e^{\ln b}$.

 b) What is the exponential growth rate, as a percentage?
 c) Estimate the monthly mobile data traffic in 2020.
 d) When will monthly mobile data traffic exceed 50 exabytes?
 e) What is the doubling time for monthly mobile data traffic?

26. a) Find an exponential function, base *e*, that fits the data, using the points (0, 1.5) and (2, 4.4). Let *x* represent the number of years since 2013.

 b) Estimate the total monthly mobile data traffic in 2019 and 2022.
 c) When will total monthly mobile data traffic exceed 50 exabytes?
 d) What is the doubling time for total monthly mobile traffic?
 e) Compare your answers with those from parts (c)–(e) of Exercise 25. Decide which exponential function seems to fit the data better, and explain why.

27. Value of Manhattan Island. Peter Minuit of the Dutch West India Company purchased Manhattan Island from the natives living there in 1626 for $24 worth of merchandise. Assuming an exponential rate of inflation of 5%, how much will Manhattan be worth in 2020?

28. Total revenue. Intel, a computer chip manufacturer, reported $1265 million in total revenue in 1986. In 2012, the total revenue was $53.3 billion. (*Source*: intel.com.) Assuming an exponential model, find the growth rate *k*, to four decimal places, and determine the revenue function *R*, with $R(t)$ in billions of dollars. Then predict the company's total revenue for 2020.

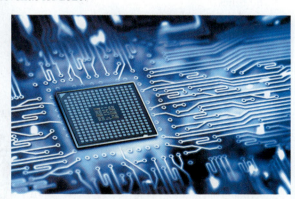

29. The U.S. Forever Stamp. The U.S. Postal Service sells the Forever Stamp, which is always valid as first-class postage on standard envelopes weighing 1 ounce or less, regardless of any subsequent increases in the first-class rate. (*Source*: U.S. Postal Service.)

 a) The cost of a first-class postage stamp was 4¢ in 1962 and 49¢ in 2014. This increase represents exponential growth. Write the function *S* for the cost of a stamp *t* years after 1962 ($t = 0$).
 b) What was the growth rate in the cost?
 c) Predict the cost of a first-class postage stamp in 2016, 2018, and 2020.

d) An advertising firm spent $4900 on 10,000 first-class postage stamps at the beginning of 2014. Knowing it will need 10,000 first-class stamps in each of the years 2015–2024, it decides to try to save money by also buying enough stamps to cover those years at the time of the 2014 purchase. Assuming there is a postage increase in each of the years 2016, 2018, and 2020 to the cost predicted in part (c), how much money will the firm save by buying the Forever Stamps in 2014?

e) Discuss the pros and cons of the purchase decision described in part (d).

30. Average salary of Major League baseball players. In 1970, the average salary of Major League baseball players was $29,303. In 2013, the average salary was $3,390,000. (*Source*: mlb.com.) Assuming exponential growth occurred, what was the growth rate to the nearest hundredth of a percent? What will the average salary be in 2020? In 2025? Round your answers to the nearest thousand.

31. Effect of advertising. Suppose that SpryBorg Inc. introduces a new computer game in Houston using television advertisements. Surveys show that $P\%$ of the target audience buys the game after x ads are broadcast, satisfying

$$P(x) = \frac{100}{1 + 49e^{-0.13x}}.$$

a) What percentage buys the game without seeing a TV ad ($x = 0$)?

b) What percentage buys the game after the ad is run 5 times? 10 times? 20 times? 30 times? 50 times? 60 times?

c) Find the rate of change, $P'(x)$.

d) Sketch a graph of the function.

32. Cost of a Hershey bar. The cost of a Hershey bar was $0.05 in 1962 and $0.99 in 2013 (in a supermarket, not in a movie theater).

a) Find an exponential function that fits the data.

b) Predict the cost of a Hershey bar in 2020 and 2025.

33. Superman comic book. In August 2014, a 1938 comic book featuring the first appearance of Superman sold at auction for a record price of $3.2 million. The comic book originally cost 10¢ ($0.10). (*Sources*: eBay; money.cnn.com.) Use the two data points (0, $0.10) and (76, $3,200,000), and assume that the value V of the comic book has grown exponentially, as given by

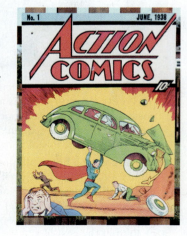

$$\frac{dV}{dt} = kV.$$

(In the summer of 2010, a family faced foreclosure on their mortgage. As they were packing, they came across some old comic books in the basement, and one of them was a copy of this first Superman comic. They sold it and saved their house.)

a) Find the function that satisfies this equation. Assume that $V_0 = \$0.10$.

b) Estimate the value of the comic book in 2020.

c) What is the doubling time for the value of the comic book?

d) After what time will the value of the comic book be $30 million, assuming there is no change in the growth rate?

34. Batman comic book. Refer to Example 6. In what year will the value of the comic book be $5 million?

35. Batman comic book. Refer to Example 6. In what year will the value of the comic book be $10 million?

Life and Physical Sciences

Population growth. *For Exercises 36–40, complete the following.*

	Population	Exponential Growth Rate, k	Doubling Time, T (in years)
36.	Mexico	3.5%/yr	_____
37.	Europe	_____	69.31
38.	Oil reserves		6.931
39.	Coal reserves	_____	17.3
40.	Alaska	2.794%/yr	_____

41. Bicentennial growth of the United States. The population of the United States in 1776 was about 2,508,000. In the country's bicentennial year, the population was about 216,000,000.

a) Assuming an exponential model, what was the growth rate of the United States through its bicentennial year?

b) Is exponential growth a reasonable assumption? Why or why not?

42. Limited population growth: human population. Seventeen adults came ashore from the British ship *HMS Bounty* in 1790 to settle on the uninhabited South Pacific island Pitcairn. The population, $P(t)$, of the island t years after 1790 can be approximated by the logistic equation

$$P(t) = \frac{3400}{17 + 183e^{-0.982t}}.$$

(*Source*: www.government.pn.)

a) Find the population of the island after 10 yr, 50 yr, and 75 yr.

b) Find the rate of change in the population, $P'(t)$.

c) Find the rate of change in the population after 10 yr, 50 yr, and 75 yr.

d) What is the limiting value for the population of Pitcairn? (The limiting value is the number to which the population gets closer and closer but never reaches.)

43. Limited population growth: tortoise population. The tortoise population, $P(t)$, in a square mile of the Mojave Desert after t years can be approximated by the logistic equation

$$P(t) = \frac{3000}{20 + 130e^{-0.214t}}$$

(*Source:* www.deserttortoise.org.)

a) Find the tortoise population after 0 yr, 5 yr, 15 yr, and 25 yr.

b) Find the rate of change in the population, $P'(t)$.

c) Find the rate of change in the population after 0 yr, 5 yr, 15, yr, and 25 yr.

d) What is the limiting value (see Exercise 42) for the population of tortoises in a square mile of the Mojave Desert?

44. Limited population growth. A lake is stocked with 400 rainbow trout. The size of the lake, the availability of food, and the number of other fish restrict population growth to a limiting value of 2500 trout. (See Exercise 42.) The population of trout in the lake after time t, in months, is approximated by

$$P(t) = \frac{2500}{1 + 5.25e^{-0.32t}}.$$

a) Find the population after 0 months, 1 month, 5 months, 10 months, 15 months, and 20 months.

b) Find the rate of change, $P'(t)$.

c) Sketch a graph of the function.

Social Sciences

45. Women college graduates. The number of women earning a bachelor's degree from a 4-yr college in the United States grew from 48,869 in 1930 to approximately 920,000 in 2010. (*Source:* National Center for Education Statistics.) Find an exponential function that fits the data, and the exponential growth rate, rounded to the nearest hundredth of a percent.

46. Hullian learning model. The Hullian learning model asserts that the probability p of mastering a task after t learning trials is approximated by

$$p(t) = 1 - e^{-kt},$$

where k is a constant that depends on the task to be learned. Suppose a new dance is taught to an aerobics class. For this particular dance, assume $k = 0.28$.

a) What is the probability of mastering the dance in 1 trial? 2 trials? 5 trials? 11 trials? 16 trials? 20 trials?

b) Find the rate of change, $p'(t)$.

c) Sketch a graph of the function.

47. Spread of infection. Spread by skin-to-skin contact or via shared towels or clothing, methicillin-resistant *Staphylococcus aureus* (MRSA) can easily infect growing numbers of students at a university. Left unchecked,

the number of cases of MRSA on a university campus t weeks after the first 9 cases occur can be modeled by

$$N(t) = \frac{568.803}{1 + 62.200e^{-0.092t}}.$$

(*Source:* Vermont Department of Health, Epidemiology Division.)

a) Find the number of infected students beyond the first 9 cases after 3 weeks, 40 weeks, and 80 weeks.

b) Find the rate at which the disease is spreading after 20 weeks.

c) Explain why an unrestricted growth model is inappropriate but a logistic equation *is* appropriate for this situation. Then use a calculator to graph the equation.

48. Diffusion of information. Pharmaceutical firms invest significantly in testing new medications. After a drug is approved by the Federal Drug Administration, it still takes time for physicians to fully accept and start prescribing it. The acceptance by physicians approaches a *limiting value* of 100%, or 1, after t months. Suppose that the percentage P of physicians prescribing a new cancer medication after t months is approximated by

$$P(t) = 100(1 - e^{-0.4t}).$$

a) What percentage of doctors are prescribing the medication after 0 months? 1 month? 2 months? 3 months? 5 months? 12 months? 16 months?

b) Find $P'(7)$, and interpret its meaning.

c) Sketch a graph of the function.

49. Spread of a rumor. The rumor "People who study math all get scholarships" spreads across a college campus. Data in the following table show the number of students N who have heard the rumor after time t, in days.

TIME, t (in days)	NUMBER, N, WHO HAVE HEARD THE RUMOR
1	1
2	2
3	4
4	7
5	12
6	18
7	24
8	26
9	28
10	28
11	29
12	30

a) Use REGRESSION to fit a logistic equation,

$$N(t) = \frac{c}{1 + ae^{-bt}},$$

to the data.

b) Estimate the limiting value of the function. At most, how many students will hear the rumor?

c) Graph the function.

d) Find the rate of change, $N'(t)$.

e) Find $\lim_{t \to \infty} N'(t)$, and explain its meaning.

SYNTHESIS

We have now studied models for linear, quadratic, exponential, and logistic growth. In the real world, understanding which is the most appropriate type of model for a given situation is an important skill. For each situation in Exercises 50–60, identify the most appropriate type of model and explain why you chose that model. List any restrictions you would place on the domain of the function.

50. The growth in value of a U.S. savings bond

51. The growth in the length of Zachary's hair following a haircut

52. The growth in sales of electric cars

53. The drop and rise of a lake's water level during and after a drought

54. The rapidly growing sales of organic foods

55. The number of manufacturing jobs that have left the United States since 1995

56. The life expectancy of the average American

57. The occupancy (number of apartments rented) of a newly opened apartment complex

58. The decrease in population of a city after its principal industry closes

59. The weight of a dog from birth to adulthood

60. The efficiency of an employee as a function of number of hours worked

61. Find an expression relating the exponential growth rate k and the *quadrupling time* T_4.

62. Find an expression relating the exponential growth rate k and the *tripling time* T_3.

63. Quantity Q_1 grows exponentially with a doubling time of 1 yr. Quantity Q_2 grows exponentially with a doubling time of 2 yr. If the initial amounts of Q_1 and Q_2 are the same, how long will it take for Q_1 to be twice the size of Q_2?

64. To what exponential growth rate per hour does a growth rate of 100% per day correspond?

65. Complete the table below, which relates growth rate k and doubling time T.

Growth Rate, k (per year)	1%	2%			14%
Doubling Time, T (in years)			15	10	

Graph $T = (\ln 2)/k$. Is this a linear relationship? Explain.

66. Describe the differences in the graphs of an exponential function and a logistic function.

The Rule of 70. *The relationship between doubling time T and growth rate k is the basis of the Rule of 70. Since*

$$T = \frac{\ln 2}{k} = \frac{0.693147}{k} = \frac{69.3147}{100k} \approx \frac{70}{100k},$$

we can estimate the length of time needed for a quantity to double by dividing the growth rate k (expressed as a percentage) into 70.

67. Estimate the time needed for an amount of money to double, if the interest rate is 7%, compounded continuously.

68. Estimate the time needed for the population in a city to double, if the growth rate is 3.5%, compounded continuously.

69. Using a calculator, find the exact doubling times for the amount of money in Exercise 67 and the population in Exercise 68.

70. Describe two situations where it would be preferable to use the Rule of 70 instead of the formula $T = \ln(2)/k$. Explain why it would be acceptable to use this rule in these situations.

71. Business: total revenue. The revenue of Red Rocks, Inc., in millions of dollars, is given by the function

$$R(t) = \frac{4000}{1 + 1999e^{-0.5t}},$$

where t is measured in years.

a) What is $R(0)$, and what does it represent?

b) Find $\lim_{t \to \infty} R(t)$. Call this value R_{max}, and explain what it means.

c) Find the value of t (to the nearest integer) for which $R(t) = 0.99R_{max}$.

Answers to Quick Checks

1. $f'(x) = 20e^{4x}; f'(x) = 4f(x)$ **2.** $N(t) = ce^{kt}$, where c is an arbitrary constant **3. (a)** $P(t) = P_0e^{0.04t}$;
(b) \$104.08; **(c)** 17.3 yr **4.** 0.69% per day
5. (a) $P(t) = 1.314e^{0.006t}$; **(b)** 1.429 billion; **(c)** 115.5 yr
6. (a) $V(t) = 100e^{0.226t}$, which is almost the same as the growth rate found in Example 6; **(b)** \$10,131,604, which is quite close to the estimate found in Example 6; **(c)** 3.07 yr;
(d) after 55.8 yr, or in 2025, again about the same as was found in Example 6 **7. (a)** 167; **(b)** 500, 3007, 3499
(c)

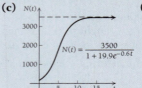

$$N(t) = \frac{3500}{1 + 19.9e^{-0.6t}}$$

(d) After 16 days, the number of people infected is growing at the rate of about 2.8 people per day.

(e) According to the model, $\lim_{t \to \infty} N(t) = 3500$, so virtually all of the town's residents will be affected.

3.4

- Find a function that satisfies $dP/dt = -kP$.
- Convert between decay rate and half-life.
- Solve applied problems involving exponential decay.

Exploratory

Using the same set of axes, graph $y_1 = e^{2x}$ and $y_2 = e^{-2x}$. Compare the graphs.

Then, using the same set of axes, graph $y_1 = 100e^{-0.06x}$ and $y_2 = 100e^{0.06x}$. Compare each pair of graphs.

Applications: Decay

In the equation of population growth, $dP/dt = kP$, the constant k is given by

$$k = (\text{Birth rate}) - (\text{Death rate}).$$

Thus, a population "grows" only when the birth rate is greater than the death rate. When the birth rate is less than the death rate, k is negative, and the population is decreasing, or "decaying," at a rate proportional to its size. For convenience in our computations, we will express such a negative value as $-k$, where $k > 0$. The equation

$$\frac{dP}{dt} = -kP, \quad \text{where } k > 0,$$

shows P to be *decreasing* as a function of time, and the solution

$$P(t) = P_0 e^{-kt}$$

shows P to be decreasing exponentially. This is called **exponential decay**. The amount present initially at $t = 0$ is again P_0.

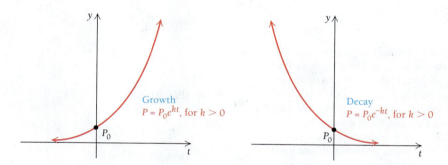

Radioactive Decay

Radioactive elements decay exponentially; that is, they disintegrate at a rate that is proportional to the amount present.

EXAMPLE 1 **Life Science: Decay.** Strontium-90 has a decay rate of 2.8% per year. The rate of change of an amount N of this radioactive isotope is given by

$$\frac{dN}{dt} = -0.028N.$$

a) Find the function that satisfies the equation. Let N_0 represent the amount present at $t = 0$.

b) Suppose that 1000 grams (g) of strontium-90 is present at $t = 0$. How much will remain after 70 yr?

c) After how long will exactly half of the 1000 g remain?

Solution

a) $N(t) = N_0 e^{-0.028t}$

b) $N(70) = 1000e^{-0.028(70)}$

$= 1000e^{-1.96}$

$\approx 140.8584209.$ Using a calculator

After 70 yr, about 140.9 g of the strontium-90 remains.

Quick Check 1 ✔

Life Science: Decay. Xenon-133 has a decay rate of 14% per day. The rate of change of an amount N of this radioactive isotope is given by

$$\frac{dN}{dt} = -0.14N.$$

a) Find the function that satisfies the equation. Let N_0 represent the amount present at $t = 0$.

b) Suppose 1000 g of xenon-133 is present at $t = 0$. How much will remain after 10 days?

c) After how long will half of the 1000 g remain?

How can scientists determine that the remains of an animal or plant have lost 30% of the carbon-14? The assumption is that the percentage of carbon-14 in the atmosphere and in living plants and animals is the same. When a plant or animal dies, its carbon-14 decays exponentially. A scientist burns the remains and uses a Geiger counter to determine the percentage of carbon-14 in the smoke. The amount by which this varies from the percentage in the atmosphere indicates how much carbon-14 was lost through decay.

The process of carbon-14 dating was developed by the American chemist Willard F. Libby in 1952. It is known that the radioactivity of a living plant measures 16 disintegrations per gram per minute. Since the half-life of carbon-14 is 5730 years, a dead plant with an activity of 8 disintegrations per gram per minute is 5730 years old, one with an activity of 4 disintegrations per gram per minute is about 11,500 years old, and so on. Carbon-14 dating can be used to determine the age of organic objects from 30,000 to 40,000 years old. Beyond such an age, it is too difficult to measure the radioactivity, and other methods are used.

c) We are asking, "At what time T will $N(T)$ be half of N_0, or $\frac{1}{2} \cdot 1000$?" The number T is called the **half-life** of strontium-90. To find T, we solve

$$500 = 1000e^{-0.028T} \qquad \text{We use 500 because } 500 = \tfrac{1}{2} \cdot 1000.$$
$$\tfrac{1}{2} = e^{-0.028T} \qquad \text{Dividing both sides by 1000}$$
$$\ln \tfrac{1}{2} = \ln e^{-0.028T} \qquad \text{Taking the natural logarithm of both sides}$$
$$\ln 1 - \ln 2 = -0.028T \qquad \text{Using the properties of logarithms}$$
$$0 - \ln 2 = -0.028T$$
$$\frac{-\ln 2}{-0.028} = T \qquad \text{Dividing both sides by } -0.028$$
$$\frac{\ln 2}{0.028} = T$$
$$25 \approx T. \qquad \text{Using a calculator}$$

Thus, the half-life of strontium-90 is about 25 yr.   1 ✔

To find a general expression relating decay rate k and half-life T, we solve

$$\tfrac{1}{2} P_0 = P_0 e^{-kT}$$
$$\tfrac{1}{2} = e^{-kT}$$
$$\ln \tfrac{1}{2} = \ln e^{-kT}$$
$$\ln 1 - \ln 2 = -kT$$
$$0 - \ln 2 = -kT$$
$$-\ln 2 = -kT$$
$$\ln 2 = kT.$$

Again, we have the following.

THEOREM 10

The *decay rate*, k, and the *half-life*, T, are related by

$$kT = \ln 2 \approx 0.693147,$$

or

$$k = \frac{\ln 2}{T} \quad \text{and} \quad T = \frac{\ln 2}{k}.$$

Thus, half-life, T, depends only on decay rate, k, and is independent of the initial population size.

The effect of half-life is shown in the radioactive decay curve at right. Note that the exponential function gets close to, but never reaches, 0 as t gets larger. Thus, in theory, a radioactive substance never completely decays.

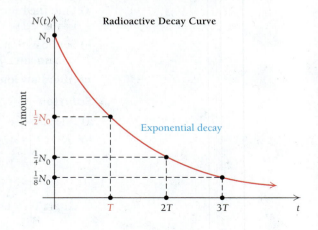

Radioactive Decay Curve

EXAMPLE 2 **Life Science: Half-life.** Plutonium-239, a common product of a functioning nuclear reactor, can be deadly to people exposed to it. Its decay rate is about 0.0028% per year. What is its half-life?

Solution We have

$$T = \frac{\ln 2}{k}$$

$$= \frac{\ln 2}{0.000028} \qquad \text{Converting 0.0028\% to decimal notation}$$

$$\approx 24{,}755.$$

Thus, the half-life of plutonium-239 is about 24,755 yr. 2 ✔

EXAMPLE 3 **Life Science: Carbon Dating.** The radioactive element carbon-14 has a half-life of 5730 yr. The percentage of carbon-14 present in the remains of plants and animals is used to determine age. Archaeologists found that the linen wrapping from one of the Dead Sea Scrolls had lost 22.3% of its carbon-14. How old was the linen wrapping?

Solution Our plan is to find an exponential equation of the form $N(t) = N_0 e^{-kt}$, replace $N(t)$ with $(1 - 0.223)N_0$, and solve for t. First, however, we find the decay rate, k:

$$k = \frac{\ln 2}{T} \approx \frac{0.693147}{5730} \approx 0.00012097, \quad \text{or} \quad 0.012097\% \text{ per year.}$$

Thus, the amount $N(t)$ that remains from an initial amount N_0 after t years is given by:

$$N(t) = N_0 e^{-0.00012097t}. \qquad \text{Remember: } k \text{ is positive, so } -k \text{ is negative.}$$

(*Note:* This equation can be used for all subsequent carbon-dating problems.)
If the linen wrapping of a Dead Sea Scroll lost 22.3% of its carbon-14 from an initial amount N_0, then 77.7% · N_0 remains. To find the age t of the wrapping, we solve the following equation for t:

$$77.7\% \ N_0 = N_0 e^{-0.00012097t}$$
$$0.777 = e^{-0.00012097t}$$
$$\ln 0.777 = \ln e^{-0.00012097t}$$
$$\ln 0.777 = -0.00012097t$$
$$\frac{\ln 0.777}{-0.00012097} = t$$
$$2086 \approx t.$$

Thus, the linen wrapping of the Dead Sea Scroll is about 2086 yr old. 3 ✔

A Business Application: Present Value

Bankers and financial planners are often asked about problems like the following.

EXAMPLE 4 **Business: Present Value.** Following the birth of their granddaughter, two grandparents want to make an initial investment, P_0, that will grow to $10,000 by the child's 20th birthday. Interest is compounded continuously at 4%. What should the initial investment be?

Quick Check 2 ✔

Life Science: Half-life.
a) The decay rate of cesium-137 is 2.3% per year. What is its half-life?
b) The half-life of barium-140 is 13 days. What is its decay rate?

In 1947, a Bedouin youth looking for a stray goat climbed into a cave at Kirbet Qumran on the shores of the Dead Sea near Jericho and came upon earthenware jars containing an incalculable treasure of ancient manuscripts, which concern the Jewish books of the Bible. Shown here are fragments of those so-called Dead Sea Scrolls, a portion of some 600 or so texts found so far. Officials date them before A.D. 70, making them the oldest biblical manuscripts by 1000 years.

Quick Check 3 ✔

Life Science: Carbon Dating.
How old is a skeleton found at an archaeological site if tests show that it has lost 60% of its carbon-14?

Solution   Using $P = P_0 e^{kt}$, we find P_0 such that

$$10{,}000 = P_0 e^{0.04 \cdot 20},$$

or

$$10{,}000 = P_0 e^{0.8}.$$

Now

$$\frac{10{,}000}{e^{0.8}} = P_0,$$

or

$$10{,}000 e^{-0.8} = P_0,$$

and, using a calculator, we have

$$P_0 = 10{,}000 e^{-0.8}$$
$$\approx 4493.29.$$

Thus, the grandparents should deposit $4493.29, which will grow to $10,000 by the child's 20th birthday.

Economists call $4493.29 the *present value* of $10,000 due 20 yr from now at 4%, compounded continuously. The process of computing present value is called **discounting**. Another way to pose this problem is to ask "What must I invest now, at 4%, compounded continuously, in order to have $10,000 in 20 years?" The answer is $4493.29, and it is the present value of $10,000 due in 20 yr.

Computing present value can be interpreted as exponential decay from the future back to the present.

Quick Check 4 ✔

Business: Present Value. Repeat Example 4 for an interest rate of 6%.

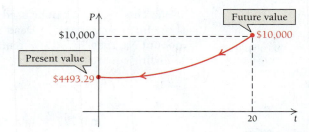

4 ✔

In general, the present value P_0 of an amount P due t years later is found by solving the following equation for P_0:

$$P_0 e^{kt} = P$$

$$P_0 = \frac{P}{e^{kt}} = P e^{-kt}.$$

THEOREM 11

The **present value** P_0 of an amount P due t years later, at interest rate k, compounded continuously, is given by

$$P_0 = P e^{-kt}.$$

Newton's Law of Cooling

Consider the following situation. A hot cup of soup, at a temperature of 200°, is placed in a 70° room.* The temperature of the soup decreases over time t, in minutes, according to the model known as **Newton's Law of Cooling**.

Newton's Law of Cooling

The temperature T of a cooling object drops at a rate that is proportional to the difference $T - C$, where C is the constant temperature of the surrounding medium. Thus,

$$\frac{dT}{dt} = -k(T - C). \qquad (1)$$

The function that satisfies equation (1) is

$$T = T(t) = ae^{-kt} + C. \qquad (2)$$

To check that $T(t) = ae^{-kt} + C$ is the solution, find dT/dt and substitute dT/dt and $T(t)$ into equation (1). This check is left to the student.

EXAMPLE 5 **Life Science: Scalding Coffee.** McDivett's Pie Shoppes, a national chain, finds that the temperature of its freshly brewed coffee is 130°. The company fears that if customers spill hot coffee on themselves, lawsuits might result. Room temperature in the restaurants is generally 72°. The temperature of the coffee cools to 120° after 4.3 min. McDivett's decides that it is safer to serve coffee at 105°. How long does it take a cup of coffee to cool to 105°?

Solution Note that C, the surrounding air temperature, is 72°. To find the value of a in equation (2) for Newton's Law of Cooling, we observe that at $t = 0$, we have $T(0) = 130°$. We solve for a as follows:

$$130 = ae^{-k \cdot 0} + 72$$
$$130 = a + 72$$
$$58 = a. \qquad \text{Noting that } a = 130 - 72, \text{ the difference between the original}$$
$$\text{temperatures}$$

Next, we find k using the fact that $T(4.3) = 120$:

$$120 = 58e^{-k \cdot (4.3)} + 72$$
$$48 = 58e^{-4.3k}$$
$$\frac{48}{58} = e^{-4.3k}$$
$$\ln \frac{48}{58} = \ln e^{-4.3k}$$
$$\ln \frac{48}{58} = -4.3k \qquad \text{Using Property P5 of logarithms}$$
$$\frac{\ln \dfrac{48}{58}}{-4.3} = k$$
$$k \approx 0.044. \qquad \text{Using a calculator}$$

---

*Assume throughout this section that all temperatures are in degrees Fahrenheit unless noted otherwise.

We now have $T(t) = 58e^{-0.044t} + 72$. To see how long it will take the coffee to cool to 105°, we set $T(t) = 105$ and solve for t:

$$105 = 58e^{-0.044t} + 72$$

$$33 = 58e^{-0.044t}$$

$$\frac{33}{58} = e^{-0.044t}$$

$$\ln \frac{33}{58} = \ln e^{-0.044t}$$

$$\ln \frac{33}{58} = -0.044t \qquad \text{Using Property P5 of logarithms}$$

$$\frac{\ln \frac{33}{58}}{-0.044} = t$$

$$t \approx 12.8 \text{ min.} \qquad \text{Using a calculator}$$

Thus, to cool to 105°, the coffee should be allowed to cool for about 13 min.   **5** ✔

Quick Check 5 ✔

Life Science: Scalding Coffee.
Repeat Example 5, but assume that the coffee is sold in an ice cream shop, where room temperature is 70°, and it cools to a temperature of 120° in 4 min.

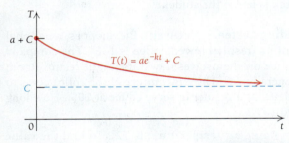

The graph of $T(t) = ae^{-kt} + C$ shows that $\lim_{t \to \infty} T(t) = C$. The temperature of the object decreases toward the temperature of the surrounding medium.

Mathematically, this model demonstrates that the object's temperature never quite reaches C. In practice, the temperature of the cooling object will get so close to that of the surrounding medium that no device could detect a difference. Let's now use Newton's Law of Cooling to solve a crime.

EXAMPLE 6   Forensics: When Was the Murder Committed?   A body is found slumped over a desk in a study. The coroner arrives at noon, immediately takes the temperature of the body, and finds it to be 94.6°. She waits 1 hr, takes the temperature again, and finds it to be 93.4°. She also notes that the temperature of the room is 70°. When was the murder committed?

Solution   Note that C, the surrounding air temperature, is 70°. To find a in $T(t) = ae^{-kt} + C$, we assume that the temperature of the body was normal when the murder occurred. Thus, $T = 98.6°$ at $t = 0$:

$$98.6 = ae^{-k \cdot 0} + 70,$$

$$a = 28.6.$$

This gives $T(t) = 28.6e^{-kt} + 70$.

To find the number of hours N since the murder was committed, we must first determine k. From the two temperature readings the coroner made, we have

$$94.6 = 28.6e^{-kN} + 70, \qquad \text{or} \quad 24.6 = 28.6e^{-kN}; \tag{3}$$

$$93.4 = 28.6e^{-k(N+1)} + 70, \qquad \text{or} \quad 23.4 = 28.6e^{-k(N+1)}. \tag{4}$$

Dividing equation (3) by equation (4), we get

$$\frac{24.6}{23.4} = \frac{28.6e^{-kN}}{28.6e^{-k(N+1)}}$$

$$= e^{-kN + k(N+1)}$$

$$= e^{-kN + kN + k} = e^k.$$

We solve this equation for k:

$$\ln \frac{24.6}{23.4} = \ln e^k \qquad \text{Taking the natural logarithm on both sides}$$

$$0.05 \approx k. \qquad \text{Using a calculator}$$

Next, we substitute $k \approx 0.05$ into equation (3) and solve for N:

$$24.6 = 28.6 e^{-0.05N}$$

$$\frac{24.6}{28.6} = e^{-0.05N}$$

$$\ln \frac{24.6}{28.6} = \ln e^{-0.05N}$$

$$\frac{\ln \dfrac{24.6}{28.6}}{-0.05} = N$$

$$3 \approx N. \qquad \text{Using a calculator}$$

Since the coroner arrived at noon, or 12 o'clock, the murder occurred at about 9:00 A.M.

6 ✔

Quick Check 6 ✔

Forensics. Repeat Example 6, assuming that the coroner arrives at 2 A.M., immediately takes the temperature of the body, and finds it to be 92.8°. She waits 1 hr, takes the temperature again, and finds it to be 90.6°. She also notes that the temperature of the room is 72°. When was the murder committed?

Section Summary

- The *decay rate*, k, and the *half-life*, T, are related by $kT = \ln 2$, or

$$k = \frac{\ln 2}{T} \quad \text{and} \quad T = \frac{\ln 2}{k}.$$

- The *present value* P_0 of an amount P due t years later, at an interest rate k, compounded continuously, is given by $P_0 = Pe^{-kt}$.

- According to *Newton's Law of Cooling*, the temperature T of a cooling object drops at a rate proportional to the difference $T - C$, when C is the surrounding room temperature. Thus, we have

$$\frac{dT}{dt} = -k(T - C), \quad \text{and} \quad T(t) = ae^{-kt} + C.$$

3.4 Exercise Set

In Exercises 1–8, find the half-life for each situation.

1. An element loses 12% of its mass every year.

2. An population of bacteria decreases by 5.75% per month.

3. A vehicle loses 0.8% of its value every month.

4. A city loses 3.9% of its population every year.

5. The value of a dollar decreases by 3% every year.

6. An element loses 1.75% of its mass every day.

7. An investment loses 1.9% of its value every week.

8. A motor home loses 0.5% of its value every week.

APPLICATIONS

Life and Physical Sciences

9. **Radioactive decay.** Iodine-131 has a decay rate of 9.6% per day. The rate of change of an amount N of iodine-131 is given by

$$\frac{dN}{dt} = -0.096N,$$

where t is the number of days since decay began.

a) Let N_0 represent the amount of iodine-131 present at $t = 0$. Find the exponential function that models the situation.

b) Suppose 500 g of iodine-131 is present at $t = 0$. How much will remain after 4 days?

c) After how many days will half of the 500 g of iodine-131 remain?

10. **Radioactive decay.** Carbon-14 has a decay rate of 0.012097% per year. The rate of change of an amount N of carbon-14 is given by

$$\frac{dN}{dt} = -0.00012097N,$$

where t is the number of years since decay began.

a) Let N_0 represent the amount of carbon-14 present at $t = 0$. Find the exponential function that models the situation.

b) Suppose 200 g of carbon-14 is present at $t = 0$. How much will remain after 800 yr?

c) After how many years will half of the 200 g of carbon-14 remain?

11. **Chemistry.** Substance A decomposes at a rate proportional to the amount of A present.

a) Write an equation that gives the amount A left of an initial amount A_0 after time t.

b) It is found that 10 lb of A will reduce to 5 lb in 3.3 hr. After how long will there be only 1 lb left?

12. **Chemistry.** Substance A decomposes at a rate proportional to the amount of A present.

a) Write an equation that gives the amount A left of an initial amount A_0 after time t.

b) It is found that 8 g of A will reduce to 4 g in 3 hr. After how long will there be only 1 g left?

Radioactive decay. *For Exercises 13–16, complete the following.*

Radioactive Substance	Decay Rate, k	Half-life, T
13. Polonium-218	_____	3 min
14. Radium-226	_____	1600 yr
15. Lead-210	3.15%/yr	_____
16. Strontium-90	2.77%/yr	_____

17. **Half-life.** Of an initial amount of 1000 g of lead-210, how much will remain after 100 yr? See Exercise 15 for the value of k.

18. **Half-life.** Of an initial amount of 1000 g of polonium-218, how much will remain after 20 min? See Exercise 13 for the value of k.

19. **Carbon dating.** How old is an ivory tusk that has lost 40% of its carbon-14?

20. **Carbon dating.** How old is a piece of wood that has lost 90% of its carbon-14?

21. **Cancer treatment.** Iodine-125 is often used to treat cancer and has a half-life of 60.1 days. In a sample, the amount of iodine-125 decreased by 25% while in storage. How long was the sample in storage?

22. **Carbon dating.** How old is a Chinese artifact that has lost 60% of its carbon-14?

23. **Carbon dating.** Recently, while digging in Chaco Canyon, New Mexico, archaeologists found corn pollen that had lost 38.1% of its carbon-14. The age of this corn pollen was evidence that Indians had been cultivating crops in the Southwest centuries earlier than previously thought. (*Source: American Anthropologist.*) What was the age of the pollen?

Chaco Canyon, New Mexico

Business and Economics

24. **Present value.** Following the birth of a child, a parent wants to make an initial investment P_0 that will grow to $30,000 by the child's 20th birthday. Interest is compounded continuously at 6%. What should the initial investment be?

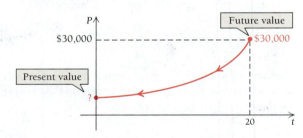

25. **Present value.** Following the birth of their child, the Irwins want to make an initial investment P_0 that will grow to $40,000 by the child's 20th birthday. Interest is compounded continuously at 5.3%. What should the initial investment be?

26. **Present value.** Desmond wants to have $15,000 available in 5 yr to pay for new siding. Interest is 4.3%, compounded continuously. How much money should be invested?

27. **Sports salaries.** An athlete signs a contract that guarantees a $9-million salary 6 yr from now. Assuming that money can be invested at 4.7%, with interest compounded continuously, what is the present value of that year's salary?

28. **Actors' salaries.** An actor signs a film contract that will pay $12 million when the film is completed 3 yr from now. Assuming that money can be invested at 4.2%, with interest compounded continuously, what is the present value of that payment?

29. **Estate planning.** Shannon has a trust fund that will yield $80,000 in 13 yr. A CPA is preparing a financial statement for Shannon and wants to take into account the present value of the trust fund in computing her net

worth. Interest is compounded continuously at 4.8%. What is the present value of the trust fund?

30. Supply and demand. The supply and demand for stereos produced by Blaster Sound, Inc., are given by

$$S(x) = \ln x \quad \text{and} \quad D(x) = \ln \frac{163,000}{x},$$

where $S(x)$ is the number of stereos that the company is willing to sell for x dollars each and $D(x)$ is the quantity that the public is willing to buy at price x. Find the equilibrium point. (See Section R.5.)

31. Salvage value. Lucas Mining estimates that the salvage value $V(t)$, in dollars, of a piece of machinery after t years is given by

$$V(t) = 40,000e^{-t}.$$

a) What did the machinery cost initially?
b) What is the salvage value after 2 yr?
c) Find the rate of change of the salvage value, and explain its meaning.

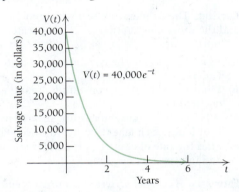

32. Salvage value. Wills Investments tracks the value of a particular photocopier over a period of years. The data in the table below show the value of the copier at time t, in years, after the date of purchase.

Time, t (in years)	Salvage Value
0	$34,000
1	22,791
2	15,277
3	10,241
4	6,865
5	4,600
6	3,084

(*Source*: International Data Corporation.)

a) Use REGRESSION to fit an exponential function $y = a \cdot b^x$ to the data. Then convert that formula to $V(t) = V_0e^{-kt}$, where V_0 is the value when the copier is purchased and t is the time, in years, from the date of purchase. (See the Technology Connection on p. 326.)
b) Estimate the salvage value of the copier after 7 yr; 10 yr.

c) After what amount of time will the salvage value be $1000?
d) After how long will the copier be worth half of its original value?
e) Find the rate of change of the salvage value, and interpret its meaning.

33. Actuarial science. An actuary works for an insurance company and calculates insurance premiums. Given an actual mortality rate (probability of death) for a given age, actuaries sometimes need to project future expected mortality rates of people of that age. For example,

$$Q(t) = (Q_0 - 0.00055)e^{0.163t} + 0.00055,$$

where t is the number of years into the future and Q_0 is the mortality rate when $t = 0$, projects future mortality rates.

a) Suppose the actual mortality rate of a group of females aged 25 is 0.014 (14 deaths per 1000). What is the future expected mortality rate of this group of females 3, 5, and 10 yr in the future?
b) Sketch the graph of the mortality function $Q(t)$ for the group in part (a) for $0 \le t \le 10$.

34. Actuarial science. Use the formula from Exercise 33.

a) Suppose the actual mortality rate of a group of males aged 25 is 0.023 (23 deaths per 1000). What is the future expected mortality rate of this group of males 3, 5, and 10 yr in the future?
b) Sketch the graph of the mortality function $Q(t)$ for the group in part (a) for $0 \le t \le 10$.
c) What is the ratio of the mortality rate for 25-year-old males 10 yr in the future to that for 25-year-old females 10 yr in the future (Exercise 33a)?

35. U.S. farms. The number N of farms in the United States has declined continually since 1950. In 1950, there were 5,650,000 farms, and in 2012, that number had decreased to 2,170,000. (*Sources*: U.S. Department of Agriculture; National Agricultural Statistics Service.)

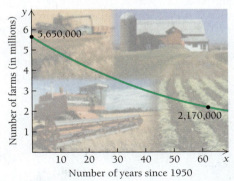

Assuming the number of farms decreased according to the exponential decay model:

a) Find the value of k, and write an exponential function that describes the number of farms after time t, where t is the number of years since 1950.
b) Estimate the number of farms in 2016 and in 2020.
c) At this decay rate, when will only 1,000,000 farms remain?

Social Sciences

36. Forgetting. In an art history class, students took a test. They were subsequently retested with an equivalent test at weekly intervals. Their average retest scores t weeks later are given in the following table.

Time, t (in weeks)	Score, y
1	84.9%
2	84.6%
3	84.4%
4	84.2%
5	84.1%
6	83.9%

a) Use REGRESSION to fit a logarithmic function $y = a + b \ln x$ to the data.
b) Use the function to predict the average test score after 8 weeks, 10 weeks, 24 weeks, and 36 weeks.
c) After how long will the test scores fall below 82%?
d) Find the rate of change of the scores, and interpret its meaning.

37. Decline in beef consumption. Annual consumption of beef per person was about 64.6 lb in 2000 and about 61.2 lb in 2008. Assuming that $B(t)$, the annual beef consumption t years after 2000, is decreasing according to the exponential decay model:

a) Find the value of k, and write the equation.
b) Estimate the consumption of beef in 2015.
c) In what year (theoretically) will the consumption of beef be 20 lb per person?

38. Population decrease of Russia. The population of Russia dropped from 150 million in 1995 to 142.5 million in 2013. (*Source: CIA–The World Factbook.*) Assume that $P(t)$, the population, in millions, t years after 1995, is decreasing according to the exponential decay model.

a) Find the value of k, and write the equation.
b) Estimate the population of Russia in 2018.
c) When will the population of Russia be 100 million?

39. Population decrease of Ukraine. The population of Ukraine dropped from 51.9 million in 1995 to 44.5 million in 2013. (*Source: CIA–The World*

Factbook.) Assume that $P(t)$, the population, in millions, t years after 1995, is decreasing according to the exponential decay model.

a) Find the value of k, and write the equation.
b) Estimate the population of Ukraine in 2018.
c) In what year will the population of Ukraine be 40 million, according to this model?

Life and Natural Sciences

40. Cooling. After warming the water in a hot tub to 100°, the heating element fails. The surrounding air temperature is 40°, and in 5 min the water temperature drops to 95°.

a) Find the value of the constant a in Newton's Law of Cooling.
b) Find the value of the constant k. Round to five decimal places.
c) What is the water temperature after 10 min?
d) How long does it take the water to cool to 41°?
e) Find the rate of change of the water temperature, and interpret its meaning.

41. Cooling. The temperature in a whirlpool bath is 102°, and the room temperature is 75°. The water cools to 90° in 10 min.

a) Find the value of the constant a in Newton's Law of Cooling.
b) Find the value of the constant k. Round to five decimal places.
c) What is the water temperature after 20 min?
d) How long does it take the water to cool to 80°?
e) Find the rate of change of the water temperature, and interpret its meaning.

42. Forensics. A coroner arrives at a murder scene at 2 A.M. He takes the temperature of the body and finds it to be 61.6°. He waits 1 hr, takes the temperature again, and finds it to be 57.2°. The body is in a freezer, where the temperature is 10°. When was the murder committed?

43. Forensics. A coroner arrives at a murder scene at 11 P.M. She finds the temperature of the body to be 85.9°. She waits 1 hr, takes the temperature again, and finds it to be 83.4°. She notes that the room temperature is 60°. When was the murder committed?

44. Prisoner-of-war protest. The initial weight of a prisoner of war is 140 lb. To protest the conditions of her imprisonment, she begins a fast. Her weight t days after her last meal is approximated by

$$W = 140e^{-0.009t}.$$

a) How much does the prisoner weigh after 25 days?
b) At what rate is the prisoner's weight changing after 25 days?

45. Political protest. A monk weighing 170 lb begins a fast to protest a war. His weight after t days is given by

$$W = 170e^{-0.008t}.$$

a) When the war ends 20 days later, how much does the monk weigh?
b) At what rate is the monk losing weight after 20 days (before any food is consumed)?

46. Atmospheric pressure. Atmospheric pressure P at altitude a is given by

$$P = P_0 e^{-0.00005a},$$

where P_0 is the pressure at sea level. Assume that $P_0 = 14.7 \text{ lb/in}^2$ (pounds per square inch).

a) Find the pressure at an altitude of 1000 ft.
b) Find the pressure at an altitude of 20,000 ft.
c) At what altitude is the pressure 14.7 lb/in^2?
d) Find the rate of change of the pressure, and interpret its meaning.

47. Satellite power. The power supply of a satellite is a radioisotope (radioactive substance). The power output P, in watts (W), decreases at a rate proportional to the amount present and P is given by

$$P = 50e^{-0.004t},$$

where t is the time, in days.

a) How much power will be available after 375 days?
b) What is the half-life of the power supply?
c) The satellite cannot operate on less than 10 W of power. How long can the satellite stay in operation?
d) How much power did the satellite have to begin with?
e) Find the rate of change of the power output, and interpret its meaning.

48. Cases of tuberculosis. The number of cases N of tuberculosis in the United States has decreased continually since 1956. In 1956, there were 69,895 cases. By 2012, this number had decreased by over 80%, to 10,521 cases.

a) Find the value of k, and write an exponential function that describes the number of tuberculosis cases after time t, where t is the number of years since 1956.
b) Estimate the number of cases in 2016 and in 2020.
c) At this decay rate, in what year will there be 5000 cases?

Modeling

For each of the scatterplots in Exercises 49–58, determine which, if any, of these functions might be used as a model for the data:

a) Quadratic: $f(x) = ax^2 + bx + c$
b) Polynomial, not quadratic
c) Exponential: $f(x) = ae^{kx}, k > 0$
d) Exponential: $f(x) = ae^{-kx}, k > 0$
e) Logarithmic: $f(x) = a + b \ln x$
f) Logistic: $f(x) = \dfrac{a}{1 + be^{-kx}}$

49.

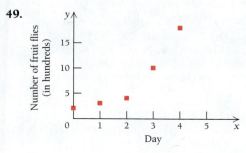

50.

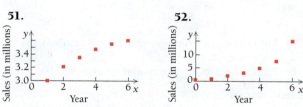

51.   **52.**

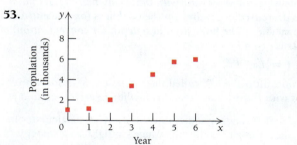

53.

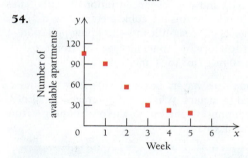

54.

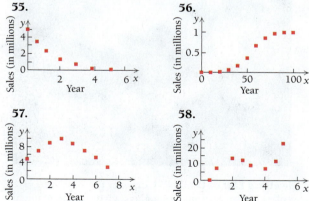

55.   **56.**

57.   **58.**

SYNTHESIS

59. A sample of an element lost 25% of its mass in 5 weeks.

a) Find the half-life.
b) After how many weeks will the sample have lost 75% of its mass?

60. A vehicle lost 15% of its value in 2 yr.

   a) Find the half-life.

   b) After how many years will the vehicle be worth 40% of its original value?

61. Economics: supply and demand elasticity. The demand, $D(x)$, and supply, $S(x)$, functions for a multipurpose printer are as follows:

$$D(x) = q = 480e^{-0.003x}$$

and

$$S(x) = q = 150e^{0.004x}.$$

   a) Find the equilibrium point. Assume that x is the price in dollars.

   b) Find the elasticity of demand when $x = \$100$.

The Beer–Lambert Law. *A beam of light enters a medium such as water or smoky air with initial intensity I_0. Its intensity is decreased depending on the thickness (or concentration) of the medium. The intensity I at a depth (or concentration) of x units is given by*

$$I = I_0 e^{-\mu x}.$$

The constant μ ("mu"), called the coefficient of absorption, varies with the medium. Use this law for Exercises 62 and 63.

62. Light through smog. Concentrations of particulates in the air due to pollution reduce sunlight. In a smoggy area, $\mu = 0.01$ and x is the concentration of particulates measured in micrograms per cubic meter (mcg/m^3). What change is more significant—dropping pollution levels from 100 mcg/m^3 to 90 mcg/m^3 or dropping them from 60 mcg/m^3 to 50 mcg/m^3? Why?

63. Light through sea water. Sea water has $\mu = 1.4$ and x is measured in meters. What would increase cloudiness more —dropping x from 2 m to 5 m or dropping x from 7 m to 10 m? Explain.

64. Newton's Law of Cooling. Consider the following exploratory situation. Fill a glass with hot tap water. Place a thermometer in the glass and measure the temperature. Check the temperature every 30 min

thereafter. Plot your data on this graph, and connect the points with a smooth curve.

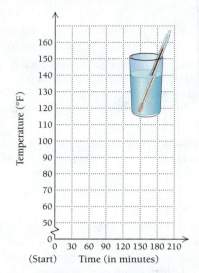

   a) What was the temperature of the water when you began?

   b) At what temperature does there seem to be a leveling off of the graph?

   c) What is the difference between your answers to parts (a) and (b)?

   d) How does the water temperature in part (b) compare with the room temperature?

   e) Find an equation that fits the data. Use this equation to check values of other data points. How do they compare?

   f) Is it ever "theoretically" possible for the temperature of the water to be the same as the room temperature? Explain.

   g) Find the rate of change of the temperature, and interpret its meaning.

65. An interest rate decreases from 8% to 7.2%. Explain why this increases the present value of an amount due 10 yr later.

66. The city of Rayburn loses half its population every 12 yr.

   a) Explain why Rayburn's population will not be zero after 2 half-lives, or 24 yr.

   b) What percentage of the original population remains after 2 half-lives?

   c) What percentage of the original population remains after 4 half-lives?

Answers to Quick Checks

1. (a) $N(t) = N_0 e^{-0.14t}$; **(b)** 246.6 g; **(c)** 4.95 days
2. (a) 30.1 yr; **(b)** 5.3% per day   **3.** 7574 yr
4. $3011.94 **5.** 11.7 min   **6.** $N = 2.2$ hr, so the murder was committed 2 hr and 12 min before the coroner arrived, at about 11:48 P.M. on the previous day.

3.5

- Differentiate functions involving a^x.
- Differentiate functions involving $\log_a x$.
- Find the future value and the rate of growth of an annuity.

The Derivatives of a^x and $\log_a x$; Application: Annuities

The Derivative of a^x

To find the derivative of a^x, for any base a, we first express a^x as a power of e. To do this, recall that $\log_b x$ is the power to which b is raised in order to get x. Thus,

$$b^{\log_b x} = x.$$

In particular, it follows that

$$e^{\log_e A} = A, \quad \text{or} \quad e^{\ln A} = A.$$

If we replace A with a^x, we have

$$e^{\ln a^x} = a^x, \quad \text{or} \quad a^x = e^{\ln a^x}. \tag{1}$$

To find the derivative of a^x, we differentiate both sides:

$$\frac{d}{dx} a^x = \frac{d}{dx} e^{\ln a^x}$$

$$= \frac{d}{dx} e^{x \ln a} \qquad \text{Using a property of logarithms}$$

$$= \frac{d}{dx} e^{(\ln a)x}$$

$$= e^{(\ln a)x} \cdot \ln a \qquad \text{Differentiating and using the Chain Rule}$$

$$= e^{\ln a^x} \cdot \ln a \qquad \text{Using a property of logarithms}$$

$$= a^x \cdot \ln a. \qquad \text{Using equation (1)}$$

Thus, we have the following theorem.

THEOREM 12

$$\frac{d}{dx} a^x = (\ln a) a^x$$

EXAMPLE 1   Differentiate: **a)** $y = 2^x$;   **b)** $y = (1.4)^x$;   **c)** $f(x) = 3^{2x}$.

Solution

a) $\dfrac{d}{dx} 2^x = (\ln 2) 2^x$ Using Theorem 12

Note that $\ln 2 \approx 0.6931$, so this equation verifies our earlier approximation (in Section 3.1) of the derivative of 2^x as $(0.6931)2^x$.

b) $\dfrac{d}{dx} (1.4)^x = (\ln 1.4)(1.4)^x$

c) Since $f(x) = 3^{2x}$ is of the form $f(x) = 3^{g(x)}$, the Chain Rule applies:

$$f'(x) = (\ln 3) 3^{2x} \cdot \frac{d}{dx} (2x)$$

$$= \ln 3 \cdot 3^{2x} \cdot 2 = 2 \ln 3 \cdot 3^{2x}. \qquad \mathbf{1} \checkmark$$

Quick Check 1 ✔

Differentiate:

a) $y = 5^x$;
b) $f(x) = 4^x$;
c) $y = 7^{2x}$.

Compare these formulas:

$$\frac{d}{dx}a^x = (\ln a)a^x \quad \text{and} \quad \frac{d}{dx}e^x = e^x.$$

The simplicity of the latter formula is a reason for the use of base e in calculus. The many applications of e in natural phenomena provide additional reasons.

THEOREM 13

$$\ln a = \lim_{h \to 0} \frac{a^h - 1}{h}$$

The Derivative of $\log_a x$

Just as the derivative of a^x is expressed in terms of $\ln a$, so too is the derivative of $\log_a x$. To find this derivative, we first express $\log_a x$ in terms of $\ln a$ using the change-of-base formula (P7 of Theorem 3) from Section 3.2:

$$\frac{d}{dx}\log_a x = \frac{d}{dx}\left(\frac{\log_e x}{\log_e a}\right) \qquad \text{Using the change-of-base formula}$$

$$= \frac{d}{dx}\left(\frac{\ln x}{\ln a}\right)$$

$$= \frac{1}{\ln a} \cdot \frac{d}{dx}(\ln x) \qquad \frac{1}{\ln a} \text{ is a constant.}$$

$$= \frac{1}{\ln a} \cdot \frac{1}{x}.$$

THEOREM 14

$$\frac{d}{dx}\log_a x = \frac{1}{\ln a} \cdot \frac{1}{x}.$$

Comparing this equation with

$$\frac{d}{dx}\ln x = \frac{1}{x},$$

we see another advantage of using base e in calculus: with base e, the constant $1/(\ln a)$ is simply 1.

EXAMPLE 2 Differentiate:

a) $y = \log_8 x$;
b) $y = \log x$;
c) $f(x) = \log_3 (x^2 + 1)$;
d) $f(x) = x^3 \log_5 x$.

Solution

a) $\dfrac{d}{dx}\log_8 x = \dfrac{1}{\ln 8} \cdot \dfrac{1}{x}$ Using Theorem 14

b) $\dfrac{d}{dx}\log x = \log_{10} x$ $\log x$ means $\log_{10} x$.

$$= \frac{1}{\ln 10} \cdot \frac{1}{x}$$

TECHNOLOGY CONNECTION

Exploratory

Using the nDeriv feature, check the results of Examples 1 and 2 graphically. Then differentiate $y = \log_2 x$, and check the result with your calculator.

Quick Check 2 ✔

Differentiate:

a) $y = \log_2 x$;
b) $f(x) = -7 \log x$;
c) $g(x) = x^6 \log x$;
d) $y = \log_8 (x^3 - 7)$.

c) Note that $f(x) = \log_3 (x^2 + 1)$ is of the form $f(x) = \log_3 (g(x))$, so the Chain Rule is required:

$$f'(x) = \frac{1}{\ln 3} \cdot \frac{1}{x^2 + 1} \cdot \frac{d}{dx}(x^2 + 1) \qquad \text{Using the Chain Rule}$$

$$= \frac{1}{\ln 3} \cdot \frac{1}{x^2 + 1} \cdot 2x$$

$$= \frac{2x}{(\ln 3)(x^2 + 1)}.$$

d) Since $f(x) = x^3 \log_5 x$ is of the form $f(x) = g(x) \cdot h(x)$, we use the Product Rule:

$$f'(x) = x^3 \cdot \frac{d}{dx} \log_5 x + \log_5 x \cdot \frac{d}{dx}x^3 \qquad \text{Using the Product Rule}$$

$$= x^3 \cdot \frac{1}{\ln 5} \cdot \frac{1}{x} + \log_5 x \cdot 3x^2 \qquad \text{Differentiating}$$

$$= \frac{x^2}{\ln 5} + 3x^2 \log_5 x, \quad \text{or} \quad x^2\left(\frac{1}{\ln 5} + 3 \log_5 x\right). \qquad \text{2 ✔}$$

Business Application: Annuities

A savings account into which deposits of equal size (called *payments*) are made on a regular basis is called an **annuity**. Many people use annuities to save for retirement or for college.

If we let p represent the amount paid into the annuity on a regular basis, r the annual interest rate, n the number of compounding periods per year (which is also the number of payments made per year), and t the time in years, then the future value of the annuity, A, is given by

$$A(t) = \frac{p\left[\left(1 + \frac{r}{n}\right)^{nt} - 1\right]}{\frac{r}{n}}.$$

EXAMPLE 3 **Value of an Annuity.** Javier deposits $150 every month into an annuity with an annual interest rate of 4.5%. Assume that interest is compounded monthly.

a) Find a function $A(t)$ that gives the value of Javier's annuity after t years.

b) What is the value of Javier's annuity after 5 years?

c) What is the rate of change in the value of Javier's annuity after 5 years?

Solution

a) We have $p = 150$, $r = 0.045$, and $n = 12$. Thus, we substitute and simplify:

$$A(t) = \frac{150\left[\left(1 + \frac{0.045}{12}\right)^{12t} - 1\right]}{\frac{0.045}{12}} \qquad \begin{array}{l}\text{Substituting 150 for } p, \text{ 0.045 for } r, \text{ and}\\ \text{12 for } n.\end{array}$$

$$= 40{,}000[(1.00375)^{12t} - 1]. \qquad \text{Simplifying}$$

b) In 5 yr, the value of Javier's annuity will be

$$A(5) = 40{,}000[(1.00375)^{12 \cdot 5} - 1]$$

$$= \$10{,}071.83.$$

Quick Check 3 ☑

Kerrie deposits $10,000 every
3 months into an annuity.
Assume an annual interest rate of
3.75%, compounded quarterly.
a) Find $A(t)$, the value of Kerrie's
   annuity after t years.
b) What is the value of Kerrie's
   annuity after 10 yr?
c) What is the rate of change in
   the value of Kerrie's annuity
   after 10 yr?

c) To find the rate of change in the value of Javier's annuity, we first find the derivative
of $A(t)$:

$$\frac{d}{dt}A(t) = \frac{d}{dt}\left(40{,}000[(1.00375)^{12t} - 1]\right)$$

$$A'(t) = 40{,}000\frac{d}{dt}\left[(1.00375)^{12t} - 1\right]$$

$$= 40{,}000(1.00375)^{12t} \cdot \ln(1.00375) \cdot 12 \qquad \text{Using } \frac{d}{dt}a^u = a^u \cdot \ln a \cdot \frac{du}{dt}$$

$$= 1796.63(1.00375)^{12t} \qquad\qquad\qquad \text{Simplifying}$$

After 5 yr, the value of Javier's annuity is changing at the rate of

$$A'(5) = 1796.63(1.00375)^{12 \cdot 5} = \$2249.01/\text{yr.} \qquad\qquad 3 ☑$$

Section Summary

- The following rules apply when we differentiate exponential and logarithmic functions for any base a:

$$\frac{d}{dx}a^x = (\ln a)a^x, \quad \text{and} \quad \frac{d}{dx}\log_a x = \frac{1}{\ln a} \cdot \frac{1}{x}.$$

- The value of an annuity is given by

$$A = \frac{p\left[\left(1 + \dfrac{r}{n}\right)^{nt} - 1\right]}{\dfrac{r}{n}},$$

where p is the amount of the regular payment, r the
annual percentage rate, and n the number of compounding periods per year.

3.5 | Exercise Set

Differentiate.

1. $y = 6^x$

2. $y = 7^x$

3. $f(x) = 8^x$

4. $f(x) = 15^x$

5. $g(x) = x^5(3.7)^x$

6. $g(x) = x^3(5.4)^x$

7. $y = 7^{x^4 + 2}$

8. $y = 4^{x^2 + 5}$

9. $y = e^{x^2}$

10. $y = e^{8x}$

11. $f(x) = 3^{x^4 + 1}$

12. $f(x) = 12^{7x - 4}$

13. $y = \log_8 x$

14. $y = \log_4 x$

15. $y = \log_{17} x$

16. $y = \log_{23} x$

17. $g(x) = \log_{32}(9x - 2)$

18. $g(x) = \log_6(5x + 1)$

19. $F(x) = \log(6x - 7)$

20. $G(x) = \log(5x + 4)$

21. $y = \log_9(x^4 - x)$

22. $y = \log_8(x^3 + x)$

23. $f(x) = 4\log_7\left(\sqrt{x} - 2\right)$

24. $g(x) = -\log_6\left(\sqrt[3]{x} + 5\right)$

25. $y = 5^x \cdot \log_2 x$

26. $y = 6^x \cdot \log_7 x$

27. $G(x) = (\log_{12} x)^5$

28. $F(x) = (\log_9 x)^7$

29. $f(x) = \dfrac{6^x}{5x - 1}$

30. $g(x) = \dfrac{7^x}{4x + 1}$

31. $y = 5^{2x^3 - 1} \cdot \log(6x + 5)$

32. $y = \log(7x + 3) \cdot 4^{2x^4 + 8}$

33. $G(x) = \log_9 x \cdot (4^x)^6$

34. $F(x) = 7^x \cdot (\log_4 x)^9$

35. $f(x) = (3x^5 + x)^5 \log_3 x$

36. $g(x) = \sqrt{x^3 - x}\,(\log_5 x)$

APPLICATIONS

Business and Economics

37. Double declining balance depreciation. An office machine is purchased for $5200. Under certain assumptions, its salvage value, *V*, in dollars, is depreciated according to a method called *double declining balance*, by basically 80% each year, and is given by

$$V(t) = 5200(0.80)^t,$$

where *t* is the time, in years, after purchase.

a) Find $V'(t)$.
b) Interpret the meaning of $V'(t)$.

38. Recycling aluminum cans. It is known that 45% of all aluminum cans distributed are recycled each year. A beverage company uses 250,000 lb of aluminum cans. After recycling, the amount of aluminum, in pounds, still in use after *t* years is given by

$$N(t) = 250,000(0.45)^t.$$

(*Source*: The Container Recycling Institute.)

a) Find $N'(t)$.
b) Interpret the meaning of $N'(t)$.

39. Recycling glass. In 2012, 34.1% of all glass containers were recycled. A beverage company used 400,000 lb of glass containers per year. After recycling, the amount of glass, in pounds, still in use after *t* years is given by

$$N(t) = 400,000(0.341)^t.$$

(*Source*: www.gpi.org.)

a) Find $N'(t)$
b) Interpret the meaning of $N'(t)$.

40. Household liability. The total financial liability, in billions of dollars, of U.S. households can be modeled by

$$L(t) = 1547(1.083)^t,$$

where *t* is the number of years after 1980. The graph of this function follows.

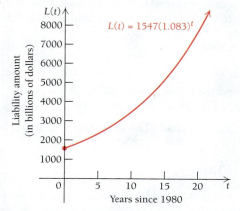

a) Using this model, predict the total financial liability of U.S. households in 2012.
b) Find $L'(25)$.
c) Interpret the meaning of $L'(25)$.

41. Small business. The number of nonfarm proprietorships, in thousands, in the United States can be modeled by

$$N(t) = 8400 \ln t - 10,500,$$

where *t* is the number of years after 1970. The graph of this function is given below.

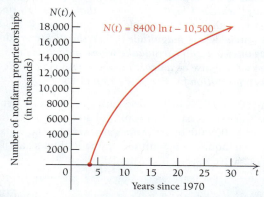

a) Using this model, predict the number of nonfarm proprietorships in the United States in 2020.
b) Find $N'(45)$.
c) Interpret the meaning of $N'(45)$.

42. Annuities. Yukiko opens a savings account to pay for her new baby's college education. She deposits $200 every month into the account at an annual interest rate of 4.2%, compounded monthly.

a) Find $A(t)$, the value of Yukiko's account after *t* years.
b) What is the value of Yukiko's account after 8 yr?
c) What is the rate of change in the value of Yukiko's account after 8 yr?

43. Annuities. Nasim opens a retirement savings account, and she deposits $1000 every 4 months into the account, which has an annual interest rate of 3.8%, compounded every 4 months.

a) Find $A(t)$, the value of Nasim's account after *t* years.
b) What is the value of Nasim's account after 12 yr?
c) What is the rate of change in the value of Nasim's account after 12 yr?

Life and Physical Sciences

44. Agriculture. Farmers wishing to avoid the use of genetically modified (GMO) seeds are increasingly concerned about inadvertently growing GMO plants as a result of pollen drifting from nearby farms. Assuming that these farmers raise their own seeds, the fractional portion of their crop that remains free of GMO plants *t* years later can be approximated by

$$P(t) = (0.98)^t.$$

a) Using this model, predict the fractional portion of the crop that will be GMO-free 10 yr after a neighboring farm begins to use GMO seeds.
b) Find $P'(15)$.
c) Interpret the meaning of $P'(15)$.

Earthquake magnitude. *The magnitude R (measured on the Richter scale) of an earthquake of intensity I is defined as*

$$R = \log \frac{I}{I_0},$$

where I_0 is a minimum intensity used for comparison. If one earthquake is 10 times as intense as another, its magnitude on the Richter scale is 1 higher; if one earthquake is 100 times as intense as another, its magnitude is 2 higher, and so on. Thus, an earthquake whose magnitude is 6 on the Richter scale is 10 times as intense as an earthquake whose magnitude is 5, and 100 times as intense as an earthquake whose magnitude is 4. Use this information for Exercises 45 and 46.

45. On January 12, 2010, a devastating earthquake struck the Caribbean nation of Haiti. It had an intensity that was 10,000,000, or 10^7, times as intense as I_0. What was this earthquake's magnitude on the Richter scale? (*Hint:* Let $I = 10^7 \cdot I_0$.)

46. Find the magnitude of an earthquake that is 5,600,000 times as intense as I_0.

If two earthquakes have magnitudes R_1 and R_2, where $R_1 > R_2$, their relative intensity is given by

$$R_1 - R_2 = \log \frac{I_1}{I_2}.$$

Thus, comparing an earthquake of magnitude 8.0 with another earthquake of magnitude 5.5, we have

$8.0 - 5.5 = \log \dfrac{I_1}{I_2}$, *or* $2.5 = \log \dfrac{I_1}{I_2}$, *and* $I_1 = 10^{2.5} I_2$. *Since*

$10^{2.5} \approx 316$, *an earthquake of magnitude 8.0 is about 316 times as intense as an earthquake of magnitude 5.5. Use this information for Exercises 47 and 48.*

47. The following table shows the magnitudes of selected large earthquakes.

EARTHQUAKE	MAGNITUDE
Sumatran-Andaman, 2004	9.2
Japan, 2011	9.0
San Francisco, 1906	8.0
Baja California, 2010	7.2
San Fernando, 1971	6.6

a) How many times more intense was the Japanese earthquake of 2011 than the Baja California earthquake of 2010?

b) How many times more intense was the Sumatran-Andaman earthquake of 2004 than the San Fernando earthquake of 1971?

48. An earthquake of magnitude 5.0 is considered "moderate," with the potential to damage unreinforced structures. How many times more intense was the San Francisco earthquake of 1906 than a "moderately damaging" earthquake with a magnitude 5.0?

49. Earthquake intensity. The intensity of an earthquake is given by

$$I = I_0 10^R,$$

where R is the magnitude on the Richter scale and I_0 is the minimum intensity, at which $R = 0$, used for comparison.

a) Find I, in terms of I_0, for an earthquake of magnitude 7 on the Richter scale.

b) Find I, in terms of I_0, for an earthquake of magnitude 8 on the Richter scale.

c) Compare your answers to parts (a) and (b).

d) Find the rate of change dI/dR.

e) Interpret the meaning of dI/dR.

50. Intensity of sound. The intensity of a sound is given by

$$I = I_0 10^{0.1L},$$

where L is the loudness of the sound measured in decibels and I_0 is the minimum intensity detectable by the human ear.

a) Find I, in terms of I_0, for the loudness of a power mower, which is 100 decibels.

b) Find I, in terms of I_0, for the loudness of a just audible sound, which is 10 decibels.

c) Compare your answers to parts (a) and (b).

d) Find the rate of change dI/dL.

e) Interpret the meaning of dI/dL.

51. Earthquake magnitude. The magnitude R (measured on the Richter scale) of an earthquake of intensity I is defined as

$$R = \log \frac{I}{I_0},$$

where I_0 is the minimum intensity (used for comparison).

a) Find the rate of change dR/dI.

b) Interpret the meaning of dR/dI.

52. Loudness of sound. The loudness L of a sound of intensity I is defined as

$$L = 10 \log \frac{I}{I_0},$$

where I_0 is the minimum intensity detectable by the human ear and L is the loudness measured in decibels.

a) Find the rate of change dL/dI.

b) Interpret the meaning of dL/dI.

53. Response to drug dosage. The response y to a dosage x of a drug can be approximated by

$$y = m \log x + b,$$

where m and b are constants. This response may be hard to measure with a number. The patient might perspire more, have an increase in temperature, or faint.

a) Find the rate of change dy/dx.
b) Interpret the meaning of dy/dx.

SYNTHESIS

Finding natural logarithms as limits. *Given that the derivative of $f(x) = a^x$ is $f'(x) = a^x(\ln a)$, in Section 3.1, we showed that*

$$f'(x) = a^x \cdot \lim_{h \to 0} \frac{a^h - 1}{h}.$$

Thus, we can define

$$\ln a = \lim_{h \to 0} \frac{a^h - 1}{h}.$$

In Exercises 54–57, use this definition to find each limit.

54. $\displaystyle\lim_{h \to 0} \frac{3^h - 1}{h}$

55. $\displaystyle\lim_{h \to 0} \frac{5^h - 1}{h}$

56. $\displaystyle\lim_{h \to 0} \frac{e^h - 1}{h}$

57. $\displaystyle\lim_{h \to 0} \frac{\sqrt{e^h} - 1}{h}$

Use the Chain Rule, implicit differentiation, and other techniques to differentiate each function given in Exercises 58–65.

58. $f(x) = 3^{(2^x)}$

59. $y = 2^{x^4}$

60. $y = x^x$, for $x > 0$

61. $y = \log_3 (\log x)$

62. $f(x) = x^{e^x}$, for $x > 0$

63. $y = a^{f(x)}$

64. $y = \log_a f(x)$, for $f(x)$ positive

65. $y = [f(x)]^{g(x)}$, for $f(x)$ positive

66. Consider the function $y = x^x$, with $x > 0$.

a) Find $\dfrac{dy}{dx}$. (*Hint:* Take the natural logarithm of both sides and differentiate implicitly.)
b) Find the minimum value of y on $(0, \infty)$.

67. In your own words, derive the formula for the derivative of $f(x) = a^x$.

68. In your own words, derive the formula for the derivative of $f(x) = \log_a x$.

Answers to Quick Checks

1. (a) $(\ln 5)5^x$; **(b)** $(\ln 4)4^x$; **(c)** $(\ln 7)(7^{2x})(2)$
2. (a) $\dfrac{1}{\ln 2} \cdot \dfrac{1}{x}$; **(b)** $\dfrac{-7}{\ln 10} \cdot \dfrac{1}{x}$; **(c)** $x^5\left(\dfrac{1}{\ln 10} + 6\log x\right)$;
(d) $\dfrac{3x^2}{(\ln 8)(x^3 - 7)}$
3. (a) $A(t) = 1{,}066{,}666.67[(1.009375)^{4t} - 1]$;
(b) \$482,615.39; **(c)** \$57,827.44/yr

3.6

A Business Application: Amortization*

- Find the regular payment needed to amortize a loan.
- Use the amortization formula to solve applied problems.
- Complete an amortization schedule.

The value of an investment typically grows (or decays) exponentially. In Section R.1, we saw that an initial amount P, invested at an annual interest rate r, compounded n times per year for t years, grows to an amount $A(t)$, given by

$$A(t) = P\left(1 + \frac{r}{n}\right)^{nt}.$$

The amount $A(t)$ is called the **compound interest future value**. The formula assumes that amount P is invested once and r does not change.

If an investor chooses to make regular, equal-size payments p into a savings account, the account is an *annuity*, as we saw in Section 3.5. The **annuity future value**, $A(t)$, is given by

$$A(t) = \frac{p\left[\left(1 + \dfrac{r}{n}\right)^{nt} - 1\right]}{\dfrac{r}{n}}.$$

In this case, the investor deposits amount p in each of the n compounding periods per year.

---

*This section does use calculus and is not relied on in later chapters.

Amortization

Most large purchases, such as a house or a car, are made with an amortized loan, which is paid off in equal payments. This payment process is called **amortization**; it is represented graphically below.

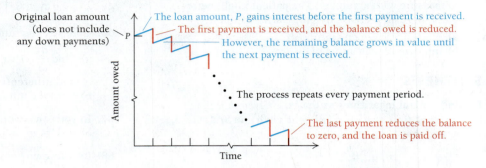

Original loan amount (does not include any down payments) ⟍ *P*

The loan amount, *P*, gains interest before the first payment is received.
The first payment is received, and the balance owed is reduced.
However, the remaining balance grows in value until the next payment is received.

Amount owed

The process repeats every payment period.

The last payment reduces the balance to zero, and the loan is paid off.

Time

Suppose you buy a new car. After making a down payment, you elect to finance the balance using a loan on which you will make monthly payments. However, the money borrowed accrues interest as long as the loan is active. We let *P* represent the loan amount and *p* represent the regular payment needed to pay off (*amortize*) the loan in the specified number of years. To determine *p*, assuming that *r*, *n*, and *t* have been agreed on, we combine the above two expressions for $A(t)$ into one **amortization formula**.

DEFINITION

A loan amount *P* is financed at an annual interest rate *r* compounded *n* times per year for *t* years. The payment amount *p* must satisfy the *amortization formula*:

$$P\left(1 + \frac{r}{n}\right)^{nt} = \frac{p\left[\left(1 + \frac{r}{n}\right)^{nt} - 1\right]}{\frac{r}{n}}.$$

EXAMPLE 1   Amortization: Car Loan.   Henri buys a new car for $20,000, with 30% down and the rest financed through an amortized loan at 7%, compounded monthly for 5 yr.

a) Find Henri's monthly loan payment.

b) Assuming that Henri makes every payment, find the total amount he will pay.

Solution

a) Henri pays 30% of $20,000, or $6000, as a down payment and finances the rest of the car's cost through an amortized loan. Since 70% of $20,000 is $14,000, we have $P = \$14,000$, $n = 12$, $t = 5$, and $r = 0.07$. These are substituted into the amortization formula

$$P\left(1 + \frac{r}{n}\right)^{nt} = \frac{p\left[\left(1 + \frac{r}{n}\right)^{nt} - 1\right]}{\frac{r}{n}} \qquad \text{Using the amortization formula}$$

$$14,000\left(1 + \frac{0.07}{12}\right)^{12 \cdot 5} = \frac{p\left[\left(1 + \frac{0.07}{12}\right)^{12 \cdot 5} - 1\right]}{\frac{0.07}{12}}$$

Quick Check 1 ✔

The Sharmas purchase a new home for $275,000. Suppose they put 40% down and finance the rest through a mortgage (an amortized home loan) at an annual rate of 4.1%, compounded monthly for 30 yr.

a) Find their monthly payment.

b) If the Sharmas make every payment, what is the total amount they will pay?

c) Find the total interest they will pay on their mortgage, assuming that they make all payments.

$$19{,}846.75363 = p(71.59290165)$$

$$p = \frac{19{,}846.75363}{71.59290165} = \$277.22. \qquad \text{Solving for } p$$

Thus, Henri's monthly car payment is $277.22.

b) If Henri makes the minimum monthly payment every month for 60 months, he will pay a total of $277.22 · 60 = $16,633.20 to cover the loan.   **1** ✔

Sometimes, a person will want to specify a payment amount p and use that to find loan amount P that is affordable.

EXAMPLE 2 **Business: Car Loan.** Chelsea wants to buy a new car and can afford to make payments of $300 per month. She qualifies for a car loan at 5%, compounded monthly for 6 years. What is the largest loan amount she can afford?

Solution We use the amortization formula and make the substitutions:

$$P\left(1 + \frac{r}{n}\right)^{nt} = \frac{p\left[\left(1 + \dfrac{r}{n}\right)^{nt} - 1\right]}{\dfrac{r}{n}} \qquad \text{Using the amortization formula}$$

$$P\left(1 + \frac{0.05}{12}\right)^{12 \cdot 6} = \frac{300\left[\left(1 + \dfrac{0.05}{12}\right)^{12 \cdot 6} - 1\right]}{\dfrac{0.05}{12}} \qquad \text{Substituting}$$

$$P(1.349017744) = 25129.27758$$

$$P = \frac{25129.27758}{1.349017744} = \$18{,}627.83. \qquad \text{Solving for } P$$

Chelsea can afford a loan of up to $18,627.83.   **2** ✔

Quick Check 2 ✔

Mr. and Mrs. Urrutia qualify for a 30-yr mortgage at 4.5%, compounded monthly. They are able to make payments of $2000 per month. What is the largest loan amount they can afford?

Amortization Schedules

To see the effect each payment has on a loan, we can set up an *amortization schedule*.

EXAMPLE 3 **Amortization Schedule.** From Example 1, Henri's car loan amount is $P = $14,000$ and his monthly payment is $277.22. Fill in the first two rows of the amortization schedule shown below.

Balance	Payment	Portion of Payment Applied to Interest	Portion of Payment Applied to Principal	New Balance
$14,000	$277.22			

Solution During the first month, interest is calculated once. Therefore, we use the simple interest formula to determine 1 month's interest on $14,000 at the annual interest rate of $r = 7\% = 0.07$, with $t = \frac{1}{12}$ representing 1 month:

$$I = Prt$$
$$= \$14{,}000 \cdot 0.07 \cdot \tfrac{1}{12}$$
$$= \$81.67.$$

Henri's payment of $277.22 is split into two parts: $81.67 pays off the first month's interest, and the remainder, $277.22 − $81.67 = $195.55, is applied toward the principal. The new balance is $14,000 − $195.55 = $13,804.45.

Balance	Payment	Portion of Payment Applied to Interest	Portion of Payment Applied to Principal	New Balance
$14,000	$277.22	$81.67	$195.55	$13,804.45
$13,804.45	$277.22			

We repeat the process: The second month's interest is calculated based on $P = \$13,804.45$. We obtain $I = \$13,804.45 \cdot 0.07 \cdot \frac{1}{12} = \$80.53$. The second payment of $277.22 is also split in two parts: $80.53 pays off the interest for the month, and the rest, $277.22 - \$80.53 = \$196.69$, is applied to the principal. The new balance is $13,804.45 - \$196.69 = \$13,607.76$.

Balance	Payment	Portion of Payment Applied to Interest	Portion of Payment Applied to Principal	New Balance
$14,000	$277.22	$81.67	$195.55	$13,804.45
$13,804.45	$277.22	$80.53	$196.69	$13,607.76

Quick Check 3 ✔

Complete the first two rows of an amortization schedule for the Sharmas' mortgage (see Quick Check 1).

Slowly, the portion of the monthly payment applied to interest decreases and the portion applied to principal increases. Eventually, very little interest accrues on the balance, and nearly all of the payment is applied to the principal. **3** ✔

TECHNOLOGY CONNECTION

Amortization Schedules on a Spreadsheet

An amortization schedule like that in Example 3 can be created with a spreadsheet. In cells A1 through E1, enter the headings shown below. In cell A2, enter 14000, and in cell B2, enter 277.22. Highlight the whole sheet and click on the button to convert all cells to dollar format:

	A	B	C	D	E
1	Balance Forward	Payment	Portion to Interest	Portion to Principal	New Balance
2	$ 14,000.00	$ 277.22			
3					

Next, in cell C2, enter =A2*0.07/12; in cell D2, enter =B2-C2; in cell E2, enter =A2-D2; and in cell A3, enter =E2. Then copy cells B2 through E2 and paste them into cells B3 through E3. Finally, copy cells A3 through E3 and paste them into cells A4 through E61. This process generates the amortization schedule for all 60 months of the loan period.

Note that the monthly portion devoted to interest decreases steadily. The final payment reduces the balance to zero. Due to rounding error, small fluctuations in the last row usually occur.

	A	B	C	D	E
1	Balance Forward	Payment	Portion to Interest	Portion to Principal	New Balance
2	$ 14,000.00	$ 277.22	$ 81.67	$ 195.55	$ 13,804.45
3	$ 13,804.45	$ 277.22	$ 80.53	$ 196.69	$ 13,607.75
4	$ 13,607.75	$ 277.22	$ 79.38	$ 197.84	$ 13,409.91
5	$ 13,409.91	$ 277.22	$ 78.22	$ 199.00	$ 13,210.92
6	$ 13,210.92	$ 277.22	$ 77.06	$ 200.16	$ 13,010.76
7	$ 13,010.76	$ 277.22	$ 75.90	$ 201.32	$ 12,809.44
8	$ 12,809.44	$ 277.22	$ 74.72	$ 202.50	$ 12,606.94
9	$ 12,606.94	$ 277.22	$ 73.54	$ 203.68	$ 12,403.26
10	$ 12,403.26	$ 277.22	$ 72.35	$ 204.87	$ 12,198.39
11	$ 12,198.39	$ 277.22	$ 71.16	$ 206.06	$ 11,992.33
12	$ 11,992.33	$ 277.22	$ 69.96	$ 207.26	$ 11,785.06
13	$ 11,785.06	$ 277.22	$ 68.75	$ 208.47	$ 11,576.59
49	$ 3,460.68	$ 277.22	$ 20.19	$ 257.03	$ 3,203.65
50	$ 3,203.65	$ 277.22	$ 18.69	$ 258.53	$ 2,945.12
51	$ 2,945.12	$ 277.22	$ 17.18	$ 260.04	$ 2,685.08
52	$ 2,685.08	$ 277.22	$ 15.66	$ 261.56	$ 2,423.52
53	$ 2,423.52	$ 277.22	$ 14.14	$ 263.08	$ 2,160.44
54	$ 2,160.44	$ 277.22	$ 12.60	$ 264.62	$ 1,895.82
55	$ 1,895.82	$ 277.22	$ 11.06	$ 266.16	$ 1,629.66
56	$ 1,629.66	$ 277.22	$ 9.51	$ 267.71	$ 1,361.95
57	$ 1,361.95	$ 277.22	$ 7.94	$ 269.28	$ 1,092.67
58	$ 1,092.67	$ 277.22	$ 6.37	$ 270.85	$ 821.82
59	$ 821.82	$ 277.22	$ 4.79	$ 272.43	$ 549.40
60	$ 549.40	$ 277.22	$ 3.20	$ 274.02	$ 275.38
61	$ 275.38	$ 277.22	$ 1.61	$ 275.61	$ (0.23)

EXERCISES

1. Suppose Henri decides to pay $300 per month. Make the necessary changes to the amortization schedule. By how many months is the length of the loan reduced, and how much interest does he save?

2. Suppose Henri decides to pay $350 per month. Make the necessary changes to the amortization schedule. By how many months is the length of the loan reduced, and how much interest does he save?

Section Summary

- *Amortization* is a process in which a loan amount P at an annual interest rate r compounded n times per year for t years is paid off (*amortized*) in equal-sized payments p, according to the formula

$$P\left(1 + \frac{r}{n}\right)^{nt} = \frac{p\left[\left(1 + \frac{r}{n}\right)^{nt} - 1\right]}{\frac{r}{n}}.$$

3.6 Exercise Set

In Exercises 1–10, find the payment amount p needed to amortize the given loan amount. Assume that a payment is made in each of the n compounding periods per year.

1. $P = \$7000$; $r = 6\%$; $t = 5$ yr, compounded monthly

2. $P = \$8000$; $r = 5\%$; $t = 8$ yr, compounded monthly

3. $P = \$12,000$; $r = 5.7\%$; $t = 6$ yr, compounded quarterly

4. $P = \$20,000$; $r = 6.2\%$; $t = 7$ yr, compounded quarterly

5. $P = \$500$; $r = 4.1\%$; $t = 1$ yr, compounded monthly

6. $P = \$800$; $r = 3.8\%$; $t = 2$ yr, compounded monthly

7. $P = \$150,000$; $r = 5.15\%$; $t = 30$ yr, compounded semiannually

8. $P = \$235,000$; $r = 4.35\%$; $t = 30$ yr, compounded semiannually

9. $P = \$75,000$; $r = 8\%$; $t = 15$ yr, compounded annually

10. $P = \$90,000$; $r = 7\%$; $t = 12$ yr, compounded annually

11. **Car loans.** Todd purchases a new Honda Accord LX for $22,150. He makes a $4000 down payment and finances the remainder through an amortized loan at an annual interest rate of 6.5%, compounded monthly for 5 yr.

   a) Find Todd's monthly car payment.
   b) Assume that Todd makes every payment for the life of the loan. Find his total payments.
   c) How much interest does Todd pay over the life of the loan?

12. **Car loans.** Katie purchases a new Jeep Wrangler Sport for $23,000. She makes a $5000 down payment and finances the remainder through an amortized loan at an annual interest rate of 5.7%, compounded monthly for 7 yr.

   a) Find Katie's monthly car payment.
   b) Assume that Katie makes every payment for the life of the loan. Find her total payments.
   c) How much interest does Katie pay over the life of the loan?

13. **Home mortgages.** The Hogansons purchase a new home for $195,000. They make a 25% down payment and finance the remainder with a 30-yr mortgage at an annual interest rate of 5.2%, compounded monthly.

   a) Find the Hogansons' monthly mortgage payment.
   b) Assume that the Hogansons make every payment for the life of the loan. Find their total payments.
   c) How much interest do the Hogansons pay over the life of the loan?

14. **Mortgages.** Andre purchases an office building for $450,000. He makes a 30% down payment and finances the remainder through a 15-yr mortgage at an annual interest rate of 4.15%, compounded monthly.

   a) Find Andre's monthly mortgage payment.
   b) Assume that Andre makes every payment for the life of the loan. Find his total payments.
   c) How much interest does Andre pay over the life of the loan?

15. **Credit cards.** Joanna uses her credit card to finance a $500 purchase. Her card charges an annual interest rate of 22.75%, compounded monthly, and assumes a 10-yr term. Assume that Joanna makes no further purchases on her credit card.

   a) Find Joanna's monthly credit card payment.
   b) Assume that Joanna makes every payment for the life of the loan. Find her total payments.
   c) How much interest does Joanna pay over the life of the loan?

16. **Credit cards.** Isaac uses his credit card to finance a $1200 purchase. His card charges an annual interest rate of 19.25%, compounded monthly, and assumes a 10-yr term. Assume that Isaac makes no further purchases on his credit card.

   a) Find Isaac's monthly credit card payment.
   b) Assume that Isaac makes every payment for the life of the loan. Find his total payments.
   c) How much interest does Isaac pay over the life of the loan?

In Exercises 17–22, complete the first two lines of an amortization schedule for each situation, using a table as shown below.

Balance	Payment	Portion of Payment Applied to Interest	Portion of Payment Applied to Principal	New Balance

17. Todd's car loan in Exercise 11

18. Katie's car loan in Exercise 12

19. The Hogansons' home loan in Exercise 13

20. Andre's office building loan in Exercise 14

21. Joanna's credit card loan in Exercise 15

22. Isaac's credit card loan in Exercise 16

23. Maximum loan amount. Desmond plans to purchase a new car. He qualifies for a loan at an annual interest rate of 5.8%, compounded monthly for 6 yr. He is willing to pay up to $300 per month. What is the largest loan he can afford?

24. Maximum loan amount. Curtis plans to purchase a new car. He qualifies for a loan at an annual interest rate of 7%, compounded monthly for 5 yr. He is willing to pay up to $200 per month. What is the largest loan he can afford?

25. Maximum loan amount. The Daleys plan to purchase a new home. They qualify for a mortgage at an annual interest rate of 4.15%, compounded monthly for 30 yr. They are willing to pay up to $1800 per month. What is the largest loan they can afford?

26. Maximum loan amount. Martina plans to purchase a new home. She qualifies for a mortgage at an annual interest rate of 4.45%, compounded monthly for 20 yr. She is willing to pay up to $2000 per month. What is the largest loan she can afford?

27. Comparing loan options. Kathy plans to finance $12,000 for a new car through an amortized loan. The lender offers two options: (1) a 5-yr term at an annual interest rate of 5.2%, compounded monthly, and (2) a 6-yr term at an annual interest rate of 5%, compounded monthly.

a) Find the monthly payments for options 1 and 2.

b) Assume that Kathy makes every monthly payment. Find her total payments for options 1 and 2.

c) Assume that Kathy intends to make every monthly payment. Which option will result in less interest paid, and how much less?

28. Comparing loan options. The Aubrys plan to finance a new home through an amortized loan of $275,000. The lender offers two options: (1) a 30-yr term at an annual interest rate of 4%, compounded monthly, and (2) a 20-yr term at an annual interest rate of 5%, compounded monthly.

a) Find the monthly payments for options 1 and 2.

b) Assume that the Aubrys make every monthly payment. Find their total payments for options 1 and 2.

c) Assume that the Aubrys intend to make every monthly payment. Which option will result in less interest paid, and by how much?

29. Comparing rates. Darnell plans to finance $200,000 for a new home through a 30-yr mortgage.

a) Find his monthly payment, assuming an annual interest rate of 5%, compounded monthly.

b) Find his monthly payment, assuming an annual interest rate of 6%, compounded monthly.

c) Assume that Darnell makes every payment. How much will he save in interest over 30 yr if he selects the 5% rate?

30. Comparing rates. The Salazars plan to finance $300,000 for a new home through a 20-yr mortgage.

a) Find their monthly payment assuming an annual interest rate of 4.25%, compounded monthly.

b) Find their monthly payment assuming an annual interest rate of 4.75%, compounded monthly.

c) Assume that the Salazars make every payment. How much will they save in interest over 30 yr if they select the 4.25% rate?

SYNTHESIS

31. Retirement planning. Dwight is 25 years old. He plans to retire at age 60 and by then wants to have saved a sum of money that will allow him to withdraw $500 per month for 25 more years (until age 85), at which time the sum will be depleted (amortized). Assume a fixed annual interest rate of 4.5%, compounded monthly for the entire duration.

a) Find the amount of money Dwight will need at age 60.

b) Find the monthly deposit Dwight needs to make in order to reach this sum, assuming that he starts saving immediately.

32. Retirement planning. Kenna is 30 years old. She plans to retire at age 55 and by then wants to have saved a sum of money that will allow her to withdraw $700 per month for 25 more years (until age 80), at which time the sum will be depleted (amortized). Assume a fixed annual interest rate of 5%, compounded monthly for the entire duration.

a) Find the amount of money Kenna will need at age 55.

b) Find the monthly deposit Kenna needs to make in order to reach this sum, assuming that she starts saving immediately.

33. Lottery winnings. Suppose you win $5,000,000 (after taxes) in a lottery. Instead of a one-time payment, you accept a structured plan of annual payments over 20 yr. Under this plan, you receive yearly payments W that assume an annual interest rate of 5% and that will accrue to $5,000,000 over 20 yr. Find the annual payment W you receive under this plan.

34. Structured settlement. Suppose you won a $700,000 (after taxes) settlement in court, to be paid in quarterly payments at an annual interest rate of 4%, compounded quarterly for 5 yr. Find your quarterly payment.

Shortening the life of a loan. *Amortization gives the borrower an advantage: by paying more than the minimum required payment, the borrower can pay off (amortize) the principal faster and save money in interest. Assume P is the loan amount or principal, r is the interest rate, n is the compounding frequency, and t is the time in years. The amortization formula is used to determine the payment amount p that will amortize the loan in t years. However, if the borrower consistently pays Q, where Q > p, the formula for the time needed to amortize the loan amount P is*

$$ t = \frac{\ln (nQ) - \ln (nQ - Pr)}{n \ln \left(1 + \dfrac{r}{n} \right)}. $$

35. Business: home loan. The Begays finance $200,000 for a 30-yr home mortgage at an annual interest rate of 5%, compounded monthly.

a) Find the monthly payment needed to amortize this loan in 30 yr.

b) Assuming that the Begays make the payment found in part (a) every month for 30 yr, find the total interest they will pay.

c) Suppose the Begays pay an extra 15% every month (thus, $Q = 1.15 \cdot p$). Find the time needed to amortize the $200,000 loan.

d) About how much total interest will the Begays pay if they pay Q every month?

e) About how much will the Begays save on interest if they pay Q rather than p every month?

36. Business: car loan. Gwen finances a $15,000 car loan at an annual interest rate of 6%, compounded monthly over a 5-yr term.

a) Find the payment amount needed to amortize this loan in 5 yr.

b) Assuming that Gwen makes the payment found in part (a) every month for 5 yr, find the total interest she will pay.

c) Suppose Gwen pays an extra 20% every month. Find the time needed to amortize the $15,000 loan.

d) About how much total interest will Gwen pay if she pays Q every month?

e) About how much will Gwen save on interest if she pays Q rather than p every month?

37. Derive the formula for t given before Exercise 35.

38. If you are shopping for a car, you will likely be offered loans at different interest rates and/or for different terms. For example, a longer term usually results in a lower monthly payment. Is increasing the term of the loan a good long-term strategy? In general, what is always a good strategy when choosing among various loan options?

TECHNOLOGY CONNECTION

39–44. Use a spreadsheet to complete the first twelve lines of an amortization schedule for each loan in Exercises 11–16.

Answers to Quick Checks

1. (a) $797.28; **(b)** $287,020.80; **(c)** $122,020.80
2. $394,722.32   **3.** First row: $165,000, $797.28, $563.75, $233.53, $164,766.47; second row: $164,766.47, $797.28, $562.95, $234.33, $164,532.14

KEY TERMS AND CONCEPTS

EXAMPLES

SECTION 3.1

An **exponential function** f is a function of the form $f(x) = a_0 \cdot a^x$, where x is any real number and a, the **base**, is any positive number other than 1.

The y-intercept of f is $(0, a_0)$; there are no x-intercepts. The domain of an exponential function is $(-\infty, \infty)$.

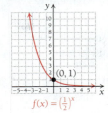

$f(x) = \left(\frac{1}{2}\right)^x$

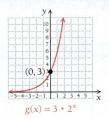

$g(x) = 3 \cdot 2^x$

The natural base e is such that

$$\frac{d}{dx} e^x = e^x,$$

where $e = \lim_{h \to 0} (1 + h)^{1/h} \approx 2.718$,

and $\frac{d}{dx} e^{f(x)} = f'(x)e^{f(x)}$.

$$\frac{d}{dx}(e^{x^2+3x}) = e^{x^2+3x} \cdot (2x + 3) = (2x + 3)e^{x^2+3x}$$

SECTION 3.2

A **logarithmic function** g is any function of the form $g(x) = \log_a x$, where a, the **base**, is a positive number other than 1:

$$y = \log_a x \quad \text{means} \quad a^y = x.$$

Logarithmic functions are inverses of exponential functions.

$\log_2 8 = 3$ means $2^3 = 8$;

$\log_3 \frac{1}{3} = -1$ means $3^{-1} = \frac{1}{3}$;

and $\log_8 64 = x$ means $8^x = 64$, so $x = 2$.

The graph of $y = \log_3 x$:

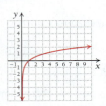

Remember that $\log_b c$ *is the exponent to which b is raised to get c.* The notation for common and natural logarithms is

$$\log c = \log_{10} c \quad \text{and} \quad \ln c = \log_e c.$$

- $\log 1000 = \log_{10} 10^3 = 3$ and $\ln e^{\sqrt{5}} = \log_e e^{\sqrt{5}} = \sqrt{5}$
- Using a calculator, $\log 50 \approx 1.69897$, and so $10^{1.69897} \approx 50$.

The following properties of logarithms allow us to manipulate expressions. We assume that M, N, and a are positive, with $a \neq 1$, and k is a real number:

$$\log_a (MN) = \log_a M + \log_a N$$

$$\log_a \frac{M}{N} = \log_a M - \log_a N$$

$$\log_a (M^k) = k \log_a M$$

$$\log_a a = 1$$

$$\log_a a^k = k$$

$$\log_a 1 = 0$$

$$\log_b M = \frac{\log_a M}{\log_a b}$$

- $\log_2 8x = \log_2 8 + \log_2 x = 3 + \log_2 x$
- $\log_3 \frac{9}{x} = \log_3 9 - \log_3 x = 2 - \log_3 x$
- $\log_4 x^5 = 5 \log_4 x$
- $\log_6 6 = 1$
- $\log_5 25 = \log_5 5^2 = 2$
- $\log_7 1 = 0$
- $\log_3 5 = \frac{\log_a 5}{\log_a 3}$. If $a = 10$, then $\log_3 5 = \frac{\log 5}{\log 3} \approx 1.465$.

| **KEY TERMS AND CONCEPTS** | **EXAMPLES** |

SECTION 3.2 (*continued*)

Logarithms are used to solve certain exponential equations.

Solve: $5e^{2t} = 80$.

We have

$$5e^{2t} = 80$$
$$e^{2t} = 16 \qquad \text{Dividing both sides by 5}$$
$$\ln e^{2t} = \ln 16 \qquad \text{Taking the natural logarithm of both sides}$$
$$2t = \ln 16 \qquad \text{Using a property of logarithms}$$
$$t = \frac{\ln 16}{2}$$
$$t \approx 1.386$$

The **natural logarithm** is defined by $f(x) = \log_e |x| = \ln |x|$, where $e \approx 2.71828$ and $x \neq 0$. When we write $y = \ln x$, we assume $x > 0$.

The derivative of the natural logarithm of x is the reciprocal of x:

$$\frac{d}{dx} \ln |x| = \frac{1}{x}, \text{ for } x \neq 0, \text{ and}$$

$$\frac{d}{dx} \ln x = \frac{1}{x}, \text{ for } x > 0.$$

The derivative of $y = \ln f(x)$ is

$$\frac{d}{dx} \ln f(x) = \frac{f'(x)}{f(x)}.$$

- $\dfrac{d}{dx}[\ln (x^2 + 8)] = \dfrac{1}{x^2 + 8} \cdot 2x = \dfrac{2x}{x^2 + 8}$

- $\dfrac{d}{dx}\left[\ln (5x) \cdot (x^3 - 7x)\right] = \ln (5x) \cdot \dfrac{d}{dx}(x^3 - 7x) + (x^3 - 7x) \cdot \dfrac{d}{dx}[\ln (5x)]$

  Using the product rule

  $$= \ln (5x) \cdot (3x^2 - 7) + (x^3 - 7x) \cdot \frac{1}{5x} \cdot 5$$

  Simplifying

  $$= (3x^2 - 7) \ln (5x) + (x^2 - 7)$$

SECTION 3.3

Because exponential functions are the only functions for which the derivative (rate of change) is directly proportional to the function value for any input value, they can model many real-world situations involving uninhibited growth.

If $\dfrac{dP}{dt} = kP$, with $k > 0$, then $P(t) = P_0 e^{kt}$, where P_0 is the initial population at $t = 0$.

Business. The balance P in an account with Turing Mutual Funds grows at a rate given by

$$\frac{dP}{dt} = 0.04P,$$

where t is time, in years. Find the function that satisfies the equation (let $P(0) = P_0$). After what period of time will an initial investment, P_0, double itself?

The function is given by
$$P = P_0 e^{0.04t}.$$

Check:

$$\frac{d}{dt} P_0 e^{0.04t} = P_0 e^{0.04t} \cdot 0.04$$
$$= 0.04 P_0 e^{0.04t}$$
$$= 0.04P$$

To find the doubling time, we set $P = 2P_0$ and solve for t:

$$2P_0 = P_0 e^{0.04t}$$
$$2 = e^{0.04t} \qquad \text{Dividing both sides by } P_0$$
$$\ln 2 = \ln (e^{0.04t}) \qquad \text{Taking the natural logarithm of both sides}$$
$$\ln 2 = 0.04t$$
$$\frac{\ln 2}{0.04} = t, \quad \text{or} \quad t \approx 17.3 \text{ yr.}$$

(continued)

KEY TERMS AND CONCEPTS	**EXAMPLES**

SECTION 3.3 (continued)

The **exponential growth rate** k and the **doubling time** T are related by

$$k = \frac{\ln 2}{T} \quad \text{and} \quad T = \frac{\ln 2}{k}.$$

Two models for limited and inhibited growth are

$$P(t) = \frac{L}{1 + be^{-kt}},$$

and

$$P(t) = L(1 - e^{-kt}),$$

where $k > 0$ and L is the limiting value.

If the exponential growth rate is 4% per year, then the doubling time is

$$T = \frac{\ln 2}{4\%} \approx \frac{0.693147}{0.04} \approx 17.3 \text{ yr.}$$

The doubling time for the number of downloads per day from iSongs is 614 days. The exponential growth rate is

$$k = \frac{\ln 2}{T} \approx \frac{0.693147}{614} \approx 0.0011289 = 0.11\% \text{ per day.}$$

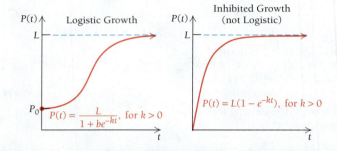

SECTION 3.4

Exponential growth is modeled by $P(t) = P_0 e^{kt}$, $k > 0$, and **exponential decay** is modeled by $P(t) = P_0 e^{-kt}$, $k > 0$.

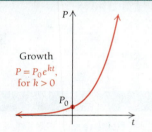 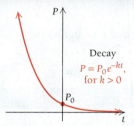

Exponential decay characterizes many real-world phenomena. One of the most common is *radioactive decay*.

Physical Science. Lead-210 has a decay rate of 3.15% per year. The rate of change of an amount N of lead-210 is given by

$$\frac{dN}{dt} = -0.0315N.$$

Find the function that satisfies the equation. How much of an 80-g sample of lead-210 will remain after 20 yr?

The function is given by $N(t) = N_0 e^{-0.0315t}$.

Check:

$$\frac{d}{dt}\left(N_0 e^{-0.0315t}\right) = N_0 e^{-0.0315t}(-0.0315)$$

$$= -0.0315 \cdot N(t).$$

The amount of an 80-g sample remaining after 20 yr is $N(20) = 80e^{-0.0315(20)}$
≈ 42.6 g.

KEY TERMS AND CONCEPTS	EXAMPLES

SECTION 3.4 (continued)

Half-life, T, and *decay rate*, k, are related by

$$k = \frac{\ln 2}{T} \quad \text{and} \quad T = \frac{\ln 2}{k}.$$

The decay equation, $P_0 = Pe^{-kt}$, can also be used to calculate the present value of P dollars due t years in the future.

The half-life of a radioactive isotope is 38 days. The decay rate is

$$T = \frac{\ln 2}{38} = \frac{0.693147}{38} \approx 0.0182 = 1.82\% \text{ per day.}$$

If a sample of an element loses 20% of it mass in 5 yr, then $0.8A = Ae^{-k \cdot 5}$, and $k = \frac{\ln 0.8}{5} \approx 0.0446$. Thus, the element's half-life is

$$T = \frac{\ln 2}{0.0446} \approx 15.54 \text{ yr.}$$

The **present value** P_0 of an amount P due t years later, at interest rate k, compounded continuously, is given by

$$P_0 = Pe^{-kt}.$$

The present value of $200,000 due 8 yr from now, at 4.6% interest, compounded continuously, is given by

$$P_0 = Pe^{-kt} = 200{,}000e^{-0.046 \cdot 8} = \$138{,}423.44.$$

SECTION 3.5

Formulas can be used to differentiate exponential and logarithmic functions for any base a, other than e:

$$\frac{d}{dx} a^x = (\ln a)a^x \text{ and } \frac{d}{dx} \log_a x = \frac{1}{\ln a} \cdot \frac{1}{x}.$$

$$\frac{d}{dx} 7^x = \ln 7 \cdot 7^x$$

$$\frac{d}{dx} \log_6 x = \frac{1}{\ln 6} \cdot \frac{1}{x}$$

An **annuity** is a savings account into which deposits of equal size (called payments) are made on a regular basis.

The **annuity future value**, A, is given by

$$A(t) = \frac{p\left[\left(1 + \frac{r}{n}\right)^{nt} - 1\right]}{\frac{r}{n}},$$

where p is the amount of each regular payment, r is the annual interest rate, n is the number of compounding periods per year (and the number of payments per year), and t is the time in years.

Matt deposits $500 every month into an annuity that has an annual interest rate of 4.2%, compounded monthly. The value, A, of Matt's annuity after t years is given by

$$A(t) = \frac{500\left[\left(1 + \frac{0.042}{12}\right)^{12t} - 1\right]}{\frac{0.042}{12}} = 142{,}857.14[(1.0035)^{12t} - 1].$$

The value of Matt's annuity after 6 yr is

$$A(6) = 142{,}857.14[(1.0035)^{12 \cdot 6} - 1] \approx \$40{,}861.44.$$

The rate of change in the value of the annuity is given by

$$A'(t) = 142{,}857.14(1.0035)^{12t}(\ln 1.0035)(12) = 5989.52(1.0035)^{12t}.$$

After 6 yr, the annuity is increasing in value at the rate of

$$A'(6) = 5989.52(1.0035)^{12 \cdot 6} = \$7702.70/\text{yr}.$$

SECTION 3.6

Amortization is a process in which a loan amount P is paid off in $n \cdot t$ equal payments over a period of t years. The loan amount P and the payment amount p are related by

$$P\left(1 + \frac{r}{n}\right)^{nt} = \frac{p\left[\left(1 + \frac{r}{n}\right)^{nt} - 1\right]}{\frac{r}{n}}.$$

Andy buys a 2014 Nissan Sentra for $25,000. He pays $5000 and finances the rest through a 5-yr loan at an annual interest rate of 6%. His monthly car payment is p:

$$20{,}000\left(1 + \frac{0.06}{12}\right)^{60} = \frac{p\left[\left(1 + \frac{0.06}{12}\right)^{60} - 1\right]}{\frac{0.06}{12}}$$

$$26977.00305 = p(69.77003051)$$

$$p = \frac{26977.00305}{69.77003051} = \$386.66.$$

CHAPTER 3
Review Exercises

These review exercises are for test preparation. They can also be used as a practice test. Answers are at the back of the book. The red bracketed section references tell you what part(s) of the chapter to restudy if your answer is incorrect.

CONCEPT REINFORCEMENT

In Exercises 1–6, match each equation in column A with the most appropriate graph in column B [3.1–3.4]

Column A

Column B

1. $P(t) = 50e^{0.03t}$

a)

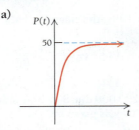

2. $P(t) = \dfrac{50}{1 + 2e^{-0.02t}}$

b)

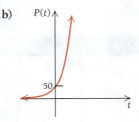

3. $P(t) = 50e^{-0.20t}$

c)

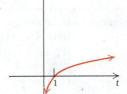

4. $P(t) = \ln t$

d)

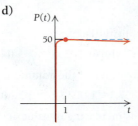

5. $P(t) = 50(1 - e^{-0.04t})$

e)

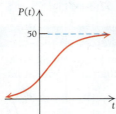

6. $P(t) = 50 + \ln t$

f)

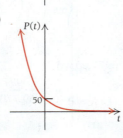

Classify each statement as either true or false.

7. The base a in the exponential function given by $f(x) = a^x$ must be greater than 1. [3.1]

8. The base a in the logarithmic function given by $g(x) = \log_a x$ must be greater than 0. [3.2]

9. If $f'(x) = c \cdot f(x)$ for $c \neq 0$ and $f(x) \neq 0$, then f must be an exponential function. [3.3]

10. With exponential growth, the doubling time depends on the size of the original population. [3.3]

11. A radioactive isotope's half-life can be used to determine the value of its decay constant. [3.4]

12. A radioactive isotope's half-life depends on how much of the substance is initially present. [3.4]

13. For any exponential function of the form $f(x) = a^x$, it follows that $f'(x) = \ln a \cdot a^x$. [3.5]

14. For any logarithmic function of the form $g(x) = \log_a x$, it follows that $g'(x) = \dfrac{1}{a} \cdot \dfrac{1}{x}$. [3.5]

15. It is possible to amortize a loan faster by paying more than the minimum amount for each payment. [3.6]

REVIEW EXERCISES

16. Find x. [3.5]

    a) $\log_2 \frac{1}{8} = x$

    b) $\log_x 9 = \frac{1}{2}$

    c) $\log_4 x = 5$

Differentiate each function.

17. $y = \ln x$ [3.2]

18. $y = e^x$ [3.1]

19. $y = \ln(x^4 + 5)$ [3.2]

20. $y = e^{2\sqrt{x}}$ [3.1]

21. $f(x) = \ln\sqrt{x}$ [3.2]

22. $f(x) = x^4 e^{3x}$ [3.1]

23. $f(x) = \dfrac{\ln x}{x^3}$ [3.2]

24. $f(x) = e^{x^2} \cdot \ln 4x$ [3.1, 3.2]

25. $f(x) = e^{4x} - \ln \dfrac{x}{4}$ [3.1, 3.2]

26. $y = \dfrac{\ln e^x}{e^x}$ [3.1, 3.2]

27. $F(x) = 9^x$ [3.5]

28. $g(x) = \log_2 x$ [3.5]

29. $y = 3^x \cdot \log_4(2x + 1)$ [3.5]

Graph each function. [3.1]

30. $f(x) = 4^x$

31. $g(x) = (\frac{1}{3})^x$

Given $\log_a 2 = 1.8301$ *and* $\log_a 7 = 5.0999,$ *find each logarithm.* [3.2]

32. $\log_a 14$

33. $\log_a \frac{2}{7}$

34. $\log_a 28$

35. $\log_a 3.5$

36. $\log_a \sqrt{7}$

37. $\log_a \frac{1}{4}$

38. Find the function Q that satisfies $dQ/dt = 7Q$, given that $Q = 25$ when $t = 0$. [3.3]

39. Life science: population growth. The population of Boomtown doubled in 16 yr. What was the growth rate of the city? Round to the nearest tenth of a percent. [3.3]

40. Business: interest compounded continuously. Suppose $8300 is invested in Noether Bond Fund, for which the interest rate is 3.8%, compounded continuously. How long will it take for the $8300 to double? Round to the nearest tenth of a year. [3.3]

41. Business: cost of a prime-rib dinner. Suppose the average cost C of a prime-rib dinner was $15.81 in 1986 and $27.95 in 2014. Assuming that the exponential growth model applies: [3.3]

a) Find the exponential growth rate to three decimal places, and write the function that models the situation.

b) What will the cost of such a dinner be in 2018? In 2024?

42. Business: cost of Oreo cookies. The average cost, C, of Oreo Cookies was $2.69/lb in 1990 and $5.14/lb in 2013. (*Source:* foodtimeline.org.) Assuming that the exponential growth model applies: [3.3]

a) Find the exponential growth rate to three decimal places, and write the function that models the situation.

b) What will the cost of 1 lb of Oreo Cookies be in 2020? In 2030?

43. Business: franchise growth. Fashionista Clothing is selling franchises throughout the United States and Canada. It is estimated that the number of franchises N will increase at the rate of 12% per year, that is,

$$\frac{dN}{dt} = 0.12N,$$

where t is the time, in years. [3.3]

a) Find the function that satisfies the equation, assuming that the number of franchises in 2014 $(t = 0)$ is 60.

b) How many franchises will there be in 2018?

c) After how long will the number of franchises be 300? Round to the nearest tenth of a year.

44. Business: franchise growth. The Coffee Casa is selling franchises throughout the southwestern United States. It is estimated that the number of franchises N will increase at the rate of 7% per year, that is,

$$\frac{dN}{dt} = 0.07N,$$

where t is the time, in years. [3.3]

a) Find the function that satisfies the equation, assuming that the number of franchises in 2010 $(t = 0)$ was 24.

b) How many franchises will there be in 2017?

c) After how long will the number of franchises be double what it was 2010? Round to the nearest tenth of a year.

45. Life science: decay rate. The decay rate of a certain radioactive substance is 13% per year. What is its half-life? Round to the nearest tenth of a year. [3.4]

46. Life science: half-life. The half-life of radon-222 is 3.8 days. What is its decay rate? Round to the nearest tenth of a percent. [3.4]

47. Life science: decay rate. A certain radioactive isotope has a decay rate of 7% per day, that is,

$$\frac{dA}{dt} = -0.07A,$$

where A is the amount of isotope present after t days. [3.4]

a) Find a function that satisfies the equation if the amount of the isotope present at $t = 0$ is 800 g.

b) After 20 days, how much of the 800 g will remain? Round to the nearest gram.

c) After how long will half of the original amount remain?

48. Social science: Hullian learning model. The probability $p(t)$ of mastering a certain assembly-line task after t learning trials is given by

$$p(t) = 1 - e^{-0.7t}. \ [3.3]$$

a) What is the probability of learning the task after 1 trial? 2 trials? 5 trials? 10 trials? 14 trials?

b) Find the rate of change, $p'(t)$.

c) Interpret the meaning of $p'(t)$.

d) Sketch a graph of the function.

49. Business: present value. Find the present value of $1,000,000 due 40 yr later at 4.2%, compounded continuously. [3.4]

50. Business: annuity. Patrice deposits $50 into a savings account every month at an annual interest rate of 4.7%, compounded monthly. [3.5]

a) How much will her account contain after 8 yr, assuming she makes no withdrawals?

b) What is the rate of change in the value of Patrice's account after 8 yr?

51. Business: car loan. Glenda buys a used Subaru Outback for $13,000. She pays 25% down and finances the rest at an annual interest rate of 6.5%, compounded monthly for 5 yr. [3.6]

  a) Find Glenda's monthly car payment.

  b) If Glenda makes every payment for the life of the loan, find the total interest she will pay.

52. Business: home mortgage. The Savards qualify for a 30-yr mortgage at an annual interest rate of 4.25%, compounded monthly. They are willing to pay up to $2500 per month as their mortgage payment. What is the most they can afford to borrow? [3.6]

53. Business: credit card. Vicki uses her credit card to buy $1200 in goods. Her credit card has an annual interest rate of 20.75%, compounded monthly over a 10-yr term. Assume that Vicki makes no additional purchases with this credit card. [3.6]

  a) Find the amount of Vicki's monthly payment.

  b) Assume that Vicki pays the amount found in part (a). In the first month, what will she pay in interest, and what will she pay toward the principal?

SYNTHESIS

54. Differentiate: $y = \dfrac{e^{2x} + e^{-2x}}{e^{2x} - e^{-2x}}$. [3.1]

55. Find the minimum value of $f(x) = x^4 \ln (4x)$. [3.2]

56. Graph: $f(x) = \dfrac{e^{1/x}}{(1 + e^{1/x})^2}$. [3.1]

57. Find $\lim\limits_{x \to 0} \dfrac{e^{1/x}}{(1 + e^{1/x})^2}$. [3.1]

58. Business: shopping on the Internet. Online sales of all types of consumer products increased at an exponential rate in recent years. Data in the following table show online retail sales in the United States, in billions of dollars. [3.3]

Years, t, after 2010	U.S. Online Retail Sales (in billions)
0	$167
1	194
2	231
3	262
4	291

(*Sources*: Internetretailer.com; www.census.gov.)

  a) Use REGRESSION to fit an exponential function $y = a \cdot b^x$ to the data. Then convert that formula to an exponential function, base e, where t is the number of years after 2010, and determine the exponential growth rate.

  b) Estimate online sales in 2016 and in 2020.

  c) After what amount of time will online sales be $1 trillion?

  d) What is the doubling time for online sales?

CHAPTER 3
Test

Differentiate.

1. $y = 2e^{3x}$

2. $y = (\ln x)^4$

3. $f(x) = e^{-x^2}$

4. $f(x) = \ln \dfrac{x}{7}$

5. $f(x) = e^x - 5x^3$

6. $f(x) = 3e^x \ln x$

7. $y = 7^x + 3^x$

8. $y = \log_{14} x$

9. Solve:

  a) $\log_2 32 = x$;

  b) $\log_8 x = \frac{1}{3}$.

Given $\log_b 2 = 0.2560$ *and* $\log_b 9 = 0.8114$, *find each of the following.*

10. $\log_b 18$

11. $\log_b 4.5$

12. $\log_b 3$

13. Find the function that satisfies $dM/dt = 6M$, if $M = 2$ at $t = 0$.

14. The doubling time for a certain bacteria population is 3 hr. What is the growth rate? Round to the nearest tenth of a percent.

APPLICATIONS

15. Business: interest compounded continuously. An investment is made at 6.931% per year, compounded continuously. What is the doubling time? Round to the nearest tenth of a year.

16. Business: cost of milk. The cost C of a gallon of milk was $3.22 in 2006. In 2013, it was $3.50. (*Source*: U.S. Department of Labor, Bureau of Labor Statistics.) Assuming that the exponential growth model applies:

  a) Find the exponential growth rate to the nearest tenth of a percent, and write the equation.

  b) Estimate the cost of a gallon of milk in 2016 and in 2020.

17. Life science: drug dosage. A dose of a drug is injected into the body of a patient. The drug amount in the body decreases at the rate of 10% per hour, that is,

$$\frac{dA}{dt} = -0.1A,$$

where A is the amount in the body and t is the time, in hours.

a) A dose of 3 cubic centimeters (cc) is administered. Assuming $A_0 = 3$, find the function that satisfies the equation.

b) How much of the initial dose of 3 cc remains after 10 hr?

c) After how long does half of the original dose remain?

18. Life science: decay rate. The decay rate of radium-226 is 4.209% per century. What is its half-life?

19. Life science: half-life. The half-life of bohrium-267 is 17 sec. What is its decay rate? Express the rate as a percentage rounded to four decimal places.

20. Business: effect of advertising. Twin City Roasters introduced a new coffee in a trial run. The firm advertised the coffee on television and found that the percentage $P(t)$ of people who bought the coffee after t ads had been run was

$$P(t) = \frac{100}{1 + 24e^{-0.28t}}.$$

a) What percentage of people bought the coffee before seeing the ad $(t = 0)$?

b) What percentage bought the coffee after the ad had been run 1 time? 5 times? 10 times? 20 times? 30 times?

c) Find the rate of change, $P'(t)$.

d) Interpret the meaning of $P'(t)$.

e) Sketch a graph of the function.

21. Business: annuity. Jill deposits $20 per week into an account for holiday shopping. The account has an annual interest rate of 3.3%, compounded weekly.

a) Find the future value of Jill's account after 1 yr (52 weeks).

b) What is the rate of change in the value of her account after 1 yr?

22. Business: amortized loan. The Langways purchase a new home for $450,000. They pay 20% down and finance the rest through a 30-yr mortgage at an annual interest rate of 3.75%, compounded monthly.

a) How much is the Langways' monthly mortgage payment?

b) Assuming that the Langways make the monthly payment found in part (a) for the life of the loan, how much will they pay in total?

c) How much interest will they pay?

23. Business: car loan. Giselle qualifies for a car loan at an annual interest rate of 5.75%, compounded monthly for

5 yr. She is willing to pay up to $400 per month in car payments. What is the maximum loan amount she can afford?

SYNTHESIS

24. Differentiate: $y = x (\ln x)^2 - 2x \ln x + 2x$.

25. Find the maximum and minimum values of $f(x) = x^4 e^{-x}$ over $[0, 10]$.

TECHNOLOGY CONNECTION

26. Graph: $f(x) = \dfrac{e^x - e^{-x}}{e^x + e^{-x}}.$

27. Find $\lim\limits_{x \to 0} \dfrac{e^x - e^{-x}}{e^x + e^{-x}}.$

28. Business: average price of a television commercial. The cost of a 30-sec television commercial that runs during the Super Bowl was increasing exponentially from 1991 to 2014. Data in the table below show costs for those years.

Years, t, after 1990	Cost of Commercial
1	$800,000
3	850,000
5	1,000,000
8	1,300,000
13	2,100,000
16	2,600,000
22	3,500,000
24	4,000,000

(*Source:* National Football League.)

a) Use REGRESSION to fit an exponential function $y = a \cdot b^x$ to the data. Then convert that formula to an exponential function, base e, where t is the number of years after 1990.

b) Estimate the cost of a commercial run during the Super Bowl in 2017 and 2020.

c) After what amount of time will the cost be $1 billion?

d) What is the doubling time of the cost of a commercial run during the Super Bowl?

e) The cost of a Super Bowl commercial in 2009 turned out to be $3 million, and in 2010 it dropped to about $2.8 million, possibly due to the decline in the world economy. Expand the table of costs, and make a scatterplot of the data. Does the cost still seem to follow an exponential function? Explain. What kind of curve seems to fit the data best? Fit that curve using REGRESSION, and predict the cost of a Super Bowl commercial in 2017 and 2020. Compare your answers to those for part (b).

The Business of Motion Picture Revenue and DVD Release

There has been increasing pressure by motion picture executives to narrow the gap between theater release of a movie and release to DVD. The executives want to reduce marketing expenses, adapt to the increasing consumption of on-demand movies, and boost decreasing DVD sales. Theater owners, on the other hand, want to protect sales. The owners are fearful that if movie executives shorten the time between theater release and DVD release, the number of ticket buyers will drop, as many people will wait for the DVD.

The table to the right gives the number of days between theater release and DVD release for 10 movies. Note that the average gap in time is about 4 months.

Let's examine the data for *Monsters University*. The table on the facing page presents the weekly estimates of gross revenue, *G*, for the movie. The total revenue, *R*, is approximated by adding each week's gross revenue to the previous week's total revenue.

Exercises

1. A movie executive wants to fit a function to the data. Make a scatterplot of the data points (t, G); then use REGRESSION to fit linear, quadratic, cubic, and exponential functions to the data, and graph each equation with the scatterplot. Which function fits best? Why?

2. Use the exponential function to predict gross revenue, *G*, for weeks 9 through 13.

MOVIE	RELEASE DATES	NUMBER OF DAYS SEPARATING RELEASE DATES
Dallas Buyers Club	Theater: Nov. 1, 2013 DVD: Feb. 4, 2014	96
12 Years a Slave	Theater: Oct. 18, 2013 DVD: Mar. 4, 2014	137
American Hustle	Theater: Dec. 13, 2013 DVD: Mar. 18, 2014	95
Lee Daniels' The Butler	Theater: Aug. 16, 2013 DVD: Jan. 14, 2014	151
Captain Phillips	Theater: Oct. 11, 2013 DVD: Jan. 21, 2014	102
Rush	Theater: Sept. 20, 2013 DVD: Jan. 28, 2014	130
Monsters University	Theater: June 21, 2013 DVD: Oct. 29, 2013	130
Man of Steel	Theater: June 14, 2013 DVD: Dec. 12, 2013	181
Star Trek: Into Darkness	Theater: May 17, 2013 DVD: Sept. 10, 2013	116
World War Z	Theater: June 21, 2013 DVD: Sept. 17, 2013	88
		Average = 122 days, or about 4 months

(*Source:* www.imdb.com.)

3. Use your predictions from Exercise 2 to compute total revenue, R, for weeks 9 through 13.

4. Use REGRESSION to fit a logistic function of the form $R(t) = c/(1 + ae^{-bt})$ to the data points (t, R) for weeks 1–8. Based on these results, what seems to be a limiting value for the total revenue from *Monsters University*?

5. Find the rate of change $R'(t)$, and explain what it means. Find $\lim_{t \to \infty} R'(t)$, and explain what it represents.

6. The DVD release for *Monsters University* was October 29, 2013. Based on the given data, do you think this release date was appropriate? What other factors might influence the release date of a DVD?

7. Now, consider the data for *Dallas Buyer's Club*, shown in the table to the right. This movie was released in selected theaters on November 1, 2013, and released nationwide (United States and Canada) on November 22, 2013. Make a scatterplot of the data points (t, G) for weeks 1 through 8, and fit linear, quadratic, cubic, and exponential functions to the data. Which function, if any, best estimates the weekly gross revenue for weeks 9 through 14?

8. Make a scatterplot of the data points (t, G) for weeks 4 through 8, and fit linear, quadratic, cubic and exponential functions to the data. Which function, if any, best estimates the weekly gross revenue for the weeks 9 through 14?

9. Using procedures similar to Exercises 2–4, find a limiting value for the total revenue of *Dallas Buyers Club*. The DVD's release date was February 4, 2014, or 14 weeks after theater release. Was the release date appropriate? Why or why not?

Revenue for *Monsters University*

WEEK IN RELEASE, t (week 1 = June 20–26, 2013)	WEEKLY REVENUE, G (current week estimates, in millions)	TOTAL REVENUE, R (or cumulative box office revenue, in millions)
1. June 26	$115.33	$115.33
2. July 2	$70.57	$185.90
3. July 9	$36.09	$221.99
4. July 16	$19.23	$241.22
5. July 23	$9.72	$250.94
6. July 30	$5.49	$256.43
7. Aug. 6	$2.66	$259.09
8. Aug. 13	$1.37	$260.46
9. Aug. 20	?	?
10. Aug. 27	?	?
11. Sept. 3	?	?
12. Sept. 10	?	?
13. Sept. 17	?	?

(*Source:* www.the-numbers.com.)

Revenue for *Dallas Buyers Club*

WEEK IN RELEASE, t (week 1 = November 1–7, 2013)	WEEKLY REVENUE, G (current week estimates, in millions)	TOTAL REVENUE, R (or cumulative box office revenue, in millions)
1. Nov. 7	$0.35	$0.35
2. Nov. 14	$0.91	$1.26
3. Nov. 21	$2.43	$3.69
4. Nov. 28	$4.01	$7.70
5. Dec. 5	$3.25	$10.95
6. Dec. 12	$2.18	$13.13
7. Dec. 19	$1.64	$14.77
8. Dec. 26	$0.68	$15.45
9. Jan. 2, 2014	?	?
10. Jan. 9	?	?
11. Jan. 16	?	?
12. Jan. 23	?	?
13. Jan. 30	?	?
14. Feb. 6 (week of DVD release)	?	?

(*Source:* www.the-numbers.com.)

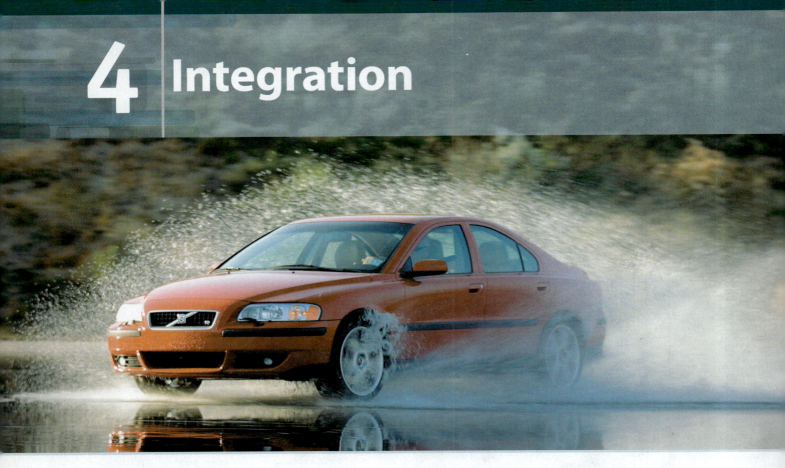

4 | Integration

What You'll Learn

4.1 Antidifferentiation
4.2 Antiderivatives as Areas
4.3 Area and Definite Integrals
4.4 Properties of Definite Integrals
4.5 Integration Techniques: Substitution
4.6 Integration Techniques: Integration by Parts
4.7 Integration Techniques: Tables

Why It's Important

When a vehicle is moving, it accumulates distance. Is it possible to determine the distance a vehicle has traveled if we know its velocity function? Similarly, for every unit of product sold, a company accumulates profit. Can we determine a company's total profit if we know its marginal-profit function? We can, using a process called *integration*, which is one of the two main branches of calculus, the other being differentiation. In this chapter, we explore some of the many practical applications of integration in science, business, and statistics.

Where It's Used

Braking Distance: Juanita's car is traveling at 40 mi/hr (58.67 ft/sec) when she applies the brakes. The velocity, t seconds after she applies the brakes, is $v(t) = -1.197t^2 + 58.67$, where $v(t)$ is in feet per second and $0 \leq t \leq 7$. How far did the car travel while Juanita was braking? (*This problem appears as Example 10 in Section 4.3.*)

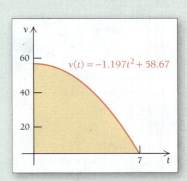

$v(t) = -1.197t^2 + 58.67$

Antidifferentiation

4.1

- Find an antiderivative of a function.
- Evaluate indefinite integrals using basic rules of antidifferentiation.
- Use initial conditions to determine an antiderivative.

Antidifferentiation is the process of differentiation in reverse. Given a function f, we determine another function F such that $\frac{d}{dx}F(x) = f(x)$. We say that F is an *antiderivative* of f.

For example, if $f(x) = 2x$, then $F(x) = x^2$ is an antiderivative of f, since $\frac{d}{dx}x^2 = 2x$. Note that $F(x) = x^2 + 1$ and $F(x) = x^2 - 10$ are also antiderivatives of $f(x) = 2x$. Thus, an antiderivative of $f(x) = 2x$ is any function that can be written in the form $F(x) = x^2 + C$, where C is any constant.

THEOREM 1

The **antiderivative** of $f(x)$ is the set of functions $F(x) + C$ such that

$$\frac{d}{dx}[F(x) + C] = f(x).$$

The constant C is called the **constant of integration**.

If two functions F and G have the same derivative, then F and G differ by a constant: $F(x) = G(x) + C$.

If F is an antiderivative of f, we write

$$\int f(x)\, dx = F(x) + C.$$

This equation is read as "the antiderivative of $f(x)$, with respect to x, is $F(x) + C$." The expression on the left side is called an **indefinite integral**. The symbol $\int$ is the *integral sign*, and $f(x)$ is the *integrand*. The symbol dx can be regarded as indicating that x is the variable of integration.

EXAMPLE 1   Determine these indefinite integrals. That is, find the antiderivative of each integrand:

a) $\int 8\, dx$;   **b)** $\int 3x^2\, dx$;   **c)** $\int e^x\, dx$;   **d)** $\int \frac{1}{x}\, dx$.

Solution   We have seen these integrands before as derivatives of other functions.

a) $\int 8\, dx = 8x + C$ *Check:* $\frac{d}{dx}(8x + C) = 8.$

b) $\int 3x^2\, dx = x^3 + C$ *Check:* $\frac{d}{dx}(x^3 + C) = 3x^2.$

c) $\int e^x\, dx = e^x + C$ *Check:* $\frac{d}{dx}(e^x + C) = e^x.$

d) $\int \frac{1}{x}\, dx = \ln|x| + C$ *Check:* $\frac{d}{dx}(\ln|x| + C) = \frac{1}{x}.$

Every antiderivative can be checked by differentiation.

The results of Example 1 suggest some rules of antidifferentiation, which are summarized in Theorem 2.

THEOREM 2   Rules of Antidifferentiation

A1. Constant Rule:

$$\int k \, dx = kx + C.$$

A2. Power Rule (where $n \neq -1$):

$$\int x^n \, dx = \frac{1}{n+1} x^{n+1} + C.$$

A3. Natural Logarithm Rule:

$$\int \frac{1}{x} \, dx = \ln |x| + C, \text{ and for } x > 0, \int \frac{1}{x} \, dx = \ln x + C.$$

A4. Exponential Rule (base e):

$$\int e^{ax} \, dx = \frac{1}{a} e^{ax} + C, \qquad a \neq 0.$$

Note that the Power Rule of Antidifferentiation can be viewed as a two-step process:

$$\int x^n \, dx = \frac{1}{n+1} x^{n+1} + C$$

①   ②

1. Raise the power by 1.
2. Divide the term by the new power.

EXAMPLE 2   Find the following indefinite integrals:

a) $\int x^7 \, dx;$ **b)** $\int x^{2.5} \, dx;$ **c)** $\int \sqrt{x} \, dx;$ **d)** $\int \frac{1}{x^3} \, dx.$

Check each answer by differentiation.

Solution

a) $\int x^7 \, dx = \dfrac{x^{7+1}}{7+1} + C = \dfrac{1}{8} x^8 + C$ $Check: \dfrac{d}{dx}\left(\dfrac{1}{8}x^8 + C\right) = \dfrac{1}{8}(8x^7) = x^7.$

b) $\int x^{2.5} \, dx = \dfrac{x^{2.5+1}}{2.5+1} + C$

$\qquad = \dfrac{1}{3.5} x^{3.5} + C$ $Check: \dfrac{d}{dx}\left(\dfrac{1}{3.5}x^{3.5} + C\right) = \dfrac{1}{3.5}(3.5x^{2.5}) = x^{2.5}.$

c) Recall that $\sqrt{x} = x^{1/2}$. Therefore,

$$\int \sqrt{x} \, dx = \int x^{1/2} \, dx = \frac{x^{(1/2)+1}}{\left(\frac{1}{2}\right)+1} + C = \frac{x^{3/2}}{\frac{3}{2}} + C$$

$$= \frac{2}{3} x^{3/2} + C.$$ $Check: \dfrac{d}{dx}\left(\dfrac{2}{3}x^{3/2} + C\right) = \dfrac{2}{3}\left(\dfrac{3}{2}x^{1/2}\right) = x^{1/2} = \sqrt{x}.$

d) Recall that $\dfrac{1}{x^3} = x^{-3}$. Therefore,

$$\int \frac{1}{x^3} \, dx = \int x^{-3} \, dx = \frac{x^{-3+1}}{-3+1} + C = -\frac{1}{2} x^{-2} + C$$

$$= -\frac{1}{2x^2} + C.$$ $Check: \dfrac{d}{dx}\left(-\dfrac{1}{2}x^{-2} + C\right) = -\dfrac{1}{2}(-2x^{-3}) = x^{-3} = \dfrac{1}{x^3}.$

Quick Check 1 ✔

Find these indefinite integrals:

a) $\int x^{10} \, dx;$

b) $\int x^{1.7} \, dx;$

c) $\int \sqrt[6]{x} \, dx;$

d) $\int \dfrac{1}{x^4} \, dx.$

1 ✔

The Power Rule of Antidifferentiation is valid for all real numbers n, except for $n = -1$. However, as we saw in Example 1(d), if $n = -1$, we have $x^{-1} = \dfrac{1}{x}$, which is the derivative of the natural logarithm function, $y = \ln |x|$. Therefore,

$$\int \frac{1}{x}\, dx = \ln |x| + C, \text{ and for } x > 0, \int \frac{1}{x}\, dx = \ln x + C.$$

In Example 3, we explore the case of $f(x) = e^{ax}$.

EXAMPLE 3 Find $\int e^{4x}\, dx$.

Solution By Rule A4 of Theorem 2, we have

$$\int e^{4x}\, dx = \frac{1}{4} e^{4x} + C.$$

This checks: $\dfrac{d}{dx}\left(\dfrac{1}{4} e^{4x} + C\right) = \dfrac{1}{4}(4e^{4x}) = e^{4x}$; multiplying $\dfrac{1}{4}$ and 4 gives 1.   **2** ✔

Two useful properties of antidifferentiation are presented in Theorem 3.

THEOREM 3   Properties of Antidifferentiation

P1. A constant factor can be moved to the front of an indefinite integral:

$$\int [c \cdot f(x)]\, dx = c \cdot \int f(x)\, dx.$$

P2. The antiderivative of a sum or difference is the sum or difference of the antiderivatives:

$$\int [f(x) \pm g(x)]\, dx = \int f(x)\, dx \pm \int g(x)\, dx.$$

In Example 4, we use the rules of antidifferentiation in conjunction with the properties of antidifferentiation. In part (b), we algebraically simplify the integrand before performing the antidifferentiation steps.

EXAMPLE 4   Find each integral. Assume $x > 0$.

a) $\int (3x^5 + 7x^2 + 8)\, dx$;   **b)** $\displaystyle\int \frac{4 + 3x + 2x^4}{x}\, dx$.

Solution

a) We antidifferentiate each term separately:

$$\int (3x^5 + 7x^2 + 8)\, dx = \int 3x^5\, dx + \int 7x^2\, dx + \int 8\, dx \quad \text{Using Antidifferentiation Property P2}$$

$$= 3\left(\tfrac{1}{6}x^6\right) + 7\left(\tfrac{1}{3}x^3\right) + 8x \quad \text{Using Antidifferentiation Property P1 and Rules A1 and A2}$$

$$= \tfrac{1}{2}x^6 + \tfrac{7}{3}x^3 + 8x + C.$$

Note the simplification of coefficients and the inclusion of just one constant of integration.

Quick Check 2 ✔

Find each antiderivative:

a) $\int e^{-3x}\, dx$;

b) $\int e^{(1/2)x}\, dx$.

b) We note that x is a common denominator and then simplify each ratio as much as possible:

$$\frac{4 + 3x + 2x^4}{x} = \frac{4}{x} + \frac{3x}{x} + \frac{2x^4}{x} = \frac{4}{x} + 3 + 2x^3.$$

Quick Check 3 ✔

Find each integral:

a) $\int (2x^4 + 3x^3 - 7x^2)\, dx;$

b) $\int \dfrac{x^2 - 7x + 2}{x^2}\, dx,\ x > 0.$

Therefore,

$$\int \frac{4 + 3x + 2x^4}{x}\, dx = \int \left(\frac{4}{x} + 3 + 2x^3 \right) dx$$
$$= 4 \ln x + 3x + \tfrac{1}{2}x^4 + C.$$

Using Antidifferentiation Properties P1 and P2 and Rules A1, A2, and A3

3 ✔

Initial Conditions

In some cases, we may be given a point that is a solution of the antiderivative, thereby allowing us to solve for C. This point is called an **initial condition.**

EXAMPLE 5   Find $\displaystyle\int (2x + 3)\, dx$ given that $F(1) = -2.$

Solution   If we specify that $F'(x) = 2x + 3$, then we have

$$F(x) = \int (2x + 3)\, dx = x^2 + 3x + C.$$

Since $F(1) = -2$, we make the substitution and solve for C:

$$-2 = (1)^2 + 3(1) + C.$$

Simplifying, we have $-2 = 4 + C$, which gives $C = -6$. Therefore, the specific antiderivative that satisfies the initial condition is

$$F(x) = x^2 + 3x - 6.$$

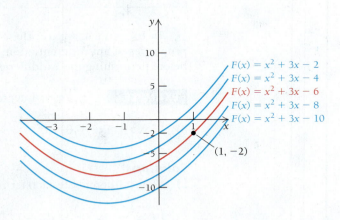

$F(x) = x^2 + 3x - 2$
$F(x) = x^2 + 3x - 4$
$F(x) = x^2 + 3x - 6$
$F(x) = x^2 + 3x - 8$
$F(x) = x^2 + 3x - 10$

$(1, -2)$

Quick Check 4 ✔

Find an antiderivative of $g(x) = e^{2x}$ such that the point $(0, 3)$ is a solution of the antiderivative.

4 ✔

The antiderivative of a function has many applications. For example, in Section 1.8, we saw that the derivative of a distance function is velocity. Therefore, if we are given a velocity function, its antiderivative is a function that describes the moving object's distance from a starting point. If information about the distance of an object at some time t is known, it provides us with an initial condition.

EXAMPLE 6 **Physical Sciences: Height of a Thrown Object.** A rock thrown upward at an initial velocity of 50 ft/sec from an initial height of 10 ft has a velocity modeled by $v(t) = -32t + 50$, where t is the number of seconds after the rock is released and v is in feet per second.

a) Determine a distance function h as a function of t (in this case, "distance" is the height of the rock).

b) Find the height and the velocity of the rock after 3 sec.

Solution

a) Since distance (height) is the antiderivative of velocity, we have

$$h(t) = \int (-32t + 50)\, dt$$
$$= -16t^2 + 50t + C.$$

The initial height, 10 ft, gives us the ordered pair $(0, 10)$ as an initial condition. We substitute 0 for t and 10 for $h(t)$, and solve for C:

$$10 = -16(0)^2 + 50(0) + C$$
$$10 = C.$$

Therefore, the distance function is given by $h(t) = -16t^2 + 50t + 10$.

b) To find the height of the rock after 3 sec, we substitute 3 for t in the distance function:

$$h(3) = -16(3)^2 + 50(3) + 10 = 16 \text{ ft.}$$

The velocity after 3 sec is

$$v(3) = -32(3) + 50 = -46 \text{ ft/sec.}$$

Thus, the rock is 16 ft above the ground, but the negative velocity indicates that it is moving downward. **5** ✔

Quick Check 5 ✔

An arrow shot directly upward has a velocity modeled by $v(t) = -32t + 125$, where t is the number of seconds after the arrow is released and v is in feet per second. Find the arrow's height and velocity after 5 sec.

EXAMPLE 7 **Life Sciences: Change in Population.** The rate of change of the population of Arizona is modeled by the exponential function $P'(t) = 0.0416e^{0.0315t}$, where t is the number of years since 1960 and $P'(t)$ is in millions of people per year. In 2010, Arizona had a population of 6.4 million. (*Source*: U.S. Census Bureau.)

a) Find the population model $P(t)$.

b) Estimate the population of Arizona in 2018.

Solution

a) To find $P(t)$, we antidifferentiate the rate-of-change model:

$$P(t) = \int 0.0416e^{0.0315t}\, dt$$
$$= \frac{0.0416}{0.0315}e^{0.0315t} + C \qquad \text{Using Antidifferentiation Property P1 and Rule A4}$$
$$= 1.321e^{0.0315t} + C. \qquad \text{Simplifying}$$

The year 2010 is 50 yr after 1960, so $(50, 6.4)$ is the initial condition. We substitute and solve for C:

$$6.4 = 1.321e^{0.0315(50)} + C$$
$$6.4 = 6.381 + C$$
$$0.019 = C.$$

Therefore, the population model is $P(t) = 1.321e^{0.0315t} + 0.019$.

Quick Check 6 ✔

Waltonville's rate of population change is modeled by $P'(t) = 34t + 16$, where t is the number of years since 1990 and $P'(t)$ is in people per year.

a) Find the population model for Waltonville if it is known that in 2000, the population was 2500.

b) Forecast the town's population in 2018.

b) The year 2018 corresponds to $t = 58$, and making the substitution gives
$$P(58) = 1.321e^{0.0315(58)} + 0.019$$
$$= 8.19.$$

According to this model, the population of Arizona in 2018 should be about 8,190,000.

6 ✔

Section Summary

- An *antiderivative* of a function f is a function F such that
$$\frac{d}{dx} F(x) = f(x),$$
where the constant C is called the *constant of integration*.
- An antiderivative is denoted by an *indefinite integral* using the integral sign, $\int$. If F is an antiderivative of f, we write
$$\int f(x)\, dx = F(x) + C.$$
We check that an antiderivative is correct by differentiating it.
- The *Constant Rule of Antidifferentiation* is $\int k\, dx = kx + C$.

- The *Power Rule of Antidifferentiation* is
$$\int x^n\, dx = \frac{1}{n+1} x^{n+1} + C, \quad \text{for } n \neq -1.$$
- The *Natural Logarithm Rule of Antidifferentiation* is
$$\int \frac{1}{x}\, dx = \ln |x| + C,$$
or, for $x > 0$,
$$\int \frac{1}{x}\, dx = \ln x + C.$$
- The *Exponential Rule* (base e) of Antidifferentiation is
$$\int e^{ax}\, dx = \frac{1}{a} e^{ax} + C, \quad \text{for } a \neq 0.$$
- An *initial condition* is an ordered pair that is a solution of a particular antiderivative of an integrand.

4.1 | Exercise Set

Find each integral.

1. $\displaystyle\int x^6\, dx$

2. $\displaystyle\int x^7\, dx$

3. $\displaystyle\int 2\, dx$

4. $\displaystyle\int 4\, dx$

5. $\displaystyle\int x^{1/4}\, dx$

6. $\displaystyle\int x^{1/3}\, dx$

7. $\displaystyle\int (x^2 + x - 1)\, dx$

8. $\displaystyle\int (x^2 - x + 2)\, dx$

9. $\displaystyle\int (2t^2 + 5t - 3)\, dt$

10. $\displaystyle\int (3t^2 - 4t + 7)\, dt$

11. $\displaystyle\int \frac{1}{x^3}\, dx$

12. $\displaystyle\int \frac{1}{x^5}\, dx$

13. $\displaystyle\int \sqrt[3]{x}\, dx$

14. $\displaystyle\int \sqrt{x}\, dx$

15. $\displaystyle\int \sqrt{x^5}\, dx$

16. $\displaystyle\int \sqrt[3]{x^2}\, dx$

17. $\displaystyle\int \frac{dx}{x^4}$

18. $\displaystyle\int \frac{dx}{x^2}$

19. $\displaystyle\int \frac{10}{x}\, dx$

20. $\displaystyle\int \frac{2}{x}\, dx$

21. $\displaystyle\int \left(\frac{3}{x} + \frac{5}{x^2} \right) dx$

22. $\displaystyle\int \left(\frac{4}{x^3} + \frac{7}{x} \right) dx$

23. $\displaystyle\int \frac{-7}{\sqrt[3]{x^2}}\, dx$

24. $\displaystyle\int \frac{5}{\sqrt[4]{x^3}}\, dx$

25. $\displaystyle\int 2e^{2x}\, dx$

26. $\displaystyle\int 4e^{4x}\, dx$

27. $\displaystyle\int e^{3x}\, dx$

28. $\displaystyle\int e^{5x}\, dx$

29. $\displaystyle\int e^{7x}\, dx$

30. $\displaystyle\int e^{6x}\, dx$

31. $\displaystyle\int 5e^{3x}\, dx$

32. $\displaystyle\int 2e^{5x}\, dx$

33. $\displaystyle\int 6e^{8x}\,dx$ **34.** $\displaystyle\int 12e^{3x}\,dx$

35. $\displaystyle\int \tfrac{2}{3}e^{-9x}\,dx$ **36.** $\displaystyle\int \tfrac{4}{5}e^{-10x}\,dx$

37. $\displaystyle\int (5x^2 - 2e^{7x})\,dx$

38. $\displaystyle\int (2x^5 - 4e^{3x})\,dx$

39. $\displaystyle\int \left(x^2 - \frac{3}{2}\sqrt{x} + x^{-4/3}\right)dx$

40. $\displaystyle\int \left(x^4 + \frac{1}{8\sqrt{x}} - \frac{4}{5}x^{-2/5}\right)dx$

41. $\displaystyle\int (3x + 2)^2\,dx$ (*Hint:* Expand first.)

42. $\displaystyle\int (x + 4)^2\,dx$

43. $\displaystyle\int \left(\frac{3}{x} - 5e^{2x} + \sqrt{x^7}\right)dx, x > 0$

44. $\displaystyle\int \left(2e^{6x} - \frac{3}{x} + \sqrt[3]{x^4}\right)dx, x > 0$

45. $\displaystyle\int \left(\frac{7}{\sqrt{x}} - \frac{2}{3}e^{5x} - \frac{8}{x}\right)dx, x > 0$

46. $\displaystyle\int \left(\frac{4}{\sqrt[5]{x}} + \frac{3}{4}e^{6x} - \frac{7}{x}\right)dx, x > 0$

Find f such that:

47. $f'(x) = x - 3, \quad f(2) = 9$

48. $f'(x) = x - 5, \quad f(1) = 6$

49. $f'(x) = x^2 - 4, \quad f(0) = 7$

50. $f'(x) = x^2 + 1, \quad f(0) = 8$

51. $f'(x) = 5x^2 + 3x - 7, \quad f(0) = 9$

52. $f'(x) = 8x^2 + 4x - 2, \quad f(0) = 6$

53. $f'(x) = 3x^2 - 5x + 1, \quad f(1) = \tfrac{7}{2}$

54. $f'(x) = 6x^2 - 4x + 2, \quad f(1) = 9$

55. $f'(x) = 5e^{2x}, \quad f(0) = \tfrac{1}{2}$

56. $f'(x) = 3e^{4x}, \quad f(0) = \tfrac{7}{4}$

57. $f'(x) = \dfrac{4}{\sqrt{x}}, \quad f(1) = -5$

58. $f'(x) = \dfrac{2}{\sqrt[3]{x}}, \quad f(1) = 1$

APPLICATIONS

Business and Economics

Credit market debt. *Since 2009, the annual rate of change in the national credit market debt can be modeled by the function*

$$D'(t) = 33.428t + 71.143,$$

where $D'(t)$ is in billions of dollars per year and t is the number of years since 2009. (Source: Based on data from federalreserve.gov/releases/g19/current.) Use the preceding information for Exercises 59 and 60.

59. Find the national credit market debt, $D(t)$, since 2009, given that $D(0) = 2555.229$.

60. What was the national credit market debt in 2013?

61. Total cost from marginal cost. Belvedere, Inc., determines that the marginal cost, C', of producing the xth thermos is given by

$$C'(x) = x^3 - 2x.$$

Find the total-cost function, C, assuming that $C(x)$ is in dollars and that fixed costs are \$7000.

62. Total cost from marginal cost. Solid Rock Industries determines that the marginal cost, C', of producing the xth climbing harness is given by

$$C'(x) = x^3 - x.$$

Find the total-cost function, C, assuming that $C(x)$ is in dollars and that fixed costs are \$6500.

63. Total revenue from marginal revenue. Eloy Chutes determines that the marginal revenue, R', in dollars per unit, from selling the xth parachute is given by

$$R'(x) = x^2 - 3.$$

a) Find the total-revenue function, R, assuming that $R(0) = 0$.

b) Why is $R(0) = 0$ a reasonable assumption?

64. Total revenue from marginal revenue. Taylor Ceramics determines that the marginal revenue, R', in dollars per unit, from selling the xth vase is given by

$$R'(x) = x^2 - 1.$$

a) Find the total-revenue function, R, assuming that $R(0) = 0$.

b) Why is $R(0) = 0$ a reasonable assumption?

65. Demand from marginal demand. Lessard & Company finds that the rate at which the quantity of flameless candles that consumers demand changes with respect to price is given by the marginal-demand function

$$D'(x) = -\frac{4000}{x^2},$$

where x is the price per candle, in dollars. Find the demand function if 1003 candles are demanded by consumers when the price is $4 per candle.

66. Supply from marginal supply. Keans Corporation finds that the rate at which a seller's quantity supplied changes with respect to price is given by the marginal-supply function

$$S'(x) = 0.24x^2 + 4x + 10,$$

where x is the price per unit, in dollars. Find the supply function if it is known that the seller will sell 121 units of the product when the price is $5 per unit.

67. Efficiency of a machine operator. The rate at which a machine operator's efficiency, E (expressed as a percentage), changes with respect to time t is given by

$$\frac{dE}{dt} = 30 - 10t,$$

where t is the number of hours the operator has been at work.

A machine operator's efficiency changes with respect to time.

a) Find $E(t)$, given that the operator's efficiency after working 2 hr is 72%; that is, $E(2) = 72$.
b) Use the answer to part (a) to find the operator's efficiency after 3 hr; after 5 hr.

68. Efficiency of a machine operator. The rate at which a machine operator's efficiency, E (expressed as a percentage), changes with respect to time t is given by

$$\frac{dE}{dt} = 40 - 10t,$$

where t is the number of hours the operator has been at work.

a) Find $E(t)$, given that the operator's efficiency after working 3 hr is 56%; that is, $E(3) = 56$.
b) Use the answer to part (a) to find the operator's efficiency after 4 hr; after 7 hr.

Social and Life Sciences

69. Heart rate. The rate of change in Trisha's pulse (in beats per minute per minute) t minutes after she stops exercising is given by

$$R'(t) = -46.964e^{-0.796t}.$$

a) Find $R(t)$, if Trisha's pulse is 78 beats per minute 2 min after she stopped exercising.
b) Find Trisha's pulse rate 4 min after she stops exercising.
c) Find the rate of change in Trisha's pulse after 4 min.
d) How can $R(t)$ and $R'(t)$ be used to find Trisha's resting pulse rate?

70. Memory. In a memory experiment, the rate at which students memorize Spanish vocabulary is found to be given by

$$M'(t) = 0.2t - 0.003t^2,$$

where $M(t)$ is the number of words memorized in t minutes.

a) Find $M(t)$ if it is known that $M(0) = 0$.
b) How many words are memorized in 8 min?

Physical Sciences

71. Physics: height of a thrown baseball. A baseball is thrown directly upward with an initial velocity of 75 ft/sec from an initial height of 30 ft. The velocity of the baseball t seconds after being released is given by

$$v(t) = -32t + 75,$$

where v is in feet per second.

a) Find the function h that gives the height (in feet) of the baseball after t seconds.
b) What are the height and the velocity of the baseball after 2 sec of flight?
c) After how many seconds does the ball reach its highest point? (*Hint*: The ball "stops" for a moment before starting its downward fall.)
d) How high is the ball at its highest point?
e) After how many seconds will the ball hit the ground?
f) What is the ball's velocity at the moment it hits the ground?

72. Physics: height of an object. A football player punts a football, which leaves his foot at a height of 3 ft above the ground and with an initial upward velocity of 70 ft/sec. The vertical velocity of the football t seconds after it is punted is given by

$$v(t) = -32t + 70,$$

where v is in feet per second.

a) Find the function h that gives the height (in feet) of the football after t seconds.
b) What are the height and the velocity of the football after 1.5 sec?
c) After how many seconds does the ball reach its highest point, and how high is the ball at this point?
d) The punt returner catches the football 5 ft above the ground. What is the vertical velocity of the ball immediately before it is caught?

General Interest

73. Population growth. The rates of change in population for two cities are as follows:

Alphaville: $P'(t) = 45$,

Betaburgh: $Q'(t) = 105e^{0.03t}$,

where t is the number of years since 2000, and both $P'(x)$ and $Q'(x)$ are measured in people per year. In 2000, Alphaville had a population of 5000, and Betaburgh had a population of 3500.

a) Determine the population models for both cities.
b) What were the populations of Alphaville and Betaburgh, to the nearest hundred, in 2010?
c) Sketch the graph of each city's population model, and estimate the year in which the two cities have the same population.

74. Comparing rates of change. Jim is offered a job that will pay him $50 on the first day, $100 on the second day, $150 on the third day, and so on; thus, the rate of change of his pay t days after starting the job is given by $J'(t) = 50$. Larry is offered the same job, but the rate of change of his pay is given by $L'(t) = e^{0.1t}$. Both $J'(t)$ and $L'(t)$ are measured in dollars per day.

a) Determine the total pay model for Jim and for Larry.
b) After 30 days, what is Jim's total pay and Larry's total pay?
c) On what day does Larry's daily pay first exceed Jim's daily pay?
d) In general, how does exponential growth compare to linear growth? Explain.

SYNTHESIS

Solve for $f(t)$.

75. $f'(t) = \sqrt{t} + \dfrac{1}{\sqrt{t}}$, $f(4) = 0$

76. $f'(t) = t^{\sqrt{3}}$, $f(0) = 8$

Solve each integral. Each can be found using rules developed in this section, but some algebra may be required.

77. $\displaystyle\int (5t + 4)^2\, t^4\, dt$

78. $\displaystyle\int (x - 1)^2 x^3\, dx$

79. $\displaystyle\int \dfrac{(t + 3)^2}{\sqrt{t}}\, dt$

80. $\displaystyle\int \dfrac{x^4 - 6x^2 - 7}{x^3}\, dx$

81. $\displaystyle\int (t + 1)^3\, dt$

82. $\displaystyle\int be^{ax}\, dx$

83. $\displaystyle\int (3x - 5)(2x + 1)^2\, dx$

84. $\displaystyle\int \sqrt[3]{64x^4}\, dx$

85. $\displaystyle\int \dfrac{x^2 - 1}{x + 1}\, dx$

86. $\displaystyle\int \dfrac{t^3 + 8}{t + 2}\, dt$

87. On a test, a student makes this statement: "The function given by $f(x) = x^2$ has a unique antiderivative." Is this a true statement? Why or why not?

Answers to Quick Checks

1. (a) $\dfrac{1}{11}x^{11} + C$; **(b)** $\dfrac{1}{2.7}x^{2.7} + C$; **(c)** $\dfrac{6}{7}\sqrt[6]{x^7} + C$;
(d) $-\dfrac{1}{3x^3} + C$ **2. (a)** $-\dfrac{1}{3}e^{-3x} + C$; **(b)** $2e^{(1/2)x} + C$
3. (a) $\dfrac{2}{5}x^5 + \dfrac{3}{4}x^4 - \dfrac{7}{3}x^3 + C$; **(b)** $x - 7\ln x - \dfrac{2}{x} + C$
4. $G(x) = \dfrac{1}{2}e^{2x} + \dfrac{5}{2}$ **5.** 225 ft, -35 ft/sec
6. (a) $P(t) = 17t^2 + 16t + 640$; **(b)** $P(28) = 14{,}416$ people

Antiderivatives as Areas

4.2

- Find the area under a graph and use it to solve real-world problems.
- Use rectangles to approximate the area under a graph.

Integral calculus studies the *accumulation* of units as the input variable increases. For example, suppose a jogger maintains a constant velocity of 5 mi/hr. As she runs, she "accumulates" distance: after 1 hr, she has run 5 mi; after 2 hr, she has run 10 mi, and so on.

Similarly, suppose a manufacturer of skateboards spends $40 per unit. That is, the total cost to produce 1 skateboard is $40, the total cost to produce 2 skateboards is $80, and so on. In this manner, the manufacturer accumulates costs as the number of skateboards produced increases.

We can view an accumulation graphically, as shown in the following example.

EXAMPLE 1

a) Emma's motor scooter travels at 15 mi/hr for 2 hr. How far has she traveled?

b) Green Leaf Skateboards determines that for the first 50 skateboards produced, its cost is $40 per skateboard. What is the total cost to produce 50 skateboards?

Solution

a) Emma's velocity is given by $v(t) = 15$. We graph v for $0 \le t \le 2$ and obtain a rectangle. This rectangle measures 2 units horizontally and 50 units vertically. Its area is width times height, which is the total distance that she has accumulated:

$$2 \text{ hr} \cdot \frac{15 \text{ mi}}{1 \text{ hr}} = 30 \text{ mi}.$$

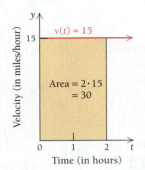
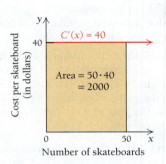

Quick Check 1 ✔

It costs Berglund Bakeries $0.85 to produce 1 dozen cinnamon rolls. Find the total cost of producing 10 dozen cinnamon rolls, and sketch a graph illustrating the total cost as an area.

b) The marginal cost is $C'(x) = 40$, for $0 \le x \le 50$. We graph C' and again obtain a rectangle that measures 50 units by 40 units. Therefore, Green Leaf Skateboards accumulates $2000 in total cost when producing 50 skateboards:

$$50 \text{ skateboards} \cdot \frac{40 \text{ dollars}}{1 \text{ skateboard}} = 2000 \text{ dollars}.$$

1 ✔

Geometry and Areas

For linear functions, we can use geometry to find the area formed by the graph of a function. Two formulas, where $b =$ base and $h =$ height, are useful.

Area of a rectangle: $A = bh$ Area of a triangle: $A = \frac{1}{2}bh$

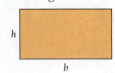
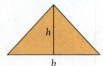

EXAMPLE 2 **Physical Sciences: Distance as Area.** The velocity of a model helicopter is given by $v(t) = 3t$, where t is in seconds and $v(t)$ is in feet per second. Use geometry to find the distance the helicopter travels between the third second and the fifth second ($3 \le t \le 5$).

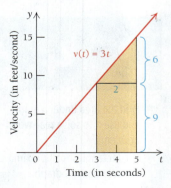

Quick Check 2 ✔

A toy boat moves with a velocity of $v(t) = \frac{1}{2}t$, where t is in minutes and $v(t)$ is in feet per minute.

a) How far does the boat travel during the first 30 min?

b) How far does the boat travel between the first hour and the second hour?

Solution The region corresponding to the interval $3 \le t \le 5$ is a trapezoid. It can be regarded as a rectangle and a triangle as shown to the right. The rectangle has area $A = (2)(9) = 18$, and the triangle has area $A = \frac{1}{2}(2)(6) = 6$. Adding these gives a total area of 24. Therefore, the helicopter travels 24 ft between the third second and the fifth second.

2 ✔

EXAMPLE 3 **Business: Total Profit as Area.** Cousland, Inc., has a marginal-profit function modeled by $P'(t) = 0.15t$, where t is in months and $P'(t)$ is in thousands of dollars per month. Graph P', and determine the total profit earned by Cousland, Inc., in 1 yr ($0 \le t \le 12$).

EXAMPLE 5 Express each of the following without using summation notation.

a) $\sum_{i=1}^{4} 3^i$ **b)** $\sum_{i=1}^{30} h(x_i)\Delta x$

Solution

a) We have

$$\sum_{i=1}^{4} 3^i = 3^1 + 3^2 + 3^3 + 3^4$$

$$= 3 + 9 + 27 + 81 = 120.$$

b) We have

$$\sum_{i=1}^{30} h(x_i)\,\Delta x = h(x_1)\,\Delta x + h(x_2)\,\Delta x + \cdots + h(x_{30})\,\Delta x.$$ 4 ✔

Quick Check 4 ✔

Express $\sum_{i=1}^{6} (i^2 + i)$ without using summation notation.

Approximating the area with rectangles becomes more accurate as we use smaller subintervals, as shown in the following figures.

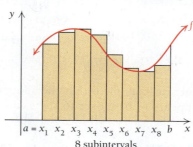

8 subintervals

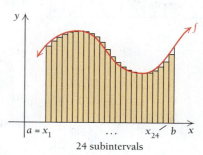

24 subintervals

In general, suppose that the interval $[a, b]$ is divided into n equally sized subintervals, each of width $\Delta x = (b - a)/n$. We construct rectangles with heights

$$f(x_1), f(x_2), \ldots, f(x_n).$$

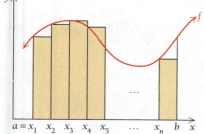

The width of each rectangle is Δx, so the first rectangle has an area of $f(x_1)\Delta x$, the second rectangle has an area of $f(x_2)\Delta x$, and so on. The area of the region under the curve is approximated by the sum of the areas of the rectangles:

$$\sum_{i=1}^{n} f(x_i)\Delta x.$$

In the examples that follow, note that we use the left endpoint of each subinterval to find $f(x)$.

EXAMPLE 6 Consider the graph of

$$f(x) = 300x - \tfrac{1}{2}x^2$$

over the interval $[0, 600]$. Use a Riemann sum to approximate the area under the graph using:

a) 6 equally sized subintervals;

b) 12 equally sized subintervals.

Solution

a) We divide $[0, 600]$ into 6 subintervals of size

$$\Delta x = \frac{600 - 0}{6} = 100,$$

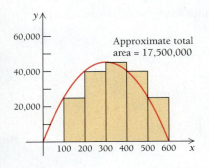

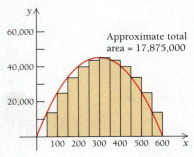

with x_i ranging from $x_1 = 0$ to $x_6 = 500$ in increments of 100. Thus, the area under the graph of f is approximately

$$\sum_{i=1}^{6} f(x_i)\,\Delta x = f(0) \cdot 100 + f(100) \cdot 100 + f(200) \cdot 100$$
$$+ f(300) \cdot 100 + f(400) \cdot 100 + f(500) \cdot 100$$
$$= 0 \cdot 100 + 25{,}000 \cdot 100 + 40{,}000 \cdot 100$$
$$+ 45{,}000 \cdot 100 + 40{,}000 \cdot 100 + 25{,}000 \cdot 100$$
$$= 17{,}500{,}000.$$

b) We divide $[0, 600]$ into 12 subintervals of size $\Delta x = (600 - 0)/12 = 50$, with x_i ranging from $x_1 = 0$ to $x_{12} = 550$ in increments of 50. This gives another approximation of the area under the graph:

$$\sum_{i=1}^{12} f(x_i)\,\Delta x = f(0) \cdot 50 + f(50) \cdot 50 + f(100) \cdot 50 + f(150) \cdot 50$$
$$+ f(200) \cdot 50 + f(250) \cdot 50 + f(300) \cdot 50 + f(350) \cdot 50$$
$$+ f(400) \cdot 50 + f(450) \cdot 50 + f(500) \cdot 50 + f(550) \cdot 50$$
$$= 0 \cdot 50 + 13{,}750 \cdot 50 + 25{,}000 \cdot 50 + 33{,}750 \cdot 50$$
$$+ 40{,}000 \cdot 50 + 43{,}750 \cdot 50 + 45{,}000 \cdot 50 + 43{,}750 \cdot 50$$
$$+ 40{,}000 \cdot 50 + 33{,}750 \cdot 50 + 25{,}000 \cdot 50 + 13{,}750 \cdot 50$$
$$= 17{,}875{,}000.$$

Note in Example 6 that the approximation of the area using $n = 12$ subintervals appears to be closer to the exact value than the approximation using $n = 6$ subintervals.

EXAMPLE 7   Use 5 subintervals to approximate the area under the graph of $f(x) = 0.1x^3 - 2.3x^2 + 12x + 25$ over the interval $[1, 16]$.

Solution   We divide $[1, 16]$ into 5 subintervals of size $\Delta x = (16 - 1)/5 = 3$, with x_i ranging from $x_1 = 1$ to $x_5 = 13$ in increments of 3, as shown below.

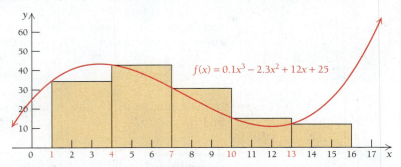

The area under the curve from 1 to 16 is approximately

$$\sum_{i=1}^{5} f(x_i)\,\Delta x = f(1) \cdot 3 + f(4) \cdot 3 + f(7) \cdot 3 + f(10) \cdot 3 + f(13) \cdot 3$$
$$= 34.8 \cdot 3 + 42.6 \cdot 3 + 30.6 \cdot 3 + 15 \cdot 3 + 12 \cdot 3$$
$$= 405.$$
5 ✔

Quick Check 5 ✔

Use 6 subintervals to approximate the area under the graph of the function in Example 7 over the interval $[0, 12]$.

Definite Integrals

The key concept being developed in this section is that the more subintervals we use, the more accurate the approximation of area becomes. As the number of subdivisions n increases, the width of each rectangle Δx decreases. If n approaches infinity, then Δx approaches 0, and the Riemann sum approaches the exact area under the graph.

The *exact* area under the graph of a continuous function $y = f(x)$ over an interval $[a, b]$ is given by a *definite integral*.

DEFINITION

Let $y = f(x)$ be continuous and nonnegative over an interval $[a, b]$. A **definite integral** is the limit as $n \to \infty$ (equivalently, $\Delta x \to 0$) of the Riemann sum of the areas of rectangles under the graph of $y = f(x)$ over $[a, b]$.

$$\text{Exact area} = \lim_{\Delta x \to 0} \sum_{i=1}^{n} f(x_i) \cdot \Delta x = \int_a^b f(x)\, dx.$$

Notice that the summation symbol becomes an integral sign (the elongated "s" is Leibniz notation representing "sum") and Δx becomes dx. The interval endpoints a and b are placed at the bottom and top, respectively, of the integral sign.

If $f(x) \geq 0$ over an interval $[a, b]$, *the definite integral represents area*. The definite integral is also defined for $f(x) < 0$. We will discuss its interpretation in Section 4.3.

We can use geometry to determine the value of some definite integrals, as the following example suggests.

EXAMPLE 8   Find the value of $\displaystyle\int_0^2 (3x + 2)\, dx$.

Solution   We sketch the graph over the interval $[0, 2]$ and note that the region is a trapezoid. Thus, we can use geometry to determine this area.

Using the method similar of Example 2, we find that the area is 10. Therefore,

$$\int_0^2 (3x + 2)\, dx = 10.$$

6 ✔

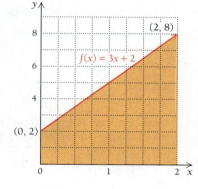

Quick Check 6 ✔

Use geometry to determine the values of these definite integrals:

a) $\displaystyle\int_0^3 (x + 1)\, dx$;

b) $\displaystyle\int_4^7 (15 - 2x)\, dx$.

Section Summary

- The area under a curve can often be interpreted as an accumulation of smaller units of area.
- Geometry can be used to find areas of regions formed by graphs of linear functions.

- *Riemann summation* uses rectangles to approximate the area under the graph of a function. The more rectangles, the better the approximation.
- The *definite integral*, $\int_a^b f(x)\, dx$, represents the exact area under the graph of a continuous function $y = f(x)$, where $f(x) \geq 0$, over an interval $[a, b]$.

4.2   Exercise Set

APPLICATIONS

Business and Economics

In Exercises 1–8, calculate total cost (disregarding any fixed costs) or total profit.

1. Total profit from marginal profit. A concert promoter sells x tickets and has a marginal-profit function given by

$$P'(x) = 2x - 150,$$

where $P'(x)$ is in dollars per ticket. This means that the rate of change of total profit with respect to the number of tickets sold, x, is $P'(x)$. Find the total profit from the sale of the first 300 tickets.

2. Total profit from marginal profit. Poyse Inc. has a marginal-profit function given by

$$P'(x) = -2x + 80,$$

where $P'(x)$ is in dollars per unit. This means that the rate of change of total profit with respect to the number of units produced, x, is $P'(x)$. Find the total profit from the production and sale of the first 40 units.

3. **Total cost from marginal cost.** Sylvie's Old World Cheeses has found that its marginal cost, in dollars per kilogram, is

$$C'(x) = -0.003x + 4.25, \quad \text{for } x \le 500,$$

where x is the number of kilograms of cheese produced. Find the total cost of producing 400 kg of cheese.

4. **Total cost from marginal cost.** Redline Roasting has found that its marginal cost, in dollars per pound, is

$$C'(x) = -0.012x + 6.50, \quad \text{for } x \le 300,$$

where x is the number of pounds of coffee roasted. Find the total cost of roasting 200 lb of coffee.

5. **Total cost from marginal cost.** Cleo's Custom Fabrics has found that its marginal cost, in dollars per yard, is

$$C'(x) = -0.007x + 12, \quad \text{for } x \le 350,$$

where x is the number of yards of fabric produced. Find the total cost of producing 200 yd of this fabric.

6. **Total cost from marginal cost.** Photos from Nature has found that its marginal cost, in cents per card, is

$$C'(x) = -0.04x + 85, \quad \text{for } x \le 1000,$$

where x is the number of cards produced. Find the total cost of producing 650 cards.

7. **Total cost from marginal cost.** Raggs, Ltd., determines that its marginal cost, in dollars per dress, is given by

$$C'(x) = -\frac{2}{25}x + 50, \quad \text{for } x \le 450.$$

Find the total cost of producing the first 200 dresses.

8. **Total cost from marginal cost.** Using the information and answer from Exercise 7, find the cost of producing the 201st dress through the 400th dress.

9. **Total profit from marginal profit.** Beuerlein Industries has found that its marginal profit, in dollars per yard, is

$$C'(x) = -0.015x + 15.50,$$

where x is the number of yards of audio cable produced. Find the total profit from producing 200 yards of audio cable.

10. **Total profit from marginal profit.** Stevens Bakery has found that its marginal profit, in dollars per wedding cake, is

$$C'(x) = -0.12x + 40,$$

where x is the number of wedding cakes produced. Find the total profit from producing 50 wedding cakes.

11. Express $\displaystyle\sum_{i=1}^{4} 2^i$ without using summation notation.

12. Express $\displaystyle\sum_{i=1}^{6} 5i$ without using summation notation.

13. Express $\displaystyle\sum_{i=0}^{10} i^2$ without using summation notation.

14. Express $\displaystyle\sum_{i=0}^{5} (-2)^i$ without using summation notation.

15. Express $\displaystyle\sum_{i=1}^{5} f(x_i)$ without using summation notation.

16. Express $\displaystyle\sum_{i=1}^{4} g(x_i)$ without using summation notation.

17. **a)** Approximate the area under the following graph of $f(x) = \dfrac{1}{x^2}$ over the interval $[1, 7]$ by computing the area of each rectangle to four decimal places and then adding.

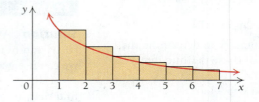

b) Approximate the area under the graph of $f(x) = \dfrac{1}{x^2}$ over the interval $[1, 7]$ by computing the area of each rectangle to four decimal places and then adding. Compare your answer to that for part (a).

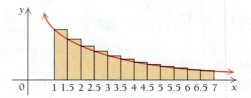

18. **a)** Approximate the area under the graph of $f(x) = x^2 + 1$ over the interval $[0, 5]$ by computing the area of each rectangle and then adding.

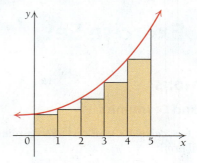

b) Approximate the area under the graph of $f(x) = x^2 + 1$ over the interval $[0, 5]$ by computing the area of each rectangle and then adding. Compare your answer to that for part (a).

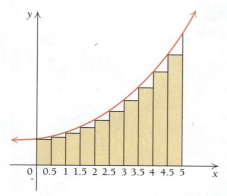

19. Total profit from marginal profit. Holcomb Hill Fitness has found that its marginal profit, $P'(x)$, in cents, is given by

$$P'(x) = -0.0006x^3 + 0.28x^2 + 55.6x, \quad \text{for } x \le 500,$$

where x is the number of members currently enrolled at the health club.

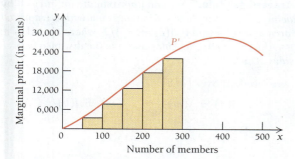

Approximate the total profit when 300 members are enrolled by computing the sum

$$\sum_{i=1}^{6} P'(x_i)\,\Delta x,$$

with $\Delta x = 50$.

20. Total cost from marginal cost. Raggs, Ltd., has found that its marginal cost, in dollars, for the xth jacket produced is given by

$$C'(x) = 0.0003x^2 - 0.2x + 50.$$

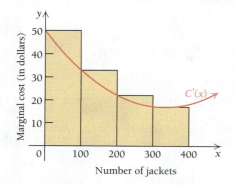

Approximate the total cost of producing 400 jackets by computing the sum

$$\sum_{i=1}^{4} C'(x_i)\,\Delta x,$$

with $\Delta x = 100$.

21. Approximate the area under the graph of

$$f(x) = 0.01x^4 - 1.44x^2 + 60$$

over the interval $[2, 10]$ using 4 subintervals.

22. Approximate the area under the graph of

$$g(x) = -0.02x^4 + 0.28x^3 - 0.3x^2 + 20$$

over the interval $[3, 12]$ using 4 subintervals.

23. Approximate the area under the graph of

$$F(x) = 0.2x^3 + 2x^2 - 0.2x - 2$$

over the interval $[-8, -3]$ using 5 subintervals.

24. Approximate the area under the graph of

$$G(x) = 0.1x^3 + 1.2x^2 - 0.4x - 4.8$$

over the interval $[-10, -4]$ using 6 subintervals.

25. Total cost from marginal cost. Ship Shape Woodworkers has found that the marginal cost of producing x feet of custom molding is given by

$$C'(x) = -0.00002x^2 - 0.04x + 45, \quad \text{for } x \le 800,$$

where $C'(x)$ is in cents. Approximate the total cost of manufacturing 800 ft of molding, using 5 subintervals over $[0, 800]$ and the left endpoint of each subinterval.

26. Total cost from marginal cost. Soulful Scents has found that the marginal cost of producing x ounces of a new fragrance is given by

$$C'(x) = 0.0005x^2 - 0.1x + 30, \quad \text{for } x \le 125,$$

where $C'(x)$ is in dollars. Use 5 subintervals over $[0, 100]$ and the left endpoint of each subinterval to approximate the total cost of producing 100 oz of the fragrance.

27. Total cost from marginal cost. Shelly's Roadside Fruit has found that the marginal cost of producing x pints of fresh-squeezed orange juice is given by

$$C'(x) = 0.000008x^2 - 0.004x + 2, \quad \text{for } x \le 350,$$

where $C'(x)$ is in dollars. Approximate the total cost of producing 270 pt of juice, using 3 subintervals over $[0, 270]$ and the left endpoint of each subinterval.

28. Total cost from marginal cost. Mangianello Paving, Inc., has found that the marginal cost, in dollars, of paving a road surface with asphalt is given by

$$C'(x) = \frac{1}{6}x^2 - 20x + 1800, \quad \text{for } x \le 80,$$

where x is measured in hundreds of feet. Use 4 subintervals over $[0, 40]$ and the left endpoint of each subinterval to approximate the total cost of paving 4000 ft of road surface.

In Exercises 29–36, use geometry to evaluate each definite integral.

29. $\int_0^2 2\, dx$

30. $\int_0^5 6\, dx$

31. $\int_2^6 3\, dx$

32. $\int_{-1}^4 4\, dx$

33. $\int_0^3 x\, dx$

34. $\int_0^5 4x\, dx$

35. $\int_0^{10} \frac{1}{2}x\, dx$

36. $\int_0^5 (2x + 5)\, dx$

SYNTHESIS

37. Show that, for any function f defined for all x_i's, and any constant k, we have

$$\sum_{i=1}^4 kf(x_i) = k\sum_{i=1}^4 f(x_i).$$

Then show that, in general,

$$\sum_{i=1}^n kf(x_i) = k\sum_{i=1}^n f(x_i),$$

for any constant k and any function f defined for all x_i's.

38. Use the following graph of $y = f(x)$ to evaluate each definite integral.

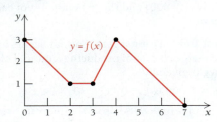

a) $\int_0^2 f(x)\, dx$

b) $\int_2^3 f(x)\, dx$

c) $\int_3^4 f(x)\, dx$

d) $\int_4^7 f(x)\, dx$

e) Use the results from parts (a)–(d) to evaluate

$$\int_0^7 f(x)\, dx.$$

39. Use geometry and the following graph of $f(x) = \frac{1}{2}x$ to evaluate each definite integral.

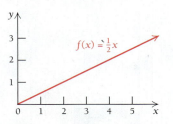

a) $\int_0^1 f(x)\, dx$

b) $\int_1^3 f(x)\, dx$

c) Find c such that $\int_0^c f(x)\, dx = 4$.

d) Find c such that $\int_0^c f(x)\, dx = 3$.

The Trapezoidal Rule. *We can approximate an integral by replacing each rectangle in a Riemann sum with a trapezoid. The area of a trapezoid is $h(c_1 + c_2)/2$, where c_1 and c_2 are the lengths of the parallel sides and h is the distance between the sides. Thus,*

Area under f over $[a, b]$

$$\approx \Delta x \frac{f(a) + f(m)}{2} + \Delta x \frac{f(m) + f(b)}{2}$$

$$\approx \Delta x \left[\frac{f(a)}{2} + f(m) + \frac{f(b)}{2} \right].$$

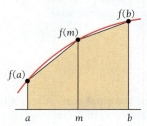

If $[a, b]$ is subdivided into n equal subintervals of length $\Delta x = (b - a)/n$, we have

Area under f over $[a, b]$

$$\approx \Delta x \left[\frac{f(x_1)}{2} + f(x_2) + f(x_3) + \cdots + f(x_n) + \frac{f(b)}{2} \right],$$

where $x_1 = a$ and

$$x_n = x_{n-1} + \Delta x \quad or \quad x_n = a + (n - 1)\Delta x.$$

40. Use the Trapezoidal Rule and the interval subdivision of Exercise 17(a) to approximate the area under the graph of $f(x) = 1/x^2$ over $[1, 7]$.

41. Use the Trapezoidal Rule and the interval subdivision of Exercise 18(a) to approximate the area under the graph of $f(x) = x^2 + 1$ over $[0, 5]$.

Simpson's Rule. *To use Simpson's Rule to approximate the area under a graph of a function f, interval [a, b] is subdivided into n equal subintervals, where n is an even number. As shown in the following graph, a parabola is fitted to the first, second, and third points, another parabola is fitted to the third, fourth, and fifth points, and so on.*

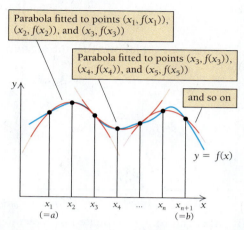

Parabola fitted to points $(x_1, f(x_1))$, $(x_2, f(x_2))$, and $(x_3, f(x_3))$

Parabola fitted to points $(x_3, f(x_3))$, $(x_4, f(x_4))$, and $(x_5, f(x_5))$

and so on

$y = f(x)$

x_1 x_2 x_3 x_4 ... x_n x_{n+1} x
$(=a)$ $(=b)$

The area under the graph of f over [a, b] is approximated by

$$\frac{b-a}{3n}(f(x_1) + 4f(x_2) + 2f(x_3) + \cdots$$
$$+ 2f(x_{n-1}) + 4f(x_n) + f(x_{n+1})),$$

where $x_1 = a$ and $x_{n+1} = b$.

42. Use Simpson's Rule and the interval subdivision of Exercise 17(a) to approximate the area under the graph of $f(x) = \dfrac{1}{x^2}$ over [1, 7].

43. Use Simpson's Rule and 4 subintervals to approximate the area under the graph of $f(x) = \sqrt{x^2 - 1}$ over [2, 4].

44. Use Simpson's Rule and 6 subintervals to approximate the area under the graph of $f(x) = \ln(x^2 + 1)$ over [1, 3].

45. When using Riemann summation to approximate the area under the graph of a function, is it necessary to construct rectangles that have their upper-left corners touching the graph? Would the method work if their upper-right corners touched the graph instead? Why or why not?

46. When using Riemann summation to approximate the area under the graph of a function, is it necessary to divide the interval $[a, b]$ into subintervals of equal width? Why or why not?

TECHNOLOGY CONNECTION

The area, A, of a semicircle of radius r is given by $A = \frac{1}{2}\pi r^2$. Using this equation, compare the answers to Exercises 47 and 48 with the exact area.

47. Approximate the area under the graph of $f(x) = \sqrt{25 - x^2}$ using 10 rectangles.

48. Approximate the area under the graph of $g(x) = \sqrt{49 - x^2}$ using 14 rectangles.

Answers to Quick Checks

1. $8.50;

$C'(x) = 0.85$

0.85

$0.85 \cdot 10 = 8.50$

10 x

2. (a) 225 ft; **(b)** 2700 ft **3.** Total profit = $8925
4. 112 **5.** 368 **6. (a)** 7.5; **(b)** 12

Area and Definite Integrals

4.3

In Sections 4.1 and 4.2, we saw examples in which the area under the graph of a function was found using the function's antiderivative. The process of using an antiderivative of a function to determine the area under the graph of the function is called *integration*.

The Fundamental Theorem of Calculus

- Find the area under the graph of a nonnegative function over a given closed interval.
- Evaluate a definite integral.
- Solve applied problems involving definite integrals.

The area under the graph of a nonnegative continuous function f over an interval $[a, b]$ is determined by an area function A, which is an antiderivative of f; that is, $\dfrac{d}{dx}A(x) = f(x)$.

We have established this fact for cases in which f is a constant or linear function by using geometry formulas for areas of a rectangle and a triangle. When the graph of f is a curve, we can approximate the area underneath the graph using a Riemann sum, which suggests a general method for calculating the area underneath the graph of *any* nonnegative continuous function f.

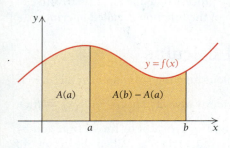

In Section 4.2, we found that the area under the graph of $f(x) = k$ is given by $A(x) = kx$ and the area under the graph of $f(x) = mx$ is given by $A(x) = \frac{1}{2}mx^2$. Note that in both cases the derivative of the area function, $A(x)$, is $f(x)$. Is this always the case?

To answer this, we let $A(x)$ represent the area under a nonnegative continuous function f over the interval $[0, x]$. To find $A'(x)$, we use the definition of derivative:

$$A'(x) = \lim_{h \to 0} \frac{A(x + h) - A(x)}{h}.$$

Since the area under f over $[0, x + h]$ is $A(x + h)$, it follows that the area under f between x and $x + h$ is $A(x + h) - A(x)$.

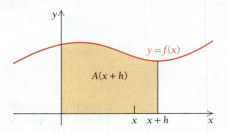
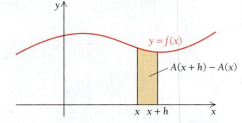

The area represented by $A(x + h) - A(x)$ can be approximated by a rectangle with width h and height $f(x)$. That is, for small values of h,

$$A(x + h) - A(x) \approx h \cdot f(x).$$

Thus, $$\frac{A(x + h) - A(x)}{h} \approx f(x). \qquad \text{Dividing both sides by } h$$

As h approaches 0, the approximation becomes more exact. Taking the limit of both sides, we have

$$\lim_{h \to 0} \frac{A(x + h) - A(x)}{h} = \lim_{h \to 0} f(x).$$

Since $\lim\limits_{h \to 0} f(x) = f(x)$, we see that $A'(x) = f(x)$. We have proved the following remarkable result.

THEOREM 4

Let f be a nonnegative continuous function over $[0, b]$, and let $A(x)$ be the area between the graph of f and the x-axis over $[0, x]$, with $0 < x < b$. Then $A(x)$ is a differentiable function of x and $A'(x) = f(x)$.

Theorem 4 answers the question posed earlier: The derivative of the area function is always the function under which the area is being calculated. We now adapt Theorem 4 for cases in which f is defined over *any* interval $[a, b]$. Referring to the graph on the left, we see that the area over $[a, b]$ is the same as the area over $[0, b]$ minus the area over $[0, a]$ or $A(b) - A(a)$.

In Section 4.1, we found that a function's antiderivatives can differ only by a constant. Thus, if F is another antiderivative of f, then $A(x) = F(x) + C$, for some constant C, and

$$\text{Area} = A(b) - A(a) = F(b) + C - (F(a) + C) = F(b) - F(a).$$

This result tells us that as long as an area is computed by substituting an interval's endpoints into an antiderivative and then subtracting, *any* antiderivative—and any choice of C—can be used. It generally simplifies computations to choose 0 as the value of C.

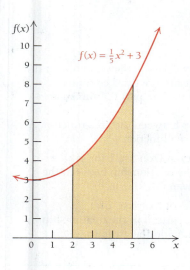

$f(x) = \frac{1}{5}x^2 + 3$

EXAMPLE 1 Find the area under the graph of $f(x) = \frac{1}{5}x^2 + 3$ over $[2, 5]$.

Solution Although making a drawing is not required, doing so helps us visualize the problem.

Note that every antiderivative of $f(x) = \frac{1}{5}x^2 + 3$ is of the form

$$F(x) = \frac{1}{15}x^3 + 3x + C. \quad \textit{Check: } \frac{d}{dx}[\frac{1}{15}x^3 + 3x + C] = \frac{1}{5}x^2 + 3 = f(x)$$

For simplicity, we set $C = 0$, so

$$\text{Area over } [2, 5] = F(5) - F(2)$$
$$= \frac{1}{15}(5)^3 + 3(5) - [\frac{1}{15}(2)^3 + 3(2)]$$
$$= \left(\frac{125}{15} + 15\right) - \left(\frac{8}{15} + 6\right)$$
$$= 16\frac{4}{5}.$$

To find the area under the graph of a nonnegative continuous function f over $[a, b]$:

1. Find any antiderivative F of f.
2. Evaluate $F(x)$ at $x = b$ and $x = a$, and compute $F(b) - F(a)$. The result is the area under the graph over $[a, b]$.

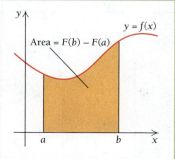

Area = $F(b) - F(a)$, $y = f(x)$

EXAMPLE 2 Find the area under the graph of $y = x^2 + 1$ over $[-1, 2]$.

Solution In this case, $f(x) = x^2 + 1$, with $a = -1$ and $b = 2$.

1. Find any antiderivative F of f. We choose the simplest one:

$$F(x) = \frac{x^3}{3} + x. \quad \text{Choosing } C = 0 \text{ as the constant term}$$

2. Substitute 2 and -1, and find the difference $F(2) - F(-1)$:

$$F(2) - F(-1) = \left[\frac{2^3}{3} + 2\right] - \left[\frac{(-1)^3}{3} + (-1)\right]$$
$$= \frac{8}{3} + 2 - \left[\frac{-1}{3} - 1\right]$$
$$= \frac{8}{3} + 2 + \frac{1}{3} + 1$$
$$= 6.$$

As a partial check, we can count the squares and parts of squares shaded on the graph to the right.

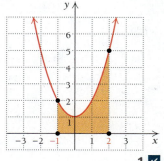

Quick Check 1 ✔

Refer to the function and graph in Example 2.

a) Calculate the area over $[0, 5]$.
b) Calculate the area over $[-2, 2]$.
c) Can you suggest a shortcut for part (b)?

1 ✔

TECHNOLOGY CONNECTION

Antiderivatives and Area

A graphing calculator can calculate the area under the graph of a function. In the Y= window, enter the function $f(x) = 2x$, for $x \geq 0$, and graph it in $[0, 10, 0, 20]$. Press **2ND** and **CALC** and then select $\int f(x)\, dx$ from the list.

For "Lower Limit," type in 0, and press **ENTER**. For "Upper Limit," let $x = 1$, and press **ENTER**. The calculator will shade in the region and report the area in the lower-left corner.

Do this for a series of x-values, and put the information in a table like that on the next page.

(continued)

Antiderivatives and Area (*continued*)

Base, x	Height, $f(x)$	Area of region, $A(x)$
1	2	1
2	4	4
3		
4		
5		
6		
7		

EXERCISES

1. **a)** Fill in the table.
   b) If $x = 20$, what is the area under the graph of f?
   c) What is the relationship between the value of x in the first column and the area in the third column?
   d) Convert your observation from part (c) into an area function $A(x)$.
   e) What is the relationship between the area function A from part (d) and the given function f?

2. Repeat parts (a) through (e) of Exercise 1 for $f(x) = 3$, and look for a pattern in the relationship between the area function A and the given function f.

3. Repeat parts (a) through (e) of Exercise 1 for $f(x) = 3x^2$, and look for a pattern in the relationship between the area function A and the given function f.

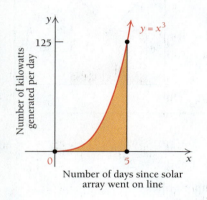

Number of days since solar array went on line

EXAMPLE 3 Let $y = x^3$ represent the number of kilowatts (kW) generated by a new community-owned solar array each day, x days after going on line. Find the area under the graph of $y = x^3$ over $[0, 5]$, and explain what this area represents.

Solution In this case, $f(x) = x^3$, $a = 0$, and $b = 5$.

1. Find any antiderivative F of f. We choose the simplest one:

$$F(x) = \frac{x^4}{4}.$$

2. Substitute 5 and 0, and find the difference $F(5) - F(0)$:

$$F(5) - F(0) = \frac{5^4}{4} - \frac{0^4}{4} = \frac{625}{4} = 156\tfrac{1}{4}.$$

The area represents the total number of kilowatts generated during the first 5 days. To see this, note that the unit of the area is the product of the units on the two axes: $(\text{kW/day})(\text{days}) = \text{kW}$. ∎

The difference $F(b) - F(a)$ is the same for all antiderivatives F of a function f whether the function is nonnegative or not. It is called the *definite integral* of f from a to b.

DEFINITION

Let f be any continuous function over $[a, b]$ and F be any antiderivative of f. Then the **definite integral** of f from a to b is

$$\int_a^b f(x)\,dx = F(b) - F(a).$$

Evaluating definite integrals is called *integration*. The numbers a and b are the **limits of integration**. Note that this use of the word *limit* indicates an endpoint of an interval, not a value that is being approached, as presented in Chapter 1.

It is often convenient to use an intermediate notation:

$$\int_a^b f(x)\, dx = [F(x)]_a^b = F(b) - F(a),$$

where $F(x)$ is an antiderivative of $f(x)$.

EXAMPLE 4 Evaluate each of the following:

a) $\int_{-1}^4 (x^2 - x)\, dx$; **b)** $\int_0^2 e^x\, dx$; **c)** $\int_2^5 \frac{1}{x}\, dx$; **d)** $\int_{-4}^{-1} \frac{1}{x}\, dx$.

Solution

a) $\int_{-1}^4 (x^2 - x)\, dx = \left[\dfrac{x^3}{3} - \dfrac{x^2}{2}\right]_{-1}^4 = \left(\dfrac{4^3}{3} - \dfrac{4^2}{2}\right) - \left(\dfrac{(-1)^3}{3} - \dfrac{(-1)^2}{2}\right)$

$$= \left(\dfrac{64}{3} - \dfrac{16}{2}\right) - \left(\dfrac{-1}{3} - \dfrac{1}{2}\right)$$

$$= \dfrac{64}{3} - 8 + \dfrac{1}{3} + \dfrac{1}{2} = 14\tfrac{1}{6}$$

b) $\int_0^2 e^x\, dx = [e^x]_0^2 = e^2 - e^0 = e^2 - 1$

c) $\int_2^5 \dfrac{1}{x}\, dx = [\ln x]_2^5$ Using the absolute value of x is unnecessary here since $x > 0$ on $[2, 5]$.

$$= \ln 5 - \ln 2$$

$$\approx 0.916$$ Using a calculator

d) $\int_{-4}^{-1} \dfrac{1}{x}\, dx = [\ln |x|]_{-4}^{-1}$

$$= \ln |-1| - \ln |-4|$$

$$= \ln 1 - \ln 4$$

$$= 0 - \ln 4$$

$$\approx -1.386$$ Using a calculator   **2** ✔

Quick Check 2 ✔

Evaluate each of the following:

a) $\int_2^4 (2x^3 - 3x)\, dx$;

b) $\int_0^{\ln 4} 2e^x\, dx$;

c) $\int_1^5 \dfrac{x - 1}{x}\, dx$.

The fact that we can express the integral of a function either as a limit of a sum or in terms of an antiderivative is so important that it has a name: the *Fundamental Theorem of Integral Calculus*.

> **The Fundamental Theorem of Integral Calculus**
>
> If a continuous function f has an antiderivative F over $[a, b]$, then
>
> $$\lim_{n \to \infty} \sum_{i=1}^n f(x_i)\Delta x = \int_a^b f(x)\, dx = F(b) - F(a).$$

It is helpful to envision taking the limit as stretching the summation sign, Σ, into something resembling an S (the integral sign) and redefining Δx as dx. Because Δx is used in the limit, dx appears in the integral notation to indicate that integration is being carried out *with respect to x*. Later we will see that more than one variable can be used.

More on Area

When we evaluate the definite integral of a nonnegative function, we get the area under the graph over an interval.

EXAMPLE 5 Find the area under the graph of $y = 1/x^2$ over $[1, 10]$.

Solution

$$\int_1^{10} \frac{dx}{x^2} = \int_1^{10} x^{-2}\, dx$$

$$= \left[\frac{x^{-2+1}}{-2+1} \right]_1^{10}$$

$$= \left[\frac{x^{-1}}{-1} \right]_1^{10} = \left[-\frac{1}{x} \right]_1^{10}$$

$$= \left(-\frac{1}{10} \right) - \left(-\frac{1}{1} \right)$$

$$= 1 - \tfrac{1}{10} = 0.9$$

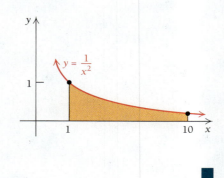

■

EXAMPLE 6 Suppose y is the profit per mile traveled and x is the number of miles traveled, in thousands. Find the area under $y = 1/x$ over the interval $[1, 4]$, and explain what this area represents.

Solution

$$\int_1^4 \frac{dx}{x} = [\ln x]_1^4 = \ln 4 - \ln 1$$

$$= \ln 4 - 0 \approx 1.3863$$

Considering the units, (dollars/mile) · miles = dollars, we see that the area represents a total profit of $1386.30 when the miles traveled increase from 1000 to 4000 miles. ■

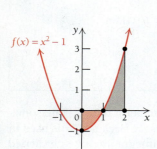

Number of miles traveled
(in thousands)

Now let's compare two similar definite integrals:

$$\int_0^2 x^2\, dx = \left[\frac{x^3}{3} \right]_0^2 \qquad\qquad \int_0^2 -x^2\, dx = \left[-\frac{x^3}{3} \right]_0^2$$

$$= \frac{2^3}{3} - \frac{0^3}{3} = \frac{8}{3} \qquad\qquad = -\frac{2^3}{3} + \frac{0^3}{3} = -\frac{8}{3}$$

The graphs of $y = x^2$ and $y = -x^2$ are reflections of each other across the x-axis. Thus, the shaded areas are the same, $\frac{8}{3}$. The integral involving $y = -x^2$ gives $-\frac{8}{3}$. This shows that for negative-valued functions, the definite integral gives us the opposite of the area between the curve and the x-axis.

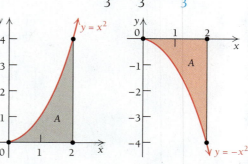

Now let's consider $f(x) = x^2 - 1$ over $[0, 2]$. Here $f(x)$ has both positive and negative values. We can find $\int_0^2 f(x)\, dx$ in two ways.

First, let's use the fact that for any a, b, c, if $a < b < c$, then

$$\int_a^c f(x)\, dx = \int_a^b f(x)\, dx + \int_b^c f(x)\, dx.$$

The area from a to b plus the area from b to c is the area from a to c.

(We will consider this property of integrals again in Section 4.4.) Note that 1 is the x-intercept in $[0, 2]$.

$$\int_0^2 (x^2 - 1)\, dx = \int_0^1 (x^2 - 1)\, dx + \int_1^2 (x^2 - 1)\, dx$$

$$= \left[\frac{x^3}{3} - x\right]_0^1 + \left[\frac{x^3}{3} - x\right]_1^2$$

$$= \underbrace{\left[\left(\frac{1^3}{3} - 1\right) - \left(\frac{0^3}{3} - 0\right)\right]}_{} + \underbrace{\left[\left(\frac{2^3}{3} - 2\right) - \left(\frac{1^3}{3} - 1\right)\right]}_{}$$

$$= \left[\frac{1}{3} - 1\right] + \left[\frac{8}{3} - 2 - \frac{1}{3} + 1\right]$$

$$= -\frac{2}{3} + \frac{4}{3} = \frac{2}{3}.$$

This shows that the area above the x-axis exceeds the area below the x-axis by $\frac{2}{3}$ unit. Now let's evaluate the original integral in another, more direct, way:

$$\int_0^2 (x^2 - 1)\, dx = \left[\frac{x^3}{3} - x\right]_0^2$$

$$= \left(\frac{2^3}{3} - 2\right) - \left(\frac{0^3}{3} - 0\right)$$

$$= \left(\frac{8}{3} - 2\right) - 0 = \frac{2}{3}.$$

The definite integral of a continuous function over an interval is the area above the x-axis minus the area below the x-axis.

EXAMPLE 7   Consider $\int_{-1}^2 (-x^3 + 3x - 1)\, dx$. Predict the sign of the result by examining the graph, and then evaluate the integral.

Solution   From the graph, it appears that there is more area below the x-axis than above. Thus, we expect that

$$\int_{-1}^2 (-x^3 + 3x - 1)\, dx < 0.$$

Evaluating the integral, we have

$$\int_{-1}^2 (-x^3 + 3x - 1)\, dx = \left[-\frac{x^4}{4} + \frac{3}{2}x^2 - x\right]_{-1}^2$$

$$= \left(-\frac{2^4}{4} + \frac{3}{2} \cdot 2^2 - 2\right) - \left(-\frac{(-1)^4}{4} + \frac{3}{2}(-1)^2 - (-1)\right)$$

$$= (-4 + 6 - 2) - \left(-\frac{1}{4} + \frac{3}{2} + 1\right) = 0 - 2\frac{1}{4}$$

$$= -2\frac{1}{4}.$$

As a partial check, we note that the result is negative, as predicted.   **3** ✔

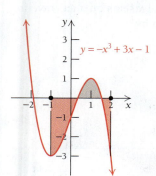

$y = -x^3 + 3x - 1$

Quick Check 3 ✔

Let $f(x) = x^4 - x^2$.

a) Graph f, and then predict the sign of the value of $\int_0^2 f(x)\, dx$ by examining the graph.

b) Evaluate this integral.

TECHNOLOGY CONNECTION

Approximating Definite Integrals

There are two methods for evaluating definite integrals with a calculator. Let's consider the function from Example 7:
$$f(x) = -x^3 + 3x - 1.$$

Method 1: fnInt

First, we select fnInt from the MATH menu. Next, we enter the function, the variable, and the endpoints of the interval over which we are integrating. The calculator returns the same value for the integral as we found in Example 7.

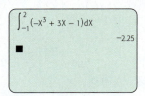

Method 2: $\int f(x)\, dx$

We first graph $y_1 = -x^3 + 3x - 1$. Then we select $\int f(x)dx$ from the CALC menu and enter the lower and upper limits of integration. The calculator shades the area and returns the value found in Example 7.

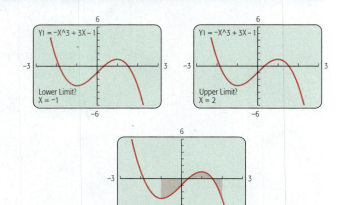

EXERCISES

Evaluate each definite integral.

1. $\displaystyle\int_{-1}^{2} (x^2 - 1)\, dx$ **2.** $\displaystyle\int_{-2}^{3} (x^3 - 3x + 1)\, dx$

3. $\displaystyle\int_{1}^{6} \frac{\ln x}{x^2}\, dx$ **4.** $\displaystyle\int_{-8}^{2} \frac{4}{(1 + e^x)^2}\, dx$

5. $\displaystyle\int_{-10}^{10} (0.002x^4 - 0.3x^2 + 4x - 7)\, dx$

Applications Involving Definite Integrals

EXAMPLE 8   **Business: Total Profit from Marginal Profit.**   Northeast Airlines determines that the marginal profit resulting from the sale of x seats on a jet traveling from Atlanta to Kansas City, in hundreds of dollars, is given by

$$P'(x) = \sqrt{x} - 6.$$

Find the total profit when 60 seats are sold.

Solution   We integrate to find $P(60)$:

$$P(60) = \int_{0}^{60} P'(x)\, dx$$

$$= \int_{0}^{60} (\sqrt{x} - 6)\, dx$$

$$= \left[\frac{2}{3}x^{3/2} - 6x\right]_{0}^{60}$$

$$\approx -50.1613. \qquad \text{Using a calculator}$$

When 60 seats are sold, Northeast's profit is $-\$5016.13$. That is, the airline will lose $\$5016.13$ on the flight.   4 ✔

Quick Check 4 ✔

Business. Referring to Example 8, find the total profit of Northeast Airlines when 140 seats are sold.

Recall that if the position coordinate at time t of a moving object is $s(t)$, then

$$s'(t) = v(t) = \text{the } \textbf{velocity} \text{ at time } t,$$

$$s''(t) = v'(t) = a(t) = \text{the } \textbf{acceleration} \text{ at time } t.$$

EXAMPLE 9   **Physical Science: Distance.**   Suppose a particle's acceleration, in m/sec², is $a(t) = 12t^2 - 6$, with initial velocity $v(0) = 5$ m/sec, and initial position (or distance) $s(0) = 10$ m.

a) Find $s(t)$.

b) Find the particle's position (distance from the initial position) after 2 sec.

Solution

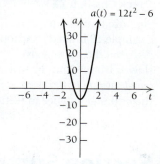
$a(t) = 12t^2 - 6$

a) We first find $v(t)$ by integrating $a(t)$:

$$v(t) = \int a(t)\, dt$$

$$= \int (12t^2 - 6)\, dt$$

$$= 4t^3 - 6t + C_1.$$

The condition $v(0) = 5$ allows us to find C_1:

$$v(0) = 4 \cdot 0^3 - 6 \cdot 0 + C_1 = 5$$

$$C_1 = 5.$$

Thus, $v(t) = 4t^3 - 6t + 5$.

Next we find $s(t)$ by integrating $v(t)$:

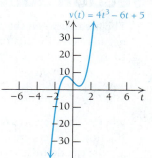
$v(t) = 4t^3 - 6t + 5$

$$s(t) = \int v(t)\, dt$$

$$= \int (4t^3 - 6t + 5)\, dt$$

$$= t^4 - 3t^2 + 5t + C_2.$$

The condition $s(0) = 10$ allows us to find C_2:

$$s(0) = 0^4 - 3 \cdot 0^2 + 5 \cdot 0 + C_2 = 10$$

$$C_2 = 10.$$

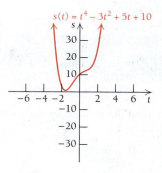
$s(t) = t^4 - 3t^2 + 5t + 10$

Thus, $s(t) = t^4 - 3t^2 + 5t + 10$.

b) After 2 sec, the particle's position is

$$s(2) = (2)^4 - 3(2)^2 + 5(2) + 10$$

$$= 24.$$

That is, it is 24 m from its initial position. ■

EXAMPLE 10   **Physical Science: Braking Distance.**   Juanita's car is traveling at 40 mi/hr (58.67 ft/sec) when she applies the brakes. The velocity, t seconds after she applies the brakes, is $v(t) = -1.197t^2 + 58.67$, where $v(t)$ is in feet per second and $0 \le t \le 7$. How far did the car travel while Juanita was braking?

Solution   The distance traveled is given by the definite integral of $v(t)$:

$$\int_0^7 (-1.197t^2 + 58.67)\, dt = \left[-\frac{1.197}{3} t^3 + 58.67t \right]_0^7$$

$$= -\frac{1.197}{3} (7)^3 + 58.67(7) - 0$$

$$= 273.83 \text{ ft.}$$

In the graph of v, the shaded area represents the distance the vehicle traveled during the 7 sec.

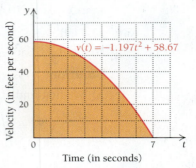

Quick Check 5 ✔

Suppose that the driver in Example 10 braked to a stop in 7 sec but did so "linearly," that is, slowed from 40 mi/hr to a stop at a constant rate of deceleration. Find the braking distance in feet.

5 ✔

Section Summary

- The area between the x-axis and the graph of the non-negative continuous function $y = f(x)$ over an interval $[a, b]$ is found by evaluating the *definite integral*

$$\int_a^b f(x)\, dx = F(b) - F(a),$$

where F is an antiderivative of f.
- If a function has areas both below and above the x-axis, the definite integral gives the net total area, or the difference between the sum of the areas above the x-axis and the sum of the areas below the x-axis.

  o If there is more area above the x-axis than below, then the definite integral will be positive.
  o If there is more area below the x-axis than above, then the definite integral will be negative.
  o If the areas above and below the x-axis are the same, then the definite integral will be 0.

4.3 | Exercise Set

Find the area under the given curve over the indicated interval.

1. $y = 4$; $[1, 3]$

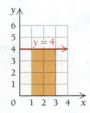

2. $y = 5$; $[1, 3]$

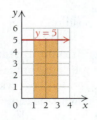

3. $y = 2x$; $[1, 3]$

4. $y = x^2$; $[0, 3]$

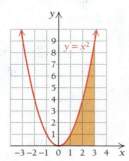

5. $y = x^2$; $[0, 5]$

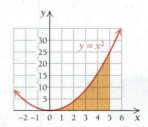

6. $y = x^3$; $[0, 2]$

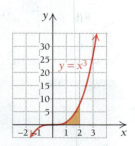

7. $y = x^3$; $[0, 1]$

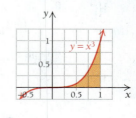

8. $y = 1 - x^2$; $[-1, 1]$

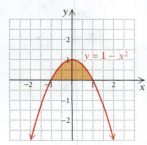

9. $y = 4 - x^2$; $[-2, 2]$

10. $y = e^x$; $[0, 2]$

11. $y = e^x$; $[0, 3]$

12. $y = \dfrac{2}{x}$; $[1, 4]$

13. $y = \dfrac{3}{x}$; $[-6, -1]$

14. $y = x^2 - 4x$; $[-4, -2]$

In each of Exercises 15–24, explain what the shaded area represents.

15.
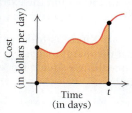

16.
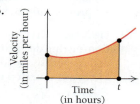

17.
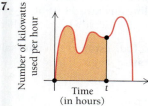

18.
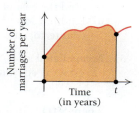

19.
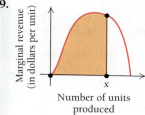

20.
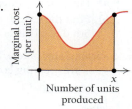

21.
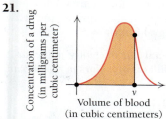

22.

23.
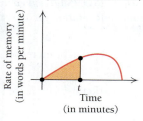

24.

Find the area under the graph of each function over the given interval.

25. $y = x^3$; $[0, 2]$

26. $y = x^4$; $[0, 1]$

27. $y = x^2 + x + 1$; $[2, 3]$

28. $y = 2 - x - x^2$; $[-2, 1]$

29. $y = 5 - x^2$; $[-1, 2]$ 30. $y = e^x$; $[-2, 3]$

31. $y = e^x$; $[-1, 5]$ 32. $y = 2x + \dfrac{1}{x^2}$; $[1, 4]$

In Exercises 33 and 34, determine visually whether $\int_a^b f(x)\,dx$ is positive, negative, or zero, and express $\int_a^b f(x)\,dx$ in terms of area A.

33. a)
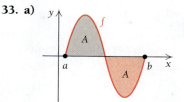

b)
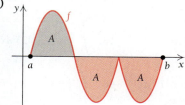

34. a)
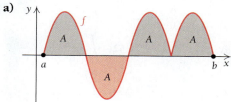

b)
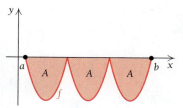

Evaluate each integral. Then state whether the result indicates that there is more area above or below the x-axis or that the areas above and below the axis are equal.

35. $\displaystyle\int_0^{1.5} (x - x^2)\,dx$ 36. $\displaystyle\int_0^2 (x^2 - x)\,dx$

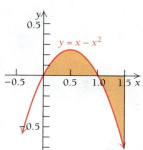

37. $\int_{-1}^{1} (x^3 - 3x)\, dx$ **38.** $\int_{0}^{b} -2e^{3x}\, dx$

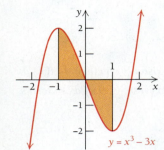

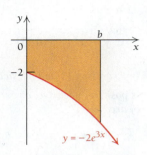

$y = x^3 - 3x$

$y = -2e^{3x}$

39–42. Check the results of each of Exercises 35–38 using a graphing calculator.

Evaluate.

43. $\int_{1}^{3} (3t^2 + 7)\, dt$ **44.** $\int_{1}^{2} (4t^3 - 1)\, dt$

45. $\int_{1}^{4} (\sqrt{x} - 1)\, dx$ **46.** $\int_{1}^{8} (\sqrt[3]{x} - 2)\, dx$

47. $\int_{-2}^{5} (2x^2 - 3x + 7)\, dx$ **48.** $\int_{-2}^{3} (-x^2 + 4x - 5)\, dx$

49. $\int_{-5}^{2} e^t\, dt$ **50.** $\int_{-2}^{3} e^{-t}\, dt$

51. $\int_{a}^{b} \frac{1}{2}x^2\, dx$ **52.** $\int_{a}^{b} \frac{1}{5}x^3\, dx$

53. $\int_{a}^{b} e^{2t}\, dt$ **54.** $\int_{a}^{b} -e^t\, dt$

55. $\int_{1}^{e} \left(x + \frac{1}{x}\right) dx$

56. $\int_{-5}^{-e} \left(x - \frac{1}{x}\right) dx$

57. $\int_{0}^{2} \sqrt{2x}\, dx$ (*Hint:* Simplify first.)

58. $\int_{0}^{27} \sqrt{3x}\, dx$

APPLICATIONS

Business and Economics

59. Business: total revenue. Sally's Sweets finds that the marginal revenue, in dollars, from the sale of x pounds of maple-coated pecans is given by

$$R'(x) = 2x^{1/6}.$$

Find the revenue when 300 lb of maple-coated pecans are produced.

60. Business: total profit. Pure Water Enterprises finds that the marginal profit, in dollars, from drilling a well that is x feet deep is given by

$$P'(x) = \sqrt[5]{x}.$$

Find the profit when a well 250 ft deep is drilled.

61. Business: increasing total profit. Laso Industries finds that the marginal profit, in dollars, from the sale of x digital control boards is given by

$$P'(x) = 2.6x^{0.1}.$$

  a) Find the cost of producing 1200 digital control boards.
  b) A customer orders 1200 digital control boards and later increases the order to 1500. Find the extra profit resulting from the increase in order size.

62. Business: increasing total cost. Kitchens-to-Please Contracting determines that the marginal cost, in dollars per square foot, of installing x square feet of kitchen countertop is given by

$$C'(x) = 4x^{1/3}.$$

  a) Find the cost of installing 50 ft² of countertop.
  b) Find the cost of installing an extra 14 ft² of countertop after 50 ft² have already been installed.

63. Accumulated sales. Raggs, Ltd., estimates that its sales are growing continuously at a rate given by

$$S'(t) = 10e^t,$$

where $S'(t)$ is in dollars per day, on day t.

  a) Find the accumulated sales for the first 5 days.
  b) Find the sales from the 2nd day through the 5th day. (This is the integral from 1 to 5.)

64. Accumulated sales. Melanie's Crafts estimates that its sales are growing continuously at a rate given by

$$S'(t) = 20e^t,$$

where $S'(t)$ is in dollars per day, on day t.

  a) Find the accumulated sales for the first 5 days.
  b) Find the sales from the 2nd day through the 5th day. (This is the integral from 1 to 5.)

Credit market debt. *The annual rate of change in the national credit market debt (in billions of dollars per year) can be modeled by the function*

$$D'(t) = 33.428t + 71.143,$$

where t is the number of years since 2009. (Source: Federal Reserve System.) Use the preceding information for Exercises 65 and 66.

65. By how much did the credit market debt increase between 2009 and 2012?

66. By how much did the credit market debt increase between 2011 and 2015?

Industrial learning curve. *A company is producing a new product, and the time required to produce each unit decreases as workers gain experience. It is determined that*

$$T(x) = 2 + 0.3\left(\frac{1}{x}\right),$$

where $T(x)$ is the time, in hours, required to produce the xth unit. Use this information for Exercises 67 and 68.

67. Find the total time required for a worker to produce units 1 through 10; units 20 through 30.

68. Find the total time required for a worker to produce units 1 through 20; units 20 through 40.

Social Sciences

Memorizing. *The rate of memorizing information initially increases. Eventually, however, a maximum rate is reached, after which it begins to decrease.*

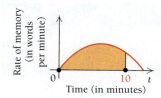

69. Suppose an experiment finds that the rate of memorizing is given by

$$M'(t) = -0.009t^2 + 0.2t,$$

where $M'(t)$ is the memory rate, in words per minute. How many words are memorized in the first 10 min (from $t = 0$ to $t = 10$)?

70. Suppose another experiment finds that the rate of memorizing is given by

$$M'(t) = -0.003t^2 + 0.2t,$$

where $M'(t)$ is the memory rate, in words per minute. How many words are memorized in the first 10 min (from $t = 0$ to $t = 10$)?

71. See Exercise 69. How many words are memorized during minutes 10–15?

72. See Exercise 70. How many words are memorized during minutes 10–17?

Life and Physical Sciences

Find s(t).

73. $v(t) = 3t^2$, $s(0) = 4$

74. $v(t) = 2t$, $s(0) = 10$

Find v(t).

75. $a(t) = 4t$, $v(0) = 20$

76. $a(t) = 6t$, $v(0) = 30$

Find s(t).

77. $a(t) = -2t + 6$, with $v(0) = 6$ and $s(0) = 10$

78. $a(t) = -6t + 7$, with $v(0) = 10$ and $s(0) = 20$

79. Physics. A particle is released as part of an experiment. Its speed t seconds after release is given by $v(t) = -0.5t^2 + 10t$, where $v(t)$ is in meters per second.

   a) How far does the particle travel during the first 5 sec?

   b) How far does it travel during the second 5 sec?

80. Physics. A particle is released during an experiment. Its speed t minutes after release is given by $v(t) = -0.3t^2 + 9t$, where $v(t)$ is in kilometers per minute.

   a) How far does the particle travel during the first 10 min?

   b) How far does it travel during the second 10 min?

81. Distance and speed. A motorcycle accelerates at a constant rate from 0 mph ($v(0) = 0$) to 60 mph in 15 sec. How far has it traveled after 15 sec? (*Hint:* Convert seconds to hours.)

82. Distance and speed. A car accelerates at a constant rate from 0 mph to 60 mph in 30 sec. How far has it traveled after 30 sec?

83. Distance and speed. A bicyclist decelerates at a constant rate from 30 km/hr to a complete stop in 45 sec.

   a) How fast is the bicyclist traveling after 20 sec?

   b) How far has the bicyclist traveled after 45 sec?

84. Distance and speed. A cheetah decelerates at a constant rate from 50 km/hr to a complete stop in 20 sec.

   a) How fast is the cheetah moving after 10 sec?

   b) How far has the cheetah traveled after 20 sec?

85. Distance. For a freely falling object, $a(t) = -32$ ft/sec^2, $v(0) =$ initial velocity $= v_0$ (in ft/sec), and $s(0) =$ initial height $= s_0$ (in ft). Find a general expression for $s(t)$ in terms of v_0 and s_0.

86. Time. A ball is thrown upward from a height of 10 ft, that is, $s(0) = 10$, at an initial velocity of 80 ft/sec, or $v(0) = 80$. How long will it take before the ball hits the ground? (See Exercise 85.)

87. Distance. A car accelerates at a constant rate from 0 to 60 mph in 30 sec. How far does the car travel during that time?

88. Distance. A motorcycle accelerates at a constant rate from 0 to 50 mph in 15 sec. How far does it travel during that time?

89. Physics. A particle starts at the origin. Its velocity, in miles per hour, after t hours is given by

$$v(t) = 3t^2 + 2t.$$

How far does it travel from the start of the 2nd hour through the end of the 5th hour (from $t = 1$ to $t = 5$)?

90. Physics. A particle starts at the origin. Its velocity, in miles per hour, after t hours is given by

$$v(t) = 4t^3 + 2t.$$

How far does it travel from the start through the end of the 3rd hour (from $t = 0$ to $t = 3$)?

SYNTHESIS

91. Total pollution. A factory is polluting a lake in such a way that the rate of pollutants entering the lake after t months is

$$N'(t) = 280t^{3/2},$$

where $N(t)$ is the total number of pounds of pollutants in the lake after t months.

a) How many pounds of pollutants enter the lake in 16 months?
b) An environmental board tells the factory that it must begin cleanup procedures after 50,000 lb of pollutants have entered the lake. After what length of time will this occur?

92. Accumulated sales. Bluetape, Inc., estimates that its sales are growing continuously at a rate given by

$$S'(t) = 0.5e^t,$$

where $S'(t)$ is in dollars per day, on day t. On what day will accumulated sales first exceed $10,000?

Evaluate.

93. $\displaystyle\int_2^3 \frac{x^2 - 1}{x - 1}\, dx$

94. $\displaystyle\int_0^1 (x + 2)^3\, dx$

95. $\displaystyle\int_4^{16} (x - 1)\sqrt{x}\, dx$

96. $\displaystyle\int_1^8 \frac{\sqrt[3]{x^2} - 1}{\sqrt[3]{x}}\, dx$

97. $\displaystyle\int_2^5 (t + \sqrt{3})(t - \sqrt{3})\, dt$

98. $\displaystyle\int_1^3 \left(x - \frac{1}{x}\right)^2 dx$

99. $\displaystyle\int_1^3 \frac{t^5 - t}{t^3}\, dt$

100. $\displaystyle\int_4^9 \frac{t + 1}{\sqrt{t}}\, dt$

Explain the error that has been made in each of Exercises 101 and 102.

101.
$$\int_1^2 (x^2 - x)\, dx = \left[\frac{1}{3}x^3 - \frac{1}{2}x^2\right]_1^2$$
$$= \left(\frac{1}{3}\cdot 2^3 - \frac{1}{2}\cdot 1^2\right)$$
$$= \frac{13}{6}$$

102.
$$\int_1^2 (\ln x - e^x)\, dx = \left[\frac{1}{x} - e^x\right]_1^2$$
$$= \left(\frac{1}{2} - e^2\right) - (1 - e^1)$$
$$= e - e^2 - \frac{1}{2}$$

TECHNOLOGY CONNECTION

Evaluate.

103. $\displaystyle\int_{-8}^{1.4} (x^4 + 4x^3 - 36x^2 - 160x + 300)\, dx$

104. $\displaystyle\int_{-2}^2 \sqrt{4 - x^2}\, dx$

105. $\displaystyle\int_0^8 x(x - 5)^4\, dx$

106. $\displaystyle\int_2^4 \frac{x^2 - 4}{x^2 - 3}\, dx$

107. Prove that $\displaystyle\int_a^b f(x)\, dx = -\int_b^a f(x)\, dx$.

Answers to Quick Checks

1. (a) $46\frac{2}{3}$, *or* $\frac{140}{3}$; **(b)** $9\frac{1}{3}$, *or* $\frac{28}{3}$; **(c)** integrate from 0 to 2, then double the result

2. (a) 102; **(b)** 6; **(c)** $4 - \ln 5 \approx 2.39$

3. (a) Positive; **(b)** $\frac{56}{15}$
4. Approximately $26,433.49
5. Approximately 205.35 ft

Properties of Definite Integrals

The Additive Property of Definite Integrals

4.4

- Use properties of definite integrals to find the area between curves.
- Solve applied problems involving definite integrals.
- Determine the average value of a function.

We have seen that the definite integral

$$\int_a^b f(x)\, dx$$

can be regarded as the area under the graph of $y = f(x) \geq 0$ over the interval $[a, b]$. Thus, if c is such that $a < c < b$, the above integral can be expressed as a sum. This **additive property of definite integrals** is stated in the following theorem.

THEOREM 5

For $a < c < b$,

$$\int_a^b f(x)\, dx = \int_a^c f(x)\, dx + \int_c^b f(x)\, dx.$$

For any number c between a and b, the integral from a to b is the integral from a to c plus the integral from c to b.

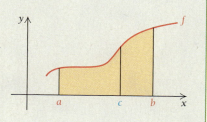

Theorem 5 is useful with piecewise-defined functions.

EXAMPLE 1   Find the area under the graph of $y = f(x)$ from -4 to 5, where

$$f(x) = \begin{cases} 9, & \text{for } x < 3, \\ x^2, & \text{for } x \geq 3. \end{cases}$$

Solution

$$\int_{-4}^5 f(x)\, dx = \int_{-4}^3 f(x)\, dx + \int_3^5 f(x)\, dx$$

$$= \int_{-4}^3 9\, dx + \int_3^5 x^2\, dx$$

$$= 9[x]_{-4}^3 + \left[\frac{x^3}{3}\right]_3^5$$

$$= 9(3 - (-4)) + \left(\frac{5^3}{3} - \frac{3^3}{3}\right)$$

$$= 95\tfrac{2}{3}$$

1 ✔

Quick Check 1 ✔

Find the area under the graph of $y = g(x)$ from -3 to 6, where

$$g(x) = \begin{cases} x^2, & \text{for } x \leq 2, \\ 8 - x, & \text{for } x > 2. \end{cases}$$

EXAMPLE 2   Evaluate each definite integral:

a) $\displaystyle\int_{-3}^4 |x|\, dx;$

b) $\displaystyle\int_0^3 |1 - x^2|\, dx.$

Solution

a) The absolute-value function $f(x) = |x|$ is defined piecewise as follows:

$$f(x) = |x| = \begin{cases} -x, & \text{for } x < 0, \\ x, & \text{for } x \geq 0. \end{cases}$$

See Example 8 in Section R.5.

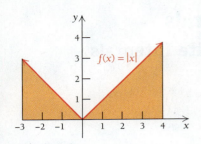

$f(x) = |x|$

Therefore,

$$\int_{-3}^{4} |x|\, dx = \int_{-3}^{0} (-x)\, dx + \int_{0}^{4} x\, dx$$

$$= \left[-\frac{x^2}{2} \right]_{-3}^{0} + \left[\frac{x^2}{2} \right]_{0}^{4}$$

$$= \left[-\frac{0^2}{2} - \left(-\frac{(-3)^2}{2} \right) \right] + \left(\frac{4^2}{2} - \frac{0^2}{2} \right)$$

$$= \frac{9}{2} + \frac{16}{2} = \frac{25}{2}.$$

As a check, this definite integral can also be evaluated using geometry, since the two regions are triangles.

b) The graph of $f(x) = |1 - x^2|$ is the graph of $y = 1 - x^2$, where any portion of the graph below the x-axis (that is, where $y < 0$) is reflected above the x-axis.

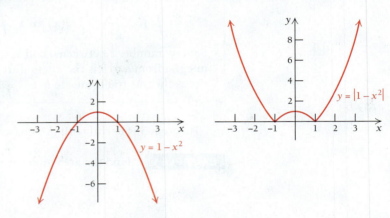

$y = 1 - x^2$ $y = |1 - x^2|$

We see that y is negative for $x < -1$ or $x > 1$. Therefore, the function is defined piecewise, as follows. Note that $-(1 - x^2) = x^2 - 1$.

$$f(x) = |1 - x^2| = \begin{cases} x^2 - 1, & \text{for } x < -1 \text{ or } x > 1, \\ 1 - x^2, & \text{for } -1 \leq x \leq 1. \end{cases}$$

See Exercises 101–108 in Section 2.1.

The definite integral of $f(x) = |1 - x^2|$ over the interval $[0, 3]$ is written as the sum of two definite integrals, of $y = 1 - x^2$ over $[0, 1]$ and of $y = x^2 - 1$ over $[1, 3]$:

$$\int_{0}^{3} |1 - x^2|\, dx = \int_{0}^{1} (1 - x^2)\, dx + \int_{1}^{3} (x^2 - 1)\, dx$$

$$= \left[x - \frac{1}{3}x^3 \right]_{0}^{1} + \left[\frac{1}{3}x^3 - x \right]_{1}^{3}$$

$$= \left[\left(1 - \frac{1}{3} \cdot 1^3 \right) - \left(0 - \frac{1}{3} \cdot 0^3 \right) \right] + \left[\left(\frac{1}{3} \cdot 3^3 - 3 \right) - \left(\frac{1}{3} \cdot 1^3 - 1 \right) \right]$$

$$= \frac{22}{3}.$$

2 ✔

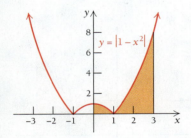

$y = |1 - x^2|$

Quick Check 2 ✔

Evaluate $\int_{0}^{7} |2x - 1|\, dx$.

The Area of a Region Bounded by Two Graphs

Suppose we want to find the area of a region bounded by the graphs of two functions, f and g, as shown below.

Note that the area of the desired region A is the area of A_2 minus that of A_1.

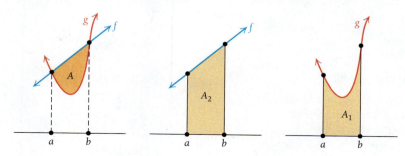

Thus,

$$A = \underbrace{\int_a^b f(x)\,dx}_{A_2} - \underbrace{\int_a^b g(x)\,dx}_{A_1},$$

or $\qquad A = \int_a^b [f(x) - g(x)]\,dx.$

In general, we have the following theorem.

THEOREM 6

Let f and g be continuous functions with $f(x) \geq g(x)$ over $[a, b]$. Then the area of the region between the two curves, from $x = a$ to $x = b$, is

$$\int_a^b [f(x) - g(x)]\,dx.$$

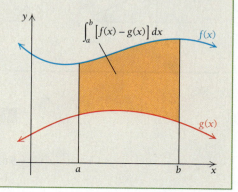

EXAMPLE 3 Find the area of the region bounded by the graphs of $f(x) = 2x + 1$ and $g(x) = x^2 + 1$.

Solution We need to know where the graphs of f and g intersect, as well as which is the upper graph. To find the points of intersection, we set $f(x)$ equal to $g(x)$ and solve.

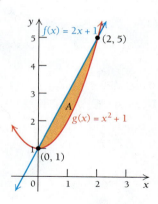

$$f(x) = g(x)$$
$$2x + 1 = x^2 + 1$$
$$0 = x^2 - 2x$$
$$0 = x(x - 2)$$
$$x = 0 \quad \text{or} \quad x = 2$$

The graphs intersect at $x = 0$ and $x = 2$. Graphing both f and g, we see that, over $[0, 2]$, $f(x) \geq g(x)$.

We now compute the area as follows:

$$\int_0^2 [(2x + 1) - (x^2 + 1)] \, dx = \int_0^2 (2x - x^2) \, dx$$

$$= \left[x^2 - \frac{x^3}{3} \right]_0^2$$

$$= \left(2^2 - \frac{2^3}{3} \right) - \left(0^2 - \frac{0^3}{3} \right)$$

$$= 4 - \frac{8}{3}$$

$$= \frac{4}{3}.$$

Quick Check 3 ✔

Find the area of the region bounded by the graphs of $y = \sqrt{x}$ and $y = \frac{1}{3}x$.

3 ✔

TECHNOLOGY CONNECTION

To find the area bounded by two graphs, such as those of $y_1 = -2x - 7$ and $y_2 = -x^2 - 4$, we graph each function and use INTERSECT from the CALC menu to determine the points of intersection.

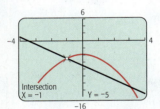

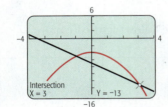

Next, we use fnInt from the MATH menu.

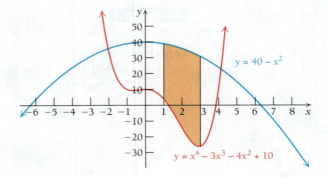

The area bounded by the two curves is about 10.7.

EXERCISES

1. Find the area of the region in Example 3 using this approach.

2. Use Graphicus to find the same area.

EXAMPLE 4 Find the area of the region bounded by

$$y = x^4 - 3x^3 - 4x^2 + 10, \qquad y = 40 - x^2, \qquad x = 1, \qquad \text{and} \qquad x = 3.$$

Solution Note that over $[1, 3]$, the upper graph is $y = 40 - x^2$, as shown below. Thus, $40 - x^2 \geq x^4 - 3x^3 - 4x^2 + 10$ over $[1, 3]$.

The limits of integration are stated and the graphs do not intersect between those limits, so we can compute the area as follows:

$$\int_1^3 [(40 - x^2) - (x^4 - 3x^3 - 4x^2 + 10)]\, dx$$

$$= \int_1^3 (-x^4 + 3x^3 + 3x^2 + 30)\, dx \qquad \text{Simplifying the integrand}$$

$$= \left[-\frac{x^5}{5} + \frac{3}{4}x^4 + x^3 + 30x \right]_1^3$$

$$= \left(-\frac{3^5}{5} + \frac{3}{4}\cdot 3^4 + 3^3 + 30\cdot 3 \right) - \left(-\frac{1^5}{5} + \frac{3}{4}\cdot 1^4 + 1^3 + 30\cdot 1 \right)$$

$$= 97.6.$$

An Environmental Application

EXAMPLE 5   **Life Science: Emission Control.**   A college student develops an engine that is believed to meet all state standards for emission control. The new engine's rate of emission is given by

$$E(t) = 2t^2,$$

where $E(t)$ is the emissions, in billions of pollution particulates per year, at time t, in years. Suppose the emission rate of a conventional engine is given by

$$C(t) = 9 + t^2.$$

The graphs of both curves are shown at the right.

a) At what point in time will the emission rates be the same?

b) What reduction in emissions results with the student's engine up to the time found in part (a)?

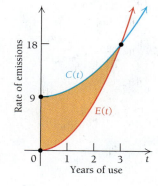

Solution
a) The rate of emission will be the same when $E(t) = C(t)$, or

$$2t^2 = 9 + t^2$$
$$t^2 - 9 = 0$$
$$(t - 3)(t + 3) = 0$$
$$t = 3 \quad \text{or} \quad t = -3.$$

Since negative time values have no meaning here, the emission rates will be the same when $t = 3$ yr.

b) The reduction in emissions is represented by the area of the shaded region in the figure above. It is the area between $C(t) = 9 + t^2$ and $E(t) = 2t^2$, from $t = 0$ to $t = 3$, and is computed as follows:

$$\int_0^3 [(9 + t^2) - 2t^2]\, dt = \int_0^3 (9 - t^2)\, dt$$

$$= \left[9t - \frac{t^3}{3} \right]_0^3$$

$$= \left(9\cdot 3 - \frac{3^3}{3} \right) - \left(9\cdot 0 - \frac{0^3}{3} \right)$$

$$= 27 - 9$$

$$= 18.$$

Over 3 yr, the engine reduces emissions by 18 billion pollution particulates.   4 ✔

Quick Check 4 ✔

Two rockets are fired upward simultaneously. The first rocket's velocity is given by $v_1(t) = 4t$; the second rocket's velocity is given by $v_2(t) = \frac{1}{10}t^2$. In both cases, t is in seconds and velocity is in feet per second.

a) After how many seconds will the velocities of the two rockets be the same?

b) How far ahead (in feet) of the second rocket is the first rocket after the number of seconds found in part (a)?

Average Value of a Continuous Function

Another important use of the area under a curve is in finding the average value of a continuous function over a closed interval.

Suppose that

$$T = f(t)$$

is the temperature at time t recorded at a weather station on a certain day. The station uses a 24-hr clock, so the domain of the temperature function is the interval $[0, 24]$. The function is continuous, as shown in the following graph.

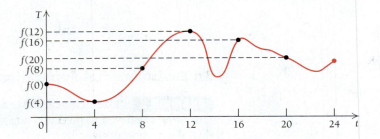

To find the average temperature for the day, we might take six temperature readings at 4-hr intervals, starting at midnight:

$$T_0 = f(0), \quad T_1 = f(4), \quad T_2 = f(8), \quad T_3 = f(12), \quad T_4 = f(16), \quad T_5 = f(20).$$

The average reading would then be the sum of these six readings divided by 6:

$$T_{av} = \frac{T_0 + T_1 + T_2 + T_3 + T_4 + T_5}{6}.$$

This computation of the average temperature has limitations. For example, suppose it is a hot summer day, and at 2:00 in the afternoon (hour 14 on the 24-hr clock), a short thunderstorm cools the air for an hour between our readings. This temporary dip would not show up in the average computed above.

What can we do? We could take 48 readings at half-hour intervals. This should give us a better result. In fact, the shorter the time between readings, the better the result should be. It seems reasonable that we might define the **average value** of T over $[0, 24]$ to be the limit, as n approaches ∞, of the average of n values:

$$\text{Average value of } T = \lim_{n \to \infty} \left(\frac{1}{n} \sum_{i=1}^{n} T_i \right) = \lim_{n \to \infty} \left(\frac{1}{n} \sum_{i=1}^{n} f(t_i) \right).$$

Note that this is not too far from our definition of an integral. All we need is to get Δt, which is $(24 - 0)/n$, or $24/n$, into the summation. We accomplish this by multiplying by 1, writing 1 as $\frac{1}{\Delta t} \cdot \Delta t$:

$$\text{Average value of } T = \lim_{n \to \infty} \left(\frac{1}{\Delta t} \cdot \frac{1}{n} \sum_{i=1}^{n} f(t_i) \, \Delta t \right)$$

$$= \lim_{n \to \infty} \left(\frac{n}{24} \cdot \frac{1}{n} \sum_{i=1}^{n} f(t_i) \, \Delta t \right) \qquad \color{red}{\Delta t = \frac{24}{n}, \text{ so } \frac{1}{\Delta t} = \frac{n}{24}}$$

$$= \frac{1}{24} \lim_{n \to \infty} \sum_{i=1}^{n} f(t_i) \, \Delta t \qquad \color{red}{\text{Using Limit Property L6}}$$

$$= \frac{1}{24} \int_0^{24} f(t) \, dt.$$

DEFINITION

Let f be a continuous function over $[a, b]$. Its **average value**, y_{av}, over $[a, b]$ is given by

$$y_{av} = \frac{1}{b - a} \int_a^b f(x)\, dx.$$

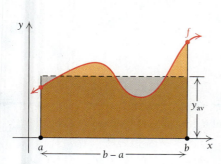

Let's consider average value another way. If we multiply both sides of

$$y_{av} = \frac{1}{b - a} \int_a^b f(x)\, dx$$

by $b - a$, we get

$$(b - a)y_{av} = \int_a^b f(x)\, dx.$$

Now the expression on the left side is the area of a rectangle of length $b - a$ and height y_{av}. The area of such a rectangle is the same as the area under the graph of $y = f(x)$ over $[a, b]$, as shown in the figure at the left.

 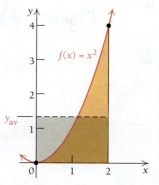

EXERCISE

1. Graph $f(x) = x^4$. Compute the average value of the function over $[0, 2]$, using the method of Example 6. Then use that value, y_{av}, and draw a graph of $y = y_{av}$ using the same set of axes. What does this line represent in comparison to the graph of $f(x) = x^4$ and its associated area?

EXAMPLE 6 Find the average value of $f(x) = x^2$ over $[0, 2]$.

Solution The average value is

$$\frac{1}{2 - 0} \int_0^2 x^2\, dx = \frac{1}{2} \left[\frac{x^3}{3} \right]_0^2$$

$$= \frac{1}{2} \left(\frac{2^3}{3} - \frac{0^3}{3} \right)$$

$$= \frac{1}{2} \cdot \frac{8}{3} = \frac{4}{3}, \quad \text{or } 1\frac{1}{3}.$$

Note that although the values of $f(x)$ increase from 0 to 4 over $[0, 2]$, we do not expect the average value to be 2 (which is half of 4), because we see from the graph that $f(x)$ is less than 2 over more than half the interval. ∎

EXAMPLE 7 Rico's speed, in miles per hour, t minutes after entering the freeway, is given by

$$v(t) = -\frac{1}{200} t^3 + \frac{3}{20} t^2 - \frac{3}{8} t + 60, \quad t \le 30.$$

From 5 min after entering the freeway to 25 min after doing so, what is Rico's average speed? How far does he travel over that time interval?

Solution Rico's average speed is

$$\frac{1}{25 - 5} \int_5^{25} \left(-\frac{1}{200} t^3 + \frac{3}{20} t^2 - \frac{3}{8} t + 60 \right) dt$$

$$= \frac{1}{20} \left[-\frac{1}{800} t^4 + \frac{1}{20} t^3 - \frac{3}{16} t^2 + 60t \right]_5^{25}$$

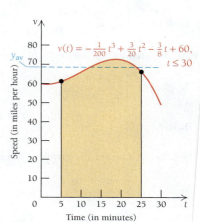

Quick Check 5 ☑

The temperature, in degrees Fahrenheit, in Minneapolis on a winter's day can be modeled by the function

$$T(x) = -0.012x^3 + 0.38x^2 - 1.99x - 10.1,$$

where x is the number of hours from midnight $(0 \le x \le 24)$. Find the average temperature in Minneapolis during this 24-hour period.

$$= \frac{1}{20}\left[\left(-\frac{1}{800} \cdot 25^4 + \frac{1}{20} \cdot 25^3 - \frac{3}{16} \cdot 25^2 + 60 \cdot 25\right)\right.$$

$$\left. - \left(-\frac{1}{800} \cdot 5^4 + \frac{1}{20} \cdot 5^3 - \frac{3}{16} \cdot 5^2 + 60 \cdot 5\right)\right]$$

$$= \frac{1}{20}\left(\frac{53,625}{32} - \frac{9625}{32}\right) = 68\frac{3}{4} \text{ mph.}$$

To find how far Rico travels over the time interval $[5, 25]$, first note that t is given in minutes, not hours. Since 25 min $-$ 5 min $=$ 20 min is $\frac{1}{3}$ hr, the distance traveled over $[5, 25]$ is

$$\tfrac{1}{3} \cdot 68\tfrac{3}{4} = 22\tfrac{11}{12} \text{ mi.} \qquad\qquad 5 ☑$$

Section Summary

- The *additive property of definite integrals* states that a definite integral can be expressed as the sum of two (or more) other definite integrals. If f is continuous over $[a, b]$ and we choose c such that $a < c < b$, then

$$\int_a^b f(x)\, dx = \int_a^c f(x)\, dx + \int_c^b f(x)\, dx.$$

- The area of a region bounded by the graphs of two functions, f and g, with $f(x) \ge g(x)$ over $[a, b]$, is

$$A = \int_a^b [f(x) - g(x)]\, dx.$$

- The *average value* of a continuous function f over $[a, b]$ is

$$y_{av} = \frac{1}{b - a} \int_a^b f(x)\, dx.$$

4.4 Exercise Set

Find the area under the graph of f over $[1, 5]$.

1. $f(x) = \begin{cases} 2x + 1, & \text{for } x \le 3, \\ 10 - x, & \text{for } x > 3 \end{cases}$

2. $f(x) = \begin{cases} x + 5, & \text{for } x \le 4, \\ 11 - \frac{1}{2}x, & \text{for } x > 4 \end{cases}$

Find the area under the graph of g over $[-2, 3]$.

3. $g(x) = \begin{cases} x^2 + 4, & \text{for } x \le 0, \\ 4 - x, & \text{for } x > 0 \end{cases}$

4. $g(x) = \begin{cases} -x^2 + 5, & \text{for } x \le 0, \\ x + 5, & \text{for } x > 0 \end{cases}$

Find the area under the graph of f over $[-6, 4]$.

5. $f(x) = \begin{cases} -x^2 - 6x + 7, & \text{for } x < 1, \\ \frac{3}{2}x - 1, & \text{for } x \ge 1 \end{cases}$

6. $f(x) = \begin{cases} -x - 1, & \text{for } x < -1, \\ -x^2 + 4x + 5, & \text{for } x \ge -1 \end{cases}$

Find the area represented by each definite integral.

7. $\displaystyle\int_0^4 |x - 3|\, dx$

8. $\displaystyle\int_{-1}^1 |3x - 2|\, dx$

9. $\displaystyle\int_0^2 |x^3 - 1|\, dx$

10. $\displaystyle\int_{-3}^4 |x^3|\, dx$

Find the area of the shaded region.

11. $f(x) = 2x + x^2 - x^3,$
$g(x) = 0$

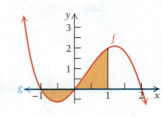

12. $f(x) = x^3 + 3x^2 - 9x - 1$
$g(x) = 4x + 3$

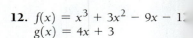
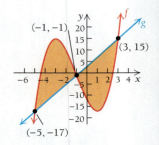

13. $f(x) = x^4 - 8x^3 + 18x^2, \quad g(x) = x + 28$

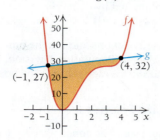

14. $f(x) = 4x - x^2, \quad g(x) = x^2 - 6x + 8$

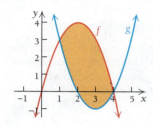

Find the area of the region bounded by the graphs of the given equations.

15. $y = x, y = x^3, x = 0, x = 1$

16. $y = x, y = x^4$

17. $y = x + 2, y = x^2$

18. $y = x^2 - 2x, y = x$

19. $y = 6x - x^2, y = x$

20. $y = x^2 - 6x, y = -x$

21. $y = 2x - x^2, y = -x$

22. $y = x^2, y = \sqrt{x}$

23. $y = x, y = \sqrt[4]{x}$

24. $y = 3, y = x, x = 0$

25. $y = 5, y = \sqrt{x}, x = 0$

26. $y = x^2, y = x^3$

27. $y = 4 - x^2, y = 4 - 4x$

28. $y = x^2 + 1, y = x^2, x = 1, x = 3$

29. $y = x^2 + 3, y = x^2, x = 1, x = 2$

30. $f(x) = x^2 - x - 5, \quad g(x) = x + 10$

31. $f(x) = x^2 - 7x + 20, \quad g(x) = 2x + 6$

32. $y = 2x^2 - x - 3, y = x^2 + x$

33. $y = 2x^2 - 6x + 5, y = x^2 + 6x - 15$

Find the average function value over the given interval.

34. $y = 2x^3; \quad [-1, 1]$

35. $y = 4 - x^2; \quad [-2, 2]$

36. $y = e^x; \quad [0, 1]$

37. $y = e^{-x}; \quad [0, 1]$

38. $y = x^2 - x + 1; \quad [0, 2]$

39. $f(x) = x^2 + x - 2; \quad [0, 4]$

40. $f(x) = mx + 1; \quad [0, 2]$

41. $f(x) = 4x + 5; \quad [0, a]$

42. $f(x) = x^n, n \neq 0; \quad [0, 1]$

43. $f(x) = x^n, n \neq 0; \quad [1, 2]$

44. $f(x) = \dfrac{n}{x}; \quad [1, 5]$

APPLICATIONS

Business and Economics

45. Total and average daily profit. Shylls, Inc., determines that its marginal revenue per day is given by

$$R'(t) = 100e^t, \quad R(0) = 0,$$

where $R(t)$ is the total accumulated revenue, in dollars, on the tth day. The company's marginal cost per day is given by

$$C'(t) = 100 - 0.2t, \quad C(0) = 0,$$

where $C(t)$ is the total accumulated cost, in dollars, on the tth day.

a) Find the total profit from $t = 0$ to $t = 10$ (the first 10 days). *Note:*

$$P(T) = R(T) - C(T) = \int_0^T [R'(t) - C'(t)]\, dt.$$

b) Find the average daily profit for the first 10 days (from $t = 0$ to $t = 10$).

46. Total and average daily profit. Great Green, Inc., determines that its marginal revenue per day is given by

$$R'(t) = 75e^t - 2t, \quad R(0) = 0,$$

where $R(t)$ is the total accumulated revenue, in dollars, on the tth day. The company's marginal cost per day is given by

$$C'(t) = 75 - 3t, \quad C(0) = 0,$$

where $C(t)$ is the total accumulated cost, in dollars, on the tth day.

a) Find the total profit from $t = 0$ to $t = 10$ (see Exercise 45).

b) Find the average daily profit for the first 10 days.

47. Accumulated sales. ProArt, Inc., estimates that its weekly online sales, $S(t)$, in hundreds of dollars, t weeks after online sales began, is given by

$$S(t) = 9e^t.$$

Find the average weekly sales for the first 5 weeks after online sales began.

48. Accumulated sales. Music Manager, Ltd., estimates that monthly revenue, $R(t)$, in thousands of dollars, attributable to its site t months after the site was launched, is given by

$$R(t) = 0.5e^t.$$

Find the average monthly revenue attributable to the Web site for its first 4 months of operation.

49. Refer to Exercise 47. Find ProArt's average weekly online sales for weeks 2 through 5 ($t = 1$ to $t = 5$).

50. Refer to Exercise 48. Find the average monthly revenue from Music Manager's Web site for months 3 through 5 ($t = 2$ to $t = 5$).

Social Sciences

51. Memorizing. In a memory experiment, Alan is able to memorize words at the rate (in words per minute) given by

$$m'(t) = -0.009t^2 + 0.2t.$$

In the same memory experiment, Bonnie is able to memorize words at the rate given by

$$M'(t) = -0.003t^2 + 0.2t.$$

a) How many more words does the person whose memorization rate is higher memorize from $t = 0$ to $t = 10$ (during the first 10 min of the experiment)?

b) Over the first 10 min of the experiment, on average, how many words per minute did Alan memorize?

c) Over the first 10 min of the experiment, on average, how many words per minute did Bonnie memorize?

52. Results of studying. Celia's score on a test, $s(t)$, after t hours of studying, is given by

$$s(t) = t^2, \quad 0 \le t \le 10,$$

Dan's score on the same test is given by

$$S(t) = 10t, \quad 0 \le t \le 10,$$

where $S(t)$ is his score after t hours of studying.

a) For $0 < t < 10$, who will have the higher test score?

b) Find the average value of $s(t)$ over $[7, 10]$, and explain what it represents.

c) Find the average value of $S(t)$ over $[6, 10]$, and explain what it represents.

d) Assuming that both students have the same study habits and are equally likely to study for any number of hours, t, in $[0, 10]$, on average, how far apart will their test scores be?

53. Results of practice. A keyboarder's speed over a 5-min interval is given by

$$W(t) = -6t^2 + 12t + 90, \quad t \text{ in } [0, 5],$$

where $W(t)$ is the speed, in words per minute, at time t.

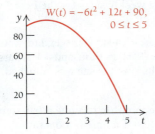

a) Find the speed at the beginning of the interval.
b) Find the maximum speed and when it occurs.
c) Find the average speed over the 5-min interval.

54. Average population. The population of the United States can be approximated by

$$P(t) = 282.3e^{0.01t},$$

where $P(t)$ is in millions and t is the number of years since 2000. (*Source:* Population Division, U.S. Census Bureau.) Find the average value of the population from 2009 to 2013.

Natural and Life Sciences

55. Average drug dose. The concentration, $C(t)$, of phenylbutazone, in micrograms per milliliter (µg/mL), in the plasma of a calf injected with this anti-inflammatory agent is approximately

$$C(t) = 42.03e^{-0.01050t},$$

where t is the number of hours after the injection and $0 \le t \le 120$. (*Source:* A. K. Arifah and P. Lees, "Pharmacodynamics and Pharmacokinetics of Phenylbutazone in Calves," *Journal of Veterinary Pharmacology and Therapeutics*, Vol. 25, 299–309 (2002).)

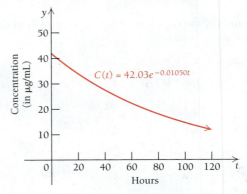

a) Given that this model is accurate for $0 \le t \le 120$, what is the initial dosage?

b) What is the average amount of phenylbutazone in the calf's body for the time between 10 and 120 hours?

56. New York temperature. For any date, the average temperature in New York can be approximated by

$$T(x) = 43.5 - 18.4x + 8.57x^2 - 0.996x^3 + 0.0338x^4,$$

where T represents the temperature in degrees Fahrenheit, $x = 1$ represents the middle of January, $x = 2$ represents the middle of February, and so on. (*Source:* Based on data from www.worldclimate.com.) Compute the average temperature in New York over the whole year to the nearest degree.

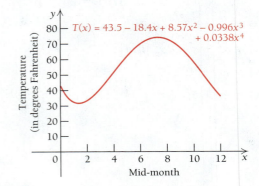

57. Outside temperature. Suppose the temperature in degrees Celsius over a 10-hr period is given by

$$f(t) = -t^2 + 5t + 40, \quad 0 \le t \le 10.$$

a) Find the average temperature.
b) Find the minimum temperature.
c) Find the maximum temperature.

58. Engine emissions. The emissions of an engine are given by

$$E(t) = 2t^2,$$

where $E(t)$ is the engine's rate of emission, in billions of pollution particulates per year, at time t, in years. Find the average emissions from $t = 1$ to $t = 5$.

SYNTHESIS

Find the area of the region bounded by the given graphs.

59. $y = x^2, y = x^{-2}, x = 5$

60. $y = e^x, y = e^{-x}, x = -2$

61. $y = x + 6, y = -2x, y = x^3$

62. $y = x^2, y = x^3, x = -1$

63. $x + 2y = 2, y - x = 1, 2x + y = 7$

64. Find the area bounded by $y = 3x^5 - 20x^3$, the x-axis, and the first coordinates of the relative maximum and minimum values of the function.

65. Find the area bounded by $y = x^3 - 3x + 2$, the x-axis, and the first coordinates of the relative maximum and minimum values of the function.

66. Life science: Poiseuille's Law. The flow of blood in a blood vessel is faster toward the center of the vessel and slower toward the outside. The speed of the blood is given by

$$V = \frac{p}{4Lv}(R^2 - r^2),$$

where R is the radius of the blood vessel, r is the distance of the blood from the center of the vessel, and p, v, and L are physical constants related to the pressure and viscosity of the blood and the length of the blood vessel. If R is constant, we can regard V as a function of r:

$$V(r) = \frac{p}{4Lv}(R^2 - r^2).$$

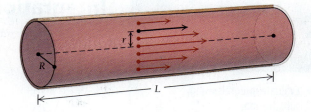

The total blood flow, Q, is given by

$$Q = \int_0^R 2\pi \cdot V(r) \cdot r \cdot dr.$$

Find Q.

67. Solve for K, given that

$$\int_1^2 [(3x^2 + 5x) - (3x + K)] \, dx = 6.$$

Find the area of the region enclosed by the given graphs.

68. $y = x^2 + 4x, y = \sqrt{16 - x^2}$

69. $y = x\sqrt{4 - x^2}, y = \dfrac{-4x}{x^2 + 1}, x = 0, x = 2$

70. $y = 2x^2 + x - 4, y = 1 - x + 8x^2 - 4x^4$

71. $y = \sqrt{1 - x^2}, y = 1 - x^2, x = -1, x = 1$

72. Consider the following functions:

$$f(x) = 3.8x^5 - 18.6x^3,$$
$$g(x) = 19x^4 - 55.8x^2.$$

a) Graph f and g in the window $[-3, 3, -80, 80]$, with Yscl $= 10$.
b) Estimate the first coordinates a, b, and c of the three points of intersection of the two graphs.
c) Find the area between the curves over $[a, b]$.
d) Find the area between the curves over $[b, c]$.

Answers to Quick Checks
1. $\frac{83}{3}$, or $27\frac{2}{3}$ **2.** $\frac{85}{2}$, or $42\frac{1}{2}$ **3.** $\frac{9}{2}$, or $4\frac{1}{2}$
4. (a) 40 sec; **(b)** $1066\frac{2}{3}$ ft **5.** Approximately $-2.5°$F

Integration Techniques: Substitution

• Evaluate integrals using substitution.
• Solve applied problems involving integration by substitution.

The following formulas provide a basis for an integration technique called **substitution**.

A. $\displaystyle\int u^r \, du = \frac{u^{r+1}}{r+1} + C, \quad \text{assuming } r \neq -1$

B. $\displaystyle\int e^u \, du = e^u + C$

C. $\displaystyle\int \frac{1}{u} \, du = \ln |u| + C; \quad \text{and} \quad \int \frac{1}{u} \, du = \ln u + C, \quad u > 0$

In the above formulas, the variable u usually represents some function of x. Recall that we solve $\int x^7 \, dx$ using the Power Rule of Antidifferentiation:

$$\int x^7 \, dx = \frac{x^{7+1}}{7+1} + C = \frac{x^8}{8} + C, \quad \text{or} \quad \frac{1}{8}x^8 + C.$$

But, what about an integral like $\int (3x - 4)^7 \, dx$? Suppose we thought the antiderivative was

$$\frac{(3x - 4)^8}{8} + C.$$

If we check by differentiating, we get

$$8 \cdot \frac{1}{8} \cdot (3x - 4)^7 \cdot 3 \cdot dx.$$

This simplifies to

$$3(3x - 4)^7, \quad not \ (3x - 4)^7.$$

To correct our antiderivative, let's make this substitution:

$$u = 3x - 4.$$

Then $du/dx = 3$ and recalling our work with differentials (Section 2.6), we have

$$du = 3 \cdot dx, \quad \text{and} \quad \frac{du}{3} = dx.$$

With *substitution*, our original integral, $\int (3x - 4)^7 \, dx$, takes the form

$$\int (3x - 4)^7 \, dx = \int u^7 \cdot \frac{du}{3} \qquad \text{Substituting } u \text{ for } 3x - 4 \text{ and } \frac{du}{3} \text{ for } dx$$

$$= \frac{1}{3} \cdot \int u^7 \, du \qquad \text{Factoring out the constant } \frac{1}{3}$$

$$= \frac{1}{3} \cdot \frac{u^8}{8} + C \qquad \text{Using formula A from above}$$

$$= \frac{1}{3 \cdot 8} \cdot (3x - 4)^8 + C = \frac{1}{24}(3x - 4)^8 + C.$$

We leave it to the student to check that this is indeed the antiderivative. Note how this procedure reverses the Chain Rule.

Throughout this section, it is important to recall that

$$\frac{dy}{dx} = f'(x) \quad \text{and} \quad dy = f'(x)\, dx.$$

EXAMPLE 1 Find dy for each function:

a) $y = f(x) = x^3$;

b) $y = f(x) = x^{2/3}$;

c) $u = g(x) = \ln x$;

d) $y = f(x) = e^{x^2}$

Solution

a) We have

$$\frac{dy}{dx} = f'(x) = 3x^2,$$

so $dy = f'(x)\, dx = 3x^2\, dx.$

b) We have

$$\frac{dy}{dx} = f'(x) = \tfrac{2}{3}x^{-1/3},$$

so $dy = f'(x)\, dx = \tfrac{2}{3}x^{-1/3}\, dx.$

c) We have

$$\frac{du}{dx} = g'(x) = \frac{1}{x},$$

so $du = g'(x)\, dx = \frac{1}{x}\, dx, \quad \text{or} \quad \frac{dx}{x}.$

d) Using the Chain Rule, we have

$$\frac{dy}{dx} = f'(x) = e^{x^2} \cdot 2x,$$

so $dy = f'(x)\, dx = 2xe^{x^2}\, dx.$ 1 ✔

Quick Check 1 ✔

Find each differential.

a) For $y = \sqrt{x}$, find dy.

b) For $u = x^2 - 3x$, find du.

c) For $y = \dfrac{1}{x^3}$, find dy.

d) For $u = 4x - 3$, find du.

So far, the dx in

$$\int f(x)\, dx$$

has played no role in integration other than to indicate the variable of integration. Now it becomes convenient to make use of dx. Consider the integral

$$\int 2xe^{x^2}\, dx.$$

If we note that $2xe^{x^2} = e^{x^2} \cdot 2x$, we see in Example 1(d) that $f(x) = e^{x^2}$ is an antiderivative of $f'(x) = 2xe^{x^2}$. How might we find such an antiderivative directly? Suppose we let $u = x^2$. Since

$$\frac{du}{dx} = 2x, \quad \text{we have} \quad du = 2x\, dx.$$

We substitute u for x^2 and du for $2x\, dx$:

$$\int 2xe^{x^2}\, dx = \int e^{x^2} 2x\, dx = \int e^u\, du.$$

Since

$$\int e^u\, du = e^u + C,$$

it follows that

$$\int 2xe^{x^2}\,dx = \int e^u\,du$$

$$= e^u + C$$

$$= e^{x^2} + C. \qquad \text{Replacing } u \text{ with } x^2$$

In effect, we have applied the Chain Rule in reverse. The procedure is referred to as *substitution*, or *change of variable*. We can check the result by differentiating. While there are many integrations that cannot be carried out using substitution, any integral that fits formula A, B, or C on p. 416 can be evaluated with this procedure.

EXAMPLE 2 Evaluate: $\int 3x^2(x^3 + 1)^{10}\,dx.$

Solution Note that $3x^2$ is the derivative of $x^3 + 1$. Thus,

$$\int 3x^2(x^3 + 1)^{10}\,dx = \int (x^3 + 1)^{10}\,3x^2\,dx \qquad \underline{\text{Substitution}}\ \begin{array}{l} u = x^3 + 1, \\ du = 3x^2\,dx \end{array}$$

$$= \int u^{10}\,du$$

$$= \frac{u^{11}}{11} + C$$

$$= \tfrac{1}{11}(x^3 + 1)^{11} + C. \qquad \text{Reversing the substitution}$$

To check, we differentiate:

$$\frac{d}{dx}\left[\tfrac{1}{11}(x^3 + 1)^{11} + C\right] = \tfrac{11}{11}(x^3 + 1)^{10} \cdot 3x^2 + 0$$

Quick Check 2 ✔

Evaluate: $\int 4x(2x^2 + 3)^3\,dx.$

$$= (x^3 + 1)^{10} \cdot 3x^2$$

$$= 3x^2(x^3 + 1)^{10}. \qquad\qquad\qquad\qquad \textbf{2} ✔$$

EXAMPLE 3 Evaluate: $\displaystyle\int \frac{2x\,dx}{1 + x^2}.$

Solution

$$\int \frac{2x\,dx}{1 + x^2} = \int \frac{du}{u} \qquad \underline{\text{Substitution}}\ \begin{array}{l} u = 1 + x^2, \\ du = 2x\,dx \end{array}$$

$$= \ln u + C \qquad \text{Remember: } \int \frac{du}{u} = \int \frac{1}{u} \cdot du.$$

Quick Check 3 ✔

Evaluate: $\displaystyle\int \frac{e^x}{1 + e^x}\,dx.$

$$= \ln(1 + x^2) + C \qquad \text{Using the absolute value is not necessary here since } 1 + x^2 > 0 \text{ for all } x. \qquad \textbf{3} ✔$$

EXAMPLE 4 Evaluate: $\displaystyle\int \frac{2x\,dx}{(1+x^2)^5}$.

Solution

$$\int \frac{2x\,dx}{(1+x^2)^5} = \int \frac{du}{u^5} \qquad \text{Substitution}\;\boxed{\begin{aligned} u &= 1+x^2, \\ du &= 2x\,dx \end{aligned}}$$

$$= \int u^{-5}\,du$$

$$= \frac{u^{-4}}{-4} + C$$

$$= -\frac{1}{4u^4} + C$$

$$= -\frac{1}{4(1+x^2)^4} + C \qquad \left.\begin{aligned}\\\\\end{aligned}\right\} \quad \text{Don't forget to reverse the substitution after integrating.}$$

4 ✔

Quick Check 4 ✔

Evaluate: $\displaystyle\int \frac{6x^2}{\sqrt{3+2x^3}}\,dx$.

EXAMPLE 5 Evaluate: $\displaystyle\int \frac{\ln(3x)\,dx}{x}$.

Solution Note that $\dfrac{d}{dx}\ln(3x) = \dfrac{1}{3x}\cdot 3 = \dfrac{1}{x}$. Thus,

$$\int \frac{\ln(3x)\,dx}{x} = \int u\,du \qquad \text{Substitution}\;\boxed{\begin{aligned} u &= \ln(3x), \\ du &= \frac{1}{x}\,dx \end{aligned}}$$

$$= \frac{u^2}{2} + C$$

$$= \frac{(\ln(3x))^2}{2} + C$$

5 ✔

Quick Check 5 ✔

Evaluate:

$$\int \frac{(\ln x)^2}{x}\,dx, \quad x > 0.$$

EXAMPLE 6 Evaluate: $\displaystyle\int xe^{x^2}\,dx$.

Solution We let $u = x^2$, so $du = 2x\,dx$. Since the integrand contains $x\,dx$, we need a "new" factor of 2. To provide this factor, we multiply by 1, in the form $\frac{1}{2}\cdot 2$:

$$\int xe^{x^2}\,dx = \frac{1}{2}\int 2xe^{x^2}\,dx$$

$$= \frac{1}{2}\int e^{x^2}(2x\,dx)$$

$$= \frac{1}{2}\int e^u\,du \qquad \text{Substitution}\;\boxed{\begin{aligned} u &= x^2, \\ du &= 2x\,dx \end{aligned}}$$

$$= \frac{1}{2}e^u + C = \frac{1}{2}e^{x^2} + C.$$

6 ✔

Quick Check 6 ✔

Evaluate: $\displaystyle\int x^2 e^{4x^3}\,dx$.

EXAMPLE 7 Evaluate: $\displaystyle\int \frac{dx}{x+3}$.

Solution

$$\int \frac{dx}{x+3} = \int \frac{du}{u} \qquad \text{Substitution}\;\boxed{\begin{aligned} u &= x+3, \\ du &= 1\,dx = dx \end{aligned}}$$

$$= \ln u + C$$

$$= \ln|x+3| + C$$

■

EXAMPLE 8 Evaluate: $\int_0^1 5x\sqrt{x^2 + 3}\, dx$. Round to the nearest thousandth.

Solution We first find the antiderivative:

$$\int 5x\sqrt{x^2 + 3}\, dx = 5 \int x\sqrt{x^2 + 3}\, dx$$

$$= \frac{5}{2} \int 2x\sqrt{x^2 + 3}\, dx \qquad \text{Note that } \frac{d}{dx}(x^2 + 3) = 2x.$$

$$= \frac{5}{2} \int \sqrt{x^2 + 3}\, 2x\, dx$$

$$= \frac{5}{2} \int \sqrt{u}\, du \qquad \boxed{\begin{array}{l} \text{Substitution} \\ u = x^2 + 3, \\ du = 2x\, dx \end{array}}$$

$$= \frac{5}{2} \int u^{1/2}\, du$$

$$= \frac{5}{2} \cdot \frac{2}{3} u^{3/2} + C$$

$$= \frac{5}{3}(x^2 + 3)^{3/2} + C. \qquad \text{Reversing the substitution before evaluating the bounds}$$

Using 0 for C, we evaluate the antiderivative:

$$\int_0^1 5x\sqrt{x^2 + 3}\, dx = \left[\frac{5}{3}(x^2 + 3)^{3/2} \right]_0^1$$

$$= \tfrac{5}{3}(1^2 + 3)^{3/2} - \frac{5}{3}(0^2 + 3)^{3/2}$$

$$= \tfrac{5}{3}\left[4^{3/2} - 3^{3/2} \right] \qquad \text{Simplifying}$$

$$\approx 4.673. \qquad \text{Using a calculator}$$

$$\boxed{7\ \checkmark}$$

Quick Check 7 ✔

Evaluate:
$\int_0^2 (x + 1)(x^2 + 2x + 3)^4\, dx$.

In some cases, after a substitution is made, a further simplification can allow us to complete an integration.

EXAMPLE 9 Evaluate: $\int \dfrac{x}{x + 2}\, dx$. Assume $x > -2$.

Solution We substitute $u = x + 2$ and $du = dx$. Note that $x = u - 2$. The substitutions are made:

$$\int \frac{x}{x + 2}\, dx = \int \frac{u - 2}{u}\, du \qquad \boxed{\begin{array}{l} \text{Substitution} \\ u = x + 2, \\ du = dx, \\ x = u - 2 \end{array}}$$

$$= \int \left(1 - \frac{2}{u} \right) du \qquad \frac{u - 2}{u} = \frac{u}{u} - \frac{2}{u} = 1 - \frac{2}{u}$$

$$= u - 2 \ln u + C_1$$

$$= x + 2 - 2 \ln (x + 2) + C_1 \qquad \text{Reversing the substitution}$$

$$= x - 2 \ln (x + 2) + C \qquad \text{Collecting constant terms:}$$

$$C = C_1 + 2 \qquad \boxed{8\ \checkmark}$$

Quick Check 8 ✔

Evaluate $\int \dfrac{x}{(x - 1)^3}\, dx$ by letting $u = x - 1$.

Strategy for Substitution

The following strategy may help with substitution:
1. Decide which rule of antidifferentiation is appropriate.
   a) If you believe it is the Power Rule, let u be the base (see Examples 2, 4, 5, and 8).
   b) If you believe it is the Exponential Rule (base e), let u be the expression in the exponent (see Example 6).
   c) If you believe it is the Natural Logarithm Rule, let u be the denominator (see Examples 3 and 7).
2. Determine du.
3. Inspect the integrand to be sure the substitution includes all factors. You may need to insert constants (see Examples 6 and 8) or make an extra substitution (see Example 9).
4. Perform the antidifferentiation.
5. Reverse the substitution. If there are bounds, use them to evaluate the integral *after* the substitution has been reversed.
6. *Always check your answer by differentiation.*

Section Summary

- Integration by *substitution* is the reverse of applying the Chain Rule.
- The substitution is reversed after the integration has been performed.

- Antiderivatives should be checked using differentiation.

4.5 Exercise Set

Evaluate. (Be sure to check by differentiating!)

1. $\displaystyle\int (8 + x^3)^5 \, 3x^2 \, dx$

2. $\displaystyle\int (x^2 - 7)^6 \, 2x \, dx$

3. $\displaystyle\int (x^2 - 6)^7 x \, dx$

4. $\displaystyle\int (x^3 + 1)^4 x^2 \, dx$

5. $\displaystyle\int (3t^4 + 2)t^3 \, dt$

6. $\displaystyle\int (2t^5 - 3)t^4 \, dt$

7. $\displaystyle\int \frac{2}{1 + 2x} \, dx$

8. $\displaystyle\int \frac{5}{5x + 7} \, dx$

9. $\displaystyle\int (\ln x)^3 \frac{1}{x} \, dx, \, x > 0$

10. $\displaystyle\int (\ln x)^7 \frac{1}{x} \, dx, \, x > 0$

11. $\displaystyle\int e^{3x} \, dx$

12. $\displaystyle\int e^{7x} \, dx$

13. $\displaystyle\int e^{x/3} \, dx$

14. $\displaystyle\int e^{x/2} \, dx$

15. $\displaystyle\int x^4 e^{x^5} \, dx$

16. $\displaystyle\int x^3 e^{x^4} \, dx$

17. $\displaystyle\int t e^{-t^2} \, dt$

18. $\displaystyle\int t^2 e^{-t^3} \, dt$

19. $\displaystyle\int \frac{1}{5 + 2x} \, dx$

20. $\displaystyle\int \frac{1}{2 + 8x} \, dx$

21. $\displaystyle\int \frac{dx}{12 + 3x}$

22. $\displaystyle\int \frac{dx}{1 + 7x}$

23. $\displaystyle\int \frac{dx}{1 - x}$

24. $\displaystyle\int \frac{dx}{4 - x}$

25. $\displaystyle\int t(t^2 - 1)^5 \, dt$

26. $\displaystyle\int t^2(t^3 - 1)^7 \, dt$

27. $\displaystyle\int (x^4 + x^3 + x^2)^7 (4x^3 + 3x^2 + 2x) \, dx$

28. $\displaystyle\int (x^3 - x^2 - x)^9 (3x^2 - 2x - 1) \, dx$

29. $\displaystyle\int \frac{e^x \, dx}{4 + e^x}$

30. $\displaystyle\int \frac{e^t \, dt}{3 + e^t}$

31. $\int \dfrac{\ln x^2}{x} \, dx$ (*Hint:* Use the properties of logarithms.)

32. $\int \dfrac{(\ln x)^2}{x} \, dx$

33. $\int \dfrac{dx}{x \ln x}, x > 1$ **34.** $\int \dfrac{dx}{x \ln x^2}, x > 1$

35. $\int x\sqrt{ax^2 + b} \, dx$ **36.** $\int \sqrt{ax + b} \, dx$

37. $\int P_0 e^{kt} \, dt$ **38.** $\int be^{ax} \, dx$

39. $\int \dfrac{x^3 \, dx}{(2 - x^4)^7}$ **40.** $\int \dfrac{3x^2 \, dx}{(1 + x^3)^5}$

41. $\int 12x\sqrt[5]{1 + 6x^2} \, dx$ **42.** $\int 5x\sqrt[4]{1 - x^2} \, dx$

Evaluate.

43. $\int_0^1 2xe^{x^2} \, dx$ **44.** $\int_0^1 3x^2 e^{x^3} \, dx$

45. $\int_0^1 x(x^2 + 1)^5 \, dx$ **46.** $\int_1^2 x(x^2 - 1)^7 \, dx$

47. $\int_0^4 \dfrac{dt}{1 + t}$ **48.** $\int_0^2 e^{4x} \, dx$

49. $\int_1^4 \dfrac{2x + 1}{x^2 + x - 1} \, dx$ **50.** $\int_1^3 \dfrac{2x + 3}{x^2 + 3x} \, dx$

51. $\int_0^b e^{-x} \, dx$ **52.** $\int_0^b 2e^{-2x} \, dx$

53. $\int_0^b me^{-mx} \, dx$ **54.** $\int_0^b ke^{-kx} \, dx$

55. $\int_0^4 (x - 6)^2 \, dx$ **56.** $\int_0^3 (x - 5)^2 \, dx$

57. $\int_0^2 \dfrac{3x^2 \, dx}{(1 + x^3)^5}$ **58.** $\int_{-1}^0 \dfrac{x^3 \, dx}{(2 - x^4)^7}$

59. $\int_0^{\sqrt{7}} 7x\sqrt[3]{1 + x^2} \, dx$ **60.** $\int_0^1 12x\sqrt[5]{1 - x^2} \, dx$

61–78. Use a graphing calculator to check the results of Exercises 43–60.

Evaluate. Use the technique of Example 9.

79. $\int \dfrac{x}{x - 5} \, dx$ **80.** $\int \dfrac{3x}{2x + 1} \, dx$

81. $\int \dfrac{x}{1 - 4x} \, dx$

82. $\int \dfrac{x + 3}{x - 2} \, dx$ (*Hint:* $u = x - 2$.)

83. $\int \dfrac{2x + 3}{3x - 2} \, dx$

84. $\int x^2(x + 1)^{10} \, dx$

85. $\int x^3(x + 2)^7 \, dx$

86. $\int x^2\sqrt{x - 2} \, dx$

APPLICATIONS

Business and Economics

87. Demand from marginal demand. Masterson Insoles, Inc., has the marginal-demand function

$$D'(x) = \dfrac{-2000x}{\sqrt{25 - x^2}},$$

where $D(x)$ is the number of units sold at x dollars per unit.

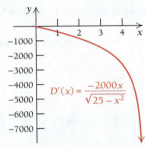

Find the demand function given that $D = 13{,}000$ when $x = \$3$ per unit.

88. Profit from marginal profit. A firm has the marginal-profit function

$$\dfrac{dP}{dx} = \dfrac{9000 - 3000x}{(x^2 - 6x + 10)^2},$$

where $P(x)$ is the profit earned at x dollars per unit.

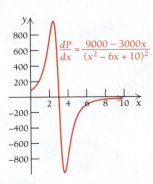

Find the total-profit function given that $P = \$1500$ at $x = \$3$.

Social Sciences

89. Marriage rate. The marriage rate in the United States is approximated by

$$M(t) = 8.3e^{-0.019t},$$

where $M(t)$ is the number of marriages per 1000 people, t years after 2000. (*Source:* Based on data from www.cdc.gov.)

a) Find the total number of marriages per 1000 people in the United States from 2000 to 2005. Note that this is given by

$$\int_0^5 M(t)\, dt.$$

b) Find the total number of marriages per 1000 people in the United States between 2005 and 2016.

90. Divorce rate. The divorce rate in the United States is approximated by

$$D(t) = 3.95e^{-0.012t}$$

where $D(t)$ is the number of divorces per 1000 people, t years after 2000. (*Source:* Based on data from www.cdc.gov.)

a) Find the total number of divorces per 1000 people in the United States from 2000 to 2008. Note that this is given by

$$\int_0^8 D(t)\, dt.$$

b) Find the total number of divorces in the United States between 2008 and 2016.

SYNTHESIS

Find the total area of the shaded region.

91.

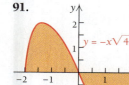

$y = -x\sqrt{4 - x^2}$

92.

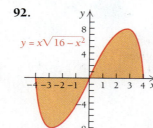

$y = x\sqrt{16 - x^2}$

Evaluate.

93. $\displaystyle\int \frac{dx}{ax + b}$

94. $\displaystyle\int 5x\sqrt{1 - 4x^2}\, dx$

95. $\displaystyle\int \frac{e^{\sqrt{t}}}{\sqrt{t}}\, dt$

96. $\displaystyle\int \frac{x^2}{e^{x^3}}\, dx$

97. $\displaystyle\int \frac{e^{1/t}}{t^2}\, dt$

98. $\displaystyle\int \frac{(\ln x)^{99}}{x}\, dx, x > 0$

99. $\displaystyle\int \frac{dx}{x(\ln x)^4}$

100. $\displaystyle\int (e^t + 2)e^t\, dt$

101. $\displaystyle\int x^2\sqrt{x^3 + 1}\, dx$

102. $\displaystyle\int \frac{t^2}{\sqrt[4]{2 + t^3}}\, dt$

103. $\displaystyle\int \frac{[(\ln x)^2 + 3(\ln x) + 4]}{x}\, dx, x > 0$

104. $\displaystyle\int \frac{x - 3}{(x^2 - 6x)^{1/3}}\, dx$

105. $\displaystyle\int \frac{t^3 \ln(t^4 + 8)}{t^4 + 8}\, dt$

106. $\displaystyle\int \frac{t^2 + 2t}{(t + 1)^2}\, dt$ (*Hint:* $u = t + 1$.)

107. $\displaystyle\int \frac{x^2 + 6x}{(x + 3)^2}\, dx$

108. $\displaystyle\int \frac{x + 3}{x + 1}\, dx$ $\left(\text{Hint: } \frac{x + 3}{x + 1} = 1 + \frac{2}{x + 1}.\right)$

109. $\displaystyle\int \frac{t - 5}{t - 4}\, dt$

110. $\displaystyle\int \frac{dx}{x(\ln x)^n}, \quad n \neq -1$

111. $\displaystyle\int \frac{dx}{e^x + 1}$ $\left(\text{Hint: } \frac{1}{e^x + 1} = \frac{e^{-x}}{1 + e^{-x}}.\right)$

112. $\displaystyle\int \frac{e^x - e^{-x}}{e^x + e^{-x}}\, dx$

113. $\displaystyle\int \frac{(\ln x)^n}{x}\, dx, x > 0, \quad n \neq -1$

114. $\displaystyle\int \frac{e^{-mx}}{1 - ae^{-mx}}\, dx$

115. $\displaystyle\int 9x(7x^2 + 9)^n\, dx, \quad n \neq -1$

116. $\displaystyle\int 5x^2(2x^3 - 7)^n\, dx, \quad n \neq -1$

117. Is the following a true statement? Why or why not?

$$\int [2f(x)]\, dx = [f(x)]^2 + C$$

Answers to Quick Checks

1. a) $dy = \dfrac{1}{2\sqrt{x}}\, dx;$ **b)** $du = (2x - 3)\, dx;$

c) $dy = -\dfrac{3}{x^4}\, dx;$ **d)** $du = 4\, dx$

2. $\frac{1}{4}(2x^2 + 3)^4 + C$ **3.** $\ln(1 + e^x) + C$

4. $2\sqrt{3 + 2x^3} + C$ **5.** $\frac{1}{3}(\ln x)^3 + C$ **6.** $\frac{1}{12}e^{4x^3} + C$

7. 16,080.8 **8.** $-\dfrac{1}{x - 1} - \dfrac{1}{2(x - 1)^2} + C$

Integration Techniques: Integration by Parts

● Evaluate integrals using the formula for integration by parts.
● Solve applied problems involving integration by parts.

Recall the Product Rule for differentiation:

$$\frac{d}{dx}(uv) = u\frac{dv}{dx} + v\frac{du}{dx}.$$

Integrating both sides with respect to x, we get

$$uv = \int u\frac{dv}{dx}\,dx + \int v\frac{du}{dx}\,dx$$

$$= \int u\,dv + \int v\,du.$$

Solving for $\int u\,dv$, we get the following theorem.

THEOREM 7 **The Integration-by-Parts Formula**

$$\int u\,dv = uv - \int v\,du$$

This equation can be used as an integration formula when the integrand is a product of two functions, and at least one of those functions can be integrated using the techniques we have already developed. For example,

$$\int xe^x\,dx$$

can be considered as

$$\int u\,dv,$$

where we let

$$u = x \quad \text{and} \quad dv = e^x\,dx.$$

In this case, differentiating u gives

$$du = dx,$$

and integrating dv gives

$$v = e^x. \qquad \text{We select } C = 0 \text{ to obtain the simplest antiderivative.}$$

Then the Integration-by-Parts Formula gives us

$$\int \overset{u}{(x)}\overset{dv}{(e^x\,dx)} = \overset{u}{(x)}\overset{v}{(e^x)} - \int \overset{v}{(e^x)}\overset{du}{(dx)}$$

$$= xe^x - e^x + C.$$

This method of integrating is called **integration by parts.** As always, to check, we differentiate the result. This check is left to the student.

Note that integration by parts, like substitution, can be a trial-and-error process. In the preceding example, if we had reversed the roles of x and e^x, we would have obtained

$$u = e^x, \qquad dv = x\,dx,$$

$$du = e^x\,dx, \qquad v = \frac{x^2}{2},$$

and
$$\int \overset{u}{(e^x)} \overset{dv}{(x\,dx)} = \overset{u}{(e^x)} \overset{v}{\left(\frac{x^2}{2}\right)} - \int \overset{v}{\left(\frac{x^2}{2}\right)} \overset{du}{(e^x\,dx)}.$$

Now the integrand on the right is *more* difficult to integrate than the one with which we began. We usually have a choice as to how to apply the Integration-by-Parts Formula. It may be that only one (or none) of the possibilities will work.

Tips on Using Integration by Parts

1. Use integration by parts when an integral can be written in the form

$$\int f(x)\,g(x)\,dx.$$

Write it as an integral of the form

$$\int u\,dv$$

by choosing a function to be $u = f(x)$, where $f(x)$ can be differentiated, and the remaining factor to be $dv = g(x)\,dx$, where $g(x)$ can be integrated.
2. Find du by differentiating and v by integrating.
3. If the resulting integral is more complicated than the original, try another choice for u and dv.
4. Check your result by differentiating.

Consider the following examples.

EXAMPLE 1 Evaluate: $\int \ln x\,dx$. Assume $x > 0$.

Solution Since we can differentiate $\ln x$, we let

$$u = \ln x \quad \text{and} \quad dv = dx.$$

Then $du = \dfrac{1}{x}\,dx$ and $v = x$.

Using the Integration-by-Parts Formula gives

$$\int \overset{u}{(\ln x)} \overset{dv}{(dx)} = \overset{u}{(\ln x)} \overset{v}{x} - \int \overset{v}{x} \overset{du}{\left(\frac{1}{x}\,dx\right)}$$

Quick Check 1 ✔

Evaluate: $\int xe^{3x}\,dx$.

$$= x\ln x - \int dx$$

$$= x\ln x - x + C. \qquad \mathbf{1} ✔$$

EXAMPLE 2 Evaluate: $\int x\ln x\,dx$.

Solution Let's examine a couple of choices, as follows.

Attempt 1: We let

$$u = x\ln x \qquad\qquad \text{and} \quad dv = dx.$$

Then $du = \left[x\left(\dfrac{1}{x}\right) + (\ln x)1\right] dx$ and $v = x$

$$= (1 + \ln x)\,dx.$$

Using the Integration-by-Parts Formula, we have

$$\int_{\overset{u}{}}^{\overset{dv}{}} (x \ln x)\, dx = \overset{u\quad v}{(x \ln x)x} - \int \overset{v}{x}(\overset{du}{(1 + \ln x)\, dx})$$

$$= x^2 \ln x - \int (x + x \ln x)\, dx.$$

This integral seems more complicated than the original. Let's try a different choice for u and dv.

Attempt 2: We let

$$u = \ln x \quad \text{and} \quad dv = x\, dx.$$

Then $\quad du = \dfrac{1}{x}\, dx \quad$ and $\quad v = \dfrac{x^2}{2}.$

Using the Integration-by-Parts Formula, we have

$$\int \overset{u\quad dv}{x \ln x\, dx} = \overset{u\quad\quad v}{\ln x \cdot \dfrac{x^2}{2}} - \int \overset{v}{\dfrac{x^2}{2}}\left(\overset{du}{\dfrac{1}{x}\, dx}\right)$$

$$= \dfrac{x^2}{2} \ln x - \dfrac{1}{2} \int x\, dx$$

$$= \dfrac{x^2}{2} \ln x - \dfrac{x^2}{4} + C.$$

This choice of u and dv allows us to evaluate the integral.

EXAMPLE 3   Evaluate: $\int x\sqrt{5x + 1}\, dx.$

Solution   We let

$$u = x \quad \text{and} \quad dv = (5x + 1)^{1/2}\, dx.$$

Then $\quad du = dx \quad$ and $\quad v = \frac{2}{15}(5x + 1)^{3/2}.$

Note that we used substitution to integrate dv:

$$\int (5x + 1)^{1/2}\, dx = \dfrac{1}{5}\int (5x + 1)^{1/2}\, 5\, dx = \dfrac{1}{5}\int w^{1/2}\, dw \quad \underline{\text{Substitution}} \quad \boxed{\begin{array}{l} w = 5x + 1, \\ dw = 5\, dx \end{array}}$$

$$v = \dfrac{1}{5} \cdot \dfrac{w^{1/2+1}}{\frac{1}{2} + 1} = \dfrac{2}{15} w^{3/2} = \dfrac{2}{15}(5x + 1)^{3/2}.$$

Using the Integration-by-Parts Formula gives us

$$\int \overset{u}{x}(\overset{dv}{\sqrt{5x + 1}\, dx}) = \overset{u\quad\quad v}{x \cdot \tfrac{2}{15}(5x + 1)^{3/2}} - \int \overset{v}{\tfrac{2}{15}(5x + 1)^{3/2}}\, \overset{du}{dx}$$

$$= \tfrac{2}{15}x(5x + 1)^{3/2} - \tfrac{2}{15} \cdot \tfrac{2}{25}(5x + 1)^{5/2} + C \quad \text{Using substitution to evaluate the integral on the right.}$$

$$= \tfrac{2}{15}x(5x + 1)^{3/2} - \tfrac{4}{375}(5x + 1)^{5/2} + C.$$

Quick Check 2 ✔

Evaluate: $\int 2x\sqrt{3x - 2}\, dx.$

This integral may also be evaluated using a substitution similar to that shown in Example 9 of Section 4.5. (See Exercise 43 at the end of this section.)

2 ✔

EXAMPLE 4 Evaluate: $\int_1^2 \ln x \, dx$.

Solution First, we find the indefinite integral (see Example 1). Next, we evaluate the definite integral:

$$\int_1^2 \ln x \, dx = [x \ln x - x]_1^2 \qquad \text{Using the result from Example 1}$$

$$= (2 \ln 2 - 2) - (1 \cdot \ln 1 - 1)$$
$$= 2 \ln 2 - 2 + 1$$
$$= 2 \ln 2 - 1 \approx 0.386.$$

Repeated Integration by Parts

In some cases, we may need to apply the Integration-by-Parts Formula more than once.

EXAMPLE 5 Evaluate $\int_0^7 x^2 e^{-x} \, dx$ to find the shaded area shown to the left.

Solution We first let

$$u = x^2 \qquad \text{and} \qquad dv = e^{-x} dx.$$

Then $du = 2x \, dx$ and $v = -e^{-x}$.

Using the Integration-by-Parts Formula gives

$$\int \overset{u}{x^2} (\overset{dv}{e^{-x} dx}) = \overset{u}{x^2} (\overset{v}{-e^{-x}}) - \int \overset{v}{-e^{-x}} (\overset{du}{2x \, dx})$$

$$= -x^2 e^{-x} + \int 2x e^{-x} \, dx. \qquad (1)$$

To evaluate the integral on the right, we use integration by parts again:

$$u = 2x \qquad \text{and} \qquad dv = e^{-x} dx.$$

so $du = 2 \, dx$ and $v = -e^{-x}$.

Using the Integration-by-Parts Formula once again, we get

$$\int \overset{u}{2x} (\overset{dv}{e^{-x} dx}) = \overset{u}{2x} (\overset{v}{-e^{-x}}) - \int \overset{v}{-e^{-x}} (\overset{du}{2 \, dx})$$

$$= -2x e^{-x} - 2e^{-x} + C. \qquad (2)$$

When we substitute equation (2) into (1), the original integral becomes

$$\int x^2 e^{-x} \, dx = -x^2 e^{-x} - 2x e^{-x} - 2e^{-x} + C \qquad \text{Substituting}$$

$$= -e^{-x}(x^2 + 2x + 2) + C. \qquad \text{Factoring simplifies the next step.}$$

We now evaluate the definite integral:

$$\int_0^7 x^2 e^{-x} dx = [-e^{-x}(x^2 + 2x + 2)]_0^7$$

$$= [-e^{-7}(7^2 + 2(7) + 2)] - [-e^{-0}(0^2 + 2(0) + 2)]$$

$$= -65e^{-7} + 2 \approx 1.94. \qquad \text{3 ✔}$$

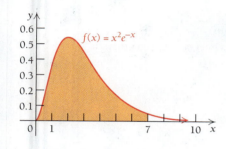

$f(x) = x^2 e^{-x}$

TECHNOLOGY CONNECTION

EXERCISE

1. Use a calculator to evaluate

$$\int_0^7 x^2 e^{-x} \, dx.$$

Quick Check 3 ✔

Evaluate: $\int_0^3 \dfrac{x}{\sqrt{x+1}} \, dx.$

Tabular Integration by Parts

In situations similar to Example 5, we have an integral,

$$\int f(x)\, g(x)\, dx,$$

for which $f(x)$ can be easily differentiated repeatedly, leading to a derivative that is eventually 0. The function $g(x)$ can also be easily integrated repeatedly. In such situations, we can use integration by parts more than once to evaluate the integral.

EXAMPLE 6   Evaluate: $\int x^3 e^x\, dx$.

Solution   We use integration by parts repeatedly, watching for patterns. First, we select $u = x^3$ and $dv = e^x\, dx$:

$$\int \overset{u}{x^3} \overset{dv}{e^x}\, dx = \overset{u}{x^3} \overset{v}{e^x} - \int \overset{v}{e^x}(\overset{du}{3x^2\, dx})$$

$$= x^3 e^x - \int 3x^2 e^x\, dx. \qquad \text{This integral is simpler than the original.} \qquad (1)$$

To solve $\int 3x^2 e^x\, dx$, we select $u = 3x^2$ and $dv = e^x\, dx$, so

$$du = 6x\, dx \quad \text{and} \quad v = e^x,$$

and
$$\int 3x^2 e^x\, dx = \overset{u}{3x^2}\overset{v}{e^x} - \int \overset{v}{e^x}(\overset{du}{6x\, dx})$$

$$= 3x^2 e^x - \int 6x e^x\, dx. \qquad \text{This integral is the simplest so far.} \qquad (2)$$

To solve $\int 6x e^x\, dx$, we select $u = 6x$ and $dv = e^x\, dx$, so

$$du = 6\, dx \quad \text{and} \quad v = e^x,$$

and
$$\int 6x e^x\, dx = \overset{u}{6x}\overset{v}{e^x} - \int \overset{v}{e^x}(\overset{du}{6\, dx})$$

$$= 6x e^x - 6 \int e^x\, dx$$

$$= 6x e^x - 6 e^x + C. \qquad (3)$$

Combining equations (1), (2), and (3), we have

$$\int x^3 e^x\, dx = x^3 e^x - \left(3x^2 e^x - \int 6x e^x\, dx \right) \qquad \text{Substituting equation (2) into equation (1)}$$

$$= x^3 e^x - 3x^2 e^x + \int 6x e^x\, dx$$

$$= x^3 e^x - 3x^2 e^x + 6x e^x - 6 e^x + C. \qquad \text{Substituting equation (3)}$$

This approach can get complicated. Using **tabular integration** can simplify the work.

$f(x)$ and Repeated Derivatives	Sign of Product	$g(x)$ and Repeated Integrals
x^3	$(+)$	e^x
$3x^2$	$(-)$	e^x
$6x$	$(+)$	e^x
6	$(-)$	e^x
0		e^x

Quick Check 4 ✔

Evaluate: $\int x^4 e^{2x}\, dx$.

We add products along the arrows, making alternating sign changes, to obtain

$$\int x^3 e^x\, dx = x^3 e^x - 3x^2 e^x + 6xe^x - 6e^x + C.$$ **4** ✔

Section Summary

- The *Integration-by-Parts Formula* makes use of the Product Rule for differentiation:

$$\int u\, dv = uv - \int v\, du.$$

- The choices for u and dv should be such that $\int v\, du$ is simpler than the original integral. If this does not occur, other choices should be made.
- Tabular integration can be used when repeated integration by parts is necessary.

4.6 Exercise Set

Evaluate using integration by parts or substitution. (Assume $u > 0$ in ln u.) Check by differentiating.

1. $\int 4xe^{4x}\, dx$

2. $\int 3xe^{3x}\, dx$

3. $\int x^3(3x^2)\, dx$

4. $\int x^2(2x)\, dx$

5. $\int xe^{5x}\, dx$

6. $\int 2xe^{4x}\, dx$

7. $\int xe^{-2x}\, dx$

8. $\int xe^{-x}\, dx$

9. $\int x^2 \ln x\, dx$

10. $\int x^3 \ln x\, dx$

11. $\int x \ln \sqrt{x}\, dx$

12. $\int x^2 \ln x^3\, dx$

13. $\int \ln (x + 5)\, dx$

14. $\int \ln (x + 4)\, dx$

15. $\int (x + 2) \ln x\, dx$

16. $\int (x + 1) \ln x\, dx$

17. $\int (x - 1) \ln x\, dx$

18. $\int (x - 2) \ln x\, dx$

19. $\int x\sqrt{x + 2}\, dx$

20. $\int x\sqrt{x + 5}\, dx$

21. $\int x^3 \ln (2x)\, dx$

22. $\int x^2 \ln (5x)\, dx$

23. $\int x^2 e^x\, dx$

24. $\int (\ln x)^2\, dx$

25. $\int x^2 e^{2x}\, dx$

26. $\int x^{-5} \ln x\, dx$

27. $\int x^3 e^{-2x}\, dx$

28. $\int x^5 e^{4x}\, dx$

29. $\int (x^4 + 4)e^{3x}\, dx$

30. $\int (x^3 - x + 1)e^{-x}\, dx$

Evaluate using integration by parts.

31. $\int_1^2 x^2 \ln x\, dx$

32. $\int_1^2 x^3 \ln x\, dx$

33. $\int_2^6 \ln (x + 8)\, dx$

34. $\int_0^5 \ln (x + 7)\, dx$

35. $\int_0^1 xe^x\, dx$

36. $\int_0^1 (x^3 + 2x^2 + 3)e^{-2x}\, dx$

37. $\int_0^8 x\sqrt{x + 1}\, dx$

38. $\int_0^{\ln 3} x^2 e^{2x}\, dx$

39. Cost from marginal cost. Larry's Lawn Chairs determines that its marginal-cost function is given by

$$C'(x) = 4x\sqrt{x + 3},$$

where x is the number of lawn chairs sold and $C'(x)$ is the marginal cost in dollars per unit. Find the total cost given that $C(13) = \$1126.40$.

40. Profit from marginal profit.   Nevin Patio Contractors determines that its marginal-profit function is given by
$$P'(x) = 1000x^2e^{-0.2x}.$$

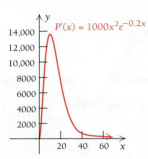

Find the total profit given that $P(0) = -\$2000$.

Life and Physical Sciences

41. Drug dosage.   Suppose an oral dose of a drug is taken. Over time, the drug is assimilated in the body and excreted in the urine. The total amount of the drug that passes through the body in T hours is given by
$$\int_0^T E(t)\, dt,$$
where E is the rate of excretion of the drug. A typical rate-of-excretion function is
$$E(t) = te^{-kt},$$
where $k > 0$ and t is the time, in hours.

a) Find a formula for
$$\int_0^T E(t)\, dt.$$

b) Find
$$\int_0^{10} E(t)\, dt, \quad \text{when } k = 0.2 \text{ mg/hr}.$$

42. Electrical energy use.   The rate at which electrical energy is used by the Ortiz family, in kilowatt-hours (kW-h) per day, is given by
$$K(t) = 10te^{-t},$$
where t is time, in hours. That is, t is in the interval $[0, 24]$.

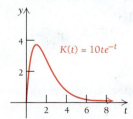

a) How many kilowatt-hours does the family use in the first T hours of a day ($t = 0$ to $t = T$)?
b) How many kilowatt-hours does the family use in the first 4 hours of the day?

SYNTHESIS

In Exercises 43 and 44, evaluate the given indefinite integral using substitution. Refer to Example 9 in Section 4.5 to review the technique.

43. Evaluate $\int x\sqrt{5x + 1}\, dx$ by letting $u = 5x + 1$ and $du = 5\, dx$ (so that $dx = \frac{1}{5}\, du$) and observing that $x = \dfrac{u - 1}{5}$. Compare your answer to that found in Example 3 of this section. Are they the same? (*Hint:* Simplify both forms of the answer into a common third form.)

44. Consider $\displaystyle\int \frac{x}{\sqrt{x - 3}}\, dx$.
a) Evaluate this integral using integration by parts.
b) Evaluate this integral using the substitution $u = x - 3$ and observing that $x = u + 3$.
c) Show algebraically that the answers from parts (a) and (b) are equivalent.

In Exercises 45 and 46, both substitution and integration by parts are used to determine the indefinite integral.

45. Evaluate $\int e^{\sqrt{x}}\, dx$ by letting $u = \sqrt{x}$. Note that $x = u^2$, so $dx = 2u\, du$. Make the substitutions and observe that the new integral (with variable u) can be evaluated using integration by parts.

46. Evaluate $\displaystyle\int \frac{1}{1 + \sqrt{x}}\, dx$ by letting $u = \sqrt{x}$ and following the procedure used in Exercise 45.

Evaluate.

47. $\displaystyle\int \sqrt{x}\ln x\, dx$

48. $\displaystyle\int \frac{te^t}{(t + 1)^2}\, dt$

49. $\displaystyle\int \frac{\ln x}{\sqrt{x}}\, dx$

50. $\displaystyle\int \frac{13t^2 - 48}{\sqrt[5]{4t + 7}}\, dt$

51. $\displaystyle\int (27x^3 + 83x - 2)\sqrt[6]{3x + 8}\, dx$

52. $\displaystyle\int x^2(\ln x)^2\, dx$

53. $\displaystyle\int x^n(\ln x)^2\, dx, \quad n \neq -1$

54. $\displaystyle\int x^n \ln x\, dx, \quad n \neq -1$

55. Verify that for any positive integer n,
$$\int x^n e^x\, dx = x^n e^x - n\int x^{n-1}e^x\, dx.$$

56. Verify that for any positive integer n,
$$\int (\ln x)^n\, dx = x(\ln x)^n - n\int (\ln x)^{n-1}\, dx.$$

57. Is the following a true statement?

$$\int f(x)g(x)\,dx = \int f(x)\,dx \cdot \int g(x)\,dx.$$

Why or why not?

58. Compare the procedures of differentiation and integration. Which seems to be the most complicated or difficult and why?

Recurring integrals. *Occasionally, integration by parts yields an integral of the form $\int u\,dv$ that is identical to the original integral. In some cases, we can then solve for $\int u\,dv$ algebraically. For example, to find $\int 2^x e^x\,dx$, we let $u = 2^x$ and $dv = e^x$, so $du = (\ln 2)2^x\,dx$ and $v = e^x$. Using integration by parts, we have*

$$\int 2^x e^x\,dx = 2^x e^x - \ln 2 \int 2^x e^x\,dx.$$

Note that $\int 2^x e^x\,dx$ appears twice. Adding $\ln 2 \int 2^x e^x\,dx$ to both sides, we have

$$\int 2^x e^x\,dx + \ln 2 \int 2^x e^x\,dx = 2^x e^x$$

$$(1 + \ln 2)\int 2^x e^x\,dx = 2^x e^x$$

$$\int 2^x e^x\,dx = \frac{2^x e^x}{1 + \ln 2} + C.$$

Use this method to evaluate the integrals in Exercises 59–62.

59. $\displaystyle\int 3^x e^x\,dx$ **60.** $\displaystyle\int 5^x e^{2x}\,dx$

61. $\displaystyle\int 10^x e^{3x}\,dx$

62. $\displaystyle\int x \ln x\,dx$ (*Hint*: let $u = x \ln x$ and $dv = dx$. Assume $x > 0$.)

TECHNOLOGY CONNECTION

63. Use a graphing calculator to evaluate

$$\int_1^{10} x^5 \ln x\,dx.$$

Answers to Quick Checks

1. $\dfrac{x}{3}e^{3x} - \dfrac{1}{9}e^{3x} + C$

2. $\dfrac{4}{9}x(3x-2)^{3/2} - \dfrac{4}{135}(3x-2)^{5/2} + C$ **3.** $2\frac{2}{3}$

4. $e^{2x}\left(\dfrac{1}{2}x^4 - x^3 + \dfrac{3}{2}x^2 - \dfrac{3}{2}x + \dfrac{3}{4}\right) + C$

4.7

- Evaluate integrals using a table of integration formulas.

Integration Techniques: Tables

Tables of Integration Formulas

You have probably noticed that, generally speaking, integration is more challenging than differentiation. Because of this, certain integral formulas have been gathered into tables. Table 1, shown below and inside the back cover of this book, is such a table. Entire books of integration formulas are available, and lengthy tables are also available online. Such tables are usually classified by the form of the integrand. The idea is to properly match the integral in question with a formula in the table. Sometimes algebra or a technique such as substitution or integration by parts may be needed in order to use a formula from a table.

TABLE 1   Integration Formulas

1. $\displaystyle\int x^n\,dx = \frac{x^{n+1}}{n+1} + C, \quad n \ne -1$

2. $\displaystyle\int \frac{dx}{x} = \ln|x| + C, \quad$ or, for $x > 0$, $\displaystyle\int \frac{dx}{x} = \ln x + C$

3. $\displaystyle\int u\,dv = uv - \int v\,du$

4. $\displaystyle\int e^x\,dx = e^x + C$

5. $\displaystyle\int e^{ax}\,dx = \frac{1}{a} \cdot e^{ax} + C$

(*continued*)

6. $\displaystyle\int xe^{ax}\,dx = \frac{1}{a^2}\cdot e^{ax}(ax-1) + C$

7. $\displaystyle\int x^n\,e^{ax}\,dx = \frac{x^n e^{ax}}{a} - \frac{n}{a}\int x^{n-1}e^{ax}\,dx + C$

8. $\displaystyle\int \ln x\,dx = x\ln x - x + C, \quad x > 0$

9. $\displaystyle\int (\ln x)^n\,dx = x(\ln x)^n - n\int (\ln x)^{n-1}\,dx + C, \quad n \neq -1, x > 0$

10. $\displaystyle\int x^n \ln x\,dx = x^{n+1}\left[\frac{\ln x}{n+1} - \frac{1}{(n+1)^2}\right] + C, \quad n \neq -1, x > 0$

11. $\displaystyle\int a^x\,dx = \frac{a^x}{\ln a} + C, \quad a > 0, a \neq 1$

12. $\displaystyle\int \frac{1}{\sqrt{x^2 + a^2}}\,dx = \ln\left|x + \sqrt{x^2 + a^2}\right| + C$

13. $\displaystyle\int \frac{1}{\sqrt{x^2 - a^2}}\,dx = \ln\left|x + \sqrt{x^2 - a^2}\right| + C$

14. $\displaystyle\int \frac{1}{x^2 - a^2}\,dx = \frac{1}{2a}\ln\left|\frac{x - a}{x + a}\right| + C$

15. $\displaystyle\int \frac{1}{a^2 - x^2}\,dx = \frac{1}{2a}\ln\left|\frac{a + x}{a - x}\right| + C$

16. $\displaystyle\int \frac{1}{x\sqrt{a^2 + x^2}}\,dx = -\frac{1}{a}\ln\left|\frac{a + \sqrt{a^2 + x^2}}{x}\right| + C$

17. $\displaystyle\int \frac{1}{x\sqrt{a^2 - x^2}}\,dx = -\frac{1}{a}\ln\left|\frac{a + \sqrt{a^2 - x^2}}{x}\right| + C$

18. $\displaystyle\int \frac{x}{a + bx}\,dx = \frac{x}{b} - \frac{a}{b^2}\ln|a + bx| + C$

19. $\displaystyle\int \frac{x}{(a + bx)^2}\,dx = \frac{a}{b^2(a + bx)} + \frac{1}{b^2}\ln|a + bx| + C$

20. $\displaystyle\int \frac{1}{x(a + bx)}\,dx = \frac{1}{a}\ln\left|\frac{x}{a + bx}\right| + C$

21. $\displaystyle\int \frac{1}{x(a + bx)^2}\,dx = \frac{1}{a(a + bx)} + \frac{1}{a^2}\ln\left|\frac{x}{a + bx}\right| + C$

22. $\displaystyle\int \sqrt{x^2 + a^2}\,dx = \frac{1}{2}\left[x\sqrt{x^2 + a^2} + a^2\ln\left|x + \sqrt{x^2 + a^2}\right|\right] + C$

23. $\displaystyle\int \sqrt{x^2 - a^2}\,dx = \frac{1}{2}\left[x\sqrt{x^2 - a^2} - a^2\ln\left|x + \sqrt{x^2 - a^2}\right|\right] + C$

24. $\displaystyle\int x\sqrt{a + bx}\,dx = \frac{2}{15b^2}(3bx - 2a)(a + bx)^{3/2} + C$

25. $\displaystyle\int x^2\sqrt{a + bx}\,dx = \frac{2}{105b^3}(15b^2x^2 - 12abx + 8a^2)(a + bx)^{3/2} + C$

26. $\displaystyle\int \frac{x\,dx}{\sqrt{a + bx}} = \frac{2}{3b^2}(bx - 2a)\sqrt{a + bx} + C$

27. $\displaystyle\int \frac{x^2\,dx}{\sqrt{a + bx}} = \frac{2}{15b^3}(3b^2x^2 - 4abx + 8a^2)\sqrt{a + bx} + C$

EXAMPLE 1 Find: $\int \dfrac{dx}{x(3-x)}$.

Solution The integral $\int \dfrac{dx}{x(3-x)}$ fits *formula 20* in Table 1:

$$\int \frac{1}{x(a+bx)}\,dx = \frac{1}{a}\ln\left|\frac{x}{a+bx}\right| + C.$$

In the given integral, $a=3$ and $b=-1$, so we have, by the formula,

$$\int \frac{dx}{x(3-x)} = \frac{1}{3}\ln\left|\frac{x}{3+(-1)x}\right| + C$$

$$= \frac{1}{3}\ln\left|\frac{x}{3-x}\right| + C.$$

We check our answer by differentiation:

$$\frac{d}{dx}\left(\frac{1}{3}\ln\left|\frac{x}{3-x}\right|\right) = \frac{1}{3}\cdot\frac{d}{dx}(\ln|x| - \ln|3-x|)$$

$$= \frac{1}{3}\left(\frac{d}{dx}\ln|x| - \frac{d}{dx}\ln|3-x|\right)$$

$$= \frac{1}{3}\left(\frac{1}{x} - \frac{1}{3-x}(-1)\right)$$

$$= \frac{1}{3}\left(\frac{1}{x} + \frac{1}{3-x}\right)$$

$$= \frac{1}{3}\left(\frac{1}{x}\cdot\left(\frac{3-x}{3-x}\right) + \frac{1}{3-x}\cdot\left(\frac{x}{x}\right)\right)$$

$$= \frac{1}{3}\left(\frac{3-x+x}{x(3-x)}\right)$$

$$= \frac{1}{3}\left(\frac{3}{x(3-x)}\right)$$

$$= \frac{1}{x(3-x)}.$$

Recalling that $\ln\left(\dfrac{a}{b}\right) = \ln a - \ln b$

Differentiating

Simplifying

Using a common denominator

Quick Check 1 ✔

Find: $\int \dfrac{2x}{(3-5x)^2}\,dx$.

1 ✔

EXAMPLE 2 Find: $\int \dfrac{5x}{7x-8}\,dx$.

Solution We first factor 5 out of the integral. The integral then fits *formula 18* in Table 1:

$$\int \frac{x}{a+bx}\,dx = \frac{x}{b} - \frac{a}{b^2}\ln|a+bx| + C.$$

In the given integral, $a=-8$ and $b=7$, so we have, by the formula,

$$\int \frac{5x}{7x-8}\,dx = 5\int \frac{x}{-8+7x}\,dx$$

$$= 5\left[\frac{x}{7} - \frac{-8}{7^2}\ln|-8+7x|\right] + C$$

$$= 5\left[\frac{x}{7} + \frac{8}{49}\ln|7x-8|\right] + C$$

$$= \frac{5x}{7} + \frac{40}{49}\ln|7x-8| + C.$$

Quick Check 2 ✔

Find: $\int \dfrac{3}{2x(7-3x)^2}\,dx$.

2 ✔

EXAMPLE 3 Find: $\int \sqrt{16x^2 + 3}\, dx$.

Solution This integral *almost* fits *formula 22* in Table 1:

$$\int \sqrt{x^2 + a^2}\, dx = \tfrac{1}{2}\left[x\sqrt{x^2 + a^2} + a^2 \ln\left|x + \sqrt{x^2 + a^2}\right|\right] + C.$$

For the coefficient of x^2 to be 1, we have to factor out 16. Then we apply *formula 22*:

$$\int \sqrt{16x^2 + 3}\, dx = \int \sqrt{16\left(x^2 + \tfrac{3}{16}\right)}\, dx \qquad \text{Factoring}$$

$$= \int 4\sqrt{x^2 + \tfrac{3}{16}}\, dx \qquad \begin{array}{l}\text{Using the properties of}\\ \text{radicals; } \sqrt{16} = 4\end{array}$$

$$= 4\int \sqrt{x^2 + \tfrac{3}{16}}\, dx \qquad \text{We have } a^2 = \tfrac{3}{16} \text{ in formula 22.}$$

$$= 4\cdot\tfrac{1}{2}\left[x\sqrt{x^2 + \tfrac{3}{16}} + \tfrac{3}{16}\ln\left|x + \sqrt{x^2 + \tfrac{3}{16}}\right|\right] + C$$

$$= 2\left[x\sqrt{x^2 + \tfrac{3}{16}} + \tfrac{3}{16}\ln\left|x + \sqrt{x^2 + \tfrac{3}{16}}\right|\right] + C.$$

Quick Check 3 ✔

Find: $\int x^2\sqrt{8 + 3x}\, dx$.

3 ✔

EXAMPLE 4 Find: $\int \dfrac{dx}{x^2 - 25}$.

Solution This integral fits *formula 14* in Table 1, with $a^2 = 25$. Then $a = 5$, and we have

$$\int \frac{1}{x^2 - a^2}\, dx = \frac{1}{2a}\ln\left|\frac{x - a}{x + a}\right| + C.$$

$$= \frac{1}{2\cdot 5}\ln\left|\frac{x - 5}{x + 5}\right| + C$$

Quick Check 4 ✔

Find: $\int \dfrac{4}{x^2 - 11}\, dx$.

$$\int \frac{dx}{x^2 - 25} = \frac{1}{10}\ln\left|\frac{x - 5}{x + 5}\right| + C.$$

4 ✔

EXAMPLE 5 Find: $\int (\ln x)^3\, dx$. Assume $x > 0$.

Solution This integral fits *formula 9* in Table 1:

$$\int (\ln x)^n\, dx = x(\ln x)^n - n\int (\ln x)^{n-1}\, dx + C, \quad n \neq -1.$$

We must apply the formula three times:

$$\int (\ln x)^3\, dx = x(\ln x)^3 - 3\int (\ln x)^2\, dx + C \qquad \text{Formula 9, with } n = 3$$

$$= x(\ln x)^3 - 3\left[x(\ln x)^2 - 2\int \ln x\, dx\right] + C \qquad \begin{array}{l}\text{Applying formula 9}\\ \text{again, with } n = 2\end{array}$$

$$= x(\ln x)^3 - 3\left[x(\ln x)^2 - 2\left(x\ln x - \int dx\right)\right] + C \qquad \begin{array}{l}\text{Applying formula 9}\\ \text{for the third time,}\\ \text{with } n = 1\end{array}$$

Quick Check 5 ✔

Find: $\int x^4 \ln x\, dx$. Assume $x > 0$.

$$= x(\ln x)^3 - 3x(\ln x)^2 + 6x\ln x - 6x + C.$$

5 ✔

The Web site www.wolframalpha.com can be used to find integrals. Use this Web site or an app to check Examples 1–5 or your answers to the exercises in the following set.

Section Summary

- We can use tables of integrals or Web sites such as www.wolframalpha.com to evaluate many integrals.

- Some algebraic simplification of the integrand may be required before the correct integral form from a table can be identified.

4.7 Exercise Set

Find each antiderivative using Table 1.

1. $\int xe^{-3x}\, dx$

2. $\int 2xe^{3x}\, dx$

3. $\int 6^x\, dx$

4. $\int \dfrac{1}{\sqrt{x^2 - 9}}\, dx$

5. $\int \dfrac{1}{25 - x^2}\, dx$

6. $\int \dfrac{1}{x\sqrt{4 + x^2}}\, dx$

7. $\int \dfrac{x}{3 - x}\, dx$

8. $\int \dfrac{x}{(1 - x)^2}\, dx$

9. $\int \dfrac{1}{x(8 - x)^2}\, dx$

10. $\int \sqrt{x^2 + 9}\, dx$

11. $\int \ln(3x)\, dx, \quad x > 0$

12. $\int \ln\left(\dfrac{4}{5}x\right) dx, \quad x > 0$

13. $\int x^4 \ln x\, dx, \quad x > 0$

14. $\int x^3 e^{-2x}\, dx$

15. $\int x^3 \ln x\, dx, \quad x > 0$

16. $\int 5x^4 \ln x\, dx, \quad x > 0$

17. $\int \dfrac{dx}{\sqrt{x^2 + 7}}$

18. $\int \dfrac{3\, dx}{x\sqrt{1 - x^2}}$

19. $\int \dfrac{10\, dx}{x(5 - 7x)^2}$

20. $\int \dfrac{2}{5x(7x + 2)}\, dx$

21. $\int \dfrac{-5}{4x^2 - 1}\, dx$

22. $\int \sqrt{9t^2 - 1}\, dt$

23. $\int \sqrt{4m^2 + 16}\, dm$

24. $\int \dfrac{3 \ln x}{x^2}\, dx$

25. $\int \dfrac{-5 \ln x}{x^3}\, dx, \quad x > 0$

26. $\int (\ln x)^4\, dx, \quad x > 0$

27. $\int \dfrac{e^x}{x^{-3}}\, dx$

28. $\int \dfrac{3}{\sqrt{4x^2 + 100}}\, dx$

29. $\int x\sqrt{1 + 2x}\, dx$

30. $\int x\sqrt{2 + 3x}\, dx$

APPLICATIONS

Business and Economics

31. Supply from marginal supply. Stellar Lawn Care introduces a new kind of lawn seeder. It finds that its marginal supply for the seeder satisfies

$$S'(x) = \dfrac{100x}{(20 - x)^2}, \quad 0 \le x \le 19,$$

where $S(x)$ is the quantity purchased when the price is x thousand dollars per seeder. Find $S(x)$, given that the company sells 2000 seeders when the price is 19 thousand dollars.

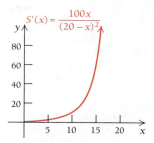

Social Sciences

32. Learning rate. The rate of change of the probability that an employee learns a task on a new assembly line is

$$p'(t) = \dfrac{1}{t(2 + t)^2},$$

where $p(t)$ is the probability of learning the task after t months. Find $p(t)$ given that $p = 0.8267$ when $t = 2$.

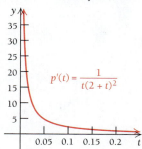

SYNTHESIS

Evaluate using Table 1 or www.wolframalpha.com.

33. $\displaystyle\int \frac{8}{3x^2 - 2x}\,dx$

34. $\displaystyle\int \frac{x\,dx}{4x^2 - 12x + 9}$

35. $\displaystyle\int \frac{dx}{x^3 - 4x^2 + 4x}$

36. $\displaystyle\int e^x\sqrt{e^{2x} + 1}\,dx$

37. $\displaystyle\int \frac{-e^{-2x}\,dx}{9 - 6e^{-x} + e^{-2x}}$

38. $\displaystyle\int \frac{\sqrt{(\ln x)^2 + 49}}{2x}\,dx$

39. Prove formula 6 in Table 1 using integration by parts.

40. Prove formula 18 in Table 1 using the substitution $u = a + bx$.

Answers to Quick Checks

1. Using formula 19: $\displaystyle\frac{6}{25(3 - 5x)} + \frac{2}{25}\ln|3 - 5x| + C$

2. Using formula 21: $\displaystyle\frac{3}{14(7 - 3x)} + \frac{3}{98}\ln\left|\frac{x}{7 - 3x}\right| + C$

3. Using formula 25:
$$\frac{2}{2835}(135x^2 - 288x + 512)(8 + 3x)^{3/2} + C$$

4. Using formula 14: $\displaystyle\frac{2}{11}\sqrt{11}\,\ln\left|\frac{x - \sqrt{11}}{x + \sqrt{11}}\right| + C$

5. Using formula 10: $\displaystyle\frac{x^5}{5}\ln x - \frac{x^2}{25} + C$

KEY TERMS AND CONCEPTS

EXAMPLES

SECTION 4.1

Antidifferentiation is the reverse of the process of differentiation. A function F is an **antiderivative** of a function f if

$$\frac{d}{dx} F(x) = f(x).$$

Antiderivatives of a function f all differ by a constant C, called the **constant of integration**.

$F(x) = x^2$ is an antiderivative of $f(x) = 2x$ because $\frac{d}{dx}(x^2) = 2x$.

$G(x) = \ln x$ is an antiderivative of $g(x) = \frac{1}{x}$ because $\frac{d}{dx}(\ln x) = \frac{1}{x}$.

Antiderivatives of $f(x) = 2x$ have the form $x^2 + C$. These antiderivatives can be expressed by $\int 2x\, dx = x^2 + C$.

An **indefinite integral** of a function is symbolized by

$$\int f(x)\, dx = F(x) + C,$$

where $f(x)$ is called the **integrand**.

In the indefinite integral $\int (x^{-4} + x + e^x)\, dx$, the integrand is $x^{-4} + x + e^x$. The

antiderivative is the set of functions $-\frac{1}{3x^3} + \frac{1}{2}x^2 + e^x + C$.

We use four **rules of antidifferentiation**:

A1. $\int k\, dx = kx + C$

A2. $\int x^n\, dx = \frac{x^{n+1}}{n+1} + C, \quad n \neq -1$

A3. $\int \frac{1}{x}\, dx = \ln |x| + C,$

   or, for $x > 0$, $\int \frac{1}{x}\, dx = \ln x + C$

A4. $\int e^{ax}\, dx = \frac{1}{a}e^{ax} + C$

- $\int 9\, dx = 9x + C \qquad$ A1

- $\int x^6\, dx = \frac{1}{7}x^7 + C \qquad$ A2

- $\int \frac{5}{x}\, dx = 5 \ln |x| + C \qquad$ A3

- $\int e^{3x}\, dx = \frac{1}{3}e^{3x} + C \qquad$ A4

There are two common **properties of indefinite integrals**:

P1. $\int [c \cdot f(x)]\, dx = c \int f(x)\, dx$

P2. $\int [f(x) \pm g(x)]\, dx$
$\quad = \int f(x)\, dx \pm \int g(x)\, dx.$

- $\int (3x^4 + 4x - 5)\, dx = \frac{3}{5}x^5 + 2x^2 - 5x + C \qquad$ A1, A2, P1, P2

- $\int (x + 4)^2\, dx = \int (x^2 + 8x + 16)\, dx$

$$= \frac{1}{3}x^3 + 4x^2 + 16x + C \qquad \text{A1, A2, P1, P2}$$

(continued)

KEY TERMS AND CONCEPTS	EXAMPLES

SECTION 4.1 (continued)

An **initial condition** is a point that is a solution of a particular antiderivative.

Find $\int (3x - 2)\, dx$ given that $(1, 4)$ is a solution of the antiderivative.

We antidifferentiate: $\int (3x - 2)\, dx = \dfrac{3}{2}x^2 - 2x + C$, so

$F(x) = \dfrac{3}{2}x^2 - 2x + C$. Given $(1, 4)$ as an initial condition, we substitute 1 for x and 4 for $F(x)$ and solve for C:

$$4 = \frac{3}{2}(1)^2 - 2(1) + C$$

$$4 = \frac{3}{2} - 2 + C$$

$$4 = -\frac{1}{2} + C$$

$$C = \frac{9}{2}.$$

The particular antiderivative that meets the initial condition is $F(x) = \dfrac{3}{2}x^2 - 2x + \dfrac{9}{2}$.

SECTION 4.2

The **area under the graph of a function** can be interpreted in a meaningful way. The units of the area are determined by multiplying the units of the input variable by the units of the output variable.

Physical Science. Suppose Josef's jogging speed, in miles per hour, is $v(t) = 6$, where t is in hours.

In 3 hr, Josef runs $3\text{ hr} \cdot 6\,\dfrac{\text{mi}}{\text{hr}} = 18$ mi, which is the area under the graph representing the velocity function.

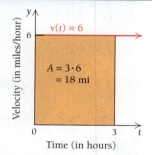

Common geometry formulas can sometimes be used to calculate the area.

Business. McLean Furniture's marginal revenue is modeled by

$$R'(x) = 0.37x,$$

where x is the number of dining room sets sold and R' is in thousands of dollars per set. The total revenue from selling 100 sets is the area under the marginal revenue function.

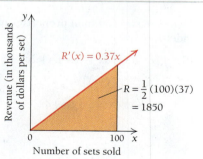

$$\text{Total revenue} = \frac{1}{2}(100 \text{ units})\left(37\,\frac{\text{thousands of dollars}}{\text{unit}}\right)$$
$$= 1850 \text{ thousand dollars, or } \$1{,}850{,}000.$$

KEY TERMS AND CONCEPTS	EXAMPLES

SECTION 4.2 (continued)

Riemann summation uses rectangles to approximate the area under a curve. The more subintervals used as bases of the rectangles, the more accurate the approximation of the area.

If the number of rectangles is allowed to approach infinity, we have a **definite integral**, which represents the exact area under the graph of a continuous and nonnegative function $f(x) \geq 0$ over an interval $[a, b]$:

$$\text{Exact area} = \int_a^b f(x)\,dx.$$

Approximate the area under the graph of $f(x) = -\frac{1}{3}x^2 + 3x$ over $[1, 7]$ using 3 equally sized subintervals.

Each subinterval will have width $\Delta x = \dfrac{7-1}{3} = 2$, with x_i ranging from $x_1 = 1$ to $x_3 = 5$ in increments of 2. The area under the curve over $[1, 7]$ is approximated as follows:

$$\sum_{i=1}^{3} f(x_i) \cdot \Delta x = f(1) \cdot 2 + f(3) \cdot 2 + f(5) \cdot 2$$

$$= \frac{8}{3} \cdot 2 + 6 \cdot 2 + \frac{20}{3} \cdot 2$$

$$= 30.666.$$

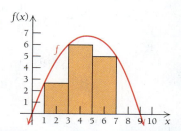

Thus, the area under f is approximately 31 square units.

SECTION 4.3

The **Fundamental Theorem of Calculus** tells us that the exact area under a continuous function f over $[a, b]$ is calculated directly using a definite integral:

$$\int_a^b f(x)\,dx = F(b) - F(a).$$

The function F is any antiderivative of f (we usually set the constant of integration equal to 0).

The exact area under the graph of $f(x) = -\frac{1}{3}x^2 + 3x$ over $[1, 7]$ is

$$\int_1^7 \left(-\frac{1}{3}x^2 + 3x\right) dx = \left[-\frac{1}{9}x^3 + \frac{3}{2}x^2\right]_1^7$$

$$= \left(-\frac{1}{9}(7)^3 + \frac{3}{2}(7)^2\right)$$

$$\quad - \left(-\frac{1}{9}(1)^3 + \frac{3}{2}(1)^2\right)$$

$$= 34.$$

The Fundamental Theorem of Calculus is true for *all* continuous functions over an interval $[a, b]$.

- If $f(x)$ is negative over $[a, b]$, then the definite integral is negative.
- If $f(x)$ has more area above the x-axis than below it over $[a, b]$, then the definite integral is positive.
- If $f(x)$ has more area below the x-axis than above it over $[a, b]$, then the definite integral is negative.
- If $f(x)$ has equal areas above and below the x-axis over $[a, b]$, then the definite integral is zero.

Evaluate the definite integral of $f(x) = x^2 - 1$ over $[-1, 1]$.

Since $f(x)$ is negative over $[-1, 1]$, the definite integral will be negative.

$$\int_{-1}^{1}(x^2 - 1)\,dx = \left[\frac{1}{3}x^3 - x\right]_{-1}^{1}$$

$$= \left(\frac{1}{3}(1)^3 - (1)\right) - \left(\frac{1}{3}(-1)^3 - (-1)\right)$$

$$= -\frac{4}{3}.$$

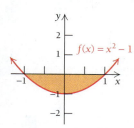

Area over $[-1, 1]$

(continued)

KEY TERMS AND CONCEPTS	EXAMPLES

SECTION 4.3 (*continued*)

Evaluate the definite integral of $f(x) = x^2 - 1$ over $[0, 3]$.

Over $[0, 3]$, f has more area above the x-axis than below, so we expect a positive result.

$$\int_0^3 (x^2 - 1)\, dx = \left[\frac{1}{3}x^3 - x\right]_0^3 = (9 - 3) - (0) = 6$$

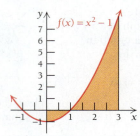

Area over $[0, 3]$

The function $g(x) = x - 2$ over $[0, 4]$ has equal areas below and above the x-axis. Therefore,

$$\int_0^4 (x - 2)\, dx = \left[\frac{1}{2}x^2 - 2x\right]_0^4 = 8 - 8 = 0.$$

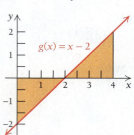

SECTION 4.4

There are many useful properties of definite integrals.

- **Additive property:** if $a < c < b$, we have

$$\int_a^b f(x)\, dx = \int_a^c f(x)\, dx + \int_c^b f(x)\, dx.$$

The additive property is useful for piecewise-defined functions, which include the absolute-value function. Since

$$f(x) = |x| = \begin{cases} -x, \text{ for } x < 0, \\ x, \text{ for } x \geq 0, \end{cases}$$

we have

$$\int_{-2}^3 |x|\, dx = \int_{-2}^0 (-x)\, dx + \int_0^3 x\, dx$$

$$= 2 + \frac{9}{2}$$

$$= \frac{13}{2}.$$

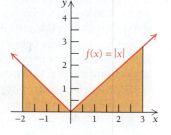

KEY TERMS AND CONCEPTS	EXAMPLES

SECTION 4.4 (*continued*)

• **Area of a region bounded by two curves:** if $f(x) \geq g(x)$ over $[a, b]$, then the area between the graphs of f and g from $x = a$ to $x = b$ is

$$A = \int_a^b [f(x) - g(x)] \, dx.$$

Let $f(x) = x^2$ and $g(x) = x + 2$. Solving $f(x) = g(x)$, we find that the curves intersect at $x = -1$ and $x = 2$. Furthermore, we see that $g(x) \geq f(x)$, on $[-1, 2]$. Therefore, the area between these curves is

$$A = \int_{-1}^2 (x + 2 - x^2) \, dx = \left[\frac{1}{2}x^2 + 2x - \frac{1}{3}x^3 \right]_{-1}^2$$

$$= \left(\frac{1}{2}(2)^2 + 2(2) - \frac{1}{3}(2)^3 \right) - \left(\frac{1}{2}(-1)^2 + 2(-1) - \frac{1}{3}(-1)^3 \right)$$

$$= \frac{9}{2}.$$

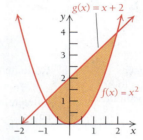

• **Average value:** the average value of $f(x)$ over $[a, b]$ is

$$y_{\text{av}} = \frac{1}{b - a} \int_a^b f(x) \, dx.$$

The average value of $y = x^3$ over $[1, 4]$ is

$$y_{\text{av}} = \frac{1}{3} \int_1^4 x^3 \, dx = \left[\frac{1}{12}x^4 \right]_1^4 = 21\frac{1}{4}.$$

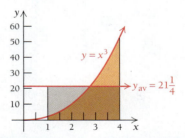

SECTION 4.5

Integration by **substitution** is the reverse of applying the Chain Rule. We choose u, determine du, and rewrite the integrand in terms of u. It may be necessary to multiply the integrand by a constant and the entire integral by the reciprocal of that constant to obtain the correct form. *Results should be checked using differentiation!*

Evaluate $\int 2x(x^2 + 1)^5 \, dx$.

We let $u = x^2 + 1$, so that $du = 2x \, dx$. Therefore, we have

$$\int 2x(x^2 + 1)^5 \, dx = \int u^5 \, du = \frac{1}{6}u^6 + C = \frac{1}{6}(x^2 + 1)^6 + C.$$

Evaluate $\int \frac{1}{3x - 2} \, dx$.

We let $u = 3x - 2$, so that $du = 3 \, dx$. We multiply the integrand by 3 and the integral by $\frac{1}{3}$ outside the integral:

$$\int \frac{1}{3x - 2} \, dx = \frac{1}{3} \int \frac{1}{3x - 2} 3 \, dx = \frac{1}{3} \int \frac{1}{u} \, du$$

$$= \frac{1}{3} \ln u + C = \frac{1}{3} \ln |3x - 2| + C.$$

(*continued*)

KEY TERMS AND CONCEPTS	EXAMPLES

SECTION 4.6

The **Integration-by-Parts Formula** makes use of the product rule for differentiation:

$$\int u \, dv = uv - \int v \, du$$

Evaluate $\int x^3 \ln x \, dx, \, x > 0$.

We let $u = \ln x$ and $dv = x^3 \, dx$. Then $du = \dfrac{1}{x} \, dx$ and $v = \dfrac{1}{4}x^4$, and we have

$$\int u \, dv = uv - \int v \, du$$

$$\int x^3 \ln x \, dx = \frac{1}{4}x^4 \ln x - \int \left(\frac{1}{4}x^4\right)\left(\frac{1}{x} \, dx\right)$$

$$= \frac{1}{4}x^4 \ln x - \frac{1}{4}\int x^3 \, dx$$

$$= \frac{1}{4}x^4 \ln x - \frac{1}{16}x^4 + C.$$

Tabular integration can be used when integration by parts has to be applied more than once to evaluate an integral.

Evaluate $\int x^3 e^{2x} \, dx$.

This will involve repeated integrations by parts. Since $f(x) = x^3$ eventually differentiates to 0 and $g(x) = e^{2x}$ is easily integrable, we use tabular integration by parts:

$f(x)$ and Repeated Derivatives	Sign of Product	$g(x)$ and Repeated Integrals
x^3	$(+)$	e^{2x}
$3x^2$	$(-)$	$\frac{1}{2}e^{2x}$
$6x$	$(+)$	$\frac{1}{4}e^{2x}$
6	$(-)$	$\frac{1}{8}e^{2x}$
0		$\frac{1}{16}e^{2x}$

We multiply along the arrows, alternate signs, and simplify when possible. The antiderivative is

$$\int x^3 e^{2x} \, dx = \frac{1}{2}x^3 e^{2x} - \frac{3}{4}x^2 e^{2x} + \frac{6}{8}xe^{2x} - \frac{6}{16}e^{2x}$$

$$= e^{2x}\left(\frac{1}{2}x^3 - \frac{3}{4}x^2 + \frac{3}{4}x - \frac{3}{8}\right) + C.$$

SECTION 4.7

Integration tables provide formulas for evaluating many general forms of integrals.

Evaluate $\displaystyle\int \frac{3}{\sqrt{x^2 + 64}} \, dx$.

We see that formula 12 from the table of integration formulas on pp. 431–432 is appropriate:

$$\int \frac{1}{\sqrt{x^2 + a^2}} \, dx = \ln\left|x + \sqrt{x^2 + a^2}\right| + C. \qquad \text{Using formula 12}$$

We factor the constant 3 outside of the integral, and we note that $a^2 = 64$, so $a = 8$:

$$\int \frac{3}{\sqrt{x^2 + 64}} \, dx = 3\int \frac{1}{\sqrt{x^2 + 64}} \, dx = 3\ln\left|x + \sqrt{x^2 + 64}\right| + C.$$

CHAPTER 4
Review Exercises

These review exercises are for test preparation. They can also be used as a practice test. Answers are at the back of the book. The red bracketed section references tell you what part(s) of the chapter to restudy if your answer is incorrect.

CONCEPT REINFORCEMENT

Classify each statement as either true or false.

1. Riemann sums are a way of approximating the area under a curve by using rectangles. [4.2]

2. If a and b are both negative, then $\int_a^b f(x)\,dx$ is negative. [4.3]

3. For any continuous function f defined over $[-1, 7]$, it follows that
$$\int_{-1}^2 f(x)\,dx + \int_2^7 f(x)\,dx = \int_{-1}^7 f(x)\,dx. \quad [4.4]$$

4. Every integral can be evaluated using integration by parts. [4.6]

5. $\int x^2 \cdot e^x \, dx = \left(\frac{1}{3}x^3\right)' e^x + C$ [4.1, 4.6]

Match each integral in column A with the corresponding antiderivative in column B. [4.1, 4.5]

Column A	Column B
6. $\int \dfrac{1}{\sqrt{x}}\,dx$	a) $\ln x + C$
7. $\int (1 + 2x)^{-2}\,dx$	b) $-x^{-1} + C$
8. $\int \dfrac{1}{x}\,dx, \quad x > 0$	c) $-(1 + x^2)^{-1} + C$
9. $\int \dfrac{2x}{1 + x^2}\,dx$	d) $-\dfrac{1}{2}(1 + 2x)^{-1} + C$
10. $\int \dfrac{1}{x^2}\,dx$	e) $2x^{1/2} + C$
11. $\int \dfrac{2x}{(1 + x^2)^2}\,dx$	f) $\ln(1 + x^2) + C$

REVIEW EXERCISES

12. Business: total cost. The marginal cost, in dollars, of producing the xth car stereo is given by
$$C'(x) = 0.004x^2 - 2x + 500.$$

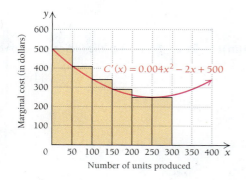

Approximate the total cost of producing 300 car stereos by computing the sum
$$\sum_{i=1}^{6} C'(x_i)\,\Delta x, \quad \text{with } \Delta x = 50. \quad [4.2]$$

Find each antiderivative. [4.1]

13. $\int 20x^4\,dx$

14. $\int (3e^{4x} + 2)\,dx$

15. $\int \left(18t^2 + 6t + \dfrac{1}{2t}\right)dt$ (assume $t > 0$)

Find the area under each curve over the indicated interval. [4.3]

16. $y = 6 - x^2; \quad [-2, 1]$

17. $y = x^2 + 4x + 4; \quad [0, 3]$

In each case, give an interpretation of what the shaded region represents. [4.2, 4.3]

18.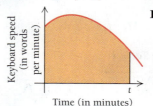

19.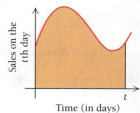

Evaluate. [4.3, 4.4]

20. $\int_a^b x^5\,dx$

21. $\int_{-1}^2 (x^3 - x^4)\,dx$

22. $\int_0^1 (e^{2x} + x)\,dx$

23. $\int_1^4 \dfrac{2}{x}\,dx$

24. $\int_{-2}^4 f(x)\,dx$, where $f(x) = \begin{cases} x + 2, & \text{for } x \le 0, \\ 2 - \frac{1}{2}\sqrt{x}, & \text{for } x > 0 \end{cases}$

25. $\int_2^4 g(x)\,dx$, for g as shown in the graph at right

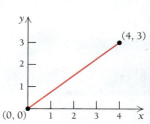

Decide whether $\int_a^b f(x)\,dx$ is positive, negative, or zero. [4.3]

26.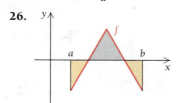

27.

443

28.

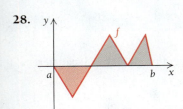

29. Find the area of the region bounded by

$$y = x^2 + 3x + 1 \text{ and } y = 6x - 1. \quad [4.4]$$

Find each antiderivative using substitution. Do not use Table 1. [4.5]

30. $\displaystyle\int x^3 e^{x^4} \, dx$

31. $\displaystyle\int \frac{24t^5}{4t^6 + 3} \, dt$

32. $\displaystyle\int \frac{\ln(4x)}{2x} \, dx$

33. $\displaystyle\int \frac{e^{2x}}{e^{2x} + 2} \, dx$

Find each antiderivative using integration by parts. Do not use Table 1. [4.6]

34. $\displaystyle\int 3x e^{4x} \, dx$

35. $\displaystyle\int \ln \sqrt[3]{x^2} \, dx$

36. $\displaystyle\int 5x^2 \ln x \, dx$

37. $\displaystyle\int x^4 e^{3x} \, dx$

Find each antiderivative using Table 1. [4.7]

38. $\displaystyle\int \frac{1}{49 - x^2} \, dx$ **39.** $\displaystyle\int x\sqrt{2 + 7x} \, dx$ **40.** $\displaystyle\int \frac{x}{7x + 1} \, dx$

41. $\displaystyle\int \frac{dx}{\sqrt{x^2 - 36}}$ **42.** $\displaystyle\int \frac{4}{x\sqrt{10 - x^2}} \, dx$

43. Business: total cost. Refer to Exercise 12. Calculate the total cost of producing 300 car stereos. [4.4]

44. Find the average value of $y = xe^{-x}$ over $[0, 2]$. [4.4]

45. A particle starts out from the origin. Its velocity in mph after t hours is $v(t) = 3t^2 + 2t$. Find the distance the particle travels during the first 4 hr (from $t = 0$ to $t = 4$). [4.3]

46. Business: total revenue. A company estimates that its revenue grows continuously at a rate given by $S'(t) = 3e^{3t}$, where t is the number of days since an innovation is introduced. Find the accumulated revenue for the first 4 days. [4.3]

Integrate using any method. [4.3–4.6]

47. $\displaystyle\int \frac{12t^2}{4t^3 + 7} \, dt$

48. $\displaystyle\int \frac{x \, dx}{\sqrt{4 + 5x}}$

49. $\displaystyle\int 5x^4 e^{x^5} \, dx$

50. $\displaystyle\int \frac{dx}{x + 9}$

51. $\displaystyle\int t^7 (t^8 + 3)^{11} \, dt$

52. $\displaystyle\int \ln(7x) \, dx$

53. $\displaystyle\int x \ln(8x) \, dx$

SYNTHESIS

Find each antiderivative. [4.5–4.7]

54. $\displaystyle\int \frac{t^4 \ln(t^5 + 3)}{t^5 + 3} \, dt$ **55.** $\displaystyle\int \frac{dx}{e^x + 2}$

56. $\displaystyle\int \frac{\ln \sqrt{x}}{x} \, dx$ **57.** $\displaystyle\int x^{91} \ln|x| \, dx$

58. $\displaystyle\int \ln\left|\frac{x - 3}{x - 4}\right| dx$ **59.** $\displaystyle\int \frac{dx}{x\,(\ln|x|)^4}$

60. $\displaystyle\int x \sqrt[3]{x + 3} \, dx$ **61.** $\displaystyle\int \frac{x^2}{2x + 1} \, dx$

TECHNOLOGY CONNECTION

62. Use a graphing calculator to approximate the area between the following curves:

$$y = 2x^2 - 2x, \quad y = 12x^2 - 12x^3. \quad [4.4]$$

CHAPTER 4
Test

1. Approximate

$$\int_0^5 (25 - x^2) \, dx$$

by computing the area of each rectangle and adding.

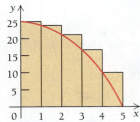

Find each antiderivative.

2. $\displaystyle\int \sqrt{3x} \, dx$

3. $\displaystyle\int 300x^5 \, dx$

4. $\displaystyle\int \left(e^{5x} + \frac{1}{x} + x^{3/8} \right) dx$ (assume $x > 0$)

Find the area under the curve over the indicated interval.

5. $y = x - x^2$; $[0, 1]$ **6.** $y = \dfrac{4}{x}$; $[1, 3]$

7. Give an interpretation of the shaded area.

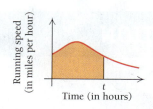

Running speed (in miles per hour)

Time (in hours)

Evaluate.

8. $\int_{-1}^{2} (4x + 5x^2)\, dx$

9. $\int_{0}^{1} e^{-2x}\, dx$

10. $\int_{e^3}^{e^5} \dfrac{dx}{x}$

11. $\int_{0}^{5} g(x)\, dx$, where $g(x) = \begin{cases} x^2, & \text{for } x \leq 2, \\ 6 - x, & \text{for } x > 2 \end{cases}$

12. Find $\int_{3}^{7} f(x)\, dx$, for f as shown in the graph.

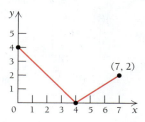

$(7, 2)$

13. Decide whether $\int_{a}^{b} f(x)\, dx$ is positive, negative, or zero.

Find each antiderivative using substitution. Assume $u > 0$ when $\ln u$ appears. Do not consult Table 1.

14. $\int \dfrac{dx}{3x + 12}$

15. $\int e^{-0.5x}\, dx$

16. $\int t^3 (t^4 + 3)^7\, dt$

Find each antiderivative using integration by parts. Do not consult Table 1.

17. $\int x e^{5x}\, dx$

18. $\int x^3 \ln x^4\, dx$

Find each antiderivative using Table 1.

19. $\int \dfrac{1}{x^2 - 81}\, dx$

20. $\int \dfrac{dx}{x(7 - x)}$

21. Find the average value of $y = 4t^3 + 2t$ over $[-1, 2]$.

22. Find the area of the region in the first quadrant bounded by $y = x$ and $y = x^6$.

23. Business: cost from marginal cost. An air conditioning company determines that the marginal cost, in dollars, for producing the xth air conditioner is given by

$$C'(x) = -0.2x + 500, \quad C(0) = 0.$$

Find the total cost of producing 250 air conditioners.

24. Social science: learning curve. A translator's speed over 4-min interval is given by

$$W(t) = -6t^2 + 12t + 90, \quad t \text{ in } [0, 4],$$

where $W(t)$ is the speed, in words per minute, at time t. How many words are translated during the second minute (from $t = 1$ to $t = 2$)?

25. A robot leaving a spacecraft has velocity given by $v(t) = -0.4t^2 + 2t$, where $v(t)$ is in kilometers per hour and t is the number of hours since the robot left the spacecraft. Find the total distance traveled during the first 4 hr.

Find each antiderivative using any method. Assume $u > 0$ when $\ln u$ appears.

26. $\int \dfrac{6}{5 + 7x}\, dx$

27. $\int x^5 e^x\, dx$

28. $\int x^5 e^{x^6}\, dx$

29. $\int \sqrt{x}\, \ln x\, dx$

30. $\int e^{3x} \sqrt{1 + e^{3x}}\, dx$

31. $\int x^4 e^{-0.1x}\, dx$

32. $\int x \ln (13x)\, dx$

SYNTHESIS

Find each antiderivative using any method.

33. $\int x^3 \sqrt{x^2 + 4}\, dx$

34. $\int \dfrac{[(\ln x)^3 - 4(\ln x)^2 + 5]}{x}\, dx, \; x > 0$

35. $\int \ln \left(\dfrac{x + 3}{x + 5} \right) dx$

36. $\int \dfrac{8x^3 + 10}{\sqrt[3]{5x - 4}}\, dx$

37. $\int \dfrac{x}{\sqrt{3x - 2}}\, dx$

38. $\int \dfrac{(x + 4)^2}{x^2}\, dx$

39. Evaluate $\int 5^x\, dx$ without using Table 1. (*Hint:* $5 = e^{\ln 5}$.)

40. Use a calculator to approximate the area between the following curves:

$$y = 3x - x^2, \quad y = 2x^3 - x^2 - 5x.$$

Business: Distribution of Wealth

Lorenz Functions and the Gini Coefficient

The distribution of wealth within a population is of great interest to many economists and sociologists. Let $y = f(x)$ represent the percentage of wealth owned by x percent of the population, with x and y expressed as decimals between 0 and 1. The assumptions are that 0% of the population owns 0% of the wealth and that 100% of the population owns 100% of the wealth. With these requirements in place, the *Lorenz function* is defined to be any continuous, increasing and concave upward function connecting the points $(0, 0)$ and $(1, 1)$, which represent the two extremes. The function is named for economist Max Otto Lorenz (1880–1962), who developed these concepts as a graduate student in 1905–1906.

If the collective wealth of a society is equitably distributed among its population, we would observe that "$x\%$ of the population owns $x\%$ of the wealth," and this is modeled by the function $f(x) = x$, where $0 \le x \le 1$. This is an example of a Lorenz function that is often called the *line of equality*.

In many societies, the distribution of wealth is not equitable. For example, the Lorenz function $f(x) = x^3$ represents a society in which a large percentage of the population owns a small percentage of the wealth. In this example, $f(0.7) = 0.7^3 = 0.343$, meaning that 70% of the population owns just 34.3% of the wealth, with the implication that the other 30% owns the remaining 65.7% of the wealth.

In the graphs below, we see the line of equality in the left-most graph, and increasingly inequitable distributions as we move to the right. Note that the area between the line of equality and the graph of the Lorenz function $f(x)$ is small if the distribution of wealth is close to equitable and is large when the distribution is very unequitable. The *Gini coefficient* (named for the Italian statistician and demographer Corrado Gini, 1884–1965) is a measure of the difference between the actual distribution of wealth in a society and the ideal distribution represented by the line of equality. It is the ratio of the area between the line of equality and the

graph of the Lorenz function to the area below the line of equality and above the x-axis, as shown below:

$$\text{Gini coefficient} = \frac{A}{A + B}.$$

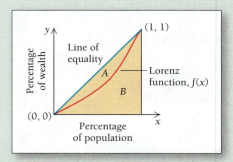

The area A is found by calculating the area between the graphs of $y = x$ and $y = f(x)$, where $y = x$ is the line of equality and f is the Lorenz function for a particular society. We observe that $A + B$ is a triangle with area $\frac{1}{2}(1)(1) = \frac{1}{2}$. Thus, the Gini coefficient can be written as an integral:

$$\text{Gini coefficient} = \frac{A}{A + B} = \frac{\int_0^1 (x - f(x))\, dx}{\left(\dfrac{1}{2}\right)}$$

$$= 2\int_0^1 (x - f(x))\, dx.$$

For the most equitable distribution of wealth, the Gini coefficient is 0, since there will be no difference (area) between the graph of the Lorenz function and the line of equality; for the most inequitable distribution of wealth, the Gini coefficient is 1. Often, the Gini coefficient is multiplied by 100 to give the *Gini index*: a Gini coefficient of 0.34 gives a Gini index of 34.

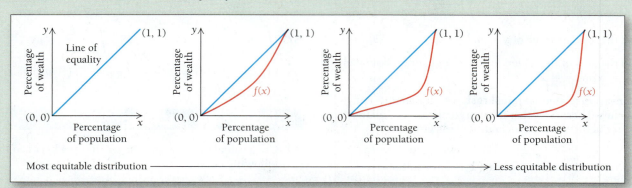

Exercises

1. Suppose the Lorenz function for a society is given by $f(x) = x^2, 0 \leq x \leq 1$.

   a) What percentage of the wealth is owned by 60% of the population?

   b) Calculate the Gini index (the value will be between 0 and 100).

2. Suppose the Lorenz function for a society is given by $f(x) = x^{3.5}, 0 \leq x \leq 1$.

   a) What percentage of the wealth is owned by 60% of the population?

   b) Calculate the Gini index.

Regression for Determining Lorenz Functions

If data exist on the distribution of wealth in a society, a Lorenz function can be determined using regression.

Exercises

3. The data in the table show the amount of wealth distributed within a population.

x	0.1	0.2	0.3	0.4	0.5	0.6	0.7	0.8	0.9
y	0.0178	0.06	0.122	0.201	0.297	0.409	0.536	0.677	0.833

Use regression to determine a power function that best fits these data. [*Note:* Entering the point $(0, 0)$ may cause an error message to appear. However, the point $(1, 1)$ should be entered along with the rest of the data.]

   a) Express the Lorenz function in the form $f(x) = x^n$. The coefficient should be 1, so you may have to do some rounding.

   b) Determine the Gini coefficient and the Gini index.

   c) What percentage of the wealth is owned by the lowest 74% of this population?

4. A fast-food chain has many hundreds of franchises nationwide. Ideally, all franchises would generate equal amounts of revenue, but in reality, some perform better than others. An internal audit reveals the following results: the lowest 30% of the franchises account for just 6% of the total revenue, the lowest 50% account for 20% of the total revenue, and the lowest 70% account for 43.5% of the total revenue. (Assume that 100% of the franchises account for 100% of the total revenue.)

   a) Use regression to determine a power function that models these data, and write the Lorenz function in the form $f(x) = x^n$. The coefficient should be 1, so you may have to do some rounding.

   b) Determine the Gini coefficient and the Gini index.

   c) What percentage of total revenue is generated by the lowest 45% of the franchises?

   d) What percentage of total revenue is generated by the top 10% of the franchises?

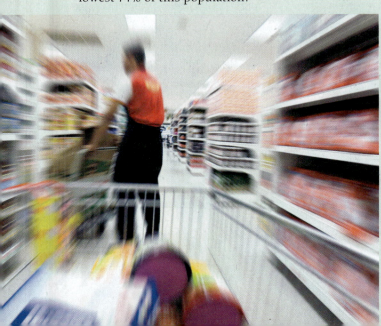

Gini Coefficient as a Function of n

Functions of the form $f(x) = x^n$, $0 \leq x \leq 1$, where $n \geq 1$, meet the criteria for Lorenz functions. We can develop a function $G(n)$ that will allow us to calculate the Gini coefficient directly, given a value of n.

$$G(n) = 2\int_0^1 (x - x^n)\, dx$$

$$= \left[2\left(\frac{1}{2}x^2 - \frac{1}{n+1}x^{n+1}\right)\right]_0^1$$

$$= 2\left(\frac{1}{2} - \frac{1}{n+1}\right)$$

$$= 1 - \frac{2}{n+1}$$

$$= \frac{n-1}{n+1}.$$

Exercises

5. Verify your results for Exercises 3 and 4 using the function $G(n) = \dfrac{n-1}{n+1}$.

6. The United States had a Gini index of 46.9 in 2013. (*Source:* www.census.gov.) Express this as a decimal: $G = 0.469$.

   a) Solve for n, and write the Lorenz function in the form $f(x) = x^n$.

   b) According to this model, what percentage of the wealth was owned by the least wealthy 55% of U.S. citizens in 2013?

7. Canada's Gini index is usually 30.0.

   a) Determine the Lorenz function.

   b) What percentage of wealth is owned by the least wealthy 55% of the citizens in Canada?

     Sometimes, data may not fit "neatly" into the $f(x) = x^n$ form, especially in cases where the distribution very heavily favors a small percentage of the population that holds most of the wealth. In these cases, an exponential function of the form $f(x) = a \cdot b^x$, $0 \leq x \leq 1$ may work better than $f(x) = x^n$, as long as the value of a is extremely small.

8. In 2010, the distribution of net worth within the United States was as given in the following table:

PERCENTAGE OF POPULATION	0.5	0.9	0.99	1
PERCENTAGE OF WEALTH	0.011	0.254	0.654	1

(*Source:* www.fas.gov, based on data from Arthur Kennickell.)

   a) According to the table, what percentage of net worth was held by the top 1%? (*Hint:* What percentage did the other 99% hold?)

   b) Use regression to fit an exponential function $g(x) = a \cdot b^x$ to these data.

   c) Determine the area between the line of equality and the graph of g over $[0, 1]$. (*Hint:* Integrate $a \cdot b^x$ using formula 11 from Table 1 in Section 4.7.)

   d) Determine the Gini coefficient and the Gini index.

   e) What percentage of the net worth was held by the lowest 30% of the population?

   f) What percentage of the net worth was held by the top 15% of the population?

5 | Applications of Integration

What You'll Learn

5.1 Consumer Surplus and Producer Surplus

5.2 Integrating Growth and Decay Models

5.3 Improper Integrals

5.4 Probability

5.5 Probability: Expected Value; The Normal Distribution

5.6 Volume

5.7 Differential Equations

Why It's Important

In this chapter, we explore a variety of applications of integration in business and economics (consumer and producer surplus), environmental science (exponential growth and decay), and probability and statistics (expected value). We also see how to use integration to find volumes of solids and to solve differential equations.

Where It's Used

Crude Oil Demand: In 2013, annual world demand for crude oil was approximately 33.3 billion barrels, and it was projected to increase by 1.5% per year. World reserves of crude oil in 2013 were approximately 1635 billion barrels. Assuming that no new oil reserves are found, when will the reserves be depleted? (*Sources*: Based on information from www.iea.gov and www.cia.gov.) (*This problem appears as Exercise 45 in Section 5.2.*)

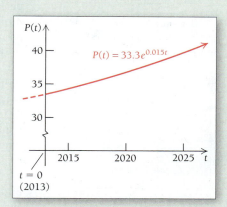

● Given demand and supply functions, find the consumer surplus and the producer surplus at the equilibrium point.

Consumer Surplus and Producer Surplus

It is convenient to consider demand and supply as functions of price. For this section, we will think of demand and supply as prices that are functions of quantity: $p = D(x)$ and $p = S(x)$. We can then use integration to calculate quantities of interest to economists, such as *consumer surplus* and *producer surplus*.

The consumer's **demand curve** is the graph of $p = D(x)$, which is the price per unit, p, that a consumer is willing to pay for x units of a product. It is usually a decreasing function since the consumer expects to pay less per unit for large quantities of the product. The producer's **supply curve** is the graph of $p = S(x)$, which is the price per unit, p, that a producer is willing to accept for selling x units. It is usually an increasing function since a higher price per unit is an incentive for the producer to make more units available for sale. The *equilibrium point*, (x_E, p_E), is the intersection of these two curves.

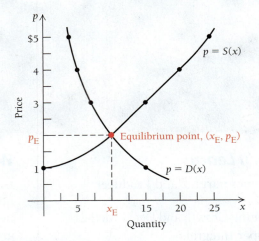

Utility is a function often considered in economics. When a consumer buys x units of a product, he or she receives a certain amount of utility, U (see Exercise 27 in Exercise Set 1.3). For example, the number of movies that you see in a month gives you a certain utility. If you see four movies, you get more utility than if you see no movies. The same notion applies to having a meal in a restaurant or paying your heating bill to warm your home.

To help explain the concepts of consumer surplus and producer surplus, we will consider the utility of seeing movies over 1 month. Suppose that, at a price of $10 per ticket, Louise will see no movies. At a price of $8.75 per ticket, she will see one movie per month, and at $7.50 per ticket, she will see two movies per month. As long as the number of movies, x, is small, Louise's demand function for movies can be modeled by $p = D(x) = 10 - 1.25x$. We want to examine the utility she receives from going to the movies.

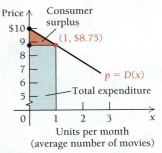

FIGURE 1

At a ticket price of $8.75, Louise sees one movie. Her total expenditure is $(1)(\$8.75) = \$8.75$, as shown by the blue region in Fig. 1. However, the area under her demand curve over $[0, 1]$ is $9.38. This is what going to one movie per month is worth to Louise. When she spends $8.75, the difference in area, represented by the orange triangle, $\$9.38 - \$8.75 = \$0.63$, can be interpreted as the extra utility she gets, but does not have to pay for, from the one movie. Economists define this amount as the *consumer surplus*. It is the extra utility that consumers enjoy when prices decrease.

Suppose Louise goes to two movies per month at $7.50 per ticket. Her total expenditure is $(2)(\$7.50) = \$15.00$, which is represented by the blue region in Fig. 2. The area under Louise's demand curve over $[0, 2]$ is $17.50. Therefore, her consumer surplus is $2.50, which measures the extra utility she received, but did not have to pay for, from the two movies.

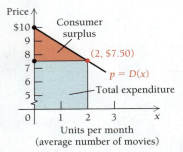

FIGURE 2

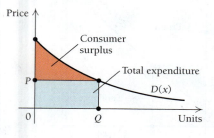

Price

Consumer surplus

Total expenditure

P

$D(x)$

0 Q Units

Suppose the graph of a demand function is a curve, as shown at the left. If Louise goes to Q movies at P dollars per movie, then her total expenditure is QP. The area under the curve over $[0, Q]$ is the total utility received:

$$\int_0^Q D(x)\, dx.$$

The *consumer surplus* is the difference between the area under the curve and the total expenditure, and it is given by

$$\int_0^Q D(x)\, dx - QP.$$

DEFINITION

Let $p = D(x)$ describe the demand function for a product. Then the **consumer surplus** for Q units of the product, at a price per unit P, is

$$\int_0^Q D(x)\, dx - QP.$$

EXAMPLE 1 **Business: Consumer Surplus.**
Arnold is shopping for high-quality sketchpads for his art class. His demand function is given by $p = D(x) = (x - 6)^2$. Find Arnold's consumer surplus if he purchases 2 sketchpads.

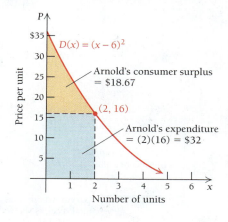

Solution When $x = 2$, we have

$$p = D(2) = (2 - 6)^2 = (-4)^2 = 16.$$

Thus, Arnold is willing to purchase 2 sketchpads at $16 per unit.

If he purchases 2 sketchpads at $16 per unit, his consumer surplus is

$$\int_0^2 (x - 6)^2\, dx - 2 \cdot 16 = \int_0^2 (x^2 - 12x + 36)\, dx - 32$$

$$= \left[\frac{x^3}{3} - 6x^2 + 36x \right]_0^2 - 32$$

$$= \left[\left(\frac{(2)^3}{3} - 6(2)^2 + 36(2) \right) - \left(\frac{(0)^3}{3} - 6(0)^2 + 36(0) \right) \right] - 32$$

$$= \left(\frac{8}{3} - 24 + 72 \right) - 0 - 32$$

$$= \$18.67. \qquad \qquad \text{1 ✔}$$

Quick Check 1 ✔

Find the consumer surplus for the demand function given by $D(x) = x^2 - 6x + 16$ when $x = 1$.

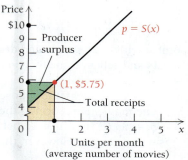

Price

$10

9 — Producer

8 — surplus

7

6 ● (1, \$5.75)

5

4 — Total receipts

0 1 2 3 4 5 x

Units per month
(average number of movies)

FIGURE 3

Let's now look at a supply curve for a movie theater, as shown in Figs. 3 and 4. Suppose the theater will not sell tickets for less than $4, but will sell one ticket for one movie at $5.75 or two tickets for two movies at $7.50 each. For small numbers of movies, x, the theater's supply curve is modeled by $p = 4 + 1.75x$. The price $5.75 is within what Louise is willing to pay for one movie, and the theater will take in $(1)(\$5.75) = \$5.75$ for selling her one ticket for one movie. The yellow region in Fig. 3 represents the total per-person cost to the theater for showing one movie, which is $4.88. Since the theater's revenue is $5.75 for selling one ticket, the difference, $\$5.75 - \$4.88 = \$0.87$, represents the profit for the theater. Economists call this the *producer surplus*.

Exploratory

Graph $D(x) = (x - 6)^2$, the demand function in Example 1, using the viewing window $[0, 6, 0, 40]$, with Yscl = 5. To find the consumer surplus at $x = 2$, we first find $D(2)$. Then we graph $y = D(2)$. What is the point of intersection of $y = D(x)$ and $y = D(2)$?

From the intersection, use DRAW to create a vertical line down to the x-axis. What does the area of the resulting rectangle represent? What does the area above the horizontal line and below the curve represent?

At a price of $7.50, the theater will sell Louise two tickets and collect (2)($7.50), or $15, in revenue. The area of the yellow region in Fig. 4 represents the total cost to the theater of selling Louise two tickets, which is $11.50. The area of the green triangle, $15.00 − $11.50 = $3.50, is the producer's surplus. It is part of the theater's profit.

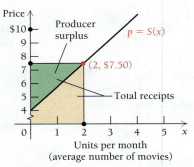

FIGURE 4

Suppose the graph of the supply function is a curve, as shown at the right. If the theater sells Louise Q tickets at a price P, the total receipts are QP. The *producer surplus* is the total receipts minus the area under the curve:

$$QP - \int_0^Q S(x)\, dx.$$

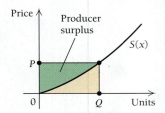

DEFINITION

Let $p = S(x)$ be a supply function for a product. Then the **producer surplus** for Q units of the product, at a price per unit P, is

$$QP - \int_0^Q S(x)\, dx.$$

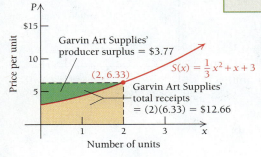

Quick Check 2 ✔

Find the producer surplus for $S(x) = \frac{1}{3}x^2 + \frac{4}{3}x + 4$ when $x = 1$.

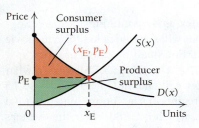

FIGURE 5

EXAMPLE 2 **Business: Producer Surplus.** Garvin Art Supplies sells artist's sketchpads, and the supply function for this product is given by $p = S(x) = \frac{1}{3}x^2 + x + 3$. Find the producer surplus when Garvin Art Supplies sells 2 sketchpads.

Solution When $x = 2$, we have $p = S(2) = \frac{1}{3}(2)^2 + (2) + 3 = 6.33$. Thus, Garvin Art Supplies is willing to sell 2 sketchpads at $6.33 per unit. If the firm sells 2 sketchpads at $6.33 per unit, the producer surplus is

$$2 \cdot 6.33 - \int_0^2 \left(\frac{1}{3}x^2 + x + 3\right) dx = 12.66 - \left[\frac{x^3}{9} + \frac{x^2}{2} + 3x\right]_0^2$$

$$= 12.66 - \left[\left(\frac{(2)^3}{9} + \frac{(2)^2}{2} + 3(2)\right) - \left(\frac{(0)^3}{9} + \frac{(0)^2}{2} + 3(0)\right)\right]$$

$$= 12.66 - \left(\frac{8}{9} + 2 + 6 - 0\right)$$

$$= \$3.77.$$

The **equilibrium point**, (x_E, p_E), in Fig. 5 is the point at which the supply and demand curves intersect. It is the ideal point at which buyers and sellers come together and purchases and sales actually occur.

Let's reconsider the example involving Louise and the movie theater. Suppose the theater charges $5.75 for one ticket, and Louise sees one movie. Since seeing the movie is worth $9.38 to Louise, she derives $9.38 − $5.75 = $3.63 in utility. To her, this is a good deal, since she paid much less than she was willing to pay. However, the theater lost potential revenue by "undercharging" her.

As Fig. 6 shows, at $7.50 per ticket, Louise's demand curve and the theater's supply curve intersect. This point is advantageous for both Louise and the theater, since she is willing to see two movies at $7.50 per ticket, while the theater can increase its revenue by selling two tickets. In other words, if the price per ticket is set too low, the theater will sell tickets but will lose revenue it could be receiving at a higher price, since Louise is willing to pay more according to the demand curve. On the other extreme, if the theater sets the price per ticket too high, it will not sell enough tickets to make a profit. The $7.50 ticket price is the best "middle ground" for both producer and consumer in this case.

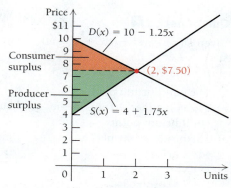

FIGURE 6

EXAMPLE 3 **Business: Equilibrium Point.** Given Arnold's demand function, $D(x) = (x - 6)^2$, and Garvin Art Supplies' supply function, $S(x) = \frac{1}{3}x^2 + x + 3$, find the equilibrium point, and explain what it represents. Assume $x \leq 5$.

Solution To find the equilibrium point, we set $D(x) = S(x)$ and solve:

$$(x - 6)^2 = \frac{1}{3}x^2 + x + 3$$

$$3(x - 6)^2 = x^2 + 3x + 9 \qquad \text{Multiplying both sides by 3}$$

$$3(x^2 - 12x + 36) = x^2 + 3x + 9$$

$$3x^2 - 36x + 108 = x^2 + 3x + 9$$

$$2x^2 - 39x + 99 = 0.$$

Factoring or using the quadratic formula, we find that $x = x_E = 3$ is a solution to this equation. Substituting 3 for x in the demand or supply function, we find that $p_E = D(3) = S(3) = 9$. The equilibrium point is $(3, 9)$. This means that Arnold is willing to purchase 3 sketchpads at $9 per unit and Garvin Art Supplies is willing to sell him 3 sketchpads at $9 per unit.

Arnold's consumer surplus is

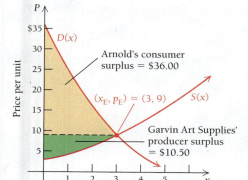

$$\int_0^3 (x - 6)^2 \, dx - 3 \cdot 9 = \int_0^3 (x^2 - 12x + 36) \, dx - 27$$

$$= \left[\frac{x^3}{3} - 6x^2 + 36x \right]_0^3 - 27$$

$$= \left[\left(\frac{(3)^3}{3} - 6(3)^2 + 36(3) \right) - \left(\frac{(0)^3}{3} - 6(0)^2 + 36(0) \right) \right] - 27$$

$$= (9 - 54 + 108) - 0 - 27$$

$$= \$36.00.$$

Garvin Art Supplies' producer surplus is

$$3 \cdot 9 - \int_0^3 \left(\frac{1}{3}x^2 + x + 3 \right) dx = 27 - \left[\frac{x^3}{9} + \frac{x^2}{2} + 3x \right]_0^3$$

$$= 27 - \left[\left(\frac{(3)^3}{9} + \frac{(3)^2}{2} + 3(3) \right) - \left(\frac{(0)^3}{9} + \frac{(0)^2}{2} + 3(0) \right) \right]$$

$$= 27 - \left(3 + \frac{9}{2} + 9 - 0 \right)$$

$$= \$10.50.$$

Quick Check 3 ✔

Given $D(x) = x^2 - 6x + 16$ and $S(x) = \frac{1}{3}x^2 + \frac{4}{3}x + 4$, find each of the following. Assume $x \le 5$.

a) The equilibrium point
b) The consumer surplus at the equilibrium point
c) The producer surplus at the equilibrium point

We saw in Example 1 that if Arnold purchases 2 sketchpads at $16 per unit, he spends $32, but would have spent up to $50.67. The difference, $18.67, is the extra utility that Arnold received. Similarly, we saw in Example 2 that when Garvin Art Supplies sells 2 sketchpads at $6.33 per unit, it receives revenue of $12.66, of which $3.77 is the consumer surplus, a contribution to its overall profit.

However, Arnold is willing to spend $9 per unit for 3 sketchpads, and Garvin Art Supplies is willing to sell 3 sketchpads at $9 per unit. At this point of equilibrium, Arnold spends more but also gains more utility, while Garvin Art Supplies collects more revenue and gains more producer surplus. Both producer and consumer benefit from this transaction.

3 ✔

Section Summary

- A *demand curve* is the graph of a function $p = D(x)$, which is the unit price p a consumer is willing to pay for x units. It is usually a decreasing function.
- A *supply curve* is the graph of a function $p = S(x)$, which is the unit price p a producer is willing to accept for x units. It is usually an increasing function.
- *Consumer surplus* for Q units at a price per unit P is

$$\int_0^Q D(x)\, dx - QP.$$

- *Producer surplus* for Q units at a price per unit P is

$$QP - \int_0^Q S(x)\, dx.$$

- The *equilibrium point*, (x_E, p_E), is the point at which the supply and demand curves intersect. The consumer surplus at the equilibrium point is

$$\int_0^{x_E} D(x)\, dx - x_E p_E.$$

The producer surplus at the equilibrium point is

$$x_E p_E - \int_0^{x_E} S(x)\, dx.$$

5.1   Exercise Set

In Exercises 1–14, $D(x)$ is the price, in dollars per unit, that consumers will pay for x units of an item, and $S(x)$ is the price, in dollars per unit, that producers will accept for x units. Find (a) the equilibrium point, (b) the consumer surplus at the equilibrium point, and (c) the producer surplus at the equilibrium point.

1. $D(x) = -3x + 7, \quad S(x) = 2x + 2$

2. $D(x) = -\frac{5}{6}x + 9, \quad S(x) = \frac{1}{2}x + 1$

3. $D(x) = (x - 3)^2, \quad S(x) = x^2 + 2x + 1$

4. $D(x) = (x - 4)^2, \quad S(x) = x^2 + 2x + 6$

5. $D(x) = (x - 8)^2, \quad S(x) = x^2$

6. $D(x) = (x - 6)^2, \quad S(x) = x^2$

7. $D(x) = 8800 - 30x, \quad S(x) = 7000 + 15x$

8. $D(x) = 1000 - 10x, \quad S(x) = 250 + 5x$

9. $D(x) = 7 - x, \text{ for } 0 \le x \le 7; \quad S(x) = 2\sqrt{x + 1}$

10. $D(x) = 5 - x, \text{ for } 0 \le x \le 5; \quad S(x) = \sqrt{x + 7}$

11. $D(x) = \dfrac{1800}{\sqrt{x + 1}}, \quad S(x) = 2\sqrt{x + 1}$

12. $D(x) = \dfrac{100}{\sqrt{x}}, \quad S(x) = \sqrt{x}$

13. $D(x) = 13 - x, \text{ for } 0 \le x \le 13; \quad S(x) = \sqrt{x + 17}$

14. $D(x) = (x - 4)^2, \quad S(x) = x^2 + 2x + 8$

APPLICATIONS

Business and Economics

15. Business: consumer and producer surplus. Beth enjoys skydiving and is willing to pay p dollars per jump for x jumps, where $p = D(x) = 7.5x^2 - 60.5x + 254$.

a) Find Beth's consumer surplus if she makes 2 jumps.

b) Suppose the supply function for Aero Skydiving Center is given by $p = S(x) = 15x + 95$. Find the producer surplus if the center sells Beth 2 jumps.

c) Find the equilibrium point and the consumer and producer surpluses at this point. Assume that Beth makes no more than 5 jumps.

d) Explain what the equilibrium point represents to both Beth and Aero Skydiving Center.

16. Business: consumer and producer surplus. Chris is enjoying a day at the Splashorama Water Park. His demand function for sliding down El Monstro Water Slide is given by $p = D(x) = 0.425x^2 - 5.7x + 22.5$, where x is the number of slides and p is the price, in dollars per slide, that he is willing to pay.

a) Find Chris's consumer surplus if he buys 3 slides.

b) Suppose the supply function for Splashorama Water Park for the El Monstro Water Slide is given by $p = S(x) = 0.3x + 5.3$. Find the producer surplus if Chris is sold 3 slides.

c) Find the equilibrium point and the consumer and producer surpluses at this point. Assume that Chris buys no more than 6 slides.

d) Explain what the equilibrium point represents to both Chris and to the Splashorama Water Park.

SYNTHESIS

For Exercises 17 and 18, follow the directions given for Exercises 1–14.

17. $D(x) = e^{-x+4.5}, \quad S(x) = e^{x-5.5}$

18. $D(x) = \sqrt{56 - x}, \quad S(x) = x$

19. Explain why both consumers and producers feel good when consumer and producer surpluses exist.

20. Research consumer and producer surpluses in an economics book. Write a brief description of what each represents.

For Exercises 21 and 22, graph each pair of demand and supply functions. Then:

a) *Find the equilibrium point using the INTERSECT feature or another feature that will allow you to find this point of intersection.*

b) *Graph $y = D(x_E)$ and identify the regions of both consumer and producer surpluses.*

c) *Find the consumer surplus.*

d) *Find the producer surplus.*

21. $D(x) = \dfrac{x+8}{x+1}, \quad S(x) = \dfrac{x^2+4}{20}$

22. $D(x) = 15 - \frac{1}{3}x, \quad S(x) = 2\sqrt[3]{x}$

23. Bungee jumping. Regina loves bungee jumping. The table shows the number of half-hours that Regina is willing to bungee jump at various prices.

Time Spent (in half-hours per month)	Price (per half-hour)
8	$25.00
7	50.00
6	75.00
5	100.00
4	125.00
3	150.00
2	175.00
1	200.00

a) Make a scatterplot of the data, and determine the type of function that you think fits best.

b) Fit that function to the data using REGRESSION.

c) If Regina goes bungee jumping for 6 half-hours per month, what is her consumer surplus?

d) At a price of $115.00 per half-hour, what is Regina's consumer surplus?

Integrating Growth and Decay Models

5.2

- Find the future value of an investment.
- Find the accumulated future value of a continuous income stream.
- Find the present value of an amount due in the future.
- Find the accumulated present value of an income stream.
- Calculate the total consumption of a natural resource.

We studied exponential growth and decay models using $P(t) = P_0 e^{kt}$ and $P(t) = P_0 e^{-kt}$ in Sections 3.3 and 3.4. In this section, we consider applications of the integrals of these functions. For convenience, we first derive formulas for evaluating these integrals. Recall that the interest rate k is expressed in decimal notation.

For the *growth* model, the formula is

$$\int_0^T P_0 e^{kt}\, dt = \left[\frac{P_0}{k} \cdot e^{kt} \right]_0^T \qquad \text{Using the substitution } u = e^{kt}$$

$$= \frac{P_0}{k} (e^{kT} - e^{k \cdot 0}) \qquad \text{Evaluating the integral}$$

$$= \frac{P_0}{k} (e^{kT} - 1).$$

Similarly, for the *decay* model, the formula is $\int_0^T P_0 e^{-kt}\, dt = \frac{P_0}{k}(1 - e^{-kT})$. Thus, we have the following integration formulas.

> **Growth formula:** $\displaystyle\int_0^T P_0 e^{kt}\, dt = \frac{P_0}{k}(e^{kT} - 1)$ (1)
>
> **Decay formula:** $\displaystyle\int_0^T P_0 e^{-kt}\, dt = \frac{P_0}{k}(1 - e^{-kT})$ (2)

We now consider several applications of these formulas to business and economics.

Future Value

Recall the basic model for the growth of an amount of money, presented in the following definition.

> **DEFINITION**
>
> If P_0 is invested for t years at interest rate k, compounded continuously (Section 3.3), then
>
> $$P(t) = P_0 e^{kt},$$ (3)
>
> where $P = P_0$ when $t = 0$. The value $P(t)$ is called the **future value** of P_0 dollars invested at interest rate k, compounded continuously, for t years.

EXAMPLE 1   Business: Future Value of an Investment.   Find the future value of $3650 invested for 3 yr at an interest rate of 5%, compounded continuously.

Solution   Using equation (3) with $P_0 = 3650$, $k = 0.05$, and $t = 3$, we have

$$P(3) = 3650e^{0.05(3)}$$
$$= 3650e^{0.15}$$
$$\approx \$4240.69. \qquad \text{Using a calculator}$$

The future value of $3650 after 3 yr will be about $4240.69.

Quick Check 1 ✔

Business: Future Value of an Investment. Find the future value of $10,000 invested for 3 yr at an interest rate of 6%, compounded continuously.

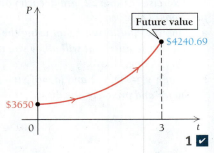

1 ✔

Accumulated Future Value of a Continuous Income Stream

Let's consider a situation involving the accumulation of future values. The owner of a parking space near a convention center receives a yearly profit of $3650 at the end of each of 4 years; this is called an *income stream*. The owner invests the $3650 at 5% interest compounded continuously. When $3650 is received at the end of the first year, it is invested for $4 - 1$, or 3 yr. The future value is $3650e^{0.05(4-1)}$, or $4240.69, as we saw in Example 1. When $3650 is received at the end of the second year, it is invested for $4 - 2$, or 2 yr. The future value of this investment is $3650e^{0.05(4-2)}$, or $4,033.87. When $3650 is received at the end of the third year, it is invested for $4 - 3$, or 1 yr. That future value is $3650e^{0.05(4-3)}$, or $3837.14. When the last $3650 is received after the fourth year, it is reinvested for $4 - 4$, or 0 yr. This amount has no time to earn interest, so its future value is $3650. The *accumulated, or total, future value of the income stream* is the sum of the four future values:

After 1st year, $t = 3$: $3650e^{0.05(3)}$ ⟶ $4,240.69

After 2nd year, $t = 2$: $3650e^{0.05(2)}$ ⟶ $4,033.87

After 3rd year, $t = 1$: $3650e^{0.05(1)}$ ⟶ $3,837.14

After 4th year, $t = 0$: $3650e^{0.05(0)}$ ⟶ $3,650.00

Total future value of the income stream = $15,761.70

Next, suppose the owner of the parking space receives the profit at a rate of $3650 per year but in 365 payments of $10 per day. Each day, when the owner gets $10, it is invested at 5%, compounded continuously, and due at the end of the fourth year. The *first* day's investment grows to

$$\frac{3650}{365} e^{0.05(4-1/365)} = 10e^{0.05(4-1/365)} \approx \$12.2124,$$

since it will be invested for only 1 day, or $1/365$ yr, less than the full 4 yr. The value of the investment on the *second* day will grow to

$$\frac{3650}{365} e^{0.05(4-2/365)} = 10e^{0.05(4-2/365)} \approx \$12.2107,$$

at the end of the fourth year, and so on, for every day in the 4-yr period. Assuming that all deposits are made into the same account, the total of the future values is

$$10e^{0.05(4-1/365)} + 10e^{0.05(4-2/365)} + \cdots + 10e^{0.05(2/365)} + 10e^{0.05(1/365)} + 10.$$

If we express 10 as $3650 \cdot \frac{1}{365}$ and let $\Delta t = \frac{1}{365}$, we have a Riemann sum, with t in years:

$$3650 \sum_{t=0}^{4} e^{0.05(4-t)} \Delta t.$$

As $\Delta t \to 0$, this Riemann sum is approximated by the definite integral

$$\int_0^4 3650e^{0.05(4-t)} \, dt = e^{0.2} \int_0^4 3650e^{-0.05t} \, dt. \tag{4}$$

Instead of receiving an income stream at the rate of $10 a day, suppose the owner receives the money *continuously* at a rate of $3650 *per year*. This means that over the course of 4 yr, profit is received at a constant rate of $3650 per year in what is called a *continuous income stream, or flow*. If at each instant the money is invested at 5%, compounded continuously, then the *accumulated future value of the continuous income stream* is approximated by the definite integral in equation (4). Let's calculate it:

$$e^{0.2} \int_0^4 3650e^{-0.05t} \, dt = -\frac{3650}{0.05} e^{0.2}(e^{-0.2} - 1) \quad \text{Using growth formula (1)}$$

$$\approx \$16,162.40 \quad \text{Using a calculator}$$

Economists call $16,162.40 the **accumulated future value of a continuous income stream**.

DEFINITION   Accumulated Future Value of a Continuous Income Stream

Let $R(t)$ represent the rate, per year, of a continuous income stream; let k be the interest rate, compounded continuously, at which the continuous income stream is invested; and let T be the number of years for which it is invested. Then the **accumulated future value of the continuous income stream** is given by

$$A = \int_0^T R(t)e^{k(T-t)}\, dt = e^{kT}\int_0^T R(t)e^{-kt}\, dt. \tag{5}$$

If $R(t)$ is a constant, it can be factored out of the integral, and the formula becomes, after simplifying,

$$A = \frac{R(t)}{k}\left(e^{kT} - 1\right). \tag{6}$$

If $R(t)$ is not a constant, then equation (6) does not apply and the integral in equation (5) must be evaluated using another integration technique.

EXAMPLE 2   Business: Insurance Settlement.   A cardiac surgeon, Sarah Maka-hone, who earns $450,000 per year, is involved in an automobile accident that prevents her from working. In a legal settlement with an insurance company, Sarah is granted a continuous income stream of $225,000 per year for 20 yr. Sarah invests the money at 3.2%, compounded continuously, in the Halmos Global Equities Fund. Find the accumulated future value of the continuous income stream.

Solution   Since this is an income stream flowing at a constant rate, we can use equation (6), with $R(t) = \$225,000$, $k = 0.032$, and $T = 20$. We have

$$A = \frac{225,000}{0.032}\left(e^{0.032(20)} - 1\right) \approx \$6,303,381.18. \qquad\qquad 2\ \checkmark$$

Present Value

We saw in Example 1 that the future value of $3650 invested for 3 yr at a continuously compounded interest rate of 5% is $4240.69. We call $3650 the **present value** of $4240.69 invested for 3 yr at interest rate 5%, compounded continuously. It answers this question: "What do we have to invest now at a certain interest to attain a certain future value?" (see Section 3.4).

In general, the present value P_0 of an amount P that results from an investment at interest rate k for t years is found by solving the growth equation for P_0:

$$P_0 e^{kt} = P$$

$$P_0 = \frac{P}{e^{kt}} = Pe^{-kt}.$$

DEFINITION

The **present value**, P_0, of an amount P due t years later, at interest rate k, compounded continuously, is given by

$$P_0 = Pe^{-kt}.$$

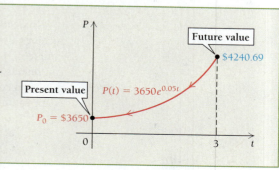

Quick Check 3 ☑

Business: Finding the Present Value of a Trust. Following the birth of a grandchild, Mira Bell wants to set up a trust fund that will be worth $120,000 on the child's 18th birthday. Mira secures an interest rate of 3.6%, compounded continuously, for the time period. What amount should she deposit to achieve her goal?

EXAMPLE 3 **Business: Finding the Present Value of a Trust.** In 10 years, Sam Bixby is going to receive $250,000 under the terms of a trust established by his aunt. If the money in the trust fund is invested at 4.8% interest, compounded continuously, what is the present value of Sam's trust?

Solution Using the equation for present value, we have

$$P_0 = 250{,}000e^{-0.048(10)} \approx \$154{,}695.85.$$

3 ☑

Accumulated Present Value of a Continuous Income Stream

To find the **accumulated present value of a continuous income stream**, when $R(t)$ is constant, we can work backward from equation (6):

$$A = \frac{R(t)}{k}(e^{kT} - 1).$$

We are looking for the principal B, the amount of a one-time deposit, at the interest rate k, that will yield the same accumulated value as the income stream. We choose B such that $Be^{kT} = A$ in equation (6). Then we solve for B:

$$Be^{kT} = \frac{R(t)}{k}(e^{kT} - 1) \qquad \text{Substituting}$$

$$\frac{Be^{kT}}{e^{kT}} = \frac{R(t)}{k}\left(\frac{e^{kT} - 1}{e^{kT}}\right) \qquad \text{Dividing both sides by } e^{kT}$$

$$B = \frac{R(t)}{k}\left(\frac{e^{kT}}{e^{kT}} - \frac{1}{e^{kT}}\right)$$

$$B = \frac{R(t)}{k}(1 - e^{-kT}). \qquad \text{Simplifying}$$

> **DEFINITION** **Accumulated Present Value of a Continuous Income Stream**
>
> Let $R(t)$ represent the rate, per year, of a continuous income stream; let k be the interest rate, compounded continuously, at which the continuous income stream is invested; and let T be the number of years over which the income stream is received.
>
> If $R(t)$ is a constant, then B, the **accumulated present value of the continuous income stream**, is given by
>
> $$B = \frac{R(t)}{k}(1 - e^{-kT}). \tag{7}$$
>
> If $R(t)$ is not a constant, the accumulated present value of the continuous income stream is given by
>
> $$B = \int_0^T R(t)e^{-kt}\, dt. \tag{8}$$

Accumulated present value is a useful tool in business decision making when evaluating a purchase, an investment, or a contract. It allows for comparison of alternatives.

EXAMPLE 4   **Business: Determining the Value of a Franchise.**   Silver Spoon, Inc., operates frozen yogurt franchises. Chris Nelson, yearning to be an entrepreneur, considers buying a franchise in his home town, Carmel, Indiana. As part of his decision to purchase, he wants to determine the accumulated present value of the income stream from the franchise over an 8-yr period. Silver Spoon tells Chris that he should expect a constant annual income stream given by

$$R_1(t) = \$275,000,$$

which Chris knows he can invest at an interest rate of 5%, compounded continuously.

However, Chris performs a linear regression on data from the annual reports of Silver Spoon, which indicates that there will be a nonconstant annual income stream of

$$R_2(t) = \$80,000t,$$

where t is the number of years the franchise is in operation.

a) Evaluate the accumulated future value of the income stream at rate $R_1(t)$. Then evaluate the accumulated present value of the income stream, and interpret the results.

b) Evaluate the accumulated future value of the income stream at rate $R_2(t)$. Then evaluate the accumulated present value of the income stream, and interpret the results.

Round all answers to the nearest ten dollars.

Solution
a) Assuming a constant income stream of $275,000 per year for 8 yr and using equation (6), we find that the accumulated *future* value is

$$A = \frac{R_1(t)}{k}(e^{kT} - 1)$$

$$= \frac{275,000}{0.05}(e^{0.05(8)} - 1) \approx \$2,705,040.$$

This gives Chris a sense of the value of the franchise over the 8-yr period. The accumulated *present* value is found by using equation (7):

$$B = \frac{R_1(t)}{k}(1 - e^{-kT})$$

$$= \frac{275,000}{0.05}(1 - e^{-0.05(8)}) \approx \$1,813,240.$$

The first result tells us that if Chris were to buy the franchise now and invest the predicted income stream at 5%, compounded continuously, he would have $2,705,040 in 8 yr. The second result tells us that $2,705,040 is worth $1,813,240 today.

b) With a nonconstant income stream, $R_2(t) = 80,000t$ per year, using equation (5), the accumulated *future* value is

$$e^{0.05(8)}\int_0^8 (80,000t)e^{-0.05t}\,dt = 80,000\,e^{0.4}\int_0^8 te^{-0.05t}\,dt.$$

To evaluate this integral, we use formula 6, from Table 1 in Chapter 4 (p. 432), with $a = -0.05$ and $x = t$:

$$\int xe^{ax}\,dx = \frac{1}{a^2}\cdot e^{ax}(ax - 1) = \frac{1}{(-0.05)^2}\cdot e^{-0.05t}(-0.05t - 1)$$

$$= 400e^{-0.05t}(-0.05t - 1) \qquad \text{Simplifying}$$

$$= -20te^{-0.05t} - 400e^{-0.05t}.$$

Then,

$$80{,}000\,e^{0.4}\int_0^8 te^{-0.05t}\,dt = 80{,}000\,e^{0.4}\Big[\left(-20(8)e^{-0.05(8)} - 400e^{-0.05(8)}\right)$$

$$- \left(-20(0)e^{-0.05(0)} - 400e^{-0.05(0)}\right)\Big]$$

$$= 80{,}000\,e^{0.4}\Big[(-160e^{-0.4} - 400e^{-0.4}) - (-400)\Big]$$

$$= 80{,}000\,e^{0.4}[-560e^{-0.4} + 400] \approx \$2{,}938{,}390.$$

This result tells us that if Chris were to buy the franchise now and invest the predicted income stream at 5%, compounded continuously, he would have $2,938,390 in 8 yr.

Using equation (8), we find that the accumulated present value is

$$\int_0^8 (80{,}000t)e^{-0.05t} = 80{,}000\int_0^8 te^{-0.05t}\,dt \approx \$1{,}969{,}660.$$

This result tells us that $2,938,390 is worth $1,969,660 at the present.

Chris's computations yield a higher accumulated present value than is claimed by Silver Spoon, suggesting that he is dealing with a reputable company. **4** ✔

> **EXAMPLE 5** **Business: Creating a College Trust.** Emma and Jake Tuttle establish a college trust fund for their new grandchild, Erica, that will yield $100,000 by her 18th birthday.

a) What lump sum do they need to deposit now, at 3.5% interest, compounded continuously, to yield $100,000?

b) They discover that the required lump sum is more than they can afford at the time, so they decide to invest a constant stream of $R(t)$ dollars per year. Find $R(t)$ such that the accumulated future value of the continuous money stream is $100,000, assuming that the interest rate is 3.5%, compounded continuously.

Solution

a) The lump sum is the *present value* of $100,000, at 3.5% interest, compounded continuously, for 18 yr:

$$P_0 = Pe^{-kt} = 100{,}000e^{-0.035(18)} \approx \$53{,}259.18.$$

b) We want $R(t)$ such that

$$100{,}000 = \frac{R(t)}{0.035}\left(e^{0.035(18)} - 1\right) \qquad \text{Using equation (6)}$$

$$0.035(100{,}000) = R(t)(e^{0.63} - 1)$$

$$\frac{3500}{(e^{0.63} - 1)} = R(t)$$

$$R(t) \approx \$3988.10.$$

A continuous money stream of $3988.10 per year, invested at 3.5%, compounded continuously for 18 yr, will yield a *future value* of $100,000. **5** ✔

Quick Check 4 ✔

Business: Determining the Value of a Franchise. Repeat Example 4, but with the following income streams:

$$R_1(t) = \$265{,}000,$$

$$R_2(t) = 75{,}000t,$$

and an interest rate of 4%, compounded continuously.

Quick Check 5 ✔

Business: Creating a College Trust. Repeat Example 5 for a yield of $50,000 and an interest rate of 4%.

EXAMPLE 6 **Business: Contract Buyout.** Gregory is playing basketball under a contract that pays him $500,000 each year for 5 yr. After 2 yr, the team offers him a buyout of his contract. How much should the team offer him? Assume an annual percentage rate of 4.75%, compounded continuously.

Solution We view the $500,000 as a continuous money stream. After 2 yr, the contract's accumulated future value, A_2, is

$$A_2 = \frac{500,000}{0.0475} (e^{0.0475(2)} - 1) \approx \$1,049,040.58.$$

If the contract were allowed to run the full 5 yr, the accumulated future value, A_5, would be

$$A_5 = \frac{500,000}{0.0475} (e^{0.0475(5)} - 1) \approx \$2,821,842.07.$$

The difference is

$$A_5 - A_2 = \$2,821,842.07 - \$1,049,040.58 = \$1,772,801.49.$$

Since the team is offering a lump sum payment to buy out the contract, Gregory should expect an amount that, if allowed to grow at 4.75%, compounded continuously for the remaining 3 yr, would yield $1,772,801.49. That is, he should receive the present value of the difference, or

$$P_0 = 1,772,801.49e^{-0.0475(3)} = \$1,537,351.39.$$ 6 ✔

Quick Check 6 ✔

Business: Contract Buyout.
Repeat Example 6 for a $400,000 contract and an interest rate of 3.2%.

Life and Physical Sciences: Consumption of Natural Resources

Another application of the integration of models of exponential growth uses

$$P(t) = P_0 e^{kt}$$

as a model of the demand for natural resources. Suppose that P_0 represents the annual amount of a natural resource (such as coal or oil) used at time $t = 0$ and that the growth rate for the use of this resource is k. Then, assuming exponential growth in demand, the amount used annually t years in the future is $P(t)$, given by

$$P(t) = P_0 e^{kt}.$$

The total amount used during an interval $[0, T]$ is then given by

$$\int_0^T P(t) \, dt = \int_0^T P_0 e^{kt} \, dt = \left[\frac{P_0}{k} e^{kt} \right]_0^T = \frac{P_0}{k} (e^{kT} - 1).$$

Consumption of a Natural Resource

Suppose $P(t)$ is the annual consumption of a natural resource in year t. If consumption of the resource is growing exponentially at growth rate k, then the total consumption of the resource after T years is given by

$$\int_0^T P_0 e^{kt} \, dt = \frac{P_0}{k} (e^{kT} - 1), \quad (9)$$

where P_0 is the annual consumption at time $t = 0$.

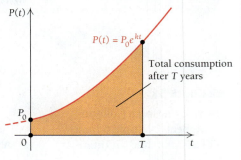

EXAMPLE 7 **Physical Science: Gold Mining.** In 2012 ($t = 0$), world production of gold was 2700 metric tons, and it was growing exponentially at the rate of 0.98% per year. (*Source:* www.dani2989.com.) If growth continues at this rate, how many tons of gold will be produced from 2012 to 2020?

Solution Using equation (9), we have

$$\int_0^8 2700e^{0.0098t}\, dt = \frac{2700}{0.0098}(e^{0.0098(8)} - 1)$$

$$= 275{,}510.2(e^{0.0784} - 1)$$

$$\approx 22{,}469.3.$$

From 2012 to 2020, approximately 22,469 metric tons of gold will be produced. ∎

EXAMPLE 8 **Physical Science: Depletion of Gold Reserves.** The world reserves of gold in 2012 were estimated to be 51,000 metric tons. (*Source:* U.S. Geological Survey, U.S. Dept. of the Interior, January 2012.) Assuming that the growth rate of 0.98% per year continues and that no new reserves are discovered, when will the world reserves of gold be depleted?

Solution Using equation (9), we want to find T such that

$$51{,}000 = \frac{2700}{0.0098}(e^{0.0098T} - 1).$$

We solve for T as follows:

$$51{,}000 = 275{,}510.2(e^{0.0098T} - 1)$$

$$0.185 \approx e^{0.0098T} - 1 \qquad \text{Dividing both sides by 275,510.2}$$

$$1.185 \approx e^{0.0098T}$$

$$\ln 1.185 \approx \ln e^{0.0098T} \qquad \text{Taking the natural logarithm of each side}$$

$$\ln 1.185 \approx 0.0098T \qquad \text{Recall that } \ln e^k = k.$$

$$17.3 \approx T. \qquad \text{Dividing both sides by 0.0098 and rounding}$$

Thus, assuming that world production of gold continues to increase at 0.98% per year and no new reserves are found, the world reserves of gold will be depleted 17.3 yr from 2012, or in 2029–2030. **7** ✔

Quick Check 7 ✔

Physical Science: Silver.

a) In 2010, world production of silver was 22,200 metric tons, and it was growing at the rate of 1.8% per year. (*Source:* www.dani2989.com.) If growth continues at this rate, how many metric tons of silver will be produced from 2010 to 2018?

b) The world reserves of silver in 2010 were 510,000 metric tons. (*Source:* U.S. Geological Survey.) Assuming that the growth rate of 1.8% per year continues and that no new reserves are discovered, when will the world reserves of silver be depleted?

Section Summary

- If P_0 dollars is invested for t years at an interest rate k, compounded continuously, the *future value* is given by $P(t) = P_0 e^{kt}$.

- If a continuous income stream $R(t)$, in dollars, is invested at an interest rate k, compounded continuously, for T years, then the *accumulated future value of the continuous income stream* is given by

$$A = e^{kT} \int_0^T R(t)e^{-kt}\, dt.$$

If $R(t)$ is constant, then

$$A = \frac{R(t)}{k}(e^{kT} - 1).$$

- To attain a future value P, the *present value*, P_0, that should be invested at an interest rate k, compounded continuously, for t years, is given by $P_0 = Pe^{-kt}$.

- If a continuous income stream $R(t)$, in dollars, is invested at an interest rate k, compounded continuously, for T years, the *accumulated present value of the continuous income stream* is given by

$$B = \int_0^T R(t)e^{-kt}\, dt.$$

If $R(t)$ is constant, then

$$B = \frac{R(t)}{k}(1 - e^{-kT}).$$

5.2 | Exercise Set

Find the future value P of each amount P_0 invested for time period t at interest rate k, compounded continuously.

1. $P_0 = \$100,000$, $t = 6$ yr, $k = 3\%$

2. $P_0 = \$55,000$, $t = 8$ yr, $k = 4\%$

3. $P_0 = \$140,000$, $t = 9$ yr, $k = 5.8\%$

4. $P_0 = \$88,000$, $t = 13$ yr, $k = 4.7\%$

Find the present value P_0 of each amount P due t years in the future and invested at interest rate k, compounded continuously.

5. $P = \$100,000$, $t = 6$ yr, $k = 3\%$

6. $P = \$100,000$, $t = 8$ yr, $k = 4\%$

7. $P = \$1,000,000$, $t = 25$ yr, $k = 6\%$

8. $P = \$2,000,000$, $t = 20$ yr, $k = 3.5\%$

Find the accumulated future value of each continuous income stream at rate R(t), for the given time T and interest rate k, compounded continuously.

9. $R(t) = \$50,000$, $T = 22$ yr, $k = 5\%$

10. $R(t) = \$125,000$, $T = 20$ yr, $k = 6\%$

11. $R(t) = \$400,000$, $T = 20$ yr, $k = 4\%$

12. $R(t) = \$50,000$, $T = 22$ yr, $k = 2.75\%$

Find the accumulated present value of each continuous income stream at rate R(t), for the given time T and interest rate k, compounded continuously.

13. $R(t) = \$250,000$, $T = 18$ yr, $k = 4\%$

14. $R(t) = \$425,000$, $T = 15$ yr, $k = 3.5\%$

15. $R(t) = \$800,000$, $T = 20$ yr, $k = 2.3\%$

16. $R(t) = \$520,000$, $T = 25$ yr, $k = 6\%$

17. $R(t) = \$5200t$, $T = 18$ yr, $k = 3.1\%$

18. $R(t) = \$6400t$, $T = 20$ yr, $k = 3.8\%$

19. $R(t) = \$(2000t + 7)$, $T = 30$ yr, $k = 4.5\%$

20. $R(t) = \$t^2$, $T = 40$ yr, $k = 4.25\%$

APPLICATIONS

Business and Economics

21. **Present value of a trust.** In 18 yr, Maggie Oaks is to receive $200,000 under the terms of a trust established by her grandparents. Assuming an interest rate of 3.8%, compounded continuously, what is the present value of Maggie's trust?

22. **Present value of a trust.** In 16 yr, Claire Beasley is to receive $180,000 under the terms of a trust established by her aunt. Assuming an interest rate of 4.2%, compounded continuously, what is the present value of Claire's trust?

23. **Salary value.** At age 35, Rochelle earns her MBA and accepts a position as vice president of an asphalt company. Assume that she will retire at the age of 65, having received an annual salary of $95,000, and that the interest rate is 5%, compounded continuously.

    a) What is the accumulated present value of her position?
    b) What is the accumulated future value of her position?

24. **Salary value.** At age 25, Del earns his CPA and accepts a position in an accounting firm. Del plans to retire at the age of 65, having received an annual salary of $125,000. Assume an interest rate of 4%, compounded continuously.

    a) What is the accumulated present value of his position?
    b) What is the accumulated future value of his position?

25. **Future value of an inheritance.** Upon the death of his uncle, David receives an inheritance of $50,000, which he invests for 16 yr at 4.3%, compounded continuously. What is the future value of the inheritance?

26. **Future value of an inheritance.** Upon the death of his aunt, Burt receives an inheritance of $80,000, which he invests for 20 yr at 3.9%, compounded continuously. What is the future value of the inheritance?

27. **Decision making.** A group of entrepreneurs is considering the purchase of a fast-food franchise. Franchise A predicts that it will bring in a constant revenue stream of $80,000 per year for 10 yr. Franchise B predicts that it will bring in a constant revenue stream of $95,000 per year for 8 yr. Based on a comparison of accumulated present values, which franchise is the better buy, assuming the interest rate is 4.1%, compounded continuously, and both franchises have the same purchase price?

28. **Decision making.** A group of entrepreneurs is considering the purchase of a fast-food franchise. Franchise A predicts that it will bring in a constant revenue stream of $120,000 per year for 8 yr. Franchise B predicts that it will bring in a constant revenue stream of $112,000 per year for 10 yr. Based on a comparison of accumulated present values, which franchise is the better buy, assuming the interest rate is 5.4%, compounded continuously, and both franchises have the same purchase price?

29. **Decision making.** An athlete attains free agency and is looking for a new team. The Bronco Crunchers offer a

salary of $100,000t$ for 8 yr, and the Doppler Radars offer a salary of $83,000t$ for 9 yr, where t is in years.

a) Based on the accumulated present values of the salaries, which team has the better offer, assuming an interest rate of 4.2%, compounded continuously?

b) What signing bonus should the team with the lower offer give to equalize the offers?

30. Capital outlay. Chrome Solutions determines that the rate of revenue coming in from a new machine is

$$R_1(t) = 8000 - 100t,$$

in dollars per year, for 8 yr, after which the machine will be replaced. The company learns that an alternative machine will yield revenue at a rate of

$$R_2(t) = 7600 - 85t.$$

a) Find the accumulated present value of the income stream from each machine at an interest rate of 5.8%, compounded continuously.

b) Find the difference in the accumulated present values.

31. Trust fund. Bob and Ann MacKenzie have a new grandchild, Brenda, and want to create a trust fund for her that will yield $250,000 on her 24th birthday.

a) What lump sum should they deposit now at 5.8%, compounded continuously, to achieve $250,000?

b) The amount in part (a) is more than they can afford, so they decide to invest a constant amount, $R(t)$ dollars per year. Find $R(t)$ such that the accumulated future value of the continuous money stream is $250,000, assuming an interest rate of 5.8%, compounded continuously.

32. Trust fund. Ted and Edith Markey have a new grandchild, Kurt, and want to create a trust fund for him that will yield $1,000,000 on his 22nd birthday.

a) What lump sum should they deposit now at 4.2%, compounded continuously, to achieve $1,000,000?

b) The amount in part (a) is more than they can afford, so they decide to invest a constant amount, $R(t)$ dollars per year. Find $R(t)$ such that the accumulated future value of the continuous money stream is $1,000,000, assuming an interest rate of 4.2%, compounded continuously.

33. Early retirement. Lauren Johnson signs a 10-yr contract as a loan officer for a bank, at a salary of $84,000 per year. After 7 yr, the bank offers her early retirement. What is the least amount the bank should offer Lauren, assuming an interest rate of 4.7%, compounded continuously?

34. Early sports retirement. Tory Johnson signs a 10-yr contract to play for a football team at a salary of $5,000,000 per year. After 6 yr, his skills deteriorate, and the team offers to buy out the rest of his contract. What is the least amount Tory should accept for the buyout, assuming an interest rate of 4.9%, compounded continuously?

35. Disability insurance settlement. A movie stuntman who receives an annual salary of $180,000 per year is injured and can no longer work. Through a settlement with an insurance company, he is granted a continuous income stream of $120,000 per year for 20 yr. The stuntman invests the money at 4%, compounded continuously.

a) Find the accumulated future value of the continuous income stream. Round your answer to the nearest $10.

b) Thinking that he might not live 20 yr, the stuntman negotiates a flat sum payment from the insurance company, which is the accumulated present value of the continuous income stream. What is that amount? Round your answer to the nearest $10.

36. Disability insurance settlement. Dale was a furnace maintenance employee who earned a salary of $70,000 per year before he was injured on the job. Through a settlement with his employer's insurance company, he is granted a continuous income stream of $40,000 per year for 25 yr. Dale invests the money at 5%, compounded continuously.

a) Find the accumulated future value of the continuous income stream. Round your answer to the nearest $10.

b) Thinking that he might not live for 25 yr, Dale negotiates a flat sum payment from the insurance company, which is the accumulated present value of the continuous stream plus $100,000. What is that amount? Round your answer to the nearest $10.

37. Lottery winnings and risk analysis. Lucky Larry wins $1,000,000 in a state lottery. The standard way in which a state pays such lottery winnings is at a constant rate of $50,000 per year for 20 yr.

a) If Lucky invests each payment from the state at 4.4%, compounded continuously, what is the accumulated future value of the income stream? Round your answer to the nearest $10.

b) What is the accumulated present value of the income stream at 4.4%, compounded continuously? This amount represents what the state has to invest at the start of its lottery payments, assuming the 4.4% interest rate holds.

c) The risk for Lucky is that he doesn't know how long he will live or what the future interest rate will be; it might drop or rise, or it could vary considerably over 20 yr. This is the *risk* he assumes in accepting payments of $50,000 a year over 20 yr. Lucky has taken a course in business calculus so he is aware of the formulas for accumulated future value and present value. He calculates the accumulated present value of the income stream for interest rates of 3%, 4%, and 5%. What values does he obtain?

d) Lucky thinks "a bird in the hand (present value) is worth two in the bush (future value)" and decides to negotiate with the state for immediate payment of his lottery winnings. He asks the state for $750,000. They offer $600,000. Discuss the pros and cons of each amount. Lucky finally accepts $675,000. Is this a good decision? Why or why not?

38. Negotiating a sports contract. Gusto Stick is a professional baseball player who has just become a free agent. His attorney begins negotiations with an interested team by asking for a contract that provides Gusto with an income stream given by $R_1(t) = 800{,}000 + 340{,}000t$, over 10 yr, where t is in years. (Round all answers to the nearest \$100.)

a) What is the accumulated future value of the offer, assuming an interest rate of 5%, compounded continuously?

b) What is the accumulated present value of the offer, assuming an interest rate of 5%, compounded continuously?

c) The team counters by offering an income stream given by $R_2(t) = 600{,}000 + 210{,}000t$. What is the accumulated present value of this counteroffer?

d) Gusto comes back with a demand for an income stream given by $R_3(t) = 1{,}000{,}000 + 250{,}000t$. What is the accumulated present value of this income stream?

e) Gusto signs a contract for the income stream in part (d) but decides to live on \$500,000 each year, investing the rest at 5%, compounded continuously. What is the accumulated future value of the remaining income, assuming an interest rate of 5%, compounded continuously?

39. Accumulated present value. The Wilkinsons want to have \$100,000 in 10 yr for a down payment on a retirement home. Find the continuous money stream, $R(t)$ dollars per year, that they need to invest at 4.33%, compounded continuously, to generate \$100,000.

40. Accumulated present value. Tania wants to have \$20,000 in 5 yr for her dream vacation. Find the continuous money stream, $R(t)$ dollars per year, that she needs to invest at 5.125%, compounded continuously, to generate \$20,000.

Life and Physical Sciences

41. Demand for natural gas. In 2013 ($t = 0$), world consumption of natural gas was approximately 117.2 trillion cubic feet and was growing exponentially at about 1.24% per year. (*Source:* U.S. Energy Information Administration.) If the demand continues to grow at this rate, how many cubic feet of natural gas will the world use from 2015 to 2025?

42. Demand for aluminum ore (bauxite). In 2013 ($t = 0$), bauxite production was approximately 232 million metric tons, and the demand was growing exponentially at a rate of 2.6% per year. (*Source:* minerals.usgs.gov.) If the demand continues to grow at this rate, how many metric tons of bauxite will the world use from 2015 to 2030?

43. Depletion of natural gas. The world reserves of natural gas were approximately 6597 trillion cubic feet in 2013. (*Source:* www.cia.com.) Assuming that the growth described in Exercise 41 continues and that no new reserves are found, when will the world reserves of natural gas be depleted?

44. Depletion of aluminum ore (bauxite). In 2013, the world reserves of bauxite were about 65 billion metric tons. (*Source:* U.S. Geological Survey summaries, Jan. 2014.) Assuming that the growth described in Exercise 42 continues and that no new reserves are discovered, when will the world reserves of bauxite be depleted?

45. Demand for and depletion of oil. In 2013, annual world demand for crude oil was approximately 33.3 billion barrels, and it was projected to increase by 1.5% per year. (*Sources:* Based on information from www.iea.gov and www.cia.gov.)

a) Assuming an exponential growth model, predict the demand in 2020.

b) World reserves of crude oil in 2013 were approximately 1635 billion barrels. Assuming that no new oil reserves are found, when will the reserves be depleted?

The model

$$\int_0^T P e^{-kt}\, dt = \frac{P}{k}\left(1 - e^{-kT}\right)$$

can be applied to calculate the buildup of a radioactive material that is being released into the atmosphere at a constant annual rate. Some of the material decays, but more continues to be released. The amount present at time T is given by the integral above, where P is the amount released per year and k is the half-life.

46. Radioactive buildup. Plutonium-239 has a decay rate of approximately 0.003% per year. Suppose plutonium-239 is released into the atmosphere for 20 yr at a constant rate of 1 lb per year. How much plutonium-239 will be present in the atmosphere after 20 yr?

47. Radioactive buildup. Cesium-137 has a decay rate of 2.3% per year. Suppose cesium-137 is released into the atmosphere for 20 yr at a rate of 1 lb per year. How much cesium-137 will be present in the atmosphere after 20 yr?

SYNTHESIS

Capitalized cost. *The capitalized cost, c, of an asset over its lifetime is the total of the initial cost and the present value of* *all maintenance expenses that will occur in the future. It is computed with the formula*

$$c = c_0 + \int_0^L m(t)e^{-kt}\,dt,$$

where c_0 is the initial cost of the asset, L is the lifetime (in years), k is the interest rate (compounded continuously), and $m(t)$ is the annual cost of maintenance. Find the capitalized cost under each set of assumptions.

48. $c_0 = \$500{,}000$, $k = 5\%$, $m(t) = \$20{,}000$, $L = 20$

49. $c_0 = \$400{,}000$, $k = 5.5\%$, $m(t) = \$10{,}000$, $L = 25$

50. $c_0 = \$600{,}000$, $k = 4\%$,
$m(t) = \$40{,}000 + \$1000e^{0.01t}$, $L = 40$

51. $c_0 = \$300{,}000$, $k = 5\%$, $m(t) = \$30{,}000 + \$500t$,
$L = 20$

52. How would you explain the concepts of present value and accumulated present value to a friend who has not studied this chapter?

53. Look up some data on rate of use and current world reserves of a natural resource not considered in this section. Predict when the world reserves for that resource will be depleted.

Answers to Quick Checks

1. $11,972.17 **2.** $6,225,857.39 **3.** $62,770.91
4. With R_1, accumulated future value is $2,498,471, and accumulated present value is $1,814,263. With R_2, accumulated future value is $2,677,864, and accumulated present value is $1,944,528. **5.** $24,337.61, $1896.75
6. $1,219,822.71 **7.** **(a)** 191,023.7 metric tons;
(b) 19.2 yr, or by 2029

5.3

- Show that an improper integral is convergent or divergent.
- Solve applied problems involving improper integrals.

Improper Integrals

Let's try to find the area of the region under the graph of $y = 1/x^2$ over $[1, \infty)$.

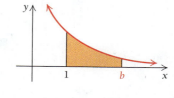

Note that this region is of infinite extent. To find the area of such a region, let's find the area over the interval from 1 to b, and then see what happens as b approaches ∞. The area under the graph over $[1, b]$ is

$$\int_1^b \frac{dx}{x^2} = \left[-\frac{1}{x}\right]_1^b$$

$$= \left(-\frac{1}{b}\right) - \left(-\frac{1}{1}\right)$$

$$= -\frac{1}{b} + 1$$

$$= 1 - \frac{1}{b}.$$

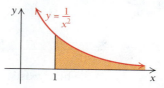

Then

$$\lim_{b \to \infty} (\text{area from 1 to } b) = \lim_{b \to \infty}\left(1 - \frac{1}{b}\right) = 1.$$

We *define* the area from 1 to infinity to be this limit.

Similar areas may not be finite. Let's try to find the area of the region under the graph of $y = 1/x$ over $[1, \infty)$.

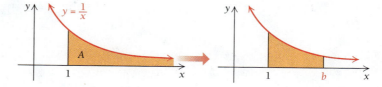

The area A, from 1 to infinity, is by definition the limit as b approaches ∞ of the area from 1 to b, so

$$A = \lim_{b \to \infty} \int_1^b \frac{dx}{x} = \lim_{b \to \infty} \left[\ln x\right]_1^b$$

$$= \lim_{b \to \infty} (\ln b - \ln 1)$$

$$= \lim_{b \to \infty} \ln b.$$

In Section 3.2, we saw that the function $y = \ln x$ is always increasing. Therefore, the limit $\lim_{b \to \infty} \ln b$ approaches infinity. The area under $y = 1/x$ from 1 to infinity is infinite.

Note that the graphs of $y = 1/x^2$ and $y = 1/x$ have similar shapes over $[1, \infty)$, but the region under $y = 1/x^2$ has a finite area and the region under $y = 1/x$ does not.

An integral such as

$$\int_a^\infty f(x)\, dx,$$

with an upper limit of infinity, is an example of an **improper integral**. Its value is defined to be the following limit, if it exists.

DEFINITION

$$\int_a^\infty f(x)\, dx = \lim_{b \to \infty} \int_a^b f(x)\, dx$$

If the limit exists, then we say that the improper integral **converges**, or is **convergent**. If the limit does not exist, then we say that the improper integral **diverges**, or is **divergent**. Thus,

$$\int_1^\infty \frac{dx}{x^2} \text{ converges,} \quad \text{and} \quad \int_1^\infty \frac{dx}{x} \text{ diverges.}$$

EXAMPLE 1   Determine whether $\displaystyle\int_0^\infty 4e^{-2x}\, dx$ is convergent or divergent, and find its value if it is convergent.

Solution   We have

$$\int_0^\infty 4e^{-2x}\, dx = \lim_{b \to \infty} \int_0^b 4e^{-2x}\, dx \qquad \textcolor{red}{\text{Using the above definition}}$$

$$= \lim_{b \to \infty} \left[\frac{4}{-2} e^{-2x}\right]_0^b$$

$$= \lim_{b \to \infty} \left[-2e^{-2x}\right]_0^b$$

$$= \lim_{b \to \infty} \left[-2e^{-2b} - (-2e^{-2 \cdot 0})\right] \qquad \textcolor{red}{\text{Evaluating}}$$

$$= \lim_{b \to \infty} (-2e^{-2b} + 2)$$

$$= \lim_{b \to \infty} \left(2 - \frac{2}{e^{2b}}\right).$$

As b approaches ∞, we know that e^{2b} approaches ∞, so

$$\frac{2}{e^{2b}} \to 0 \quad \text{and} \quad \left(2 - \frac{2}{e^{2b}}\right) \to 2.$$

Thus, $\displaystyle\int_0^\infty 4e^{-2x}\, dx = \lim_{b\to\infty} \left(2 - \frac{2}{e^{2b}}\right) = 2.$

The integral is convergent.

1 ✔

Quick Check 1 ✔

Determine whether $\displaystyle\int_2^\infty \frac{2}{x^3}\, dx$ is convergent or divergent, and find its value if it is convergent.

Following are definitions of two other types of improper integrals.

> **DEFINITIONS**
>
> **1.** $\displaystyle\int_{-\infty}^b f(x)\, dx = \lim_{a\to-\infty} \int_a^b f(x)\, dx$
>
> **2.** $\displaystyle\int_{-\infty}^\infty f(x)\, dx = \int_{-\infty}^c f(x)\, dx + \int_c^\infty f(x)\, dx,$
>
> where c can be any real number.

In order for $\int_{-\infty}^\infty f(x)\, dx$ to converge, both $\int_{-\infty}^c f(x)\, dx$ and $\int_c^\infty f(x)\, dx$ must converge.

Applications of Improper Integrals

In Section 5.2, we learned that the accumulated present value of a continuous money flow (income stream) of P dollars per year, at a constant rate, from now until T years in the future can be found by integration:

$$\int_0^T Pe^{-kt}\, dt = \frac{P}{k}\left(1 - e^{-kT}\right),$$

where k is the interest rate and interest is compounded continuously. Suppose that the money flow is to continue perpetually (forever). Under this assumption, the accumulated present value of the money flow is

$$\int_0^\infty Pe^{-kt}\, dt = \lim_{T\to\infty} \int_0^T Pe^{-kt}\, dt$$

$$= \lim_{T\to\infty} \frac{P}{k}\left(1 - e^{-kT}\right)$$

$$= \lim_{T\to\infty} \frac{P}{k}\left(1 - \frac{1}{e^{kT}}\right) = \frac{P}{k}.$$

> **THEOREM 1**
>
> The **accumulated present value** of a continuous money flow into an investment at the constant rate of P dollars per year perpetually is given by
>
> $$\int_0^\infty Pe^{-kt}\, dt = \frac{P}{k},$$
>
> where k is the annual interest rate, compounded continuously.

Quick Check 2

Find the accumulated present value of an investment for which there is a perpetual continuous money flow of $10,000 per year. Assume that the interest rate is 6%, compounded continuously.

EXAMPLE 2 Business: Accumulated Present Value. Find the accumulated present value of an investment for which there is a perpetual continuous money flow of $2000 per year. Assume that the interest rate is 5%, compounded continuously.

Solution   The accumulated present value is 2000/0.05, or $40,000.   **2** ✔

TECHNOLOGY CONNECTION

Exploratory

We can explore the situation of Example 2 with a calculator. To evaluate $\int_0^\infty 2000e^{-0.05x}\,dx$, we first consider

$$\int_0^t 2000e^{-0.05x}\,dx = \frac{2000}{0.05}\left(1 - e^{-0.05t}\right).$$

Then we examine what happens as t approaches infinity. Graph

$$f(x) = \frac{2000}{0.05}\left(1 - e^{-0.05x}\right)$$

using the window $[0, 10, 0, 50000]$, with Xscl $= 1$ and Yscl $= 5000$. On the same set of axes, graph $y = 40,000$. Then change the viewing window to $[0, 50, 0, 50000]$, with Xscl $= 10$ and Yscl $= 5000$, and finally to $[0, 100, 0, 50000]$. What happens as x gets larger? What is the significance of 40,000?

When an amount P of radioactive material is being released into the atmosphere annually, the total amount that has been released up to time T is given by

$$\int_0^T Pe^{-kt}\,dt = \frac{P}{k}\left(1 - e^{-kT}\right).$$

As T approaches infinity (the radioactive material is released forever), the buildup of radioactive material approaches a limiting value P/k. It is no wonder that scientists and environmentalists are so concerned about radioactive waste. The radioactivity is "here to stay."

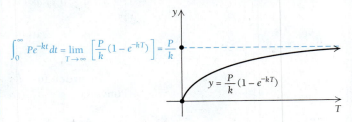

$$\int_0^\infty Pe^{-kt}\,dt = \lim_{T\to\infty}\left[\frac{P}{k}\left(1 - e^{-kT}\right)\right] = \frac{P}{k}$$

$$y = \frac{P}{k}\left(1 - e^{-kT}\right)$$

Section Summary

- An *improper integral* has infinity as one or both of its bounds and is evaluated using

$$\int_a^\infty f(x)\,dx = \lim_{b\to\infty}\int_a^b f(x)\,dx,$$

$$\int_{-\infty}^b f(x)\,dx = \lim_{a\to-\infty}\int_a^b f(x)\,dx,$$

and $\int_{-\infty}^\infty f(x)\,dx = \int_{-\infty}^c f(x)\,dx + \int_c^\infty f(x)\,dx,$

where c is any real number.

- The *accumulated present value* of a continuous money flow into an investment at the rate of P dollars per year perpetually is given by

$$\int_0^\infty Pe^{-kt}\,dt = \frac{P}{k},$$

where k is the annual interest rate, compounded continuously.

5.3 | Exercise Set

Determine whether each improper integral is convergent or divergent, and find its value if it is convergent.

1. $\int_5^\infty \dfrac{dx}{x^2}$

2. $\int_3^\infty \dfrac{dx}{x^2}$

3. $\int_3^\infty \dfrac{dx}{x}$

4. $\int_4^\infty \dfrac{dx}{x}$

5. $\int_0^\infty 3e^{-3x}\,dx$

6. $\int_0^\infty 4e^{-4x}\,dx$

7. $\int_1^\infty \dfrac{dx}{x^3}$

8. $\int_1^\infty \dfrac{dx}{x^4}$

9. $\int_0^\infty \dfrac{dx}{2 + x}$

10. $\int_0^\infty \dfrac{4\,dx}{3 + x}$

11. $\int_{2}^{\infty} 4x^{-2}\, dx$

12. $\int_{2}^{\infty} 7x^{-2}\, dx$

13. $\int_{0}^{\infty} e^{x}\, dx$

14. $\int_{0}^{\infty} e^{2x}\, dx$

15. $\int_{3}^{\infty} x^{2}\, dx$

16. $\int_{5}^{\infty} x^{4}\, dx$

17. $\int_{0}^{\infty} xe^{x}\, dx$

18. $\int_{1}^{\infty} \ln x\, dx$

19. $\int_{0}^{\infty} me^{-mx}\, dx,\ m > 0$

20. $\int_{0}^{\infty} Qe^{-kt}\, dt,\ k > 0$

21. $\int_{\pi}^{\infty} \dfrac{dt}{t^{1.001}}$

22. $\int_{1}^{\infty} \dfrac{2t}{t^{2}+.1}\, dt$

23. $\int_{-\infty}^{\infty} t\, dt$

24. $\int_{1}^{\infty} \dfrac{3x^{2}}{(x^{3}+1)^{2}}\, dx$

25. Find the area, if it is finite, of the region under the graph of $y = 1/x^{2}$ over $[\frac{1}{2}, \infty)$.

26. Find the area, if it is finite, of the region under the graph of $y = 1/x$ over $[2, \infty)$.

27. Find the area, if it is finite, of the region under the graph of $y = 2xe^{-x^{2}}$ over $[0, \infty)$.

28. Find the area, if it is finite, of the region under the graph of $y = 1/\sqrt{(3x-2)^{3}}$ over $[6, \infty)$.

APPLICATIONS

Business and Economics

29. Total profit from marginal profit. Myna's Fashions determines that its marginal profit, in dollars, from producing x shawls is given by

$$P'(x) = 200{,}000e^{-0.032x}.$$

Suppose it were possible for this firm to make infinitely many shawls. What would its total profit be?

30. Total profit from marginal profit. Find the total profit in Exercise 29 if

$$P'(x) = 200{,}000x^{-1.032}, \quad \text{where } x \geq 1.$$

31. Total cost from marginal cost. Barton Novelties determines that its marginal cost, in dollars, for producing x keyholders is given by

$$C'(x) = 3{,}600{,}000x^{-1.8}, \quad \text{where } x \geq 1.$$

Suppose it were possible for this company to make infinitely many keyholders. What would the total cost be?

32. Total production. A firm determines that it can produce tires at the rate

$$r(t) = (2 \times 10^{6})e^{-0.42t},$$

where t is in years. Assuming that the firm endures forever (it never gets tired), how many tires can it make?

33. Accumulated present value. Find the accumulated present value of an investment for which there is a perpetual continuous money flow of $3600 per year at an interest rate of 5%, compounded continuously.

34. Accumulated present value. Find the accumulated present value of an investment for which there is a perpetual continuous money flow of $3500 per year at an interest rate of 4%, compounded continuously.

35. Accumulated present value. Find the accumulated present value of an investment for which there is a perpetual continuous money flow of $5000 per year, assuming continuously compounded interest at a rate of 3.7%.

36. Accumulated present value. Find the accumulated present value of an investment for which there is a perpetual continuous money flow of $2000e^{-0.01t}$ per year, assuming continuously compounded interest at a rate of 5%.

Capitalized cost. *The capitalized cost, c, of an asset for an unlimited lifetime is the total of the initial cost and the present value of all maintenance expenses that will occur in the future. It is computed with the formula*

$$c = c_{0} + \int_{0}^{\infty} m(t)e^{-kt}\, dt,$$

where c_{0} is the initial cost of the asset, k is the interest rate (compounded continuously), and $m(t)$ is the annual cost of maintenance. Find the capitalized cost under each set of assumptions.

37. $c_{0} = \$500{,}000, \quad k = 5\%, \quad m(t) = \$20{,}000$

38. $c_{0} = \$700{,}000, \quad k = 5\%, \quad m(t) = \$30{,}000$

Life and Physical Sciences

39. Radioactive buildup. Plutonium has a decay rate of 0.003% per year. Suppose a nuclear accident causes plutonium to be released into the atmosphere perpetually at the rate of 1 lb per year. What is the limiting value of the radioactive buildup?

40. Radioactive buildup. Cesium-137 has a decay rate of 2.3% per year. Suppose a nuclear accident causes cesium-137 to be released into the atmosphere perpetually at the rate of 1 lb per year. What is the limiting value of the radioactive buildup?

Radioactive implant treatments. *In the treatment of prostate cancer, radioactive implants are often used. The implants are left in the patient and never removed. The amount of energy that is transmitted to the body from the implant is measured in rem units and is given by*

$$E = \int_{0}^{a} P_{0}e^{-kt}\, dt,$$

where k is the decay constant for the radioactive material, t is the number of years since the implant, a is the time (in years) until the rem measurement is made, and P_{0} is the initial rate at which energy is transmitted. (Source: www.cancer.gov.) Use this information for Exercises 41 and 42.

41. Suppose the treatment uses iodine-125, which has a half-life of 60.1 days.

    a) Find the decay rate, k, of iodine-125.

    b) How much energy (measured in rems) is transmitted in the first month if the initial rate of transmission is 10 rems per year?

    c) What is the total amount of energy that the implant will transmit to the body?

42. Suppose the treatment uses palladium-103, which has a half-life of 16.99 days.

    a) Find the decay rate, k, of palladium-103.

    b) How much energy (measured in rems) is transmitted in the first month if the initial rate of transmission is 10 rems per year?

    c) What is the total amount of energy that the implant will transmit to the body?

SYNTHESIS

Determine whether each improper integral is convergent or divergent, and find its value if it is convergent.

43. $\displaystyle\int_0^\infty \frac{dx}{x^{2/3}}$

44. $\displaystyle\int_1^\infty \frac{dx}{\sqrt{x}}$

45. $\displaystyle\int_0^\infty \frac{dx}{(x+1)^{3/2}}$

46. $\displaystyle\int_{-\infty}^0 e^{2x}\,dx$

47. $\displaystyle\int_0^\infty xe^{-x^2}\,dx$

48. $\displaystyle\int_{-\infty}^\infty xe^{-x^2}\,dx$

Life science: drug dosage. *Suppose an oral dose of a drug is taken. Over time, the drug is assimilated in the body and excreted through the urine. The total amount of the drug that has passed through the body in T hours is given by*

$$\int_0^T E(t)\,dt,$$

where $E(t)$ is the rate of excretion of the drug. A typical rate-of-excretion function is $E(t) = te^{-kt}$, where $k > 0$ and t is the time, in hours. Use this information for Exercises 49 and 50.

49. Find $\int_0^\infty E(t)\,dt$, and interpret the answer. That is, what does the integral represent?

50. A physician prescribes a dosage of 100 mg. Find k.

51. Consider the functions

$$y = \frac{1}{x^2} \quad \text{and} \quad y = \frac{1}{x}.$$

Suppose you go to a paint store to buy paint to cover the region under each graph over $[1, \infty)$. Discuss whether you could be successful, and explain why or why not.

52. Suppose you own a building that yields a continuous series of rental payments and you decide to sell the building. Explain how you would use the concept of the accumulated present value of a perpetual continuous money flow to determine a fair selling price.

Another form of improper integral occurs when a vertical asymptote appears at one of the bounds of integration. For example, to find the area under $f(x) = \dfrac{1}{\sqrt{x}}$ over $[0, 4]$, we note that f is not defined at $x = 0$ and that there is a vertical asymptote as x approaches 0 from the right. In such a case, we integrate over $[a, 4]$ and find the limit as a approaches 0 from the right. Use this technique for Exercises 53 and 54.

53. Find $\displaystyle\int_0^4 \frac{1}{\sqrt{x}}\,dx.$

54. Find $\displaystyle\int_4^5 \frac{3}{\sqrt[3]{x-5}}\,dx.$

55. Find and explain the error in the following calculation:

$$\int_{-1}^2 \frac{1}{x^2}\,dx = \left[-\frac{1}{x}\right]_{-1}^2 = \left(-\frac{1}{2}\right) - \left(-\frac{1}{(-1)}\right) = -\frac{3}{2}.$$

Approximate each integral.

56. $\displaystyle\int_1^\infty \frac{4}{1+x^2}\,dx$

57. $\displaystyle\int_1^\infty \frac{6}{5+e^x}\,dx$

58. $\displaystyle\int_{-\infty}^\infty \frac{1}{1+x^2}\,dx$

59. Graph the function E and shade the area under the curve for the situation in Exercises 49 and 50.

Answers to Quick Checks

    1. Convergent; $\frac{1}{4}$ **2.** $166,666.67

5.4

- Verify certain properties of probability density functions.
- Solve applied problems involving probability density functions.

A desire to calculate odds in games of chance gave rise to the theory of probability.

Probability

A number from 0 to 1 that represents the likelihood that an event will occur is referred to as the event's **probability**. A probability of 0 means that the event is impossible, and a probability of 1 means that the event is certain to occur. In this section, we will see that integration is a useful tool for calculating probabilities.

Experimental and Theoretical Probability

There are two types of probability, *experimental* and *theoretical*.

If we toss a coin a high number of times—say, 1000—and count the number of times we get heads, we can determine the probability of the coin landing heads up. If it lands heads up 503 times, we calculate the probability of it landing heads up to be

$$\frac{503}{1000}, \quad \text{or} \quad 0.503.$$

This is an **experimental** determination of probability. Such a determination of probability is discovered by the observation and study of data and is quite common and very useful. For example, here are some probabilities that have been determined experimentally:

1. The lifetime probability of being struck by lightning is about $\frac{1}{6250}$. (*Source*: www.noaa.gov.)
2. A person who is released from prison has a 52% probability of returning to prison within 3 years. (*Source*: www.crimeinamerica.net.)
3. The probability that a basketball player will make a free throw is based on his or her past success rate (number of free throws made divided by total number of free throw attempts).

If we consider tossing a coin and reason that we are just as likely to get heads as tails, we would *calculate* the probability of it landing heads up to be $\frac{1}{2}$, or 0.5. This is a **theoretical** determination of probability. Here, for example, are some probabilities that have been determined theoretically, using mathematics:

1. In any random group of 30 people, the probability that two of them have the same birthday (excluding year) is 0.706. (See Exercise 42.)
2. The probability of winning the Powerball lottery is $\dfrac{1}{175,223,510}$. (*Source*: powerball.com.)
3. The probability that two cards drawn from a standard deck of 52 cards are both kings is $\frac{1}{221}$.

In summary, experimental probabilities are determined from observations and data. Theoretical probabilities are determined by mathematical reasoning. There are situations in which it is easier to determine one type of probability than the other. For example, it would be quite difficult to determine the theoretical probability of catching a cold.

In the discussion that follows, we consider primarily theoretical probability.

EXAMPLE 1 What is the probability of drawing an ace from a well-shuffled deck of cards?

Solution There are 52 possible cards that could be drawn, and each card has the same chance of being drawn. Since there are 4 aces, the probability of drawing an ace is $\frac{4}{52}$ or $\frac{1}{13}$, or about 7.7%. 1 ✔

Quick Check 1 ✔

What is the probability of drawing each of the following from a well-shuffled deck of cards?

a) a three
b) a heart
c) the three of hearts

Quick Check 2 ✔

Assume that a jar has the same assortment of balls as in Example 2 and 1 ball is randomly selected.

a) What is the probability that it is black or yellow?

b) What is the probability that it is not green?

EXAMPLE 2   A jar contains 7 black balls, 6 yellow balls, 4 green balls, and 3 red balls, all the same size and weight. The jar is shaken well, and 1 ball is randomly selected.

a) What is the probability that the ball is red?

b) What is the probability that it is white?

Solution

a) There are 20 balls, of which 3 are red; so the probability of selecting a red ball is $\frac{3}{20}$.

b) There are no white balls, so the probability of selecting a white ball is $\frac{0}{20}$, or 0.

2 ✔

Below is a table of probabilities for the situation in Example 2. Note that the sum of these probabilities is 1. We are certain that we will select a black, yellow, green, or red ball, so the probability of that event is 1. We can arrange the data from the table into what is called a **relative frequency graph**, or **histogram**, which shows the proportion of times that each event occurs (the probability of each event). If we assign a width of 1 to each rectangle in this graph, then the sum of the areas of the rectangles is 1.

Color	Probability
Black (B)	$\frac{7}{20}$
Yellow (Y)	$\frac{6}{20}$
Green (G)	$\frac{4}{20}$
Red (R)	$\frac{3}{20}$

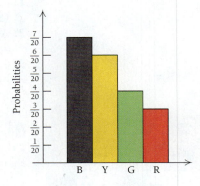

Continuous Random Variables

Suppose that we throw a dart at a number line in such a way that it always lands in the interval $[3, 5]$. Let x be the number the dart hits. There is an infinite number of possibilities for x. Note that x can be observed (or measured) repeatedly and its possible values comprise an interval of real numbers. Such a variable is an example of a **continuous random variable**.

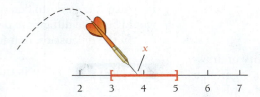

Suppose we throw the dart a large number of times and it lands 43% of the time in the subinterval $[3.6, 4.8]$. We can then conclude that the experimental probability of the dart landing in $[3.6, 4.8]$ is 0.43.

Let's consider some other examples of continuous random variables.

If x is the temperature at any point on Earth, with −128.6°F the lowest temperature ever recorded (at Vostok Station, Antarctica, in 1983) and 134°F the highest ever recorded (at Death Valley, California, in 1913), then x is a continuous random variable distributed over $[-128.6, 134]$. (*Source:* www.ncdc.noaa.gov.)

[-128.6, 134]

If buses traveling from Philadelphia to New York City require at least 2 hr and at most 5 hr for the trip and x is the number of hours a bus takes to make the trip, then x is a continuous random variable distributed over $[2, 5]$.

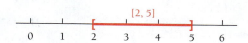

[2, 5]

There may be a function $y = f(x)$ such that the area under its graph over the subinterval $[2, 5]$ gives the probability that a particular trip time appears in the subinterval. For example, suppose we have a constant function $f(x) = \frac{1}{3}$ that gives us these probabilities. Look at its graph. The area under the graph is $3 \cdot \frac{1}{3}$, or 1.

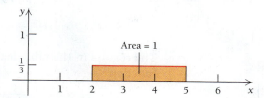

Area = 1

Using this graph, the probability that a trip takes 4 hr to 5 hr is the area that lies over the subinterval $[4, 5]$. That is,

$$P([4, 5]) = \tfrac{1}{3} = 33\tfrac{1}{3}\%.$$

The probability that a trip takes between 2 hr and 4.5 hr is $\frac{5}{6}$, or $83\frac{1}{3}\%$. This is the area of the rectangle over $[2, 4.5]$.

Note that when $f(x) = \frac{1}{3}$, any interval between the numbers 2 and 5 of width 1 has probability $\frac{1}{3}$. This does not happen for all functions.

Suppose instead that

$$f(x) = \tfrac{3}{117}x^2.$$

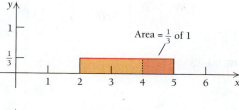

Area = $\frac{1}{3}$ of 1

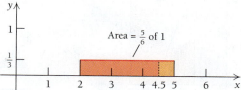

Area = $\frac{5}{6}$ of 1

The area under the graph of f from 4 to 5 is given by the definite integral over $[4, 5]$ and yields the probability that a trip takes 4 hr to 5 hr.

We have

$$P([4, 5]) = \int_4^5 f(x)\, dx$$

$$= \int_4^5 \frac{3}{117}x^2\, dx$$

$$= \frac{1}{117}\left[x^3\right]_4^5$$

$$= \frac{1}{117}(5^3 - 4^3) = \frac{61}{117} \approx 0.52.$$

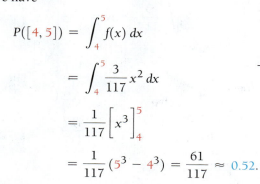

$P([3, 4]) = 0.32$

$P([4, 5]) = 0.52$

$P([2, 3]) = 0.16$

Thus, according to this model, there is a probability of 0.52 that a bus trip takes 4 hr to 5 hr. The function *f* is called a *probability density function*. Its integral over any sub-interval gives the probability that *x* "lands" in that subinterval.

Similar calculations for the bus trip example are shown in the following table.

Trip Time	Probability That a Trip Time Occurs during the Interval
2 hr to 3 hr	$P([2,3]) = \int_2^3 \frac{3}{117}x^2\,dx \approx 0.16$
3 hr to 4 hr	$P([3,4]) = \int_3^4 \frac{3}{117}x^2\,dx \approx 0.32$
4 hr to 5 hr	$P([4,5]) = \int_4^5 \frac{3}{117}x^2\,dx \approx 0.52$
2 hr to 5 hr	$P([2,5]) = \int_2^5 \frac{3}{117}x^2\,dx = 1.00$

The results in the table lead us to the following definition of a probability density function.

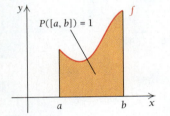

FIGURE 1

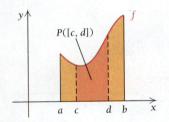

FIGURE 2

DEFINITION

Let *x* be a continuous random variable. A function *f* is said to be a **probability density function** for *x* if:

1. For all *x* in the domain of *f*, we have $f(x) \geq 0$.
2. The area under the graph of *f* is 1 (see Fig. 1).
3. For any subinterval $[c, d]$ in the domain of *f* (see Fig. 2), the probability that *x* will be in that subinterval is given by

$$P([c,d]) = \int_c^d f(x)\,dx.$$

EXAMPLE 3   Verify that Property 2 of the definition of a probability density function holds for

$$f(x) = \frac{3}{117}x^2, \quad \text{for } 2 \leq x \leq 5.$$

Solution   We need to show that the area under *f* over $[2, 5]$ is 1.

$$\int_2^5 \frac{3}{117}x^2\,dx = \frac{3}{117}\left[\frac{1}{3}x^3\right]_2^5$$

$$= \frac{1}{117}\left[x^3\right]_2^5$$

$$= \frac{1}{117}(5^3 - 2^3)$$

$$= \frac{117}{117} = 1$$

Quick Check 3 ✔

Assume that *x* is a continuous random variable. Verify that $g(x) = \frac{3}{14}\sqrt{x}$, for $1 \leq x \leq 4$, is a probability density function.

3 ✔

EXAMPLE 4   **Business: Life of a Product.**   Luminox Corp. produces compact fluorescent bulbs and determines that the life t of a bulb is from 3 to 6 yr, with the probability density function for t given by

$$f(t) = \frac{24}{t^3}, \quad \text{for } 3 \le t \le 6.$$

a) Verify Property 2 of the definition of a probability density function.

b) Find the probability that a bulb will last no more than 4 yr.

c) Find the probability that a bulb will last at least 4 yr and at most 5 yr.

Solution

a) We must show that $\int_3^6 f(t)\, dt = 1$. We have

$$\int_3^6 \frac{24}{t^3}\, dt = -12\left[\frac{1}{t^2}\right]_3^6 \qquad \text{Integrating: } 24\left[\frac{t^{-2}}{-2}\right] = -12\left[\frac{1}{t^2}\right]$$

$$= -12\left(\frac{1}{6^2} - \frac{1}{3^2}\right) \qquad \text{Substituting}$$

$$= -12\left(-\frac{3}{36}\right) = 1. \qquad \text{Simplifying}$$

b) The probability that a bulb will last no more than 4 yr is

$$P(3 \le t \le 4) = \int_3^4 \frac{24}{t^3}\, dt$$

$$= -12\left[\frac{1}{t^2}\right]_3^4$$

$$= -12\left(\frac{1}{4^2} - \frac{1}{3^2}\right) \qquad \text{Substituting}$$

$$= -12\left(-\frac{7}{144}\right) = \frac{7}{12} \approx 0.58. \qquad \text{Simplifying}$$

c) The probability that a bulb will last at least 4 yr and at most 5 yr is

$$P(4 \le t \le 5) = \int_4^5 \frac{24}{t^3}\, dt$$

$$= -12\left[\frac{1}{t^2}\right]_4^5$$

$$= -12\left(\frac{1}{5^2} - \frac{1}{4^2}\right) \qquad \text{Substituting}$$

$$= -12\left(-\frac{9}{400}\right)$$

$$= \frac{27}{100} = 0.27. \qquad \text{Simplifying} \qquad 4 ✔$$

Quick Check 4 ✔

The time between arrivals of subway trains at a station is modeled by the probability density function

$$h(x) = \frac{10}{x^2}, \quad \text{for } 5 \le x \le 10,$$

where x is in minutes. Find the probability that:

a) the time between trains is between 5 and 7 minutes;

b) the time between trains is between 8 and 10 minutes.

Constructing Probability Density Functions

Suppose that $f(x)$ is nonnegative over $[a, b]$ and that

$$\int_a^b f(x)\, dx = K.$$

Multiplying both sides by $1/K$ gives

$$\frac{1}{K}\int_a^b f(x)\, dx = \frac{1}{K} \cdot K = 1, \quad \text{or} \quad \int_a^b \frac{1}{K} \cdot f(x)\, dx = 1.$$

Thus, when we multiply $f(x)$ by $1/K$, we have a function whose area over $[a, b]$ is 1. Such a function satisfies the definition of a probability density function.

EXAMPLE 5   Find k such that

$$f(x) = kx^2$$

is a probability density function over the interval $[1, 4]$. Then write the probability density function.

Solution   Note that for $k \geq 0$, we have $kx^2 \geq 0$. For f to be a probability density function, we must also have $\int_1^4 kx^2\, dx = 1$. Since

$$\int_1^4 kx^2\, dx = k\left[\tfrac{1}{3}x^3\right]_1^4 = k\left[\tfrac{64}{3} - \tfrac{1}{3}\right] = k \cdot 21,$$

we have $21k = 1$, so $k = \tfrac{1}{21}$. Thus, the probability density function is

$$f(x) = \tfrac{1}{21}x^2, \quad \text{for } 1 \leq x \leq 4. \qquad\qquad \boxed{5\;✔}$$

Quick Check 5 ✔

Find k such that $g(x) = \dfrac{k}{x}$ is a probability density function over the interval $[2, 7]$. Then write the probability density function.

Uniform Distributions

Consider the probability density function given by $f(x) = \dfrac{1}{3}$ over $[2, 5]$, as shown below.

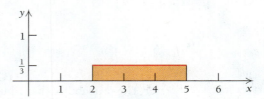

The length of the rectangle is the length of the interval $[2, 5]$, which is 3, and the area is 1.

Generalizing, the length of the rectangle shown at the left is the length of the interval $[a, b]$, which is $b - a$. In order for the shaded area to be 1, the height of the rectangle must be $1/(b - a)$. Thus, $f(x) = 1/(b - a)$.

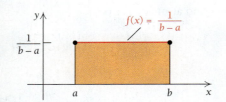

DEFINITION

A continuous random variable x is said to be **uniformly distributed** over an interval $[a, b]$ if it has a probability density function f given by

$$f(x) = \frac{1}{b - a}, \quad \text{for } a \leq x \leq b.$$

EXAMPLE 6   **Business: Quality Control.**   BlastOut Inc. produces sirens used for tornado warnings. The maximum loudness, L, of the sirens ranges from 70 to 100 decibels. The probability density function for L is

$$f(L) = \tfrac{1}{30}, \quad \text{for } 70 \leq L \leq 100.$$

Business: Quality Control.
The probability density function for the weight x, in pounds, of bags of feed sold at a feedstore is

$f(x) = \frac{1}{8}$, for $45 \le x \le 53$.

A bag is selected at random. Find the probability the bag weighs between 47.5 and 50.25 lb.

A siren is selected at random off the assembly line. Find the probability that its maximum loudness is from 70 to 92 decibels.

Solution The probability is

$$P(70 \le L \le 92) = \int_{70}^{92} \frac{1}{30} \, dL = \frac{1}{30}[L]_{70}^{92}$$

$$= \frac{1}{30}(92 - 70) = \frac{22}{30} = \frac{11}{15} \approx 0.73.$$ **6** ✔

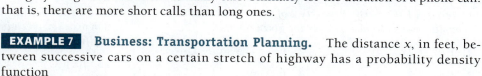

Exponential Distributions

The duration of a phone call, the distance between successive cars on a highway, and the amount of time required to learn a task are all examples of *exponentially distributed* random variables. That is, their probability density functions are exponential.

> **DEFINITION**
>
> A continuous random variable is **exponentially distributed** if it has a probability density function of the form
>
> $$f(x) = ke^{-kx}, \quad \text{over } [0, \infty).$$

To see that $f(x) = 2e^{-2x}$ is such a probability density function, note that

$$\int_{0}^{\infty} 2e^{-2x}\,dx = \lim_{b \to \infty} \int_{0}^{b} 2e^{-2x}\,dx = \lim_{b \to \infty} [-e^{-2x}]_{0}^{b} = \lim_{b \to \infty}\left(\frac{-1}{e^{2b}} - (-1)\right) = 1.$$

The general case,

$$\int_{0}^{\infty} ke^{-kx}\,dx = 1,$$

can be verified in a similar way.

Why is it reasonable to assume that the distance between cars is exponentially distributed? Part of the reason is that there are many more cases in which traffic is highly congested and involves many cars. Similarly for the duration of a phone call: that is, there are more short calls than long ones.

EXAMPLE 7 **Business: Transportation Planning.** The distance x, in feet, between successive cars on a certain stretch of highway has a probability density function

$$f(x) = ke^{-kx}, \quad \text{for } 0 \le x < \infty,$$

where $k = 1/a$ and a is the average distance between successive cars over some period of time. A transportation planner determines that the average distance between cars on a certain stretch of highway is 166 ft. What is the probability that the distance between two successive cars, chosen at random, is 50 ft or less?

Solution We first determine k:

$$k = \frac{1}{166}$$
$$\approx 0.006024.$$

The probability density function for x is

$$f(x) = 0.006024e^{-0.006024x}, \quad \text{for } 0 \le x < \infty.$$

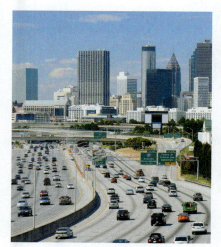

A transportation planner can determine the probabilities that cars are certain distances apart.

Quick Check 7 ✔

The response time for a para-medic unit has the probability density function

$$f(t) = 0.05e^{-0.05t},$$
for $0 \le t < \infty$,

where t is in minutes. Find the probability of each response time:

a) between 5 and 10 min;
b) between 8 and 20 min;
c) between 1 and 45 min;
d) more than 60 min.

The probability that the distance between the cars is 50 ft or less is

$$P(0 \le x \le 50) = \int_0^{50} 0.006024e^{-0.006024x} \, dx$$

$$= \left[\frac{0.006024}{-0.006024} e^{-0.006024x} \right]_0^{50}$$

$$= \left[-e^{-0.006024x} \right]_0^{50}$$

$$= (-e^{-0.006024 \cdot 50}) - (-e^{-0.006024 \cdot 0})$$

$$= -e^{-0.301200} + 1$$

$$= 1 - e^{-0.301200}$$

$$= 1 - 0.739930 \approx 0.260.$$

7 ✔

Section Summary

- The *probability* of an event is a number between 0 and 1, with 0 meaning that the event is impossible and 1 meaning that the event is certain.
- Probabilities are determined *experimentally* (by conducting trials) or *theoretically* (by mathematical reasoning).
- A *continuous random variable* is a quantity that can be observed (or measured) and whose possible values comprise an interval of real numbers.
- If x is a continuous random variable, then f is a *probability density function* for x if it meets the following criteria:
  (1) For all x in $[a, b]$, $f(x) \ge 0$.

(2) The area under the graph of f over $[a, b]$ is 1; that is, $\int_a^b f(x) \, dx = 1$.
(3) The probability that x is within the subinterval $[c, d]$ is given by $P([c, d]) = \int_c^d f(x) \, dx$.

- A continuous random variable x is *uniformly distributed* over $[a, b]$ if its probability density function has the form $f(x) = \dfrac{1}{b - a}$.

- A continuous random variable x is *exponentially distributed* over $[0, \infty)$ if its probability density function has the form $f(x) = ke^{-kx}$.

5.4 Exercise Set

In Exercises 1–12, verify Property 2 of the definition of a probability density function over the given interval.

1. $f(x) = 2x, \quad [0, 1]$

2. $f(x) = \frac{1}{4}x, \quad [1, 3]$

3. $f(x) = 3, \quad [0, \frac{1}{3}]$

4. $f(x) = \frac{1}{5}, \quad [3, 8]$

5. $f(x) = \frac{3}{26}x^2, \quad [1, 3]$

6. $f(x) = \frac{3}{64}x^2, \quad [0, 4]$

7. $f(x) = \frac{1}{x}, \quad [1, e]$

8. $f(x) = \frac{1}{e - 1}e^x, \quad [0, 1]$

9. $f(x) = \frac{1}{3}x^2, \quad [-2, 1]$

10. $f(x) = \frac{3}{2}x^2, \quad [-1, 1]$

11. $f(x) = 3e^{-3x}, \quad [0, \infty)$

12. $f(x) = 4e^{-4x}, \quad [0, \infty)$

Find k such that each function is a probability density function over the given interval. Then write the probability density function.

13. $f(x) = kx, \quad [1, 4]$

14. $f(x) = kx, \quad [2, 5]$

15. $f(x) = kx^2, \quad [-1, 1]$

16. $f(x) = kx^2, \quad [-2, 2]$

17. $f(x) = k, \quad [3, 9]$

18. $f(x) = k, \quad [1, 4]$

19. $f(x) = k(2 - x), \quad [0, 2]$

20. $f(x) = k(4 - x), \quad [0, 4]$

21. $f(x) = \frac{k}{x}, \quad [1, 2]$

22. $f(x) = \frac{k}{x}, \quad [1, 3]$

23. $f(x) = ke^x, \quad [0, 3]$

24. $f(x) = ke^x, \quad [0, 2]$

25. A dart is thrown at a number line in such a way that it always lands in $[0, 10]$. Let x represent the number the dart hits. Suppose the probability density function for x is given by

$$f(x) = \tfrac{1}{50}x, \quad \text{for } 0 \leq x \leq 10.$$

Find $P(2 \leq x \leq 6)$, the probability that the dart lands in $[2, 6]$.

26. In Exercise 25, suppose the dart always lands in $[0, 5]$, and the probability density function for x is given by

$$f(x) = \tfrac{3}{125}x^2, \quad \text{for } 0 \leq x \leq 5.$$

Find $P(1 \leq x \leq 4)$, the probability that the dart lands in $[1, 4]$.

27. A number x is selected at random from $[4, 20]$. The probability density function for x is given by

$$f(x) = \tfrac{1}{16}, \quad \text{for } 4 \leq x \leq 20.$$

Find the probability that a number selected is in the subinterval $[9, 20]$.

28. A number x is selected at random from $[5, 29]$. The probability density function for x is given by

$$f(x) = \tfrac{1}{24}, \quad \text{for } 5 \leq x \leq 29.$$

Find the probability that a number selected is in the subinterval $[14, 29]$.

APPLICATIONS

Business and Economics

29. Transportation planning. Refer to Example 7. A transportation planner determines that the average distance between cars on a certain highway is 100 ft. Find the probability that the distance between two successive cars, chosen at random, is at most 40 ft.

30. Transportation planning. Refer to Example 7. A transportation planner determines that the average distance between cars on a certain highway is 200 ft. Find the probability that the distance between two successive cars, chosen at random, is at most 10 ft.

31. Duration of a phone call. A telephone company determines that the duration t, in minutes, of a phone call is an exponentially distributed random variable with a probability density function

$$f(t) = 2e^{-2t}, \quad 0 \leq t < \infty.$$

Find the probability that a phone call will last no more than 5 min.

32. Duration of a phone call. Referring to Exercise 31, find the probability that a phone call will last no more than 2 min.

33. Time to failure. The *time to failure*, t, in hours, of a machine is often exponentially distributed with a probability density function

$$f(t) = ke^{-kt}, \quad 0 \leq t < \infty,$$

where $k = 1/a$ and a is the average amount of time that will pass before a failure occurs. Suppose the average amount of time that will pass before a failure occurs is

100 hr. What is the probability that a failure will occur in 50 hr or less?

34. Reliability of a machine. The *reliability* of the machine (the probability that it will work) in Exercise 33 is defined as

$$R(T) = 1 - \int_0^T 0.01e^{-0.01t}\, dt,$$

where $R(T)$ is the reliability at time T. Write $R(T)$ without using an integral.

Life and Physical Sciences

35. Wait time for 911 calls. The wait time before a 911 call is answered in the state of California has a probability density function $f(t) = 0.23e^{-0.23t}$, for $0 \leq t < \infty$, where t is in seconds. (*Source:* California Government Code.)

   a) The state standard is that 90% of 911 calls are to be answered within 10 sec. Verify that this standard is met using the probability density function f.

   b) What is the probability that a 911 call is answered within 15 to 25 sec after being made?

36. Emergency room wait times. The wait time at an emergency room has the probability density function $f(t) = 0.116e^{-0.116t}$, for $0 \leq t < \infty$, where t is in hours. (*Source:* www.pressganey.com.)

   a) Find the probability that a wait time is at most 1 hr.

   b) In 2009, half of all emergency room patients waited up to 6 hr. Verify this using the probability density function f.

Social Sciences

37. Time in a maze. In a psychology experiment, the time t, in seconds, that it takes a rat to learn its way through a maze is an exponentially distributed random variable with the probability density function

$$f(t) = 0.02e^{-0.02t}, \quad 0 \leq t < \infty.$$

Find the probability that a rat will learn its way through a maze in 150 sec or less.

The time that it takes a rat to learn its way through a maze is an exponentially distributed random variable.

38. Time in a maze. Using the equation in Exercise 37, find the probability that a rat will learn its way through the maze in 50 sec or less.

39. Use your answer to Exercise 37 to find the probability that a rat requires more than 150 sec to learn its way through the maze.

SYNTHESIS

40. If $f(x) = x^3$ is a probability density function over $[0, b]$, what is b?

41. If $f(x) = 12x^2$ is a probability density function over $[-a, a]$, what is a?

42. The birthday problem. Assume that birthdays are uniformly distributed throughout the year and that February 29 is omitted from consideration. The probability of at least one shared birthday (month and day only) among n randomly chosen people is

$$P(n) = 1 - \left(\frac{364 \cdot 363 \cdot 362 \cdot \cdots \cdot (366 - n)}{365^{n-1}} \right).$$

For example, in a group of 10 people the probability of at least one shared birthday is

$$P(10) = 1 - \left(\frac{364 \cdot 363 \cdot 362 \cdot \cdots \cdot 356}{365^9} \right) = 0.117.$$

a) Verify the claim made at the start of this section that the probability that two people in a group of 30 have the same birthday is about 70%.

b) How many people are required so that the probability that two of them share a birthday exceeds 50%?

c) How many people are in your calculus class? What is the probability of at least one shared birthday among you and your classmates? Test your calculation experimentally.

43. The wait times at an urgent care center are exponentially distributed. There is a 30% probability that a patient will have to wait up to 1 hr to see a doctor.

a) Find k, and then write the probability density function f.

b) Find the probability that a patient will have to wait between 90 min and 3 hr for a doctor.

44. The elapsed time between the arrivals of cars at a rural intersection is exponentially distributed. There is a 20% probability that 10 min will pass between the arrivals of cars at the intersection.

a) Find k, and then write the probability density function f.

b) Find the probability that two cars will come to the intersection within 5 min of one another.

45. The graph of f is a probability density function.

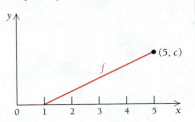

a) Find c.
b) Find f.
c) Find $P(2 \le x \le 3)$.

46. The graph of f is a probability density function.

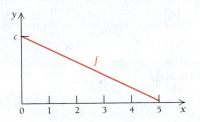

a) Find c.
b) Find f.
c) Find $P(1 \le x \le 4)$.

47. Find c such that $f(x) = cxe^{2x}$, for $1 \le x \le 2$, is a probability density function.

48. Find c such that $f(x) = cx\sqrt{1 + x}$, for $0 \le x \le 1$, is a probability density function.

TECHNOLOGY CONNECTION

49–60. Verify Property 2 of the definition of a probability density function for each of the functions in Exercises 1–12.

Answers to Quick Checks

1. (a) $\frac{4}{52}$ or $\frac{1}{13}$; **(b)** $\frac{13}{52}$ or $\frac{1}{4}$; **(c)** $\frac{1}{52}$

2. (a) $\frac{13}{20}$; **(b)** $\frac{16}{20}$ or $\frac{4}{5}$

3. $\int_1^4 \frac{3}{14}\sqrt{x}\, dx = \frac{3}{14}\left[\frac{2}{3}x^{3/2}\right]_1^4 = \frac{3}{14}\left(\frac{16}{3} - \frac{2}{3}\right) = 1$;

$g(x) \ge 0$ on $[1, 4]$ **4. (a)** 0.57; **(b)** 0.25

5. $k = 0.7982$; $g(x) = \dfrac{0.7982}{x}$, for $2 \le x \le 7$

6. 0.34375 **7. (a)** 0.172; **(b)** 0.302; **(c)** 0.846; **(d)** 0.050

5.5

● Find $E(x)$, $E(x^2)$, the mean, the variance, and the standard deviation.

● Evaluate normal distribution probabilities using a table.

● Calculate percentiles for a normal distribution.

Probability: Expected Value; The Normal Distribution

Expected Value

Let's again consider a dart thrown at a number line so that it lands somewhere in the interval $[3, 5]$. We assume a uniform distribution, meaning that it is equally likely that the dart will land anywhere in the interval.

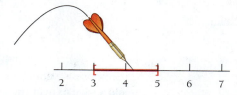

Suppose we throw the dart at the interval 100 times and keep track of the numbers it hits. We can calculate the arithmetic mean, or average, $\bar{x}$ (read "x bar") of these numbers:

$$\bar{x} = \frac{x_1 + x_2 + x_3 + \cdots + x_{100}}{100} = \frac{\sum\limits_{i=1}^{100} x_i}{100} = \sum_{i=1}^{100} x_i \cdot \frac{1}{100}.$$

Assuming the x_i's are uniformly distributed over $[3, 5]$, as shown to the left,

$$\sum_{i=1}^{n} x_i \cdot \frac{1}{n}, \quad \text{or} \quad \sum_{i=1}^{n} \left(x_i \cdot \frac{1}{2} \right) \frac{2}{n}, \qquad \textcolor{red}{\text{Note that the interval width is 2.}}$$

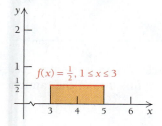

$f(x) = \frac{1}{2}, 1 \le x \le 3$

is analogous to

$$\int_{3}^{5} x \cdot f(x)\, dx,$$

where $f(x) = 1/2$ is a probability density function for x. Because the width of $[3, 5]$ is 2, we can regard $2/n$ as Δx. The probability density function gives a "weight" to x. We add the values of

$$\left(x_i \cdot \frac{1}{2} \right) \left(\frac{2}{n} \right)$$

when we find $\sum\limits_{i=1}^{n} \left(x_i \cdot \frac{1}{2} \right) \left(\frac{2}{n} \right)$. Similarly, we add all values of

$$(x \cdot f(x))(\Delta x)$$

when we find $\int_{3}^{5} x \cdot f(x)\, dx$:

$$\int_{3}^{5} x \cdot \frac{1}{2}\, dx = \frac{1}{2} \left[\frac{x^2}{2} \right]_{3}^{5} = \frac{1}{4} [5^2 - 3^2] = 4.$$

This result, representing the average value of the distribution, is not surprising, since the distribution is uniform and centered around 4. The average of 100 dart throws may not be exactly 4, but as $n \to \infty$, we will have $\bar{x} \to 4$.

Suppose we use the probability density function $f(x) = \frac{1}{4}x$ over the interval $[1, 3]$. As we can see from the graph, this function gives more "weight" to the right

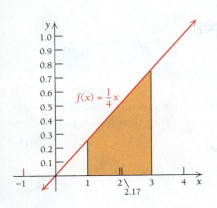

$f(x) = \frac{1}{4}x$

side of the interval than to the left. Perhaps more points are awarded if a dart lands closer to 3 than to 1. Then we have

$$\int_1^3 x \cdot f(x)\, dx = \int_1^3 x \cdot \frac{1}{4}x\, dx$$

$$= \frac{1}{4}\int_1^3 x^2\, dx$$

$$= \frac{1}{4}\left[\frac{x^3}{3}\right]_1^3$$

$$= \frac{1}{12}(3^3 - 1^3) = \frac{26}{12} \approx 2.17.$$

Suppose we continue to throw the dart and compute averages. The more throws, the closer we expect the average to approach 2.17.

DEFINITION

Let x be a continuous random variable over $[a, b]$ with probability density function f. The **expected value** of x is defined by

$$E(x) = \int_a^b x \cdot f(x)\, dx.$$

The concept of the expected value of a random variable can be generalized to functions of the random variable. Suppose $y = g(x)$ is a function of the random variable x. Then we have the following.

DEFINITION

The **expected value** of $g(x)$ is defined by

$$E(g(x)) = \int_a^b g(x) \cdot f(x)\, dx,$$

where f is a probability density function for x.

EXAMPLE 1   Given the probability density function

$$f(x) = \tfrac{1}{2}x - 2, \quad \text{over } [4, 6],$$

find $E(x)$ and $E(x^2)$.

Solution

$$E(x) = \int_4^6 x \cdot \left(\tfrac{1}{2}x - 2\right) dx = \int_4^6 \left(\tfrac{1}{2}x^2 - 2x\right) dx$$

$$= \left[\tfrac{1}{6}x^3 - x^2\right]_4^6$$

$$= \left[\tfrac{1}{6}(6^3) - 6^2\right] - \left[\tfrac{1}{6}(4^3) - 4^2\right]$$

$$= \tfrac{16}{3}$$

Quick Check 1 ✔

Given the probability density function

$$f(x) = \tfrac{1}{2} - \tfrac{1}{8}x, \quad \text{over } [0, 4],$$

find $E(x)$ and $E(x^2)$.

$$E(x^2) = \int_4^6 x^2 \cdot \left(\tfrac{1}{2}x - 2 \right) dx = \int_4^6 \left(\tfrac{1}{2}x^3 - 2x^2 \right) dx$$

$$= \left[\tfrac{1}{8}x^4 - \tfrac{2}{3}x^3 \right]_4^6$$

$$= \left[\tfrac{1}{8}(6^4) - \tfrac{2}{3}(6^3) \right] - \left[\tfrac{1}{8}(4^4) - \tfrac{2}{3}(4^3) \right]$$

$$= \tfrac{86}{3}$$

1 ✔

> **DEFINITION**
>
> The **mean, μ,** of a continuous random variable x is defined to be $E(x)$. That is,
>
> $$\mu = E(x) = \int_a^b x \cdot f(x) \, dx,$$
>
> where f is a probability density function for x defined over $[a, b]$. (The symbol μ is the lowercase Greek letter *mu*.)

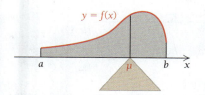

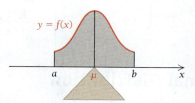

We can get a physical idea of the mean of a random variable by pasting the graph of the probability density function on cardboard and cutting out the area under the curve over the interval $[a, b]$. If we find a balance point on the x-axis, that balance point is the mean, μ.

Variance and Standard Deviation

Because two different distributions can have the same mean, it is useful to have a second statistic that serves as a measure of how the data in a distribution are spread out. Statistics that provide such a measure are the *variance* and (especially) the *standard deviation* of a distribution. A full derivation of how these statistics were developed is beyond the scope of this book.

> **DEFINITION**
>
> The **variance, σ^2,** of a continuous random variable x, defined on $[a, b]$, with probability density function f, is
>
> $$\sigma^2 = E(x^2) - \mu^2$$
>
> $$= E(x^2) - [E(x)]^2$$
>
> $$= \int_a^b x^2 \cdot f(x) \, dx - \left[\int_a^b x \cdot f(x) \, dx \right]^2.$$
>
> The **standard deviation, σ,** of a continuous random variable is defined as
>
> $$\sigma = \sqrt{\text{variance}}.$$
>
> (The symbol σ is the lowercase Greek letter *sigma*.)

EXAMPLE 2   Given the probability density function

$$f(x) = \tfrac{1}{2}x - 2, \quad \text{over } [4, 6],$$

find the mean, the variance, and the standard deviation.

Solution From Example 1, we have $E(x) = \frac{16}{3}$ and $E(x^2) = \frac{86}{3}$. Thus,

$$\text{mean} = \mu = E(x) = \frac{16}{3};$$

$$\begin{aligned}
\text{variance} = \sigma^2 &= E(x^2) - [E(x)]^2 \\
&= \frac{86}{3} - \left(\frac{16}{3}\right)^2 = \frac{86}{3} - \frac{256}{9} \\
&= \frac{258}{9} - \frac{256}{9} = \frac{2}{9};
\end{aligned}$$

$$\begin{aligned}
\text{standard deviation} = \sigma &= \sqrt{\frac{2}{9}} \\
&= \frac{1}{3}\sqrt{2} \approx 0.47.
\end{aligned}$$

2 ✔

Loosely speaking, the standard deviation is a measure of how closely bunched the graph of f is, that is, how far the points on f are, on average, from the line $x = \mu$, as indicated below.

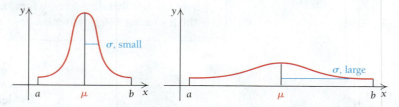

The Normal Distribution

Suppose the average score on a test is 70. Usually there are about as many scores above the average as there are below the average; and the farther away from the average a particular score is, the fewer people who received that score. Scores on a test are an example of a random variable that is often *normally* distributed.

Consider the function given by

$$g(x) = e^{-x^2/2}, \quad \text{over } (-\infty, \infty).$$

This function has the set of real numbers as its domain. Its graph is the bell-shaped curve shown below. We can find function values using a calculator:

x	$g(x)$
0	1
1	0.6
2	0.1
3	0.01
−1	0.6
−2	0.1
−3	0.01

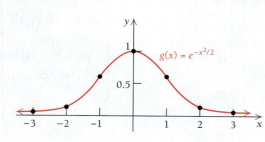

This function has an antiderivative, but that antiderivative has no basic integration formula. Nevertheless, it can be shown that the improper integral converges over the interval $(-\infty, \infty)$ to a number given by

$$\int_{-\infty}^{\infty} e^{-x^2/2}\, dx = \sqrt{2\pi}.$$

That is, although an elementary expression for the antiderivative cannot be found, there is a numerical value for the improper integral evaluated over the set of real numbers. Note that since the area is not 1, the function g is not a probability density function. However, the following function is:

$$f(x) = \frac{1}{\sqrt{2\pi}} e^{-x^2/2}, \quad \text{over } (-\infty, \infty).$$

> **DEFINITION**
>
> A continuous random variable x has a **standard normal distribution** if its probability density function is
>
> $$f(x) = \frac{1}{\sqrt{2\pi}} e^{-x^2/2}, \quad \text{over } (-\infty, \infty).$$

TECHNOLOGY CONNECTION

Exploratory

Use a graphing calculator to approximate

$$\int_{-b}^{b} \frac{1}{\sqrt{2\pi}} e^{-x^2/2}\, dx$$

for $b = 10, 100$, and 1000. What does this suggest about

$$\int_{-\infty}^{\infty} \frac{1}{\sqrt{2\pi}} e^{-x^2/2}\, dx?$$

This is a way to verify part of the assertion that

$$f(x) = \frac{1}{\sqrt{2\pi}} e^{-x^2/2}$$

is a probability density function. Use a similar approximation procedure to show that the mean is 0 and the standard deviation is 1.

This standard normal distribution has a mean of 0 and a standard deviation of 1. Its graph follows.

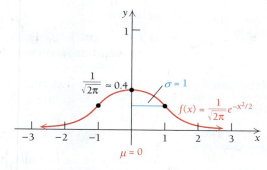

The general case is defined as follows.

> **DEFINITION**
>
> A continuous random variable x is **normally distributed** with mean μ and standard deviation σ if its probability density function is given by
>
> $$f(x) = \frac{1}{\sigma\sqrt{2\pi}} e^{-(1/2)[(x-\mu)/\sigma]^2}, \quad \text{over } (-\infty, \infty).$$

The graph of any normal distribution is a transformation of the graph of the standard normal distribution. This can be visualized as translating the graph of a normal distribution along the x-axis and adjusting how tightly clustered the graph is about the mean. Some examples follow.

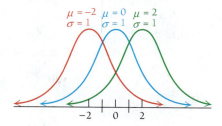

Normal distributions with same standard deviations but different means

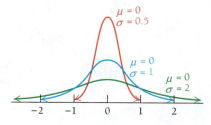

Normal distributions with same means but different standard deviations

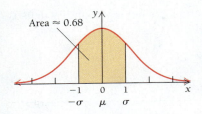

Area ≈ 0.68

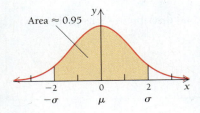

Area ≈ 0.95

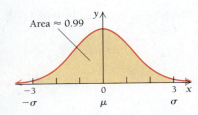

Area ≈ 0.99

The 68-95-99 Rule: In a standard normal distribution, about 68% of the data is within one standard deviation of the mean, about 95% of the data is within two standard deviations of the mean, and about 99% of the data is within three standard deviations of the mean.

The normal distribution is extremely important in statistics; it underlies much of the research in the behavioral and social sciences. Because of this, tables of approximate values of the definite integral of the standard normal distribution have been prepared using numerical approximation methods like the Trapezoidal Rule given in Exercise Set 4.2. Table A at the back of the book (p. 000) is such a table. It contains values of

$$P(0 \leq x \leq z) = \int_0^z \frac{1}{\sqrt{2\pi}} e^{-x^2/2} \, dx.$$

The symmetry of the graph of this function about the mean allows many types of probabilities to be computed from the table. Some involve addition or subtraction of areas.

EXAMPLE 3   Let x be a continuous random variable with a standard normal distribution. Using Table A at the back of the book, find each of the following.

a) $P(0 \leq x \leq 1.68)$ b) $P(-0.97 \leq x \leq 0)$

c) $P(-2.43 \leq x \leq 1.01)$ d) $P(0.90 \leq x \leq 1.74)$

e) $P(-1.98 \leq x \leq -0.42)$ f) $P(x \geq 0.61)$

Solution

a) $P(0 \leq x \leq 1.68)$ is the area bounded by the standard normal curve and the lines $x = 0$ and $x = 1.68$. We look this up in Table A by going down the left column to 1.6, then moving to the right to the column headed 0.08. There we read 0.4535. Thus,

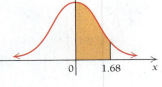

$$P(0 \leq x \leq 1.68) = 0.4535.$$

b) Because of the symmetry of the graph,

$$P(-0.97 \leq x \leq 0)$$
$$= P(0 \leq x \leq 0.97) \quad \text{Using symmetry}$$
$$= 0.3340.$$

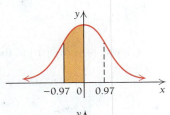

c) $P(-2.43 \leq x \leq 1.01)$
$$= P(-2.43 \leq x \leq 0) + P(0 \leq x \leq 1.01)$$
$$= P(0 \leq x \leq 2.43) + P(0 \leq x \leq 1.01)$$
$$= 0.4925 + 0.3438$$
$$= 0.8363$$

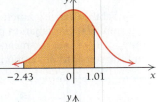

d) $P(0.90 \leq x \leq 1.74)$
$$= P(0 \leq x \leq 1.74) - P(0 \leq x \leq 0.90)$$
$$= 0.4591 - 0.3159$$
$$= 0.1432$$

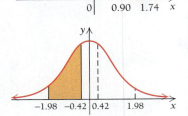

e) $P(-1.98 \leq x \leq -0.42)$
$$= P(0.42 \leq x \leq 1.98)$$
$$= P(0 \leq x \leq 1.98) - P(0 \leq x \leq 0.42)$$
$$= 0.4761 - 0.1628$$
$$= 0.3133$$

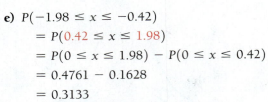

Quick Check 3 ✔

Let x be a continuous random variable with a standard normal distribution. Using Table A, find each of the following.

a) $P(-1 \leq x \leq \frac{1}{2})$
b) $P(x \leq -0.77)$
c) $P(x \geq \frac{2}{5})$

f) $P(x \geq 0.61)$

$= P(x \geq 0) - P(0 \leq x \leq 0.61)$

$= 0.5000 - 0.2291$

Because of the symmetry about the line $x = 0$, half the area is on each side of the line, and since the entire area is 1, we have $P(x \geq 0) = 0.5000$.

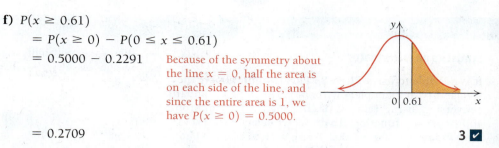

$= 0.2709$

3 ✔

For most normal distributions, $\mu \neq 0$ and $\sigma \neq 1$. It would be a hopeless task to make tables for all values of the mean μ and the standard deviation σ. For any normal distribution, the transformation

$$z = \frac{x - \mu}{\sigma}$$

standardizes the distribution, since $(x - \mu)/\sigma$ always tells us how many standard deviations x is from μ. Subtracting μ from all x-values and then dividing by σ preserves the order of the x-values while permitting the use of Table A. Such converted values are called *z-scores*, or *z-values*.

$$P(a \leq x \leq b) = P\left(\frac{a - \mu}{\sigma} \leq z \leq \frac{b - \mu}{\sigma}\right),$$

and this last probability can be found using Table A.

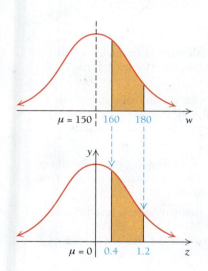

EXAMPLE 4   The weights, w, of the members of a cross-country team are normally distributed with a mean, μ, of 150 lb and a standard deviation, σ, of 25 lb. Find the probability that the weight of a student on the team is between 160 lb and 180 lb.

Solution   We first standardize the weights:

180 is standardized to $\dfrac{b - \mu}{\sigma} = \dfrac{180 - 150}{25} = 1.2;$

160 is standardized to $\dfrac{a - \mu}{\sigma} = \dfrac{160 - 150}{25} = 0.4.$

These z-values measure the distance of w from μ in terms of σ.

Then we have

$$P(160 \leq w \leq 180) = P(0.4 \leq z \leq 1.2)$$
$$= P(0 \leq z \leq 1.2) - P(0 \leq z \leq 0.4)$$
$$= 0.3849 - 0.1554$$
$$= 0.2295.$$

Now we can use Table A.

Quick Check 4 ✔

Referring to Example 4, find the probability of each of the following.

a) The weight of a randomly selected member of the team is below 165 lb.
b) The weight of a randomly selected member of the team is between 135 lb and 155 lb.
c) The weight of a randomly selected member of the team is above 175 lb.

Thus, the probability that the weight of a randomly selected member of the team is 160 lb to 180 lb is 0.2295. That is, about 23% of the team members weigh between 160 lb and 180 lb.

4 ✔

Percentiles of a Normal Distribution

Suppose you take an exam and score better than 85% of all students taking that exam. We say that your score is in the 85th *percentile*. For the standard normal distribution,

TECHNOLOGY CONNECTION

Statistics on a Calculator

It is possible to use a TI-83/84 Plus to approximate the probability in Example 4 without performing a standard conversion, using Table A, or entering the normal probability density function. To do so, we select an appropriate window, $[0, 300, -0.002, 0.02]$, with Xscl $= 50$ and Yscl $= 0.01$. Next, we use the ShadeNorm command, which we find by pressing **2ND** **DISTR** **▷** **ENTER**. We then enter values as shown and press **ENTER**. (If necessary, the ClearDraw option on the DRAW menu can be used to clear the graph.)

Left endpoint of interval   Right endpoint of interval

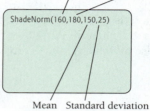 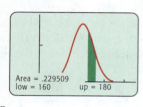

Mean   Standard deviation

The area is shaded and given as 0.229509, or about 23%.
Alternatively, the probability can be calculated directly by pressing **2ND** and **DISTR** and selecting normalcdf. The values are entered as shown:

Mean   Standard deviation

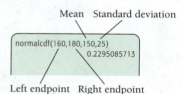

Left endpoint   Right endpoint

Open-ended intervals can be handled by selecting a "distant" value for the open end. For example, if we want to

know the probability that a member of the cross-country team weighs less than 145 lb, we can use 0 as the left endpoint:

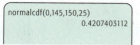

If we want to know the probability that a team member weighs more than 160 lb, we can use 300 as the right endpoint:

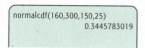

For cases involving an open-ended (infinite) bound, a rule of thumb is to set that bound at a value approximately five standard deviations below or above the mean. This will give areas that are accurate to over six decimal places.

EXERCISES

1. The annual rainfall in Crosleyville is normally distributed with mean $\mu = 150$ mm and standard deviation $\sigma = 25$ mm.

   a) What is the probability that the annual rainfall is between 125 mm and 170 mm?

   b) What is the probability that the annual rainfall is greater than 200 mm?

2. **Test scores.** All the high school students in a state take a standardized math test, and the scores are normally distributed with $\mu = 68$ and $\sigma = 5.3$.

   a) What is the probability that a randomly selected student scored between 65 and 75?

   b) What is the probability that the student scored above 70?

the **percentile** for each z-value is the area under the standard normal curve to the left of z, multiplied by 100.

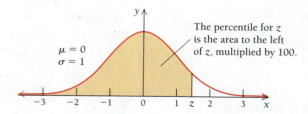

EXAMPLE 5   For the standard normal distribution, with $\mu = 0$ and $\sigma = 1$, determine the percentile corresponding to each of the following z-values.

a) $z = 0$

b) $z = 2.25$

c) $z = -1.75$

Solution

a) The percentile corresponding to $z = 0$ is the area under the curve shown at the right from $-\infty$ to 0 (to the left of 0). This is half of the total area of the standard normal distribution. Thus, a z-value of 0 corresponds to the 50th percentile: a score exactly at the mean is higher than 50% of all other scores.

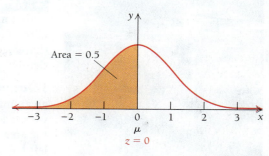

b) Table A shows that the area from 0 to 2.25 is 0.4878. We add this to 0.5, so the total area from $-\infty$ to 2.25 is 0.9878. A z-value of 2.25 corresponds to the 98.78th percentile.

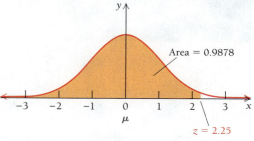

Quick Check 5 ✔

For the standard normal distribution, with $\mu = 0$ and $\sigma = 1$, determine the percentile corresponding to each of the following z-values.

a) $z = -1$
b) $z = 0.25$
c) $z = 2.8$

c) We use Table A to determine the area from 0 to 1.75, which is 0.4599. By symmetry, the area between -1.75 and 0 is also 0.4599. Since the area from $-\infty$ to 0 is 0.5, to find the area from $-\infty$ to -1.75, we subtract: $0.5 - 0.4599 = 0.0401$. Therefore, a z-value of -1.75 corresponds to the 4th percentile (rounded). **5** ✔

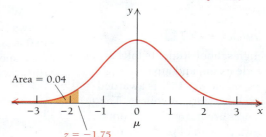

EXAMPLE 6 A large chemistry class takes an exam, and the scores are normally distributed; the mean score is $\mu = 72$, and the standard deviation is $\sigma = 4.5$. The professor gives any student scoring in the top 10% an A. What is the minimum score needed to get an A?

Solution The top 10% corresponds to the 90th percentile. We first determine the z-value that corresponds to an area of 0.9. From Table A, we see that an area of 0.4 occurs when $z = 1.28$ (with rounding). Therefore, when we include the area 0.5 for the left half of the distribution, we see that a z-value of 1.28 corresponds to the 90th percentile.

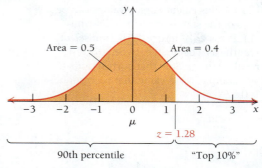

Quick Check 6 ✔

Speeds along a stretch of highway are normally distributed and have a mean of $\mu = 59$ mph with a standard deviation of $\sigma = 8$. The police will issue a speeding ticket to any driver whose speed is in the top 2% of this distribution. What is the minimum speed that will get a driver a ticket?

Next, we use the transformation formula to determine the x-value (test score) that corresponds to $z = 1.28$:

$$1.28 = \frac{x - 72}{4.5}$$

$$5.76 = x - 72 \quad \text{Multiplying both sides by 4.5}$$

$$x = 77.76.$$

Rounding gives a score of 78 as the minimum score needed to get an A. **6** ✔

EXAMPLE 7   **Business: Quality Control.**   Bottles of cola are to contain a volume with a mean of 591 mL, but some variation is expected. Any bottle at or below the 20th percentile of the volume distribution is rejected. Suppose we know that a bottle that contains 593 mL of cola is in the 60th percentile. What is the smallest volume that will be accepted? Assume that the volumes are normally distributed.

Solution   We are not given the standard deviation, but we can determine it from the given information. Since we know that a bottle with 593 mL is in the 60th percentile, we need to determine a z-value that corresponds to that percentile. Table A shows that the area from 0 to about 0.25 is 0.10, which gives 0.6 when added to the 0.5 from the left half of the distribution. Thus, $z = 0.25$ corresponds to the 60th percentile. We use the transformation formula to solve for σ, with $x = 593$ and $\mu = 591$:

$$0.25 = \frac{593 - 591}{\sigma} \qquad \text{It is important to remember that } z = \frac{x - \mu}{\sigma}$$

$$0.25 = \frac{2}{\sigma}$$

$$\sigma = \frac{2}{0.25} = 8. \qquad \text{Solving for } \sigma$$

We now determine the z-value that corresponds to the 20th percentile. Table A shows that the area from 0 to 0.84 is about 0.3, so, by symmetry, the area from -0.84 to 0 is also about 0.3. Therefore, the area to the left of -0.84 is 0.2. The 20th percentile corresponds to a z-value of -0.84. We now solve for the x-value that corresponds to this z-value:

$$-0.84 = \frac{x - 591}{8}$$

$$-6.72 = x - 591 \qquad \text{Multiplying both sides by 8}$$

$$x = 584.3. \qquad \text{Adding 591 and rounding}$$

We conclude that any bottle containing 584 mL of cola or less is rejected and any bottle containing more than 584 mL is accepted.   **7 ✔**

Quick Check 7 ✔

High school students take a state-wide exam, and anyone scoring in the top 15% is awarded a scholarship for college. The scores are normally distributed, with a mean of $\mu = 79$. Your friend scored 87, which was in the 80th percentile, and did not get a scholarship. Your score was 91. Did you get a scholarship?

TECHNOLOGY CONNECTION

Percentiles on a Calculator

On the T1-83 Plus and TI-84 Plus calculators, z-values can be determined for percentiles of the standard normal distribution ($\mu = 0$ and $\sigma = 1$). Press **2ND** and (**DISTR**) and select InvNorm. Enter the percentile as a decimal between 0 and 1, and press **ENTER**. The result is the z-value that corresponds to the given percentile. For example, the z-value that corresponds to the 70th percentile is 0.524.

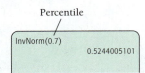

Percentile

For any mean and standard deviation, the x-value can be found directly by entering the percentile, followed by the given mean and standard deviation, separated by commas. For example, if the mean is $\mu = 90$ and the standard

deviation is $\sigma = 12$, then the x-value that corresponds to the 70th percentile is 96.293.

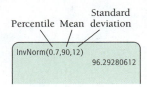

Percentile Mean Standard deviation

EXERCISES

Use InvNorm to find the x-values for the given percentiles.

1. 50th percentile, $\mu = 0, \sigma = 1$

2. 25th percentile, $\mu = 0, \sigma = 1$

3. 95th percentile, $\mu = 5, \sigma = 3$

4. 3rd percentile, $\mu = 10, \sigma = 4$

5. 58.45th percentile, $\mu = 100, \sigma = 15$

Section Summary

Assume that x is a continuous random variable and f is a probability density function for x over $[a, b]$.

- The *expected value* of x is

$$E(x) = \int_a^b x \cdot f(x)\, dx.$$

- If g is a function of x on the interval $[a, b]$, then the expected value of $g(x)$ is

$$E(g(x)) = \int_a^b g(x) \cdot f(x)\, dx.$$

- The mean, μ, is the expected value:

$$\mu = E(x) = \int_a^b x \cdot f(x)\, dx.$$

- The variance, σ^2, of x is

$$\sigma^2 = E(x^2) - \mu^2$$

$$= \int_a^b x^2 f(x)\, dx - \left[\int_a^b x \cdot f(x)\, dx \right]^2.$$

- The *standard deviation* is the square root of the variance:

$$\sigma = \sqrt{\text{variance}}.$$

It is used to describe the "spread" of the data.

- The *standard normal distribution* of x is defined by the probability density function

$$f(x) = \frac{1}{\sqrt{2\pi}} e^{-x^2/2},$$

over $(-\infty, \infty)$, where the mean $\mu = 0$ and the standard deviation $\sigma = 1$.

- The general case of a normally distributed random variable x with mean μ and standard deviation σ has the probability density function

$$f(x) = \frac{1}{\sigma\sqrt{2\pi}} e^{-(1/2)[(x-\mu)/\sigma]^2}, \quad \text{over } (-\infty, \infty).$$

- For a normal distribution, data values x are converted into z-values by the transformation formula:

$$z = \frac{x - \mu}{\sigma}.$$

- For the standard normal distribution, the *percentile* for a z-value is the area under the standard normal curve to the left of z, multiplied by 100.

5.5   Exercise Set

For each probability density function, over the given interval, find $E(x)$, $E(x^2)$, the mean, the variance, and the standard deviation.

1. $f(x) = \frac{1}{5}, \quad [3, 8]$

2. $f(x) = \frac{1}{4}, \quad [3, 7]$

3. $f(x) = \frac{1}{8}x, \quad [0, 4]$

4. $f(x) = \frac{2}{9}x, \quad [0, 3]$

5. $f(x) = \frac{2}{3}x, \quad [1, 2]$

6. $f(x) = \frac{1}{4}x, \quad [1, 3]$

7. $f(x) = \frac{3}{2}x^2, \quad [-1, 1]$

8. $f(x) = \frac{1}{3}x^2, \quad [-2, 1]$

9. $f(x) = \frac{1}{\ln 4} \cdot \frac{1}{x}, \quad [0.8, 3.2]$

10. $f(x) = \frac{1}{\ln 5} \cdot \frac{1}{x}, \quad [1.5, 7.5]$

Let x be a continuous random variable with a standard normal distribution. Using Table A, find each of the following.

11. $P(0 \le x \le 0.36)$

12. $P(0 \le x \le 2.13)$

13. $P(-1.37 \le x \le 0)$

14. $P(-2.01 \le x \le 0)$

15. $P(-2.94 \le x \le 2.00)$

16. $P(-1.89 \le x \le 0.45)$

17. $P(1.35 \le x \le 1.45)$

18. $P(0.76 \le x \le 1.45)$

19. $P(-2.45 \le x \le -1.24)$

20. $P(-1.27 \le x \le -0.58)$

21. $P(x \ge 3.01)$

22. $P(x \ge 1.01)$

23. a) $P(-2 \le x \le 2)$
    b) What percentage of the area is from -2 to 2?

24. a) $P(-1 \le x \le 1)$
    b) What percentage of the area is from -1 to 1?

Let x be a continuous random variable that is normally distributed with mean $\mu = 22$ and standard deviation $\sigma = 5$. Using Table A, find each of the following.

25. $P(22 \le x \le 27)$

26. $P(24 \le x \le 30)$

27. $P(18 \le x \le 26)$

28. $P(19 \le x \le 25)$

29. $P(20.3 \le x \le 27.5)$

30. $P(17.2 \le x \le 21.7)$

31. $P(x \ge 20)$

32. $P(x \le 22.5)$

33–54. *Use a graphing calculator to verify the solutions to Exercises 11–32.*

55. Find the *z*-value that corresponds to each percentile for a standard normal distribution.

a) 30th percentile   **b)** 50th percentile
c) 95th percentile

56. In a normal distribution with $\mu = 60$ and $\sigma = 7$, find the *x*-value that corresponds to the

a) 35th percentile   **b)** 75th percentile

57. In a normal distribution with $\mu = -15$ and $\sigma = 0.4$, find the *x*-value that corresponds to the

a) 46th percentile   **b)** 92nd percentile

58. In a normal distribution with $\mu = 0$ and $\sigma = 4$, find the *x*-value that corresponds to the

a) 50th percentile   **b)** 84th percentile

APPLICATIONS

Business and Economics

59. **Mail orders.** The number of orders, *N*, received daily by an online vendor of used Blu-ray discs is normally distributed with mean 250 and standard deviation 20. The company has to hire extra help or pay overtime on those days when the number of orders received is 300 or higher. On what percentage of days will the company have to hire extra help or pay overtime?

60. **Bread baking.** The number of loaves of bread, *N*, baked each day by Fireside Bakers is normally distributed with mean 1000 and standard deviation 50. The bakery pays bonuses to its employees on days when at least 1100 loaves are baked. On what percentage of days will the bakery have to pay a bonus?

Manufacturing. *In an automotive body-welding line, delays encountered during the process can be modeled by various probability distributions. (Source: R. R. Inman, "Empirical Evaluation of Exponential and Independence Assumptions in Queueing Models of Manufacturing Systems," Production and Operations Management, Vol. 8, 409–432 (1999).)*

61. The processing time for the robogate has a normal distribution with mean 38.6 sec and standard deviation 1.729 sec. Find the probability that the next operation of the robogate will take 40 sec or less.

62. The processing time for the automatic piercing station has a normal distribution with mean 36.2 sec and standard deviation 2.108 sec. Find the probability that the next operation of the piercing station will take between 35 and 40 sec.

63. **Test score distribution.** In 2013, combined SAT scores were normally distributed with mean 1498 and standard deviation 348. Find the combined SAT scores that correspond to these percentiles. (*Source:* www.collegeboard.com.)

a) 35th percentile
b) 60th percentile
c) 92nd percentile

General Interest

64. **Test score distribution.** The scores on a biology test are normally distributed with mean 65 and standard deviation 20. A score from 80 to 89 is a B. What is the probability of getting a B?

65. **Test score distribution.** In a large class, student test scores had a mean of $\mu = 76$ and a standard deviation $\sigma = 7$.

a) The top 12% of students got an A. Find the minimum score needed to get an A (round to the appropriate integer).
b) The top 75% of students passed. Find the minimum score needed to pass (round to the appropriate integer).

66. **Average temperature.** Las Vegas, Nevada, has an average daily high temperature of 104 degrees in July, with a standard deviation of 4.5 degrees. (*Source:* www.weatherspark.com.)

a) In what percentile is a temperature of 112 degrees?
b) What temperature would be at the 67th percentile?
c) What temperature would be in the top 0.5% of all July temperatures for this location?

67. **Heights of basketball players.** Players in the National Basketball Association have a mean height of 6 ft 7 in. (79 in.) (*Source:* www.apbr.org.) If a basketball player who is 7 ft 2 in. (86 in.) tall is in the top 1% of players by height, in what percentile is a 6 ft 11 in. (83 in.) player?

68. **Bowling scores.** At the time this book was written, the bowling scores, *S*, of author Marv Bittinger (shown below) were normally distributed with mean 201 and standard deviation 23.

a) Find the probability that one of Marv's scores is from 185 to 215.
b) Find the probability that one of his scores is from 160 to 175.
c) Find the probability that one of his scores is greater than 200.
d) Marv's best score is 299. Find the percentile that corresponds to this score, and explain what that number represents.

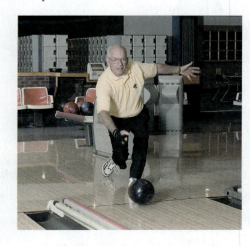

SYNTHESIS

For each probability density function, over the given interval, find E(x), E(x²), the mean, the variance, and the standard deviation.

69. $f(x) = \dfrac{1}{b-a}$, over $[a, b]$

70. $f(x) = \dfrac{3a^3}{x^4}$, over $[a, \infty)$

Median. *Let x be a continuous random variable over $[a, b]$ with probability density function f. Then the median of the x-values is that number m for which*

$$\int_a^m f(x)\, dx = \frac{1}{2}.$$

Find each median.

71. $f(x) = \frac{1}{2}x, [0, 2]$

72. $f(x) = \frac{3}{2}x^2, [-1, 1]$

73. $f(x) = ke^{-kx}, [0, \infty)$

74. Business: coffee production. Suppose the amount of coffee beans loaded into a vacuum-packed bag has a mean weight of μ ounces, which can be adjusted on the filling machine. Also, the amount dispensed is normally distributed with $\sigma = 0.2$ oz. What should μ be set at to ensure that only 1 bag in 50 will have less than 16 oz?

75. Business: does thy cup overflow? Suppose the mean amount of cappuccino, μ, dispensed by a vending machine can be set. If a cup holds 8.5 oz and the amount dispensed is normally distributed with $\sigma = 0.3$ oz, what should μ be set at to ensure that only 1 cup in 100 will overflow?

76. Explain why a normal distribution may not apply if you are analyzing the distribution of weights of students in a classroom.

77. A professor gives an easy test worth 100 points. The mean is 94, and the standard deviation is 5. Is it possible to apply a normal distribution to this situation? Why or why not?

78. Approximate the integral

$$\int_{-\infty}^{\infty} e^{-x^2}\, dx.$$

Answers to Quick Checks

1. $E(x) = \frac{4}{3}, E(x^2) = \frac{8}{3}$ **2.** $\mu = \frac{4}{3}, \sigma^2 = \frac{8}{9}$,

$\sigma = \dfrac{\sqrt{8}}{3} \approx 0.943$ **3. (a)** 0.533; **(b)** 0.221; **(c)** 0.345

4. (a) 0.726; **(b)** 0.305; **(c)** 0.159 **5. (a)** 15.9th percentile; **(b)** about the 59.9th percentile; **(c)** 99.7th percentile **6.** About 75.4 mph **7.** Yes, you were in the 89.7th percentile.

5.6

Volume

Volume by Disks

- Find the volume of a solid of revolution using disks.
- Find the volume of a solid of revolution using shells.

Consider the graph of $y = f(x)$ in Fig. 1. If the upper half-plane is rotated around the x-axis, then each point on the graph has a circular path, and the whole graph sweeps out a certain surface, called a *surface of revolution*.

The plane region bounded by the graph of $y = f(x)$, the x-axis, $x = a$, and $x = b$ sweeps out a *solid of revolution*. To calculate the volume of this solid, we first approximate it as a finite sum of thin right circular cylinders, or **disks** (Fig. 2). We divide the interval $[a, b]$ into equally sized subintervals, each of length Δx. Thus, the height h of each disk is Δx (Fig. 3). The radius of each disk is $f(x_i)$, where x_i is the right-hand endpoint of the subinterval that determines that disk.

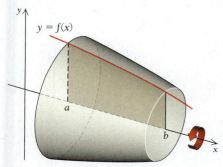

FIGURE 1

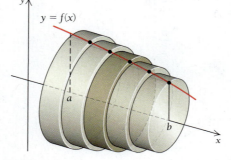

FIGURE 2

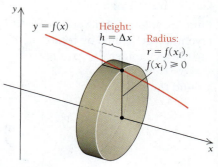

FIGURE 3

Since the volume of a right circular cylinder is given by

$$V = \pi r^2 h, \quad \text{or} \quad \text{Volume} = \text{area of the base} \cdot \text{height},$$

each of the approximating disks has volume

$$\pi[f(x_i)]^2 \, \Delta x. \qquad \text{Squaring allows for cases where } f(x) < 0$$

The volume of the solid of revolution is approximated by the sum of the volumes of all the disks:

$$V \approx \sum_{i=1}^{n} \pi[f(x_i)]^2 \, \Delta x.$$

The actual volume is the limit of this sum as the thickness of the disks, Δx, approaches zero, and the number of disks approaches infinity:

$$V = \lim_{n \to \infty} \sum_{i=1}^{n} \pi[f(x_i)]^2 \, \Delta x = \pi \int_a^b [f(x)]^2 \, dx.$$

That is, the volume is the value of the definite integral of the function $y = \pi[f(x)]^2$ from a to b.

THEOREM 2   Volume by Disks

For a continuous function f defined on $[a, b]$, the **volume, V, of the solid of revolution** obtained by rotating the area under the graph of f from a to b around the x-axis is given by

$$V = \pi \int_a^b [f(x)]^2 \, dx.$$

EXAMPLE 1   Find the volume of the solid of revolution generated by rotating the area under the graph of $y = \sqrt{x}$ from $x = 0$ to $x = 1$ around the x-axis.

Solution

$$V = \pi \int_a^b [f(x)]^2 \, dx$$

$$= \pi \int_0^1 [\sqrt{x}]^2 \, dx \qquad \text{Substituting}$$

$$= \pi \int_0^1 x \, dx$$

$$= \pi \left[\frac{x^2}{2} \right]_0^1$$

$$= \frac{\pi}{2} \left[x^2 \right]_0^1$$

$$= \frac{\pi}{2} (1^2 - 0^2) \qquad \text{Evaluating}$$

$$= \frac{\pi}{2}$$

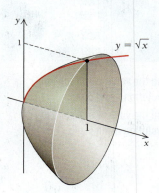

Quick Check 1 ✔

Find the volume of the solid of revolution generated by rotating the area under the graph of $y = x^3$ from $x = 0$ to $x = 2$ around the x-axis.

1 ✔

EXAMPLE 2 Find the volume of the solid of revolution generated by rotating the region under the graph of

$$y = e^x$$

from $x = -1$ to $x = 2$ around the x-axis.

Solution

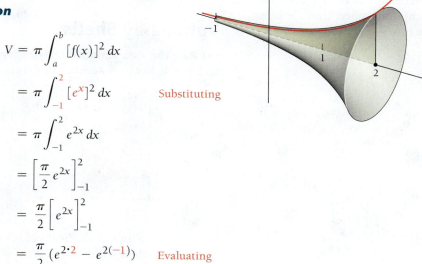

$$V = \pi \int_a^b [f(x)]^2 \, dx$$

$$= \pi \int_{-1}^2 [e^x]^2 \, dx \qquad \text{Substituting}$$

$$= \pi \int_{-1}^2 e^{2x} \, dx$$

$$= \left[\frac{\pi}{2} e^{2x} \right]_{-1}^2$$

$$= \frac{\pi}{2} \left[e^{2x} \right]_{-1}^2$$

$$= \frac{\pi}{2} (e^{2 \cdot 2} - e^{2(-1)}) \qquad \text{Evaluating}$$

$$= \frac{\pi}{2} (e^4 - e^{-2})$$

$$\approx 85.55 \qquad\qquad\qquad\qquad 2\ ✔$$

Quick Check 2 ✔

Find the volume of the solid of revolution generated by rotating the area under the graph of $y = \dfrac{1}{x}$ from $x = 1$ to $x = 3$ around the x-axis.

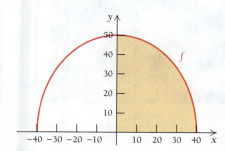

EXAMPLE 3 **Business: Water Storage.** The interior of a university's water storage tank has the shape of the solid of revolution generated by rotating the area under the graph of

$$f(x) = 50\sqrt{1 - \frac{x^2}{40^2}}$$

from $x = -40$ ft to $x = 40$ ft around the x-axis. What is the volume of this tank?

Solution The graph of f is shown at left. Rotating the graph of f about the x-axis gives the shape of the tank. The actual tank has this shape turned on its side.

If we rotate the portion of the graph, from $x = 0$ to $x = 40$, we will get half of the solid. This has the advantage of using 0 as a bound of integration.

The volume for $0 \leq x \leq 40$ is

$$V = \pi \int_0^{40} \left[50\sqrt{1 - \frac{x^2}{40^2}} \right]^2 dx$$

$$= \pi \int_0^{40} \left[2500 \left(1 - \frac{x^2}{1600} \right) \right] dx \qquad \text{Simplifying}$$

$$= \pi \int_0^{40} \left[2500 - \frac{25}{16} x^2 \right] dx \qquad \text{Multiplying by 2500}$$

$$= \pi \left[2500x - \frac{25}{48} x^3 \right]_0^{40} \qquad \text{Antidifferentiating}$$

$$= \frac{200{,}000}{3} \pi.$$

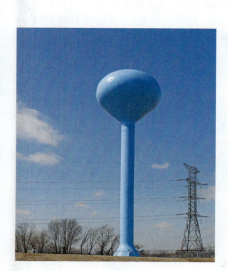

Quick Check 3 ✔

A tepee is a cone with a height of 15 ft at its center and a circular base with a radius of 8 ft. Determine the volume contained within this tepee. (*Hint:* Rotate the line $y = \frac{8}{15}x$ for $x = 0$ to $x = 15$ around the x-axis.)

Multiplying this result by 2 gives the tank's volume:

$$\frac{400,000}{3}\pi \approx 418,879 \text{ ft}^3.$$

Since 1 ft³ holds 7.48 gal, this tank holds over 3.13 million gallons of water.   **3** ✔

Volume by Shells

Let R be the area bounded by the graph of $y = f(x)$, the x-axis, and the lines $x = a$ and $x = b$. Rotate R around the y-axis, as shown below.

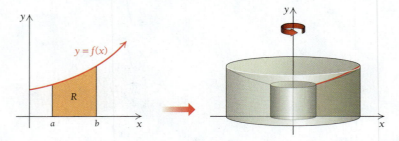

To find the volume of this solid of revolution, we divide $[a, b]$ into equally sized subintervals, each of width Δx. A rectangle, similar to those used in a Riemann sum, is drawn above each subinterval. When the area R is rotated around the y-axis, each rectangle sweeps out a **shell**, as shown below.

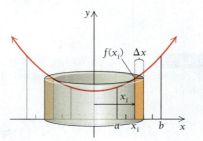

Explain how this amphitheater could be interpreted as a solid of revolution.

Let $f(x_i)$ be the height of the shell swept out by the rectangle above the ith subinterval. The radius of this shell is x_i, and its thickness is Δx. If we cut and flatten this shell, we can view it as a rectangular solid. Its width is Δx, its height is $f(x_i)$, and its length is the circumference, or $2\pi x_i$.

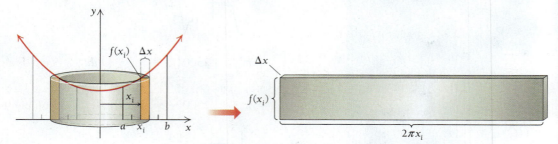

Therefore, the volume of the ith shell is

$$V_i = (\text{length})(\text{height})(\text{width})$$
$$= (2\pi x_i)(f(x_i))(\Delta x).$$

The volume of the solid of revolution is approximated by the sum of all such shells:

$$V \approx \sum_{i=1}^{n} (2\pi x_i)(f(x_i))(\Delta x).$$

As Δx approaches 0, the number of subdivisions, n, approaches ∞, and the volume of the solid of revolution is given by

$$V = \lim_{n \to \infty} \sum_{i=1}^{n} (2\pi x_i)(f(x_i))(\Delta x) = 2\pi \int_{a}^{b} x \cdot f(x) \, dx.$$

THEOREM 3   Volume by Shells

Let R be an area bounded by the graph of $y = f(x)$, the x-axis, and the lines $x = a$ and $x = b$. If R is rotated around the y-axis, the volume of the solid of revolution is given by

$$V = 2\pi \int_{a}^{b} x \cdot f(x) \, dx.$$

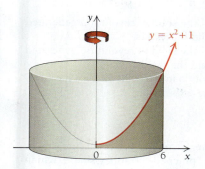

EXAMPLE 4   Find the volume of the solid of revolution formed by rotating the area under $y = x^2 + 1$ over $[0, 6]$ around the y-axis.

Solution   We have

$$V = 2\pi \int_{a}^{b} x \cdot f(x) \, dx$$

$$= 2\pi \int_{0}^{6} x(x^2 + 1) \, dx \qquad \text{Substituting}$$

$$= 2\pi \int_{0}^{6} (x^3 + x) \, dx \qquad \text{Multiplying}$$

$$= 2\pi \left[\frac{x^4}{4} + \frac{x^2}{2} \right]_{0}^{6}$$

$$= 2\pi \left[\left(\frac{6^4}{4} + \frac{6^2}{2} \right) - \left(\frac{0^4}{4} + \frac{0^2}{2} \right) \right] \qquad \text{Evaluating}$$

$$= 684\pi. \qquad\qquad\qquad\qquad\qquad\qquad\qquad 4 ✔$$

Quick Check 4 ✔

Find the volume of the solid of revolution formed by rotating the area under $y = \sqrt{x}$ over $[1, 4]$ around the y-axis.

Section Summary

- *Volume by disks:* If f is continuous over $[a, b]$, then the volume of the solid of revolution formed by rotating the area under the graph of f from a to b around the x-axis is given by

$$V = \pi \int_{a}^{b} [f(x)]^2 \, dx.$$

- *Volume by shells:* If f is continuous over $[a, b]$, then the volume of the solid of revolution formed by rotating the area under the graph of f from a to b around the y-axis is given by

$$V = 2\pi \int_{a}^{b} x \cdot f(x) \, dx.$$

5.6 Exercise Set

Find the volume generated by rotating the area bounded by the graphs of each set of equations around the x-axis.

1. $y = x, x = 0, x = 1$

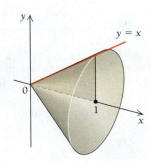

2. $y = x, x = 0, x = 2$

3. $y = 2x, x = 1, x = 3$

4. $y = \sqrt{x}, x = 1, x = 4$

5. $y = e^x, x = -2, x = 5$

6. $y = e^x, x = -3, x = 2$

7. $y = \dfrac{1}{x}, x = 1, x = 4$ **8.** $y = \dfrac{1}{x}, x = 1, x = 3$

9. $y = \dfrac{2}{\sqrt{x}}, x = 4, x = 9$ **10.** $y = \dfrac{1}{\sqrt{x}}, x = 1, x = 4$

11. $y = 5, x = 1, x = 3$ **12.** $y = 4, x = 1, x = 3$

13. $y = x^2, x = 0, x = 2$

14. $y = x + 1, x = -1, x = 2$

15. $y = 2\sqrt{x}, x = 1, x = 2$

16. $y = \sqrt{1 + x}, x = 2, x = 10$

17. $y = \sqrt{4 - x^2}, x = -2, x = 2$

18. $y = \sqrt{1 - x^2}, x = -1, x = 1$

Find the volume generated by rotating the area bounded by the graphs of each set of equations around the y-axis.

19. $y = 2x, x = 0, x = 3$ **20.** $y = 3x, x = 0, x = 5$

21. $y = x^2, x = 0, x = 3$ **22.** $y = x^3, x = 0, x = 3$

23. $y = \dfrac{1}{x}, x = 1, x = 5$ **24.** $y = \dfrac{1}{x^2}, x = 2, x = 4$

25. $y = x^2 + 3, x = 1, x = 2$

26. $y = x^3 + 1, x = 2, x = 5$

27. $y = \sqrt{x}, x = 4, x = 9$

28. $y = \sqrt[3]{x}, x = 1, x = 8$

29. $y = \sqrt{x^2 + 1}, x = 0, x = 3$

30. $y = \dfrac{1}{x^2 + 1}, x = 1, x = 7$

31. Let R be the area bounded by the graph of $y = 9 - x^2$ and the x-axis over $[0, 3]$.

   a) Find the volume of the solid of revolution generated by rotating R around the x-axis.

   b) Find the volume of the solid of revolution generated by rotating R around the y-axis.

   c) Explain why the solids in parts (a) and (b) do not have the same volume.

32. Let R be the area bounded by the graph of $y = 9 - x$ and the x-axis over $[0, 9]$.

   a) Find the volume of the solid of revolution generated by rotating R around the x-axis.

   b) Find the volume of the solid of revolution generated by rotating R around the y-axis.

   c) Explain why the solids in parts (a) and (b) have the same volume.

APPLICATIONS

33. Cooling tower volume. Cooling towers at nuclear power plants have a "pinched" chimney shape (which promotes cooling within the tower) formed by rotating a hyperbola around an axis. The function

$$y = 50\sqrt{1 + \frac{x^2}{22,500}}, \quad \text{for } -250 \le x \le 150,$$

where x and y are in feet, describes the shape of such a tower (laying on its side). Determine the volume of the tower by rotating the area bounded by the graph of y around the x-axis.

34. Volume of a football. A regulation football used in the National Football League (NFL) is 11 in. from tip to tip and 7 in. in diameter at its thickest (the regulations allow for slight variation in these dimensions). (*Source:* NFL.) The shape of a football can be modeled by the function

$$f(x) = -0.116x^2 + 3.5, \quad \text{for } -5.5 \leq x \leq 5.5,$$

where x is in inches. Find the volume of an NFL football by rotating the area bounded by the graph of f around the x-axis.

35. Volume of a hogan. A *hogan* is a circular shelter used by Native Americans in the Four Corners region of the southwestern United States. The volume of a hogan can be approximated if the graph of $y = -0.02x^2 + 12$, for $0 \leq x \leq 15$, where x and y are in feet and the x-axis represents ground level, is rotated around the y-axis. Find the volume.

36. Volume of a domed stadium. The volume of a stadium with a domed roof can be approximated if the graph of $y = -0.00025x^2 + 130$, for $0 \leq x \leq 400$, where x and y are in feet and the x-axis represents ground level, is rotated around the y-axis. Find the volume.

SYNTHESIS

37. Using volume by disks, prove that the volume of a sphere of radius r is $\frac{4}{3}\pi r^3$.

38. Using volume by shells, prove that the volume of a right circular cone of height h and radius r is $V = \frac{1}{3}\pi r^2 h$.

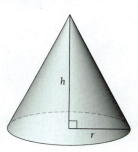

Find the volume generated by rotating about the x-axis the area bounded by the graphs of each set of equations and the x-axis.

39. $y = \sqrt{\ln x}, x = e, x = e^3$

40. $y = \sqrt{xe^{-x}}, x = 1, x = 2$

In Exercises 41 and 42, the first quadrant is the region of the xy-plane in which $x \geq 0$ and $y \geq 0$.

41. Find the volume of the solid of revolution generated by rotating the area in the first quadrant between $y = -\frac{1}{3}x^3 + 3x$ and the x-axis around the y-axis.

42. Find the volume of the solid of revolution generated by rotating the area in the first quadrant between $y = x - x^3$ and the x-axis around the y-axis.

43. Let R be the area between $y = x + 1$ and the x-axis over $[0, a]$, where $a > 0$.

 a) Find the volumes of the solids generated by rotating R, with $a = 1$, around the x-axis and around the y-axis. Which volume is larger?

 b) Find the volumes of the solids generated by rotating R, with $a = 2$, around the x-axis and around the y-axis. Which volume is larger?

 c) Find the value of a for which the two solids of revolution have the same volume.

44. Let R be the area between $y = kx$, the x-axis, and the line $x = 1$. Assume $k > 0$.

 a) Find the value of k for which the volumes of the solids formed by rotating R around the x-axis and around the y-axis are the same.

 b) Show that the answer to part (a) is true if the limits of integration are $[a, b]$, where $0 < a < b$.

45. Consider the function $y = 1/x$ over the interval $[1, \infty)$. We showed in Section 5.3 that the area under the curve is not finite; that is,

$$\int_1^\infty \frac{1}{x}\, dx$$

diverges. Find the volume of the solid of revolution formed by rotating the area under the graph of $y = 1/x$ over $[1, \infty)$ around the x-axis. That is, find

$$\int_1^\infty \pi \left[\frac{1}{x}\right]^2 dx.$$

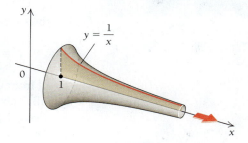

This solid is sometimes referred to as *Gabriel's horn.*

46. Paradox of Gabriel's horn or the infinite paint can.
Though we cannot prove it here, the surface area of
Gabriel's horn (see Exercise 45) is given by

$$S = \int_{1}^{\infty} \frac{2\pi}{x} \sqrt{1 + \frac{1}{x^4}}\, dx.$$

Show that the surface area of Gabriel's horn is infinite.
The paradox is that the volume of the horn is finite, but
the surface area is infinite. This is like a can of paint that
has a finite volume but, when full, does not hold enough
paint to paint the outside of the can.

Answers to Quick Checks

1. $\dfrac{128\pi}{7}$ **2.** $\dfrac{2\pi}{3}$ **3.** 320π ft^3 **4.** $\dfrac{124\pi}{5}$

Differential Equations

5.7

- Determine general and particular solutions of a differential equation.
- Verify that a function is a solution of a differential equation.
- Solve differential equations using the uninhibited growth model.
- Solve differential equations using separation of variables.
- Solve applied problems using separation of variables.

In Chapter 3, we studied an important equation:

$$\frac{d}{dt} P(t) = k \cdot P(t) \qquad \text{or} \qquad \frac{dP}{dt} = kP.$$

We saw that the solution of this equation is

$$P(t) = P_0 e^{kt}, \tag{1}$$

where P_0 is the initial quantity at time $t = 0$. Equations such as $dP/dt = kP$ are called
differential equations.

> **DEFINITION**
>
> A **differential equation** is an equation that includes a derivative and has a function as a solution.

Solving Certain Differential Equations

In differential equations, P and $P(t)$ are often used interchangeably, as are dP/dt and
$P'(t)$. We will also use the notation y' for a derivative. That is, if $y = f(x)$, then

$$y' = \frac{dy}{dx} = f'(x).$$

Certain differential equations are solved by finding the antiderivative. For example,
the differential equation

$$\frac{dy}{dx} = f(x), \qquad \text{or} \qquad y' = f(x),$$

has the solution

$$y = \int f(x)\, dx = F(x) + C, \text{ where } \frac{d}{dx} F(x) = f(x).$$

Recall from Chapter 4 that an indefinite integral results in a set of functions, so there
are infinitely many solutions. The set of all functions that solve a differential equation
is called the **general solution**.

The motion of waves can be represented by differential equations.

EXAMPLE 1   Find the general solution of $y' = 2x$.

Solution   We find the solution by integrating both sides of the equation:

$$y = \int 2x\, dx = x^2 + C. \quad \text{As a check, note that } \frac{d}{dx}(x^2 + C) = 2x.$$

1 ✔

The solution in Example 1, $y = x^2 + C$, is a general solution because substituting *all* values of C in it gives *all* of the solutions. Substituting a specific value of C gives a **particular solution** of the differential equation. For example, the following are particular solutions of $y' = 2x$:

$$y = x^2 + 3, \quad y = x^2, \quad y = x^2 - 3.$$

The graph shows the curves of these particular solutions. The general solution can be regarded as the set of all particular solutions.

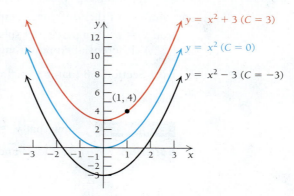

Knowing the value of the function at a specific point often allows us to find a particular solution from the general solution. For example, if we are told that the solution in Example 1 must include the ordered pair $(1, 4)$, then a particular solution is $y = x^2 + 3$, since $4 = 1^2 + 3$. The requirement that the ordered pair $(1, 4)$ must be included is called an **initial condition**.

Verifying Solutions

To verify that a function is a solution of a differential equation, we find the necessary derivatives and substitute. For example, we can verify that $P(t) = P_0 e^{kt}$ is a solution of the differential equation $dP/dt = kP$. To do so, we find the derivative of P, which is $P'(t) = P_0 k e^{kt}$. We then substitute $P_0 k e^{kt}$ for dP/dt and $P_0 e^{kt}$ for P. Since

$$P_0 k e^{kt} = k \cdot P_0 e^{kt}$$

is true, we know that $P(t) = P_0 e^{kt}$ is a solution of $dP/dt = kP$.

EXAMPLE 2   Consider the differential equation $y' = 2xy$.

a) Show that $y = e^{x^2}$ is a solution of this differential equation.
b) Show that $y = Ce^{x^2}$ is a solution, where C is a constant.

Solution   For both parts (a) and (b), we find y' and then substitute.

a) If $y = e^{x^2}$, then $y' = 2xe^{x^2}$. Substituting, we have

$$\frac{y' - 2xy = 0}{2xe^{x^2} - 2x \cdot e^{x^2} \;?\; 0}$$
$$0 \;\big|\; 0 \quad \text{TRUE}$$

Thus, $y = e^{x^2}$ is a solution of $y' - 2xy = 0$.

b) If $y = Ce^{x^2}$, then $y' = 2Cxe^{x^2}$. Substituting, we have

$$\frac{y' - 2xy = 0}{2Cxe^{x^2} - 2x \cdot Ce^{x^2} \;?\; 0}$$
$$0 \;\big|\; 0 \quad \text{TRUE}$$

Quick Check 2 ✔

Let $y' - 2y = 0$.

a) Show that $y = e^{2x}$ is a solution of this differential equation.

b) Show that $y = Ce^{2x}$ is a solution.

2 ✔

In Example 2, $y = Ce^{x^2}$ is the general solution of $y' - 2xy = 0$, meaning that all solutions of this differential equation can be written in the form $y = Ce^{x^2}$. The solution $y = e^{x^2}$ is a particular solution, where $C = 1$.

A differential equation may include second- or higher-order derivatives. Recall from Section 1.8 that if $y = f(x)$, then $y' = \dfrac{d}{dx} f(x) = f'(x), y'' = \dfrac{d}{dx} f'(x) = f''(x)$, and so on.

> **EXAMPLE 3**   Show that $y = 4e^x + 5e^{3x}$ is a solution of $y'' - 4y' + 3y = 0$.

Solution   We first find y' and y'':

$$y' = \frac{d}{dx}(4e^x + 5e^{3x}) = 4e^x + 15e^{3x};$$

$$y'' = \frac{d}{dx}y' = 4e^x + 45e^{3x}.$$

Then we substitute in the differential equation, as follows:

$$\frac{y'' - 4y' + 3y = 0}{(4e^x + 45e^{3x}) - 4(4e^x + 15e^{3x}) + 3(4e^x + 5e^{3x}) \;\big|\; 0}$$
$$4e^x + 45e^{3x} - 16e^x - 60e^{3x} + 12e^x + 15e^{3x} \;?\; 0$$
$$0 \;\big|\; 0 \quad \text{TRUE}$$

Quick Check 3 ✔

Show that $y = x^2 + x$ is a solution of $y' + \dfrac{1}{x}y = 3x + 2$.

Since a true statement, $0 = 0$, results, we know that $y = 4e^x + 5e^{3x}$ is a solution of the differential equation.

3 ✔

The Uninhibited Growth Model

Recall from Section 3.3 that $dP/dt = kP$, with $k > 0$, is called the **uninhibited growth model** and is used for situations in which growth continues forever, unaffected by outside constraints. The equation $dP/dt = kP$ is read "the rate of change of P is directly proportional to the amount P." Thus, as P increases, the rate of change dP/dt also increases. In simple terms, the larger P is, the faster P grows. The **constant of proportionality** k is also referred to as the **continuous growth rate**.

> **EXAMPLE 4**   **Business: Growth of a Savings Account.**   Donald deposits $1000 into a savings account that earns interest at a rate of 4.5%, compounded continuously. Find the particular solution of the differential equation that describes the growth in value of this account.

Solution   We let $A(t)$ represent the value of Donald's account after t years. The interest rate of 4.5% is expressed as $k = 0.045$, and at $t = 0$, we have $A_0 = \$1000$. We have

$$\frac{dA}{dt} = kA$$

$$A(t) = A_0 e^{kt} \qquad \text{Using equation (1)}$$

$$A(t) = 1000 e^{0.045t} \qquad \text{Substituting}$$

Thus, the particular solution is $A(t) = 1000 e^{0.045t}$. The graph of $A(t)$ shows the relationship between the value of A at time t and the corresponding rate of change of A. Note that as $A(t)$ increases in value, so does the rate $A'(t)$.

Quick Check 4 ✔

Find the particular solution of the differential equation $dA/dt = kA$, when $k = 0.032$ and $A_0 = \$2000$.

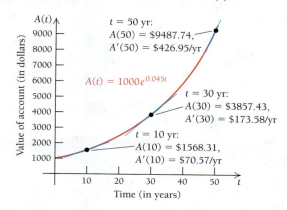

4 ✔

Separable Differential Equations

Consider the differential equation from Example 2:

$$\frac{dy}{dx} = 2xy. \tag{2}$$

We treat dy/dx as a quotient, as in Sections 2.6 and 4.5. Multiplying equation (2) by dx and then by $1/y$, we get

$$\frac{1}{y}\, dy = 2x\, dx, \quad \text{or} \quad \frac{dy}{y} = 2x\, dx, \quad y \neq 0. \tag{3}$$

We have *separated the variables*, meaning that all the expressions involving one variable are on one side of the equation and all the expressions involving the other variable are on the other, with all expressions held together by multiplication. A differential equation in which the variables can be separated is called a *separable differential equation*.

> **DEFINITION**
>
> A **separable differential equation** is a differential equation that can be written in the form
>
> $$f(y)\, dy = g(x)\, dx.$$

Equation (2) is an example of a separable differential equation, and equation (3) is equivalent to equation (2), but with the variables separated. To find a solution, we integrate both sides of equation (3):

$$\int \frac{dy}{y} = \int 2x\, dx$$

$$\ln |y| = x^2 + C_1.$$

At the 1968 Olympic Games in Mexico City, Bob Beamon made a world-record long jump of 29 ft, $2\frac{1}{2}$ in. Many believed that the jump's record length was due to the altitude, which was 7400 ft. Using differential equations for analysis, M. N. Bearley refuted the altitude theory in "The Long Jump Miracle of Mexico City" (*Mathematics Magazine*, Vol. 45, 241–246 (November 1972)). Bearley argues that the world-record jump was a result of Beamon's exceptional speed (9.5 sec in the 100-yd dash) and the fact that he left the take-off board in perfect position.

We use only one constant because the two antiderivatives differ by, at most, a constant. Recall that the definition of logarithms says that if $\log_a b = t$, then $b = a^t$. Since $\ln |y| = \log_e |y| = x^2 + C_1$, we have

$$|y| = e^{x^2+C_1}, \quad \text{or} \quad y = \pm e^{x^2} \cdot e^{C_1}.$$

Thus, the solution of equation (2) is

$$y = Ce^{x^2}, \quad \text{where } C = \pm e^{C_1}.$$

As a check, we have $y' = \dfrac{dy}{dx} = 2Cxe^{x^2}$. Substituting y' and y into equation (2), we have

$$\frac{dy}{dx} = 2xy$$

$$2Cxe^{x^2} \overset{?}{=} 2x \cdot Ce^{x^2}$$
$$2Cxe^{x^2} \;\Big|\; 2Cxe^{x^2} \quad \text{TRUE}$$

EXAMPLE 5   Find the particular solution of $3y^2 \dfrac{dy}{dx} + x = 0$, if $y = 5$ when $x = 0$.

Solution   We first separate the variables:

$$3y^2 \frac{dy}{dx} = -x \qquad \text{Adding } -x \text{ to both sides}$$

$$3y^2 \, dy = -x \, dx. \qquad \text{Multiplying both sides by } dx$$

We then integrate both sides:

$$\int 3y^2 \, dy = \int -x \, dx$$

$$y^3 = -\frac{x^2}{2} + C$$

$$y^3 = C - \frac{x^2}{2}. \tag{4}$$

We are given that $y = 5$ when $x = 0$, so we substitute to find C:

$$(5)^3 = C - \frac{(0)^2}{2} \qquad \text{Substituting}$$

$$125 = C.$$

Quick Check 5 ✔

Solve $2\sqrt{y}\,\dfrac{dy}{dx} - 3x = 0$, if $y = 9$ when $x = 0$.

Using this value for C and then taking the cube root of both sides of equation (4), we have the particular solution:

$$y = \sqrt[3]{125 - \frac{x^2}{2}}.$$

5 ✔

EXAMPLE 6   Solve: $y' = x - xy$.

Solution   Before separating variables, we replace y' with dy/dx:

$$\frac{dy}{dx} = x - xy.$$

Then we separate the variables:

$$dy = (x - xy) \, dx$$
$$dy = x(1 - y) \, dx$$
$$\frac{dy}{1 - y} = x \, dx.$$

Next, we integrate both sides:

$$\int \frac{dy}{1-y} = \int x \, dx$$

$$-\ln|1-y| = \frac{x^2}{2} + C_1$$

$$\ln|1-y| = -\frac{x^2}{2} - C_1$$

$$|1-y| = e^{-x^2/2 - C_1} \qquad \text{Writing an equivalent exponential equation}$$

$$1-y = \pm e^{-x^2/2 - C_1}$$

$$1-y = \pm e^{-x^2/2} e^{-C_1}. \qquad \text{Recalling that } a^{x+y} = a^x a^y$$

Since C_1 is an arbitrary constant, $\pm e^{-C_1}$ is also an arbitrary constant. Thus, we can replace $\pm e^{-C_1}$ with C_2. Since $\pm e^{-C_1}$ is never zero, C_2 cannot be zero.

$$1-y = C_2 e^{-x^2/2}$$

$$-y = C_2 e^{-x^2/2} - 1$$

$$y = 1 - C_2 e^{-x^2/2}. \qquad \text{Multiplying both sides by } -1$$

Quick Check 6 ✔

Solve

$$(1 + x^2)y' = xy.$$

If we replace $-C_2$ with C, we can replace the subtraction with addition. The general solution is

$$y = 1 + Ce^{-x^2/2}, \quad \text{where } C \neq 0. \qquad \qquad \text{6} ✔$$

TECHNOLOGY CONNECTION 〜

Exploratory

Solutions to Differential Equations

At www.wolframalpha.com, we can solve differential equations and graph families of solutions. When given an initial condition, we can find the particular solution. For example, to solve $y' = x - xy$ (Example 6), we key in y'=x-xy and press **ENTER**. Initial conditions are entered in the form y(a)=b and separated from the differential equation by a comma.

An Application to Economics: Elasticity

EXAMPLE 7 Suppose the elasticity of demand for a certain product (see Section 2.7) is 1 for all prices $x > 0$. That is, $E(x) = 1$ for $x > 0$. Find the demand function, $q = D(x)$.

Solution Recall that elasticity of demand is given by

$$E(x) = -\frac{xD'(x)}{D(x)}.$$

Since $E(x) = 1$ for all $x > 0$,

$$1 = -\frac{xD'(x)}{D(x)} \qquad \text{Setting } E(x) \text{ equal to 1}$$

$$1 = -\frac{x}{q} \cdot \frac{dq}{dx} \qquad \text{Letting } D(x) = q \text{ and } D'(x) = \frac{dq}{dx}$$

$$\frac{q}{x} = -\frac{dq}{dx} \qquad \text{Multiplying both sides by } \frac{q}{x}$$

$$\frac{dx}{x} = -\frac{dq}{q} \qquad \text{Separating variables}$$

$$\int \frac{dx}{x} = -\int \frac{dq}{q} \qquad \text{Integrating both sides}$$

$$\ln x = -\ln q + C_1 \qquad \text{Both the price, } x, \text{ and the quantity, } q, \text{ are assumed to be positive.}$$

$$\ln x + \ln q = C_1$$

$$\ln(xq) = C_1 \qquad \text{Using a property of logarithms}$$

$$xq = e^{C_1}. \qquad \text{Rewriting as an exponential equation}$$

Quick Check 7 ✔

Find the demand function
$q = D(x)$ if the elasticity of
demand is $E(x) = x$.

We let $C = e^{C_1}$ so that $xq = C$:

$$q = \frac{C}{x} \quad \text{and} \quad x = \frac{C}{q}.$$

This result characterizes demand functions for which the elasticity is always 1.   **7** ✔

An Application to Psychology: Reaction to a Stimulus

In psychology, one model of stimulus-response asserts that the rate of change dR/dS of the reaction R with respect to a stimulus S is inversely proportional to the intensity of the stimulus. That is,

$$\frac{dR}{dS} = \frac{k}{S},$$

where k is a positive constant. This is known as the *Weber-Fechner Law*. To solve this equation, we separate variables and then integrate both sides:

$$dR = k \cdot \frac{dS}{S}$$

$$\int dR = k \int \frac{dS}{S}$$

$$R = k \ln S + C. \qquad \text{We assume } S > 0. \tag{5}$$

Let S_0 be the lowest level of a stimulus that can be detected. This is the *threshold value*, or the *detection threshold*. For example, the detection threshold for light is the flame of a candle 30 miles away on a clear, dark night. If S_0 is the lowest level of the stimulus that can be detected, we can assume that $R(S_0) = 0$. Substituting this condition into equation (5), we have

$$0 = k \ln S_0 + C, \quad \text{or} \quad -k \ln S_0 = C.$$

Replacing C in equation (5) with $-k \ln S_0$ gives us

$$R = k \ln S - k \ln S_0 \qquad \text{As a check, note that } \frac{dR}{dS} = \frac{k}{S}.$$

$$= k(\ln S - \ln S_0)$$

$$= k \ln \frac{S}{S_0}. \qquad \text{This is the solution of the original differential equation.}$$

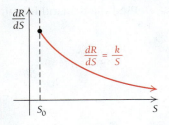

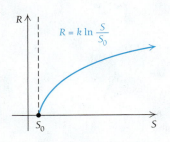

Look at the graphs of dR/dS and R. Note that as the stimulus and the response get larger, the rate of change, dR/dS, decreases. For example, suppose a lamp has a 50-watt bulb in it. If the bulb were suddenly changed to a 100-watt one, you would probably be very aware of the difference. That is, your response would be strong. If the bulb were then changed to a 150-watt one, your response would increase, but not as dramatically as it did for the change from 50 to 100 watts. A change from a 150-watt to a 200-watt bulb would cause even less increase in response, and so on.

The table lists some other detection thresholds you may find interesting.

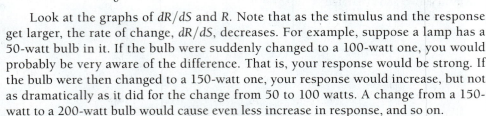

STIMULUS	DETECTION THRESHOLD
Sound	The tick of a watch from 20 feet away in a quiet room
Taste	Water diluted with sugar in the ratio of 1 teaspoon to 2 gallons
Smell	One drop of perfume diffused into the volume of three average-size rooms
Touch	The wing of a bee dropped on your cheek at a distance of 1 centimeter (about $\frac{3}{8}$ of an inch)

Section Summary

- A *differential equation* is an equation that includes a derivative and has a function as a solution.
- The *general solution* of a differential equation of the form $y' = f(x)$ is a set of functions $y = F(x) + C$, where $\dfrac{d}{dx} F(x) = f(x)$. For many differential equations, the general solution is of the form $y = Cf(x)$ where C is any constant.
- A requirement that a solution of a differential equation include a specific ordered pair is called an *initial condition*. If an initial condition is given, then a *particular solution* in which the value of C is determined may be found.

- The differential equation $dP/dt = kP$ is called the *uninhibited growth model*, and its general solution is $P(t) = P_0 e^{kt}$, where P_0 is the initial quantity at time $t = 0$. The value k is the *constant of proportionality*, also known as the *continuous growth rate*.
- A *separable differential equation* is a differential equation in which the variables can be separated to opposite sides of the equals sign, in the form $f(y)dy = g(x)\,dx$.
- To solve a separable differential equation, the variables are separated, then each side is integrated, and, if possible, the result is solved for one of the variables.

5.7 | Exercise Set

In Exercises 1–6, find the general solution and three particular solutions.

1. $y' = 10x^2$

2. $y' = 5x^6$

3. $y' = 2e^{-x} + x$

4. $y' = e^{4x} - x + 2$

5. $y' = \dfrac{4}{x} - \dfrac{1}{x^2}$

6. $y' = \dfrac{3}{x} + x^2 - x^4$

7. Show that $y = x \ln x + 3x - 2$ is a solution of

$$y'' - \frac{1}{x} = 0.$$

8. Show that $y = x \ln x - 5x + 7$ is a solution of

$$y'' - \frac{1}{x} = 0.$$

9. Show that $y = e^x + 3xe^x$ is a solution of

$$y'' - 2y' + y = 0.$$

10. Show that $y = -2e^x + xe^x$ is a solution of

$$y'' - 2y' + y = 0.$$

11. Let $y' + 4y = 0$.

a) Show that $y = e^{-4x}$ is a solution of this differential equation.

b) Show that $y = Ce^{-4x}$ is a solution, where C is a constant.

12. Let $y' - 3x^2 y = 0$.

a) Show that $y = e^{x^3}$ is a solution of this differential equation.

b) Show that $y = Ce^{x^3}$ is a solution, where C is a constant.

13. Let $y'' - y' - 30y = 0$.

a) Show that $y = e^{6x}$ is a solution of this differential equation.

b) Show that $y = e^{-5x}$ is a solution.

c) Show that $y = C_1 e^{6x} + C_2 e^{-5x}$ is a solution, where C_1 and C_2 are constants.

14. Let $y'' - 7y' - 44y = 0$.

a) Show that $y = e^{11x}$ is a solution of this differential equation.

b) Show that $y = e^{-4x}$ is a solution.

c) Show that $y = C_1 e^{11x} + C_2 e^{-4x}$ is a solution, where C_1 and C_2 are constants.

In Exercises 15–22, (a) find the general solution of each differential equation, and (b) check the solution by substituting into the differential equation.

15. $\dfrac{dM}{dt} = 0.05M$

16. $\dfrac{dC}{dt} = 0.66C$

17. $\dfrac{dR}{dt} = 0.35R$

18. $\dfrac{dV}{dt} = 1.33V$

19. $\dfrac{dG}{dt} = 0.005G$

20. $\dfrac{dh}{dt} = 0.023h$

21. $\dfrac{dR}{dt} = R$

22. $\dfrac{dQ}{dt} = 2Q$

In Exercises 23–34, (a) find the particular solution of each differential equation as determined by the initial condition, and (b) check the solution by substituting into the differential equation.

23. $y' = x^2 + 2x - 3$; $\quad y = 4$ when $x = 0$

24. $y' = 3x^2 - x + 5$; $\quad y = 6$ when $x = 0$

25. $f'(x) = x^{2/3} - x$; $\quad f(1) = -6$

26. $f'(x) = x^{2/5} + x$; $\quad f(1) = -7$

27. $\dfrac{dB}{dt} = 0.03B$, where $B(0) = 500$

28. $\dfrac{dG}{dt} = 0.75G$, where $G(0) = 2000$

29. $\dfrac{dS}{dt} = 0.12S$, where $S = 750$ when $t = 0$

30. $\dfrac{dL}{dt} = 0.68L$, where $L = 1200$ when $t = 0$

31. $\dfrac{dT}{dt} = 0.015T$, where $T = 50$ when $t = 0$

32. $\dfrac{dP}{dt} = 0.024P$, where $P = 32$ when $t = 0$

33. $\dfrac{dM}{dt} = M$, where $M = 6$ when $t = 0$

34. $\dfrac{dN}{dt} = 3N$, where $N = 3.5$ when $t = 0$

Solve by separating variables.

35. $\dfrac{dy}{dx} = 4x^3y$

36. $\dfrac{dy}{dx} = 5x^4y$

37. $3y^2 \dfrac{dy}{dx} = 8x$

38. $3y^2 \dfrac{dy}{dx} = 5x$

39. $\dfrac{dy}{dx} = \dfrac{2x}{y}$

40. $\dfrac{dy}{dx} = \dfrac{x}{2y}$

41. $\dfrac{dy}{dx} = \dfrac{6}{y}$

42. $\dfrac{dy}{dx} = \dfrac{7}{y^2}$

Solve for y.

43. $y' = 3x + xy$; $\quad y = 5$ when $x = 0$

44. $y' = 2x - xy$; $\quad y = 9$ when $x = 0$

45. $y' = 5y^{-2}$; $\quad y = 3$ when $x = 2$

46. $y' = 7y^{-2}$; $\quad y = 3$ when $x = 1$

In Exercises 47–52, (a) write a differential equation that models the situation, and (b) find the general solution. If an initial condition is given, find the particular solution. Recall that when y is directly proportional to x, we have $y = kx$, and when y is inversely proportional to x, we have $y = k/x$,

where k is the constant of proportionality. In these exercises, let $k = 1$.

47. The rate of change of y with respect to x is directly proportional to the square of y.

48. The rate of change of y with respect to x is directly proportional to the cube of y.

49. The rate of change of y with respect to x is inversely proportional to the cube of y.

50. The rate of change of y with respect to x is inversely proportional to the square root of y.

51. The rate of change of y with respect to x is directly proportional to the product of x and y, and $y = 3$ when $x = 2$.

52. The rate of change of y with respect to x is directly proportional to the quotient of x divided by y, and $y = 2$ when $x = -1$.

APPLICATIONS

Business and Economics

53. Growth of an account. Debra deposits $A_0 = \$500$ into an account that earns interest at a rate of 3.75%, compounded continuously.

  a) Write the differential equation that represents $A(t)$, the value of Debra's account after t years.
  b) Find the particular solution of the differential equation from part (a).
  c) Find $A(5)$ and $A'(5)$.
  d) Find $A'(5)/A(5)$, and explain what this number represents.

54. Growth of an account. Jennifer deposits $A_0 = \$1200$ into an account that earns 4.2% compounded continuously.

  a) Write the differential equation that represents $A(t)$, the value of Jennifer's account after t years.
  b) Find the particular solution of the differential equation from part (a).
  c) Find $A(7)$ and $A'(7)$.
  d) Find $A'(7)/A(7)$, and explain what this number represents.

55. Capital expansion. *Domar's capital expansion model is*

$$\frac{dI}{dt} = hkI,$$

where I is the investment, h is the investment productivity (constant), k is the marginal productivity to the consumer (constant), and t is the time.

  a) Use separation of variables to solve the differential equation.
  b) Rewrite the solution in terms of the condition $I_0 = I(0)$.

56. Total profit from marginal profit. Hanna's Hat Company's marginal profit, P, as a function of its total cost, C, is given by

$$\frac{dP}{dC} = \frac{-200}{(C+3)^{3/2}}.$$

a) Find the profit function, $P(C)$, if $P = \$10$ when $C = \$61$.

b) At what cost will the firm break even ($P = 0$)?

57. Stock growth. The value of a share of stock of Leslie's Designs, Inc., is modeled by

$$\frac{dV}{dt} = k(L - V),$$

where V is the value of the stock, in dollars, after t months; k is a constant; $L = \$24.81$, the *limiting value* of the stock; and $V(0) = 20$. Find the solution of the differential equation in terms of t and k.

58. Utility. The reaction R in pleasure units by a consumer receiving S units of a product can be modeled by the differential equation

$$\frac{dR}{dS} = \frac{k}{S+1},$$

where k is a positive constant.

a) Use separation of variables to solve the differential equation.

b) Rewrite the solution in terms of the initial condition $R(0) = 0$.

c) Explain why the condition $R(0) = 0$ is reasonable.

Elasticity. *Find the demand function $q = D(x)$, given each set of elasticity conditions.*

59. $E(x) = \dfrac{4}{x}$; $q = 2$ when $x = 4$

60. $E(x) = \dfrac{x}{200 - x}$; $q = 190$ when $x = 10$

61. $E(x) = 2$, for all $x > 0$

62. $E(x) = n$, for some constant n and all $x > 0$

Life and Physical Sciences

63. Population growth. The city of New River had a population of 17,000 in 2002 ($t = 0$) with a continuous growth rate of 1.75% per year.

a) Write the differential equation that represents $P(t)$, the population of New River after t years.

b) Find the particular solution of the differential equation from part (a).

c) Find $P(10)$ and $P'(10)$.

d) Find $P'(10)/P(10)$, and explain what this number represents.

64. Population growth. An initial population of 70 bacteria is growing continuously at a rate of 2.5% per hour.

a) Write the differential equation that represents $P(t)$, the population of bacteria after t hours.

b) Find the particular solution of the differential equation from part (a).

c) Find $P(24)$ and $P'(24)$.

d) Find $P'(24)/P(24)$, and explain what this number represents.

65. Population growth. Before 1859, rabbits did not exist in Australia. That year, a settler released 24 rabbits into the wild. Without natural predators, the growth of the Australian rabbit population can be modeled by the uninhibited growth model $dP/dt = kP$, where $P(t)$ is the population of rabbits t years after 1859. (*Source:* www .dpi.vic.gov.au/agriculture.)

a) When the rabbit population was estimated to be 8900, its rate of growth was about 2630 rabbits per year. Use this information to find k, and then find the particular solution of the differential equation.

b) Find the rabbit population in 1900 ($t = 41$) and the rate at which it was increasing in that year.

c) Without using a calculator, find $P'(41)/P(41)$.

66. Population growth. Suppose 30 sparrows are released into a region where they have no natural predators. The growth of the region's sparrow population can be modeled by the uninhibited growth model $dP/dt = kP$, where $P(t)$ is the population of sparrows t years after their initial release.

a) When the sparrow population is estimated at 12,500, its rate of growth is about 1325 sparrows per year. Use this information to find k, and then find the particular solution of the differential equation.

b) Find the number of sparrows after 70 yr.

c) Without using a calculator, find $P'(70)/P(70)$.

67. Exponential growth.

a) Use separation of variables to solve the differential-equation model of uninhibited growth,

$$\frac{dP}{dt} = kP.$$

b) Rewrite the solution of part (a) in terms of the condition $P_0 = P(0)$.

Social Sciences

68. The Brentano-Stevens Law. The validity of the Weber–Fechner Law has been the subject of great debate among psychologists. An alternative model,

$$\frac{dR}{dS} = k \cdot \frac{R}{S},$$

where k is a positive constant, has been proposed. Find the general solution of this equation. (This model has also been referred to as the *Power Law of Stimulus–Response*.)

SYNTHESIS

69. The amount of money, $A(t)$, in Ina's savings account after t years is modeled by the differential equation $dA/dt = 0.0418A$.

    a) What is the continuous growth rate?

    b) Find the particular solution, $A(t)$, if Ina's account is worth $3479.02 after 2 yr.

    c) Find the amount that Ina deposited initially.

70. The amount of money, $A(t)$, in John's savings account after t years is modeled by the differential equation $dA/dt = 0.0325A$.

    a) What is the continuous growth rate?

    b) Find the particular solution, $A(t)$, if John's account is worth $2582.58 after 1 yr.

    c) Find the amount that John deposited initially.

Solve.

71. $\dfrac{dy}{dx} = 5x^4y^2 + x^3y^2$

72. $e^{-1/x} \cdot \dfrac{dy}{dx} = x^{-2} \cdot y^2$

73. Explain the difference between a constant rate of growth and a constant percentage rate of growth.

74. What function is also its own derivative? Write a differential equation for which this function is a solution. Are there any other solutions to this differential equation? Why or why not?

75. Charlie deposited a sum of money in a savings account. After 1 yr, the account was worth $4467.90, and after 3 yr, the account was worth $4937.80.

    a) Use regression to find an exponential function of the form $y = ae^{bt}$ that models this situation.

    b) Write a differential equation in the form $dA/dt = kA$, including the initial condition at time $t = 0$, to model this situation.

76. Solve $dy/dx = 5/y$. Graph the particular solutions for $C_1 = 5, C_2 = -200,$ and $C_3 = 100$.

77. Solve $dy/dx = 2/y^2$. Graph the particular solutions for $C_1 = 0, C_2 = 4,$ and $C_3 = -10$.

Answers to Quick Checks

1. $y = x^3 - \frac{1}{2}x^2 + C$ **2. (a)** $y' = 2e^{2x}$, so $2e^{2x} - 2e^{2x} = 0$;   **(b)** $y' = 2Ce^{2x}$, so $2Ce^{2x} - C(2e^{2x}) = 0$ **3.** We have $y' = 2x + 1$, so

$(2x + 1) + \dfrac{1}{x}(x^2 + x) = 2x + 1 + x + 1 = 3x + 2.$

4. $A(t) = 2000e^{0.032t}$ **5.** $y = \left(\frac{9}{8}x^2 + 27\right)^{2/3}$

6. $y = C\sqrt{1 + x^2}$ **7.** $q = Ce^{-x}$

CHAPTER 5   Summary

KEY TERMS AND CONCEPTS

SECTION 5.1

If $p = D(x)$ is the demand function for a product, then the **consumer surplus** for Q units, at a price per unit P, is given by

$$\int_0^Q D(x)\, dx - QP.$$

If $p = S(x)$ is the supply function for a product, then the **producer surplus** for Q units, at a price per unit P, is given by

$$QP - \int_0^Q S(x)\, dx.$$

The **equilibrium point** (x_E, p_E) is the point at which the supply and demand curves intersect.

EXAMPLES

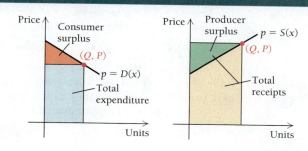

Let $p = D(x) = 12 - 1.5x$ be a demand function and $p = S(x) = 4 + 0.5x$ be a supply function. The two curves intersect at $(4, 6)$, the equilibrium point. At this point, the consumer surplus is

$$\int_0^4 (12 - 1.5x)\, dx - (4)(6) = 36 - 24 = \$12,$$

and the producer surplus is

$$(4)(6) - \int_0^4 (4 + 0.5x)\, dx = 24 - 20 = \$4.$$

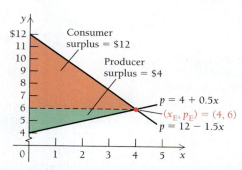

SECTION 5.2

The **future value** of P_0 dollars invested at interest rate k for t years, compounded continuously, is given by $P = P_0 e^{kt}$.

The future value of $6000 invested at 3.75%, compounded continuously, for 5 yr is

$$P = 6000 e^{0.0375(5)} = \$7237.38.$$

(continued)

KEY TERMS AND CONCEPTS	EXAMPLES

SECTION 5.2 (continued)

The amount P_0 is called the **present value**. If the future value P is known, then $P_0 = Pe^{-kt}$.

Sue wants to have \$15,000 in 4 yr to make a down payment on a house. She opens a savings account that offers 4.5% interest, compounded continuously. The present value is the amount she needs to deposit now to have \$15,000 in 4 yr:

$$P = 15{,}000e^{-0.045(4)} = \$12{,}529.05$$

The **accumulated future value of a continuous income stream** is given by

$$A = e^{kT}\int_0^T R(t)e^{-kt}\,dt,$$

where $R(t)$ is the rate of the continuous income stream, k is the interest rate, and T is the number of years. If $R(t)$ is a constant function, then

$$A = \frac{R(t)}{k}\left(e^{kT} - 1\right).$$

A baseball pitcher signs a 6-yr contract and will be paid \$1,150,000 per year. The money will be invested at 5%, compounded continuously, for the 6-yr term. The accumulated future value is

$$A = \frac{1{,}150{,}000}{0.05}\left(e^{0.05(6)} - 1\right) = \$8{,}046{,}752.57.$$

The **accumulated present value of a continuous income stream** is given by

$$B = \int_0^T R(t)e^{-kt}\,dt.$$

If $R(t)$ is a constant, then

$$B = \frac{R(t)}{k}\left(1 - e^{-kT}\right).$$

If $R(t)$ is not constant, the definite integral must be solved by an appropriate integration technique.

The accumulated present value of the pitcher's contract is

$$B = \int_0^6 1{,}150{,}000e^{-0.05t}\,dt$$

$$= \frac{1{,}150{,}000}{0.05}\left(1 - e^{-0.05(6)}\right) = \$5{,}961{,}180.92.$$

Consumption of a natural resource can be modeled by

$$\int_0^T P_0 e^{kt}\,dt = \frac{P_0}{k}\left(e^{kT} - 1\right),$$

where $P(t) = P_0 e^{kt}$ is the annual consumption of the natural resource in year t and consumption is growing exponentially at growth rate k.

Canada's diamond mines produce diamonds according to the model $P(t) = 2.5e^{0.272t}$, where t is the number of years since 2000, and $P(t)$ is in millions of carats. (*Source*: USGS *Mineral Commodities Summaries*.) Using this model, we can forecast the total production of diamonds between 2000 and 2016:

$$\int_0^{16} 2.5e^{0.272t}\,dt = \frac{2.5}{0.272}\left(e^{0.272(16)} - 1\right) = 704.353 \text{ million carats.}$$

KEY TERMS AND CONCEPTS   **EXAMPLES**

SECTION 5.3

An integral with infinity as a bound is called an **improper integral**. All improper integrals are evaluated as limits:

$$\int_a^\infty f(x)\,dx = \lim_{b \to \infty} \int_a^b f(x)\,dx,$$

$$\int_{-\infty}^b f(x)\,dx = \lim_{a \to -\infty} \int_a^b f(x)\,dx,$$

$$\int_{-\infty}^\infty f(x)\,dx = \int_{-\infty}^c f(x)\,dx + \int_c^\infty f(x)\,dx.$$

If the limit exists, the improper integral is **convergent**. Otherwise, it is **divergent**.

$$\int_1^\infty \frac{1}{x^3}\,dx = \lim_{b \to \infty} \int_1^b \frac{1}{x^3}\,dx$$

$$= \lim_{b \to \infty} \left[-\frac{1}{2x^2} \right]_1^b$$

$$= \lim_{b \to \infty} \left[-\frac{1}{2(b)^2} - \left(-\frac{1}{2(1)^2} \right) \right]$$

$$= 0 - \left(-\frac{1}{2} \right) = \frac{1}{2}$$

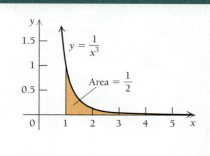

$$\int_{-\infty}^0 e^{4x}\,dx = \lim_{a \to -\infty} \int_a^0 e^{4x}\,dx$$

$$= \lim_{a \to -\infty} \left[\frac{1}{4} e^{4x} \right]_a^0$$

$$= \lim_{a \to -\infty} \left[\frac{1}{4}\left(e^{4(0)} - e^{4(a)} \right) \right]$$

$$= \frac{1}{4}(1 - 0) = \frac{1}{4}$$

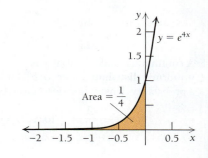

The **accumulated present value** of a continuous money flow into an investment at the rate of P dollars per year perpetually is

$$\int_0^\infty Pe^{-kt}\,dt = \frac{P}{k},$$

where k is the continuously compounded interest rate.

An investment of $5000 per year perpetually at 7%, compounded continuously, has a present value of $\dfrac{5000}{0.07} = 71{,}428.57$.

SECTION 5.4

In probability, a **continuous random variable** is a quantity that can be observed (or measured) repeatedly and whose possible values comprise an interval of real numbers.

A function f is a **probability density function** for a continuous random variable x if it meets the following conditions:

- For all x in its domain, $f(x) \geq 0$.
- The area under the graph of f is 1.
- For any subinterval $[c, d]$ in the domain of f, the probability that x will be in that subinterval is $P([c, d]) = \int_c^d f(x)\,dx$.

The function $f(x) = \frac{2}{9}x$, for $0 \leq x \leq 3$, is a probability density function since

- $f(x) \geq 0$ for all x in $[0, 3]$.

- $\displaystyle\int_0^3 \frac{2}{9}x\,dx = \left[\frac{1}{9}x^2 \right]_0^3 = \frac{3^2}{9} - 0 = 1.$

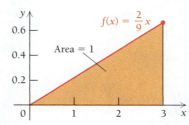

(continued)

KEY TERMS AND CONCEPTS	EXAMPLES

SECTION 5.4 (continued)

A probability density function is always stated with its domain.

The probability that x is between 1.5 and 2.3 is

$$\int_{1.5}^{2.3} \frac{2}{9} x \, dx = \left[\frac{1}{9} x^2 \right]_{1.5}^{2.3} = \frac{(2.3)^2}{9} - \frac{(1.5)^2}{9} \approx 0.338.$$

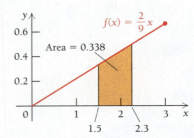

A continuous random variable is **uniformly distributed** over $[a, b]$ if it has a probability density function f given by

$$f(x) = \frac{1}{b - a}, \quad \text{for } a \le x \le b.$$

Helicopter tours over Hoover Dam last from 45 to 55 min, with the times uniformly distributed. If x is the time a tour lasts, in minutes, the probability density function f is given by

$$f(x) = \frac{1}{10}, \quad \text{for } 45 \le x \le 55.$$

The probability that a flight lasts between 48 and 53.5 min is

$$\int_{48}^{53.5} \frac{1}{10} \, dx = \left[\frac{1}{10} x \right]_{48}^{53.5} = \frac{1}{10} (53.5 - 48) = 0.55.$$

A continuous random variable is **exponentially distributed** if it has a probability density function f of the form

$$f(x) = ke^{-kx}, \quad \text{over } [0, \infty).$$

The time x (in minutes) between shoppers entering a store is modeled by the probability density function

$$f(x) = 3e^{-3x}, \quad \text{for } 0 \le x < \infty.$$

The probability that the time between shoppers is 30 seconds or less is

$$\int_{0}^{0.5} 3e^{-3x} \, dx = [-e^{-3x}]_{0}^{0.5} = (-e^{-1.5} - (-1)) \approx 0.78.$$

SECTION 5.5

If x is a continuous random variable over $[a, b]$ with probability density function f:

the **mean** (μ) is the expected value of x,

$$\mu = E(x) = \int_{a}^{b} x \cdot f(x) \, dx;$$

the **variance** (σ^2) of x is

$$\sigma^2 = E(x^2) - \mu^2$$

$$= \int_{a}^{b} x^2 \cdot f(x) \, dx - \left[\int_{a}^{b} x \cdot f(x) \, dx \right]^2;$$

and the **standard deviation** (σ) is the square root of the variance,

$$\sigma = \sqrt{\text{variance}}.$$

Consider the probability density function $f(x) = \frac{2}{9}x$ over $[0, 3]$.

• Its mean is

$$\mu = \int_{0}^{3} x \cdot \frac{2}{9} x \, dx = \int_{0}^{3} \frac{2}{9} x^2 \, dx = 2.$$

• Its variance is

$$\sigma^2 = \left[\int_{0}^{3} x^2 \cdot \frac{2}{9} x \, dx \right] - \mu^2 = 4.5 - 4 = 0.5.$$

• Its standard deviation is

$$\sigma = \sqrt{0.5} \approx 0.71.$$

KEY TERMS AND CONCEPTS	**EXAMPLES**

SECTION 5.5 (*continued*)

A continuous random variable x has a **standard normal distribution** if it has a probability density function f given by

$$f(x) = \frac{1}{\sqrt{2\pi}} e^{-x^2/2}, \quad \text{over } (-\infty, \infty),$$

with $\mu = 0$ and $\sigma = 1$.

Tables or calculators are used to determine areas within the standard normal distribution.

To convert an x-value to a z-value for use with the standard normal distribution, we use the transformation formula

$$z = \frac{x - \mu}{\sigma}.$$

Weights of packages of gourmet coffee are normally distributed with mean $\mu = 3$ oz and standard deviation $\sigma = 0.5$. The probability a packet of coffee has a weight between 2.75 oz and 3.15 oz is

$$P\left(\frac{2.75 - 3}{0.5} \le x \le \frac{3.15 - 3}{0.5}\right) = P(-0.5 \le z \le 0.3)$$

$$= 0.309 = 30.9\%.$$

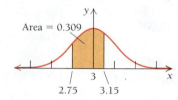

For the standard normal distribution, the *percentile* corresponding to a z-value is the area under the standard normal curve to the left of z, multiplied by 100.

In the standard normal distribution, the area to the left of $z = 1.5$ is 0.933. Therefore, a z-value of 1.5 corresponds to the 93.3rd percentile.

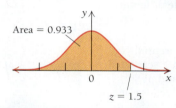

SECTION 5.6

Volume by disks: If f is continuous over $[a, b]$, the **volume of the solid of revolution** generated by rotating the area under the graph of f from a to b around the x-axis is given by

$$V = \pi \int_a^b [f(x)]^2 \, dx.$$

Volume by shells: If f is continuous over $[a, b]$, the volume of the solid of revolution generated by rotating the area under the graph of f from a to b around the y-axis is given by

$$V = 2\pi \int_a^b x f(x) \, dx.$$

The volume of the solid formed by rotating the area under the graph of $f(x) = \frac{1}{4}x^2$ from $x = -1$ to $x = 3$ around the x-axis is

$$V = \pi \int_{-1}^{3} \left(\frac{1}{4}x^2\right)^2 dx = \pi \int_{-1}^{3} \left(\frac{1}{16}x^4\right) dx = \frac{\pi}{16}\left[\frac{x^5}{5}\right]_{-1}^{3}$$

$$= \frac{\pi}{16}\left[\frac{243}{5} - \left(-\frac{1}{5}\right)\right] = \frac{244}{80}\pi = \frac{61}{20}\pi \approx 9.582.$$

The volume of the solid of revolution formed by rotating the area under the graph of $f(x) = -\frac{1}{2}x^2 + 9$ from $x = 1$ to $x = 4$ around the y-axis is

$$V = 2\pi \int_1^4 x\left(-\frac{1}{2}x^2 + 9\right) dx$$

$$= 2\pi \int_1^4 \left(-\frac{1}{2}x^3 + 9x\right) dx$$

$$= 2\pi \left[-\frac{1}{8}x^4 + \frac{9}{2}x^2\right]_1^4$$

$$= 2\pi \left[\left(-\frac{1}{8}(4)^4 + \frac{9}{2}(4)^2\right) - \left(-\frac{1}{8}(1)^4 + \frac{9}{2}(1)^2\right)\right]$$

$$= 2\pi \left[(-32 + 72) - \left(-\frac{1}{8} + \frac{9}{2}\right)\right]$$

$$= \frac{285}{4}\pi \approx 223.838.$$

(*continued*)

KEY TERMS AND CONCEPTS	EXAMPLES

SECTION 5.7

A **differential equation** is an equation that includes a derivative and has a function as a solution.

The equation $y' + 3y = 3x^2 + 2x$ is a differential equation since it involves the derivative y. The function $y = x^2$ is a solution of this differential equation since $y' = 2x$ and

$$y' + 3y = 3x^2 + 2x$$
$$2x + 3x^2 = 3x^2 + 2x.$$

The **general solution** of a differential equation of the form $y' = f(x)$ is a set of functions $F(x) + C$, where $\frac{d}{dx}F(x) = f(x)$. For many differential equations, the general solution is of the form $y = Cf(x)$ where C is any constant.

The general solution of $y' = 4x^2$ is

$$y = \int 4x^2\,dx = \frac{4}{3}x^3 + C.$$

The general solution of $y' = 7y$ is

$$y = Ce^{7x}.$$

A **separable differential equation** is a differential equation for which the variables can be separated to opposite sides. That is, a separable differential equation in x and y can be written in the form $f(y)dy = g(x)dx$.

To solve a separable differential equation, the variables are separated, each side is integrated, and, if possible, the result is solved for one of the variables.

The differential equation $y' = \dfrac{x^2}{y}$ can be solved by separating variables:

$$\frac{dy}{dx} = \frac{x^2}{y}$$
$$y\,dy = x^2\,dx$$
$$\int y\,dy = \int x^2\,dx$$
$$\frac{y^2}{2} = \frac{x^3}{3} + C_1$$
$$y^2 = \frac{2}{3}x^3 + C, \quad \text{where } C = 2C_1.$$

Therefore, the general solution is $y = \pm\sqrt{\frac{2}{3}x^3 + C}$.

If an initial condition is known, then a *particular solution* may be determined by solving for C.

If $(0, 2)$ is an initial condition, we can solve the general solution for C. We choose the positive root since the output, $y = 2$, is positive:

$$2 = \sqrt{\frac{2}{3}(0)^2 + C}$$
$$2 = \sqrt{C}$$
$$4 = C.$$

Therefore, the particular solution that passes through the point $(0, 2)$ is $y = \sqrt{\frac{2}{3}x^3 + 4}$.

CHAPTER 5
Review Exercises

These review exercises are for test preparation. They can also be used as a practice test. Answers are at the back of the book. The red bracketed section references tell you what part(s) of the chapter to restudy if your answer is incorrect.

CONCEPT REINFORCEMENT

Match each term in column A with the most appropriate graph in column B.

Column A   **Column B**

1. Consumer surplus   a)
   [5.1]

2. Producer surplus   b)
   [5.1]

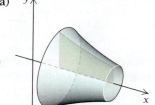

$$f(x) = \frac{1}{\sqrt{2\pi}} e^{-x^2/2}$$

3. Exponential distribution   c)
   [5.4]

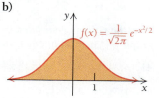

$\frac{1}{b-a}$ a b x

4. Standard normal distribution   d) Price
   [5.5]

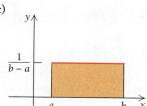

Supply, p_E, Demand, x_E, Units

5. Uniform distribution   e) Price
   [5.4]

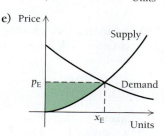

Supply, p_E, Demand, x_E, Units

6. Solid of revolution   f)
   [5.6]

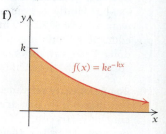

$f(x) = ke^{-kx}$

Classify each statement as either true or false.

7. The consumer surplus is the extra amount a consumer pays as a tax on a purchased product. [5.1]

8. The accumulated present value of an investment is the value of the investment as a tax-deductible gift to a nonprofit charity. [5.2]

9. If an integral has $-\infty$ or ∞ as one of the limits of integration, it is an improper integral. [5.3]

10. If f is a probability density function over $[a, b]$, then $f(x) \geq 0$ for all x in $[a, b]$. [5.4]

11. If f is a probability density function over $[a, b]$ and x is a continuous random variable over $[a, b]$, then the mean value of $f(x)$ is always $(b - a)/2$. [5.5]

12. If $y = f(x)$ is a solution of $y' = x^2 + 3x$, then $y = f(x) + C$ is also a solution. [5.7]

REVIEW EXERCISES

Let $D(x) = (x - 6)^2$ be the price, in dollars per unit, that consumers are willing to pay for x units of an item, and $S(x) = x^2 + 12$ be the price, in dollars per unit, that producers are willing to accept for x units.

13. Find the equilibrium point. [5.1]

14. Find the consumer surplus at the equilibrium point. [5.1]

15. Find the producer surplus at the equilibrium point. [5.1]

16. **Business: future value.** Find the future value of $5000, at an annual percentage rate of 3.2%, compounded continuously, for 7 yr. [5.2]

17. **Business: present value.** Find the present value of $10,000 due in 5 yr, at an interest rate of 4.3%, compounded continuously. [5.2]

18. **Business: future value of a continuous income stream.** Find the accumulated future value of $2500 per year, at 4.25% compounded continuously, for 8 yr. [5.2]

19. **Business: present accumulated value of a trust.** The DeMars family welcomes a new baby, and the parents want to have $250,000 in 18 yr for the child's college education. Find the continuous money stream, at $R(t)$ dollars per year, they need to invest at 4.75% compounded continuously, to generate $250,000. [5.2]

20. **Business: early retirement.** Cal Earl signs a 7-yr contract as a session drummer for a major recording company. His contract gives him a salary of $150,000 per year. After 3 yr, the company offers to buy out the remainder of his contract. What is the least amount Cal should accept, if the interest rate is 4.15%, compounded continuously? [5.2]

21. **Physical science: iron ore consumption.** In 2013 ($t = 0$), world production of iron ore was estimated at 2.95 billion metric tons, and production was growing exponentially at the rate of 0.7% per year. (*Source:* minerals.usgs.gov.) If production continues to grow at this rate, how much iron ore will be produced from 2015 to 2025? [5.2]

22. **Physical science: depletion of iron ore.** World reserves of iron ore in 2013 were estimated to be 81 billion metric tons. (*Source:* minerals.usgs.gov.) Assuming that the growth rate in Exercise 21 continues and no new reserves are discovered, when will world reserves of iron ore be depleted? [5.2]

Determine whether each improper integral is convergent or divergent, and find its value if it is convergent. [5.3]

23. $\int_1^\infty \frac{1}{x^2}\, dx$

24. $\int_1^\infty e^{4x}\, dx$

25. $\int_0^\infty e^{-2x}\, dx$

26. Find k such that $f(x) = k/x^3$ is a probability density function over $[1, 2]$. Then write the probability density function. [5.4]

27. **Business: waiting time.** Sharif arrives at a random time at a doctor's office where the waiting time t to see a doctor is no more than 25 min. The probability density function for t is $f(t) = \frac{1}{25}$, for $0 \le t \le 25$. Find the probability that Sharif will have to wait no more than 15 min to see a doctor. [5.4]

Given the probability density function

$$f(x) = 6x(1 - x), \quad \text{over } [0, 1],$$

find each of the following. [5.5]

28. $E(x^2)$

29. $E(x)$

30. The mean

31. The variance

32. The standard deviation

Let x be a continuous random variable with a standard normal distribution. Using Table A, find each of the following. [5.5]

33. $P(0 \le x \le 1.85)$

34. $P(-1.74 \le x \le 1.43)$

35. $P(-2.08 \le x \le -1.18)$

36. $P(x \ge 0)$

37. **Business: pizza sales.** The number of pizzas sold daily at Benito's Pizzeria is normally distributed with mean $\mu = 90$ and standard deviation $\sigma = 20$. What is the probability that at least 100 pizzas are sold during a day? [5.5]

38. **Business: distribution of revenue.** Benito's Pizzeria has daily mean revenues that are normally distributed, with $\mu = \$5500$ and $\sigma = \$425$. What is the lowest amount in the top 5% of daily revenues? [5.5]

Find the volume generated by rotating the area bounded by the graphs of each set of equations around the x-axis. [5.6]

39. $y = x^3, x = 1, x = 2$

40. $y = \frac{1}{x + 2}, x = 0, x = 1$

Find the volume generated by rotating the area bounded by the graphs of each set of equations around the y-axis.

41. $y = 12 - x^2, x = 0, x = 2$

42. $y = e^{x^2}, x = 1, x = 3$

Solve each differential equation. [5.7]

43. $\frac{dy}{dx} = 11x^{10}y$

44. $\frac{dy}{dx} = \frac{2}{y}$

45. $\frac{dy}{dx} = 4y; \quad y = 5 \text{ when } x = 0$

46. $\frac{dv}{dt} = 5v^{-2}; \quad v = 4 \text{ when } t = 3$

47. $y' = \frac{3x}{y}$

48. $y' = 8x - xy$

49. **Economics: elasticity.** Find the demand function $q = D(x)$, given the elasticity condition

$$E(x) = \frac{x}{100 - x}; \quad q = 70 \text{ when } x = 30. [5.7]$$

50. **Business: stock growth.** The growth rate of the stock of Greenwich Corp., in dollars per month, can be modeled by

$$\frac{dV}{dt} = k(L - V),$$

where V is the value of a share, in dollars, after t months; k is a constant; $L = \$36.37$, the limiting value of the stock; and $V(0) = 30$. Express the solution of the differential equation in terms of t and k. [5.7]

SYNTHESIS

51. The function $f(x) = \frac{1}{3}x^2$ is a probability density function over $[0, c]$. Find c. [5.4]

Determine whether each improper integral is convergent or divergent, and find its value if it is convergent. [5.3]

52. $\displaystyle\int_{-\infty}^{0} x^4 e^{-x^5} \, dx$

53. $\displaystyle\int_{0}^{\infty} \frac{dx}{(x+1)^{4/3}}$

54. $\displaystyle\int_{1}^{\infty} \frac{\ln x}{x^2} \, dx.$ [5.3]

55. $\displaystyle\int_{-\infty}^{\infty} \frac{2}{1 + 4x^2} \, dx.$ [5.3]

CHAPTER 5
Test

Let $D(x) = (x - 7)^2$ be the price, in dollars per unit, that consumers are willing to pay for x units of an item, and let $S(x) = x^2 + x + 4$ be the price, in dollars per unit, that producers are willing to accept for x units. Find:

1. The equilibrium point

2. The consumer surplus at the equilibrium point

3. The producer surplus at the equilibrium point

4. Business: future value. Find the future value of $12,000 invested for 10 yr at an annual percentage rate of 3.75%, compounded continuously.

5. Business: future value of a continuous income stream. Find the accumulated future value of $8000 per year, at an interest rate of 3.78%, compounded continuously, for 6 yr.

6. Physical science: demand for potash. In 2011 ($t = 0$), world production of potash was approximately 37 million metric tons, and demand was increasing at the rate of 9.8% a year. (*Source: U.S. Energy Information Administration.*) If the demand continues to grow at this rate, how much potash will be produced from 2016 to 2030?

7. Physical science: depletion of potash. See Exercise 6. The world reserves of potash in 2011 were approximately 9500 million metric tons. (*Source: U.S. Geological Survey.*) Assuming the demand for potash continues to grow at the rate of 9.8% per year and no new reserves are discovered, when will world reserves be depleted?

8. Business: accumulated present value of a continuous income stream. Bruce Kent wants to have $25,000 in 5 yr for a down payment on a house. Find the amount he needs to save, at $R(t)$ dollars per year, at 4.125%, compounded continuously, to achieve the desired future value.

9. Business: contract buyout. Guy Laplace signs a 6-yr contract to play professional hockey at a salary of $475,000 per year. After 2 yr, his team offers to buy out the remainder of his contract. What is the least amount Guy should accept, if the going interest rate is 5.1%, compounded continuously?

10. Business: future value of a noncontinuous income stream. Sonia signs a contract that will pay her an income of $R(t) = 100,000 + 10,000t$, where t is in years and $0 \le t \le 8$. If she invests this money at 5%, compounded continuously, what is the future value of the income stream?

Determine whether each improper integral is convergent or divergent, and find its value if it is convergent.

11. $\displaystyle\int_{1}^{\infty} \frac{dx}{x^5}$

12. $\displaystyle\int_{0}^{\infty} \frac{4}{1 + 3x} \, dx$

13. Find k such that $f(x) = kx^3$ is a probability density function over $[0, 2]$. Then write the probability density function.

14. Business: times of telephone calls. A telephone company determines that the length of a phone call, t, in minutes, is an exponentially distributed random variable with probability density function

$$f(t) = 2e^{-2t}, \quad 0 \le t < \infty.$$

Find the probability that a phone call will last between 1 min and 2 min.

Given the probability density function $f(x) = \frac{1}{4}x$ over $[1, 3]$, find each of the following.

15. $E(x)$

16. $E(x^2)$

17. The mean

18. The variance

19. The standard deviation

Let x be a continuous random variable with a standard normal distribution. Using Table A, find each of the following.

20. $P(0 \le x \le 1.3)$

21. $P(-2.31 \le x \le -1.05)$

22. $P(-1.61 \le x \le 1.76)$

23. Business: price distribution. The price per pound p of wild salmon at various stores in a certain city is normally distributed with mean $\mu = \$16$ and standard deviation $\sigma = \$2.50$. What is the probability that the price at a randomly selected store is at least $\$17.25$ per pound?

24. Business: price distribution. If the price per pound p of wild salmon is normally distributed with mean $\mu = \$12$ and standard deviation $\sigma = \$2.50$, what is the lowest price in the top 15% of salmon prices?

Find the volume generated by rotating the area bounded by each set of equations around the x-axis.

25. $y = \dfrac{1}{\sqrt{x}}, \quad x = 1, x = 5$

26. $y = \sqrt{2 + x}, \quad x = 0, x = 1$

27. Find the volume generated by rotating the area bounded by $y = x^2 + x$, $x = 1$, $x = 4$, and the x-axis, around the y-axis.

28. Business: grain storage. A grain silo is a circular cylinder with a conical roof. The roof can be modeled by rotating the area under the graph of $y = 30 - 0.5x$, where $0 \le x \le 10$, around the y-axis. Find the volume of the silo, assuming the floor of the cylinder lies on the x-axis and both x and y are measured in feet.

Solve each differential equation.

29. $\dfrac{dy}{dx} = 8x^7 y$

30. $\dfrac{dy}{dx} = \dfrac{9}{y}$

31. $\dfrac{dy}{dt} = 6y; \quad y = 11$ when $t = 0$

32. $y' = 5x^2 - x^2 y$

33. $\dfrac{dv}{dt} = 2v^{-3}$

34. $y' = 4y + xy$

35. Economics: elasticity. Find the demand function $q = D(x)$, given the elasticity condition

$$E(x) = 4 \quad \text{for all } x > 0.$$

36. Business: stock growth. The growth rate of Fabric Industries stock, in dollars per month, can be modeled by

$$\frac{dV}{dt} = k(L - V),$$

where V is the value of a share, in dollars, after t months; $L = \$36$, the limiting value of the stock; k is a constant; and $V(0) = 0$.

a) Express the solution $V(t)$ in terms of L and k.

b) If $V(6) = 18$, determine k to the nearest hundredth.

c) Rewrite $V(t)$ in terms of t and k using the value of k found in part (b).

d) Use the equation in part (c) to find $V(12)$, the value of the stock after 12 months.

e) In how many months will the value be $\$30$?

SYNTHESIS

37. If $f(x) = x^3$ is a probability density function over $[0, b]$, what is b?

38. Determine whether the following improper integral is convergent or divergent, and find its value if it is convergent:

$$\int_{-\infty}^{0} x^3 e^{-x^4} \, dx.$$

TECHNOLOGY CONNECTION

39. Approximate the integral to three decimal places.

$$\int_{-\infty}^{\infty} \frac{2}{1 + 3x^2} \, dx.$$

Curve Fitting and Volumes of Containers

Consider the urn or vase shown at the right. How could we estimate volume? One way would be to fill the container with a liquid and then pour the liquid into a measuring device.

Another way, using calculus and a graphing calculator, would be to turn the urn on its side, as shown below, take a series of vertical measurements of the radius of the urn, use REGRESSION, generate a solid of revolution by rotating the regression curve around the *x*-axis, and use integration to find the volume.

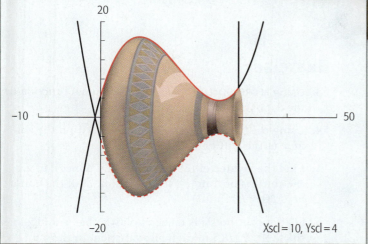

The following table is a table of values for the red curve.

x (in centimeters)	y (in centimeters)
0	4
2	10
7	17
12	16
17	10
22	5
25	3.5
31	7

Exercises

1. Using REGRESSION, fit a cubic polynomial function to the data.

2. Using the function found in Exercise 1, find the volume of the urn over $[0, 31]$. (*Hint:* If the function in Exercise 1 is Y1, find the volume by using the **VARS** key to enter πY1^2 as Y2. Then use the CALC option to integrate.)

Now consider the bottle shown at the right. To find the bottle's volume in a similar manner, we turn it on its side, use a measuring device to take vertical measurements of the radii, and proceed as we did with the urn.

The table of measurements is as follows.

x (in inches)	y (in inches)
0	1.125
1	1.275
2	1.250
3	1.275
4	1.275
5	1.125
6	1.000
7	0.875
8	0.750
9	0.500
10	0.500

Exercises

3. Using REGRESSION, fit a quartic polynomial function to the data. Label this function y_1.

4. Using the function found in Exercise 3, find the volume of the bottle in cubic inches.

5. Find the volume of the bottle in fluid ounces, using the conversion $1 \text{ in}^3 = 0.55424$ fl oz. Will the bottle hold 20 oz of fluid?

6. Suppose the bottle is made of plastic that is 0.015-in. thick. Add 0.015 to each measurement in the table, and find a regression curve for these adjusted radii. Fit a quartic polynomial function to the data, and label this curve y_2.

7. To determine the volume of the bottle's base, extend the x-axis 0.015 unit to the left of 0. The total volume of the plastic that forms the bottle can then be found by first integrating $y_3 = \pi(y_2)^2$ over $[-0.015, 0]$ to find the volume of the base and then integrating $y_4 = \pi(y_2 - y_1)^2$ over $[0, 10]$ to find the volume of the side. What is the total volume of plastic used to form the bottle?

6 | Functions of Several Variables

What You'll Learn

6.1 Functions of Several Variables
6.2 Partial Derivatives
6.3 Maximum–Minimum Problems
6.4 An Application: The Least-Squares Technique
6.5 Constrained Optimization
6.6 Double Integrals

Why It's Important

Functions with more than one input variable are called *functions of several variables*. We introduce these functions in this chapter and learn to differentiate them to find *partial derivatives*. Then we use such functions and their partial derivatives to find regression lines and solve maximum–minimum problems. Finally, we consider the integration of functions of several variables and some applications.

Where It's Used

Monthly Car Payments: Ashley wants to buy a 2014 Nissan Leaf hybrid and finance $20,000 of the cost through a loan. If the lender offers an APR of 5.0% for 7 yr, what will Ashley's monthly payment be? (*This problem appears as Exercise 19 in Section 6.1.*)

Monthly Payment per $1000 Borrowed

ANNUAL PERCENTAGE RATE, r	TERM, t (in years)				
	4	5	6	7	8
0.05	$23.03	$18.87	$16.10	$14.13	$12.66
0.055	$23.26	$19.10	$16.34	$14.37	$12.90
0.06	$23.49	$19.33	$16.57	$14.61	$13.14
0.065	$23.71	$19.57	$16.81	$14.85	$13.39
0.07	$23.95	$19.80	$17.05	$15.09	$13.63
0.075	$24.18	$20.04	$17.29	$15.34	$13.88

Functions of Several Variables

- Find a function value for a function of several variables.
- Solve applied problems involving functions of several variables.

Suppose Sunny Press, a small publisher of children's books, sells x copies of a book at a profit of \$4 per book. Then its total profit P is given by

$$P(x) = 4x.$$

This is a function of one variable.

Suppose Sunny Press sells x copies of one book at a profit of \$4 per book and y copies of a second book at a profit of \$6 per book. Then its total profit P is a function of the *two* variables x and y, and is given by

$$P(x, y) = 4x + 6y.$$

This function assigns to each input pair (x, y) a unique output number, $4x + 6y$.

DEFINITION

A **function of two variables**, f, assigns to each input pair, (x, y), exactly one output number, $f(x, y)$.

We can regard a function of two variables as a machine that has two inputs. Its domain can be viewed as a set of pairs (x, y) in the plane. When such a function is given by a formula, the domain normally consists of all ordered pairs (x, y) that are meaningful replacements in the formula.

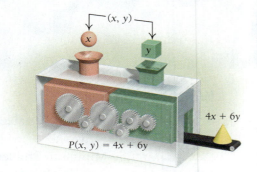

EXAMPLE 1   For the above profit function, $P(x, y) = 4x + 6y$, find $P(25, 10)$.

Solution $P(25, 10)$ is found by substituting 25 for x and 10 for y:

$$P(25, 10) = 4 \cdot 25 + 6 \cdot 10 \qquad \text{Substituting}$$
$$= 100 + 60$$
$$= \$160.$$

This result means that by selling 25 copies of the first book and 10 copies of the second, Sunny Press will make a profit of \$160.   **1** ✔

Quick Check 1 ✔

Bellview Crafts' cost function for x mugs and y vases is given by

$$C(x, y) = 6.5x + 7.25y.$$

Find $C(10, 15)$, and explain what this number represents.

The following are examples of **functions of several variables**, that is, functions of two or more variables. If there are n variables, then there are n inputs for such a function.

EXAMPLE 2   **Business: Monthly Payment on an Amortized Loan.**   Large purchases are often financed with an amortized loan (see Section 3.6). Borrowers like to know how much they can expect to pay per month for every thousand dollars borrowed. The monthly payment P depends on the annual percentage rate (APR), r, and the term of the loan, t (in years). The function P of the two variables r and t is given by

$$P(r, t) = \frac{1000r\left(1 + \dfrac{r}{12}\right)^{12t}}{12\left(1 + \dfrac{r}{12}\right)^{12t} - 12}.$$

How much per month can a borrower expect to pay per thousand dollars borrowed at an APR of 4.5% for a 6-yr term?

Solution   We let $r = 0.045$ and $t = 6$ and evaluate $P(0.045, 6)$:

$$P(0.045, 6) = \frac{1000(0.045)\left(1 + \dfrac{0.045}{12}\right)^{12(6)}}{12\left(1 + \dfrac{0.045}{12}\right)^{12(6)} - 12} = \$15.87.$$

The monthly payment is $15.87 per thousand dollars borrowed.   **2** ✔

Quick Check 2 ✔

Determine the monthly payment per thousand dollars borrowed at an APR of 7.25% for a term of 8 yr.

EXAMPLE 3   **Business: Payment Tables.**   The formula in Example 2 is used to generate a table of payments that shows borrowers the combined effects of r, the APR, and the term t (in years).

Monthly Payment per $1000 Borrowed

ANNUAL PERCENTAGE RATE, r	TERM, t (in years)				
	4	5	6	7	8
0.05	$23.03	$18.87	$16.10	$14.13	$12.66
0.055	$23.26	$19.10	$16.34	$14.37	$12.90
0.06	$23.49	$19.33	$16.57	$14.61	$13.14
0.065	$23.71	$19.57	$16.81	$14.85	$13.39
0.07	$23.95	$19.80	$17.05	$15.09	$13.63
0.075	$24.18	$20.04	$17.29	$15.34	$13.88

a) Suppose Sherry can borrow $3000 at an APR of 5.5% for a 7-yr term. Find her monthly payment.

b) Sherry is also considering borrowing the $3000 from another lender at an APR of 6.5% for a 5-yr term. Find her monthly payment.

Solution   The monthly payment per $1000 borrowed is read directly from the table above.

Quick Check 3 ✔

a) What is the monthly payment per thousand dollars borrowed at an APR of 7% for a term of 5 yr?

b) How much less per month would the payment be with the same APR but a term of 6 yr?

a) For every $1000 borrowed, we see that $P(0.055, 7) = \$14.37$. Thus, Sherry's monthly payment in this case will be $(\$14.37)(3) = \$43.11$.

b) From the table, we have $P(0.065, 5) = \$19.57$ per $1000 borrowed. Thus, Sherry's monthly payment will be $(\$19.57)(3) = \$58.71$.

   If Sherry chooses option (a), she will pay $(\$43.11)(84) = \$3621.24$ to pay off the loan. If she chooses option (b), she will pay $(\$58.71)(60) = \$3522.60$. Although the monthly payments will be higher with option (b), the total payment is lower, mostly because of the shorter term.   **3** ✔

EXAMPLE 4   **Business: Total Cost.**   The total cost to Bradshaw Steel, in hundreds of dollars, for producing a spool of rolled steel is given by

$$C(x, y, z, w) = 4x^2 + 5y + z - \ln(w + 1),$$

where x dollars are spent for labor, y dollars for raw materials, z dollars for advertising, and w dollars for machinery. This is a function of four variables (all in thousands of dollars). Find $C(3, 2, 1, 10)$.

Solution   We substitute 3 for x, 2 for y, 1 for z, and 10 for w:

$$C(3, 2, 0, 10) = 4 \cdot 3^2 + 5 \cdot 2 + 1 - \ln(10 + 1)$$
$$= 4 \cdot 9 + 10 + 1 - 2.397895$$
$$\approx \$44.6 \text{ hundred, or } \$4460.$$

EXAMPLE 5 **Business: Cost of Storage Equipment.** Terrell Petroleum purchases a storage tank that costs C_1 dollars and has capacity V_1. Later it wishes to replace the tank with a new one that costs C_2 dollars and has capacity V_2. Industrial economists have found that in such cases, the cost of the new piece of equipment can be estimated by the function of three variables

$$C_2 = \left(\frac{V_2}{V_1}\right)^{0.6} C_1.$$

Suppose that, for $45,000, Terrell Petroleum buys a 10,000-gal tank. Later it decides to buy a replacement tank with double that capacity. Estimate the cost of the new tank.

Quick Check 4 ✔

a) Repeat Example 5 assuming that the company buys a tank with a capacity of 2.75 times that of the original.

b) What is the percentage increase in cost for this tank compared to the cost of the original tank?

Solution We substitute 20,000 for V_2, 10,000 for V_1, and 45,000 for C_1:

$$C_2 = \left(\frac{20,000}{10,000}\right)^{0.6} (45,000)$$
$$= 2^{0.6}(45,000)$$
$$\approx \$68,207.25.$$

Note that a 100% increase in capacity was achieved by about a 52% increase in cost. This is independent of any increase in the costs of labor, management, or other equipment resulting from the purchase of the tank.

4 ✔

Quick Check 5 ✔

The cost, in dollars per copy, of printing Q copies of a book that contains P photographs and has R pages is given by

$$C(Q, P, R) = \frac{4.57QP}{R^{1.88}}.$$

Find the cost, in dollars per copy, of printing 500 copies of a book that contains 25 photographs and has 175 pages.

EXAMPLE 6 **Social Science: The Gravity Model.** As the populations of two cities grow, the number of telephone calls between the cities increases, much like the gravitational pull increases between two growing objects in space. The average number of telephone calls per day between two cities is given by

$$N(d, P_1, P_2) = \frac{2.8P_1P_2}{d^{2.4}},$$

where d is the distance, in miles, between the cities and P_1 and P_2 are their populations. The cities of Dallas and Fort Worth are 30 mi apart, and their populations are, respectively, 1,241,000 and 778,000. (*Sources:* Population Division, U.S. Census Bureau, 2013 estimates, and Rand McNally.) Find the average number of calls per day between the two cities.

Solution We evaluate the function with the aid of a calculator:

$$N(30, 1,241,000, 778,000) = \frac{2.8(1,241,000)(778,000)}{30^{2.4}}$$
$$\approx 770,581,000. \qquad \text{Rounding to the nearest thousand}$$

5 ✔

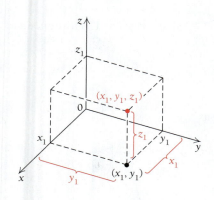

Geometric Interpretations

Visually, a function of two variables, $z = f(x, y)$, can be thought of as matching a point (x_1, y_1) in the xy-plane with the number z_1 on a number line. Thus, to graph a function of two variables, we need a three-dimensional coordinate system. The axes are generally placed as shown to the left. The line z, called the z-axis, is perpendicular to the xy-plane at the origin.

To help visualize this, think of looking into the corner of a room, where the floor is the xy-plane and the z-axis is the intersection of the two walls. To plot a point (x_1, y_1, z_1), we locate the point (x_1, y_1) in the xy-plane and move up or down in space according to the value of z_1.

EXAMPLE 7   Plot these points:

$$P_1(2, 3, 5),$$
$$P_2(2, -2, -4),$$
$$P_3(0, 5, 2),$$

and $$P_4(2, 3, 0).$$

Solution   The solution is shown at the left.

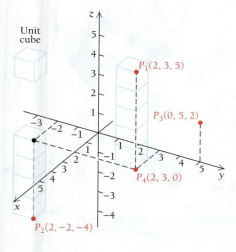

Unit cube

The *graph* of a function of two variables, $z = f(x, y)$, consists of ordered triples (x_1, y_1, z_1), where $z_1 = f(x_1, y_1)$. This graph takes the form of a **surface**. The **domain** of such a function is the set of all points in the xy-plane for which f is defined.

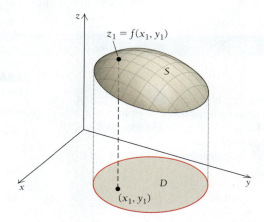

EXAMPLE 8   Find the domain of each two-variable function.

a) $f(x, y) = x^2 + y^2$

b) $g(x, y) = \sqrt{1 - x^2 - y^2}$

c) $h(x, y) = x^2 + y^2 + \dfrac{1}{x^2 + y^2}$

Solution

a) Since we can square any real number and add any two squares, f is defined for all x and all y. Therefore, the domain of f is

$$D = \{(x, y) \mid -\infty < x < \infty, \quad -\infty < y < \infty\}.$$

The graph of f is a surface called an *elliptic paraboloid*. A satellite dish is an example of an elliptic paraboloid: the weak incoming signals bounce off the interior surface of the paraboloid and collect at a single point, called the *focus*, thus amplifying the signal.

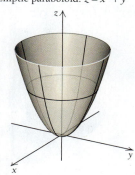

Elliptic paraboloid: $z = x^2 + y^2$

b) For $g(x, y)$ to exist, we must have $1 - x^2 - y^2 \geq 0$, or $x^2 + y^2 \leq 1$. The domain of g is

$$D = \{(x, y) \,|\, x^2 + y^2 \leq 1\}.$$

The graph of g is a surface called a *hemisphere,* of radius 1. Its domain is a filled-in circle of radius 1 on the xy-plane. We can think of the domain of g as the "shadow" it casts on the xy-plane.

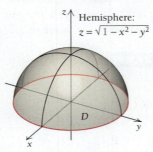

Hemisphere:
$z = \sqrt{1 - x^2 - y^2}$

D

Quick Check 6 ✔

Find the domain of each two-variable function.

a) $f(x, y) = \dfrac{x + y}{x - y}$

b) $g(x, y) = \dfrac{1}{x - 2} + \dfrac{2}{3 + y}$

c) $h(x, y) = \ln (y - x^3)$

c) Since zero cannot be in the denominator, we must have $x^2 + y^2 \neq 0$. Therefore, x and y cannot be 0 simultaneously. The domain of h is

$$D = \{(x, y) \,|\, (x, y) \neq (0, 0)\}.$$

The graph of h is shown at right.   **6** ✔

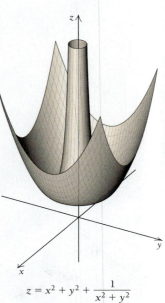

$z = x^2 + y^2 + \dfrac{1}{x^2 + y^2}$

TECHNOLOGY CONNECTION

Exploratory

A useful and inexpensive app is Quick Graph, a graphing calculator that creates visually appealing 3D graphs of functions of two variables. It has full graphing interactivity, with touch-based zoom and scroll features.

Some functions and their graphs are presented here.

EXAMPLE 1 Graph:
$(1 - \sqrt{x^2 + y^2})^2 + z^2 = 0.2$.
This is entered as follows:

(1-sqrt(x^2+y^2))^2+z^2=0.2

The graph is shown at the right.

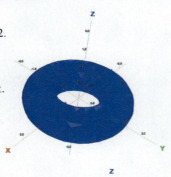

EXAMPLE 2 Graph:
$|(2x^2 + 2y^2)^{0.25}| + \sqrt{|z|} = 1$.
This is entered as follows:

abs((2x^2+2y^2)^0.25)+(abs(z))^0.5=1

The graph is shown at the right.

EXAMPLE 3 Graph: $z = e^{-4(x^2+y^2)}$.
This is entered as follows:

z=e^(-4(x^2+y^2))

The graph is shown at the right.

EXAMPLE 4 Graph: $(xy)^2 + (yz)^2 + (zx)^2 = xyz$.
This is entered as follows:

(xy)^2+(yz)^2+(xz)^2=xyz

The graph is shown at the right.

(continued)

EXAMPLE 5   Graph: $4x^2 + 2y^2 + z^2 = 1$.
This is entered as follows:

4x^2+2y^2+z^2=1

The graph is shown at the right.

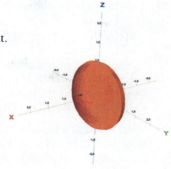

EXAMPLE 6   Graph: $z = -8xe^{-4(x^2+y^2)}$.
This is entered as follows:

z=-8xe^(-4(x^2+y^2))

The graph is shown at
the right.

EXERCISE

Use Quick Graph or another 3D graphing utility to graph the
functions in Exercises 1–12 below.

Section Summary

- A *function of two variables* assigns to each input pair, (x, y), exactly one output, $f(x, y)$.
- A function of two variables generates points (x, y, z), where $z = f(x, y)$.

- The graph of a function of two variables is a *surface* and requires a three-dimensional coordinate system.
- The *domain* of a function of two variables is the set of points in the xy-plane for which the function is defined.

6.1 | Exercise Set

1. For $f(x, y) = x^2 - 3xy$, find $f(0, -2)$, $f(2, 3)$, and $f(10, -5)$.

2. For $f(x, y) = (y^2 + 2xy)^3$, find $f(-2, 0)$, $f(3, 2)$, and $f(-5, 10)$.

3. For $f(x, y) = 3^x + 7xy$, find $f(0, -2)$, $f(-2, 1)$, and $f(2, 1)$.

4. For $f(x, y) = \log_{10}(x + y) + 3x^2$, find $f(3, 7)$, $f(1, 99)$, and $f(2, -1)$.

5. For $f(x, y) = \ln x + y^3$, find $f(e, 2)$, $f(e^2, 4)$, and $f(e^3, 5)$.

6. For $f(x, y) = 2^x - 3^y$, find $f(0, 2)$, $f(3, 1)$, and $f(2, 3)$.

7. For $f(x, y, z) = x^2 - y^2 + z^2$, find $f(-1, 2, 3)$ and $f(2, -1, 3)$.

8. For $f(x, y, z) = 2^x + 5zy - x$, find $f(0, 1, -3)$ and $f(1, 0, -3)$.

In Exercises 9–14, determine the domain of each function of two variables.

9. $f(x, y) = x^2 + 4x + y^2$ 10. $g(x, y) = \ln(x^2 - y)$

11. $f(x, y) = \sqrt{y - 3x}$ 12. $h(x, y) = xe^{\sqrt{y}}$

13. $g(x, y) = \dfrac{1}{y + x^2}$ 14. $k(x, y) = \dfrac{1}{x} + \dfrac{y}{x - 1}$

APPLICATIONS

Business and Economics

15. **Price–earnings ratio.** The *price–earnings ratio* of a stock is given by

$$R(P, E) = \frac{P}{E},$$

where P is the price of the stock and E is the earnings per share. For the quarter ending in January 2014, the price per share of Hewlett-Packard stock was $29.00, and the earnings per share were $0.74. (*Source:* yahoo.finance.com.) Find the price–earnings ratio. Use decimal notation rounded to the nearest hundredth.

16. **Yield.** The *yield* of a stock is given by

$$Y(D, P) = \frac{D}{P},$$

where D is the dividend per share of stock and P is the price per share. On April 1, 2014, the price per share of Texas Instruments stock was $47.53, and the dividend per share was $1.20. (*Source:* yahoo.finance.com.) Find the yield. Use percent notation rounded to the nearest hundredth of a percent.

17. Cost of storage equipment. Consider the cost model in Example 5. For $100,000, Tonopah Storage buys a storage tank that has a capacity of 80,000 gal. Later it replaces the tank with a new one that has triple that capacity. Estimate the cost of the new tank.

18. Savings and interest. A sum of $1000 is deposited in a savings account for which interest is compounded monthly. The future value A is a function of the annual percentage rate r and the term t, in months, and is given by

$$A(r, t) = 1000\left(1 + \frac{r}{12}\right)^{12t}.$$

a) Determine $A(0.05, 10)$.
b) What is the interest earned for the rate and term in part (a)?
c) How much more interest can be earned over the same term as in part (a) if the APR is increased to 5.75%?

19. Monthly car payments. Ashley wants to buy a 2014 Nissan Leaf hybrid and finance $20,000 of the cost through a loan. Use the table in Example 3 to answer the following questions.

a) SouthBank will lend Ashley $20,000 at an APR of 5.0% for 7 yr. What will her monthly payment be?
b) Find Ashley's total payments, assuming that she pays the amount found in part (a) every month for the full term of the loan.
c) Valley Credit Union offers a 5-yr term on a loan of $20,000. What is the highest APR that Ashley can accept if she wants to pay less overall than what she would with the loan from SouthBank?
d) If she accepts Valley Credit Union's offer of a 5-yr term with an APR of 6.5% and makes every monthly payment, how much less overall will she pay?

20. Monthly car payments. Kim is shopping for a car. She will finance $10,000 through a lender. Use the table in Example 3 to answer the following questions.

a) One lender offers Kim an APR of 6% for a 6-yr term. What would Kim's monthly payment be?
b) A competing lender offers an APR of 5.5% but for a 7-yr term. What would Kim's monthly payment be?
c) Assume that Kim makes the minimum payment each month for the entire term of the loan. Calculate her total payments for both options described in parts (a) and (b). Which option costs Kim less overall?

Life and Physical Sciences

21. Poiseuille's Law. The speed of blood in a vessel is given by

$$V(L, p, R, r, v) = \frac{p}{4Lv}(R^2 - r^2),$$

where R is the radius of the vessel, r is the distance of the blood from the center of the vessel, L is the length of the blood vessel, p is the blood pressure, and v is the viscosity of the blood. Find $V(1, 100, 0.0075, 0.0025, 0.05)$.

22. Wind speed of a tornado. Under certain conditions, the *wind speed S*, in miles per hour, of a tornado at a distance d feet from its center can be approximated by the function

$$S(a, d, V) = \frac{aV}{0.51d^2},$$

where a is a constant that depends on certain atmospheric conditions and V is the approximate volume of the tornado, in cubic feet. Approximate the wind speed 100 ft from the center of a tornado when its volume is 1,600,000 ft^3 and $a = 0.78$.

23. Body surface area. The Mosteller formula for approximating the surface area S, in square meters (m^2), of a human is given by

$$S(h, w) = \frac{\sqrt{hw}}{60},$$

where h is the person's height in centimeters and w is the person's weight in kilograms. (*Source:* www.halls.md.) Use the Mosteller approximation to estimate the surface area of a person whose height is 165 cm and whose weight is 80 kg.

24. Body surface area. The Haycock formula for approximating the surface area S, in square meters (m^2), of a human is given by

$$S(h, w) = 0.024265h^{0.3964}w^{0.5378},$$

where h is the person's height in centimeters and w is the person's weight in kilograms. (*Source:* www.halls.md.) Use the Haycock approximation to estimate the surface area of a person whose height is 165 cm and whose weight is 80 kg.

General Interest

25. Goals against average. A hockey goalie's goals against average A is a function of the number of goals allowed g and the number of minutes played m and is given by

$$A(g, m) = \frac{60g}{m}.$$

a) Find the goals against average of a goalie who allows 35 goals while playing 820 min. Round A to the nearest hundredth.

b) A goalie gave up 124 goals during the season and had a goals against average of 3.75. How many minutes did he play? (Round to the nearest integer.)

c) State the domain for A.

26. Dewpoint. The *dewpoint* is the temperature at which moisture in the air condenses into liquid (dew). It is a function of air temperature t and relative humidity h. The table below shows the dewpoints for select values of t and h.

AIR TEMPERATURE, t (°F)	RELATIVE HUMIDITY (%)				
	20	40	60	80	100
70	29	44	55	63	70
80	35	53	65	73	80
90	43	62	74	83	90
100	52	71	84	93	100

a) What is the dewpoint when the air temperature is 80°F with a relative humidity of 60%?

b) What is the dewpoint when the air temperature is 90°F with a relative humidity of 40%?

c) The air feels humid when the dewpoint reaches about 60. If the air temperature is 100°F, at what approximate relative humidity will the air feel humid?

d) Explain why the dewpoint is equal to the air temperature when the relative humidity is 100%.

SYNTHESIS

27. For the tornado described in Exercise 22, if the wind speed measures 200 mph, how far from the center was the measurement taken?

28. According to the Mosteller formula in Exercise 23, if a person's weight drops 19%, by what percentage does his or her surface area change?

29. Explain the difference between a function of two variables and a function of one variable.

30. Find some examples of functions of several variables not considered in the text, even some that may not have formulas.

General Interest

Wind chill temperature. *Because wind speed enhances the loss of heat from the skin, we feel colder when there is wind than when there is not. The* wind chill temperature *is what the temperature would have to be with no wind in order to give the same chilling effect. The wind chill temperature, W, is given by*

$$W(v, T) = 91.4 - \frac{(10.45 + 6.68\sqrt{v} - 0.447v)(457 - 5T)}{110},$$

where T is the temperature measured by a thermometer, in degrees Fahrenheit, and v is the speed of the wind, in miles per hour. Find the wind chill temperature in each case. Round to the nearest degree.

31. $T = 30°F, v = 25$ mph

32. $T = 20°F, v = 20$ mph

33. $T = 20°F, v = 40$ mph

34. $T = -10°F, v = 30$ mph

35. Use a computer graphics program such as *Maple* or *Mathematica*, an Internet site such as www.wolframalpha.com, a graphing calculator, or an app such as Quick Graph to view the graph of each function given in Exercises 1–8.

Use a 3D graphics program to generate the graph of each function.

36. $f(x, y) = y^2$

37. $f(x, y) = x^2 + y^2$

38. $f(x, y) = (x^4 - 16x^2)e^{-y^2}$

39. $f(x, y) = 4(x^2 + y^2) - (x^2 + y^2)^2$

40. $f(x, y) = x^3 - 3xy^2$

41. $f(x, y) = \dfrac{1}{x^2 + 4y^2}$

- Find the partial derivatives of a given function.
- Evaluate partial derivatives.
- Find the four second-order partial derivatives of a function in two variables.

Partial Derivatives

Finding Partial Derivatives

Consider the function f given by

$$z = f(x, y) = x^2y^3 + xy + 4y^2.$$

Suppose we fix y at 3. Then

$$f(x, 3) = x^2(3^3) + x(3) + 4(3^2) = 27x^2 + 3x + 36.$$

Note that we now have a function of only one variable. Taking the first derivative with respect to x, we have

$$54x + 3.$$

In general, without replacing y with a specific number, we can consider y fixed. Then f becomes a function of x alone, and we can calculate its derivative with respect to x. This is called the *partial derivative of f with respect to x*, denoted by

$$\frac{\partial f}{\partial x} \quad \text{or} \quad \frac{\partial z}{\partial x}.$$

Now, let's again consider the function

$$z = f(x, y) = x^2y^3 + xy + 4y^2.$$

The color blue indicates the variable x when we fix y and treat it as a constant. The expressions y^3, y, and y^2 are then also treated as constants. We have

$$\frac{\partial f}{\partial x} = \frac{\partial}{\partial x}(x^2 y^3 + xy + 4y^2)$$
$$= 2xy^3 + (1)y + 0$$
$$= 2xy^3 + y.$$

Similarly, we find $\partial f/\partial y$ or $\partial z/\partial y$ by fixing x (treating it as a constant) and calculating the derivative with respect to y. From

$$z = f(x, y) = x^2y^3 + xy + 4y^2, \qquad \text{The color blue indicates the variable.}$$

we get

$$\frac{\partial f}{\partial y} = \frac{\partial}{\partial y}(x^2 y^3 + xy + 4y^2)$$
$$= x^2(3y^2) + x(1) + 8y$$
$$= 3x^2y^2 + x + 8y.$$

A definition of partial derivatives is as follows.

> **DEFINITION**
>
> For $z = f(x, y)$, the **partial derivatives with respect to x and y** are
>
> $$\frac{\partial z}{\partial x} = \lim_{h \to 0} \frac{f(x + h, y) - f(x, y)}{h} \quad \text{and} \quad \frac{\partial z}{\partial y} = \lim_{h \to 0} \frac{f(x, y + h) - f(x, y)}{h}.$$

We can find partial derivatives of functions of any number of variables. Since the earlier theorems for finding derivatives apply, we rarely need to use the definition to find a partial derivative.

EXAMPLE 1 For $w = x^2 - xy + y^2 + 2yz + 2z^2 + z$, find

$$\frac{\partial w}{\partial x}, \quad \frac{\partial w}{\partial y}, \quad \text{and} \quad \frac{\partial w}{\partial z}.$$

Solution In order to find $\partial w/\partial x$, we regard x as the variable and treat y and z as constants. From

$$w = x^2 - xy + y^2 + 2yz + 2z^2 + z,$$

we get

$$\frac{\partial w}{\partial x} = 2x - y.$$

To find $\partial w/\partial y$, we regard y as the variable and treat x and z as constants. We get

$$\frac{\partial w}{\partial y} = -x + 2y + 2z.$$

Quick Check 1 ✔

For $u = x^2y^3z^4$, find

$$\frac{\partial u}{\partial x}, \frac{\partial u}{\partial y}, \text{ and } \frac{\partial u}{\partial z}.$$

To find $\partial w/\partial z$, we regard z as the variable and treat x and y as constants. We get

$$\frac{\partial w}{\partial z} = 2y + 4z + 1.$$

1 ✔

We often write $f_x(x, y)$ or f_x for the partial derivative of f with respect to x and $f_y(x, y)$ or f_y for the partial derivative of f with respect to y. Similarly, if $z = f(x, y)$, then z_x represents the partial derivative of z with respect to x, and z_y represents the partial derivative of z with respect to y.

EXAMPLE 2 For $f(x, y) = 3x^2y + xy^2$, find $f_x(x, y)$ and $f_y(x, y)$.

Quick Check 2 ✔

For $f(x, y) = 7x^3y^2 - \dfrac{x}{y}$, find $f_x(x, y)$ and $f_y(x, y)$.

Solution We have

$$\begin{aligned} f_x(x, y) &= 6xy + y^2, &&\text{\color{red}Treating } y \text{ and } y^2 \text{ as constants} \\ f_y(x, y) &= 3x^2 + 2xy. &&\text{\color{red}Treating } x^2 \text{ and } x \text{ as constants} \end{aligned}$$

2 ✔

For the function in Example 2, let's evaluate f_x at $(2, -3)$:

$$\begin{aligned} f_x(2, -3) &= 6 \cdot 2 \cdot (-3) + (-3)^2 \\ &= -27. \end{aligned}$$

If we use the notation $\partial f/\partial x = 6xy + y^2$, where $f = 3x^2y + xy^2$, the value of the partial derivative at $(2, -3)$ can be written as

$$\begin{aligned} \left. \frac{\partial f}{\partial x} \right|_{(2,-3)} &= 6 \cdot 2 \cdot (-3) + (-3)^2 \\ &= -27. \end{aligned}$$

TECHNOLOGY CONNECTION 〜

Exploratory

Consider finding values of a partial derivative of $f(x, y) = 3x^3y + 2xy$ using a calculator that finds derivatives of functions of one variable. How can you find $f_x(-4, 1)$? Then how can you find $f_y(2, 6)$?

EXAMPLE 3 For $f(x, y) = e^{xy} + y \ln x$, find f_x and f_y.

Solution We have

$$\begin{aligned} f_x &= y \cdot e^{xy} + y \cdot \frac{1}{x} \\ &= ye^{xy} + \frac{y}{x}, \end{aligned}$$

Quick Check 3 ✔

For $g(x, y) = \ln(x^2 + xy^4)$, find g_x and g_y.

and

$$\begin{aligned} f_y &= x \cdot e^{xy} + 1 \cdot \ln x \\ &= xe^{xy} + \ln x. \end{aligned}$$

3 ✔

The Geometric Interpretation of Partial Derivatives

The roof of this building is a smooth continuous surface. The slope at a point on the surface depends on the direction in which the tangent line is oriented.

The graph of a function of two variables $z = f(x, y)$ is a surface S, which might have a graph similar to the one shown to the right, where each input pair (x, y) in the domain D has only one output, $z = f(x, y)$.

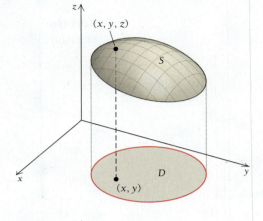

Now suppose we hold x fixed at the value a. The set of all points for which $x = a$ is a plane parallel to the yz-plane; thus, when x is fixed at a, y and z vary along that plane, as shown to the right. The plane $x = a$ in the figure cuts the surface along the curve C_1. The partial derivative f_y gives the slope of tangent lines to this curve, in the positive y-direction.

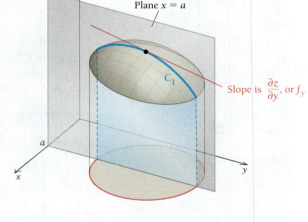

Similarly, if we hold y fixed at the value b, we obtain a curve C_2, as shown to the right. The partial derivative f_x gives the slope of tangent lines to this curve, in the positive x-direction.

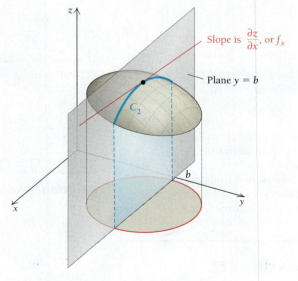

An Economics Application: The Cobb–Douglas Production Function

One model of production that is frequently applied in business and economics is the *Cobb–Douglas production function*:

$$p(x, y) = Ax^a y^{1-a}, \quad \text{for} \quad A > 0 \quad \text{and} \quad 0 < a < 1,$$

where p is the number of units produced with x units of labor and y units of capital. (Capital is the cost of machinery, buildings, tools, and other supplies.) The partial derivatives

$$\frac{\partial p}{\partial x} \quad \text{and} \quad \frac{\partial p}{\partial y}$$

are called, respectively, the *marginal productivity of labor* and the *marginal productivity of capital*.

EXAMPLE 4   MyTell Cellular has the following production function for a smartphone:

$$p(x, y) = 50x^{2/3} y^{1/3},$$

where p is the number of units produced with x units of labor and y units of capital.

a) Find the number of units produced with 125 units of labor and 64 units of capital.

b) Find the marginal productivities.

c) Evaluate the marginal productivities at $x = 125$ and $y = 64$.

Solution

a) $p(125, 64) = 50(125)^{2/3}(64)^{1/3} = 50(25)(4) = 5000$ units

b) Marginal productivity of labor $= \dfrac{\partial p}{\partial x} = p_x = 50\left(\dfrac{2}{3}\right)x^{-1/3}y^{1/3} = \dfrac{100y^{1/3}}{3x^{1/3}}$

Marginal productivity of capital $= \dfrac{\partial p}{\partial y} = p_y = 50\left(\dfrac{1}{3}\right)x^{2/3}y^{-2/3} = \dfrac{50x^{2/3}}{3y^{2/3}}$

c) For 125 units of labor and 64 units of capital, we have

Marginal productivity of labor $= p_x(125, 64)$

$$= \frac{100(64)^{1/3}}{3(125)^{1/3}} = \frac{100(4)}{3(5)} = 26\tfrac{2}{3},$$

Marginal productivity of capital $= p_y(125, 64)$

$$= \frac{50(125)^{2/3}}{3(64)^{2/3}} = \frac{50(25)}{3(16)} = 26\tfrac{1}{24}. \qquad \text{4} \; ✔$$

> **Quick Check 4** ✔
>
> A publisher's production function for textbooks is given by $p(x, y) = 72x^{0.8}y^{0.2}$, where p is the number of books produced, x is units of labor, and y is units of capital. Determine the marginal productivities at $x = 90$ and $y = 50$.

Let's interpret the marginal productivities of Example 4. To visualize the marginal productivity of labor, suppose capital is fixed at 64 units. Then a one-unit change in labor, from 125 to 126, will cause production to increase by about $26\tfrac{2}{3}$ units. To visualize the marginal productivity of capital, suppose the amount of labor is fixed at 125 units. Then a one-unit change in capital from 64 to 65 will cause production to increase by about $26\tfrac{1}{24}$ units.

A Cobb–Douglas production function is consistent with the law of diminishing returns. That is, if one input (either labor or capital) is held fixed while the other increases infinitely, then production will eventually increase at a decreasing rate. With such functions, it also turns out that if a certain maximum production is possible, then the expense of more labor, for example, may be required for that maximum output to be attainable.

Higher-Order Partial Derivatives

Consider

$$z = f(x, y) = 3xy^2 + 2xy + x^2.$$

Then $\quad \dfrac{\partial z}{\partial x} = \dfrac{\partial f}{\partial x} = 3y^2 + 2y + 2x.$

Suppose we continue and find the first partial derivative of $\partial z/\partial x$ with respect to y. This will be a **second-order partial derivative** of the original function z, denoted by

$$\frac{\partial}{\partial y}\left(\frac{\partial z}{\partial x}\right) = \frac{\partial}{\partial y}\left(\frac{\partial f}{\partial x}\right) = \frac{\partial}{\partial y}(3y^2 + 2y + 2x) = 6y + 2.$$

The notation $\dfrac{\partial}{\partial y}\left(\dfrac{\partial z}{\partial x}\right)$ is often expressed as

$$\frac{\partial^2 z}{\partial y\, \partial x} \quad \text{or} \quad \frac{\partial^2 f}{\partial y\, \partial x}.$$

We can also denote the preceding partial derivative using the notation f_{xy}:

$$f_{xy} = 6y + 2.$$

Note that in the notation f_{xy}, x and y are in the order (left to right) in which the differentiation is done, but in

$$\frac{\partial^2 f}{\partial y\, \partial x},$$

the order of x and y is reversed. In each case, the differentiation with respect to x is done first, followed by differentiation with respect to y.

Notation for the four second-order partial derivatives is as follows.

DEFINITION   **Second-Order Partial Derivatives**

1. $\dfrac{\partial^2 z}{\partial x\, \partial x} = \dfrac{\partial^2 f}{\partial x\, \partial x} = \dfrac{\partial^2 z}{\partial x^2} = \dfrac{\partial^2 f}{\partial x^2} = f_{xx}$ Take the partial derivative with respect to x, and then with respect to x again.

2. $\dfrac{\partial^2 z}{\partial y\, \partial x} = \dfrac{\partial^2 f}{\partial y\, \partial x} = f_{xy}$ Take the partial derivative with respect to x, and then with respect to y.

3. $\dfrac{\partial^2 z}{\partial x\, \partial y} = \dfrac{\partial^2 f}{\partial x\, \partial y} = f_{yx}$ Take the partial derivative with respect to y, and then with respect to x.

4. $\dfrac{\partial^2 z}{\partial y\, \partial y} = \dfrac{\partial^2 f}{\partial y\, \partial y} = \dfrac{\partial^2 z}{\partial y^2} = \dfrac{\partial^2 f}{\partial y^2} = f_{yy}$ Take the partial derivative with respect to y, and then with respect to y again.

EXAMPLE 5   For

$$z = f(x, y) = x^2 y^3 + x^4 y + xe^y,$$

find the four second-order partial derivatives.

Solution

a) $\dfrac{\partial^2 f}{\partial x^2} = f_{xx} = \dfrac{\partial}{\partial x}(2xy^3 + 4x^3 y + e^y)$ Differentiate f with respect to x.

$\qquad\qquad = 2y^3 + 12x^2 y$ Differentiate f_x with respect to x.

b) $\dfrac{\partial^2 f}{\partial y\,\partial x} = f_{xy} = \dfrac{\partial}{\partial y}(2xy^3 + 4x^3y + e^y)$ Differentiate f with respect to x.

$\qquad\qquad\quad = 6xy^2 + 4x^3 + e^y$ Differentiate f_x with respect to y.

c) $\dfrac{\partial^2 f}{\partial x\,\partial y} = f_{yx} = \dfrac{\partial}{\partial x}(3x^2y^2 + x^4 + xe^y)$ Differentiate f with respect to y.

$\qquad\qquad\quad = 6xy^2 + 4x^3 + e^y$ Differentiate f_y with respect to x.

d) $\dfrac{\partial^2 f}{\partial y^2} = f_{yy} = \dfrac{\partial}{\partial y}(3x^2y^2 + x^4 + xe^y)$ Differentiate f with respect to y.

$\qquad\qquad\quad = 6x^2y + xe^y$ Differentiate f_y with respect to y.   **5** ✔

Quick Check 5 ✔

For

$$z = g(x, y)$$
$$= 6x^2 + 3xy^4 - y^2,$$

find the four second-order partial derivatives.

Comparing parts (b) and (c) of Example 5, we see that

$$\frac{\partial^2 f}{\partial y\,\partial x} = \frac{\partial^2 f}{\partial x\,\partial y} \quad \text{and} \quad f_{xy} = f_{yx}.$$

Although this will be true for virtually all functions that we consider in this text, it is *not* true for all functions. One function for which it is not true is given in Exercise 69. In Section 6.3, we will see how higher-order partial derivatives are used in applications to find extrema for functions of two variables.

Section Summary

- For $z = f(x, y)$, the *partial derivatives with respect to* x *and* y are, respectively:

$$\frac{\partial z}{\partial x} = \lim_{h\to 0} \frac{f(x+h, y) - f(x, y)}{h} \quad \text{and}$$

$$\frac{\partial z}{\partial y} = \lim_{h\to 0} \frac{f(x, y+h) - f(x, y)}{h}.$$

- Simpler notations for partial derivatives are $f_x(x, y)$ or f_x, $f_y(x, y)$ or f_y, and z_x for $\dfrac{\partial z}{\partial x}$ and z_y for $\dfrac{\partial z}{\partial y}$.

- For a surface $z = f(x, y)$ and a point (x_0, y_0, z_0) on this surface, the partial derivative of f with respect to x gives the slope of the tangent line at (x_0, y_0, z_0) in the positive x-direction. Similarly, the partial derivative of f with respect to y gives the slope of the tangent line at (x_0, y_0, z_0) in the positive y-direction.

- For $z = f(x, y)$, the *second-order partial derivatives* are

$$f_{xx} = \frac{\partial^2 f}{\partial x^2}, f_{xy} = \frac{\partial^2 f}{\partial y\,\partial x}, f_{yx} = \frac{\partial^2 f}{\partial x\,\partial y}, \text{and} f_{yy} = \frac{\partial^2 f}{\partial y^2}.$$

Often (but not always), $f_{xy} = f_{yx}$.

6.2   Exercise Set

Find $\dfrac{\partial z}{\partial x}, \dfrac{\partial z}{\partial y}, \dfrac{\partial z}{\partial x}\bigg|_{(-2,-3)}$, and $\dfrac{\partial z}{\partial y}\bigg|_{(0,-5)}$.

1. $z = 7x - 5y$ **2.** $z = 2x - 3y$

3. $z = 3x^2 - 2xy + y$ **4.** $z = 2x^3 + 3xy - x$

Find $f_x(x, y), f_y(x, y), f_x(-2, 4), \text{and } f_y(4, -3)$.

5. $f(x, y) = 2x - 5xy$ **6.** $f(x, y) = 5x + 7y$

Find $f_x, f_y, f_x(-2, 1), \text{and } f_y(-3, -2)$.

7. $f(x, y) = \sqrt{x^2 + y^2}$

8. $f(x, y) = \sqrt{x^2 - y^2}$

Find f_x and f_y.

9. $f(x, y) = e^{3x-2y}$ **10.** $f(x, y) = e^{2x-y}$

11. $f(x, y) = e^{xy}$ **12.** $f(x, y) = e^{2xy}$

13. $f(x, y) = x \ln(x - y)$

14. $f(x, y) = y \ln(x + 2y)$

15. $f(x, y) = x \ln(xy)$

16. $f(x, y) = y \ln(xy)$

17. $f(x, y) = \dfrac{x}{y} - \dfrac{y}{3x}$

18. $f(x, y) = \dfrac{x}{y} + \dfrac{y}{5x}$

19. $f(x, y) = 4(3x + y - 8)^2$

20. $f(x, y) = 3(2x + y - 5)^2$

Find $\dfrac{\partial f}{\partial b}$ *and* $\dfrac{\partial f}{\partial m}$.

21. $f(b, m) = m^3 + 4m^2 b - b^2 + (2m + b - 5)^2$
$\qquad + (3m + b - 6)^2$

22. $f(b, m) = 5m^2 - mb^2 - 3b + (2m + b - 8)^2$
$\qquad + (3m + b - 9)^2$

Find f_x, f_y, *and* f_λ. *(The symbol* λ *is the Greek letter lambda.)*

23. $f(x, y, \lambda) = 5xy - \lambda(2x + y - 8)$

24. $f(x, y, \lambda) = 9xy - \lambda(3x - y + 7)$

25. $f(x, y, \lambda) = x^2 + y^2 - \lambda(10x + 2y - 4)$

26. $f(x, y, \lambda) = x^2 - y^2 - \lambda(4x - 7y - 10)$

Find the four second-order partial derivatives.

27. $f(x, y) = 2xy$

28. $f(x, y) = 5xy$

29. $f(x, y) = 7xy^2 + 5xy - 2y$

30. $f(x, y) = 3x^2 y - 2xy + 4y$

31. $f(x, y) = x^5 y^4 + x^3 y^2$

32. $f(x, y) = x^4 y^3 - x^2 y^3$

Find f_{xx}, f_{xy}, f_{yx}, *and* f_{yy}. *(Remember,* f_{yx} *means to differentiate with respect to y and then with respect to x.)*

33. $f(x, y) = 2x - 3y$

34. $f(x, y) = 3x + 5y$

35. $f(x, y) = e^{2xy}$

36. $f(x, y) = e^{xy}$

37. $f(x, y) = x + e^y$

38. $f(x, y) = y - e^x$

39. $f(x, y) = y \ln x$

40. $f(x, y) = x \ln y$

APPLICATIONS

Business and Economics

41. The Cobb–Douglas model. Riverside Appliances has the following production function for a certain product:

$$p(x, y) = 1800x^{0.621} y^{0.379},$$

where p is the number of units produced with x units of labor and y units of capital.

a) Find the number of units produced with 2500 units of labor and 1700 units of capital.

b) Find the marginal productivities.

c) Evaluate the marginal productivities at $x = 2500$ and $y = 1700$.

d) Interpret the meanings of the marginal productivities found in part (c).

42. The Cobb–Douglas model. Lincolnville Sporting Goods has the following production function for a certain product:

$$p(x, y) = 2400x^{2/5} y^{3/5},$$

where p is the number of units produced with x units of labor and y units of capital.

a) Find the number of units produced with 32 units of labor and 1024 units of capital.

b) Find the marginal productivities.

c) Evaluate the marginal productivities at $x = 32$ and $y = 1024$.

d) Interpret the meanings of the marginal productivities found in part (c).

Nursing facilities. *A study of Texas nursing homes found that the annual profit P (in dollars) of profit-seeking, independent nursing homes in urban locations is modeled by*

$$P(w, r, s, t) = 0.007955w^{-0.638} r^{1.038} s^{0.873} t^{2.468},$$

*where w is the average hourly wage of nurses and aides (in dollars), r is the occupancy rate (as a percentage), s is the total square footage of the facility, and t is the Texas Index of Level of Effort (TILE), a number between 1 and 11 that measures state Medicaid reimbursement. (Source: K. J. Knox, E. C. Blankmeyer, and J. R. Stutzman, "Relative Economic Efficiency in Texas Nursing Facilities," **Journal of Economics and Finance**, Vol. 23, 199–213 (1999).) Use this information for Exercises 43 and 44.*

43. A profit-seeking, independent Texas nursing home in an urban setting has nurses and aides with an average hourly wage of \$20 an hour, a TILE of 8, an occupancy rate of 70%, and 400,000 ft^2 of space.

a) Estimate the nursing home's annual profit.

b) Find the four partial derivatives of P.

c) Interpret the meaning of the partial derivatives found in part (b).

44. The change in P due to a change in w when the other variables are held constant is approximately

$$\Delta P \approx \frac{\partial P}{\partial w} \Delta w.$$

Use the values of w, r, s, and t in Exercise 43 and assume that the nursing home gives its nurses and aides a small raise so that the average hourly wage is now \$20.25 an hour. By approximately how much does profit change?

Life and Physical Sciences

Temperature–humidity heat index. *In summer, higher humidity interacts with the outdoor temperature, making a person feel hotter because of reduced heat loss from the skin. The temperature–humidity index,* T_h, *is what the temperature would have to be with no humidity in order to give the same heat effect. One index often used is given by*

$$T_h = 1.98T - 1.09(1 - H)(T - 58) - 56.9,$$

where T is the air temperature, in degrees Fahrenheit, and H is the relative humidity, expressed as a decimal. Find the temperature–humidity index in each case. Round to the nearest tenth of a degree.

45. $T = 85°F$ and $H = 60\%$

46. $T = 90°F$ and $H = 90\%$

47. $T = 90°F$ and $H = 100\%$

48. $T = 78°F$ and $H = 100\%$

Use the equation for T_h given above for Exercises 49 and 50.

49. Find $\dfrac{\partial T_h}{\partial H}$, and interpret its meaning.

50. Find $\dfrac{\partial T_h}{\partial T}$, and interpret its meaning.

51. Body surface area. The Mosteller formula for approximating the surface area, S, in m^2, of a human is

$$S = \frac{\sqrt{hw}}{60},$$

where h is the person's height in centimeters and w is the person's weight in kilograms. (*Source:* www.halls.md.)

a) Compute $\dfrac{\partial S}{\partial h}$.

b) Compute $\dfrac{\partial S}{\partial w}$.

c) The change in S due to a change in w when h is constant is approximately

$$\Delta S \approx \frac{\partial S}{\partial w}\Delta w.$$

Use this formula to approximate the change in someone's surface area given that the person is 170 cm tall, weighs 80 kg, and loses 2 kg.

52. Body surface area. The Haycock formula for approximating the surface area, S, in m^2, of a human is

$$S = 0.024265h^{0.3964}w^{0.5378},$$

where h is the person's height in centimeters and w is the person's weight in kilograms. (*Source:* www.halls.md.)

a) Compute $\dfrac{\partial S}{\partial h}$.

b) Compute $\dfrac{\partial S}{\partial w}$.

c) The change in S due to a change in w when h is constant is approximately

$$\Delta S \approx \frac{\partial S}{\partial w}\Delta w.$$

Use this formula to approximate the change in someone's surface area given that the person is 170 cm tall, weighs 80 kg, and loses 2 kg.

Social Sciences

Reading ease. *The following formula is used by psychologists and educators to predict the reading ease, E, of a passage of words:*

$$E = 206.835 - 0.846w - 1.015s,$$

where w is the number of syllables in a 100-word section and s is the average number of words per sentence. Use this information for Exercises 53–56.

53. Find E when $w = 146$ and $s = 5$

54. Find E when $w = 180$ and $s = 6$

55. Find $\dfrac{\partial E}{\partial w}$.

56. Find $\dfrac{\partial E}{\partial s}$.

SYNTHESIS

Find f_x and f_t.

57. $f(x, t) = \dfrac{x^2 + t^2}{x^2 - t^2}$

58. $f(x, t) = \dfrac{x^2 - t}{x^3 + t}$

59. $f(x, t) = \dfrac{2\sqrt{x} - 2\sqrt{t}}{1 + 2\sqrt{t}}$

60. $f(x, t) = \sqrt[4]{x^3 t^5}$

61. $f(x, t) = 6x^{2/3} - 8x^{1/4}t^{1/2} - 12x^{-1/2}t^{3/2}$

62. $f(x, t) = \left(\dfrac{x^2 + t^2}{x^2 - t^2}\right)^5$

In Exercises 63 and 64, find f_{xx}, f_{xy}, f_{yx}, and f_{yy}.

63. $f(x, y) = \dfrac{x}{y^2} - \dfrac{y}{x^2}$

64. $f(x, y) = \dfrac{xy}{x - y}$

65. Do some research on the Cobb–Douglas production function, and explain how it was developed.

66. Explain the meaning of the first partial derivatives of a function of two variables in terms of slopes of tangent lines.

67. Consider $f(x, y) = \ln(x^2 + y^2)$. Show that f is a solution of the partial differential equation

$$\frac{\partial^2 f}{\partial x^2} + \frac{\partial^2 f}{\partial y^2} = 0.$$

68. Consider $f(x, y) = x^3 - 5xy^2$. Show that f is a solution of the partial differential equation

$$xf_{xy} - f_y = 0.$$

69. Consider the function f defined as follows:

$$f(x, y) = \begin{cases} \dfrac{xy(x^2 - y^2)}{x^2 + y^2}, & \text{for } (x, y) \neq (0, 0), \\ 0, & \text{for } (x, y) = (0, 0). \end{cases}$$

a) Find $f_x(0, y)$ by evaluating the limit

$$\lim_{h \to 0} \frac{f(h, y) - f(0, y)}{h}.$$

b) Find $f_y(x, 0)$ by evaluating the limit

$$\lim_{h \to 0} \frac{f(x, h) - f(x, 0)}{h}.$$

c) Now find and compare $f_{yx}(0, 0)$ and $f_{xy}(0, 0)$.

Answers to Quick Checks

1. $\dfrac{\partial u}{\partial x} = 2xy^3 z^4$, $\dfrac{\partial u}{\partial y} = 3x^2 y^2 z^4$, $\dfrac{\partial u}{\partial z} = 4x^2 y^3 z^3$

2. $f_x(x, y) = 21x^2 y^2 - \dfrac{1}{y}$, $f_y(x, y) = 14x^3 y + \dfrac{x}{y^2}$

3. $g_x = \dfrac{2x + y^4}{x^2 + xy^4}$, $g_y = \dfrac{4xy^3}{x^2 + xy^4} = \dfrac{4y^3}{x + y^4}$

4. $p_x(90, 50) = 51.21$ textbooks/unit of labor, $p_y(90, 50) = 23.05$ textbooks/unit of capital

5. $g_{xx} = 12$, $g_{yy} = 36xy^2 - 2$, $g_{xy} = 12y^3$, $g_{yx} = 12y^3$

6.3 Maximum–Minimum Problems

We will now find maximum and minimum values of functions of two variables.

● Find relative extrema of a function of two variables.

DEFINITION

A function f of two variables:

1. has a **relative maximum** at (a, b) if

$$f(x, y) \leq f(a, b)$$

for all points (x, y) in a region containing (a, b);

2. has a **relative minimum** at (a, b) if

$$f(x, y) \geq f(a, b)$$

for all points (x, y) in a region containing (a, b).

This definition is illustrated in Figs. 1 and 2. A relative maximum (or minimum) may not be an "absolute" maximum (or minimum), as illustrated in Fig. 3.

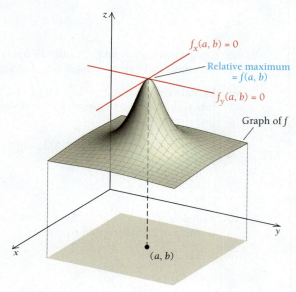

FIGURE 1

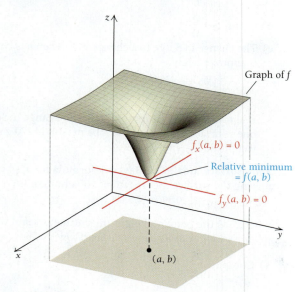

FIGURE 2

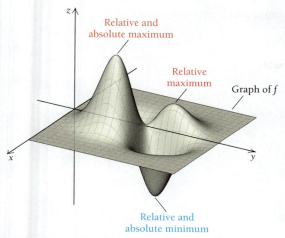

Relative and
absolute maximum

Relative
maximum

Graph of *f*

Relative and
absolute minimum

FIGURE 3

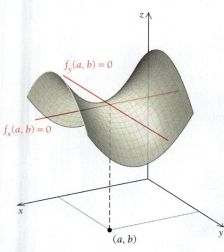

$f_y(a, b) = 0$

$f_x(a, b) = 0$

(a, b)

FIGURE 4

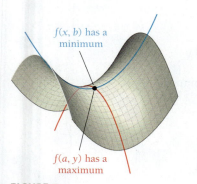

$f(x, b)$ has a
minimum

$f(a, y)$ has a
maximum

FIGURE 5

Determining Maximum and Minimum Values

Suppose a differentiable function f has a relative maximum or minimum value at some point (a, b) inside its domain. If we fix y at the value b, then $f(x, b)$ can be regarded as a function of x. Because a relative maximum or minimum occurs at (a, b), we know that $f(x, b)$ achieves a maximum or minimum at (a, b) and $f_x = 0$. Similarly, if we fix x at a, then $f(a, y)$ can be regarded as a function of y that achieves a relative extremum at (a, b), and thus $f_y = 0$. In short, since an extremum exists at (a, b), we must have

$$f_x(a, b) = 0 \quad \text{and} \quad f_y(a, b) = 0. \tag{1}$$

We call a point (a, b) at which both partial derivatives are 0 a **critical point**. This is comparable to a critical value for functions of one variable. Thus, one strategy for finding relative maximum or minimum values is to solve a system of equations like (1) to find critical points. Just as for functions of one variable, this strategy does *not* guarantee that we will have a relative maximum or minimum value. We have argued only that *if* a continuous, differentiable function f has a maximum or minimum value at (a, b), *then* both its partial derivatives must be 0 at that point. Look back at Figs. 1 and 2 and then at Fig. 4, which illustrates a case in which the partial derivatives are 0 but the function does not have a relative maximum or minimum value at (a, b).

In Fig. 4, suppose we fix y at a value b. Then $f(x, b)$ has a minimum at a, but f does not (see the blue curve in Fig. 5). Similarly, if we fix x at a, then $f(a, y)$ has a maximum at b, but f does not (see the red curve in Fig. 5). The point $(a, b, f(a, b))$ is called a **saddle point**. At this point, $f_x(a, b) = 0$ and $f_y(a, b) = 0$ [the point (a, b) is a critical point], but f does not attain a relative maximum or minimum value at (a, b).

A test that uses first- and second-order partial derivatives to find relative maximum and minimum values is stated below. We will not prove this theorem.

THEOREM 1   **The *D*-Test**

If f is a differentiable function of x and y, to find the relative maximum and minimum values of f:

1. Find f_x, f_y, f_{xx}, f_{yy}, and f_{xy}.
2. Solve the system of equations $f_x = 0$, $f_y = 0$. Let (a, b) represent a solution.
3. Evaluate D, where $D = f_{xx}(a, b) \cdot f_{yy}(a, b) - [f_{xy}(a, b)]^2$.
4. Then
   a) f has a maximum at (a, b) if $D > 0$ and $f_{xx}(a, b) < 0$.
   b) f has a minimum at (a, b) if $D > 0$ and $f_{xx}(a, b) > 0$.
   c) f has neither a maximum nor a minimum at (a, b) if $D < 0$.
      In this case, f has a *saddle point* at (a, b). See Figs. 4 and 5.
   d) This test is not applicable if $D = 0$.

The *D*-test is somewhat analogous to the Second Derivative Test (Section 2.2) for functions of one variable. Saddle points are analogous to critical values at which concavity changes and there are no relative maximum or minimum values.

A relative maximum or minimum *may or may not be an absolute maximum or an absolute minimum*. Testing for absolute maximum or minimum values is rather complicated. We will restrict our attention to finding *relative* maximum or minimum values. In most of the applications we will consider, relative maximum or minimum values turn out to be absolute as well.

The shape of a perfect tent. To give a tent roof the maximum strength possible, designers draw the fabric into a series of three-dimensional shapes that, viewed in profile, resemble a horse's saddle and that mathematicians call an *anticlastic curve*. Two people with a stretchy piece of fabric such as Spandex can duplicate the shape, as shown above. One person pulls up and out on two diagonal corners; the other person pulls down and out on the other two corners. The opposing tensions draw each point of the fabric's surface into rigid equilibrium. The more pronounced the curve, the stiffer the surface.

Quick Check 1

Find the relative maximum and minimum values of

$$f(x, y) = x^2 + xy + y^2 - 3x.$$

EXAMPLE 1   Find the relative maximum and minimum values of

$$f(x, y) = x^2 + xy + 2y^2 - 7x.$$

Solution

1. Find f_x, f_y, f_{xx}, f_{yy}, and f_{xy}:

$$f_x = 2x + y - 7, \qquad f_y = x + 4y,$$
$$f_{xx} = 2; \qquad\qquad f_{yy} = 4;$$
$$f_{xy} = 1.$$

2. Solve the system of equations $f_x = 0, f_y = 0$:

$$2x + y - 7 = 0, \qquad\qquad\qquad (2)$$
$$x + 4y = 0. \qquad\qquad\qquad (3)$$

Solving equation (3) for x, we get $x = -4y$. Substituting $-4y$ for x in equation (2) and solving, we get

$$2(-4y) + y - 7 = 0$$
$$-8y + y - 7 = 0$$
$$-7y = 7$$
$$y = -1.$$

To find x when $y = -1$, we substitute -1 for y in equation (2) or equation (3). We choose equation (3):

$$x + 4(-1) = 0$$
$$x = 4.$$

Thus, $(4, -1)$ is the only critical point, and $f(4, -1)$ is our candidate for a maximum or minimum value.

3. We check to see whether $f(4, -1)$ is a maximum or minimum value:

$$D = f_{xx}(4, -1) \cdot f_{yy}(4, -1) - [f_{xy}(4, -1)]^2$$
$$= 2 \cdot 4 - [1]^2 \qquad \text{Substituting}$$
$$= 7.$$

4. Thus, $D = 7$ and $f_{xx}(4, -1) = 2$. Since $D > 0$ and $f_{xx}(4, -1) > 0$, it follows from the D-test that f has a relative minimum at $(4, -1)$. That minimum value is found as follows:

$$f(4, -1) = 4^2 + 4(-1) + 2(-1)^2 - 7 \cdot 4$$
$$= 16 - 4 + 2 - 28$$
$$= -14. \qquad \text{This is the relative minimum.}$$

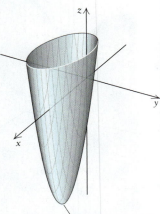

Relative minimum: $(4, -1, -14)$

$z = f(x, y) = x^2 + xy + 2y^2 - 7x$

1 ✔

EXAMPLE 2   Find the relative maximum and minimum values of

$$f(x, y) = xy - x^3 - y^2.$$

Solution

1. Find f_x, f_y, f_{xx}, f_{yy}, and f_{xy}:

$$f_x = y - 3x^2, \qquad f_y = x - 2y,$$
$$f_{xx} = -6x; \qquad f_{yy} = -2;$$
$$f_{xy} = 1.$$

2. Solve the system of equations $f_x = 0, f_y = 0$:

$$y - 3x^2 = 0, \tag{4}$$
$$x - 2y = 0. \tag{5}$$

Solving equation (4) for y, we get $y = 3x^2$. Substituting $3x^2$ for y in equation (5) and solving, we get

$$x - 2(3x^2) = 0$$
$$x - 6x^2 = 0$$
$$x(1 - 6x) = 0. \quad \text{Factoring}$$

Setting each factor equal to 0 and solving, we have

$$x = 0 \quad or \quad 1 - 6x = 0$$
$$x = 0 \quad or \quad x = \tfrac{1}{6}.$$

To find y when $x = 0$, we substitute 0 for x in equation (4) or equation (5):

$$0 - 2y = 0 \quad \text{Substituting into equation (5)}$$
$$-2y = 0$$
$$y = 0.$$

Thus, $(0, 0)$ is a critical point, and $f(0, 0)$ is one candidate for a maximum or minimum value. To look for another candidate, we substitute $\tfrac{1}{6}$ for x in equation (4) or equation (5):

$$\tfrac{1}{6} - 2y = 0 \quad \text{Substituting into equation (5)}$$
$$-2y = -\tfrac{1}{6}$$
$$y = \tfrac{1}{12}.$$

Thus, $(\tfrac{1}{6}, \tfrac{1}{12})$ is another critical point, and $f(\tfrac{1}{6}, \tfrac{1}{12})$ is another candidate for a maximum or minimum value.

3–4. We check both $(0, 0)$ and $(\tfrac{1}{6}, \tfrac{1}{12})$ to see whether they yield maximum or minimum values.

$$\text{For } (0, 0): \quad D = f_{xx}(0, 0) \cdot f_{yy}(0, 0) - [f_{xy}(0, 0)]^2$$
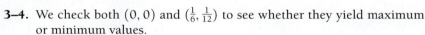
$$= (-6 \cdot 0) \cdot (-2) - [1]^2 \quad \begin{array}{l}\text{Substituting and}\\\text{using step 1}\end{array}$$
$$= -1.$$

Since $D < 0$, it follows that $f(0, 0)$ is neither a maximum nor a minimum value, but a saddle point.

$$\text{For } (\tfrac{1}{6}, \tfrac{1}{12}): \quad D = f_{xx}(\tfrac{1}{6}, \tfrac{1}{12}) \cdot f_{yy}(\tfrac{1}{6}, \tfrac{1}{12}) - [f_{xy}(\tfrac{1}{6}, \tfrac{1}{12})]^2$$
$$= (-6 \cdot \tfrac{1}{6}) \cdot (-2) - [1]^2 \quad \text{Substituting and using step 1}$$
$$= -1(-2) - 1$$
$$= 1.$$

Thus, $D = 1$ and $f_{xx}(\tfrac{1}{6}, \tfrac{1}{12}) = -1$. Since $D > 0$ and $f_{xx}(\tfrac{1}{6}, \tfrac{1}{12}) < 0$, it follows that f has a relative maximum at $(\tfrac{1}{6}, \tfrac{1}{12})$; that maximum value is

$$f(\tfrac{1}{6}, \tfrac{1}{12}) = \tfrac{1}{6} \cdot \tfrac{1}{12} - (\tfrac{1}{6})^3 - (\tfrac{1}{12})^2$$
$$= \tfrac{1}{72} - \tfrac{1}{216} - \tfrac{1}{144} = \tfrac{1}{432}. \quad \text{This is the relative maximum.} \qquad 2 ✔$$

TECHNOLOGY CONNECTION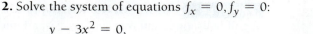

Exploratory

Examine the graph of the equation in Example 2 with a 3D graphing utility to visualize the relative maximum and the saddle point.

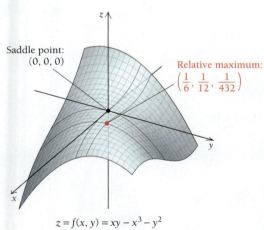

Saddle point: $(0, 0, 0)$

Relative maximum: $\left(\tfrac{1}{6}, \tfrac{1}{12}, \tfrac{1}{432}\right)$

$z = f(x, y) = xy - x^3 - y^2$

Quick Check 2 ✔

Find the critical points of

$$g(x, y) = x^3 + y^2 - 3x - 4y + 3.$$

Then use the D-test to determine if each point has a relative maximum or a relative minimum or is a saddle point.

EXAMPLE 3 **Business: Maximizing Profit.** Fly Straight, Inc. produces two kinds of golf balls, one kind that sells for $3 and one that sells for $2. The total revenue, in thousands of dollars, from the sale of x thousand balls at $3 each and y thousand at $2 each is given by

$$R(x, y) = 3x + 2y.$$

The company determines that the total cost, in thousands of dollars, of producing x thousand of the $3 ball and y thousand of the $2 ball is given by

$$C(x, y) = 2x^2 - 2xy + y^2 - 9x + 6y + 7.$$

How many balls of each type must be produced and sold to maximize profit?

Solution The total profit $P(x, y)$ is given by

$$
\begin{aligned}
P(x, y) &= R(x, y) - C(x, y) \\
&= 3x + 2y - (2x^2 - 2xy + y^2 - 9x + 6y + 7) \quad \text{Substituting} \\
P(x, y) &= -2x^2 + 2xy - y^2 + 12x - 4y - 7. \quad \text{Simplifying}
\end{aligned}
$$

1. Find P_x, P_y, P_{xx}, P_{yy}, and P_{xy}:

$$P_x = -4x + 2y + 12, \qquad P_y = 2x - 2y - 4,$$
$$P_{xx} = -4; \qquad\qquad\quad P_{yy} = -2;$$
$$P_{xy} = 2.$$

2. Solve the system of equations $P_x = 0, P_y = 0$:

$$-4x + 2y + 12 = 0, \tag{6}$$
$$2x - 2y - 4 = 0. \tag{7}$$

Adding the left and right sides of equations (6) and (7), we get

$$-2x + 8 = 0$$
$$-2x = -8$$
$$x = 4.$$

To find y when $x = 4$, we substitute 4 for x in equation (6) or equation (7):

$$2 \cdot 4 - 2y - 4 = 0 \quad \text{\color{red}Substituting into equation 7}$$
$$-2y + 4 = 0$$
$$-2y = -4$$
$$y = 2.$$

Thus, $(4, 2)$ is the only critical point, and $P(4, 2)$ is a candidate for a maximum or minimum value.

3. We must check to see whether $P(4, 2)$ is a maximum or minimum value:

$$
\begin{aligned}
D &= P_{xx}(4, 2) \cdot P_{yy}(4, 2) - [P_{xy}(4, 2)]^2 \\
&= (-4)(-2) - 2^2 \quad \text{Substituting} \\
&= 4.
\end{aligned}
$$

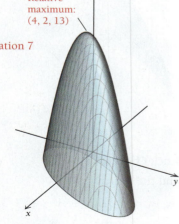

Relative maximum: $(4, 2, 13)$

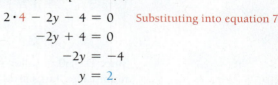

$z = P(x, y) = -2x^2 + 2xy - y^2 + 12x - 4y - 7$

4. Thus, $D = 4$ and $P_{xx}(4, 2) = -4$. Since $D > 0$ and $P_{xx}(4, 2) < 0$, it follows that P has a relative maximum at $(4, 2)$. Therefore, to maximize profit, the company must produce and sell 4 thousand of the $3 golf balls and 2 thousand of the $2 golf balls. The maximum profit will be

$$P(4, 2) = -2 \cdot 4^2 + 2 \cdot 4 \cdot 2 - 2^2 + 12 \cdot 4 - 4 \cdot 2 - 7 = 13,$$

or $13 thousand.

Quick Check 3 ✔

Repeat Example 3 using the same cost function and assuming that the company's total revenue, in thousands of dollars, comes from the sale of x thousand balls at $3.50 each and y thousand at $2.75 each.

3 ✔

Section Summary

- A two-variable function f has a *relative maximum* at (a, b) if $f(x, y) \leq f(a, b)$ for all points in a region containing (a, b) and has a *relative minimum* at (a, b) if $f(x, y) \geq f(a, b)$ for all points in a region containing (a, b).

- The *D-test* is used to determine if a *critical point* has a relative minimum or a relative maximum or is a *saddle point*.

6.3 | Exercise Set

Find the relative maximum and minimum values.

1. $f(x, y) = x^2 + xy + 3y^2 + 11x$

2. $f(x, y) = x^2 + xy + y^2 - 5y$

3. $f(x, y) = 2xy - x^3 - y^2$

4. $f(x, y) = 4xy - x^3 - 2y^2$

5. $f(x, y) = x^3 + y^3 - 6xy$

6. $f(x, y) = x^3 + y^3 - 3xy$

7. $f(x, y) = x^2 + y^2 - 4x + 2y - 5$

8. $f(x, y) = x^2 + 2xy + 2y^2 - 6y + 2$

9. $f(x, y) = x^2 + y^2 + 8x - 10y$

10. $f(x, y) = 4y + 6x - x^2 - y^2$

11. $f(x, y) = 4x^2 - y^2$

12. $f(x, y) = x^2 - y^2$

13. $f(x, y) = e^{x^2 + y^2 + 1}$

14. $f(x, y) = e^{x^2 - 2x + y^2 - 4y + 2}$

APPLICATIONS

Business and Economics

In Exercises 15–22, assume that relative maximum and minimum values are absolute maximum and minimum values.

15. Maximizing profit. Safe Shades produces two kinds of sunglasses; one kind sells for $17, and the other for $21. The total revenue in thousands of dollars from the sale of x thousand sunglasses at $17 each and y thousand at $21 each is given by

$$R(x, y) = 17x + 21y.$$

The company determines that the total cost, in thousands of dollars, of producing x thousand of the $17 sunglasses and y thousand of the $21 sunglasses is given by

$$C(x, y) = 4x^2 - 4xy + 2y^2 - 11x + 25y - 3.$$

How many of each type of sunglasses must be produced and sold to maximize profit?

16. Maximizing profit. A concert promoter produces two kinds of souvenir shirt. Total revenue from the sale of x thousand shirts at $18 each and y thousand at $25 each is given by

$$R(x, y) = 18x + 25y.$$

The company determines that the total cost, in thousands of dollars, of producing x thousand of the $18 shirt and y thousand of the $25 shirt is

$$C(x, y) = 4x^2 - 6xy + 3y^2 + 20x + 19y - 12.$$

How many of each type of shirt must be produced and sold to maximize profit?

17. Maximizing profit. McLeod Corp. finds that its profit, P, in millions of dollars, is given by

$$P(a, p) = 2ap + 80p - 15p^2 - \tfrac{1}{10}a^2 p - 80,$$

where a is the amount spent on advertising, in millions of dollars, and p is the price charged per unit, in dollars. Find the maximum value of P and the values of a and p at which it occurs.

18. Maximizing profit. Humphrey's Medical Supply finds that its profit, P, in millions of dollars, is given by

$$P(a, n) = -5a^2 - 3n^2 + 48a - 4n + 2an + 290,$$

where a is the amount spent on advertising, in millions of dollars, and n is the number of items sold, in thousands. Find the maximum value of P and the values of a and n at which it occurs.

19. Minimizing the cost of a container. ProHauling Services is designing an open-top, rectangular container that will have a volume of 320 ft^3. The cost of making the bottom of the container is $5 per square foot, and the cost of the sides is $4 per square foot. Find the dimensions of the container that will minimize total cost. (*Hint:* Make a substitution using the formula for volume.)

20. Two-variable revenue maximization. Rad Designs sells two kinds of sweatshirts that compete with one another. Their demand functions are expressed by the following relationships:

$$q_1 = 78 - 6p_1 - 3p_2, \tag{1}$$

$$q_2 = 66 - 3p_1 - 6p_2, \tag{2}$$

where p_1 and p_2 are the prices of the sweatshirts, in multiples of \$10, and q_1 and q_2 are the quantities of the sweatshirts demanded, in hundreds of units.

a) Find a formula for the total-revenue function, R, in terms of the variables p_1 and p_2. [*Hint:* $R = p_1q_1 + p_2q_2$; then substitute expressions from equations (1) and (2) to find $R(p_1, p_2)$.]

b) What prices p_1 and p_2 should be charged for each product in order to maximize total revenue?

c) How many units will be demanded?

d) What is the maximum total revenue?

21. Two-variable revenue maximization. Repeat Exercise 20, using

$$q_1 = 64 - 4p_1 - 2p_2$$

and

$$q_2 = 56 - 2p_1 - 4p_2.$$

Life and Physical Sciences

22. Temperature. A flat metal plate is mounted on a coordinate plane. The temperature of the plate, in degrees Fahrenheit, at point (x, y) is given by

$$T(x, y) = x^2 + 2y^2 - 8x + 4y.$$

Find the minimum temperature and where it occurs. Is there a maximum temperature?

SYNTHESIS

In Exercises 23–26, find the relative maximum and minimum values as well as any saddle points.

23. $f(x, y) = e^x + e^y - e^{x+y}$

24. $f(x, y) = xy + \dfrac{2}{x} + \dfrac{4}{y}$

25. $f(x, y) = 2y^2 + x^2 - x^2y$

26. $S(b, m) = (m + b - 72)^2 + (2m + b - 73)^2 + (3m + b - 75)^2$

27. Is a cross-section of an anticlastic curve always a parabola? Why or why not?

28. Explain the difference between a relative minimum and an absolute minimum of a function of two variables.

TECHNOLOGY CONNECTION

Use a 3D graphics program to graph each of the following functions. Then estimate any relative extrema.

29. $f(x, y) = \dfrac{-5}{x^2 + 2y^2 + 1}$

30. $f(x, y) = x^3 + y^3 + 3xy$

31. $f(x, y) = \dfrac{3xy(x^2 - y^2)}{x^2 + y^2}$

32. $f(x, y) = \dfrac{y + x^2y^2 - 8x}{xy}$

Answers to Quick Checks

1. $(2, -1, -3)$, relative minimum **2.** $(1, 2, -3)$, relative minimum; $(-1, 2, 1)$, saddle point **3.** Maximum profit is \$17.031 thousand when $x = \$4.625$ thousand and $y = \$3$ thousand

6.4 An Application: The Least-Squares Technique

Objectives:
• Find a regression line.
• Solve applied problems involving regression lines.
• Find an exponential regression curve.

We have regularly used a graphing calculator to perform regression. In this section, we develop an understanding of regression by using the method for finding the minimum value for a function of two variables. An equation found by regression provides a model of data from which predictions can be made. For example, in business, one might want to predict future sales on the basis of past data. In environmental studies, one might want to predict future demand for natural gas on the basis of past usage. Suppose we wish to find a linear equation,

$$y = mx + b,$$

to fit some data. To determine this equation is to determine the values of m and b. Let's do this with an example.

EXAMPLE 1 Sky Blue Car Rentals offers hybrid (gas–electric) vehicles and charts its revenue as shown in Fig. 1 and the table below. We can use the data to predict the company's revenue for the year 2021.

YEARLY REVENUE OF SKY BLUE CAR RENTALS

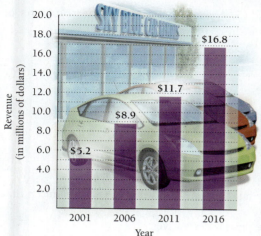

FIGURE 1

Year, x	2001	2006	2011	2016	2021
Yearly Revenue, y (in millions of dollars)	5.2	8.9	11.7	16.8	?

Solution We plot these points and try to draw a line through them that fits. There are several ways in which this might be done (see Figs. 2 and 3). Each would give a different estimate of the company's total revenue for 2021.

Note that the years for which revenue is given follow 5-yr increments. Thus, computations can be simplified if we use the data points $(1, 5.2)$, $(2, 8.9)$, $(3, 11.7)$, and $(4, 16.8)$, as plotted in Fig. 3, where each horizontal unit represents 5 years and $x = 1$ is 2001.

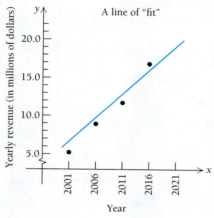

FIGURE 2

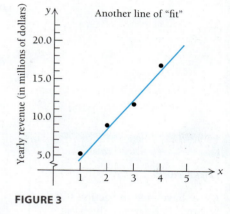

FIGURE 3

To determine the equation of the line that "best" fits the data, we note that for each data point there is a *deviation*, or error, between its y-value and the y-value at the point on the line directly above or below the data point. Those deviations, in this case, $y_1 - 5.2$, $y_2 - 8.9$, $y_3 - 11.7$, and $y_4 - 16.8$, are positive or negative, depending on the location of the line (see Fig. 4).

We wish to fit these data points with a line,

$$y = mx + b,$$

such that the values of m and b minimize the y-deviations. To minimize the deviations, we use the *least-squares assumption*.

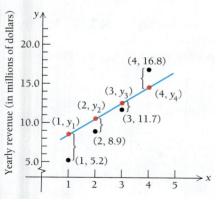

FIGURE 4

The Least-Squares Assumption

The line of best fit is the line for which the sum of the squares of the y-deviations is a minimum. This is called the **regression line**.

Note that squaring each y-deviation gives us a series of nonnegative terms we can add.

Using the least-squares assumption with the yearly revenue data, we want to minimize

$$(y_1 - 5.2)^2 + (y_2 - 8.9)^2 + (y_3 - 11.7)^2 + (y_4 - 16.8)^2. \tag{1}$$

Also, since the points $(1, y_1)$, $(2, y_2)$, $(3, y_3)$, and $(4, y_4)$ must be solutions of $y = mx + b$, it follows that

$$y_1 = m(1) + b = m + b,$$
$$y_2 = m(2) + b = 2m + b,$$
$$y_3 = m(3) + b = 3m + b,$$
$$y_4 = m(4) + b = 4m + b.$$

Substituting $m + b$ for y_1, $2m + b$ for y_2, $3m + b$ for y_3, and $4m + b$ for y_4 in equation (1), we form a function of two variables:

$$S(m, b) = (m + b - 5.2)^2 + (2m + b - 8.9)^2 + (3m + b - 11.7)^2 + (4m + b - 16.8)^2.$$

To find the regression line for the given data, we must find the values of m and b that minimize $S(m, b)$.

To apply the D-test, we first find the partial derivatives:

$$\frac{\partial S}{\partial b} = 2(m + b - 5.2) + 2(2m + b - 8.9) + 2(3m + b - 11.7) + 2(4m + b - 16.8)$$

$$= 20m + 8b - 85.2,$$

and

$$\frac{\partial S}{\partial m} = 2(m + b - 5.2) + 2(2m + b - 8.9)2 + 2(3m + b - 11.7)3 + 2(4m + b - 16.8)4$$

$$= 60m + 20b - 250.6.$$

We set these derivatives equal to 0 and solve the resulting system:

$$20m + 8b - 85.2 = 0, \qquad 5m + 2b = 21.3,$$
$$\text{or}$$
$$60m + 20b - 250.6 = 0; \qquad 15m + 5b = 62.65.$$

It can be shown that the solution of this system is

$$b = 1.25, \qquad m = 3.76. \qquad \text{(See the next Technology Connection.)}$$

We leave it to the student to complete the D-test to verify that $(1.25, 3.76)$ does, in fact, yield the minimum of S. There is no need to compute $S(1.25, 3.76)$.

The values of m and b are enough to determine $y = mx + b$. The regression line is

$$y = 3.76x + 1.25. \qquad \text{\color{red}Substituting for } m \text{ and } b$$

The graph of this "best-fit" regression line together with the data points is shown in Fig. 5. Compare it to Figs. 2, 3, and 4.

Using this regression line, we can now predict Sky Blue Car Rentals' yearly revenue in 2021:

$$y = 3.76(5) + 1.25 = 20.05.$$

The yearly revenue in 2021 is predicted to be about $20.05 million.

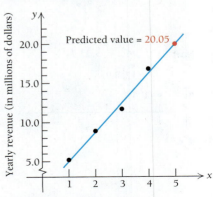

FIGURE 5

Quick Check 1 ✔

Use the method of least squares to determine the regression line for the data points $(1, 25)$, $(2, 48)$, $(3, 76.7)$, and $(4, 104.8)$.

The method of least squares is demonstrated here with only four data points in order to simplify the explanation. Most researchers would want more than four data points to get a "good" regression line. Furthermore, making predictions too far into the future from any mathematical model may not be valid. The further into the future a prediction is made, the more skeptical one should be about the prediction.

1 ✔

TECHNOLOGY CONNECTION

Solving Linear Systems Using Matrices

Let's explore a method of solution called *reduced row echelon form* (rref). We can use this method to solve the system of equations discussed above:

$$5m + 2b = 21.3,$$
$$15m + 5b = 62.65.$$

From this system, we can write the matrix

$$\begin{bmatrix} 5 & 2 & 21.3 \\ 15 & 5 & 62.65 \end{bmatrix}.$$

The first column is called the *m*-column (because the entries are the coefficients of the variable *m*), the second column is called the *b*-column, and the final column is the constants column. When entering the numbers of the system of equations into a matrix, it is crucial that the *m* and *b* terms are in the correct positions to the left of the equal signs and the constants are to the right of the equal signs.

When a matrix is in reduced row echelon form, it has the following appearance:

$$\begin{bmatrix} 1 & 0 & p \\ 0 & 1 & q \end{bmatrix}.$$

With this form, the system is considered solved, as we can rewrite this matrix as

$$1x + 0y = p,$$
$$0x + 1y = q.$$

Therefore,

$$x = p \text{ and } y = q.$$

On a calculator, select MATRIX, and under EDIT, select [A]. With this setting, matrix [A] has 2 rows and 3 columns, so it is of size 2 × 3. Enter these values, pressing **ENTER** after each one. Then enter the values of the matrix into the matrix field, pressing **ENTER** after each entry. Finally, press **2ND** and QUIT to exit. Matrix [A] is now stored in the calculator's memory.

To convert matrix [A] into its equivalent reduced row echelon form, press MATRIX, and under MATH, scroll down to rref. Press **ENTER**. Now press MATRIX again, and under NAMES, select [A], and press **ENTER**. The result will be the reduced row echelon form equivalent to the original matrix [A]:

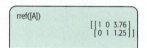

rref([A])

$$\begin{bmatrix} 1 & 0 & 3.76 \\ 0 & 1 & 1.25 \end{bmatrix}$$

Therefore, $m = 3.76$ and $b = 1.25$.

EXERCISES

Use the reduced row echelon form of a matrix to solve each system of equations with a calculator.

1. $2x + 6y = 14$
   $x - 5y = -17$

2. $3x + y = 7$
   $10x + 3y = 11$

3. $2x + y + 7 = 0$
   $x = 6 - y$

4. $x + 3y - 2z = 7$
   $2x - y + 3z = 7$
   $5x + 4y - z = 18$

Exponential Regression

This process can be expanded for other forms of regression that we often consider in this book. In the following example, we show how an exponential regression model can be developed from a set of data.

EXAMPLE 2 The table below gives the yearly revenue of Helman Insurance for 2012 through 2015.

Number of years since 2012, x	0	1	2	3
Revenue, y (in thousands of dollars)	150	235	480	815

Find the exponential model that best fits the data, and use it to predict the revenue of Helman Insurance in 2018.

Solution　We plot the given data points and note that they appear to lie on an exponential curve.

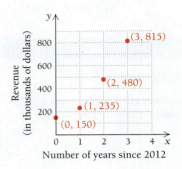

Number of years since 2012

To find a function of the form $y = a \cdot b^x$, we first take the natural logarithm of both sides:

$$\ln y = \ln(a \cdot b^x)$$
$$\ln y = \ln a + \ln b^x \qquad \text{Recalling that } \ln(uv) = \ln u + \ln v$$
$$\ln y = \ln a + x \ln b \qquad \text{Recalling that } \ln(u^x) = x \ln u$$

If we relabel $\ln y$ as $\tilde{y}$ (read "y-tilde"), $\ln b$ as $\tilde{b}$, and $\ln a$ as $\tilde{a}$, the equation $\ln y = \ln a + x \ln b$ can be treated as a linear equation in terms of x and $\tilde{y}$:

$$\tilde{y} = \tilde{b}x + \tilde{a}.$$

We find the natural logarithms of the given y-values:

x	0	1	2	3
y	150	235	480	815
$\tilde{y} = \ln y$	$\ln 150 \approx 5.0106$	$\ln 235 \approx 5.4596$	$\ln 480 \approx 6.1738$	$\ln 815 \approx 6.7032$

Plotting the ordered pairs $(x, \tilde{y})$, we see that they appear to have a linear relationship.

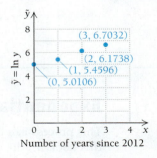

Number of years since 2012

The method outlined in Example 1 yields a regression line, $\tilde{y} = \tilde{b}x + \tilde{a}$, using the values in the table. We find that

$$\tilde{b} \approx 0.579 \text{ and } \tilde{a} \approx 4.968.$$

Since $\ln b = \tilde{b}$, we have $b = e^{\tilde{b}} = e^{0.579}$, and since $\ln a = \tilde{a}$, we have $a = e^{\tilde{a}} = e^{4.968} = 143.74$. The exponential regression model is

$$y = a \cdot b^x$$
$$= 143.74 \cdot \left(e^{0.579}\right)^x \qquad \text{Substituting}$$
$$= 143.74e^{0.579x}.$$

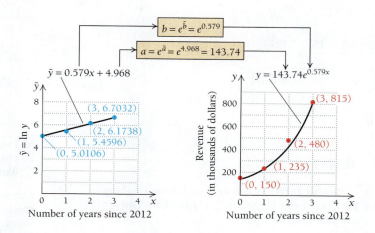

Quick Check 2 ✔

Use the method outlined in Example 2 to find an exponential regression curve for the data points $(0, 3)$, $(2, 5)$, $(4, 9)$, and $(6, 20)$.

Using this model, the estimated revenue of Helman Insurance in 2018 is found by substituting 6 for x:

$$y = 143.74e^{0.579(6)} \approx 4637.85.$$

Helman Insurance can expect a revenue of about \$4,637,850 in 2018, if growth continues at the rate it did between 2012 and 2015.   **2 ✔**

Section Summary

- Regression is a technique for determining a continuous function that "best fits" a set of data points.
- According to the least-squares assumption, the line of best fit is that for which the sum of the squares of the y-deviations is a minimum.
- For linear regression, calculus is used on a function of two variables to find the values m and b that determine the *regression line* $y = mx + b$, the line of best fit.

- To do exponential regression, we find $\tilde{y} = \ln y$ and then find a regression line $\tilde{y} = \tilde{b}x + \tilde{a}$. The *exponential regression curve* is of the form $y = ae^{bx}$, where $a = e^{\tilde{a}}$ and $b = e^{\tilde{b}}$.

6.4   Exercise Set

In Exercises 1–4, find the regression line for each data set.

1.

x	1	2	4	5
y	1	3	3	4

2.

x	1	3	5
y	2	4	7

3.

x	1	2	3	5
y	0	1	3	4

4.

x	1	2	4
y	3	5	8

In Exercises 5–8, find an exponential regression curve for each data set.

5.

x	0	1	2
y	10	19	42

6.

x	1	2	3	4
y	8	25	72	225

7.

x	1	3	7
y	8	4	1.5

8.

x	2	4	6	8
y	13	7	3.7	1

All of the following exercises can be done with a graphing calculator if your instructor so directs. The calculator can also be used to check your work.

APPLICATIONS

Business and Economics

9. Labor force. The minimum hourly wage in the United States has grown over the years, as shown in the table below.

NUMBER OF YEARS, x, SINCE 1997	MINIMUM HOURLY WAGE (dollars)
0	$5.15
10	5.85
11	6.55
12	7.25
18	10.10

(*Source:* U.S. Dept. of Labor.)

a) For the data in the table, find the regression line, $y = mx + b$.
b) Use the regression line to predict the minimum hourly wage in 2020 and 2025.

10. Football ticket prices. Ticket prices for NFL football games have experienced steady growth, as shown in the table below.

NUMBER OF YEARS, x, SINCE 2009	AVERAGE TICKET PRICE (dollars)
0	$73.18
1	76.47
2	77.36
3	79.09
4	81.54

(*Source:* Team Marketing Report.)

a) Find the regression line, $y = mx + b$.
b) Use the regression line to predict the average ticket price for an NFL game in 2020 and 2025.

Life and Physical Sciences

11. Life expectancy of women. Consider the data in the following table showing the average life expectancy of women in various years.

NUMBER OF YEARS, x, SINCE 1990	LIFE EXPECTANCY OF WOMEN, y (years)
0	78.8
10	79.5
13	80.1
17	80.4
21	81.1

(*Source:* www.ssa.gov.)

a) Find the regression line, $y = mx + b$.
b) Use the regression line to predict the life expectancy of women in 2020 and 2025.

12. Life expectancy of men. Consider the following data showing the average life expectancy of men in various years.

NUMBER OF YEARS, x, SINCE 1990	LIFE EXPECTANCY OF MEN, y (years)
0	71.8
10	74.1
13	74.8
17	75.4
21	76.3

(*Source:* www.ssa.gov.)

a) Find the regression line, $y = mx + b$.
b) Use the regression line to predict the life expectancy of men in 2020 and 2025.

General Interest

13. Grade predictions. A professor wants to predict students' final examination scores on the basis of their midterm test scores. An equation was determined on the basis of data consisting of the scores of three students who took the same course with the same instructor the previous semester (see the following table).

MIDTERM SCORE, x	FINAL EXAM SCORE, y
70%	75%
60	62
85	89

a) Find the regression line, $y = mx + b$. (*Hint:* The y-deviations are $70m + b - 75$, $60m + b - 62$, and so on.)
b) The midterm score of a student was 81%. Use the regression line to predict the student's final exam score.

14. Predicting the world record in the high jump. It has been established that most world records in track and field can be modeled by a linear function. The table below shows world high-jump records for various years.

NUMBER OF YEARS, x, SINCE 1912	WORLD RECORD IN HIGH JUMP, y (in inches)
0 (George Horme)	78.0
44 (Charles Dumas)	84.5
61 (Dwight Stones)	90.5
77 (Javier Sotomayer)	96.0
81 (Javier Sotomayer)	96.5

(*Source*: www.topendsports.com.)

a) Find the regression line, $y = mx + b$.
b) Use the regression line to predict the world record in the high jump in 2020 and in 2050.
c) Does your answer in part (b) for 2050 seem realistic? Explain why extrapolating so far into the future could be a problem.

15. Population. The data in the following table give the population of Detroit since 1970 (see Exercise 18, Section R.6). (*Source*: www.census.gov.)

NUMBER OF YEARS, x, SINCE 1970	POPULATION (in millions)
0	1.5
10	1.2
20	1
30	0.95
40	0.71

a) Find the exponential regression curve, $y = ae^{kx}$.
b) Use the regression curve to estimate the population of Detroit in 2020 and 2025.

16. Stock prices. The data in the following table give the price of one share of Starbucks stock on January 1 of various years (see Exercise 20, Section R.6). (*Sources*: yahoo.finance and NASDAQ.)

NUMBER OF YEARS, x, SINCE 2010	PRICE OF ONE SHARE OF STARBUCKS STOCK ON JANUARY 1
0	$20.59
1	$30.21
2	$46.61
3	$55.39
4	$76.17

a) Find the exponential regression curve, $y = ae^{kx}$.
b) Use the regression curve to estimate the price of one share of Starbucks stock on January 1 in 2016 and 2020.

17. How would you explain the concept of linear regression to a friend?

18. Discuss the idea of linear regression with a professor from another discipline in which regression is used. Explain how it is used in that field.

TECHNOLOGY CONNECTION

19. General interest: predicting the world record for running the mile. Note that x represents the actual year in following table.

YEAR, x	WORLD RECORD, y (in minutes:seconds)
1875 (Walter Slade)	4:24.5
1894 (Fred Bacon)	4:18.2
1923 (Paavo Nurmi)	4:10.4
1937 (Sidney Wooderson)	4:06.4
1942 (Gunder Hägg)	4:06.2
1945 (Gunder Hägg)	4:01.4
1954 (Roger Bannister)	3:59.6
1964 (Peter Snell)	3:54.1
1967 (Jim Ryun)	3:51.1
1975 (John Walker)	3:49.4
1979 (Sebastian Coe)	3:49.0
1980 (Steve Ovett)	3:48.40
1985 (Steve Cram)	3:46.31
1993 (Noureddine Morceli)	3:44.39

(*Sources*: USA Track & Field and infoplease.com.)

a) Find the regression line, $y = mx + b$, that fits the data in the table. (*Hint*: Convert each time to decimal notation; for instance, $4{:}24.5 = 4\frac{24.5}{60} = 4.4083$.)

b) Use the regression line to predict the world record in the mile in 2015 and 2020.

c) In July 1999, Hicham El Guerrouj set the current (as of July 2014) world record of 3:43.13 for the mile. (*Sources*: USA Track & Field and infoplease.com.) How does this compare with what is predicted by the regression line?

20. Power regression. The data in the following table show the percentage of net worth (as a decimal) held by the given percentage of the population (also as a decimal) (see the Extended Technology Application, Chapter 4). (*Source*: http://www.fas.org/sgp/crs/misc/RL33433.pdf, based on data from Arthur Kennickel, 2010.)

PERCENTAGE OF POPULATION, x	PERCENTAGE OF NET WORTH HELD, y
0.5	0.011
0.9	0.254
0.99	0.654
1	1

The data can be modeled by a power regression curve of the form $y = ax^b$. Use a method similar to that in Example 2 to find values for a and b.

a) Take the natural logarithm of both sides of $y = ax^b$, and use laws of logarithms to simplify the right side.

b) Relabel $\ln y$ as $\tilde{y}$, $\ln a$ as $\tilde{a}$, and $\ln x$ as $\tilde{x}$, and find a regression line that relates $\tilde{y}$ as a function of $\tilde{x}$.

c) Use the results from part (b) to write the power regression curve, $y = ax^b$.

d) Predict the percentage of the net worth held by 80% of the population.

Constrained Optimization

In Section 6.3, we discussed a method for determining maximum and minimum values of a two-variable function, $z = f(x, y)$. If restrictions are placed on the input variables x and y, we can determine the maximum and minimum values on a surface subject to the restrictions. This process is called **constrained optimization**.

- Find minimum and maximum values using Lagrange multipliers.
- Solve constrained optimization problems involving Lagrange multipliers.
- Use the Extreme Value Theorem for Two-Variable Functions to find absolute minimum and maximum values.

Path Constraints: Lagrange Multipliers

Imagine that you are hiking up a mountain. If there are no constraints on your movement, you may seek out the mountain's summit—its "maximum point." The figure at the right shows a relief map of a mountaintop; its unconstrained maximum point occurs at ●, labeled with a spot elevation of 6903 ft. A hiking trail, marked as a black dashed line, bypasses the summit. If you were constrained to this path, you would not reach the summit. You could, however, achieve a maximum elevation along the path. This constrained maximum point is approximated at M.

(*Source*: USGS maps at www.mytopo.com.)

In many applications modeled by two-variable functions, constraints on the input variables are necessary. If the input variables are related to one another by an equation, that equation is called a **constraint**.

Let's return to a problem from Chapter 2: A hobby store has 20 ft of fencing to fence off a rectangular electric-train area in a corner of its display room. The two sides up against the wall require no fence. What dimensions of the rectangle will maximize the area?

We maximize the function

$$A = xy$$

subject to the condition, or *constraint*, $x + y = 20$. Note that A is a function of two variables.

When we solved this earlier, we first solved the constraint for y:

$$y = 20 - x.$$

We then substituted $20 - x$ for y to obtain

$$A(x, y) = x(20 - x) = 20x - x^2,$$

which is a function of one variable. Next, we found a maximum value using Maximum–Minimum Principle 1 (see Section 2.4). By itself, the function of two variables

$$A(x, y) = xy$$

has no maximum value. This can be checked using the *D*-test. With the constraint $x + y = 20$, however, the function does have a maximum. We see this in the following graph.

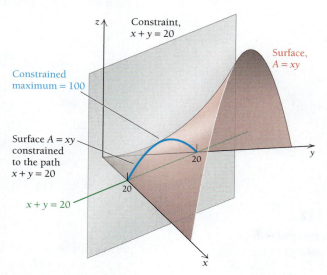

It can be difficult to solve a constraint for one variable. The method outlined below allows us to proceed without doing so.

The Method of Lagrange Multipliers

To find a maximum or minimum value of a function $f(x, y)$ subject to the constraint $g(x, y) = 0$:

1. Form a new function, called the **Lagrange function**:

$$F(x, y, \lambda) = f(x, y) - \lambda g(x, y).$$

The variable λ (lambda) is called a **Lagrange multiplier**.

2. Find the first partial derivatives F_x, F_y, and F_λ.

3. Solve the system

$$F_x = 0, \qquad F_y = 0, \quad \text{and} \quad F_\lambda = 0.$$

Let (a, b, λ) represent a solution of this system. We then must determine whether (a, b) yields a maximum or minimum of the function f.

The method of Lagrange multipliers can be extended to functions of three (or more) variables.

We illustrate the method of Lagrange multipliers by solving the electric-train area problem.

EXAMPLE 1   Find the extremum of

$$A(x, y) = xy$$

subject to the constraint $x + y = 20$.

Solution   Note first that $x + y = 20$ is equivalent to $x + y - 20 = 0$.

1. We form the Lagrange function F, given by

$$F(x, y, \lambda) = xy - \lambda \cdot (x + y - 20).$$

2. We find the first partial derivatives:

$$F_x = y - \lambda,$$
$$F_y = x - \lambda,$$
$$F_\lambda = -(x + y - 20).$$

3. We set each derivative equal to 0 and solve the resulting system:

$$y - \lambda = 0, \tag{1}$$
$$x - \lambda = 0, \tag{2}$$
$$-(x + y - 20) = 0, \quad \text{or} \quad x + y - 20 = 0. \tag{3}$$

From equations (1) and (2), it follows that $x = y = \lambda$. Substituting x for y in equation (3), we get

$$x + x - 20 = 0$$
$$2x = 20$$
$$x = 10.$$

Thus, $y = x = 10$. The extreme value of A subject to the constraint occurs at $(10, 10)$ and is

$$A(10, 10) = 10 \cdot 10 = 100.$$

Test points can be used to show that this value is a maximum. We can choose values for x and y such that $x + y = 20$ and find A: for example, $A(11, 9) = 99$, $A(9, 11) = 99$, $A(8, 12) = 96$, and so on. These results support our assertion that $A(10, 10) = 100$ is a maximum value.   **1** ✔

Quick Check 1 ✔

Find the extremum of $A(x, y) = xy$ subject to the constraint $x + 2y = 30$. State whether it is a maximum or a minimum value.

EXAMPLE 2   Find the extremum of

$$f(x, y) = 3xy$$

subject to the constraint

$$2x + y = 8.$$

Note: f might be interpreted, for example, as a production function with a budget constraint $2x + y = 8$.

Solution   We first express $2x + y = 8$ as $2x + y - 8 = 0$.

1. Next we form the Lagrange function F, given by

$$F(x, y, \lambda) = 3xy - \lambda(2x + y - 8).$$

2. We find the first partial derivatives:

$$F_x = 3y - 2\lambda,$$
$$F_y = 3x - \lambda,$$
$$F_\lambda = -(2x + y - 8).$$

3. We set each derivative equal to 0 and solve the resulting system:

$$3y - 2\lambda = 0, \tag{4}$$
$$3x - \lambda = 0, \tag{5}$$
$$-(2x + y - 8) = 0, \quad \text{or} \quad 2x + y - 8 = 0. \tag{6}$$

Solving equation (5) for λ, we get

$$\lambda = 3x.$$

Substituting for λ in equation (4), we get

$$3y - 2 \cdot 3x = 0, \quad \text{or} \quad 3y = 6x, \quad \text{or} \quad y = 2x. \tag{7}$$

Substituting $2x$ for y in equation (6), we get

$$2x + 2x - 8 = 0$$
$$4x = 8$$
$$x = 2.$$

Then, using equation (7), we have

$$y = 2 \cdot 2 = 4.$$

The extreme value of $f(x, y)$ subject to the constraint occurs at $(2, 4)$ and is

$$f(2, 4) = 3 \cdot 2 \cdot 4 = 24.$$

Test points can be used to support our assertion that this value is a maximum. **2 ✔**

TECHNOLOGY CONNECTION

Exploratory

Use a 3D graphics program to graph both $f(x, y) = 3xy$ and $2x + y = 8$. Then check the results of Example 2 visually.

Quick Check 2 ✔

Find the extremum of $g(x, y) = x^2 + y^2$ subject to the constraint $3x - y = 1$. State whether it is a maximum or a minimum value.

EXAMPLE 3 **Business: The Beverage-Can Problem.** The standard beverage can holds 12 fl. oz and has a volume of 21.66 in^3. What dimensions yield the minimum surface area? Find the minimum surface area.

Solution We want to minimize s, the surface area of a right circular cylinder, given by

$$s(h, r) = 2\pi rh + 2\pi r^2$$

subject to the volume constraint

$$\pi r^2 h = 21.66,$$

or $\quad \pi r^2 h - 21.66 = 0.$

Note that $s(h, r)$ does not have a minimum without the constraint.

1. We form the Lagrange function S, given by

$$S(h, r, \lambda) = 2\pi rh + 2\pi r^2 - \lambda(\pi r^2 h - 21.66).$$

2. We find the first partial derivatives:

$$\frac{\partial S}{\partial h} = 2\pi r - \lambda \pi r^2,$$

$$\frac{\partial S}{\partial r} = 2\pi h + 4\pi r - 2\lambda \pi rh,$$

$$\frac{\partial S}{\partial \lambda} = -(\pi r^2 h - 21.66).$$

3. We set each derivative equal to 0 and solve the resulting system:

$$2\pi r - \lambda \pi r^2 = 0, \tag{8}$$
$$2\pi h + 4\pi r - 2\lambda \pi rh = 0, \tag{9}$$
$$-(\pi r^2 h - 21.66) = 0, \quad \text{or} \quad \pi r^2 h - 21.66 = 0. \tag{10}$$

We can solve equation (8) for r:

$$\pi r(2 - \lambda r) = 0$$

$$\pi r = 0 \quad \text{or} \quad 2 - \lambda r = 0$$

$$r = 0 \quad \text{or} \quad r = \frac{2}{\lambda}. \qquad \text{We assume } \lambda \neq 0.$$

Since $r = 0$ cannot be a solution to the original problem, we solve for h by substituting $2/\lambda$ for r in equation (9):

$$2\pi h + 4\pi \cdot \frac{2}{\lambda} - 2\lambda\pi \cdot \frac{2}{\lambda} \cdot h = 0$$

$$2\pi h + \frac{8\pi}{\lambda} - 4\pi h = 0$$

$$\frac{8\pi}{\lambda} - 2\pi h = 0$$

$$\frac{8\pi}{\lambda} = 2\pi h,$$

so

$$\frac{4}{\lambda} = h.$$

Since $h = 4/\lambda$ and $r = 2/\lambda$, it follows that $h = 2r$. Substituting $2r$ for h in equation (10) yields

$$\pi r^2(2r) - 21.66 = 0 \qquad \text{Using equation (1)}$$

$$2\pi r^3 - 21.66 = 0$$

$$2\pi r^3 = 21.66$$

$$\pi r^3 = 10.83$$

$$r^3 = \frac{10.83}{\pi}$$

$$r = \sqrt[3]{\frac{10.83}{\pi}} \approx 1.51 \text{ in.}$$

Thus, when $r = 1.51$ in., we have $h = 3.02$ in. The surface area is then a minimum and is approximately

$$2\pi(1.51)(3.02) + 2\pi(1.51)^2, \quad \text{or about } 42.98 \text{ in}^2.$$

Test points can be used to support our assertion that this value is a minimum. **3** ✔

Quick Check 3 ✔

Repeat Example 3 for a right circular cylinder with no top and a volume of 500 mL.

The actual dimensions of a standard 12-oz beverage can are $r = 1.25$ in. and $h = 4.875$ in. A natural question is, "Why don't beverage companies make cans using the dimensions found in Example 3?" To do this would mean an enormous cost for retooling. New machines, including vending machines, would have to be designed. Market research has also shown that a can with the dimensions found in Example 3 is not as comfortable to hold and might not be accepted by consumers. A partial response to the desire to save aluminum has been found in recycling and in manufacturing cans with bevelled edges. These cans require less aluminum. As a result of many engineering advances, the amount of aluminum required to make 1000 cans has been reduced over the years from 36.5 lb to 28.1 lb.

Closed and Bounded Regions: The Extreme-Value Theorem

In Examples 1, 2, and 3, all constraints were equations. Constraints may also be inequalities. Multiple constraints on the input variables x and y can be plotted on the xy-plane to form a *region of feasibility*, which contains all (x, y) pairs that satisfy all the constraints simultaneously. If the constraints form a closed and bounded region (*closed* meaning it includes the boundaries, and *bounded* meaning it has finite area, with no portions tending to infinity), then the Extreme-Value Theorem can be adapted for the two-variable function.

THEOREM   Extreme-Value Theorem for Two-Variable Functions

If $f(x, y)$ is continuous for all (x, y) within a region of feasibility that is closed and bounded, then f has both an absolute maximum value and an absolute minimum value over that region.

Recall from Section 2.4 that a continuous function f defined on a closed interval $[a, b]$ must have an absolute maximum value and an absolute minimum value over $[a, b]$. In such cases, it is possible that the absolute maximum or minimum value occurs at an endpoint of the interval. Similarly, absolute maximum or minimum values may occur on the boundaries of a region of feasibility.

EXAMPLE 4   Business: Maximizing Revenue. Kim creates embroidered knit caps; one style has a script x on the front, the other a script y on the front. She sells them to raise funds for the Math Club. She determines that her weekly revenue is modeled by the two-variable function

$$R(x, y) = -x^2 - xy - y^2 + 20x + 22y - 25,$$

where x is the number of x-caps produced and sold and y is the number of y-caps produced and sold. Kim spends 2 hr working on each x-cap, and 4 hr on each y-cap, and she works no more than 40 hr per week. How many of each style should she produce and sell in order to maximize her weekly revenue? Assume $x \geq 0$ and $y \geq 0$.

Solution   The number of hours that Kim works is a constraint:

$$2x + 4y \leq 40.$$

Along with $x \geq 0$ and $y \geq 0$, we sketch these constraints and shade the region of feasibility.

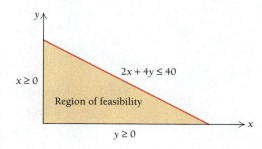

The region of feasibility is closed and bounded. Since R is continuous for all x and y in this region, by the Extreme-Value Theorem for Two-Variable Functions, R will have an absolute maximum value.

First, we check for possible critical points within the interior of the region using techniques from Section 6.3. Differentiating R with respect to x and to y, we have

$$R_x = -2x - y + 20,$$
$$R_y = -x - 2y + 22.$$

Setting these expressions equal to 0, we solve the system for x and y:

$$-2x - y + 20 = 0,$$
$$-x - 2y + 22 = 0;$$
$$\left. \begin{array}{l} 2x + y = 20 \\ x + 2y = 22 \end{array} \right\} \quad \text{After simplification}$$

The solution of this system is $x = 6$ and $y = 8$. However, this point lies outside the region of feasibility, since $2(6) + 4(8) = 44$, which is greater than 40. Thus, this point is discarded (see the explanation below).

The boundaries must also be checked for possible critical points. In the process, the points at which two constraints intersect will also be considered as critical points.

● Along the y-axis, we have $x = 0$ for $0 \leq y \leq 10$:

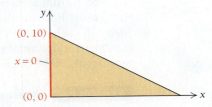

To find possible critical points, we substitute $x = 0$ into R:

$$R(0, y) = 0^2 - 0 \cdot y - y^2 + 20 \cdot 0 + 22y - 25$$
$$= -y^2 + 22y - 25.$$

The first derivative is $R_y(0, y) = -2y + 22$. Setting $-2y + 22$ equal to 0, we obtain $y = 11$. However, $y = 11$ lies outside the region of feasibility, and is ignored. However, the two endpoints of the interval, $(0, 0)$ and $(0, 10)$, are considered as possible critical points.

● Along the x-axis, we have $y = 0$ for $0 \leq x \leq 20$:

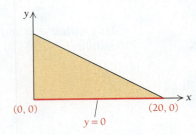

To find possible critical points, we substitute $y = 0$ into R:

$$R(x, 0) = -x^2 - x \cdot 0 - 0^2 + 20x + 22 \cdot 0 - 25$$
$$= -x^2 + 20x - 25.$$

The first derivative is $R_x(x, 0) = -2x + 20$. Setting $-2x + 20$ equal to 0, we obtain $x = 10$. This lies within the region of feasibility, so it is a critical point. The two endpoints of the interval, $(0, 0)$ and $(20, 0)$, are also considered as possible critical points.

● Along the line $2x + 4y = 40$, we have $0 \leq x \leq 20$ and $0 \leq y \leq 10$. The endpoints, $(0, 10)$ and $(20, 0)$, are critical points, identified earlier.

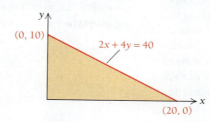

To check for critical points along the line $2x + 4y = 40$, we use the method of Lagrange multipliers. The constraint is written as $2x + 4y - 40 = 0$, and we have

$$F(x, y, \lambda) = R(x, y) - \lambda(2x + 4y - 40)$$
$$= -x^2 - xy - y^2 + 20x + 22y - 25 - 2\lambda x - 4\lambda y + 40\lambda.$$

The first derivatives are

$$F_x = -2x - y + 20 - 2\lambda$$
$$F_y = -x - 2y + 22 - 4\lambda$$
$$F_\lambda = -2x - 4y + 40.$$

We set each partial derivative equal to 0:

$$-2x - y + 20 - 2\lambda = 0 \tag{11}$$
$$-x - 2y + 22 - 4\lambda = 0 \tag{12}$$
$$-2x - 4y + 40 = 0.$$

We solve equations (11) and (12) for λ:

$$\lambda = -x - \tfrac{1}{2}y + 10 \quad \text{and} \quad \lambda = -\tfrac{1}{4}x - \tfrac{1}{2}y + \tfrac{11}{2}.$$

Because both equations include the term $-\tfrac{1}{2}y$, let's equate their right-hand sides and solve for x:

$$-x - \tfrac{1}{2}y + 10 = -\tfrac{1}{4}x - \tfrac{1}{2}y + \tfrac{11}{2}$$
$$-\tfrac{3}{4}x = -\tfrac{9}{2}$$
$$x = 6.$$

We now substitute 6 for x in the constraint:

$$2(6) + 4y = 40$$
$$12 + 4y = 40$$
$$4y = 28$$
$$y = 7.$$

Since $(6, 7)$ is in the region of feasibility, it is a critical point.

We now have six critical points, which we evaluate in the revenue function $R(x, y)$:

Critical point (x, y)	$R(x, y)$
$(0, 0)$	-25
$(0, 10)$	95
$(10, 0)$	75
$(20, 0)$	-25
$(6, 7)$	122

Quick Check 4 ✔

Repeat Example 4, using the same revenue function but assuming that each x-cap requires 4 hr to create, each y-cap requires 2 hr to create, and Kim is willing to work at most 36 hr per week. How many of each style of cap should she produce to maximize her weekly revenue?

Thus, revenue is maximized when Kim produces 6 of the x-caps and 7 of the y-caps, for a weekly revenue of $122. If there were no constraints, the maximum weekly revenue of $123 would occur at $x = 6$ and $y = 8$. Kim might think it is not worth working an extra 4 hr for one more dollar in revenue.

4 ✔

Section Summary

- If input variables x and y for a function $f(x, y)$ are related by another equation, that equation is a *constraint*.
- *Constrained optimization* is a method of determining maximum and minimum points on a surface represented by $z = f(x, y)$, subject to given restrictions (constraints) on the input variables x and y.
- The *method of Lagrange multipliers* allows us to find a maximum or minimum value of $f(x, y)$ subject to the constraint $g(x, y) = 0$.

- If the constraints are inequalities, the set of points that satisfy all the constraints simultaneously is called the *region of feasibility*.
- If the region of feasibility is closed and bounded and the surface $z = f(x, y)$ is continuous over the region, then the *Extreme-Value Theorem* guarantees that f will have both an absolute maximum and an absolute minimum value over that region.
- Critical points may be located at vertices, along a boundary, or in the interior of a region of feasibility.

6.5 | Exercise Set

Find the extremum of $f(x, y)$ subject to the given constraint, and state whether it is a maximum or a minimum.

1. $f(x, y) = xy$; $3x + y = 10$

2. $f(x, y) = 2xy$; $4x + y = 16$

3. $f(x, y) = x^2 + y^2$; $2x + y = 10$

4. $f(x, y) = x^2 + y^2$; $x + 4y = 17$

5. $f(x, y) = 4 - x^2 - y^2$; $x + 2y = 10$

6. $f(x, y) = 3 - x^2 - y^2$; $x + 6y = 37$

7. $f(x, y) = 2y^2 - 6x^2$; $2x + y = 4$

8. $f(x, y) = 2x^2 + y^2 - xy$; $x + y = 8$

9. $f(x, y, z) = x^2 + y^2 + z^2$; $y + 2x - z = 3$

10. $f(x, y, z) = x^2 + y^2 + z^2$; $x + y + z = 2$

Use the method of Lagrange multipliers to solve each of the following.

11. Of all numbers whose sum is 50, find the two that have the maximum product.

12. Of all numbers whose sum is 70, find the two that have the maximum product.

13. Of all numbers whose difference is 6, find the two that have the minimum product.

14. Of all numbers whose difference is 4, find the two that have the minimum product.

15. Of all points (x, y, z) that satisfy $x + 2y + 3z = 13$, find the one that minimizes
$$(x - 1)^2 + (y - 1)^2 + (z - 1)^2.$$

16. Of all points (x, y, z) that satisfy $3x + 4y + 2z = 52$, find the one that minimizes
$$(x - 1)^2 + (y - 4)^2 + (z - 2)^2.$$

17. Find the point on the line $3x + y = 6$ that is closest to the origin.

18. Find the point on the line $-2x + 5y = 10$ that is closest to the origin.

APPLICATIONS

Business and Economics

19. **Maximizing typing area.** A standard piece of printer paper has a perimeter of 39 in. Find the dimensions of the paper that will give the most area. What is that area? Does standard $8\frac{1}{2} \times 11$ in. paper have maximum area?

20. **Maximizing room area.** A carpenter is building a rectangular room with a fixed perimeter of 80 ft. What are the dimensions of the largest room that can be built? What is its area?

21. Minimizing surface area. An oil drum of standard size has a volume of 200 gal, or 27 ft³. What dimensions yield the minimum surface area? Find the minimum surface area.

Do these drums appear to be made in such a way as to minimize surface area?

22. Juice-can problem. A large juice can has a volume of 99 in³. What dimensions yield the minimum surface area? Find the minimum surface area.

23. Maximizing total sales. Total sales, S, of Cre-Tech are given by

$$S(L, M) = ML - L^2,$$

where M is the cost of materials and L is the cost of labor. Find the maximum value of this function subject to the budget constraint

$$M + L = 90.$$

24. Maximizing total sales. Total sales, S, of Sea Change, Inc., are given by

$$S(L, M) = ML - L^2,$$

where M is the cost of materials and L is the cost of labor. Find the maximum value of this function subject to the budget constraint

$$M + L = 70.$$

25. Minimizing construction costs. Denney Construction is planning to build a warehouse whose interior volume is to be 252,000 ft³. Construction costs per square foot are estimated as follows:

Walls: $3.00
Floor: $4.00
Ceiling: $3.00

a) The total cost of the building is $C(x, y, z)$, where x is the length, y is the width, and z is the height, all in feet. Find a formula for $C(x, y, z)$.

b) What dimensions of the building will minimize the total cost? What is the minimum cost?

26. Minimizing the costs of container construction. Northside Nursery is designing a 12-ft³ shipping crate with a square bottom and top. The cost of the top and the sides is $2 per square foot, and the cost of the bottom is $3 per square foot. What dimensions will minimize the cost of the crate?

27. Minimizing total cost. Each unit of a product can be made on either machine A or machine B. The nature of the machines makes their cost functions differ:

Machine A: $C(x) = 10 + \dfrac{x^2}{6},$

Machine B: $C(y) = 200 + \dfrac{y^3}{9}.$

Total cost is given by $C(x, y) = C(x) + C(y)$. How many units should be made on each machine in order to minimize total costs if $x + y = 10{,}100$ units are required?

28. Minimizing distance and cost. A highway passes by the small town of Las Cienegas. From Las Cienegas, the highway is 5 miles to the north and 3 miles to the east. Assume that the highway is straight as it passes through this region. The town wants to build an access road at a cost of $250,000 per mile to connect to the highway. What is the shortest possible distance (to three decimal places) from Las Cienegas to the highway, and what would be the minimum cost, to the nearest dollar, of constructing such a road?

29. Minimizing distance and cost. From the center of Bridgeton, a water main runs northwest at a 45° angle relative to due north and due west. A new house is being built at a point 2 miles west and 1 mile north of the center of town. The cost to connect the house to the water main is $8,000 per mile. What is the shortest possible distance (to three decimal places) from the house to the water main, and what would be the minimum cost, to the nearest dollar, of connecting the house to the water main?

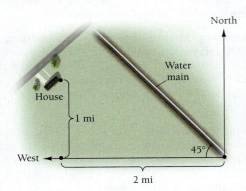

In Exercises 30–33, find the absolute maximum and minimum values of each function, subject to the given constraints.

30. $f(x, y) = x^2 + y^2 - 2x - 2y$; $x \geq 0, y \geq 0, x \leq 4$, and $y \leq 3$

31. $g(x, y) = x^2 + 2y^2$; $-1 \leq x \leq 1$ and $-1 \leq y \leq 2$

32. $h(x, y) = x^2 + y^2 - 4x - 2y + 1$; $x \geq 0, y \geq 0$, and $x + 2y \leq 5$

33. $k(x, y) = -x^2 - y^2 + 4x + 4y$; $0 \leq x \leq 3, y \geq 0$, and $x + y \leq 6$

34. Business: minimizing costs with constraints. Farmer Frank grows two crops: celery and lettuce. He has determined that the cost of planting these crops is modeled by

$$C(x, y) = x^2 + 3xy + 3.5y^2 - 775x - 1600y + 250,000,$$

where x is the number of acres of celery and y is the number of acres of lettuce. Suppose Farmer Frank has 300 acres available for planting and must plant more acres of lettuce than of celery. Find the number of acres of celery and of lettuce he should plant to minimize the cost, and state the cost.

35. Business: maximizing profits with constraints. A manufacturer of decorative end tables produces two models, basic and large. Its weekly profit function is modeled by

$$P(x, y) = -x^2 - 2y^2 - xy + 140x + 210y - 4300,$$

where x is the number of basic models sold each week and y is the number of large models sold each week. The warehouse can hold at most 90 tables. Assume that x and y must be nonnegative. How many of each model should be produced to maximize weekly profit, and what will the maximum profit be?

SYNTHESIS

Find the indicated maximum or minimum values of $f(x, y)$ subject to the given constraint.

36. Minimum: $f(x, y) = xy$; $x^2 + y^2 = 9$

37. Minimum: $f(x, y) = 2x^2 + y^2 + 2xy + 3x + 2y$; $y^2 = x + 1$

38. Maximum: $f(x, y, z) = x + y + z$; $x^2 + y^2 + z^2 = 1$

39. Maximum: $f(x, y, z) = x^2 y^2 z^2$; $x^2 + y^2 + z^2 = 2$

40. Maximum: $f(x, y, z) = x + 2y - 2z$; $x^2 + y^2 + z^2 = 4$

41. Maximum: $f(x, y, z, t) = x + y + z + t$; $x^2 + y^2 + z^2 + t^2 = 1$

42. Minimum: $f(x, y, z) = x^2 + y^2 + z^2$; $x - 2y + 5z = 1$

43. Economics: the Law of Equimarginal Productivity. Suppose $p(x, y)$ represents the production of a two-product firm. The company produces x units of the first product at a cost of c_1 each and y units of the second product at a cost of c_2 each. The budget constraint, B, is

$$B = c_1 x + c_2 y.$$

Use the method of Lagrange multipliers to find the value of λ in terms of p_x, p_y, c_1, and c_2. The resulting equation holds for any production function p and is called the *Law of Equimarginal Productivity*.

44. Business: maximizing production. A computer company has the following Cobb–Douglas production function for a certain product:

$$p(x, y) = 800x^{3/4} y^{1/4},$$

where x is labor and y is capital, both measured in dollars. Suppose the company can make a total investment in labor and capital of $1,000,000. How should it allocate the investment between labor and capital in order to maximize production?

45. Discuss the difference between solving a maximum–minimum problem using the method of Lagrange multipliers and the method of Section 6.3.

46. Write a brief report on the life and work of the mathematician Joseph Louis Lagrange (1736–1813).

47. Find the point on the parabola $y = x^2 + 2x - 5$ that is closest to the origin.

48. Find the point on the circle $x^2 + y^2 = 1$ that is closest to the point $(2, 1)$.

49–56. Use a 3D graphics program to graph both equations in each of Exercises 1–8. Then visually check the results that you found analytically.

Answers to Quick Checks

1. $A = \dfrac{225}{2}$ at $x = 15, y = \dfrac{15}{2}$, maximum   **2.** $g = \dfrac{1}{10}$ at $x = \dfrac{3}{10}, y = -\dfrac{1}{10}$, minimum   **3.** $r \approx 5.419$ cm, $h \approx 5.419$ cm, $s \approx 276.76$ cm^2 **4.** $x = 5, y = 8$, maximum revenue $= \$122$

Double Integrals

So far in this chapter, we have discussed functions of two variables and their partial derivatives. In this section, we consider integration of a function of two variables, in a process called *iterated integration*.

The following is an example of a *double integral*:

$$\int_3^6 \int_{-1}^2 10xy^2 \, dx \, dy, \quad \text{or} \quad \int_3^6 \left(\int_{-1}^2 10xy^2 \, dx \right) dy.$$

Evaluating a double integral is somewhat similar to "undoing" a second partial derivative. We first evaluate the inside integral, with respect to the innermost differential (here dx), and treat the other variable(s) (here y) as constant(s):

$$\int_{-1}^2 10xy^2 \, dx = 10y^2 \left[\frac{x^2}{2} \right]_{-1}^2 = 5y^2[x^2]_{-1}^2 = 5y^2[2^2 - (-1)^2] = 15y^2.$$

Color indicates the variable of integration.

Then we evaluate the outside integral, associated with the differential dy:

$$\int_3^6 15y^2 \, dy = 15 \left[\frac{y^3}{3} \right]_3^6$$
$$= 5[y^3]_3^6$$
$$= 5(6^3 - 3^3)$$
$$= 945.$$

More precisely, the given double integral is called a **double iterated integral**. The word "iterate" means "to do again."

If dx and dy, as well as the limits of integration, are interchanged, we have

$$\int_{-1}^2 \int_3^6 10xy^2 \, dy \, dx.$$

We first evaluate the inside, y-integral, treating x as a constant:

$$\int_3^6 10xy^2 \, dy = 10x \left[\frac{y^3}{3} \right]_3^6 \qquad \text{Treating } y \text{ as the variable}$$
$$= \frac{10x}{3} \left[y^3 \right]_3^6$$
$$= \frac{10}{3} x(6^3 - 3^3) = 630x.$$

Then we evaluate the outside, x-integral:

$$\int_{-1}^2 630x \, dx = 630 \left[\frac{x^2}{2} \right]_{-1}^2$$
$$= 315[x^2]_{-1}^2$$
$$= 315[2^2 - (-1)^2] = 945.$$

Note that we get the same result as we did before.

> **DEFINITION**
>
> If $f(x, y)$ is defined over the rectangular region R bounded by $a \le x \le b$ and $c \le y \le d$, then the **double integral** of $f(x, y)$ over R is given by
>
> $$\int_c^d \int_a^b f(x, y)\, dx\, dy \quad \text{or} \quad \int_a^b \int_c^d f(x, y)\, dy\, dx.$$

In a more technical definition of the double integral, Riemann sums are used. However, for the functions in this text, the above definition is sufficient.

Sometimes double integrals are defined over a nonrectangular region, in which case the bounds of integration may contain variables.

EXAMPLE 1 Evaluate

$$\int_0^1 \int_{x^2}^x xy^2\, dy\, dx.$$

Solution We first evaluate the inside integral with respect to y, treating x as a constant:

$$\int_{x^2}^x xy^2\, dy = x\left[\frac{y^3}{3}\right]_{x^2}^x$$

$$= \frac{1}{3}x[x^3 - (x^2)^3]$$

$$= \frac{1}{3}(x^4 - x^7).$$

Then we evaluate the outside integral:

$$\frac{1}{3}\int_0^1 (x^4 - x^7)\, dx = \frac{1}{3}\left[\frac{x^5}{5} - \frac{x^8}{8}\right]_0^1$$

$$= \frac{1}{3}\left[\left(\frac{1^5}{5} - \frac{1^8}{8}\right) - \left(\frac{0^5}{5} - \frac{0^8}{8}\right)\right] = \frac{1}{40}.$$

Quick Check 1 ✔

Evaluate

$$\int_0^4 \int_{\frac{1}{2}x}^{\sqrt{x}} 2xy\, dy\, dx.$$

Thus, $\displaystyle\int_0^1 \int_{x^2}^x xy^2\, dy\, dx = \frac{1}{40}.$

1 ✔

The Geometric Interpretation of Multiple Integrals

Suppose region D in the xy-plane is bounded by the graphs of $y_1 = g(x)$ and $y_2 = h(x)$ and the lines $x_1 = a$ and $x_2 = b$. We want the volume, V, of the solid between D and the surface $z = f(x, y)$. We can think of the solid as composed of many vertical columns, one of which is shown in Fig. 1 in red. The volume of this column can be thought of as $l \cdot w \cdot h$, or $z \cdot \Delta y \cdot \Delta x$. Integrating such columns in the y-direction, we obtain

$$\int_{y_1}^{y_2} z\, dy, \quad \text{so} \quad \left[\int_{y_1}^{y_2} z\, dy\right]\Delta x$$

can be pictured as a "slab," or slice, outlined in blue in Fig. 1. Integrating such slices in the x-direction, we obtain the entire volume:

$$V = \int_a^b \left[\int_{y_1}^{y_2} z \, dy \right] dx,$$

This can be thought of as a collection of slices that fills the volume.

or

$$V = \int_a^b \int_{g(x)}^{h(x)} z \, dy \, dx,$$

where $z = f(x, y)$.

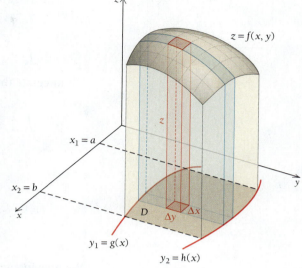

FIGURE 1

In Example 1, the region of integration D is the plane region between the graphs of $y = x^2$ and $y = x$, as shown in Figs. 2 and 3.

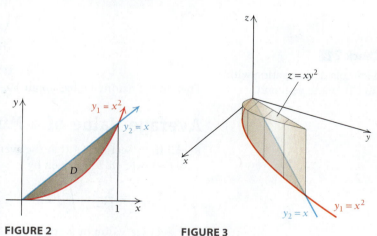

FIGURE 2   **FIGURE 3**

When we evaluated the double integral in Example 1, we found the volume of the solid based on D and capped by the surface $z = xy^2$, as shown in Fig. 3.

EXAMPLE 2 **Business: Demographics and Vehicle Ownership.** The density of privately owned vehicles in Leatonville is given by $p(x, y) = \frac{1}{8}xy$, where x is miles in the east–west direction, y is miles in the north–south direction, and p is the number of privately owned vehicles per square mile, in thousands. If the city limits are as shown in the figure to the left, what is the total number of privately owned vehicles in Leatonville?

Solution If we choose to integrate with respect to y first and x second, the iterated integral is

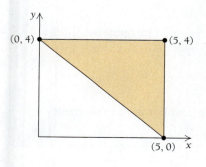

$$\int_a^b \int_{g_1(x)}^{g_2(x)} \frac{1}{8}xy \, dy \, dx.$$ See below for clarification of $g_1(x)$ and $g_2(x)$.

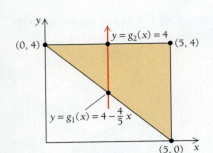

$y = g_2(x) = 4$

$(0, 4)$ $(5, 4)$

$y = g_1(x) = 4 - \frac{4}{5}x$

$(5, 0)$ x

Since we are integrating with respect to y first, a helpful visual method for determining the bounds of integration is to draw an arrow in the positive y-direction, intersecting the region representing the city. The arrow enters the region at $g_1(x) = 4 - \frac{4}{5}x$ and exits the region through $g_2(x) = 4$. The bounds for the outer integral, with respect to x, are constants: the region extends from $x = 0$ to $x = 5$.

The diagonal boundary is a line with slope $-\frac{4}{5}$ and a y-intercept of 4. Therefore, $g_1(x) = 4 - \frac{4}{5}x$. The horizontal boundary is $g_2(x) = 4$. The integral is thus

$$\int_0^5 \int_{4-\frac{4}{5}x}^4 \frac{1}{8}xy \, dy \, dx.$$

We integrate the inside integral first, with respect to y:

$$\int_{4-\frac{4}{5}x}^4 \frac{1}{8}xy \, dy = \frac{1}{8}x\left[\frac{1}{2}y^2\right]_{4-\frac{4}{5}x}^4$$

$$= \frac{1}{8}x\left[\frac{1}{2}(4)^2 - \frac{1}{2}\left(4 - \frac{4}{5}x\right)^2\right]$$

$$= \frac{1}{8}x\left(\frac{16}{5}x - \frac{8}{25}x^2\right) \qquad \text{Simplifying}$$

$$= \frac{2}{5}x^2 - \frac{1}{25}x^3.$$

We now integrate with respect to x:

$$\int_0^5 \left(\frac{2}{5}x^2 - \frac{1}{25}x^3\right)dx = \left[\frac{2}{15}x^3 - \frac{1}{100}x^4\right]_0^5$$

$$= \left(\frac{2}{15}(5)^3 - \frac{1}{100}(5)^4\right) - 0$$

$$= \frac{125}{12} \approx 10.417.$$

Therefore, Leatonville has about 10,417 privately owned vehicles.   **2 ✔**

Quick Check 2 ✔

Repeat Example 2, integrating with respect to x first and y second.

Average Value of a Multivariable Function

Recall from Section 4.4 that the average value of a continuous single-variable function $y = f(x)$ over $[a, b]$ is given by

$$y_{av} = \frac{1}{b - a}\int_a^b f(x) \, dx.$$

The average value of a continuous two-variable function $z = f(x, y)$ is found in a similar way.

> **DEFINITION**
>
> Let $z = f(x, y)$ be a continuous function over a region of integration R in the xy-plane. The **average value**, z_{av}, over R is given by
>
> $$z_{av} = \frac{1}{A(R)} \iint_R f(x, y) \, dy \, dx,$$
>
> where $A(R)$ is the area of the region of integration and R is used to determine the limits of integration.

EXAMPLE 3 **Business: Average Vehicle Ownership.** Find the average number of privately owned vehicles per square mile in Leatonville (see Example 2).

Solution Since the region of integration R is a triangle with vertices $(5, 0), (5, 4)$, and $(0, 4)$, its area can be found using the formula for the area of a triangle, $A = \frac{1}{2}bh$. Thus, we have

$$A(R) = \tfrac{1}{2}(5)(4) = 10.$$

Therefore, the average number of privately owned vehicles per square mile in the city is

$$p_{av} = \frac{1}{A(R)} \int_0^5 \int_{4-\frac{4}{5}x}^4 \tfrac{1}{8}xy \, dy \, dx$$

$$\approx \frac{1}{10} \cdot 10.417 \qquad \textcolor{red}{\text{Using the result from Example 2}}$$

$$\approx 1.042.$$

Leatonville has an average of about 1042 privately owned vehicles per square mile.   **3** ✔

Quick Check 3 ✔

Find the average value of $g(x, y) = 2x + y$, where $0 \le x \le 4$ and $-2 \le y \le 1$.

Section Summary

- The *double integral* of a two-variable function $f(x, y)$ over a rectangular region R bounded by $a \le x \le b$ and $c \le y \le d$ is written

  $$\int_c^d \int_a^b f(x, y) \, dx \, dy \qquad \text{or} \qquad \int_a^b \int_c^d f(x, y) \, dy \, dx.$$

- If the region of integration is not rectangular, the double integral may have variables in its bounds.

- The *average value* of a continuous two-variable function $z = f(x, y)$ over a region R is given by

  $$z_{av} = \frac{1}{A(R)} \iint_R f(x, y) \, dy \, dx,$$

  where $A(R)$ is the area of the region of integration and R is used to determine the limits of integration.

6.6 Exercise Set

Evaluate.

1. $\displaystyle\int_0^4 \int_0^3 3x \, dx \, dy$

2. $\displaystyle\int_0^3 \int_0^2 2y \, dx \, dy$

3. $\displaystyle\int_{-1}^3 \int_1^2 x^2 y \, dy \, dx$

4. $\displaystyle\int_1^4 \int_{-2}^1 x^3 y \, dy \, dx$

5. $\displaystyle\int_{-4}^{-1} \int_1^3 (x + 5y) \, dx \, dy$

6. $\displaystyle\int_0^5 \int_{-2}^{-1} (3x + y) \, dx \, dy$

7. $\displaystyle\int_0^1 \int_x^1 xy \, dy \, dx$

8. $\displaystyle\int_{-1}^1 \int_x^2 (x^2 + y) \, dy \, dx$

9. $\displaystyle\int_0^1 \int_{x^2}^x (x + y) \, dy \, dx$

10. $\displaystyle\int_0^2 \int_0^x e^{x+y} \, dy \, dx$

11. $\displaystyle\int_0^1 \int_1^{e^x} \frac{1}{y} \, dy \, dx$

12. $\displaystyle\int_0^1 \int_{-1}^x (x^2 + y^2) \, dy \, dx$

13. $\displaystyle\int_0^2 \int_0^x (x + y^2) \, dy \, dx$

14. $\displaystyle\int_1^3 \int_0^x 2e^{x^2} \, dy \, dx$

15. Find the volume of the solid capped by the surface $z = 1 - y - x^2$ over the region bounded on the xy-plane by $y = 1 - x^2, y = 0, x = 0$, and $x = 1$, by evaluating the integral

$$\int_0^1 \int_0^{1-x^2} (1 - y - x^2) \, dy \, dx.$$

16. Find the volume of the solid capped by the surface $z = x + y$ over the region bounded on the xy-plane by $y = 1 - x, y = 0, x = 0$, and $x = 1$, by evaluating the integral

$$\int_0^1 \int_0^{1-x} (x + y) \, dy \, dx.$$

17. Find the average value of $f(x, y) = 2x - y$, where $0 \le x \le 2$ and $2 \le y \le 3$.

18. Find the average value of $g(x, y) = 4 - x - y$, where $-1 \le x \le 1$ and $-2 \le y \le 3$.

19. Find the average value of $f(x, y) = x^2 y$, where the region of integration is a triangle with vertices $(0, 0), (6, 0)$, and $(6, 3)$.

20. Find the average value of $g(x, y) = 2 + xy$, where the region of integration is a triangle with vertices $(0, 0)$, $(4, 4)$, and $(4, -4)$.

21. Life sciences: population. The population density of fireflies in a field is given by $p(x, y) = \frac{1}{100}x^2 y$, where $0 \leq x \leq 30$ and $0 \leq y \leq 20$, x and y are in yards, and p is the number of fireflies per square yard.

    a) Determine the total population of fireflies in this field.
    b) Determine the average number of fireflies per square yard of the field.

22. Life sciences: population. The population density of Gladstone City is given by $p(x, y) = 2x^2 + 5y$, where x and y are in miles and p is the number of people per square mile, in thousands. The city limits are as shown in the graph below.

    a) Determine the city's population.
    b) Determine the average number of people per square mile of the city.

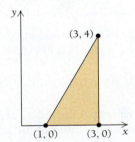

SYNTHESIS

A triple iterated integral such as

$$\int_r^s \int_c^d \int_a^b f(x, y, z)\, dx\, dy\, dz$$

is evaluated in much the same way as a double iterated integral. We first evaluate the inside x-integral, treating y and z as constants. Then we evaluate the middle y-integral, treating z as a constant. Finally, we evaluate the outside z-integral. Evaluate the following triple integrals.

23. $\displaystyle\int_0^1 \int_1^3 \int_{-1}^2 (2x + 3y - z)\, dx\, dy\, dz$

24. $\displaystyle\int_0^2 \int_1^4 \int_{-1}^2 (8x - 2y + z)\, dx\, dy\, dz$

25. $\displaystyle\int_0^1 \int_0^{1-x} \int_0^{2-x} xyz\, dz\, dy\, dx$

26. $\displaystyle\int_0^2 \int_{2-y}^{6-2y} \int_0^{\sqrt{4-y^2}} z\, dz\, dx\, dy$

27. Describe the geometric meaning of the double integral of a function of two variables.

28. Explain how Exercise 1 can be answered without finding any antiderivatives.

A function $z = f(x, y)$ is a joint probability density function over a region R in the xy-plane if $f(x, y) \geq 0$ for all x and y in R and if $\displaystyle\iint_R f(x, y)\, dx\, dy = 1$. Use this information in Exercises 29 and 30.

29. Suppose a dart is thrown at a region R in the xy-plane for which $1 \leq x \leq 2$ and $3 \leq y \leq 5$, with a joint probability density function given by

$$f(x, y) = x^2 - 3x + \tfrac{1}{3}xy - \tfrac{1}{3}y + 2.$$

    a) Verify that $\displaystyle\iint_R f(x, y)\, dx\, dy = 1$.
    b) Find the probability that the dart lands at a point in R for which $1 \leq x \leq 2$ and $3 \leq y \leq 4$.
    c) Find the probability that the dart lands at a point in R for which $1 \leq x \leq 2$ and $4 \leq y \leq 5$.
    d) Find the probability that the dart lands at a point in R for which $y \leq x + 2$.

30. Suppose a dart is thrown at a square region R for which $-1 \leq x \leq 1$ and $-1 \leq y \leq 1$, and the joint probability density function g is uniform over R.

    a) Find g.
    b) Find the probability that the dart lands at a point where both x and y are non-negative.
    c) Use the result from part (b) to find the probability that the dart lands at a point where at least one of x and y is negative.
    d) Explain why the probability of the dart landing on the origin is 0.

TECHNOLOGY CONNECTION

31. Use a calculator that does multiple integration to confirm your answers to Exercises 1, 5, 9, and 13.

Answers to Quick Checks

1. $\frac{16}{3}$ **2.** $\displaystyle\int_0^4 \int_{5 - \frac{5}{4}y}^5 \tfrac{1}{8}xy\, dx\, dy = 10.417$, or 10,417 vehicles

3. $\frac{7}{2}$

KEY TERMS AND CONCEPTS

EXAMPLES

SECTION 6.1

A **function of two variables** assigns to each input pair, (x, y), exactly one output number, $f(x, y)$.

Business. Ken's Millinery produces two styles of hat. The first style costs $5.25 per unit to produce, and the second costs $7.50 per unit to produce. If x is the number of units of the first style of hat and y is the number of units of the second, the cost function C is given by

$$C(x, y) = 5.25x + 7.50y.$$

If the company produces 30 units of the first style and 45 units of the second, the total cost of producing the hats is

$$C(30, 45) = 5.25(30) + 7.50(45) = \$495.$$

The graph of a two-variable function is a **surface**; graphing such a function requires a three-dimensional coordinate system. Points on the surface are expressed as ordered triples (x, y, z), where $z = f(x, y)$.

The graph of $g(x, y) = \sqrt{4 - x^2 - y^2}$ is a hemisphere of radius 2. Examples of points on the surface of this hemisphere are $(0, 0, 2)$, $(2, 0, 0)$, $(1, 1, \sqrt{2})$, and $(-\frac{1}{2}, -\frac{2}{3}, 1.818)$.

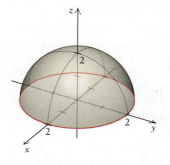

The **domain** of a two-variable function is the set of points in the xy-plane for which f is defined.

- The function $f(x, y) = x^3 + y^3 - 3x - 27y - 2$ is defined for all x and for all y. Therefore, the domain of f is
$$D = \{(x, y) \,|\, -\infty < x < \infty, -\infty < y < \infty\}.$$

- The function $g(x, y) = \sqrt{4 - x^2 - y^2}$ is defined provided that $4 - x^2 - y^2 \geq 0$, or $x^2 + y^2 \leq 4$. Therefore, the domain of g is
$$D = \{(x, y) \,|\, x^2 + y^2 \leq 4\}.$$

- For the cost function $C(x, y) = 5.25x + 7.50y$ considered above, x and y represent quantities of hats. Thus, they cannot be negative. Therefore, the domain for C is
$$D = \{(x, y) \,|\, x \geq 0, y \geq 0\}.$$

SECTION 6.2

Let f be a function of two variables, x and y. The **partial derivative of f with respect to x** is defined as

$$\frac{\partial f}{\partial x} = \lim_{h \to 0} \frac{f(x + h, y) - f(x, y)}{h}.$$

The variable y is treated as a constant during this differentiation.

The **partial derivative of f with respect to y** is defined as

$$\frac{\partial f}{\partial y} = \lim_{h \to 0} \frac{f(x, y + h) - f(x, y)}{h}.$$

The variable x is treated as a constant during this differentiation.

Let $f(x, y) = x^2 + 2xy^3 + \sqrt{y}$.

The partial derivative of f with respect to x is

$$\frac{\partial f}{\partial x} = 2x + 2y^3.$$

The partial derivative of f with respect to y is

$$\frac{\partial f}{\partial y} = 6xy^2 + \frac{1}{2\sqrt{y}}.$$

(continued)

KEY TERMS AND CONCEPTS	EXAMPLES

SECTION 6.2 (continued)

Other common notations for partial derivatives are f_x for the partial derivative of f with respect to x and f_y for the partial derivative of f with respect to y.

The partial derivative $\partial f/\partial x$ is interpreted as the slope of the tangent line at a point (x, y, z) on the surface representing the graph of f in the positive x-direction. Similarly, $\partial f/\partial y$ is interpreted as the slope of the tangent line at a point (x, y, z) on the surface in the positive y-direction.

When $x = 2$ and $y = 1$, the slope of the tangent line in the positive x-direction at $(2, 1, 13)$ on the surface representing the graph of f is

$$f_x(2, 1) = 2(2) + 2(1)^3 = 6.$$

The slope of the tangent line at that point on the surface in the positive y-direction is

$$f_y(2, 1) = 6(2)(1)^2 + \frac{1}{2\sqrt{(1)}} = 12.5.$$

For functions of many variables, the partial derivative with respect to one of the variables is found by treating all the other variables as constants and differentiating using normal techniques.

Let $w(x, y, z) = 3x^2 y^3 z^7$. The partial derivatives of w are

$$w_x = 6xy^3 z^7,$$
$$w_y = 9x^2 y^2 z^7,$$
$$w_z = 21x^2 y^3 z^6.$$

Let f be a function of two variables, x and y. Its **second-order partial derivatives** are

$$f_{xx} = \frac{\partial^2 f}{\partial x^2}, \quad f_{xy} = \frac{\partial^2 f}{\partial y\, \partial x},$$

$$f_{yx} = \frac{\partial^2 f}{\partial x\, \partial y}, \quad \text{and} \quad f_{yy} = \frac{\partial^2 f}{\partial y^2}.$$

For the functions considered in this text, $f_{xy} = f_{yx}$.

Let $f(x, y) = x^2 + 2xy^3 + \sqrt{y}$. Its first partial derivatives are

$$f_x = \frac{\partial f}{\partial x} = 2x + 2y^3 \quad \text{and} \quad f_y = \frac{\partial f}{\partial y} = 6xy^2 + \frac{1}{2\sqrt{y}}.$$

Its second-order partial derivatives are

$$f_{xx} = 2, \quad f_{xy} = 6y^2, \quad f_{yx} = 6y^2, \quad \text{and} \quad f_{yy} = 12xy - \frac{1}{4\sqrt{y^3}}.$$

SECTION 6.3

If f is a function of two variables, x and y, it has a **relative maximum** at (a, b) if

$$f(x, y) \leq f(a, b)$$

for all points (x, y) in a region containing (a, b). Similarly, f has a **relative minimum** at (a, b) if

$$f(x, y) \geq f(a, b)$$

for all points (x, y) in a region containing (a, b).

For a differentiable function f, a **critical point** occurs at (a, b) if both partial derivatives of f at (a, b) are 0; that is, if $f_x(a, b) = 0$ and $f_y(a, b) = 0$.

Let $f(x, y) = x^3 + y^3 - 3x - 27y - 2$.

The first partial derivatives are $f_x(x, y) = 3x^2 - 3$ and $f_y(x, y) = 3y^2 - 27$. When $f_x = 0$, we have $x = \pm 1$. Similarly, when $f_y = 0$, we have $y = \pm 3$. This gives four critical points:

$$(1, 3, -58), \text{ where } f(1, 3) = -58,$$
$$(1, -3, 50), \text{ where } f(1, -3) = 50,$$
$$(-1, 3, -54), \text{ where } f(-1, 3) = -54,$$

and $\quad (-1, -3, 54), \text{ where } f(-1, -3) = 54.$

The four second-order partial derivatives are $f_{xx} = 6x$, $f_{yy} = 6y$, and $f_{xy} = f_{yx} = 0$. Therefore, $D = (6x)(6y) - 0^2 = 36xy$.

KEY TERMS AND CONCEPTS	EXAMPLES

SECTION 6.3 (*continued*)

The **D-test** is used to determine whether critical points are relative maxima or minima. If (a, b) is a critical point, then

$$D = f_{xx}(a, b) \cdot f_{yy}(a, b) - [f_{xy}(a, b)]^2.$$

And:

1. If $D > 0$ and $f_{xx}(a, b) < 0$, then f has a **maximum** at (a, b).
2. If $D > 0$ and $f_{xx}(a, b) > 0$, then f has a **minimum** at (a, b).
3. If $D < 0$, then f has a **saddle point** at (a, b).
4. The D-test is not applicable if $D = 0$.

At $(1, 3)$, we have $D = 108 > 0$, and $f_{xx}(1, 3) = 6 > 0$. Therefore, $(1, 3, -58)$ is a relative minimum.

At $(1, -3)$ and at $(-1, 3)$, we have $D = -108 < 0$. Therefore, $(1, -3, 50)$ and $(-1, 3, -54)$ are saddle points.

At $(-1, -3)$, we have $D = 108 > 0$, and $f_{xx}(-1, -3) = -6 < 0$. Therefore, $(-1, -3, 54)$ is a relative maximum.

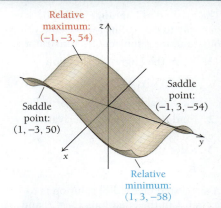

SECTION 6.4

A curve of best fit for a set of data points is called a **regression curve**. The **method of least squares** uses partial derivatives to determine this curve.

A researcher obtains the data points $(1, 3)$, $(4, 5)$, and $(6, 7)$. Find the regression line.

Using the method of least squares, we must minimize

$$(y_1 - 3)^2 + (y_2 - 5)^2 + (y_3 - 7)^2,$$

where
$$y_1 = m(1) + b = m + b,$$
$$y_2 = m(4) + b = 4m + b,$$
$$y_3 = m(6) + b = 6m + b.$$

After substitution, we have a two-variable function:

$$S(m, b) = (m + b - 3)^2 + (4m + b - 5)^2 + (6m + b - 7)^2.$$

The partial derivatives are (after simplification):

$$S_m = 106m + 22b - 130,$$
$$S_b = 22m + 6b - 30.$$

Setting the partial derivatives equal to 0 and solving the system, we get

$$m = \frac{15}{19}, \quad b = \frac{40}{19}.$$

Therefore, the regression line is $y = \frac{15}{19}x + \frac{40}{19}$.

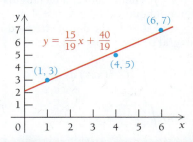

SECTION 6.5

If the input variables x and y of a two-variable function $f(x, y)$ are themselves related by an equation $g(x, y) = 0$, then $g(x, y) = 0$ is a **constraint**. The process of determining maximum and minimum values of f subject to the constraint g is called **constrained optimization**.

Maximize $f(x, y) = xy$, subject to the constraint $x + 2y = 1$.

We can substitute $x = 1 - 2y$ into f:

$$f(1 - 2y, y) = (1 - 2y)y = y - 2y^2.$$

Differentiating f with respect to y, we get $f_y = 1 - 4y$. Setting this equal to 0, we get $y = \frac{1}{4}$. Therefore, $x = 1 - 2\left(\frac{1}{4}\right) = \frac{1}{2}$. Since $f_{yy} = -4 < 0$, the function f has a maximum value of $\frac{1}{8}$ at $x = \frac{1}{2}$ and $y = \frac{1}{4}$, which lies on the line given by the constraint.

(continued)

KEY TERMS AND CONCEPTS	EXAMPLES

SECTION 6.5 (continued)

The method of **Lagrange multipliers** is one way to determine an extreme value of a function f subject to a constraint g.

1. Form the **Lagrange function**

$$F(x, y, \lambda) = f(x, y) - \lambda g(x, y).$$

The variable λ is a *Lagrange multiplier*.

2. Find the first partial derivatives F_x, F_y, and F_λ.

3. Solve the system

$$F_x = 0, \quad F_y = 0, \quad \text{and} \quad F_\lambda = 0.$$

A suggested method of solution is to isolate λ in the equations $F_x = 0$ and $F_y = 0$, substitute for λ, and solve the resulting equation for x or y. Make another substitution into $F_\lambda = 0$, and determine the values of x and y.

Find the extremum of $f(x, y) = xy$, subject to the constraint $x + 2y = 1$.

1. We write the constraint as $x + 2y - 1 = 0$ and form the Lagrange function:

$$F(x, y, \lambda) = xy - \lambda(x + 2y - 1).$$

2. Differentiating F with respect to its three input variables, we have

$$F_x = y - \lambda, \quad F_y = x - 2\lambda, \quad \text{and} \quad F_\lambda = -x - 2y + 1.$$

3. We set all three expressions equal to 0. From $F_x = 0$ and $F_y = 0$, we isolate λ:

$$\lambda = y \text{ and } \lambda = \tfrac{1}{2}x.$$

By substitution, we have $y = \tfrac{1}{2}x$. Substituting $\tfrac{1}{2}x$ for y in F_λ and setting the expression equal to 0 gives

$$-x - 2\left(\tfrac{1}{2}x\right) + 1 = 0.$$

Solving for x, we have $x = \tfrac{1}{2}$. Therefore, $y = \tfrac{1}{2}\left(\tfrac{1}{2}\right) = \tfrac{1}{4}$, and f has a constrained maximum value of $\tfrac{1}{8}$ at $\left(\tfrac{1}{2}, \tfrac{1}{4}\right)$.

Choosing test points (x, y) such that $x + 2y = 1$ and evaluating f with them supports the assertion that this value is a maximum.

If $f(x, y)$ is continuous over a region of feasibility that is closed and bounded, then the **Extreme-Value Theorem** guarantees the existence of an absolute maximum value and an absolute minimum value of f over that region.

Find the absolute maximum and minimum values of $f(x, y) = x^2 + y^2 - 2x - 2y$, subject to the constraints $0 \le x \le 2$ and $0 \le y \le 3$.

Since f is continuous on $0 \le x \le 2$ and on $0 \le y \le 3$, we check for critical points at all boundaries and vertices and in the interior. We find that f has an absolute maximum of 3 at both $(0, 3)$ and $(2, 3)$ and an absolute minimum of -2 at $(1, 1)$.

SECTION 6.6

If $f(x, y)$ is defined over a rectangular region R bounded by $a \le x \le b$ and $c \le y \le d$, then the **double iterated integral** of $f(x, y)$ over R is

$$\int_c^d \int_a^b f(x, y)\, dx\, dy$$

or $\displaystyle \int_a^b \int_c^d f(x, y)\, dy\, dx$.

If the region is not rectangular, then the bounds of integration may contain variables.

Iterated integrals are evaluated by first integrating the inside integral, indicated by the innermost differential, and then integrating the outer integral.

To evaluate

$$\int_0^2 \int_1^3 x^2 y\, dy\, dx,$$

the inside integral is integrated first. We integrate with respect to y, treating x as a constant:

$$\int_1^3 x^2 y\, dy = x^2 \left[\tfrac{1}{2} y^2\right]_1^3 = x^2 \left(\tfrac{9}{2} - \tfrac{1}{2}\right) = 4x^2.$$

We then integrate the result with respect to x:

$$\int_0^2 4x^2\, dx = \left[\tfrac{4}{3} x^3\right]_0^2 = \tfrac{4}{3}(8 - 0) = \tfrac{32}{3}.$$

The **average value** of a continuous two-variable function $z = f(x, y)$ over a region of integration R is given by

$$z_{av} = \frac{1}{A(R)} \iint_R f(x, y)\, dy\, dx,$$

where $A(R)$ is the area of the region of integration.

The average value of $f(x, y) = 2xy$ over a region R that is a triangle with vertices $(-2, 0)$, $(2, 0)$, and $(2, 8)$ is

$$z_{av} = \frac{1}{A(R)} \int_{-2}^2 \int_0^{2x+4} 2xy\, dy\, dx$$

$$= \tfrac{1}{16} \cdot \tfrac{256}{3} = \tfrac{16}{3}.$$

CHAPTER 6
Review Exercises

These review exercises are for test preparation. They can also be used as a practice test. Answers are at the back of the book. The red bracketed section references tell you what part(s) of the chapter to restudy if your answer is incorrect.

CONCEPT REINFORCEMENT

Match each expression in column A with an equivalent expression in column B. Assume that $z = f(x, y)$.   [6.2, 6.6]

Column A

1. $\dfrac{\partial z}{\partial x}$

2. $\dfrac{\partial z}{\partial y}$

3. $\dfrac{\partial}{\partial x}(5x^3y^7)$

4. $\dfrac{\partial}{\partial y}(5x^3y^7)$

5. $\dfrac{\partial^2 z}{\partial x\, \partial y}$

6. $\dfrac{\partial^2 z}{\partial y\, \partial x}$

7. $\displaystyle\int_2^3 \int_0^1 2xy^3\, dx\, dy$

8. $\displaystyle\int_2^3 \int_0^1 2xy^3\, dy\, dx$

Column B

a) $\displaystyle\int_2^3 \tfrac{1}{2}x\, dx$

b) f_{yx}

c) f_{xy}

d) $\displaystyle\int_2^3 y^3\, dy$

e) f_x

f) $15x^2y^7$

g) $35x^3y^6$

h) f_y

REVIEW EXERCISES

Given $f(x, y) = e^y + 3xy^3 + 2y$, find each of the following.
[6.1, 6.2]

9. $f(2, 0)$ 10. f_x 11. f_y

12. f_{xy} 13. f_{yx} 14. f_{xx}

15. f_{yy}

16. State the domain of $f(x, y) = \dfrac{2}{x - 1} + \sqrt{y - 2}.$ [6.1]

Given $z = 2x^3 \ln y + xy^2$, find each of the following.   [6.2]

17. $\dfrac{\partial z}{\partial x}$ 18. $\dfrac{\partial z}{\partial y}$ 19. $\dfrac{\partial^2 z}{\partial x\, \partial y}$

20. $\dfrac{\partial^2 z}{\partial y \partial x}$ 21. $\dfrac{\partial^2 z}{\partial x^2}$ 22. $\dfrac{\partial^2 z}{\partial y^2}$

Find the relative maximum and minimum values.   [6.3]

23. $f(x, y) = x^3 - 6xy + y^2 + 6x + 3y - \frac{1}{5}$

24. $f(x, y) = x^2 - xy + y^2 - 2x + 4y$

25. $f(x, y) = 3x - 6y - x^2 - y^2$

26. $f(x, y) = x^4 + y^4 + 4x - 32y + 80$

27. Consider the data in the following table regarding enrollment in colleges and universities during a recent 3-year period.   [6.4]

YEAR, x	ENROLLMENT, y (in millions)
1	7.2
2	8.0
3	8.4

a) Find the regression line, $y = mx + b$.

b) Use the regression line to predict enrollment in the fourth year.

28. Consider the data in the table below regarding workers' average monthly out-of-pocket premium for health insurance for a family.   [6.4]

a) Find the regression line, $y = mx + b$.

b) Use the regression line to predict workers' average monthly out-of-pocket premium for health insurance for a family in 2020.

c) Find the exponential regression curve, $y = ae^{kx}$.

d) Use the exponential regression curve to estimate the average monthly out-of-pocket premium in 2020.

NUMBER OF YEARS, x, SINCE 1999	WORKERS' AVERAGE MONTHLY OUT-OF-POCKET PREMIUM FOR HEALTH INSURANCE FOR A FAMILY
0	$129
6	$226
12	$360
13	$380

(*Source:* kff.org.)

29. Find the extremum of
$$f(x, y) = x^2 - 2xy + 2y^2 + 20$$
subject to the constraint $2x - 6y = 15$. State whether it is a maximum or a minimum.   [6.5]

30. Find the extremum of $f(x, y) = 6xy$ subject to the constraint $2x + y = 20$. State whether it is a maximum or a minimum.   [6.5]

31. Find the absolute maximum and minimum values of $f(x, y) = x^2 - y^2$ subject to the constraints $-1 \le x \le 3$ and $-1 \le y \le 2$.   [6.5]

Evaluate.   [6.6]

32. $\displaystyle\int_0^1 \int_1^2 x^2 y^3\, dy\, dx$

33. $\displaystyle\int_0^1 \int_{x^2}^x (x - y)\, dy\, dx$

34. Business: demographics.   The density of students living in a region near a university is modeled by

$$p(x, y) = 9 - x^2 - y^2,$$

where x and y are in miles and p is the number of students per square mile, in hundreds. Assume the university is located at $(0, 0)$ in the following graph representing the region.   [6.6]

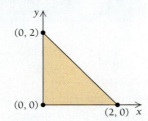

a) Find the number of students who live in the region.

b) Find the average number of students per square mile of the region.

SYNTHESIS

35. Evaluate $\displaystyle\int_0^2 \int_{1-2x}^{1-x} \int_0^{\sqrt{2-x^2}} z \, dz \, dy \, dx.$ [6.6]

36. Business: minimizing surface area.   Suppose a beverage could be packaged in either a cylindrical container or a rectangular container with a square top and bottom. Each container is designed to have the minimum surface area for its shape. If we assume a volume of 26 in^3, which container would have the smaller surface area and by how much?   [6.3, 6.5]

37. Nature uses spheres, such as oranges and grapefruits, to "package" juices. How does the surface area of a spherical container compare to that of a cylindrical container of the same volume (see Exercise 36)?

TECHNOLOGY CONNECTION

38. Use a 3D graphics program to graph

$$f(x, y) = x^2 + 4y^2.$$ [6.1]

CHAPTER 6
Test

Given $f(x, y) = e^x + 2x^3y + y$, find each of the following.

1. $f(-1, 2)$

2. $\dfrac{\partial f}{\partial x}$

3. $\dfrac{\partial f}{\partial y}$

4. $\dfrac{\partial^2 f}{\partial x^2}$

5. $\dfrac{\partial^2 f}{\partial x \, \partial y}$

6. $\dfrac{\partial^2 f}{\partial y \, \partial x}$

7. $\dfrac{\partial^2 f}{\partial y^2}$

Find the relative maximum and minimum values.

8. $f(x, y) = x^2 - xy + y^3 - x$

9. $f(x, y) = x^3 + y^3 - 3x - 12y$

10. Business: predicting total sales.   Consider the data in the following table regarding the total sales of a company during the first three years of operation.

Year, x	Sales, y (in millions)
1	$10
2	15
3	19

a) Find the regression line, $y = mx + b$.

b) Use the regression line to estimate sales in the fourth year.

c) Find the exponential regression curve, $y = ae^{kx}$.

d) Use the exponential regression curve to estimate sales in the fourth year.

11. Find the maximum value of

$$f(x, y) = 6xy - 4x^2 - 3y^2$$

subject to the constraint $x + 3y = 19$.

12. Evaluate

$$\int_0^3 \int_1^3 4x^3y^2 \, dx \, dy.$$

13. Evaluate $\displaystyle\int_{-1}^2 \int_{x^2}^{x+2} xy \, dy \, dx.$

SYNTHESIS

14. Business: maximizing production.   Southwest Appliances has the following Cobb–Douglas production function for a certain product:

$$p(x, y) = 50x^{2/3}y^{1/3},$$

where x is labor and y is capital, both measured in dollars. Suppose Southwest can make a total investment in labor and capital of $600,000. How should it allocate the investment between labor and capital in order to maximize production?

15. Find f_x and f_t:

$$f(x, t) = \frac{x^2 - 2t}{x^3 + 2t}.$$

TECHNOLOGY CONNECTION

16. Use a 3D graphics program to graph

$$f(x, y) = x - \tfrac{1}{2}y^2 - \tfrac{1}{3}x^3.$$

Minimizing Employees' Travel Time in a Building

If employees spend considerable time moving between offices, a building designed to minimize travel time can save a company money.

For a multilevel building with a square base, one design concern is minimizing travel time between the most remote points. We will make use of Lagrange multipliers to help design such a building.

Let's assume that each floor has a square grid of hallways, as shown in the figure at the lower right. Suppose that you are standing at point P in the top northeast corner of the twelfth floor of this building. How long will it take to reach the most remote point at the southwest corner on the first floor—that is, point Q?

Let's call the time t. We find a formula for t in two steps:

1. You are to go from the twelfth floor to the first floor. This is a move in a vertical direction.
2. You need to cross horizontally from one corner of the building to the other corner.

Let h represent the height of point P above the ground floor and a represent the speed at which you can travel in a vertical direction (elevator speed). Thus, vertical time is given by h/a.

The horizontal time is the time it takes to go across one level, by way of the square grid of hallways (from R to Q in the figure). If each floor is a square with side of length k, then the distance from R to Q is $2k$. If the walking speed is b, then the horizontal time is given by $2k/b$.

Thus, the time it takes to go from P to Q is a function of h and k, given by

$$t(h, k) = \text{vertical time} + \text{horizontal time}$$
$$= \frac{h}{a} + \frac{2k}{b},$$

where a and b, the elevator speed and walking speed, are constants.

What if we must choose between two (or more) building plans with the same total floor area, but with different dimensions?

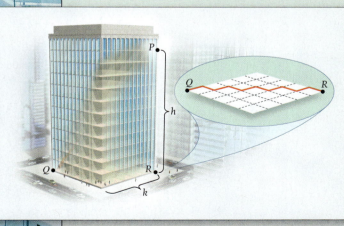

Will the travel time be the same? Or will it differ for the two buildings? First, what is the total floor area of a given building? Suppose the building has n floors, each a square with sides k feet long. Then the total floor area is given by

$$A = nk^2.$$

If h is the height of point P and c is the height of each floor—that is, the distance from the carpeting on one floor to the carpeting on the floor above—then $n = 1 + h/c$, and we have the constraint

$$A = (1 + h/c)k^2.$$

Let's return to the problem of two buildings with the same total floor area, but with different dimensions, and see what happens to the travel time, $t(h, k)$.

Exercises

1. Use the TABLE feature of a calculator or spreadsheet software to complete the table below. For each case in the table, let the elevator speed $a = 10$ ft/sec, the walking speed $b = 4$ ft/sec, and the height of each floor $c = 15$ ft. Each case in the table covers two situations, with the total floor area essentially the same for a particular case.

2. Do different dimensions, with a fixed floor area, yield different travel times?

In Exercises 3–5, assume that you are finding the dimensions of a multilevel building with a square base that will minimize travel time t between the most remote points in the building. Each floor has a square grid of hallways. The height of point P is h, and the length of a side of each floor is k. The elevator speed is 10 ft/sec and the average speed of a person walking is 4 ft/sec. The total floor area of the building is 40,000 ft^2. The height of each floor is 12 ft.

3. Use the information given to find a formula for the function $t(h, k)$.

4. Find a formula for the constraint.

5. Use the method of Lagrange multipliers to find the dimensions of the building that will minimize travel time t between the most remote points in the building.

6. Use a 3D graphics program to graph the equation in Exercise 3 subject to the constraint in Exercise 4. Then visually check the results you found analytically.

Case	Building	n	k	A	h	$t(h, k)$
1	B1	2	40	3200	15	21.5
	B2	3	32.66	3200	30	19.4
2	B1	2	60	7200		
	B2	3	48.99			
3	B1	4	40			
	B2	5	35.777			
4	B1	5	60			
	B2	10	42.426			
5	B1	5	150			
	B2	10	106.066			
6	B1	10	40			
	B2	17	30.679			
7	B1	10	80			
	B2	17	61.357			
8	B1	17	40			
	B2	26	32.344			
9	B1	17	50			
	B2	26	40.43			
10	B1	26	77			
	B2	50	55.525			

Cumulative Review

1. Write an equation of the line with slope -4 and containing the point $(-7, 1)$.

2. Find the slope and the y-intercept of $2x - 6y = 9$.

3. For $f(x) = x^2 - 5$, find $f(x + h)$.

4. a) Graph:
$$f(x) = \begin{cases} 5 - x, & \text{for } x \neq 2, \\ -3, & \text{for } x = 2. \end{cases}$$
   b) Find $\lim\limits_{x \to 2} f(x)$.
   c) Find $f(2)$.
   d) Is f continuous at 2?

5. a) Graph: $g(x) = \begin{cases} 9 - x^2, & x \leq 1, \\ x + 7, & x > 1. \end{cases}$
   b) Find $\lim\limits_{x \to 1} g(x)$.
   c) Find $g(1)$.
   d) Is g continuous at 1?

Find each limit, if it exists. If a limit does not exist, state that fact.

6. $\lim\limits_{x \to 1} \sqrt{x^3 + 8}$

7. $\lim\limits_{x \to -4} \dfrac{x^2 - 16}{x + 4}$

8. $\lim\limits_{x \to 3} \dfrac{4}{x - 3}$

9. $\lim\limits_{x \to \infty} \dfrac{12x - 7}{3x + 2}$

10. $\lim\limits_{x \to \infty} \dfrac{2x^3 - x}{8x^5 - x^2 + 1}$

11. If $f(x) = x^2 + 3$, find $f'(x)$ by determining
$$\lim\limits_{h \to 0} \dfrac{f(x + h) - f(x)}{h}.$$

For exercises 12–14, refer to the following graph of $y = h(x)$.

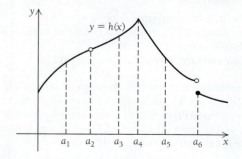

12. Identify the input values for which h has no limit.

13. Identify the input values for which h is discontinuous.

14. Identify the input values for which the derivative of h does not exist.

Differentiate.

15. $y = -9x + 3$

16. $y = x^2 - 7x + 3$

17. $y = x^{1/4}$

18. $f(x) = x^{-6}$

19. $f(x) = \sqrt[3]{2x^5 - 8}$

20. $f(x) = \dfrac{5x^3 + 4}{2x - 1}$

21. $y = \ln(x^2 + 5)$

22. $y = e^{\ln x}$

23. $y = e^{3x} + x^2$

24. $y = e^{\sqrt{x - 3}}$

25. $f(x) = \ln(e^x - 4)$

26. For $y = x^2 - \dfrac{2}{x}$, find d^2y/dx^2.

27. **Business: average cost.** Doubletake Clothing finds that the cost, in dollars, of producing x pairs of jeans is given by $C(x) = 320 + 9\sqrt{x}$. Find the rate at which the average cost is changing when 100 pairs of jeans have been produced.

28. Differentiate implicitly to find dy/dx if $x^3 + x/y = 7$.

29. Find an equation of the tangent line to the graph of $y = e^x - x^2 - 3$ at the point $(0, -2)$.

30. Find the x-value(s) at which the tangent line to $f(x) = x^3 - 2x^2$ has a slope of -1.

Sketch the graph of each function. List and label the coordinates of any extrema and points of inflection. State where the function is increasing or decreasing, where it is concave up or concave down, and where any asymptotes occur.

31. $f(x) = x^3 - 3x + 1$

32. $f(x) = 2x^2 - x^4 - 3$

33. $f(x) = \dfrac{8x}{x^2 + 1}$

34. $f(x) = \dfrac{8}{x^2 - 4}$

Find the absolute maximum and minimum values, if they exist, over the indicated interval. If no interval is indicated, consider the entire real number line.

35. $f(x) = 3x^2 - 6x - 4$

36. $f(x) = -5x + 1$

37. $f(x) = \frac{1}{3}x^3 - x^2 - 3x + 5$; $\quad [-2, 0]$

38. **Business: maximizing profit.** Detailed Clothing's total revenue and total cost, in dollars, from the sale of x custom sweatshirts are given by
$$R(x) = 4x^2 + 11x + 110,$$
$$C(x) = 4.2x^2 + 5x + 10.$$

Find the number of sweatshirts, x, that must be produced and sold in order to maximize profit.

39. Business: minimizing inventory costs. An electronics store sells 450 sets of earbuds each year. It costs $4 to store a set for a year. When placing an order, there is a fixed cost of $1 plus $0.75 for each set. How many times per year should the store reorder earbuds, and in what lot size, in order to minimize inventory costs?

40. Let $y = 3x^2 - 2x + 1$. Use differentials to find the approximate change in y when $x = 2$ and $\Delta x = 0.05$.

41. Business: exponential growth. Frieda's Frozen Yogurt is experiencing growth of 10% per year in the number, N, of franchises it owns; that is,

$$\frac{dN}{dt} = 0.1N,$$

where N is the number of franchises and t is the time, in years, from 2011.

a) Given that there were 8000 franchises in 2011, find the solution of the equation, assuming that $N_0 = 8000$ and $k = 0.1$.

b) Predict the number of franchises in 2019.

c) What is the doubling time for the number of franchises?

42. Economics: elasticity of demand. Consider the demand function given by

$$q = D(x) = 240 - 20x,$$

where q is the quantity of coffee mugs demanded at a price of x dollars per mug.

a) Find the elasticity.

b) Find the elasticity at $x = \$2$, and state whether the demand is elastic or inelastic.

c) Find the elasticity at $x = \$9$, and state whether the demand is elastic or inelastic.

d) At a price of $2, will a small increase in price cause total revenue to increase or decrease?

e) Find the value of x for which the total revenue is a maximum.

43. Business: approximating cost overage. A square plot of ground measures 75 ft by 75 ft, with a tolerance of ± 4 in. Landscapers are going to cover the plot with grass sod. Each square of sod costs $8 and measures 3 ft by 3 ft.

a) Use differentials to estimate the change in area when the measurement tolerance is taken into acount.

b) How many extra squares of sod should the landscapers bring to the job, and how much extra will this cost?

44. Gail deposits $250 every month into an annuity at an annual interest rate of 3.35%, compounded monthly.

a) Find a function $A(t)$ that gives the value of Gail's annuity after t years.

b) Find the value of Gail's annuity after 3 yr.

c) What is the rate of change in the value of Gail's annuity after 3 yr?

45. Garth purchases a new Toyota Highlander for $31,050. He makes a down payment of $7500 and finances the rest of the cost through a 6-yr amortized loan at an annual interest rate of 5.25%, compounded monthly.

a) Find Garth's monthly payment.

b) Assume that Garth makes every payment for the life of the loan. Find the total of his payments.

c) How much total interest will Garth pay?

d) Fill in the two rows of Garth's amortization schedule below.

Balance	Payment	Portion of Payment Applied to Interest	Portion of Payment Applied to Principal	New Balance
$23,550				

Evaluate.

46. $\displaystyle\int 3x^5 \, dx$

47. $\displaystyle\int_{-1}^{0} (2e^x + 1) \, dx$

48. $\displaystyle\int \frac{x}{(7 - 3x)^2} \, dx$ (Use Table 1 on pp. 431–432.)

49. $\displaystyle\int x^3 e^{x^4} \, dx$

50. $\displaystyle\int (x + 3) \ln x \, dx$

51. $\displaystyle\int \frac{75}{x} \, dx$

52. $\displaystyle\int_{0}^{1} 3\sqrt{x} \, dx$

53. Find the area under the graph of $y = x^2 + 3x$ over the interval $[1, 5]$.

54. Business: present value. Find the present value of $250,000 due in 30 yr at 6%, compounded continuously.

55. Business: value of a fund. Leigh Ann wants to have $50,000 saved in 10 yr.

a) She makes a one-time deposit at an APR of 5.45%, compounded continuously. Find the amount she should deposit.

b) Calculate the interest earned in part (a).

56. Business: contract buyout. An athlete has an 8-yr contract that pays her $200,000 per year. She invests the money at an APR of 4.85%, compounded continuously. After 5 yr, the team offers her a buyout of the contract. What is the lowest amount she should accept?

57. Determine whether the following improper integral is convergent or divergent, and calculate its value if it is convergent:

$$\int_{3}^{\infty} \frac{1}{x^7} \, dx.$$

58. Economics: supply and demand. Demand and supply functions are given by

$$p = D(x) = (x - 20)^2$$

and

$$p = S(x) = x^2 + 10x + 50,$$

where p is the price per unit, in dollars, when x units are sold. Find the equilibrium point and the consumer's surplus.

59. Find the volume of the solid of revolution generated by rotating the region under the graph of $y = e^{-x}$, from $x = 0$ to $x = 5$, around the x-axis.

60. Find the volume of the solid of revolution generated by rotating the region under the graph of $y = (x^2 + 1)^2$, from $x = 0$ to $x = 2$, around the y-axis.

61. Consider the data in the following table.

Age of Business (in years)	1	3	5
Profit (in tens of thousands of dollars)	4	7	12

a) Find the regression line, $y = mx + b$.
b) Use the regression line to predict the profit when the business is 10 years old.
c) Find the exponential regression curve, $y = ae^{bx}$.
d) Use the exponential regression curve to predict the profit when the business is 10 years old.

Given $f(x, y) = e^y + 4x^2y^3 + 3x$, find each of the following.

62. f_x **63.** f_{yy}

64. Find the relative maximum and minimum values of $f(x, y) = x^2 + y^2 - xy - 8x + 7y + 9$.

65. Maximize $f(x, y) = 4x + 2y - x^2 - y^2 + 4$, subject to the constraint $x + 2y = 9$.

66. Evaluate $\displaystyle\int_0^3 \int_{-1}^2 e^x \, dy \, dx$.

67. Business: demographics. The number of shoppers, in hundreds per square mile, who shop at Los Arcos Mall is modeled by $f(x, y) = 10 - x - y^2$, where x is miles from the mall toward the east and y is miles from the mall toward the north. The graph below shows a shaded region to the northeast of the mall, which is at $(0, 0)$. Find the total number of mall shoppers in the region.

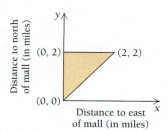

68. Solve the differential equation $dy/dx = xy$.

69. Solve the differential equation $y' + 4xy = 3x$, where $y = 3$ when $x = -1$.

70. Suppose the rate of change of y with respect to x is directly proportional to the square root of y. Assume $y \geq 0$.

a) Write a differential equation that models this situation.
b) Find the general solution of the equation from part (a).

71. A jar contains 12 yellow candies, 15 red candies, and 20 green candies. A single candy is chosen at random.

a) Find the probability that a yellow candy is chosen.
b) Find the probability that a green candy is not chosen.
c) Find the probability that a red or green candy is chosen.
d) Find the probability that a blue candy is chosen.

72. Business: distribution of weights. The weight, in pounds, of a box of a certain cereal is uniformly distributed between 2 lb and 2.25 lb. Find the probability that a randomly chosen box of this cereal weighs between 2.07 lb and 2.1 lb.

73. Business: wait times. The wait time t, in minutes, between customers at Teri's Book Nook can be modeled by the probability density function $f(t) = 0.45e^{-0.45t}$, for $0 \leq t < \infty$.

a) What is the probability that the wait time is at most 2 min?
b) What is the probability that the wait time is greater than 5 min?

74. Let $f(x) = \dfrac{c}{x^2}$ be a probability density function over the interval $[1, 3]$.

a) Find c.
b) Find the mean μ and the standard deviation σ.
c) Find $P([1.5, 2.5])$.

75. Business: distribution of salaries. The salaries paid by a large corporation are normally distributed with a mean $\mu = \$75,000$ and a standard deviation $\sigma = \$6,000$.

a) Find the probability that a randomly chosen employee earns between $72,000 and $85,000 per year.
b) An executive of the corporation earns $90,000 per year. In what percentile of the salaries does this salary place him?
c) A new employee insists on a salary that is in the top 2% of salaries. What is the minimum salary that this employee would accept?

Review of Basic Algebra

- Manipulate exponential expressions.
- Multiply and factor algebraic expressions.
- Solve equations, inequalities, and applied problems.

This appendix covers most of the algebraic topics essential to a study of calculus. It can be used in conjunction with Chapter R or as needed throughout the book.

Exponential Notation

Let's review the meaning of an expression

$$a^n,$$

where a is any real number and n is an integer; that is, n is a number in the set $\{\ldots, -3, -2, -1, 0, 1, 2, 3, \ldots\}$. The number a is the **base** and n is the **exponent**.

> For an integer n greater than 1,
> $$a^n = \underbrace{a \cdot a \cdot a \cdots a}_{n \text{ factors}}.$$

In other words, a^n is the product of n **factors**, each of which is a.

EXAMPLE 1 Write equivalent expressions without exponents.

a) 4^3 **b)** $(-2)^5$ **c)** $(-2)^4$ **d)** -2^4 **e)** $\left(\dfrac{1}{2}\right)^3$

Quick Check 1 ✔

Write equivalent expressions without exponents.

a) 3^5

b) $(-5)^2$

c) $\left(\frac{2}{3}\right)^4$

Solution

a) $4^3 = 4 \cdot 4 \cdot 4 = 64$

b) $(-2)^5 = (-2)(-2)(-2)(-2)(-2) = -32$

c) $(-2)^4 = (-2)(-2)(-2)(-2) = 16$

d) $-2^4 = -(2^4) = -(2)(2)(2)(2) = -16$ The base is 2 , not -2.

e) $\left(\dfrac{1}{2}\right)^3 = \dfrac{1}{2} \cdot \dfrac{1}{2} \cdot \dfrac{1}{2} = \dfrac{1}{8}$

1 ✔

We define an exponent of 1 as follows:

> $a^1 = a$, for any real number a.

In other words, any real number to the first power is that number itself.
We define an exponent of 0 as follows:

> $a^0 = 1$, for any nonzero real number a.

EXAMPLE 2 Write equivalent expressions without exponents.

a) $(-2x)^0$ **b)** $(-2x)^1$ **c)** $\left(\dfrac{1}{2}\right)^0$ **d)** $-e^0$ **e)** e^1 **f)** $\left(\dfrac{1}{2}\right)^1$

Quick Check 2 ☑

Write equivalent expressions without exponents.

a) $\left(\frac{2}{5}\right)^0$

b) $\left(\frac{2}{5}\right)^1$

c) -4^0

Solution

a) $(-2x)^0 = 1$ b) $(-2x)^1 = -2x$ c) $\left(\frac{1}{2}\right)^0 = 1$

d) $-e^0 = -1$ e) $e^1 = e$ f) $\left(\frac{1}{2}\right)^1 = \frac{1}{2}$

2 ☑

The meaning of a negative integer as an exponent is as follows:

$$a^{-n} = \left(\frac{1}{a}\right)^n = \frac{1}{a^n}, \text{ for any nonzero real number } a.$$

That is, any nonzero real number a to the $-n$ power is (the reciprocal of a)^n, or equivalently, the reciprocal of a^n.

EXAMPLE 3 Write equivalent expressions without negative exponents.

a) 2^{-5} b) $\left(\frac{3}{4}\right)^{-2}$ c) $2x^{-5}$ d) e^{-k} e) $\frac{1}{t^{-1}}$

Quick Check 3 ☑

Write equivalent expressions without negative exponents.

a) 3^{-2}

b) $\frac{1}{4^{-3}}$

c) $8x^{-5}$

Solution

a) $2^{-5} = \frac{1}{2^5} = \frac{1}{2 \cdot 2 \cdot 2 \cdot 2 \cdot 2} = \frac{1}{32}$ b) $\left(\frac{3}{4}\right)^{-2} = \left(\frac{4}{3}\right)^2 = \frac{4^2}{3^2} = \frac{16}{9}$

c) $2x^{-5} = \frac{2}{x^5}$ d) $e^{-k} = \frac{1}{e^k}$

e) $\frac{1}{t^{-1}} = \left(\frac{1}{t}\right)^{-1} = t^1 = t$

3 ☑

Properties of Exponents

Note the following:

$$\begin{aligned} b^5 \cdot b^{-3} &= (b \cdot b \cdot b \cdot b \cdot b) \cdot \frac{1}{b \cdot b \cdot b} \\ &= \frac{b \cdot b \cdot b}{b \cdot b \cdot b} \cdot b \cdot b \\ &= 1 \cdot b \cdot b = b^2. \end{aligned}$$

We can obtain the same result by adding the exponents. This is true in general.

THEOREM 1

For any nonzero real number a and any integers n and m,

$$a^n \cdot a^m = a^{n+m}.$$

(To multiply when the bases are the same, add the exponents.)

EXAMPLE 4   Multiply:

a) $x^5 \cdot x^6$ **b)** $x^{-5} \cdot x^6$ **c)** $2x^{-3} \cdot 5x^{-4}$ **d)** $r^2 \cdot r$ **e)** $x^3 \cdot x^7 \cdot x^{-2}$

Quick Check 4 ✔

Multiply.

a) $t^5 \cdot t^3$
b) $4x^2 \cdot 7x^{-5}$
c) $y^4 y^{-9} y$

Solution

a) $x^5 \cdot x^6 = x^{5+6} = x^{11}$

b) $x^{-5} \cdot x^6 = x^{-5+6} = x$

c) $2x^{-3} \cdot 5x^{-4} = 10x^{-3+(-4)} = 10x^{-7}$, or $\dfrac{10}{x^7}$

d) $r^2 \cdot r = r^{2+1} = r^3$

e) $x^3 \cdot x^7 \cdot x^{-2} = x^{3+7+(-2)} = x^8$

4 ✔

Note the following:

$$b^5 \div b^2 = \frac{b^5}{b^2} = \frac{b \cdot b \cdot b \cdot b \cdot b}{b \cdot b}$$

$$= \frac{b \cdot b}{b \cdot b} \cdot b \cdot b \cdot b$$

$$= 1 \cdot b \cdot b \cdot b = b^3.$$

We can obtain the same result by subtracting the exponents. This is true in general.

THEOREM 2

For any nonzero real number a and any integers n and m,

$$\frac{a^n}{a^m} = a^{n-m}.$$

(To divide when the bases are the same, subtract the exponent in the denominator from the exponent in the numerator.)

EXAMPLE 5   Divide:

a) $\dfrac{a^3}{a^2}$ **b)** $\dfrac{x^7}{x^7}$ **c)** $\dfrac{e^3}{e^{-4}}$ **d)** $\dfrac{e^{-4}}{e^{-1}}$ **e)** $\dfrac{c^2 c^8}{c^{12}}$

Quick Check 5 ✔

Divide.

a) $\dfrac{d^7}{d^2}$

b) $\dfrac{t^3}{t^{-5}}$

c) $\dfrac{u^5 u^{12}}{u^3}$

Solution

a) $\dfrac{a^3}{a^2} = a^{3-2} = a^1 = a$

b) $\dfrac{x^7}{x^7} = x^{7-7} = x^0 = 1$

c) $\dfrac{e^3}{e^{-4}} = e^{3-(-4)} = e^{3+4} = e^7$

d) $\dfrac{e^{-4}}{e^{-1}} = e^{-4-(-1)} = e^{-4+1} = e^{-3}$, or $\dfrac{1}{e^3}$

e) $\dfrac{c^2 c^8}{c^{12}} = c^{2+8-12} = c^{-2}$, or $\dfrac{1}{c^2}$

5 ✔

Note the following:

$$(b^2)^3 = b^2 \cdot b^2 \cdot b^2 = b^{2+2+2} = b^6.$$

We can obtain the same result by multiplying the exponents. The other results in Theorem 3 can be similarly motivated.

> **THEOREM 3**
>
> For any nonzero real numbers a and b, and any integers n and m,
>
> $$(a^n)^m = a^{nm}, \qquad (ab)^n = a^n b^n, \quad \text{and} \quad \left(\frac{a}{b}\right)^n = \frac{a^n}{b^n}.$$

EXAMPLE 6 Simplify:

a) $(x^{-2})^3$ 　　 b) $(e^x)^2$ 　　 c) $(2x^4 y^{-5} z^3)^{-3}$ 　　 d) $\left(\dfrac{x^2}{p^4 q^5}\right)^3$

Solution

a) $(x^{-2})^3 = x^{-2 \cdot 3} = x^{-6}, \quad \text{or} \quad \dfrac{1}{x^6}$ 　　　　 b) $(e^x)^2 = e^{2x}$

c) $(2x^4 y^{-5} z^3)^{-3} = 2^{-3}(x^4)^{-3}(y^{-5})^{-3}(z^3)^{-3}$

$$= \frac{1}{2^3} x^{-12} y^{15} z^{-9}, \quad \text{or} \quad \frac{y^{15}}{8x^{12} z^9}$$

d) $\left(\dfrac{x^2}{p^4 q^5}\right)^3 = \dfrac{(x^2)^3}{(p^4 q^5)^3} = \dfrac{x^6}{(p^4)^3 (q^5)^3}$

$$= \frac{x^6}{p^{12} q^{15}}$$

Quick Check 6 ✔

Simplify.

a) $(x^5)^7$

b) $(z^{-2})^8$

c) $\left(\dfrac{4x}{y^2 z^{-1}}\right)^4$

6 ✔

Multiplication

The distributive law is important when multiplying. This law is as follows.

> **The Distributive Law**
>
> For any numbers A, B, and C,
>
> $$A(B + C) = AB + AC.$$
>
> Because subtraction can be regarded as addition of an additive inverse, it follows that
>
> $$A(B - C) = AB - AC.$$

EXAMPLE 7 Multiply:

a) $3(x - 5)$ 　　 b) $P(1 + r)$ 　　 c) $(x - 5)(x + 3)$ 　　 d) $(a + b)(a + b)$

Solution

a) $3(x - 5) = 3 \cdot x - 3 \cdot 5 = 3x - 15$ 　　　　 b) $P(1 + r) = P \cdot 1 + P \cdot r = P + Pr$

c) $(x - 5)(x + 3) = (x - 5)x + (x - 5)3$

$$= x \cdot x - 5x + 3x - 5 \cdot 3$$

$$= x^2 - 2x - 15$$

Quick Check 7 ✔

Multiply.

a) $-4(x - 3)$

b) $x(y + 3z)$

c) $(x + 1)(2x - 7)$

d) $(a + b)(a + b) = (a + b)a + (a + b)b$

$$= a \cdot a + ba + ab + b \cdot b$$

$$= a^2 + 2ab + b^2$$

7 ✔

The following formulas, which are obtained using the distributive law, are also useful. All three are used in Example 8, which follows.

$$(A + B)^2 = A^2 + 2AB + B^2$$
$$(A - B)^2 = A^2 - 2AB + B^2$$
$$(A - B)(A + B) = A^2 - B^2$$

EXAMPLE 8 Multiply:

a) $(x + h)^2$ **b)** $(2x - t)^2$ **c)** $(3c + d)(3c - d)$

Solution
a) $(x + h)^2 = x^2 + 2xh + h^2$
b) $(2x - t)^2 = (2x)^2 - 2(2x)t + t^2 = 4x^2 - 4xt + t^2$
c) $(3c + d)(3c - d) = (3c)^2 - d^2 = 9c^2 - d^2$ **8** ✔

Quick Check 8 ✔
Multiply.
a) $(2x + 3)^2$
b) $(3x - 1)^2$
c) $(y - 5z)(y + 5z)$

Factoring

Factoring is the reverse of multiplication. That is, to factor an expression, we find an equivalent expression that is a product. Always remember to look first for a common factor.

EXAMPLE 9 Factor:

a) $P + Pi$ **b)** $2xh + h^2$ **c)** $x^2 - 6xy + 9y^2$
d) $x^2 - 5x - 14$ **e)** $6x^2 + 7x - 5$ **f)** $x^2 - 9t^2$

Solution
a) $P + Pi = P \cdot 1 + P \cdot i = P(1 + i)$ ⎫
b) $2xh + h^2 = h(2x + h)$ ⎬ Using the distributive law
c) $x^2 - 6xy + 9y^2 = (x - 3y)^2$
d) $x^2 - 5x - 14 = (x - 7)(x + 2)$ Finding factors of -14 whose sum is -5
e) $6x^2 + 7x - 5 = (2x - 1)(3x + 5)$ Considering ways of factoring the first term—for example, $(2x \ )(3x \ )$; then looking for factors of -5 that yield the desired product
f) $x^2 - 9t^2 = (x - 3t)(x + 3t)$ Using the formula $(A - B)(A + B) = A^2 - B^2$ **9** ✔

Quick Check 9 ✔
Factor.
a) $2m^2n + 4mn^2$
b) $x^2 - 10x + 25$
c) $3x^2 + 10x + 3$
d) $16t^2 - 81u^2$

Some expressions with four terms can be factored by first looking for a common binomial factor. This is called **factoring by grouping**.

EXAMPLE 10 Factor:

a) $t^3 + 6t^2 - 2t - 12$ **b)** $x^3 - 7x^2 - 4x + 28$

Solution
a) $t^3 + 6t^2 - 2t - 12 = t^2(t + 6) - 2(t + 6)$ Factoring the first two terms and then the second two terms

$\qquad\qquad\qquad\qquad = (t + 6)(t^2 - 2)$ Factoring out the common binomial factor, $(t + 6)$

b) $x^3 - 7x^2 - 4x + 28 = x^2(x - 7) - 4(x - 7)$ Factoring the first two terms and then the second two terms

$\qquad\qquad\qquad\qquad = (x - 7)(x^2 - 4)$ Factoring out the common binomial factor, $(x - 7)$

$\qquad\qquad\qquad\qquad = (x - 7)(x - 2)(x + 2)$ Using $(A - B)(A + B) = A^2 - B^2$ **10** ✔

Quick Check 10 ✔
Factor.
a) $u^3 + 4u^2 + 3u + 12$
b) $x^3 + 2x^2 - x - 2$

Solving Equations

Basic to the solution of many equations are the *Addition Principle* and the *Multiplication Principle*. We can add (or subtract) the same number on both sides of an equation and obtain an equivalent equation, that is, a new equation that has the same solutions as the original equation. We can also multiply (or divide) by a nonzero number on both sides of an equation and obtain an equivalent equation.

The Addition Principle	The Multiplication Principle
For any real numbers a, b, and c, $a = b$ is equivalent to $a + c = b + c$.	For any real numbers a, b, and c, with $c \neq 0$, $a = b$ is equivalent to $a \cdot c = b \cdot c$.

When solving a linear equation, we use these principles and other properties of real numbers to get the variable alone on one side. Then it is easy to determine the solution.

EXAMPLE 11   Solve: $-\frac{5}{6}x + 10 = \frac{1}{2}x + 2$.

Solution   We first multiply by 6 on both sides to clear the fractions:

$$6\left(-\frac{5}{6}x + 10\right) = 6\left(\frac{1}{2}x + 2\right) \qquad \text{Using the Multiplication Principle}$$
$$6\left(-\frac{5}{6}x\right) + 6 \cdot 10 = 6\left(\frac{1}{2}x\right) + 6 \cdot 2 \qquad \text{Using the distributive law}$$
$$-5x + 60 = 3x + 12 \qquad \text{Simplifying}$$
$$60 = 8x + 12 \qquad \text{Using the Addition Principle: adding } 5x \text{ on both sides}$$
$$48 = 8x \qquad \text{Adding } -12, \text{ or subtracting 12, on both sides}$$
$$\frac{1}{8} \cdot 48 = \frac{1}{8} \cdot 8x \qquad \text{Multiplying by } \frac{1}{8} \text{ on both sides}$$
$$6 = x.$$

Quick Check 11 ✔

Solve:
$3x - 15 = \frac{1}{2}x + 4$.

The variable is now alone on one side, and we see that 6 is the solution. You can check by substituting 6 into the original equation.

11 ✔

Another principle for solving equations is the *Principle of Zero Products*.

The Principle of Zero Products

For any numbers a and b, if $ab = 0$, then $a = 0$ or $b = 0$; and if $a = 0$ or $b = 0$, then $ab = 0$.

To solve an equation using this principle, we must have 0 on one side and the other side must be a product. Solutions are then obtained by setting each factor equal to 0 and solving the resulting equations.

EXAMPLE 12   Solve: $3x(x - 2)(5x + 4) = 0$.

Solution   We have

$$3x(x - 2)(5x + 4) = 0$$
$$3x = 0 \quad or \quad x - 2 = 0 \quad or \quad 5x + 4 = 0 \qquad \text{Using the Principle of Zero Products}$$
$$\tfrac{1}{3} \cdot 3x = \tfrac{1}{3} \cdot 0 \quad or \qquad x = 2 \quad or \qquad 5x = -4 \qquad \text{Solving each equation separately}$$
$$x = 0 \quad or \qquad x = 2 \quad or \qquad x = -\tfrac{4}{5}.$$

Quick Check 12 ✔

Solve:
$2x(3x - 1)(x + 5) = 0$.   The solutions are 0, 2, and $-\frac{4}{5}$.

12 ✔

Note that the Principle of Zero Products applies *only* when a product is 0.

EXAMPLE 13 Solve: $4x^3 = x$.

Solution We have

$$4x^3 = x$$

$$4x^3 - x = 0 \qquad \text{Adding } -x \text{ to both sides to get 0 on one side}$$

$$x(4x^2 - 1) = 0 \quad \Big\} \quad \text{Factoring}$$

$$x(2x - 1)(2x + 1) = 0$$

$$x = 0 \quad or \quad 2x - 1 = 0 \quad or \quad 2x + 1 = 0 \qquad \text{Using the Principle of Zero Products}$$

$$x = 0 \quad or \quad 2x = 1 \quad or \quad 2x = -1$$

$$x = 0 \quad or \quad x = \tfrac{1}{2} \quad or \quad x = -\tfrac{1}{2}.$$

Quick Check 13 ✔

Solve: $9x^3 = 4x$.

The solutions are 0, $\tfrac{1}{2}$, and $-\tfrac{1}{2}$.

13 ✔

Rational Equations

Expressions like the following are polynomials in one variable:

$$x^2 - 4, \qquad x^3 + 7x^2 - 8x + 9, \qquad t - 19.$$

The **least common multiple**, **LCM**, of two polynomials is found by factoring and using each factor the greatest number of times that it occurs in any one factorization.

EXAMPLE 14 Find the LCM: $x^2 + 2x + 1$, $5x^2 - 5x$, and $x^2 - 1$.

Solution

Quick Check 14 ✔

Find the LCM:
$x^2 + 2x + 1$, $x^2 - 9$,
and $x^2 + 4x + 3$.

$$x^2 + 2x + 1 = (x + 1)(x + 1);$$

$$5x^2 - 5x = 5x(x - 1); \quad \Big\} \quad \text{Factoring}$$

$$x^2 - 1 = (x + 1)(x - 1)$$

$$\text{LCM} = 5x(x + 1)(x + 1)(x - 1)$$

14 ✔

A **rational expression** is a ratio of polynomials. Each of the following is a rational expression:

$$\frac{x^2 - 6x + 9}{x^2 - 4}, \qquad \frac{x - 2}{x - 3}, \qquad \frac{a + 7}{a^2 - 16}, \qquad \frac{5}{5t - 15}.$$

A **rational equation** is an equation containing one or more rational expressions. Here are some examples:

$$\frac{2}{3} - \frac{5}{6} = \frac{1}{x}, \qquad x + \frac{6}{x} = 5, \qquad \frac{2x}{x - 3} - \frac{6}{x} = \frac{18}{x^2 - 3x}.$$

To solve a rational equation, we first clear the equation of fractions by multiplying both sides by the LCM of all the denominators. The resulting equation might have solutions that are *not* solutions of the original equation. Thus, we must check all possible solutions in the original equation.

EXAMPLE 15 Solve: $\dfrac{2x}{x - 3} - \dfrac{6}{x} = \dfrac{18}{x^2 - 3x}$.

Solution Note that $x^2 - 3x = x(x - 3)$. The LCM of the denominators is $x(x - 3)$. We multiply by $x(x - 3)$.

$$x(x - 3)\left(\frac{2x}{x - 3} - \frac{6}{x}\right) = x(x - 3)\left(\frac{18}{x^2 - 3x}\right) \qquad \text{Multiplying by the LCM on both sides}$$

$$x(x - 3) \cdot \frac{2x}{x - 3} - x(x - 3) \cdot \frac{6}{x} = x(x - 3)\left(\frac{18}{x^2 - 3x}\right) \qquad \text{Using the distributive law}$$

$$2x^2 - 6(x - 3) = 18 \qquad \text{Simplifying}$$

$$2x^2 - 6x + 18 = 18$$

$$2x^2 - 6x = 0$$

$$2x(x - 3) = 0$$

$$2x = 0 \quad or \quad x - 3 = 0$$

$$x = 0 \quad or \quad x = 3$$

The numbers 0 and 3 are possible solutions. But each makes a denominator in the original equation 0. Since division by 0 is not allowed, we see that $x \neq 0$ and $x \neq 3$. We can also carry out a check, as follows.

Check

For 0:
$$\frac{2x}{x-3} - \frac{6}{x} = \frac{18}{x^2 - 3x}$$
$$\frac{2(0)}{0-3} - \frac{6}{0} \overset{?}{\,\big|\,} \frac{18}{0^2 - 3(0)}$$
$$0 - \frac{6}{0} \,\Big|\, \frac{18}{0} \qquad \text{UNDEFINED;}$$
$$\qquad\qquad \text{FALSE}$$

For 3:
$$\frac{2x}{x-3} - \frac{6}{x} = \frac{18}{x^2 - 3x}$$
$$\frac{2(3)}{3-3} - \frac{6}{3} \overset{?}{\,\big|\,} \frac{18}{3^2 - 3(3)}$$
$$\frac{6}{0} - 2 \,\Big|\, \frac{18}{0} \qquad \text{UNDEFINED;}$$
$$\qquad\qquad \text{FALSE}$$

The equation has *no solution*.

EXAMPLE 16   Solve: $\dfrac{x^2}{x-2} = \dfrac{4}{x-2}$.

Solution   Note first that $x \neq 2$. The LCM of the denominators is $x - 2$. We multiply both sides by $x - 2$.

$$(x-2) \cdot \frac{x^2}{x-2} = (x-2) \cdot \frac{4}{x-2}$$
$$x^2 = 4 \qquad \text{Simplifying}$$
$$x^2 - 4 = 0$$
$$(x+2)(x-2) = 0$$
$$x = -2 \quad or \quad x = 2 \qquad \text{Using the Principle of Zero Products}$$

Check

For 2:
$$\frac{x^2}{x-2} = \frac{4}{x-2}$$
$$\frac{2^2}{2-2} \overset{?}{\,\big|\,} \frac{4}{2-2}$$
$$\frac{4}{0} \,\Big|\, \frac{4}{0} \qquad \text{UNDEFINED;}$$
$$\qquad \text{FALSE}$$

For -2:
$$\frac{x^2}{x-2} = \frac{4}{x-2}$$
$$\frac{(-2)^2}{-2-2} \overset{?}{\,\big|\,} \frac{4}{-2-2}$$
$$\frac{4}{-4} \,\Big|\, \frac{4}{-4}$$
$$-1 \,\Big|\, -1 \qquad \text{TRUE}$$

The number -2 is a solution, but 2 is not (it results in division by 0).   **15** ✔

Quick Check 15 ✔

Solve: $\dfrac{t^2}{t+3} = \dfrac{9}{t+3}$.

Square Roots and Equations of the Type $x^2 = a$

When we raise a number to the second power, we say that we have *squared* the number. Sometimes, we may need to find the number that was squared. We call this process *finding the square root* of a number.

> **DEFINITION**
>
> The number c is a **square root** of a if $c^2 = a$. The positive square root of a is written $\sqrt{a}$.

For example, 5 is a square root of 25 since $5^2 = 25$, and -5 is also a square root of 25 since $(-5)^2 = 25$. Thus, we can write $\sqrt{25} = 5$ and $\sqrt{25} = -5$. However, -4 does <u>not</u> have a real-number square root since there is no real number c such that $c^2 = -4$. That is, $\sqrt{-4}$ is not a real number.

Properties of the Square Root

- Every positive real number a has two real-number square roots. For $a > 0$, the square roots of a are $\sqrt{a}$ and $-\sqrt{a}$. We often represent the two square roots using the plus-minus symbol: $\pm\sqrt{a}$.
- The number 0 has just one square root, 0 itself.
- Negative numbers do not have real-number square roots.

Consider the equation $x^2 = 25$. The solutions are the two square roots of 25, so $x = \pm 5$. In general, we conclude that if $x^2 = a$, then its solutions are $x = \pm\sqrt{a}$. This is the *Principle of Square Roots*.

The Principle of Square Roots

The solutions of $x^2 = a$ are $x = \pm\sqrt{a}$.

- When $a > 0$, the solutions are two real numbers.
- When $a = 0$, the solution is 0.
- When $a < 0$, there are no real-number solutions.

EXAMPLE 17 Use the Principle of Square Roots to solve each equation.

a) $x^2 = 81$ **b)** $x^2 = 7$ **c)** $x^2 = -16$ **d)** $3x^2 = 12$ **e)** $(2x + 1)^2 = 36$

Solution
a) We have

$$x^2 = 81$$
$$x = \pm\sqrt{81} \qquad \text{Using the Principle of Square Roots}$$
$$x = \pm 9. \qquad \text{Simplifying}$$

b) We have

$$x^2 = 7$$
$$x = \pm\sqrt{7} \qquad \text{Using the Principle of Square Roots}$$
$$x \approx \pm 2.646. \qquad \text{Using a calculator}$$

c) Since $-16 < 0$, there is no real-number solution of $x^2 = -16$.

d) We have

$$3x^2 = 12$$
$$\tfrac{1}{3} \cdot 3x = \tfrac{1}{3} \cdot 12 \qquad \text{Multiplying both sides by } \tfrac{1}{3}$$
$$x^2 = 4 \qquad \text{Simplifying}$$
$$x = \pm\sqrt{4} \qquad \text{Using the Principle of Square Roots}$$
$$x = \pm 2. \qquad \text{Simplifying}$$

The solutions are 2 and -2.

Quick Check 16 ✔

Use the Principle of Square Roots to solve each equation.

a) $x^2 = 121$
b) $t^2 = 13$
c) $4(3 - 5r)^2 = 100$

e) We have

$$(2x + 1)^2 = 36$$
$$2x + 1 = 6 \quad or \quad 2x + 1 = -6 \qquad \text{Using the Principle of Square Roots}$$
$$2x = 5 \qquad\qquad 2x = -7 \qquad \text{Simplifying}$$
$$x = \tfrac{5}{2} \qquad\qquad x = -\tfrac{7}{2}.$$

The solutions are $\tfrac{5}{2}$ and $-\tfrac{7}{2}$.

16 ✔

Solving Inequalities

Two inequalities are **equivalent** if they have the same solutions. For example, the inequalities $x > 4$ and $4 < x$ are equivalent. Principles for solving inequalities are similar to those for solving equations. We can add the same number to both sides of an inequality. We can also multiply on both sides by the same nonzero number, but if that number is negative, we must reverse the inequality sign. The following are the inequality-solving principles.

The Inequality-Solving Principles

For any real numbers a, b, and c,

$$a < b \quad \text{is equivalent to} \quad a + c < b + c.$$

For any real numbers a, b, and any *positive* number c,

$$a < b \quad \text{is equivalent to} \quad ac < bc.$$

For any real numbers a, b, and any *negative* number c,

$$a < b \quad \text{is equivalent to} \quad ac > bc.$$

Similar statements hold for $\leq$ and $\geq$.

EXAMPLE 18   Solve: $17 - 8x \geq 5x - 4$.

Solution   We have

$$17 - 8x \geq 5x - 4$$
$$-8x \geq 5x - 21 \qquad \text{Adding } -17 \text{ to both sides}$$
$$-13x \geq -21 \qquad \text{Adding } -5x \text{ to both sides}$$
$$-\tfrac{1}{13}(-13x) \leq -\tfrac{1}{13}(-21) \qquad \text{Multiplying both sides by } -\tfrac{1}{13} \text{ and } \textit{reversing} \text{ the inequality sign}$$
$$x \leq \tfrac{21}{13}.$$

Any number less than or equal to $\tfrac{21}{13}$ is a solution.   **17** ✔

Quick Check 17 ✔
Solve:
$3x + 4 < 5x - 2$.

Applications

To solve applied problems, we first translate to mathematical language, usually an equation. Then we solve the equation and check to see whether the solution of the equation is a solution of the problem.

EXAMPLE 19   **Life Science: Weight Gain.**   After a 5% gain in weight, a grizzly bear weighs 693 lb. What was its original weight?

Solution   We first translate to an equation:

$$\underbrace{\textit{Original weight}}_{w} + \underbrace{\textit{5% of the}}_{+\ 0.05} \underbrace{\textit{original weight}}_{\cdot\quad w} = 693$$
$$= 693.$$

Now we solve the equation:

$$1 \cdot w + 0.05 \cdot w = 693$$
$$w(1 + 0.05) = 693$$
$$1.05w = 693$$
$$w = \frac{693}{1.05} = 660.$$

Check: $600 + 5\% \cdot 660 = 660 + 0.05 \cdot 660 = 660 + 33 = 693.$

The original weight of the bear was 660 lb.   **18** ✔

Quick Check 18 ✔

Including a sales tax of 6%, the cost of a set of dinner plates is \$132.50. What is the price of the plates without the tax?

Quick Check 19 ✔

Shoe Express determines that its total revenue, in dollars, from the sale of x pairs of shoes is given by

$$85x + 120.$$

How many pairs of shoes must the store sell to ensure that its total revenue will be more than $30,000?

EXAMPLE 20   **Business: Total Sales.**   Raggs, Ltd., a clothing firm, determines that its total revenue, in dollars, from the sale of x suits is given by

$$200x + 50.$$

Determine the number of suits that the firm must sell to ensure that its total revenue will be more than $70,050.

Solution   We translate to an inequality and solve:

$$200x + 50 > 70{,}050$$
$$200x > 70{,}000 \qquad \text{Adding } -50 \text{ to both sides}$$
$$x > 350. \qquad \text{Multiplying both sides by } \tfrac{1}{200}$$

Thus the company's total revenue will exceed $70,050 when it sells more than 350 suits.   **19** ✔

A | Exercise Set

Express as an equivalent expression without exponents.

1. 5^3

2. 7^2

3. $(-7)^2$

4. $(-5)^3$

5. $(1.01)^2$

6. $(1.01)^3$

7. $\left(\dfrac{1}{2}\right)^4$

8. $\left(\dfrac{1}{4}\right)^3$

9. $(6x)^0$

10. $(6x)^1$

11. $-t^1$

12. $-t^0$

13. $\left(\dfrac{1}{3}\right)^0$

14. $\left(\dfrac{1}{3}\right)^1$

Express as an equivalent expression without negative exponents.

15. 3^{-2}

16. 4^{-2}

17. $\left(\dfrac{1}{2}\right)^{-3}$

18. $\left(\dfrac{7}{2}\right)^{-2}$

19. $\dfrac{1}{3^{-2}}$

20. $\dfrac{1}{4^{-1}}$

21. 10^{-1}

22. 10^{-4}

23. e^{-b}

24. t^{-k}

25. b^{-1}

26. h^{-1}

Multiply.

27. $x^2 \cdot x^3$

28. $t^3 \cdot t^4$

29. $x^{-7} \cdot x$

30. $x^5 \cdot x$

31. $5x^2 \cdot 7x^3$

32. $4t^3 \cdot 2t^4$

33. $x^{-4} \cdot x^7 \cdot x$

34. $x^{-3} \cdot x \cdot x^3$

35. $e^{-t} \cdot e^t$

36. $e^k \cdot e^{-k}$

Divide.

37. $\dfrac{x^5}{x^2}$

38. $\dfrac{x^7}{x^3}$

39. $\dfrac{x^2}{x^5}$

40. $\dfrac{x^3}{x^7}$

41. $\dfrac{e^k}{e^k}$

42. $\dfrac{t^k}{t^k}$

43. $\dfrac{e^t}{e^4}$

44. $\dfrac{e^k}{e^3}$

45. $\dfrac{t^6}{t^{-8}}$

46. $\dfrac{t^5}{t^{-7}}$

47. $\dfrac{t^{-9}}{t^{-11}}$

48. $\dfrac{t^{-11}}{t^{-7}}$

49. $\dfrac{ab(a^2b)^3}{ab^{-1}}$

50. $\dfrac{x^2y^3(xy^3)^2}{x^{-3}y^2}$

Simplify.

51. $(t^{-2})^3$

52. $(t^{-3})^4$

53. $(e^x)^4$

54. $(e^x)^5$

55. $(2x^2y^4)^3$

56. $(2x^2y^4)^5$

57. $(3x^{-2}y^{-5}z^4)^{-4}$

58. $(5x^3y^{-7}z^{-5})^{-3}$

59. $(-3x^{-8}y^7z^2)^2$

60. $(-5x^4y^{-5}z^{-3})^4$

61. $\left(\dfrac{cd^3}{2q^2}\right)^4$

62. $\left(\dfrac{4x^2y}{a^3b^3}\right)^3$

Multiply.

63. $5(x - 7)$

64. $x(1 + t)$

65. $(x - 5)(x - 2)$

66. $(x - 4)(x - 3)$

67. $(a - b)(a^2 + ab + b^2)$

68. $(x^2 - xy + y^2)(x + y)$

69. $(2x + 5)(x - 1)$

70. $(3x + 4)(x - 1)$

71. $(a - 2)(a + 2)$

72. $(3x - 1)(3x + 1)$

73. $(5x + 2)(5x - 2)$

74. $(t - 1)(t + 1)$

75. $(a - h)^2$

76. $(a + h)^2$

77. $(5x + t)^2$

78. $(7a - c)^2$

79. $5x(x^2 + 3)^2$

80. $-3x^2(x^2 - 4)(x^2 + 4)$

Use the following equation for Exercises 81–84.

$$\begin{aligned}(x + h)^3 &= (x + h)(x + h)^2 \\ &= (x + h)(x^2 + 2xh + h^2) \\ &= (x + h)x^2 + (x + h)2xh + (x + h)h^2 \\ &= x^3 + x^2h + 2x^2h + 2xh^2 + xh^2 + h^3 \\ &= x^3 + 3x^2h + 3xh^2 + h^3\end{aligned}$$

81. $(a + b)^3$

82. $(a - b)^3$

83. $(x - 5)^3$

84. $(2x + 3)^3$

Factor.

85. $x - xt$

86. $x + xh$

87. $x^2 + 6xy + 9y^2$

88. $x^2 - 10xy + 25y^2$

89. $x^2 - 2x - 15$

90. $x^2 + 8x + 15$

91. $x^2 - x - 20$

92. $x^2 - 9x - 10$

93. $49x^2 - t^2$

94. $9x^2 - b^2$

95. $36t^2 - 16m^2$

96. $25y^2 - 9z^2$

97. $a^3b - 16ab^3$

98. $2x^4 - 32$

99. $a^8 - b^8$

100. $36y^2 + 12y - 35$

101. $10a^2x - 40b^2x$

102. $x^3y - 25xy^3$

103. $2 - 32x^4$

104. $2xy^2 - 50x$

105. $9x^2 + 17x - 2$

106. $6x^2 - 23x + 20$

107. $x^3 + 8$ (Hint: See Exercise 68.)

108. $a^3 - 27$ (Hint: See Exercise 67.)

109. $y^3 - 64t^3$

110. $m^3 + 1000p^3$

111. $3x^3 - 6x^2 - x + 2$

112. $5y^3 + 2y^2 - 10y - 4$

113. $x^3 - 5x^2 - 9x + 45$

114. $t^3 + 3t^2 - 25t - 75$

Solve.

115. $-7x + 10 = 5x - 11$

116. $-8x + 9 = 4x - 70$

117. $5x - 17 - 2x = 6x - 1 - x$

118. $5x - 2 + 3x = 2x + 6 - 4x$

119. $x + 0.8x = 216$

120. $x + 0.5x = 210$

121. $x + 0.08x = 216$

122. $x + 0.05x = 210$

123. $2x(x + 3)(5x - 4) = 0$

124. $7x(x - 2)(2x + 3) = 0$

125. $x^2 + 1 = 2x + 1$

126. $2t^2 = 9 + t^2$

127. $t^2 - 2t = t$

128. $6x - x^2 = x$

129. $6x - x^2 = -x$

130. $2x - x^2 = -x$

131. $9x^3 = x$

132. $16x^3 = x$

133. $(x - 3)^2 = x^2 + 2x + 1$

134. $(x - 5)^2 = x^2 + x + 3$

135. $\dfrac{4x}{x + 5} + \dfrac{20}{x} = \dfrac{100}{x^2 + 5x}$

136. $\dfrac{x}{x + 1} + \dfrac{3x + 5}{x^2 + 4x + 3} = \dfrac{2}{x + 3}$

137. $\dfrac{50}{x} - \dfrac{50}{x - 2} = \dfrac{4}{x}$

138. $\dfrac{60}{x} = \dfrac{60}{x - 5} + \dfrac{2}{x}$

139. $0 = 2x - \dfrac{250}{x^2}$

140. $5 - \dfrac{35}{x^2} = 0$

141. $x^2 = 64$

142. $x^2 = 144$

143. $u^2 = 30$

144. $w^2 = 97$

145. $4t^2 = 49$

146. $100k^2 = 169$

147. $9x^2 = 20$

148. $25y^2 = 3$

149. $5z^2 = 4$

150. $3b^2 = 144$

151. $(3x - 2)^2 = 1$

152. $(6x + 5)^2 = 400$

153. $(2 - 5x)^2 = 10$

154. $(1 - 4y)^2 = 2$

155. $3 - x \le 4x + 7$

156. $x + 6 \le 5x - 6$

157. $5x - 5 + x > 2 - 6x - 8$

158. $3x - 3 + 3x > 1 - 7x - 9$

159. $-7x < 4$

160. $-5x \ge 6$

161. $5x + 2x \le -21$

162. $9x + 3x \ge -24$

163. $2x - 7 < 5x - 9$

164. $10x - 3 \ge 13x - 8$

165. $8x - 9 < 3x - 11$

166. $11x - 2 \ge 15x - 7$

167. $8 < 3x + 2 < 14$

168. $2 < 5x - 8 \le 12$

169. $3 \le 4x - 3 \le 19$

170. $9 \le 5x + 3 < 19$

171. $-7 \le 5x - 2 \le 12$

172. $-11 \le 2x - 1 < -5$

APPLICATIONS

Business and Economics

173. Investment increase. An investment is made at $8\frac{1}{2}\%$, compounded annually. It grows to \$705.25 at the end of 1 yr. How much was invested originally?

174. Investment increase. An investment is made at 7%, compounded annually. It grows to \$856 at the end of 1 yr. How much was invested originally?

175. Total revenue. Sunshine Products determines that the total revenue, in dollars, from the sale of x flowerpots is $3x + 1000$. Determine the number of flowerpots that must be sold so that the total revenue will be more than $22,000.

176. Total revenue. Beeswax Inc. determines that the total revenue, in dollars, from the sale of x candles is $5x + 1000$. Determine the number of candles that must be sold so that the total revenue will be more than $22,000.

Life and Physical Sciences

177. Weight gain. After a 6% gain in weight, an elk weighs 508.8 lb. What was its original weight?

178. Weight gain. After a 7% gain in weight, a deer weighs 363.8 lb. What was its original weight?

Social Sciences

179. Population increase. After a 2% increase, the population of Burnside City is 826,200. What was the city's former population?

180. Population increase. After a 3% increase, the student population of Glen Oaks College is 5356. What was the former student population?

General Interest

181. Grade average. To get a B in a course, a student's average must be greater than or equal to 80% (at least 80%) and less than 90%. On the first three tests, Claudia scores 78%, 90%, and 92%. Determine the scores on the fourth test that will guarantee her a B.

182. Grade average. To get a C in a course, a student's average must be greater than or equal to 70% and less than 80%. On the first three tests, Horace scores 65%, 83%, and 82%. Determine the scores on the fourth test that will guarantee him a C.

183. Auditorium seating. The seats at Ardon Auditorium are arranged in a square, with n rows, each containing n seats. If the auditorium can seat 324 people, how many people can be seated in one row?

184. Tiling a room. The conference room at the Fireside Hotel is square, and the floor is covered by 5625 equal-sized square tiles. How many tiles run along each edge of the room?

Answers to Quick Checks

1. (a) 243; (b) 25, (c) $\frac{16}{81}$ 2. (a) 1; (b) $\frac{2}{5}$; (c) -1

3. (a) $\frac{1}{9}$; (b) 64; (c) $\dfrac{8}{x^5}$ 4. (a) t^8; (b) $\dfrac{28}{x^3}$; (c) $\dfrac{1}{y^4}$

5. (a) d^5; (b) t^8; (c) u^{14} 6. (a) x^{35}; (b) $\dfrac{1}{z^{16}}$;

(c) $\dfrac{256x^4z^4}{y^8}$ 7. (a) $-4x + 12$; (b) $xy + 3xz$;

(c) $2x^2 - 5x - 7$ 8. (a) $4x^2 + 12x + 9$;

(b) $9x^2 - 6x + 1$; (c) $y^2 - 25z^2$ 9. (a) $2mn(m + 2n)$;

(b) $(x - 5)^2$; (c) $(x + 3)(3x + 1)$;

(d) $(4t + 9u)(4t - 9u)$ 10. (a) $(u + 4)(u^2 + 3)$;

(b) $(x + 2)(x - 1)(x + 1)$ 11. $x = \frac{38}{5}$ 12. $0, \frac{1}{3}, -5$

13. $0, \frac{2}{3}, -\frac{2}{3}$ 14. $(x + 1)^2(x^2 - 9)$ 15. $t = 3$

16. (a) $x = \pm 11$; (b) $t = \pm\sqrt{13}$; (c) $r = -\frac{2}{5}, \frac{8}{5}$

17. $x > 3$ 18. $125 19. $x \geq 352$

Regression and Microsoft Excel

- Use Microsoft Excel to perform regression

Using Excel 2013

We can use Microsoft Excel to enter and plot data and to find lines of best fit using regression. Suppose we are given the following data:

x	0	2	4	5	6
y	3	4.7	6	6.8	8

Step 1: We open a blank workbook and enter the data into two columns, as shown in Fig. 1.

Step 2: We highlight the columns of data (see Fig. 2). Then within the Insert tab, we select Scatter in Charts. Next, we choose the first option that appears, with the markers shown as distinct points.

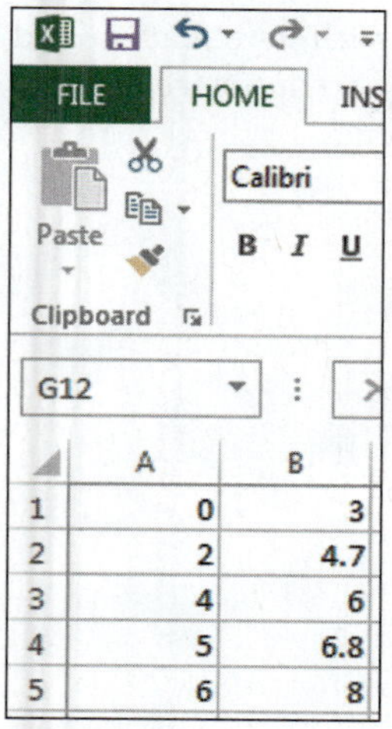

FIGURE 1

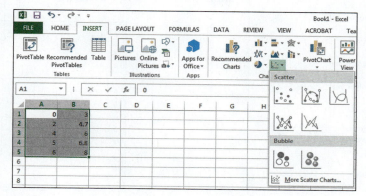

FIGURE 2

Step 3: A scatterplot of the data appears, as shown in Fig. 3. We click on ⊞ near the upper-right corner of the graph. We select Trendline, then click on the pointer and choose More Options… (see Fig. 4).

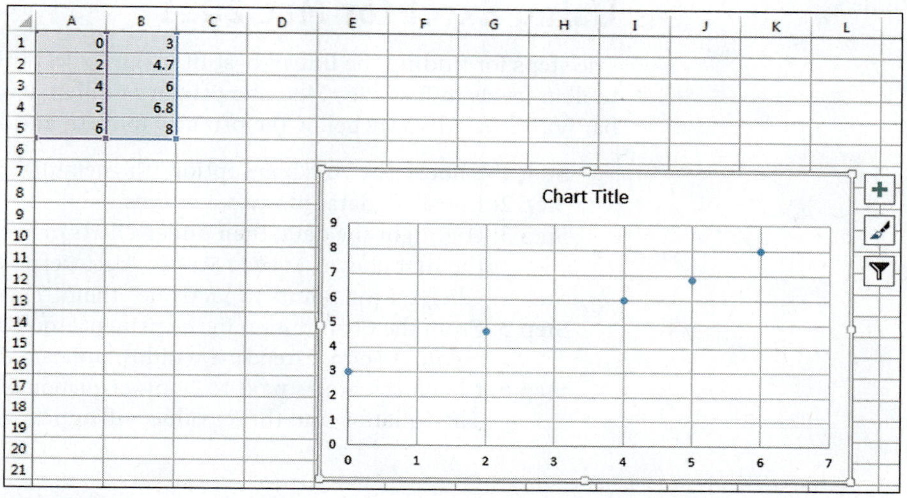

FIGURE 3

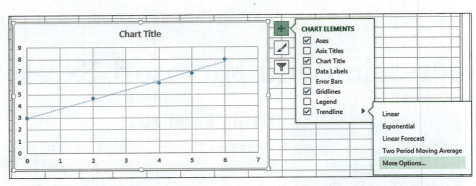

FIGURE 4

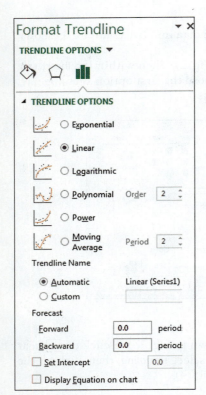

FIGURE 5

Step 4: A Format Trendline box opens (Fig. 5), with Linear selected as the default. At the bottom, we check the box next to Display Equation ON CHART, as well as the box next to Display R-squared VALUE ON CHART.

The line of best fit is now displayed on the scatterplot, along with its equation and the R^2 value, as shown in Fig. 6. The R^2 value is called the *squared correlation coefficient.* An R^2 value close to 1 indicates that the data have a strong linear trend.

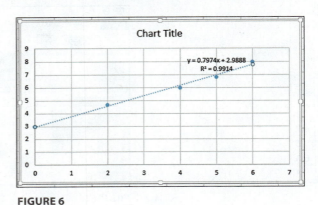

FIGURE 6

In this book, regression is also used to fit exponential and polynomial functions to data. The student can experiment with the various regression options under Format Trendline.

Using Excel for Mac 2011

The steps for finding the line of best fit using Excel for Mac 2011 are given below. There are three toolbar levels in Excel for Mac: the primary toolbar across the top of the screen, a secondary toolbar with icons directly below the primary toolbar, and a tertiary toolbar below the secondary one.

Step 1: Choose the Workbook option (the default).

Step 2: Enter the data into two columns.

Step 3: Highlight the data; then under Charts in the tertiary toolbar, select Scatter. Next, select the first choice, Marked Scatter. A scatterplot of the data appears on the screen.

Step 4: Click on the points to "activate" them. They will appear as X's.

Step 5: From the Chart option in the primary toolbar, select Add Trendline from the drop-down menu. A Format Trendline window appears. Select Options.

Step 6: Check the boxes next to Display Equation on chart and Display R-squared value on chart. The equation and the R^2 value will appear on the scatterplot, along with the trendline.

B   Exercise Set

1. Use Excel to find the line of best fit for the following data.

x	4	6	8	10	12
y	15	22	27	33	44

2. Use Excel to find the line of best fit for the following data.

x	−5	−1	3	8
y	2	10	19	35

3. Use Excel to find a quadratic function (polynomial of power 2) that best fits the data in Exercise 2.

Areas for a Standard Normal Distribution

Entries in the table represent area under the curve between $z = 0$ and a positive value of z. Because of the symmetry of the curve, area under the curve between $z = 0$ and a negative value of z are found in a similar manner.

Area = Probability
$$= P(0 \le x \le z)$$
$$= \int_0^z \frac{1}{\sqrt{2\pi}} e^{-x^2/2}\, dx$$

z	0.00	0.01	0.02	0.03	0.04	0.05	0.06	0.07	0.08	0.09
0.0	.0000	.0040	.0080	.0120	.0160	.0199	.0239	.0279	.0319	.0359
0.1	.0398	.0438	.0478	.0517	.0557	.0596	.0636	.0675	.0714	.0753
0.2	.0793	.0832	.0871	.0910	.0948	.0987	.1026	.1064	.1103	.1141
0.3	.1179	.1217	.1255	.1293	.1331	.1368	.1406	.1443	.1480	.1517
0.4	.1554	.1591	.1628	.1664	.1700	.1736	.1772	.1808	.1844	.1879
0.5	.1915	.1950	.1985	.2019	.2054	.2088	.2123	.2157	.2190	.2224
0.6	.2257	.2291	.2324	.2357	.2389	.2422	.2454	.2486	.2517	.2549
0.7	.2580	.2611	.2642	.2673	.2704	.2734	.2764	.2794	.2823	.2852
0.8	.2881	.2910	.2939	.2967	.2995	.3023	.3051	.3078	.3106	.3133
0.9	.3159	.3186	.3212	.3238	.3264	.3289	.3315	.3340	.3365	.3389
1.0	.3413	.3438	.3461	.3485	.3508	.3531	.3554	.3577	.3599	.3621
1.1	.3643	.3665	.3686	.3708	.3729	.3749	.3770	.3790	.3810	.3830
1.2	.3849	.3869	.3888	.3907	.3925	.3944	.3962	.3980	.3997	.4015
1.3	.4032	.4049	.4066	.4082	.4099	.4115	.4131	.4147	.4162	.4177
1.4	.4192	.4207	.4222	.4236	.4251	.4265	.4279	.4292	.4306	.4319
1.5	.4332	.4345	.4357	.4370	.4382	.4394	.4406	.4418	.4429	.4441
1.6	.4452	.4463	.4474	.4484	.4495	.4505	.4515	.4525	.4535	.4545
1.7	.4554	.4564	.4573	.4582	.4591	.4599	.4608	.4616	.4625	.4633
1.8	.4641	.4649	.4656	.4664	.4671	.4678	.4686	.4693	.4699	.4706
1.9	.4713	.4719	.4726	.4732	.4738	.4744	.4750	.4756	.4761	.4767
2.0	.4772	.4778	.4783	.4788	.4793	.4798	.4803	.4808	.4812	.4817
2.1	.4821	.4826	.4830	.4834	.4838	.4842	.4846	.4850	.4854	.4857
2.2	.4861	.4864	.4868	.4871	.4875	.4878	.4881	.4884	.4887	.4890
2.3	.4893	.4896	.4898	.4901	.4904	.4906	.4909	.4911	.4913	.4916
2.4	.4918	.4920	.4922	.4925	.4927	.4929	.4931	.4932	.4934	.4936
2.5	.4938	.4940	.4941	.4943	.4945	.4946	.4948	.4949	.4951	.4952
2.6	.4953	.4955	.4956	.4957	.4959	.4960	.4961	.4962	.4963	.4964
2.7	.4965	.4966	.4967	.4968	.4969	.4970	.4971	.4972	.4973	.4974
2.8	.4974	.4975	.4976	.4977	.4977	.4978	.4979	.4979	.4980	.4981
2.9	.4981	.4982	.4982	.4983	.4984	.4984	.4985	.4985	.4986	.4986
3.0	.4987	.4987	.4987	.4988	.4988	.4989	.4989	.4989	.4990	.4990

Answers

CHAPTER R

Technology Connection, p. 6

1–20. Left to the student

Exercise Set R.1, p. 10

1.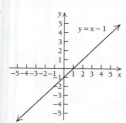

3.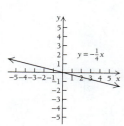

5.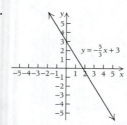

7.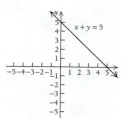

9.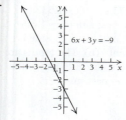

11.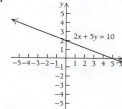

13.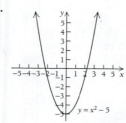

15.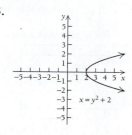

17.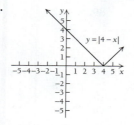

19.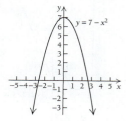

21.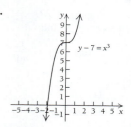

23. 344.4 mg   **25.** About 27.25 mi/hr, or mph   **27. (a)** About 1.8 million, about 4.2 million, about 5.4 million, about 6.1 million; **(b)** about 39 and 82; **(c)** about 63; **(d)** ✎
29. (a) $102,800.00; **(b)** $102,819.60; **(c)** $102,829.54; **(d)** $102,839.46; **(e)** $102,839.56   **31. (a)** $33,745.92; **(b)** $33,784.87; **(c)** $33,804.75; **(d)** $33,824.68; **(e)** $33,824.90
33. $550.86   **35.** $97,881.97   **37. (a)** 2007–2012; **(b)** 2005; **(c)** 2012; **(d)** 2002   **39. (a)** $88,382.67; **(b)** $42,000; $46,382.67   **41.** 5.43%   **43.** 3.82%   **45. (a)** Western: 4.5%, Commonwealth: 4.52%; **(b)** Commonwealth   **47.** $4.1\overline{2}\%$

49.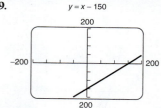

51.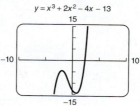

53.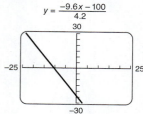

55.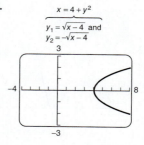

Technology Connection, p. 16

1. 951; 42,701   **2.** 21,813

Technology Connection, p. 17

1. 6; 3.99; 150; $-1.\overline{5}$, or $-\frac{14}{9}$ **2.** -21.3; -18.39; -117.3; $3.2\overline{5}$, or $\frac{293}{90}$ **3.** -75; -65.466; -420.6; $1.6\overline{8}$, or $\frac{76}{45}$

Technology Connection, p. 20

1–3. Left to the student

Exercise Set R.2, p. 21

1. Yes   **3.** Yes   **5.** Yes   **7.** Yes   **9.** Yes   **11.** Yes
13. Yes   **15.** No   **17.** Yes

A ✎ indicates that the exercise asks for a written interpretation or explanation; answers will vary.

19. (a)

x	5.1	5.01	5.001	5
$f(x)$	17.4	17.04	17.004	17

(b) $f(4) = 13, f(3) = 9, f(-2) = -11, f(k) = 4k - 3,$
$f(1 + t) = 4t + 1, f(x + h) = 4x + 4h - 3$

21. $g(-1) = -2, g(0) = -3, g(1) = -2, g(5) = 22,$
$g(u) = u^2 - 3, g(a + h) = a^2 + 2ah + h^2 - 3,$ and
$\dfrac{g(a + h) - g(a)}{h} = 2a + h, h \neq 0$ **23. (a)** $f(4) = \dfrac{1}{49},$
$f(0) = \dfrac{1}{9}, f(a) = \dfrac{1}{(a + 3)^2}, f(t + 4) = \dfrac{1}{(t + 7)^2},$
$f(x + h) = \dfrac{1}{(x + h + 3)^2},$ and $\dfrac{f(x + h) - f(x)}{h} =$
$\dfrac{-2x - h - 6}{(x + h + 3)^2(x + 3)^2}, h \neq 0$ **(b)** Take an input, square it, add
six times the input, add 9, and then take the reciprocal of the result.

25. (a) $f(x) = 4x + 2$
(b)
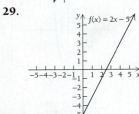

27. (a) $h(x) = x^2 + x$
(b)
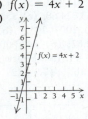

29.
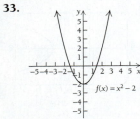

31.
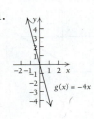

33.
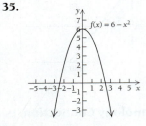

35.
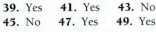

37.
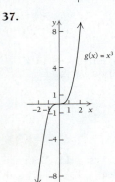

39. Yes **41.** Yes **43.** No
45. No **47.** Yes **49.** Yes
51. No

53. (a)
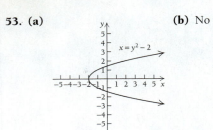
(b) No

55. $\dfrac{f(x + h) - f(x)}{h} = 2x + h - 3, h \neq 0$

57. $f(-1) = 3, f(1) = -2$ **59.** $f(0) = 17, f(10) = 6$

61.
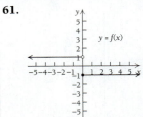

63.
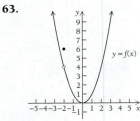

65.
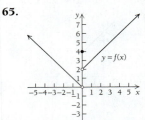

67.
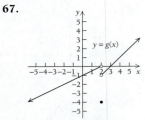

69.
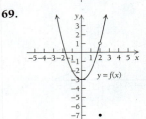

71. $530.80 **73. (a)** 1.818 m^2;
(b) 2.173 m^2; **(c)** 1.537 m^2
75. (a) Yes; a unique "scale of
impact" number is assigned to
each event. **(b)** The inputs are
the events; the outputs are the
scale of impact numbers.

77. $y = \pm\sqrt{\dfrac{x + 5}{2}}$; this is *not* a function.

79. $y = 2 \cdot \sqrt[3]{x}$; this is a function. **81.**

83.

X	Y1
-3	.6
-2	ERR:
-1	-1
0	-.75
1	-1
2	ERR:
3	.6
X = -3	

85. Left to the student

87. (a) $f(x) = 5(x + 2); g(x) = 5x + 2$
(b)
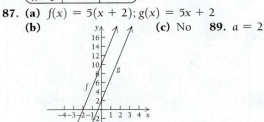
(c) No **89.** $a = 2$

Technology Connection, p. 28

1.

X	Y1
-3	.2
-2.5	.4444
-2	ERR:
-1.5	-.5714
-1	-.3333
-.5	-.2667
0	-.25

X = -3

2. Answers may vary.

Technology Connection, p. 30

1. Domain $= \mathbb{R}$; range $= [-4, \infty)$
2. Domain $= \mathbb{R}$; range $= \mathbb{R}$
3. Domain $= \{x \mid x \text{ is a real number and } x \neq -1\}$;
range $= \{y \mid y \text{ is a real number and } y \neq 0\}$
4. Domain $= \mathbb{R}$; range $= [-8, \infty)$
5. Domain $= [-4, \infty)$; range $= [0, \infty)$
6. Domain $= [-3, 3]$; range $= [0, 3]$
7. Domain $= [-3, 3]$; range $= [-3, 0]$
8. Domain $= \mathbb{R}$; range $= \mathbb{R}$

Exercise Set R.3, p. 31

1. $(-1, 3)$ **3.** $(0, 5)$ **5.** $(-9, -5]$ **7.** $[x, x + h]$
9. $(-\infty, q]$ **11.** $[-2, 2]$

13. $(6, 20]$

15. $(-3, \infty)$

17. $(-2, 3]$

19. $[12.5, \infty)$

21. (a) 3; (b) $\{-3, -1, 1, 3, 5\}$; (c) 3; (d) $\{-2, 0, 2, 3, 4\}$
23. (a) 4; (b) $\{-5, -3, 1, 2, 3, 4, 5\}$; (c) $\{-5, -3, 4\}$;
(d) $\{-3, 2, 4, 5\}$ **25.** (a) -1; (b) $[-2, 4]$; (c) 3; (d) $[-3, 3]$
27. (a) -2; (b) $[-4, 2]$; (c) -2; (d) $[-3, 3]$ **29.** (a) 3;
(b) $[-3, 3]$; (c) about -1.4 and 1.4; (d) $[-5, 4]$ **31.** (a) 1;
(b) $[-5, 5)$; (c) $[3, 5)$; (d) $\{-2, -1, 0, 1, 2\}$
33. $\{x \mid x \text{ is a real number and } x \neq 2\}$ **35.** $\{x \mid x \geq 0\}$ **37.** $\mathbb{R}$
39. $\{x \mid x \text{ is a real number and } x \neq 2\}$ **41.** $\mathbb{R}$
43. $\{x \mid x \text{ is a real number and } x \neq 3.5\}$ **45.** $\{x \mid x \geq -\frac{4}{5}\}$
47. $\mathbb{R}$ **49.** $\{x \mid x \text{ is a real number and } x \neq 5, x \neq -5\}$ **51.** $\mathbb{R}$
53. $\{x \mid x \text{ is a real number and } x \neq 5, x \neq 1\}$ **55.** $[-1, 2]$

57. (a) $A(t) = 5000\left(1 + \frac{0.031}{2}\right)^{2t}$; (b) $\{t \mid t \geq 0\}$

59. (a) $[25, 102]$; (b) approximately $[0, 455]$; (c) ✏️
61. (a) $[0, 70]$; (b) $[8, 75]$ **63.** ✏️ **65.** $(-\infty, 0) \cup (0, \infty)$;
$[0, \infty)$; $\left\{\frac{1}{6}\right\}$; $(-\infty, 0) \cup (0, \infty)$; $[0, \infty)$

Technology Connection, p. 35

1. The line slants up from left to right, intersects the y-axis at
$(0, 1)$, and is steeper than $y = 10x + 1$. **2.** The line slants
up from left to right, passes through the origin, and is less steep
than $y = \frac{2}{31}x$. **3.** The line slants down from left to right, passes through the origin, and is steeper than $y = -10x$. **4.** The
line slants down from left to right, intersects the y-axis at
$(0, -1)$, and is less steep than $y = -\frac{5}{32}x - 1$.

Technology Connection, p. 38

1. The graph of y_2 is a shift 3 units up of the graph of y_1, and y_2
has y-intercept $(0, 3)$. The graph of y_3 is a shift 4 units down
of the graph of y_1, and y_3 has y-intercept $(0, -4)$. The graph

of $y = x - 5$ is a shift 5 units down of the graph of $y = x$, and
$\cdot$ $y = x - 5$ has y-intercept $(0, -5)$. All lines are parallel.
2. For any x-value, the y_2-value is 3 more than the y_1-value,
and the y_3-value is 4 less than the y_1-value.

Exercise Set R.4, p. 44

1.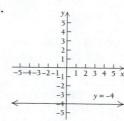

3.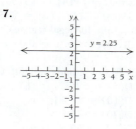

5.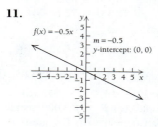

7.

9.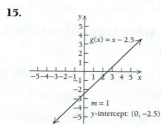

11.

13.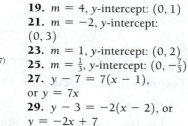

15.

17.

19. $m = 4$, y-intercept: $(0, 1)$
21. $m = -2$, y-intercept: $(0, 3)$
23. $m = 1$, y-intercept: $(0, 2)$
25. $m = \frac{1}{3}$, y-intercept: $\left(0, -\frac{7}{3}\right)$
27. $y - 7 = 7(x - 1)$, or $y = 7x$
29. $y - 3 = -2(x - 2)$, or $y = -2x + 7$
31. $y - 0 = -5(x - 5)$, or $y = -5x + 25$
33. $y + 6 = \frac{1}{2}(x - 0)$, or $y = \frac{1}{2}x - 6$
35. $y - 8 = 0 \cdot (x - 4)$, or $y = 8$ **37.** $-\frac{4}{7}$ **39.** $-\frac{1}{4}$
41. Undefined slope **43.** $-\frac{4}{13}$ **45.** 0 **47.** 4 **49.** 2
51. $y = -\frac{4}{7}x - \frac{1}{7}$ **53.** $y = -\frac{1}{4}x - \frac{23}{4}$ **55.** $x = 3$
57. $y = -\frac{4}{13}x - \frac{29}{52}$ **59.** $y = 3$ **61.** $\frac{2}{25}$, or 8% **63.** 3.5%
65. (a) $I(s) = 0.0057s$; (b) 18 cartridges
67. (a) $C(x) = 80x + 45{,}000$; (b) $R(x) = 450x$;

(c) $P(x) = 370x - 45{,}000$;
(d) profit of $\$1{,}065{,}000$; **(e)** 122 pairs

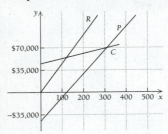

69. (a) $V(t) = 5200 - 512.5t$;
(b) $V(0) = \$5200$, $V(1) = \$4687.50$, $V(2) = \$4175$,
$V(3) = \$3662.50$, $V(4) = \$3150$, $V(7) = \$1612.50$,
$V(8) = \$1100$ **71.** $V(3) = \$25{,}200$ **73.** About 91%
(don't round up, for legal reasons!)   **75.** $\$232.50$/yr
77. 1.8 billion/yr   **79.** $t \approx 0.02$ sec   **81. (a)** $B(W) = 0.025W$;
(b) $0.025 = 2.5\%$, so $M = 2.5\% \cdot W$, and the weight of the
brain is 2.5% of the body weight; **(c)** 4 lb
83. (a) $D(5) = 6$ ft,   **(b)** 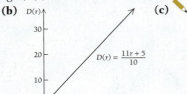   **(c)**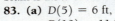
    $D(10) = 11.5$ ft,
    $D(20) = 22.5$ ft,
    $D(50) = 55.5$ ft,
    $D(65) = 72$ ft

85. (a) $y = 0.057x - 113.85$; **(b)** approximately 94.8%;
(c) 2015; **(d)**   **87. (a)** $N = 1.02P$; **(b)** $N = 204{,}000$
people; **(c)** $P = 360{,}000$ people   **89.** $y = 25$
91. (a) Graph III; **(b)** graph IV; **(c)** graph I; **(d)** graph II
93. Answers may vary.

Technology Connection, p. 50

1–2. Left to the student

Technology Connection, p. 51

1. (a) 2 and 4; **(b)** 2 and −5; **(c)** and **(d)** left to the student

Technology Connection, p. 53

1. −5 and 2   **2.** −4 and 6   **3.** −2 and 1   **4.** 0, −1.414,
and 1.414 (approx.)   **5.** 0 and 700   **6.** −2.079, 0.463, and
3.116 (approx.)   **7.** −3.096, −0.646, 0.646, and 3.096 (approx.)
8. −1 and 1   **9.** −0.387 and 1.721 (approx.)   **10.** 6.133
11. −2, −1.414, 1, and 1.414 (approx.)   **12.** −3, −1, 2, and 3

Technology Connection, p. 54

Left to the student

Technology Connection, p. 55

1. Left to the student   **2.** For $x = 3$, y_1 is ERR and y_2 is 6.

Technology Connection, p. 56

1–8. Left to the student

Technology Connection, p. 57

1–2. Left to the student

Technology Connection, p. 63

1. Approximately $(14, 266)$

Exercise Set R.5, p. 64

1.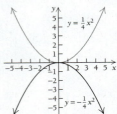

3.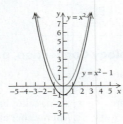

5.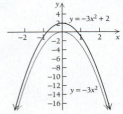

7.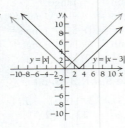

9.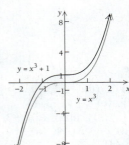

11.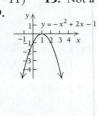

13. Parabola with vertex at $(-2, -11)$ **15.** Not a parabola
17.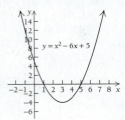

19.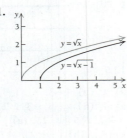

21.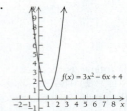

23.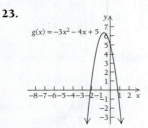

25.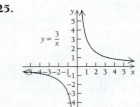

27.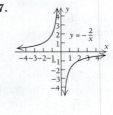

29.

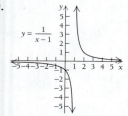

31.

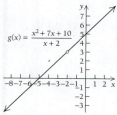

33.

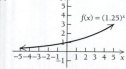

35.

37.

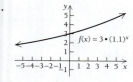

39.

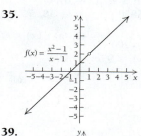

41.

43.

45. $1 \pm \sqrt{3}$, or $-0.732, 2.732$ **47.** $-3 \pm \sqrt{10}$, or $-6.162,$
0.162 **49.** $\dfrac{1 \pm \sqrt{2}}{2}$, or $-0.207, 1.207$ **51.** $\dfrac{-4 \pm \sqrt{10}}{3}$, or
$-2.387, -0.279$ **53.** $\dfrac{-7 \pm \sqrt{13}}{2}$, or $-5.303, -1.697$

55. $\sqrt[5]{x}$ **57.** $\sqrt[3]{y^2}$ **59.** $\dfrac{1}{\sqrt[5]{t^2}}$ **61.** $\dfrac{1}{\sqrt[3]{b}}$ **63.** $\dfrac{1}{\sqrt{x^2 - 3}}$

65. $x^{3/2}$ **67.** $a^{3/5}$ **69.** x^3 **71.** $t^{-5/2}$ **73.** $(x^2 + 7)^{-1/2}$
75. 27 **77.** 16 **79.** $\{x | x \neq 5\}$ **81.** $\{x | x \neq 2, x \neq 3\}$
83. $\{x | x \geq -\frac{4}{5}\}$ **85.** $\{x | x \leq 7\}$ **87.** $(50, 500)$; $x = \$50$,
$q = 500$ items **89.** $(5, 1)$; price is $500, and quantity is 1000.
91. $(1, 4)$; price is $1, and quantity is 400. **93.** $(2, 3)$; price is
$2000, and quantity is 3000. **95.** 140 tickets

97. **(a)** 166 mi, **(b)**
176 mi,
184 mi;

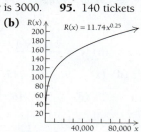

99. **(a)** $10.43\ \mu g/m^3$, **(b)**
$10.20\ \mu g/m^3$,
$10.03\ \mu g/m^3$;

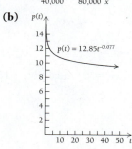

101. 24 cities; 47 cities **103.** **105.** $-1.831, -0.856, 3.188$
107. $1.489, 5.673$ **109.** $-2, 3$ **111.** $[-1, 2]$
113. Approximately $(10.33, 8.83)$; produce 8830 units at a price
of \$10.33 each

Technology Connection, p. 72

1. **(a)** $y = 2.7x + 63.8$ **(b)** 93.5; **(c)**

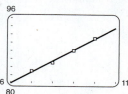

Technology Connection, p. 74

1. **(a)** $y = 0.00112141561x^4 - 0.1159429698x^3 +$
$3.670845719x^2 - 32.77404732x +$
2.857997859 **(b)**

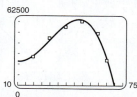

2. **(a)** $y = -62.8327x^2 + 5417.8404x - 57264.7856$

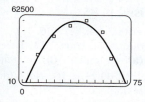

(b) $y = -1.6519x^3 + 145.6606x^2 - 2658.3088x + 36491.7730$

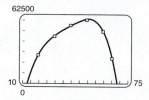

(c) $y = -0.0771x^4 + 11.3952x^3 - 639.2276x^2 + 17037.1915x - 135483.9938$

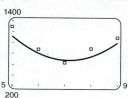

(d) quartic function; **(e)** \$38,853, \$58,887
3. $y = 93.2857x^2 - 1336x + 5460.8286$

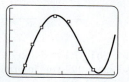

Exercise Set R.6, p. 75

1. Linear   **3.** Quadratic, $a < 0$ **5.** Quadratic, $a < 0$
7. Quadratic, $a < 0$ **9. (a)** $y = -1.5x + 65.4$;
(b) 47.4 million   **11. (a)** $y = -44\frac{1}{3}x^2 + 188\frac{2}{3}x$; **(b)** 45.3 mg;
(c) 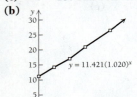   **13. (a)** $y = 0.144x^2 - 4.63x + 60$; **(b)** 188.5 ft;
(c) 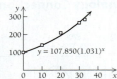   **15.** Answers will vary.   **17. (a)** $y = 11.421(1.020)^x$;
(b)   **(c)** 30.7 million

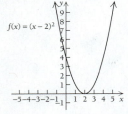

19. (a) $y = 107.850(1.031)^x$ **(b)**

(c) \$365.74 **(d)**   **21.**   **23.**
25. (a) $y = -1.5964x + 66.2607$; **(b)** 47.10 million **(c)** The
answers are very close, giving essentially the same information.
(d) $y = 66.332 \cdot (0.975)^x$, 48.95 million; **(e)**

Chapter R Review Exercises, p. 85

1. (d)   **2.** (b)   **3.** (f)   **4.** (a)   **5.** (e)   **6.** (g)   **7.** (c)
8. True   **9.** False   **10.** True   **11.** True   **12.** False
13. False   **14.** True   **15. (a)** 4.1 million; **(b)** 32, 85;
(c)   **16.** \$5017.60   **17.** \$1340.24   **18.** Not a function.
One input, Elizabeth, has three outputs.
19. (a) $f(3) = -6$; **(b)** $f(-5) = -30$; **(c)** $f(a) = -a^2 + a$;
(d) $f(x + h) = -x^2 - 2xh - h^2 + x + h$
20.   **21.**

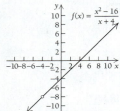

22.   **23.**

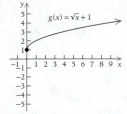

24. Not a function   **25.** Function   **26.** Function
27. Not a function   **28. (a)** $f(2) = 1$; **(b)** $[-4, 4]$;
(c) $x = -3$; **(d)** $[-1, 3]$
29. (a) $f(-1) = 1$, **(b)**
$f(1.5) = 4$,
$f(6) = 3$

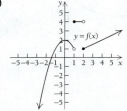

30. (a) $[-2, 5]$; **(b)** $(-1, 3]$; **(c)** $(-\infty, a)$
31. (a) $[-4, 5)$;
(b) $(2, \infty)$;
32. (a) $f(-3) = -2$; **(b)** $\{-3, -2, -1, 0, 1, 2, 3\}$;
(c) $-1, 3$; **(d)** $\{-2, 1, 2, 3, 4\}$ **33. (a)** $(-\infty, 5) \cup (5, \infty)$;
(b) $[-6, \infty)$ **34.** Slope, -3; y-intercept, 2
35. $y + 5 = \frac{1}{4}(x - 8)$, or $y = \frac{1}{4}x - 7$ **36.** -3
37. $-\$350$ per year   **38.** 75 pages per day
39. $A = \dfrac{7}{200}V$ **40. (a)** $C(x) = 0.50x + 4000$; **(b)** $R(x) = 10x$;
(c) $P(x) = 9.5x - 4000$ **(d)** 422 CDs

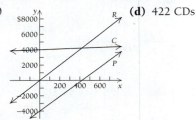

41. (a)   **(b)**

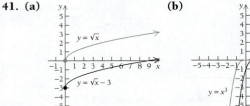

42. (a)   **(b)**

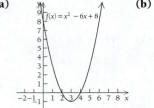 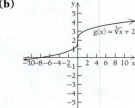

Vertex at $(3, -1)$

(c)   **(d)**

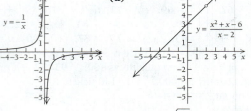

43. (a) $x = 1, x = 3$; **(b)** $x = \dfrac{2 \pm \sqrt{10}}{2}$ **44. (a)** $x^{4/5}$;
(b) t^4; **(c)** $m^{-2/3}$; **(d)** $(x^2 - 9)^{-1/2}$ **45. (a)** $\sqrt[5]{x^2}$
(b) $\dfrac{1}{\sqrt[5]{m^3}}$; **(c)** $\sqrt{x^2 - 5}$; **(d)** $\sqrt{3t}$ **46.** $[\frac{9}{2}, \infty)$

47. (a)   **(b)**

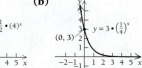

48. $(3, 16)$; price = \$3, quantity = 1600 units
49. About 3.3 hr

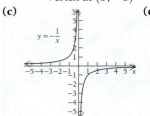

50. (a) $y = 0.2x + 160$ **(b)**
(c) 173.4 beats/min

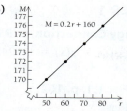

51. (a)

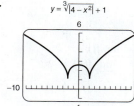

(b) Data fit a quadratic function.
(c) $y = 2.0x^2 - 89.8\overline{4}x + 870$;
(d) About $-\$25.33$; **(e)** ✎

52. (a) 525,375 lb; **(b)** $4.31/lb
53.

$f(x) = x^3 - 9x^2 + 27x + 50$

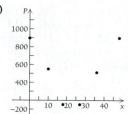

54.

$y = \sqrt[3]{|4 - x^2|} + 1$

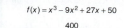

Zero: $x = -1.25$;
domain: $\mathbb{R}$; range: $\mathbb{R}$

Zero: none; domain: $\mathbb{R}$;
range: $[1, \infty)$

55. $(-1.21, 2.36)$ **56. (a)** $y = 0.2x + 160$;
(b) 173.4 beats/min; **(c)** ✎
57. (a) $y = 1.86x^2 - 84.18x + 943.86$; **(b)** $95.46; **(c)** ✎
58. (a) $y = 37.58x + 294.48$;
$y = -0.59x^2 + 74.61x - 117.72$;
$y = 0.02x^3 - 2.60x^2 + 125.71x - 439.65$;
$y = 0.003x^4 - 0.324x^3 + 11.46x^2 - 88.51x + 507.84$

(b)
$y_1 = 37.58x + 294.48$
$y_2 = -0.59x^2 + 74.61x - 117.72$
$y_3 = 0.02x^3 - 2.60x^2 + 125.71x - 439.65$
$y_4 = 0.003x^4 - 0.324x^3 + 11.46x^2 - 88.51x + 507.84$

(c) ✎

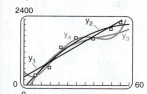

Chapter R Test, p. 88

1. [R.1] $750 **2.** [R.2] **(a)** $f(-3) = -4$;
(b) $f(a + h) = -a^2 - 2ah - h^2 + 5$ **3.** [R.4] Slope, $\frac{4}{5}$;
y-intercept, $-\frac{2}{3}$ **4.** [R.4] $y - 7 = \frac{1}{4}(x + 3)$, or $y = \frac{1}{4}x + \frac{31}{4}$
5. [R.4] $-\frac{1}{2}$ **6.** [R.4] $-\$700/yr$ **7.** [R.4] $\frac{1}{2}$ lb/bag
8. [R.4] $F = \frac{2}{3}W$ **9.** [R.4] **(a)** $C(x) = 0.08x + 8000$;
(b) $R(x) = 0.50x$; **(c)** $P(x) = 0.42x - 8000$; **(d)** 19,048 cards
10. [R.5] $(3, 25)$; $x = \$3$, $q = 25$ thousand units
11. [R.2] Yes **12.** [R.2] No **13.** [R.3] **(a)** $f(1) = -4$;
(b) $\mathbb{R}$ **(c)** $x = \pm 3$; **(d)** $[-5, \infty)$

14. [R.5]

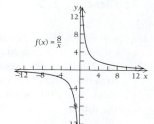

$f(x) = \frac{8}{x}$

15. [R.5] $t^{-1/2}$
16. [R.5] $\dfrac{1}{\sqrt[5]{t^3}}$

17. [R.5]

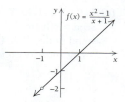

$f(x) = \frac{x^2 - 1}{x + 1}$

18. [R.5] $(-\infty, -7) \cup (-7, 2) \cup (2, \infty)$ **19.** [R.5] $(-2, \infty)$
20. [R.3] $[c, d)$
21. [R.2]

$f(x) = \begin{cases} x^2 + 2, & \text{for } x \geq 0 \\ x^2 - 2, & \text{for } x < 0 \end{cases}$

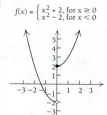

22. [R.5]

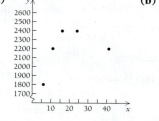

$f(x) = \frac{1}{2} \cdot (3)^x$

$(0, \frac{1}{2})$

23. [R.6] **(a)**

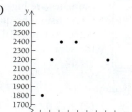

(b) yes;

(c) $y = -1.94x^2 + 102.74x + 1253.49$; **(d)** 2589.9 calories;
(e) ✎ **24.** [R.5] $\frac{1}{16}$ **25.** [R.5] Domain: $\left(-\infty, \frac{5}{3}\right]$;
zero: $x = -798\frac{2}{3}$ **26.** [R.5] Answers will vary. One possibility
is $(x + 3)(x - 1)(x - 4) = 0$. **27.** [R.4] $\frac{51}{7}$
28. [R.5] Zeros: $\pm \sqrt{8} \approx \pm 2.828$,
$\pm \sqrt{10} \approx \pm 3.162$;
domain: $\mathbb{R}$; range: $[-1, \infty)$

$y = \sqrt[3]{|9 - x^2|} - 1$

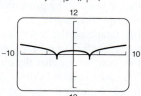

29. [R.6] **(a)** $y = -1.51x^2 + 79.98x + 1436.93$;
(b) 2480.4 calories; **(c)** ✎

Extended Technology Application, p. 91

1. (a) $y = 0.121x - 0.465$

(b)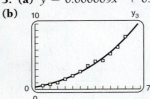

(c) \$7.40, \$8.01. Not reasonable estimates. The data are increasing, but these prices seem lower than the trend of the curve indicates. **(d)** 2119; seems too far in the future.

2. (a) $y = 0.002x^2 + 0.024x + 0.420$

(b)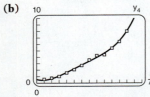

(c) \$10.43, \$11.90. Yes, the curve seems to better follow the trend in the data. **(d)** 2043; seems more reasonable.

3. (a) $y = 0.000009x^3 + 0.00082x^2 + 0.0426x + 0.347$

(b)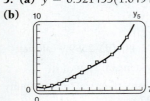

(c) \$9.05, \$10.43. Yes, the curve seems to follow the trend in the data, but the resulting ticket prices are virtually the same as those found with the quadratic function. **(d)** 2046; seems reasonable.

4. (a) $y = 0.000002x^4 - 0.00023x^3 + 0.0098x^2 - 0.067x + 0.558$

(b)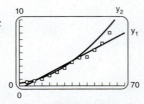

(c) \$10.15, \$13.02. Yes, but the estimates are higher than those found using the quadratic or cubic function. **(d)** 2028; seems too soon.

5. (a) $y = 0.521453(1.049459)^x$

(b)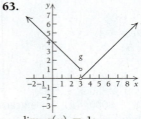

(c) \$12.02, \$15.30. Yes, but the curve rises much faster as x increases. **(d)** 2025, seems too soon.

6. (a) If the scatterplot and the linear function are graphed on the same axes, the predicted prices seem lower than the trend of the curve indicates.

(b) Graphing the scatterplot, the linear function, and the quadratic function, we see that the quadratic function seems to better follow the trend of the data.

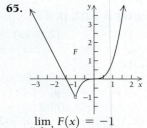

(c) The exponential function seems to fit the data more closely than the quadratic function does. The trend of the data seems to suggest an exponential model.

CHAPTER 1

Technology Connection, p. 95

1. 5 **2.** -4 **3.** $g(x) = 475.24, 492.1, 493.81, 494.19, 495.9, 513.24$ **4.** 494 **5.** -1

Exercise Set 1.1, p. 102

1. -2 **3.** The limit, as x approaches 4, of $f(x)$ **5.** The limit, as x approaches 5 from the left, of $F(x)$ **7.** $\lim\limits_{x \to 2^+}$ **9.** $\lim\limits_{x \to 5}$ **11.** 1; 2; does not exist **13.** 4; 2; does not exist **15.** 4 **17.** 0 **19.** 0 **21.** 4 **23.** 3 **25.** -1 **27.** Does not exist **29.** 0 **31.** 1 **33.** 1 **35.** 2 **37.** Does not exist **39.** 1 **41.** -1 **43.** 2 **45.** Does not exist **47.** 0 **49.** 2

51.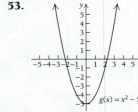

$\lim\limits_{x \to 0} f(x) = 0;$
$\lim\limits_{x \to -2} f(x) = 2$

53.

$\lim\limits_{x \to 0} g(x) = -5;$
$\lim\limits_{x \to -1} g(x) = -4$

55.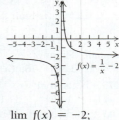

$\lim\limits_{x \to -1} G(x) = 1;$
$\lim\limits_{x \to -2} G(x)$ does not exist

57.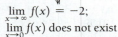

$\lim\limits_{x \to \infty} f(x) = -2;$
$\lim\limits_{x \to 0} f(x)$ does not exist

59.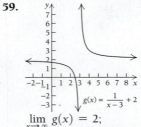

$\lim\limits_{x \to \infty} g(x) = 2;$
$\lim\limits_{x \to 3} g(x)$ does not exist

61.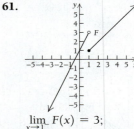

$\lim\limits_{x \to 1^-} F(x) = 3;$
$\lim\limits_{x \to 1^+} F(x) = 1;$
$\lim\limits_{x \to 1} F(x)$ does not exist

63.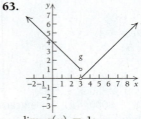

$\lim\limits_{x \to 3^-} g(x) = 1;$
$\lim\limits_{x \to 3^+} g(x) = 0;$
$\lim\limits_{x \to 3} g(x)$ does not exist

65.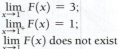

$\lim\limits_{x \to -1} F(x) = -1$

67. 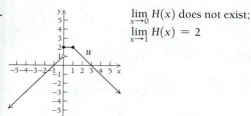 $\lim\limits_{x\to 0} H(x)$ does not exist;

$\lim\limits_{x\to 1} H(x) = 2$

69. $3.50, $3.50, $3.50 **71.** $4.00, $4.50, limit does not exist
73. $1.19, $1.40, limit does not exist **75.** Limit does not exist.
77. 10%, 15%, limit does not exist **79.** 25%, limit does not exist
81. 15%, 25%, limit does not exist **83.** 3 **85.** -1
87. Limit does not exist; 2 **89.** Limit does not exist; $\frac{1}{6}$

Technology Connection, p. 109

1. 53 **2.** 2.8284 **3.** 0.25 **4.** 0.16667

Exercise Set 1.2, p. 113

1. False **3.** True **5.** False **7.** False **9.** 3 **11.** -3
13. 4 **15.** 15 **17.** -4 **19.** 6 **21.** $\frac{3}{2}$ **23.** $\frac{13}{4}$ **25.** -12
27. $\frac{1}{10}$ **29.** Limit does not exist. **31.** 3 **33.** Limit does
not exist. **35.** 0 **37.** Not continuous **39.** Continuous
41. Not continuous **43. (a)** -1, 2, limit does not exist;
(b) -1; **(c)** no, the limit does not exist; **(d)** 3; **(e)** 3;
(f) yes, $\lim\limits_{x\to -2} f(x) = f(-2)$ **45. (a)** 2; **(b)** does not exist;
(c) no, $k(-1)$ does not exist; **(d)** -2; **(e)** -2; **(f)** yes,
$\lim\limits_{x\to 3} k(x) = k(3)$ **47. (a)** 3; **(b)** 1; **(c)** limit does not exist;
(d) 1; **(e)** no, $\lim\limits_{x\to 3} G(x)$ does not exist; **(f)** yes, $\lim\limits_{x\to 0} G(x) = G(0)$;
(g) yes, $\lim\limits_{x\to 2.9} G(x) = G(2.9)$ **49.** Yes; $\lim\limits_{x\to 4} g(x) = g(4)$
51. No; $\lim\limits_{x\to 0} G(x)$ does not exist and $G(0)$ does not exist.
53. Yes; $\lim\limits_{x\to 4} f(x) = f(4)$ **55.** No; $\lim\limits_{x\to 3} F(x)$ does not exist.
57. No; $g(4)$ does not exist. **59.** No; $\lim\limits_{x\to 2} G(x)$ does not equal
$G(2)$. **61.** Yes; $\lim\limits_{x\to 4} G(x) = G(4)$ **63.** No; $g(5)$ does not
exist and $\lim\limits_{x\to 5} g(x)$ does not exist. **65.** No; $\lim\limits_{x\to 2} G(x)$ does not
exist and $G(2)$ does not exist. **67.** Yes; $g(x)$ is continuous over
$(-4, 4)$ because $\lim\limits_{x\to a} g(x) = g(a)$ for all a in $(-4, 4)$. **69.** No;
$G(x)$ is not continuous at $x = 1$. **71.** Yes; $g(x)$ is continuous at
each real number because $\lim\limits_{x\to a} g(x) = g(a)$ for all real numbers a.
73. 30, 25, does not exist **75.** 120, 120, 120 **77.** $k = 2$
79. 0.5 **81.** 0.5 **83.** 0.378, or $\frac{1}{\sqrt{7}}$ **85.** 0

Exercise Set 1.3, p. 122

1. (a) $10x + 5h$; **(b)** 60, 55, 50.5, 50.05 **3. (a)** $-10x - 5h$;
(b) $-60, -55, -50.5, -50.05$ **5. (a)** $2x + h - 1$;
(b) 11, 10, 9.1, 9.01 **7. (a)** $\dfrac{-9}{x\cdot(x+h)}$;
(b) $-\frac{9}{35}, -\frac{3}{10}, -\frac{6}{17}, -\frac{60}{167}$ **9. (a)** 2; **(b)** 2, 2, 2, 2
11. (a) $36x^2 + 36xh + 12h^2$; **(b)** 1308, 1092, 918.12, 901.8012
13. (a) $2x + h - 4$; **(b)** 8, 7, 6.1, 6.01 **15. (a)** $2x + h - 3$;
(b) 9, 8, 7.1, 7.01 **17.** About 0.275%/yr, about 0.16%/yr,
about 0.211%/yr **19.** About 0.35%/yr, about 0.26%/yr,
about 0.3%/yr **21.** About -0.025%/yr, about -0.12%/yr,
about -0.078%/yr **23.** About 0.05%/yr, about 0.72%/yr, about
0.422%/yr **25.** 1.268 quadrillion BTUs/yr, 1.187 quadrillion BTUs/yr,
-0.255 quadrillion BTUs/yr **27. (a)** 70 pleasure units/unit of
product, 39 pleasure units/unit of product, 29 pleasure units/

unit of product, 23 pleasure units/unit of product; **(b)**
29. (a) $27.16; **(b)** $29.56; **(c)** $2.40; **(d)** $1.20, which
means the average price of a ticket increases by $1.20/yr.
31. -116; Payton County lost 116 people per year on average
from the 5th to the 8th year after the last census. **33.** The
average cost of production of between 300 and 305 holders is
$19.80 per unit. **35. (a)** 1.0 lb/month; **(b)** 0.54 lb/month;
(c) 0.77 lb/month; **(d)** 0.67 lb/month; **(e)** growth rate is
greatest in the first 3 months. **37. (a)** Approximately
1.49 hectares/g; **(b)** 1.09 represents the average growth rate,
in hectares/g, of home range with respect to body weight when
the mammal grows from 200 to 300 g. **39. (a)** 1.25 words/
min, 1.25 words/min, 0.625 word/min, 0 words/min, 0 words/
min; **(b)** 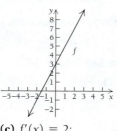 **41. (a)** 256 ft; **(b)** 128 ft/sec
43. (a) 125 million people/yr for both countries; **(b)**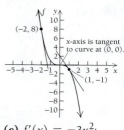
(c) A: 290 million people/yr, -40 million people/yr,
-50 million people/yr, 300 million people/yr,
B: 125 million people/yr in all intervals; **(d)**
45. (a) 2006–07; **(b)** 2003–04, 2008–09; **(c)** about $7970
for public and about $20,130 for private **47.** $2ax + b + ah$
49. $4x^3 + 6x^2h + 4xh^2 + h^3$
51. $5ax^4 + 10ax^3h + 10ax^2h^2 + 5axh^3 + ah^4 + 4bx^3 +$
$6bx^2h + 4bxh^2 + bh^3$ **53.** $\dfrac{1}{(1 - x - h)(1 - x)}$
55. $\dfrac{2}{\sqrt{2x + 2h + 1} + \sqrt{2x + 1}}$

Technology Connection, p. 131

$y = -\frac{3}{4}x - 3; y = -12x - 12$

Technology Connection, p. 134

1–8. Left to the student

Exercise Set 1.4, p. 135

1. (a) and **(b)**

(c) $f'(x) = x$;
(d) $-2, 0, 1$

3. (a) and **(b)**

(c) $f'(x) = -4x$;
(d) 8, 0, -4

5. (a) and **(b)**

(c) $f'(x) = -3x^2$;
(d) $-12, 0, -3$

7. (a) and **(b)** All tangent lines
are identical to the graph of the
original function.

(c) $f'(x) = 2$;
(d) 2, 2, 2

A-10   Answers

9. (a) and **(b)** All tangent lines are identical to the graph of the original function.

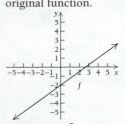

(c) $f'(x) = \dfrac{3}{4}$;

(d) $\dfrac{3}{4}, \dfrac{3}{4}, \dfrac{3}{4}$

11. (a) and **(b)**

(c) $f'(x) = 2x + 1$;
(d) $-3, 1, 3$

13. (a) and **(b)**

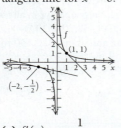

(c) $f'(x) = -10x - 2$;
(d) $18, -2, -12$
17. (a) $y = 12x + 16$; **(b)** $y = 0$; **(c)** $y = 48x - 128$
19. (a) $y = -2x + 4$; **(b)** $y = -2x - 4$;
(c) $y = -0.0002x + 0.04$ **21. (a)** $y = -6x - 4$;
(b) $y = -1$; **(c)** $y = 6x - 16$ **23.** $f'(x) = m$
25. $x_0, x_3, x_4, x_6, x_{12}$ **27.** x_1, x_2, x_3, x_4 **29–34.** Answers will vary. **35.** $x = 1, 2, 3, 4, 5, 6, 7, 8, 9, 10, 11, 12$
37. Increase: 16 points/day on Dec. 13; decrease: -129 points/day on Dec. 11 **39.** ✏ **41.** $f'(x) = 4x^3$ **43.** $f'(x) = 5x^4$
45. $f'(x) = \dfrac{1}{2\sqrt{x}}$ **47.** $f'(x) = -\dfrac{1}{2\sqrt{x^3}}$, or $\dfrac{-1}{2x\sqrt{x}}$
49. (a) f is not differentiable at $x = -3$; **(b)** 1 **51. (a)** $x = 3$;
(b) $k'(0) = -1, k'(1) = -1, k'(4) = 1, k'(10) = 1$
53. $f'(x)$ is not defined at $x = -1$. **55. (a)** $\lim\limits_{x\to 2} F(x) = 5$,
$F(2) = 5$; therefore, $\lim\limits_{x\to 2} F(x) = F(2)$. **(b)** No, the graph has a corner there **57.** $m = 11, b = -18$ **59–63.** Left to the student **65.** $f'(x)$ does not exist for $x = 5$.

15. (a) and **(b)** There is no tangent line for $x = 0$.

(c) $f'(x) = -\dfrac{1}{x^2}$;
(d) $-\dfrac{1}{4}$, does not exist, -1

Technology Connection, p. 140

1. $152, -76, -100, -180$ **2.** $36, 0, 12, 0, 43.47$
3. $-2.31, 3.69, 0.81$

Technology Connection, p. 144

1. Tangent line is horizontal at $\left(2, \dfrac{8}{3}\right)$.

Exercise Set 1.5, p. 145

1. $8x^7$ **3.** -0.5 **5.** 0 **7.** $30x^9$ **9.** $-8x^{-9}$ **11.** $-15x^{-6}$
13. $4x^3 - 7$ **15.** $\dfrac{2}{\sqrt{x}}$ **17.** $0.7x^{-0.3}$ **19.** $-1.6x^{-2/3}$

21. $-\dfrac{24}{x^5}$ **23.** $\dfrac{3}{4}$ **25.** $\dfrac{1}{4\sqrt[4]{x^3}} + \dfrac{3}{x^2}$ **27.** $-\dfrac{10}{3}\sqrt[3]{x^2}$
29. $10x - 7$ **31.** $0.36x^{0.2}$ **33.** $\dfrac{3}{4}$ **35.** $-\dfrac{12}{5x^7}$
37. $-\dfrac{4}{x^2} - \dfrac{3}{5}x^{-2/5}$ **39.** 7 **41.** $\dfrac{1}{2}x^{1/2}$ **43.** $-0.02x + 0.4$
45. $-\dfrac{3}{4}x^{-7/4} - 2x^{-1/3} + \dfrac{5}{4}x^{1/4} - \dfrac{8}{x^5}$ **47.** $\dfrac{1}{7} - \dfrac{7}{x^2}$ **49.** $\dfrac{1}{4}$
51. -5 **53.** $\dfrac{1}{12}$ **55.** $-\dfrac{3}{640}$ **57. (a)** $y = \dfrac{3}{2}x - \dfrac{3}{2}$;
(b) $y = \dfrac{31}{4}x - 17$; **(c)** $y = \dfrac{107}{6}x - \dfrac{165}{2}$ **59. (a)** $y = -\dfrac{2}{3}x + \dfrac{1}{3}$;
(b) $y = \dfrac{2}{3}x + \dfrac{1}{3}$; **(c)** $y = \dfrac{1}{3}x + \dfrac{4}{3}$ **61.** $(0, 4)$ **63.** $(0, -2)$
65. $(0.3, 7.55)$ **67.** $(20, 54)$ **69.** None **71.** The tangent line is horizontal at *all* points of the graph.
73. $\left(-1, -55\tfrac{1}{3}\right), \left(11, 111\tfrac{1}{3}\right)$ **75.** $(\sqrt{2}, 1 - 4\sqrt{2})$, or approximately $(1.41, -4.66)$; $(-\sqrt{2}, 1 + 4\sqrt{2})$, or approximately $(-1.41, 6.66)$ **77.** $(3, 0)$ **79.** $(2.5, 8.75)$
81. $(50, 75)$ **83.** $(1 + \sqrt{6}, -\tfrac{11}{3} - 3\sqrt{6})$, or approximately $(3.45, -11.02)$; $(1 - \sqrt{6}, -\tfrac{11}{3} + 3\sqrt{6})$, or approximately $(-1.45, 3.68)$ **85. (a)** $C'(r) = 6.28$; **(b)** 6.28 cm;
(c) ✏ **87. (a)** $w'(t) = 1.82 - 0.1192t + 0.002274t^2$;
(b) about 21 lb; **(c)** about 0.86 lb/month
89. (a) $R'(v) = -\dfrac{6000}{v^2}$; **(b)** 75 beats/min;
(c) -0.94 beat/min per mL **91. (a)** $\dfrac{dP}{dt} = 4000t$;
(b) 300,000 people; **(c)** 40,000 people/yr; **(d)** ✏
93. (a) $V' = \dfrac{0.61}{\sqrt{h}}$; **(b)** 244 mi; **(c)** 0.0031 mi/ft; **(d)** ✏
95. $(2, \infty)$ **97.** $(0, 4)$, $\left(-\sqrt{\tfrac{2}{3}}, -\tfrac{40}{9}\right)$, and $\left(\sqrt{\tfrac{2}{3}}, -\tfrac{40}{9}\right)$
99. $f'(x) = 5x^4 + 3x^2$, which is greater than or equal to 0 for all x.
101. $k'(x) = -\dfrac{2}{x^3}$, which is less than 0 for all x **103.** $2x + 1$
105. $3x^2 - 1$ **107.** $\dfrac{\sqrt{7}}{2\sqrt{x}}$ **109.** $1 - \dfrac{1}{x^2}$ **111.** ✏
113.

$y = \dfrac{5x^2 + 8x - 3}{3x^2 + 2}$

$(-0.346, -2.191)$,
$(1.929, 2.358)$
115.

$f'(1) = 1$
117.

$f'(1) = 1.2$

Technology Connection, p. 152

1. (c) **2–5.** Left to the student

Technology Connection, p. 153

1–2. Left to the student

Exercise Set 1.6, p. 154

1. $13x^{12}$ **3.** $12x + 7$ **5.** $18x^5 - 60x^4$ **7.** $\frac{15}{2}x^{3/2} + 4x$

9. $18x^2 + 14x - 18$ **11.** $\frac{9\sqrt{t}}{2} - \frac{1}{2\sqrt{t}} + 2$ **13.** $2x$, for $x \neq 0$

15. $18x^5 - 2x$, for $x \neq 0$ **17.** $8x + 2$, for $x \neq \frac{1}{2}$

19. 1, for $t \neq -4$ **21.** $40x^3 - 21x^2 - 26x + 13$

23. $\dfrac{-2x(5x^3 - 3x - 15)}{(2x^3 + 3)^2}$ **25.** $-\frac{105}{2}x^{3/2} - 6x + 42x^{-1/2} + 4$

27. $\dfrac{3}{(3 - t)^2} + 15t^2$ **29.** $50x - 40$ **31.** $2x(x^2 - 4)(3x^2 - 4)$

33. $-36x^{-2} - 120x^{-3} + 144x^{-4} - 72x^{-5}$

35. $3t^2 - 1 + \dfrac{6}{t^2}$ **37.** $\dfrac{x^4 + 3x^2 + 2x}{(x^2 + 1)^2} + 12x^2$

39. $\dfrac{(x^{1/2} + 3)(\frac{1}{3}x^{-2/3}) - (x^{1/3} - 7)(\frac{1}{2}x^{-1/2})}{(\sqrt{x} + 3)^2}$, or

$\dfrac{6 - \sqrt{x} + 21x^{1/6}}{6x^{2/3}(\sqrt{x} + 3)^2}$ **41.** $\dfrac{-2x^{-1}}{(x + x^{-1})^2}$, or $\dfrac{-2x}{(x^2 + 1)^2}$, for $x \neq 0$

43. $\dfrac{-1}{(t - 4)^2}$ **45.** $\dfrac{5x^2 - 6x + 5}{(x^2 - 1)^2}$

47. $\dfrac{(t^2 - 2t + 4)(-2t + 3) - (-t^2 + 3t + 5)(2t - 2)}{(t^2 - 2t + 4)^2}$, or

$\dfrac{-t^2 - 18t + 22}{(t^2 - 2t + 4)^2}$ **49.** (a) $y = 2$; (b) $y = \frac{1}{2}x + 2$

51. (a) $y = x + 5$; (b) $y = \frac{21}{4}x - \frac{21}{4}$ **53.** $-\$0.0925$/belt

55. $-\$0.0153$/belt **57.** $\$0.0772$/belt **59.** $\$1.64$/vase

61. (a) $P'(t) = 57.6t^{0.6} - 104$; (b) $\$461.4$ billion/yr;

(c) 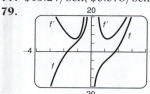 **63.** (a) $T'(t) = \dfrac{-4(t^2 - 1)}{(t^2 + 1)^2}$; (b) $100.2°$F;

(c) $-0.48°$F/hr **65.** $30t^2 + 10t - 15$

67. $3x^2\left(\dfrac{x^2 + 1}{x^2 - 1}\right) + \dfrac{-4x}{(x^2 - 1)^2}(x^3 - 8)$, or

$\dfrac{3x^6 - 4x^4 - 3x^2 + 32x}{(x^2 - 1)^2}$ **69.** $\dfrac{-x^6 + 4x^3 - 24x^2}{(x^4 - 3x^3 - 5)^2}$

71. (a) $f'(x) = \dfrac{-2x}{(x^2 - 1)^2}$; (b) $g'(x) = \dfrac{-2x}{(x^2 - 1)^2}$; (c)

73. **75.** (a) Definition of derivative; (b) adding and subtracting the same quantity is the same as adding 0; (c) the limit of a sum is the sum of the limits; (d) factoring; (e) the limit of a product is the product of the limits and $\lim\limits_{h \to 0} f(x + h) = f(x)$; (f) definition of derivative; (g) using Leibniz notation

77. $\$13.24$/belt; $\$0.078$/belt

79.

There are no points at which the tangent line is horizontal.

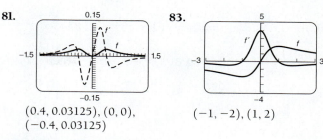

81.

(0.4, 0.03125), (0, 0), (−0.4, 0.03125)

83.

(−1, −2), (1, 2)

Exercise Set 1.7, p. 163

1. $8x - 12$ **3.** $-55(7 - x)^{54}$ **5.** $\dfrac{-1}{2\sqrt{1 - x}}$ **7.** $\dfrac{3x}{\sqrt{3x^2 - 4}}$

9. $-400x(4x^2 + 1)^{-51}$ **11.** $4(x - 4)^7(2x + 3)^5(7x - 6)$

13. $\dfrac{-8}{(4x + 5)^3}$ **15.** $\dfrac{4x(5x + 14)}{(7 - 5x)^4}$ **17.** $15x^2(3 + x^3)^4 -$

$28x^6(1 + x^7)^3$ **19.** $4x - 400$ **21.** $\dfrac{2}{3\sqrt[3]{(2x - 1)^2}} - 8 + 2x$

23. $-5(2x - 3)^3(10x - 3)$

25. $(5x + 2)^3(2x - 3)^7(120x - 28)$,

or $4(5x + 2)^3(2x - 3)^7(30x - 7)$ **27.** $\dfrac{2x(5x - 1)}{\sqrt{4x - 1}}$

29. $\dfrac{2x - 5}{4(x^2 - 5x + 2)^{3/4}}$ **31.** $\dfrac{44(3x - 1)^3}{(5x + 2)^5}$

33. $\dfrac{13}{2(2x + 3)^{1/2}(5 - x)^{3/2}}$

35. $200(2x^3 - 3x^2 + 4x + 1)^{99}(3x^2 - 3x + 2)$

37. $\dfrac{-5(7x - 2)^4}{(3x - 1)^6}$ **39.** $\dfrac{-1}{(x - 1)^{3/2}(x + 1)^{1/2}}$

41. $\dfrac{(5x - 4)^6(120x + 107)}{(6x + 1)^4}$ **43.** $\dfrac{(3x - 4)^{1/4}(138x - 19)}{(2x + 1)^{1/3}}$

45. $\dfrac{-45}{u^4}, 2, \dfrac{-90}{(2x + 1)^4}$ **47.** $50u^{49}, 12x^2 - 4x,$

$50(4x^3 - 2x^2)^{49}(12x^2 - 4x)$ **49.** $2u, 3x^2, 6x^2(x^3 + 1)$

51. $3x^2(10x^3 + 13)$ **53.** $\dfrac{-3x}{\sqrt{34 - 3x^2}}$

55. $\dfrac{3(-6t - 11)}{(5 + 3t)^2(6 + 3t)^2}$, or $\dfrac{-6t - 11}{3(t + 2)^2(3t + 5)^2}$ **57.** $y = 0$

59. $y = 4x - 3$ **61.** (a) $\dfrac{-64(6x + 1)}{(2x - 5)^3}$; (b) $\dfrac{-64(6x + 1)}{(2x - 5)^3}$;

(c) **63.** -216 **65.** $4(13)^{-2/3}$, or about 0.72

67. $f'(x) = 6[2x^3 + (4x - 5)^2]^5[6x^2 + 8(4x - 5)]$

69. $f'(x) = \dfrac{1}{2\sqrt{x^2 + \sqrt{1 - 3x}}}\left(2x - \dfrac{3}{2\sqrt{1 - 3x}}\right)$

71. $\$1,000,000$/airplane

73. $P'(x) = \dfrac{500(2x - 0.1)}{\sqrt{x^2 - 0.1x}} - \dfrac{4000x}{3(x^2 + 2)^{2/3}}$

75. (a) $37.04x^3 - 255.81x^2 + 574.48x - 309.12$ (b)

(c) $\$1929.24$ billion/yr **77.** (a) $\dfrac{da}{dr} = 5000\left(1 + \dfrac{r}{4}\right)^{19}$; (b)

79. (a) $P(t) = 2t^2 + 400.8t + 80.08$; (b) $\$592.80$/month

81. (a) $D(c) = 4.25c + 106.25, c(w) = \dfrac{95w}{43.2} \approx 2.199w$

(b) 4.25 mg/unit of creatine clearance; (c) 2.199 units of creatine clearance/kg; (d) 9.35 mg/kg; (e)

83. $1 + \dfrac{1}{2\sqrt{x}} + \dfrac{1}{2\sqrt{x+\sqrt{x}}}\cdot\left(1+\dfrac{1}{2\sqrt{x}}\right)$ **85.** $\frac{1}{27}x^{-26/27}$

87. $\dfrac{6x^7 + 32x^5 + 5x^4}{(x^3 + 6x + 1)^{2/3}}$ **89.** $3x^2(x^2+1)^{3/2} + 3x^4(x^2+1)^{1/2}$,

or $(3x^2 + 6x^4)\sqrt{1+x^2}$ **91.** $\dfrac{3(x^2 - x - 1)^2(x^2 + 4x - 1)}{(x^2+1)^4}$

93. $\dfrac{6\sqrt{t}+1}{4\sqrt{t}\sqrt{3t+\sqrt{t}}}$

95. **(a)** $\dfrac{d}{dx}[f(x)]^3 = \dfrac{d}{dx}[f(x)^2\cdot f(x)]$

$= f(x)^2\cdot f'(x) + 2f(x)\cdot f'(x)\cdot f(x)$
$= f(x)^2\cdot f'(x) + 2f(x)^2\cdot f'(x)$
$= 3f(x)^2\cdot f'(x);$

(b) $\dfrac{d}{dx}[f(x)]^4 = \dfrac{d}{dx}[f(x)^3\cdot f(x)]$

$= f(x)^3\cdot f'(x) + 3f(x)^2\cdot f'(x)\cdot f(x)$
$= f(x)^3\cdot f'(x) + 3f(x)^3\cdot f'(x) = 4f(x)^3\cdot f'(x)$

97.

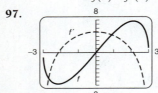

$(-2.14476, -7.728)$,
$(2.14476, 7.728)$

99. $\dfrac{4 - 2x^2}{\sqrt{4-x^2}}$

Exercise Set 1.8, p. 172

1. $12x^2$ **3.** $24x^2$ **5.** 8 **7.** 0 **9.** $\dfrac{12}{x^5}$ **11.** $\dfrac{-1}{4x^{3/2}}$

13. $6x - \dfrac{10}{x^3}$ **15.** $\dfrac{-4}{25x^{9/5}}$ **17.** $\dfrac{12}{x^4}$

19. $14(x^2 + 3x)^5(13x^2 + 39x + 27)$
21. $10(3x^2 + 2x + 1)^3(81x^2 + 54x + 11)$

23. $\dfrac{3(x^2+2)}{4(x^2+1)^{5/4}}$ **25.** $\dfrac{3}{4\sqrt{x}}$ **27.** $\dfrac{45x^4 - 54x^2 - 3}{16(x^3-x)^{5/4}}$
29. $\frac{4}{3}x^{-2/3} + \frac{1}{4}x^{-3/2}$ **31.** $24x^{-5} + 6x^{-4}$ **33.** $60x^2 + 6x$

35. $\dfrac{44}{(2x-3)^3}$ **37.** $120x$ **39.** $720x$

41. $-2520x^{-8} + \dfrac{1760}{243}x^{-14/3}$ **43.** 0

45. **(a)** $v(t) = -20t + 2$; **(b)** $a(t) = -20$; **(c)** $v(1) = -18$
m/sec, $a(1) = -20$ m/sec² **47.** **(a)** $v(t) = 3$; **(b)** $a(t) = 0$;
(c) $v(2) = 3$ mi/hr, $a(2) = 0$ mi/hr²; **(d)** 🖉 **49.** **(a)** 144 ft;
(b) 96 ft/sec; **(c)** 32 ft/sec² **51.** $v(2) = 19.62$ m/sec, $a(2) =$
9.81 m/sec² **53.** **(a)** The velocity is greatest at time 0. The
slope of a tangent line is greatest there. **(b)** The acceleration is
negative, since the slopes of tangent lines are decreasing.
55. **(a)** $(7, 11)$; **(b)** $(2, 4), (7, 11), (13, 15)$; **(c)** $(0, 2), (11, 13)$;
(d) $(4, 7)$; **(e)** 🖉 **57.** **(a)** $146,000$/month, $84,000$/month,
$-$4000$/month; **(b)** $-$68,000$/month², $-$56,000$/month²,
$-$32,000$/month²; **(c)** 🖉 **59.** **(a)** 11.34, 1.98, 0.665;

(b) $-0.789, -0.0577, -0.0112$; **(c)** 🖉

61. $\dfrac{6}{(1-x)^4}$ **63.** $\dfrac{3x^{1/2} - 1}{2x^{3/2}(x^{1/2}-1)^3}$

65. $k(k-1)(k-2)(k-3)(k-4)x^{k-5}$

67. $f'(x) = \dfrac{3}{(x+2)^2}, f''(x) = \dfrac{-6}{(x+2)^3}, f'''(x) = \dfrac{18}{(x+2)^4},$

$f^4(x) = \dfrac{-72}{(x+2)^5}$ **69.** $s''(t) = 9.81$ m/sec²; the value of the
gravitational constant for Earth · **71.** 🖉 **73.** **(a)** Graph I;
(b) Graph IV; **(c)** Graph II; **(d)** Graph III **75.** 5 **77.** $\dfrac{5}{2}$

79. -6 **81.** $\dfrac{17}{8}$ **83.** 2

85.

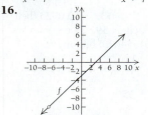

$v(t)$ switches at $t = -1.29$
and $t = 1.29$.

87.

$v(t)$ switches at $t = 0.604$
and $t = -1.104$.

Chapter 1 Review Exercises, p. 181

1. False **2.** False **3.** True **4.** False **5.** True **6.** True
7. False **8.** True **9.** (e) **10.** (c) **11.** (a) **12.** (f)
13. (b) **14.** (d)
15. (a)

$x\to -7^-$	$f(x)$
-7.1	-10.1
-7.01	-10.01
-7.001	-10.001

$x\to -7^+$	$f(x)$
-6.9	-9.9
-6.99	-9.99
-6.999	-9.999

(b) $\lim\limits_{x\to -7^-} f(x) = -10;\ \lim\limits_{x\to -7^+} f(x) = -10;\ \lim\limits_{x\to -7} f(x) = -10$

16.

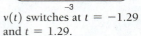

17. $\lim\limits_{x\to -7} \dfrac{x^2 + 4x - 21}{x + 7} = \lim\limits_{x\to -7} \dfrac{(x+7)(x-3)}{x+7} =$
$\lim\limits_{x\to -7} (x - 3) = -10$ **18.** -4 **19.** 10 **20.** -12 **21.** 3
22. -4 **23.** -4 **24.** Continuous at $x = 1$, since $\lim\limits_{x\to 1} g(x) = g(1)$
25. Does not exist **26.** -2 **27.** Not continuous, since

$\lim\limits_{x\to -2} g(x)$ does not exist **28.** $-\dfrac{2}{3}$ **29.** 0 **30.** Not defined
at $x = -2$, since g is discontinuous at $x = -2$ **31.** Not defined
at $x = -2$, since there is a corner there **32.** -2 **33.** -3
34. Yes, since $\lim\limits_{x\to -2} f(x) = f(-2)$ **35.** 2
36. $-6x + 3h, h \neq 0$ **37.** $6x^2 + 6xh + 2h^2, h \neq 0$
38. $y = x - 1$ **39.** $(4, 5)$ **40.** $(5, -108)$ **41.** $45x^4$

42. $\dfrac{8}{3}x^{-2/3}$ **43.** $\dfrac{24}{x^9}$ **44.** $6x^{-3/5}$ **45.** $0.7x^6 - 12x^3 - 3x^2$

46. $\frac{5}{2}x^5 + 32x^3 - 2$ **47.** $(x^3 - 5)\left(\dfrac{1}{2\sqrt{x}} + 4\right) +$

$(\sqrt{x} + 4x)(3x^2)$ **48.** $\dfrac{-x^2 + 16x + 8}{(8-x)^2}$

49. $2(5 - x)(2x - 1)^4(-7x + 26)$ **50.** $35x^4(x^5 - 3)^6$

51. $\dfrac{x(11x + 4)}{(4x+2)^{1/4}}$ **52.** $-48x^{-5}$ **53.** $3x^5 - 60x + 26$

54. **(a)** $P'(t) = 100t$; **(b)** 30,000 people; **(c)** 2000 people/yr

55. $(3, 5), (7, 8), (9, 13), (15, 18)$ **56.** $(7, 8), (15, 18)$
57. $(0, 3), (8, 9)$ **58.** $(5, 7), (13, 15)$ **59. (a)** $v(t) = 1 + 4t^3$;
(b) $a(t) = 12t^2$; **(c)** $v(2) = 33$ ft/sec, $a(2) = 48$ ft/sec^2
60. (a) $A_C(x) = 5x^{-1/2} + 100x^{-1}, A_R(x) = 40$,
$A_P(x) = 40 - 5x^{-1/2} - 100x^{-1}$;
(b) average cost is dropping at approximately \$1.33 per lamp.
61. $\dfrac{d}{dx}(f \circ g)(x) = -4(1 - 2x), \dfrac{d}{dx}(g \circ f) = -4x$
62. $\dfrac{-9x^4 - 4x^3 + 9x + 2}{2\sqrt{1 + 3x}(1 + x^3)^2}$ **63.** $\dfrac{1}{243}x^{-242/243}$ **64.** -0.25
65. $\frac{1}{6}$ **66.**

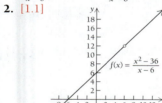

$(-1.7137, 37.445), (0, 0), (1.7137, -37.445)$

Chapter 1 Test, p. 183

1. [1.1] **(a)**

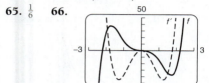

$x \to 6^-$	$f(x)$
5.9	11.9
5.99	11.99
5.999	11.999

$x \to 6^+$	$f(x)$
6.1	12.1
6.01	12.01
6.001	12.001

(b) $\lim\limits_{x \to 6^-} f(x) = 12$; $\lim\limits_{x \to 6^+} f(x) = 12$; $\lim\limits_{x \to 6} f(x) = 12$
2. [1.1]

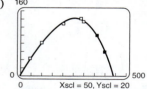

$f(x) = \dfrac{x^2 - 36}{x - 6}$

3. [1.2]
$$\lim_{x \to 6} \frac{x^2 - 36}{x - 6} = \lim_{x \to 6} \frac{(x + 6)(x - 6)}{x - 6} = \lim_{x \to 6} (x + 6) = 12$$
4. [1.1] Does not exist   **5.** [1.1] 0   **6.** [1.1] Does not exist
7. [1.1] 2   **8.** [1.1] 4   **9.** [1.1] 1   **10.** [1.1] 1
11. [1.1] 2   **12.** [1.4] 0   **13.** [1.4] 0
14. [1.2] $-5, -3, -2, 1, 4$ **15.** [1.4] $-5, -3, -2, -1, 1, 3, 4$
16. [1.2] Continuous   **17.** [1.2] Not continuous,
since $\lim\limits_{x \to 3} f(x)$ does not exist   **18.** [1.1, 1.2] **(a)** Does not exist;
(b) 1; **(c)** no   **19.** [1.1, 1.2] 3   **20.** [1.1, 1.2] 6
21. [1.1, 1.2] $\frac{1}{8}$ **22.** [1.1, 1.2] Does not exist, since
$$\lim_{x \to 0^-} \frac{1}{x} \neq \lim_{x \to 0^+} \frac{1}{x}$$ **23.** [1.3] $4x + 3 + 2h$
24. [1.4] $y = \frac{3}{4}x + 2$ **25.** [1.5] $(0, 0), (2, -4)$
26. [1.5] $23x^{22}$ **27.** [1.5] $\frac{4}{3}x^{-2/3} + \frac{5}{2}x^{-1/2}$ **28.** [1.5] $\dfrac{10}{x^2}$
29. [1.5] $\frac{5}{4}x^{1/4}$ **30.** [1.5] $-1.0x + 0.61$ **31.** [1.5]
$x^2 - 2x + 2$ **32.** $(3\sqrt{x} + 1)(2x - 1) + (x^2 - x)\dfrac{1}{3\sqrt[3]{x^2}}$
33. [1.6] $\dfrac{5}{(5 - x)^2}$ **34.** [1.7] $(x + 3)^3(7 - x)^4(-9x + 13)$
35. [1.7] $-5(x^5 - 4x^3 + x)^{-6}(5x^4 - 12x^2 + 1)$
36. [1.6, 1.7] $\dfrac{2x^2 + 5}{\sqrt{x^2 + 5}}$ **37.** [1.8] $24x$

38. [1.5] **(a)** $M'(t) = -0.003t^2 + 0.2t$; **(b)** 9; **(c)** 1.7 words/min
39. [1.6] **(a)** $A_R = 50, A_C = x^{-1/3} + 750x^{-1}$,
$A_P = 50 - x^{-1/3} - 750x^{-1}$; **(b)** average cost is dropping at
approximately \$11.74 per speaker.   **40.** [1.7] $24x^5 - 1$
41. [1.7] $6(x^2 - x)^2(2x - 1)$ **42.** [1.8] (a)
43. [1.6, 1.7] $\dfrac{-1 - 9x}{2(1 - 3x)^{2/3}(1 + 3x)^{5/6}}$ **44.** [1.2] 27
45. [1.5]

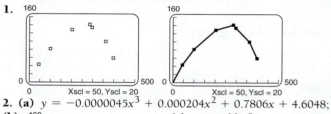

$(1.0836, 25.1029)$ and
$(2.9503, 8.6247)$

46. [1.5] 0.5

Extended Technology Application, p. 185

1.

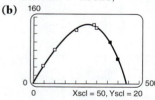

Xscl = 50, Yscl = 20   Xscl = 50, Yscl = 20

2. (a) $y = -0.0000045x^3 + 0.000204x^2 + 0.7806x + 4.6048$;
(b)

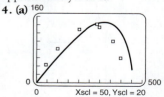

Xscl = 50, Yscl = 20

(c) acceptable fit;
(d) about 441 ft;

(e) $dy/dx = -0.0000135x^2 + 0.000408x + 0.7806$;
(f) approximately $(256, 142)$; at about 256 ft from home plate,
the ball reached its maximum height of approximately 142 ft.
3. (a) $y = -0.0000000024x^4 - 0.0000026x^3 - 0.00026x^2$
$+ 0.8150x + 4.3026$;
(b)

(c) acceptable fit;
(d) about 440 ft;

Xscl = 50, Yscl = 20

(e) $dy/dx = -0.0000000096x^3 - 0.0000078x^2 -$
$0.00053x + 0.815$; **(f)** approximately $(257, 142)$; at about 257 ft
from home plate, the ball reached its maximum height of
approximately 142 ft.
4. (a)

Xscl = 50, Yscl = 20

(b) 450 ft;
(c) $\dfrac{dy}{dx} = \dfrac{303.75 - 0.003x^2}{\sqrt{202,500 - x^2}}$;
(d) approximately $(318, 152)$;
at about 318 ft from home plate,
the ball reached its maximum
height of approximately 152 ft.
5. The two models are very similar. The main difference seems
to be that the maximum height is reached further from home
plate with the model in Exercise 4.   **6.** 466 ft, 442 ft, 430 ft
7. The estimate of the reporters is way off. Even with a low
trajectory, the ball would at best have traveled a horizontal
distance of about 526 ft.   **8. (a)** 523 ft; **(b)** 490 ft, 450 ft;
(c) 464 ft (rounded)

CHAPTER 2

Technology Connection, p. 198

1.

$f(x) = 2 - (x-1)^{2/3}$

2. $f'(x) = -\frac{2}{3}(x-1)^{-1/3}$

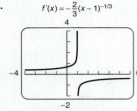

The derivative is not defined at $(1, 2)$.

Technology Connection, p. 200

1. Relative maximum at $(-1, 42)$; relative minimum at $\left(\frac{5}{2}, -\frac{175}{4}\right)$

Exercise Set 2.1, p. 201

1. Relative minimum at $(-3, -12)$

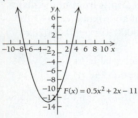

3. Relative maximum at $\left(-\frac{3}{4}, \frac{25}{8}\right)$

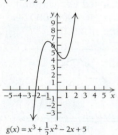

5. Relative minimum at $(-2, -13)$

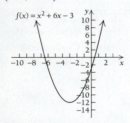

7. Relative minimum at $\left(\frac{2}{3}, \frac{113}{27}\right)$; relative maximum at $\left(-1, \frac{13}{2}\right)$

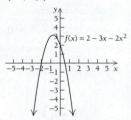

9. Relative minimum at $(2, -4)$; relative maximum at $(0, 0)$

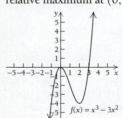

11. No relative extrema exist.

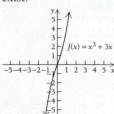

13. No relative extrema exist.

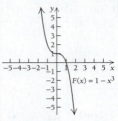

15. Relative minimum at $(4, -22)$; relative maximum at $(0, 10)$

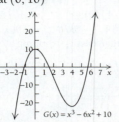

17. Relative maximum at $\left(\frac{3}{4}, \frac{27}{256}\right)$

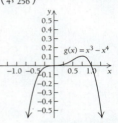

19. No relative extrema exist.

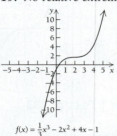

21. Relative minima at $\left(-\frac{\sqrt{10}}{2}, -\frac{27}{4}\right)$ and $\left(\frac{\sqrt{10}}{2}, -\frac{27}{4}\right)$; relative maximum at $(0, 12)$

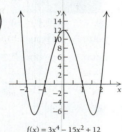

23. No relative extrema exist.

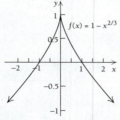

25. Relative maximum at $(0, 1)$

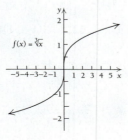

27. Relative minimum at $(0, -8)$

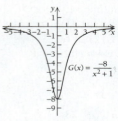

29. Relative minimum at $(-1, -2)$; relative maximum at $(1, 2)$

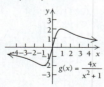

31. No relative extrema exist.

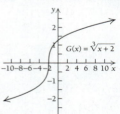

33. Relative minimum at $(-1, 2)$

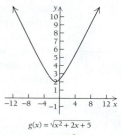

$g(x) = \sqrt{x^2 + 2x + 5}$

35–83. Left to the student.   **85.** ✎

87. Relative maximum at $(2.108, 52{,}644.383)$; relative minimum at $(3.897, 52{,}335.73)$

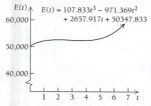

$E(t) = 107.833t^3 - 971.369t^2 + 2657.917t + 50347.833$

89. $16.08°S$

91. Increasing on $(-1, \infty)$, decreasing on $(-\infty, -1)$; relative minimum at $x = -1$. Graph left to the student.

93. Increasing on $(-\infty, 1)$, decreasing on $(1, \infty)$; relative maximum at $x = 1$. Graph left to the student.

95. Increasing on $(-4, 2)$, decreasing on $(-\infty, -4)$ and $(2, \infty)$; relative minimum at $x = -4$, relative maximum at $x = 2$. Graph left to the student.

97.

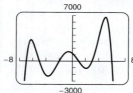

$f(x) = -x^6 - 4x^5 + 54x^4 + 160x^3 - 641x^2 - 828x + 1200$

Relative minima at $(-3.683, -2288.03)$ and $(2.116, -1083.08)$; relative maxima at $(-6.262, 3213.8)$ and $(-0.559, 1440.06)$ and $(5.054, 6674.12)$

99.

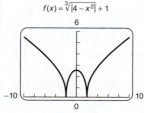

$f(x) = \sqrt[3]{|4 - x^2|} + 1$

Relative minima at $(-2, 1)$ and $(2, 1)$; relative maximum at $(0, 2.587)$

101.

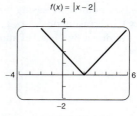

$f(x) = |x - 2|$

Relative minimum at $(2, 0)$; increasing on $(2, \infty)$; decreasing on $(-\infty, 2)$; f' does not exist at $x = 2$

103.

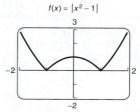

$f(x) = |x^2 - 1|$

Relative maximum at $(0, 1)$; relative minima at $(-1, 0)$ and $(1, 0)$; increasing on $(-1, 0)$ and $(1, \infty)$; decreasing on $(-\infty, -1)$ and $(0, 1)$; f' does not exist at $x = -1$ and $x = 1$

105.

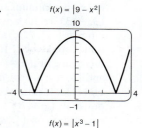

$f(x) = |9 - x^2|$

Relative maximum at $(0, 9)$; relative minima at $(-3, 0)$ and $(3, 0)$; increasing on $(-3, 0)$ and $(3, \infty)$; decreasing on $(-\infty, -3)$ and $(0, 3)$; f' does not exist at $x = -3$ and $x = 3$

107.

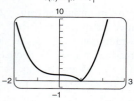

$f(x) = |x^3 - 1|$

Relative minimum at $(1, 0)$; increasing on $(1, \infty)$; decreasing on $(-\infty, 1)$; f' does not exist at $x = 1$

109. ✎   **111.** ✎

Technology Connection, p. 211

Left to the student

Technology Connection, p. 212

1. Relative minimum at $(1, -1)$; inflection points at $(0, 0)$, $(0.553, -0.512)$, $(1.447, -0.512)$, and $(2, 0)$

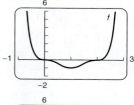

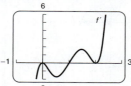

Technology Connection, p. 217

1–9. Left to the student

Exercise Set 2.2, p. 217

1. Relative maximum is $f(0) = 4$.   **3.** Relative minimum is $f\left(-\frac{1}{2}\right) = -\frac{5}{4}$.   **5.** Relative maximum is $f\left(\frac{3}{8}\right) = -\frac{7}{16}$.   **7.** Relative minimum is $f(2) = -17$; relative maximum is $f(-2) = 15$.

9.

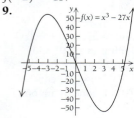

$f(x) = x^3 - 27x$

Relative minimum at $(3, -54)$, relative maximum at $(-3, 54)$; inflection point at $(0, 0)$; increasing on $(-\infty, -3)$ and $(3, \infty)$, decreasing on $(-3, 3)$; concave down on $(-\infty, 0)$, concave up on $(0, \infty)$

11.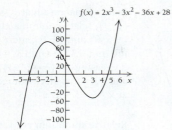

Relative minimum at $(3, -53)$, relative maximum at $(-2, 72)$; inflection point at $\left(\frac{1}{2}, \frac{19}{2}\right)$; increasing on $(-\infty, -2)$ and $(3, \infty)$, decreasing on $(-2, 3)$; concave down on $\left(-\infty, \frac{1}{2}\right)$, concave up on $\left(\frac{1}{2}, \infty\right)$

13.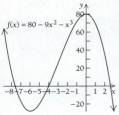

Relative minimum at $(-6, -28)$, relative maximum at $(0, 80)$; inflection point at $(-3, 26)$; increasing on $(-6, 0)$, decreasing on $(-\infty, -6)$ and $(0, \infty)$; concave up on $(-\infty, -3)$, concave down on $(-3, \infty)$

15.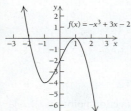

Relative minimum at $(-1, -4)$, relative maximum at $(1, 0)$; inflection point at $(0, -2)$; increasing on $(-1, 1)$, decreasing on $(-\infty, -1)$ and $(1, \infty)$; concave up on $(-\infty, 0)$, concave down on $(0, \infty)$

17.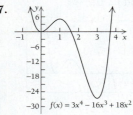

Relative minima at $(0, 0)$ and $(3, -27)$, relative maximum at $(1, 5)$; inflection points at $(0.451, 2.321)$ and $(2.215, -13.358)$; increasing on $(0, 1)$ and $(3, \infty)$, decreasing on $(-\infty, 0)$ and $(1, 3)$; concave up on $(-\infty, 0.451)$ and $(2.215, \infty)$, concave down on $(0.451, 2.215)$

19.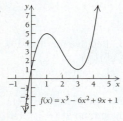

Relative minima at $\left(-\sqrt{3}, -9\right)$ and $\left(\sqrt{3}, -9\right)$, relative maximum at $(0, 0)$; inflection points at $(-1, -5)$ and $(1, -5)$; increasing on $\left(-\sqrt{3}, 0\right)$ and $\left(\sqrt{3}, \infty\right)$, decreasing on $\left(-\infty, -\sqrt{3}\right)$ and $\left(0, \sqrt{3}\right)$; concave up on $(-\infty, -1)$ and $(1, \infty)$, concave down on $(-1, 1)$

21.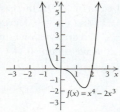

Relative minimum at $(3, 1)$, relative maximum at $(1, 5)$; inflection point at $(2, 3)$; increasing on $(-\infty, 1)$ and $(3, \infty)$, decreasing on $(1, 3)$; concave down on $(-\infty, 2)$, concave up on $(2, \infty)$

23.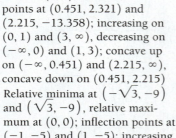

Relative minimum at $\left(\frac{3}{2}, -\frac{27}{16}\right)$; inflection points at $(0, 0)$ and $(1, -1)$; increasing on $\left(\frac{3}{2}, \infty\right)$, decreasing on $\left(-\infty, \frac{3}{2}\right)$; concave down on $(0, 1)$, concave up on $(-\infty, 0)$ and $(1, \infty)$

25.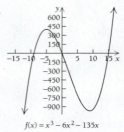

Relative minimum at $(9, -972)$, relative maximum at $(-5, 400)$; inflection point at $(2, -286)$; increasing on $(-\infty, -5)$ and $(9, \infty)$, decreasing on $(-5, 9)$; concave down on $(-\infty, 2)$, concave up on $(2, \infty)$

27.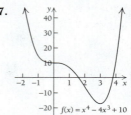

Relative minimum at $(3, -17)$; inflection points at $(0, 10)$ and $(2, -6)$; increasing on $(3, \infty)$, decreasing on $(-\infty, 3)$; concave down on $(0, 2)$, concave up on $(-\infty, 0)$ and $(2, \infty)$

29.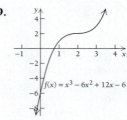

No relative extrema; inflection point at $(2, 2)$; increasing on $(-\infty, \infty)$; concave down on $(-\infty, 2)$, concave up on $(2, \infty)$

31.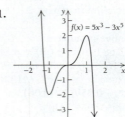

Relative minimum at $(-1, -2)$, relative maximum at $(1, 2)$; inflection points at $(-0.707, -1.237)$, $(0, 0)$, and $(0.707, 1.237)$; increasing on $(-1, 1)$, decreasing on $(-\infty, -1)$ and $(1, \infty)$; concave down on $(-0.707, 0)$ and $(0.707, \infty)$, concave up on $(-\infty, -0.707)$ and $(0, 0.707)$

33.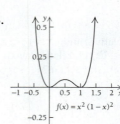

Relative minima at $(0, 0)$ and $(1, 0)$, relative maximum at $\left(\frac{1}{2}, \frac{1}{16}\right)$; inflection points at $(0.211, 0.028)$ and $(0.789, 0.028)$; increasing on $\left(0, \frac{1}{2}\right)$ and $(1, \infty)$, decreasing on $(-\infty, 0)$ and $\left(\frac{1}{2}, 1\right)$; concave down on $(0.211, 0.789)$, concave up on $(-\infty, 0.211)$ and $(0.789, \infty)$

35.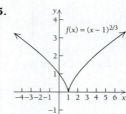

Relative minimum at $(1, 0)$; no inflection points; increasing on $(1, \infty)$, decreasing on $(-\infty, 1)$; concave down on $(-\infty, 1)$ and $(1, \infty)$

37.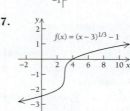

No relative extrema; inflection point at $(3, -1)$; increasing on $(-\infty, \infty)$; concave up on $(-\infty, 3)$, concave down on $(3, \infty)$

39.

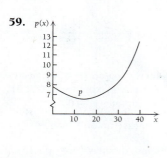

Relative maximum at $(4, 5)$; no inflection points; increasing on $(-\infty, 4)$, decreasing on $(4, \infty)$; concave up on $(-\infty, 4)$ and $(4, \infty)$

41.

Relative minimum at $\left(\sqrt{\frac{1}{2}}, -\frac{1}{2}\right)$, relative maximum at $\left(-\sqrt{\frac{1}{2}}, \frac{1}{2}\right)$; inflection point at $(0, 0)$; increasing on $\left(-1, -\sqrt{\frac{1}{2}}\right)$ and $\left(\sqrt{\frac{1}{2}}, 1\right)$, decreasing on $\left(-\sqrt{\frac{1}{2}}, \sqrt{\frac{1}{2}}\right)$; concave up on $(0, 1)$, concave down on $(-1, 0)$

43.

Relative minimum at $(-1, -4)$, relative maximum at $(1, 4)$; inflection points at $\left(-\sqrt{3}, -2\sqrt{3}\right)$ and $(0, 0)$ and $\left(\sqrt{3}, 2\sqrt{3}\right)$; increasing on $(-1, 1)$, decreasing on $(-\infty, -1)$ and $(1, \infty)$; concave up on $\left(-\sqrt{3}, 0\right)$ and $\left(\sqrt{3}, \infty\right)$, concave down on $\left(-\infty, -\sqrt{3}\right)$ and $\left(0, \sqrt{3}\right)$

45.

Relative minimum at $(0, -4)$; inflection points at $\left(-\sqrt{\frac{1}{3}}, -3\right)$ and $\left(\sqrt{\frac{1}{3}}, -3\right)$; decreasing on $(-\infty, 0)$, increasing on $(0, \infty)$; concave down on $\left(-\infty, -\sqrt{\frac{1}{3}}\right)$ and $\left(\sqrt{\frac{1}{3}}, \infty\right)$, concave up on $\left(-\sqrt{\frac{1}{3}}, \sqrt{\frac{1}{3}}\right)$

47–55. Left to the student

57.

59.

61. $(3.26, 18.36)$; the rate of change of the amount of rainfall is decreasing the fastest at this point   **63.** ✏️   **65.** Left to the student   **67.** False   **69.** True   **71.** True   **73.** True

75. $f(x) = 4x - 6x^{2/3}$

Relative maximum at $(0, 0)$; relative minimum at $(1, -2)$

77. $f(x) = x^2(1 - x)^3$

Relative minimum at $(0, 0)$; relative maximum at $(0.4, 0.035)$

79. $f(x) = (x - 1)^{2/3} - (x + 1)^{2/3}$

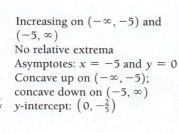

Relative maximum at $(-1, 1.587)$; relative minimum at $(1, -1.587)$

Technology Connection, p. 221

1. Vertical asymptotes: $x = -7$ and $x = 4$
2. Vertical asymptote: $x = 0$

Technology Connection, p. 224

1. Horizontal asymptote: $y = 3$ **2.** Horizontal asymptote: $y = 0$

Exercise Set 2.3, p. 230

1. $x = 2$ **3.** $x = -5$ and $x = 5$ **5.** $x = -1, x = 0$, and $x = 1$ **7.** $x = -4$ **9.** No vertical asymptotes   **11.** $y = \frac{3}{4}$
13. $y = 0$ **15.** $y = 4$ **17.** No horizontal asymptote
19. $y = 4$ **21.** $y = \frac{1}{2}$

23.

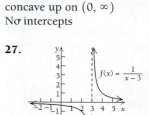

Decreasing on $(-\infty, 0)$ and $(0, \infty)$
No relative extrema
Asymptotes: $x = 0$ and $y = 0$
Concave down on $(-\infty, 0)$; concave up on $(0, \infty)$
No intercepts

25.

Increasing on $(-\infty, 5)$ and $(5, \infty)$
No relative extrema
Asymptotes: $x = 5$ and $y = 0$
Concave up on $(-\infty, 5)$; concave down on $(5, \infty)$
y-intercept: $\left(0, \frac{2}{5}\right)$

27.

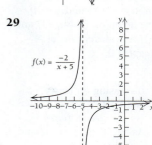

Decreasing on $(-\infty, 3)$ and $(3, \infty)$
No relative extrema
Asymptotes: $x = 3$ and $y = 0$
Concave down on $(-\infty, 3)$; concave up on $(3, \infty)$
y-intercept: $\left(0, -\frac{1}{3}\right)$

29

Increasing on $(-\infty, -5)$ and $(-5, \infty)$
No relative extrema
Asymptotes: $x = -5$ and $y = 0$
Concave up on $(-\infty, -5)$; concave down on $(-5, \infty)$
y-intercept: $\left(0, -\frac{2}{5}\right)$

31.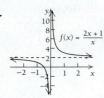

Decreasing on $(-\infty, 0)$ and $(0, \infty)$
No relative extrema
Asymptotes: $x = 0$ and $y = 2$
Concave down on $(-\infty, 0)$; concave
up on $(0, \infty)$
x-intercept: $\left(-\frac{1}{2}, 0\right)$

$f(x) = \frac{2x+1}{x}$

33.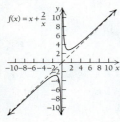

$f(x) = x + \frac{2}{x}$

Increasing on $\left(-\infty, -\sqrt{2}\right)$
and $\left(\sqrt{2}, \infty\right)$; decreasing on
$\left(-\sqrt{2}, 0\right)$ and $\left(0, \sqrt{2}\right)$
Relative minimum at
$\left(\sqrt{2}, 2\sqrt{2}\right)$; relative maximum
at $\left(-\sqrt{2}, -2\sqrt{2}\right)$;
Asymptotes: $x = 0$ and $y = x$
Concave down on $(-\infty, 0)$;
concave up on $(0, \infty)$
No intercepts

35.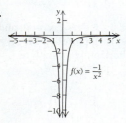

$f(x) = \frac{-1}{x^2}$

Decreasing on $(-\infty, 0)$; increasing
on $(0, \infty)$
No relative extrema
Asymptotes: $x = 0$ and $y = 0$
Concave down on $(-\infty, 0)$ and
$(0, \infty)$
No intercepts

37.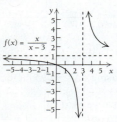

$f(x) = \frac{x}{x-3}$

Decreasing on $(-\infty, 3)$ and $(3, \infty)$
No relative extrema
Asymptotes: $x = 3$ and $y = 1$
Concave down on $(-\infty, 3)$;
concave up on $(3, \infty)$
Intercept: $(0, 0)$

39.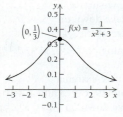

$\left(0, \frac{1}{3}\right)$ $f(x) = \frac{1}{x^2 + 3}$

Increasing on $(-\infty, 0)$; decreasing
on $(0, \infty)$
Relative maximum at $\left(0, \frac{1}{3}\right)$
Asymptote: $y = 0$
Concave down on $(-1, 1)$; con-
cave up on $(-\infty, -1)$ and $(1, \infty)$
Inflection points: $\left(-1, \frac{1}{4}\right)$ and $\left(1, \frac{1}{4}\right)$
y-intercept: $\left(0, \frac{1}{3}\right)$

41.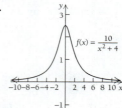

$f(x) = \frac{x+3}{x^2 - 9}$

Decreasing on $(-\infty, -3)$, $(-3, 3)$,
and $(3, \infty)$
No relative extrema
Asymptotes: $x = 3$ and $y = 0$
Concave down on $(-\infty, -3)$ and
$(-3, 3)$; concave up on $(3, \infty)$
y-intercept: $\left(0, -\frac{1}{3}\right)$

43.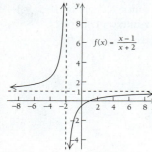

$f(x) = \frac{x-1}{x+2}$

Increasing on $(-\infty, -2)$
and $(-2, \infty)$
No relative extrema
Asymptotes: $x = -2$ and
$y = 1$
Concave up on $(-\infty, -2)$;
concave down on $(-2, \infty)$
x-intercept: $(1, 0)$;
y-intercept: $\left(0, -\frac{1}{2}\right)$

45.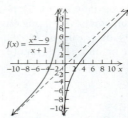

$f(x) = \frac{x^2 - 9}{x+1}$

Increasing on $(-\infty, -1)$ and
$(-1, \infty)$
No relative extrema
Asymptotes: $x = -1$ and
$y = x - 1$
Concave up on $(-\infty, -1)$;
concave down on $(-1, \infty)$
x-intercepts: $(-3, 0)$ and $(3, 0)$;
y-intercept: $(0, -9)$

47.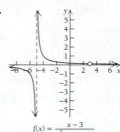

$f(x) = \frac{x-3}{x^2 + 2x - 15}$

Decreasing on $(-\infty, -5)$, $(-5, 3)$,
and $(3, \infty)$
No relative extrema
Asymptotes: $x = -5$ and $y = 0$
Concave down on $(-\infty, -5)$;
concave up on $(-5, 3)$ and $(3, \infty)$
y-intercept: $\left(0, \frac{1}{5}\right)$

49.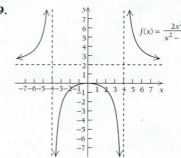

$f(x) = \frac{2x^2}{x^2 - 16}$

Increasing on
$(-\infty, -4)$ and
$(-4, 0)$; decreasing
on $(0, 4)$ and $(4, \infty)$
Relative maximum at
$(0, 0)$
Asymptotes: $x = -4$,
$x = 4$, and $y = 2$
Concave up on
$(-\infty, -4)$ and $(4, \infty)$;
concave down on
$(-4, 4)$
x- and y-intercept:
$(0, 0)$

51.

$f(x) = \frac{10}{x^2 + 4}$

Increasing on $(-\infty, 0)$; decreasing on $(0, \infty)$
Relative maximum at $(0, 2.5)$
Asymptote: $y = 0$
Concave up on $\left(-\infty, -\frac{2}{\sqrt{3}}\right)$ and $\left(\frac{2}{\sqrt{3}}, \infty\right)$, or
approximately $(-\infty, -1.1547)$ and $(1.1547, \infty)$; concave down
on $\left(-\frac{2}{\sqrt{3}}, \frac{2}{\sqrt{3}}\right)$, or approximately $(-1.1547, 1.1547)$

Inflection points: $\left(-\frac{2}{\sqrt{3}}, \frac{15}{8}\right)$ and $\left(\frac{2}{\sqrt{3}}, \frac{15}{8}\right)$, or approximately
$(-1.1547, 1.875)$ and $(1.1547, 1.875)$
y-intercept: $(0, 2.5)$

53.

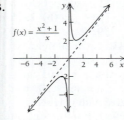

$f(x) = \dfrac{x^2+1}{x}$

Increasing on $(-\infty, -1)$ and $(1, \infty)$; decreasing on $(-1, 0)$ and $(0, 1)$
Relative maximum at $(-1, -2)$; relative minimum at $(1, 2)$
Asymptotes: $x = 0$ and $y = x$
Concave down on $(-\infty, 0)$; concave up on $(0, \infty)$
No intercepts

55.

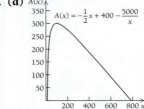

$f(x) = \dfrac{x^2 - 16}{x+4}$

Increasing on $(-\infty, -4)$ and $(-4, \infty)$
No relative extrema
No asymptotes
No concavity
x-intercept: $(4, 0)$; y-intercept: $(0, -4)$

57. (a) $A(x) = 3x + \dfrac{80}{x}$;

(c) Slant asymptote: $y = 3x$. As x, the number of units produced, increases, average cost approaches $3x$.

(b)

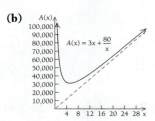

$A(x) = 3x + \dfrac{80}{x}$

59. (d)

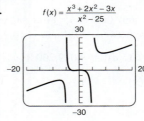

$A(x) = -\dfrac{1}{2}x + 400 - \dfrac{5000}{x}$

61. (a) $0.85, $0.73, $0.52;
(b) 27.6 yr after 1990;
(c) 0

63. (a)

n	9	6	3	1	2/3	1/3
E	4.00	6.00	12.00	36.00	54.00	108.00

(b) $\lim\limits_{n \to 0} E(n) = \infty$. The pitcher gives up one or more runs but gets no one out (0 innings pitched).

65. ✏️   **67.** Does not exist   **69.** ∞

71.

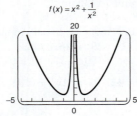

$f(x) = x^2 + \dfrac{1}{x^2}$

73.

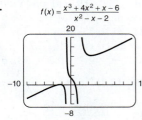

$f(x) = \dfrac{x^3 + 4x^2 + x - 6}{x^2 - x - 2}$

75.

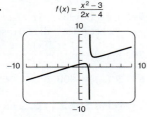

$f(x) = \dfrac{x^3 + 2x^2 - 3x}{x^2 - 25}$

77.

$f(x) = \dfrac{x^2 - 3}{2x - 4}$

(a) $(-1.732, 0)$ and $(1.732, 0)$;
(b) $(0, 0.75)$; **(c)** vertical: $x = 2$; slant: $y = 0.5x + 1$

79. Asymptote: $y = x^2 - 6$

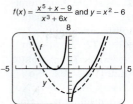

$f(x) = \dfrac{x^5 + x - 9}{x^3 + 6x}$ and $y = x^2 - 6$

81. $f(x) = \dfrac{-2x}{x - 2}$

83. $g(x) = \dfrac{x^2 - 2}{x^2 - 1}$

85. $h(x) = \dfrac{x - 9}{x^2 + x - 6}$

Technology Connection, p. 236

1. On $[-2, 1]$, absolute minimum is -8 at $x = -2$, and absolute maximum is 2.185 at $x = -0.333$; on $[-1, 2]$, absolute minimum is 1 at $x = -1$ and $x = 1$, and absolute maximum is 4 at $x = 2$

Technology Connection, p. 237

1. Absolute minimum: -4 at $x = 2$; no absolute maximum

Technology Connection, p. 239

1. No absolute maximum; absolute minimum: 6.325 at $x = 0.316$

Exercise Set 2.4, p. 240

1. (a) 55 mph   **(b)** 5 mph   **(c)** 25 mpg
3. Absolute maximum: $5\frac{1}{4}$ at $x = \frac{1}{2}$; absolute minimum: 3 at $x = 2$ **5.** Absolute maximum: $6\frac{1}{2}$ at $x = -1$; absolute minimum: 3 at $x = -2$ **7.** Absolute maximum: $5\frac{1}{2}$ at $x = -1$; absolute minimum: 2 at $x = -2$ **9.** Absolute maximum: 6 at $x = 1$; absolute minimum: 2 at $x = -1$ **11.** Absolute maximum: 15 at $x = -2$; absolute minimum: -13 at $x = 5$
13. Absolute maximum: -5 for $-1 \le x \le 1$; absolute minimum: -5 for $-1 \le x \le 1$ **15.** Absolute maximum: 24 for $4 \le x \le 13$; absolute minimum: 24 for $4 \le x \le 13$ **17.** Absolute maximum: 10 at $x = -1$; absolute minimum: 1 at $x = 2$ **19.** Absolute maximum: 4 at $x = 1$; absolute minimum: -23 at $x = 4$ **21.** Absolute maximum: 2 at $x = -1$; absolute minimum: -110 at $x = -5$ **23.** Absolute maximum: 513 at $x = -8$; absolute minimum: -511 at $x = 8$ **25.** Absolute maximum: 10 at $x = 0$; absolute minimum: -22 at $x = 4$ **27.** Absolute maximum: $\frac{27}{256}$ at $x = \frac{3}{4}$; absolute minimum: -2 at $x = -1$
29. Absolute maximum: 13 at $x = -2$ and $x = 2$; absolute minimum: 4 at $x = -1$ and $x = 1$ **31.** Absolute maximum: 1 at $x = 0$; absolute minimum: -3 at $x = -8$ and $x = 8$ **33.** Absolute maximum: -4 at $x = -2$; absolute minimum: $-\frac{17}{2}$ at $x = -8$ **35.** Absolute maximum: $\frac{4}{5}$ at $x = -2$ and $x = 2$; absolute minimum: 0 at $x = 0$ **37.** Absolute maximum: 3 at $x = 26$; absolute minimum: -1 at $x = -2$ **39–47.** Left to the student
49. Absolute maximum: 225 at $x = 15$ **51.** Absolute minimum: 70 at $x = 10$ **53.** Absolute maximum: $21\frac{1}{3}$ at $x = 2$ **55.** Absolute maximum: 900 at $x = 30$
57. Absolute maximum: $\frac{10\sqrt{5}}{3}$ at $x = -\sqrt{5}$; absolute minimum: $-\frac{10\sqrt{5}}{3}$ at $x = \sqrt{5}$ **59.** Absolute maximum: 5700 at $x = 2400$
61. Absolute maximum: $5\frac{13}{27}$ at $x = \frac{5}{3}$ **63.** Absolute maximum: 2000 at $x = 20$; absolute minimum: 0 at $x = 0$ and $x = 30$ **65.** Absolute minimum: 24 at $x = 6$ **67.** Absolute

minimum: 108 at $x = 6$　**69.** Absolute maximum: 3 at $x = 1$; absolute minimum: $-\frac{3}{8}$ at $x = -\frac{1}{2}$　**71.** Absolute maximum: $2\sqrt[3]{3}$ at $x = 8$; absolute minimum: 0 at $x = 0$　**73.** No absolute maximum or minimum　**75.** Absolute maximum: -1 at $x = 1$; absolute minimum: -5 at $x = -1$　**77.** No absolute maximum; absolute minimum: -5 at $x = -1$　**79.** Absolute maximum; 19 at $x = -2$; no absolute minimum　**81.** No absolute maximum; absolute minimum: 0 at $x = 0$　**83.** No absolute maximum or minimum　**85.** No absolute maximum; absolute minimum: -1 at $x = -1$ and $x = 1$　**87–95.** Left to the student　**97.** 1430 units; after 25 yr of employment　**99.** 2011　**101.** 2001; about 23,858,000 barrels

103. (a) $P(x) = -\frac{1}{2}x^2 + 400x - 5000$;　(b) 400 items
105. About 1.26 at $x = \frac{1}{9}$ cc, or about 0.11 cc

107. Absolute maximum: 3 at $x = 1$; absolute minimum: -5 at $x = -3$

109. Absolute maxima: 1 at $x = 0$ and 1 at $x = 2$; absolute minimum: -15 at $x = -4$

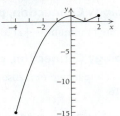

111. (a)

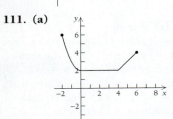

(b) Absolute maximum: 6 at $x = -2$; (c) $f(x) = 2$ over $[0, 4]$

113. Absolute maximum: $3\sqrt{6}$ at $x = 3$; absolute minimum: -2 at $x = -2$

115. Minimum: \$20,000 at $x = 7$ "quality units"

117. 2003–2004; at about 640,000 barrels/yr
119. (a) 2.755 billion barrels in 1981; (b) 0.1014 billion barrels/yr; 0.1234 billion barrels/yr　**121.** No absolute maximum; absolute minimum: 0 at $x = 1$
123. (a) $P(t) = t + 8.857$; $P(7) = 15.857$ mm Hg;
(b) $P(t) = 0.117t^4 - 1.520t^3 + 6.193t^2 - 7.018t + 10.009$; $P(7) = 24.86$ mm Hg; $P(0.765) = 7.62$ mm Hg is the smallest contraction

Technology Connection, p. 253

1–4. Left to the student

Exercise Set 2.5, p. 254

1. Maximum $Q = 1225$; $x = 35, y = 35$　**3.** Minimum product $= -9$; $x = 3, y = -3$　**5.** Maximum $Q = 4$; $x = 2, y = \sqrt{2}$, or approximately 1.414　**7.** Minimum $Q = 6$; $x = 2, y = 1$　**9.** Maximum $Q = \frac{1}{3}$; $x = \frac{2}{3}, y = \frac{1}{2}$
11. Maximum area $= 4800$ yd^2 ($x = 40$ yd, $y = 120$ yd)
13. Maximum area $= 45,000$ yd^2 (150 yd by 300 yd)
15. Maximum area $= 110.25$ ft^2; $x = 10.5$ ft; $y = 10.5$ ft
17. Dimensions: $13\frac{1}{3}$ in. by $13\frac{1}{3}$ in. by $3\frac{1}{3}$ in.; maximum volume: $592\frac{16}{27}$ in^3　**19.** Dimensions: 4 ft by 4 ft by 2 ft; minimum surface area $= 48$ ft^2　**21.** 2.08 yd by 4.16 yd by 1.387 yd
23. \$1048; 46 units　**25.** \$19; 70 units　**27.** \$5481; 1667 units　**29.** (a) $R(x) = x(280 - 0.4x)$;
(b) $P(x) = -x^2 + 280x - 5000$; (c) 140 refrigerators;

(d) \$14,600;　(e) \$224/refrigerator　**31.** \$201 per day
33. 1 fewer officer　**35.** (a) $q(x) = -\frac{12}{41}x + \frac{53}{41}$; (b) 221% of the January 2001 price　**37.** \$11.50　**39.** 62.55 ft by 79.93 ft; \$1439　**41.** 6.98 in. by 10.47 in.　**43.** Order 20 times/yr; lot size is 10.　**45.** Order 12 times/yr; lot size is 30.
47. Order 12 times/yr; lot size is 30.　**49.** $r \approx 5.03$ cm, $h \approx 5.03$ cm　**51.** $r \approx 4.08$ cm, $h \approx 7.65$ cm　**53.** 6 yd by 8 yd (8-yd side is opposite the side shared with neighbor)
55. $x \approx 2.76$ ft, $y \approx 4.92$ ft　**57–109.** Left to the student
111. $\sqrt[3]{0.1}$, or approximately 0.4642　**113.** 9%　**115.** S is 3.25 mi downshore from A.　**117.** Left to the student

119. (a) $A(x) = 8 + \dfrac{20}{x} + \dfrac{x^2}{100}$　(b) $C'(x) = \dfrac{3x^2}{100} + 8$ and

$A'(x) = \dfrac{x}{50} - \dfrac{20}{x^2}$;　(c) minimum $= \$11$/unit at $x_0 = 10$

units; $C'(10) = \$11$/unit;　(d) they are equal.

121. Minimum: $6 - 4\sqrt{2}$, or approximately 0.343, at $x = 2 - \sqrt{2} \approx 0.586$ and $y = -1 + \sqrt{2} \approx 0.414$

123. Order $\sqrt{\dfrac{aQ}{2b}}$ times; lot size: $\sqrt{\dfrac{2bQ}{a}}$

125. $x = 25, y = 16.67$; $Q = 416.67$
127. $x \approx 2.24, y \approx 2.74$; $Q \approx 45.93$

Technology Connection, p. 261

1. $P'(50) = \$140$/unit; $P(51) - P(50) = \$217$/unit

Exercise Set 2.6, p. 265

1. (a) $P(x) = -0.5x^2 + 46x - 10$;
(b) $R(20) = \$800, C(20) = \$90, P(20) = \$710$;
(c) $R'(x) = 50 - x$; $C'(x) = 4$; $P'(x) = 46 - x$;
(d) $R'(20) = \$30, C'(20) = \$4, P'(20) = \$26$
3. (a) \$226,800; (b) \$5994; (c) \$5960; (d) \$238,720
5. (a) \$1799; (b) \$235.88; (c) \$75.40;
(d) $R(71) = \$1874.40, R(72) = \$1949.80$, $R(73) = \$2025.20$　**7.** (a) \$3685.03; (b) \$276.63;
(c) \$138.20; (d) \$3823.23　**9.** If the price increases from \$1000 to \$1001, sales will decrease by 100 units.
11. \$3.21; \$3.20　**13.** \$2; \$2
15. (a) $P(x) = -0.01x^2 + 1.4x - 100$; (b) $-\$0.21$; $-\$0.20$
17. (a) $dS/dp = 0.021p^2 - p + 150$; (b) 3547 units; (c) ✏
(d) ✏　**19.** (a) 630 units, 980 units, 1430 units, 630 units; (b) $M'(t) = -4t + 100$; (c) ✏ (d) ✏
21. \$526.75 billion　**23.** ✏　**25.** About \$0.28 paid in taxes per dollar earned　**27.** 0.1206; 0.12　**29.** 0.2816; 0.28
31. -0.1667; -0.2　**33.** 6; 6　**35.** 5.100　**37.** 10.100

39. 10.017　**41.** $\dfrac{3}{2\sqrt{3x - 2}}\, dx$　**43.** $9x^2\sqrt{2x^3 + 1}\, dx$

45. $\dfrac{x^4 + 8x^2 - 4x + 3}{(x^2 + 3)^2}\, dx$　**47.** $(4x^3 - 6x^2 + 10x + 3)\, dx$

49. $-22,394.88$　**51.** 7.2　**53.** 657.00　**55.** -0.01345 m^2
57. Ticket price will increase more between 2014 and 2016.

59. $A'(x) = \dfrac{x \cdot C'(x) - C(x)}{x^2}$　**61.** $dV = 4,019,200$ ft^3

63. $R'(x) = 400 - 2x$　**65.** $R'(x) = 3$　**67.** ✏

Exercise Set 2.7, p. 273

1. (a) $E(x) = \dfrac{x}{400 - x}$; (b) $\dfrac{5}{11}$, inelastic; (c) \$200

3. (a) $E(x) = \dfrac{x}{50 - x}$; **(b)** 11.5, elastic; **(c)** \$25

5. (a) $E(x) = 1$; **(b)** 1, unit elasticity; **(c)** total revenue is independent of x.   **7. (a)** $E(x) = \dfrac{x}{2(600 - x)}$; **(b)** 0.10,

inelastic; **(c)** \$400   **9. (a)** $E(x) = \dfrac{2x}{x + 3}$; **(b)** 0.5, inelastic;

(c) \$3   **11. (a)** $E(x) = \dfrac{6x^2 - 300x}{50,000 + 300x - 3x^2}$;

(b) 0.2, inelastic; **(c)** 0.6, inelastic; **(d)** 1.38, elastic;
(e) \$114.98 per barrel; **(f)** about 51.55 billion barrels;

(g) increase   **13. (a)** $E(x) = \dfrac{3x^3}{2(200 - x^3)}$; **(b)** $\dfrac{81}{346}$;

(c) increase   **15. (a)** $E(x) = \dfrac{10x}{180 - 10x}$; **(b)** 0.8; **(c)** \$9

(d) ✎   **17.** ✎

Exercise Set 2.8, p. 279

1. $\dfrac{9x^2}{2y}$; 4.5   **3.** $\dfrac{-3x^2}{4y}$; 3   **5.** $\dfrac{-x}{y}$; $\dfrac{-1}{\sqrt{3}}$ **7.** $\dfrac{-3y}{2x}$; 4.5

9. $\dfrac{4x^2 - 2y^3}{3xy^2}$; $-\dfrac{7}{3}$ **11.** $\dfrac{2 - y}{x + 2y}$; $-\dfrac{4}{3}$ **13.** $\dfrac{12x^2 + 5}{4y^3 + 3}$; $-\dfrac{17}{29}$

15. $\dfrac{y + x}{3y - x}$ **17.** $\dfrac{x}{y}$ **19.** $\dfrac{5x^4}{3y^2}$ **21.** $\dfrac{-3xy^2 - 2y}{4x^2y + 3x}$

23. $\dfrac{-2}{2p + 1}$ **25.** $\dfrac{-p}{3x}$ **27.** $\dfrac{2 - 2xp - p}{x^2 + x - 1}$ **29.** $\dfrac{-p - 4}{x + 3}$

31. 0; $-\dfrac{3}{2\sqrt{6}}$; $-\dfrac{9}{4}$ **33.** \$200/day, \$50/day, \$150/day

35. \$36,000/day, \$72,000/day, $-$\$36,000/day
37. -1.18 sales/day   **39.** $-23,389$ mi^2/yr
41. Decreasing by 0.0256 m^2/month

43. (a) $\dfrac{dV}{dt} = 952.38R\,\dfrac{dR}{dt}$; **(b)** 0.0143 mm/sec^2

45. $-2\frac{1}{12}$ ft/sec   **47.** About 494.8 cm^3/week   **49.** $\dfrac{-y^3}{x^3}$

51. $\dfrac{2x}{y(x^2 + 1)^2}$, or $\dfrac{x(1 - y^2)}{y(1 + x^2)}$

53. $\dfrac{5x^4 - 3(x - y)^2 - 3(x + y)^2}{3(x + y)^2 - 3(x - y)^2 - 5y^4}$, or $\dfrac{6y^2 - 5x^4 + 6x^2}{y(5y^3 - 12x)}$

55. $\dfrac{-6(y^2 - xy + x^2)}{(2y - x)^3}$ **57.** $\dfrac{2x(y^3 - x^3)}{y^5}$ **59.** ✎

61. $x^4 = y^2 + x^6$

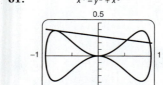

63. $x^3 = y^2(2 - x)$

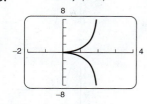

Chapter 2 Review Exercises, p. 288

1. (g)   **2.** (e)   **3.** (f)   **4.** (a)   **5.** (b)   **6.** (d)   **7.** (c)
8. False   **9.** False   **10.** True   **11.** False   **12.** True
13. False   **14.** False   **15.** True   **16.** True

17. Relative maximum: $\dfrac{25}{4}$ at $x = -\dfrac{3}{2}$

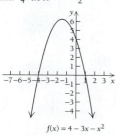

$f(x) = 4 - 3x - x^2$

18. Relative minima: 2 at $x = -1$ and 2 at $x = 1$; relative maximum: 3 at $x = 0$

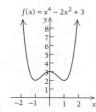

$f(x) = x^4 - 2x^2 + 3$

19. Relative minimum: -4 at $x = 1$; relative maximum: 4 at $x = -1$

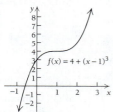

$f(x) = \dfrac{-8x}{x^2 + 1}$

20. No relative extrema

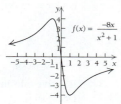

$f(x) = 4 + (x - 1)^3$

21. Relative minimum: $\dfrac{76}{27}$ at $x = \dfrac{1}{3}$; relative maximum: 4 at $x = -1$

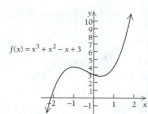

$f(x) = x^3 + x^2 - x + 3$

22. Relative minimum: 0 at $x = 0$

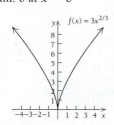

$f(x) = 3x^{2/3}$

23. Relative maximum: 17 at $x = -1$; relative minimum: -10 at $x = 2$

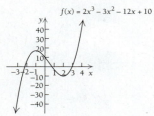

$f(x) = 2x^3 - 3x^2 - 12x + 10$

24. Relative maximum: 4 at $x = -1$; relative minimum: 0 at $x = 1$

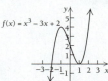

$f(x) = x^3 - 3x + 2$

25.

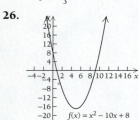

$f(x) = \frac{1}{3}x^3 + 3x^2 + 9x + 2$

No relative extrema
Inflection point at $(-3, -7)$
Increasing on $(-\infty, \infty)$
Concave down on $(-\infty, -3)$; concave up on $(-3, \infty)$

26.

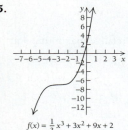

$f(x) = x^2 - 10x + 8$

Relative minimum: -17 at $x = 5$
Decreasing on $(-\infty, 5)$; increasing on $(5, \infty)$
Concave up on $(-\infty, \infty)$

27.

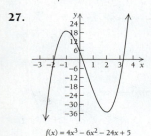

$f(x) = 4x^3 - 6x^2 - 24x + 5$

Relative minimum: -35 at $x = 2$;
relative maximum: 19 at $x = -1$
Inflection point at $\left(\frac{1}{2}, -8\right)$
Increasing on $(-\infty, -1)$ and $(2, \infty)$; decreasing on $(-1, 2)$
Concave down on $\left(-\infty, \frac{1}{2}\right)$; concave up on $\left(\frac{1}{2}, \infty\right)$

28.

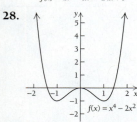

$f(x) = x^4 - 2x^2$

Relative minima: -1 at $x = -1$ and -1 at $x = 1$;
relative maximum: 0 at $x = 0$
Inflection points at $\left(-\sqrt{\frac{1}{3}}, -\frac{5}{9}\right)$ and $\left(\sqrt{\frac{1}{3}}, -\frac{5}{9}\right)$
Increasing on $(-1, 0)$ and $(1, \infty)$;
decreasing on $(-\infty, -1)$ and $(0, 1)$
Concave up on $\left(-\infty, -\sqrt{\frac{1}{3}}\right)$ and $\left(\sqrt{\frac{1}{3}}, \infty\right)$;
concave down on $\left(-\sqrt{\frac{1}{3}}, \sqrt{\frac{1}{3}}\right)$

29.

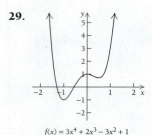

$f(x) = 3x^4 + 2x^3 - 3x^2 + 1$

Relative minima: -1 at $x = -1$ and $\frac{11}{16}$ at $x = \frac{1}{2}$;
relative maximum: 1 at $x = 0$
Inflection points at $(-0.608, -0.147)$ and $(0.274, 0.833)$
Increasing on $(-1, 0)$ and $\left(\frac{1}{2}, \infty\right)$;
decreasing on $(-\infty, -1)$ and $\left(0, \frac{1}{2}\right)$
Concave down on $(-0.608, 0.274)$; concave up on $(-\infty, -0.608)$ and $(0.274, \infty)$

30.

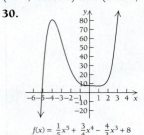

$f(x) = \frac{1}{5}x^5 + \frac{3}{4}x^4 - \frac{4}{3}x^3 + 8$

Relative minimum: $\frac{457}{60}$ at $x = 1$;
relative maximum: $\frac{1208}{15}$ at $x = -4$
Inflection points at $(-2.932, 53.701)$, $(0, 8)$, and $(0.682, 7.769)$
Increasing on $(-\infty, -4)$ and $(1, \infty)$; decreasing on $(-4, 1)$
Concave down on $(-\infty, -2.932)$ and $(0, 0.682)$; concave up on $(-2.932, 0)$ and $(0.682, \infty)$

31.

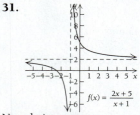

$f(x) = \frac{2x + 5}{x + 1}$

No relative extrema
Decreasing on $(-\infty, -1)$ and $(-1, \infty)$
Concave down on $(-\infty, -1)$; concave up on $(-1, \infty)$
Asymptotes: $x = -1$ and $y = 2$
x-intercept: $\left(-\frac{5}{2}, 0\right)$; y-intercept: $(0, 5)$

32.

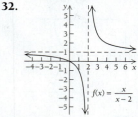

$f(x) = \frac{x}{x - 2}$

No relative extrema
Decreasing on $(-\infty, 2)$ and $(2, \infty)$
Concave down on $(-\infty, 2)$; concave up on $(2, \infty)$
Asymptotes: $x = 2$ and $y = 1$
x-intercept: $(0, 0)$; y-intercept: $(0, 0)$

33.

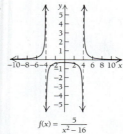

$$f(x) = \frac{5}{x^2 - 16}$$

Relative maximum at $\left(0, -\frac{5}{16}\right)$
Decreasing on $(0, 4)$ and $(4, \infty)$; increasing on $(-\infty, -4)$ and $(-4, 0)$
Concave down on $(-4, 4)$; concave up on $(-\infty, -4)$ and $(4, \infty)$
Asymptotes: $x = -4$, $x = 4$, and $y = 0$
y-intercept: $\left(0, -\frac{5}{16}\right)$

34.

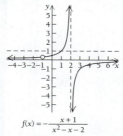

$$f(x) = -\frac{x+1}{x^2 - x - 2}$$

No relative extrema
Increasing on $(-\infty, -1)$, $(-1, 2)$, and $(2, \infty)$
Concave up on $(-\infty, -1)$ and $(-1, 2)$; concave down on $(2, \infty)$
Asymptotes: $x = 2$ and $y = 0$
y-intercept: $\left(0, \frac{1}{2}\right)$

35.

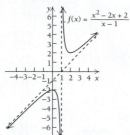

$$f(x) = \frac{x^2 - 2x + 2}{x - 1}$$

Relative minimum at $(2, 2)$; relative maximum at $(0, -2)$
Decreasing on $(0, 1)$ and $(1, 2)$; increasing on $(-\infty, 0)$ and $(2, \infty)$
Concave down on $(-\infty, 1)$; concave up on $(1, \infty)$
Asymptotes: $x = 1$ and $y = x - 1$
y-intercept: $(0, -2)$

36.

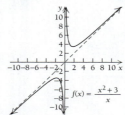

$$f(x) = \frac{x^2 + 3}{x}$$

Relative minimum at $\left(\sqrt{3}, 2\sqrt{3}\right)$; relative maximum at $\left(-\sqrt{3}, -2\sqrt{3}\right)$
Decreasing on $\left(-\sqrt{3}, 0\right)$ and $\left(0, \sqrt{3}\right)$; increasing on $\left(-\infty, -\sqrt{3}\right)$ and $\left(\sqrt{3}, \infty\right)$
Concave down on $(-\infty, 0)$; concave up on $(0, \infty)$
Asymptotes: $x = 0$ and $y = x$
No intercepts

37. Absolute maximum: 66 at $x = 3$; absolute minimum: 2 at $x = 1$ **38.** Absolute maximum: $75\frac{23}{27}$ at $x = \frac{16}{3}$; absolute minima: 0 at $x = 0$ and $x = 8$ **39.** No absolute maxima; absolute minimum: $10\sqrt{2}$ at $x = 5\sqrt{2}$ **40.** No absolute maxima; absolute minima: 0 at $x = -1$ and $x = 1$ **41.** $Q = 12$ when $x = 4$ and $y = 3$ **42.** $Q = -1$ when $x = -1$ and $y = -1$ **43.** Maximum profit is \$451 when 30 units are produced and sold. **44.** 10 ft by 10 ft by 25 ft **45.** Order 12 times per year with a lot size of 30 **46. (a)** \$108; **(b)** \$1/dinner; **(c)** \$109 **47.** $\Delta y = -0.335$, $dy = -0.35$
48. (a) $(6x^2 + 1)\, dx$; **(b)** 0.25 **49.** 9.111

50. $dV = \pm 240{,}000$ ft^3 **51. (a)** $E(x) = \dfrac{5x}{400 - 5x}$, or $\dfrac{x}{80 - x}$
(b) 0.14, inelastic; **(c)** 1.67, elastic; **(d)** increase; **(e)** \$40
52. $\dfrac{-3y - 2x^2}{2y^2 + 3x}; \dfrac{4}{5}$ **53.** -1.75 ft/sec
54. \$600/day, \$450/day, \$150/day **55.** No maximum; absolute minimum: 0 at $x = 3$ **56.** Absolute maxima: 4 at $x = 2$ and $x = 6$; absolute minimum: -2 at $x = -2$
57. $\dfrac{3x^5 - 2(x - y)^3 - 2(x + y)^3}{2(x + y)^3 - 2(x - y)^3 - 3y^5}$ **58.** Relative maximum at $(0, 0)$; relative minima at $(-9, -9477)$ and $(15, -37{,}125)$
59. $f(x) = \dfrac{3x + 3}{x + 2}$ (answers may vary) **60.** Relative maxima at $(-1.714, 37.445)$; relative minimum at $(1.714, -37.445)$
61. Relative maximum at $(0, 1.08)$; relative minima at $(-3, -1)$ and $(3, -1)$
62. (a) Linear: $y = 6.998187602x - 124.6183581$
Quadratic: $y = 0.0439274846x^2 + 2.881202838x - 53.51475166$
Cubic: $y = -0.0033441547x^3 + 0.4795643605x^2 - 11.35931622x + 5.276985809$
Quartic: $y = -0.00005539834x^4 + 0.0067192294x^3 - 0.0996735857x^2 - 0.8409991942x - 0.246072967$
(b) The quartic function best fits the data. **(c)** The domain is $[26, 102]$. Very few women outside of the age range from 26 to 102 years old develop breast cancer. **(d)** Maximum: 466 per 100,000 women at $x = 79.0$ years old

Chapter 2 Test, p. 290

1. $[2.1, 2.2]$ Relative minimum: -9 at $x = 2$
Decreasing on $(-\infty, 2)$ increasing on $(2, \infty)$

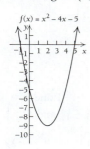

$$f(x) = x^2 - 4x - 5$$

2. [2.1, 2.2] Relative minimum: 2 at $x = -1$; relative maximum: 6 at $x = 1$
Decreasing on $(-\infty, -1)$ and $(1, \infty)$; increasing on $(-1, 1)$

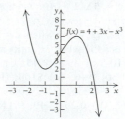

3. [2.1, 2.2] Relative minimum: -4 at $x = 2$
Decreasing on $(-\infty, 2)$; increasing on $(2, \infty)$

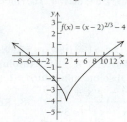

4. [2.1, 2.2] Relative maximum: 4 at $x = 0$
Increasing on $(-\infty, 0)$; decreasing on $(0, \infty)$

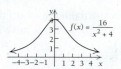

5. [2.3]

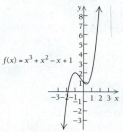

Relative maximum: 2 at $x = -1$; relative minimum: $\frac{22}{27}$ at $x = \frac{1}{3}$
Inflection point: $\left(-\frac{1}{3}, \frac{38}{27}\right)$

6. [2.3]

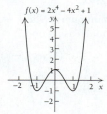

Relative maximum: 1 at $x = 0$;
relative minima: -1 at $x = -1$ and $x = 1$
Inflection points: $\left(-\sqrt{\frac{1}{3}}, -\frac{1}{9}\right)$ and $\left(\sqrt{\frac{1}{3}}, -\frac{1}{9}\right)$

7. [2.3]

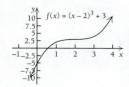

No relative extrema
Inflection point: $(2, 3)$

8. [2.3]

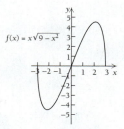

Relative maximum: $\frac{9}{2}$ at $x = \sqrt{\frac{9}{2}}$;
relative minimum: $-\frac{9}{2}$ at $x = -\sqrt{\frac{9}{2}}$
Inflection point: $(0, 0)$

9. [2.3]

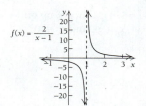

No relative extrema
Asymptotes: $x = 1$ and $y = 0$

10. [2.3]

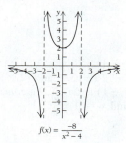

Relative minimum: 2 at $x = 0$
Asymptotes: $x = -2$, $x = 2$, and $y = 0$

11. [2.3]

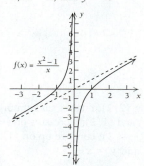

No relative extrema
Asymptotes: $x = 0$ and $y = x$

12. [2.3]

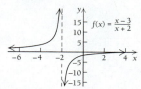

No relative extrema
Asymptotes: $x = -2$ and $y = 1$
13. [2.4] Absolute maximum: 9 at $x = 3$; no absolute minimum
14. [2.4] Absolute maximum: 2 at $x = -1$; absolute minimum: -1 at $x = -2$ **15.** [2.4] Absolute maximum: 28.49 at $x = 4.3$; no absolute minimum **16.** [2.4] Absolute maximum: 7 at $x = -1$; absolute minimum: 3 at $x = 1$ **17.** [2.4] There

are no absolute extrema. **18.** [2.4] Absolute minimum: $-\frac{13}{12}$ at $x = \frac{1}{6}$ **19.** [2.4] Absolute minimum: 48 at $x = 4$

20. [2.5] $Q = \dfrac{125{,}000}{27}$ when $x = \dfrac{50}{3}$ and $y = \dfrac{50}{3}$

21. [2.5] $Q = 50$ for $x = 5$ and $y = -5$

22. [2.5] Maximum profit: \$24,980; 500 units

23. [2.5] Dimensions: 40 in. by 40 in. by 10 in; maximum volume: 16,000 in^3 **24.** [2.5] Order 35 times per year; lot size, 35

25. [2.6] $\Delta y = 1.01$; $f'(x)\,\Delta x = 1$ **26.** [2.6] 7.0714

27. [2.6] **(a)** $\dfrac{x}{\sqrt{x^2 + 3}}\,dx$; **(b)** 0.00756

28. [2.7] **(a)** $E(x) = \dfrac{2x}{x + 4}$; **(b)** 0.4, inelastic; **(c)** 1.5,

elastic; **(d)** decrease; **(e)** \$4 **29.** [2.8] $\dfrac{-x^2}{y^2}$; $-\dfrac{1}{4}$

30. [2.6] $dV = \pm 1413$ cm^3 **31.** [2.8] -0.96 ft/sec

32. [2.4] Absolute maximum: $\dfrac{2^{2/3}}{3} \approx 0.529$ at $x = \sqrt[3]{2}$;

absolute minimum: 0 at $x = 0$ **33.** [2.5] 10,000 units

34. [2.4] Absolute minimum: 0 at $x = 0$; relative maximum: 25.103 at $x = 1.084$; relative minimum: 8.625 at $x = 2.95$

35. [2.4] Relative minimum: -0.186 at $x = 0.775$; relative maximum: 0.186 at $x = -0.775$ **36.** [2.5] $Q = 1185.19$ when $x = 2.58$ and $y = 13.33$

37. [2.1, 2.2] **(a)** Linear: $y = -0.7707142857x + 12691.60714$
Quadratic: $y = -0.9998904762x^2 + 299.1964286x + 192.9761905$
Cubic: $y = 0.000084x^3 - 1.037690476x^2 + 303.3964286x + 129.9761905$
Quartic: $y = -0.000001966061x^4 + 0.0012636364x^3 - 1.256063636x^2 + 315.8247403x + 66.78138528$
(b) Since the number of bowling balls sold cannot be negative, the domain is $[0, 300]$. This is supported by both the quadratic model and the raw data. The cubic and quartic models can also be used but are more complicated. **(c)** Based on the quadratic function, the maximum value is 22,575 bowling balls. The company should spend \$150,000 on advertising.

Extended Technology Application, p. 293

1. (a)

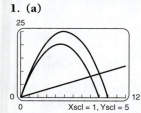

(b) 4500; **(c)** 20,250

2. (a)

(b) 60,000; **(c)** 90,000

3. (a)

(b) 50,000; **(c)** 25,000

4. (a)

(b) 400,000; **(c)** 400,000

5. (a)

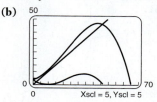

(b) 30,513; **(c)** 205,923

6. (a) $y = -0.0011P^3 + 0.0715P^2 - 0.0338P + 4$

(b)

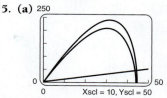

(c) 33,841

CHAPTER 3

Technology Connection, p. 298

Left to the student

Exercise Set 3.1, p. 304

1.

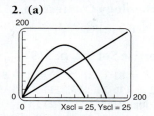

3.

5.

7.

9.

11. e^x **13.** $2e^{2x}$ **15.** $6e^x$ **17.** $-7e^{-7x}$ **19.** $15e^{5x}$

21. $7e^{-x}$ **23.** $-\frac{5}{2}e^{-5x}$ **25.** $-\dfrac{4x}{3} \cdot e^{x^2}$ **27.** $-2e^{2x}$

29. $3x^2 - 10e^{2x}$ **31.** $5x^4 \cdot e^{2x} + 2x^5 \cdot e^{2x}$ **33.** $\dfrac{2e^{2x}(x - 2)}{x^5}$

35. $x^2 \cdot e^x$ **37.** $\dfrac{e^x(x - 4)}{x^5}$ **39.** $(-2x + 8)e^{-x^2 + 8x}$

41. $xe^{x^2/2}$ **43.** $\dfrac{e^{\sqrt{x-7}}}{2\sqrt{x - 7}}$ **45.** $\dfrac{e^x}{2\sqrt{e^x - 1}}$ **47.** $3x^2 - xe^x$

49. $3e^{-3x}$ **51.** ke^{-kx} **53.** $(4x^2 + 3x)e^{x^2 - 7x}(2x - 7) + (8x + 3)e^{x^2 - 7x}$, or $(8x^3 - 22x^2 - 13x + 3)e^{x^2 - 7x}$

55.

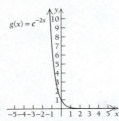

57.

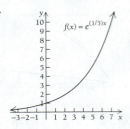

59.

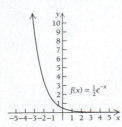

No critical values
No inflection points
Decreasing on $(-\infty, \infty)$
Concave up on $(-\infty, \infty)$

61.

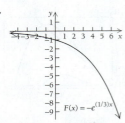

No critical values
No inflection points
Decreasing on $(-\infty, \infty)$
Concave down on $(-\infty, \infty)$

63.

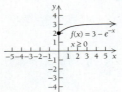

65–73. Left to the student
75. -6 **77.** $y = 2x + 1$
79. Left to the student
81. (a) $184.5 billion, $222.7 billion;   **(b)** $17.3 billion/yr, $20.9 billion/yr

83. (a) $C'(t) = 50e^{-t}$; **(b)** $50 million/yr; **(c)** $916,000/yr; **(d)** $100, 0$ **85. (a)** $113,000$; **(b)**

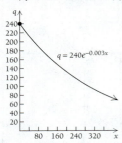

(c) $q'(x) = -0.72e^{-0.003x}$; **(d)** 🖉   **87. (a)** $29,287.17; **(b)** $1595.80/yr; **(c)** in about 12.7 yr   **89. (a)** $V(t) = 14,450(0.85)^t$; **(b)** $-$2348.40/yr$ **(c)** $20,000; **(d)** 4.27 yr
91. (a) 0 ppm, 3.7 ppm, 5.4 ppm, 4.5 ppm, 0.05 ppm;
(b)

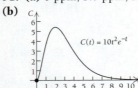

(c) $C'(t) = 10te^{-t}(2 - t)$;
(d) 5.4 ppm at $t = 2$ hr;
(e) 🖉

93. $15e^{3x}(e^{3x} + 1)^4$ **95.** $\dfrac{(x^2 - 2x + 1)e^x}{(x^2 + 1)^2}$

97. $e^{x/2}\left(\dfrac{x}{2\sqrt{x - 1}}\right)$ **99.** $\dfrac{4}{(e^x + e^{-x})^2}$

101. Equilibrium point: $(166.16, 145,787)$; supply is increasing at a rate of about 583 units per dollar, and demand is decreasing at a rate of about 437 units per dollar.   **103.** 4; 2.86797; 2.73200; 2.71964; 2.71855   **105.** 🖉

107.

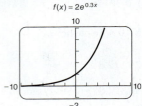

Relative minimum at $(0, 0)$;
relative maximum at $(2, 0.5413)$

109.

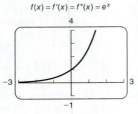

111.

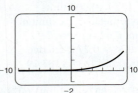

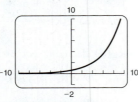

113.

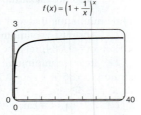

Technology Connection, p. 309

Left to the student

Technology Connection, p. 311

1. Graph is left to the student; function values are 1000, 5, 0.699, 3.
2. Left to the student; graph is obtained by entering $y = \log x/\log 2$.

Technology Connection, p. 312

1. $t = 6.907755$ **2.** $x = -4.094345$ **3.** $t = 74.893307$
4. $x = 46.219585$ **5.** $x = 38.744475$

Technology Connection, p. 313

Left to the student

Exercise Set 3.2, p. 319

1. $3^4 = 81$ **3.** $27^{1/3} = 3$ **5.** $a^K = J$ **7.** $b^{-w} = V$
9. 2   **11.** 2   **13.** 243   **15.** $\frac{1}{2}$ **17.** $\ln p = t$
19. $\log_{10}1000 = 3$ **21.** $\log_{10} 0.01 = -2$ **23.** $\log_Q T = n$
25. -1.609 **27.** $\frac{3}{2}$ **29.** 4.317   **31.** 4.382   **33.** -1.6094
35. 2.3863   **37.** 4   **39.** -0.2231 **41.** 0.3863
43. 11.512915   **45.** -4.006334 **47.** -4.509860
49. $t \approx 4.382$ **51.** $t \approx 2.267$ **53.** $t \approx 4.605$
55. $t \approx 140.671$ **57.** $-\dfrac{9}{x}$ **59.** $\dfrac{7}{x}$ **61.** $x^5 + 6(\ln x)x^5 - x^3$
63. $\dfrac{1}{x}$ **65.** $\dfrac{1}{x}$ **67.** $x^4 + 5x^4 \ln (3x)$ **69.** $x^3 + 4x^3 \ln |6x|$
71. $\dfrac{1 - 5\ln x}{x^6}$ **73.** $\dfrac{1 - 2\ln |3x|}{x^3}$ **75.** $\dfrac{2}{x}$ **77.** $\dfrac{2(3x + 1)}{3x^2 + 2x - 1}$
79. $\dfrac{x^2 + 7}{x(x^2 - 7)}$ **81.** $\dfrac{2e^x}{x} + 2e^x \ln x$ **83.** $\dfrac{e^x}{e^x + 1}$

85. $\dfrac{4(\ln x)^3}{x}$ **87.** $\dfrac{1}{x\ln(8x)}$ **89.** $\dfrac{\ln(5x)+\ln(3x)}{x}$

91. $y=8.455x-11.94$ **93.** $y=0.732x-0.990$

95. (a) 2000 units; **(b)** $N'(a)=\dfrac{500}{a}$, $N'(10)=50$ units per
$1000 spent on advertising; **(c)** minimum is 2000 units;
(d) ✏️   **97.** 86 days   **99. (a)** $R(x)=x(53.5-8\ln x)$;
(b) $R'(x)=45.5-8\ln x$; **(c)** ✏️   **101. (a)** $P'(x)=$
$1.7-0.3\ln x$; **(b)** ✏️ **(c)** 289.069   **103. (a)** 78%;
(b) 45.8%; **(c)** 13.6%; **(d)** 17.4%; **(e)** $S'(t)=\dfrac{-20}{t+1}$;
(f) maximum $=78\%$; **(g)** ✏️   **105. (a)** 2.44 ft/s;
(b) 3.39 ft/s; **(c)** $v'(p)=\dfrac{0.37}{p}$; **(d)** ✏️   **107.** $t=\dfrac{\ln(P/P_0)}{k}$

109. $\dfrac{4[\ln(x+5)]^3}{x+5}$ **111.** $\dfrac{-1}{1-t}-\dfrac{1}{1+t}$, or $\dfrac{-2}{(1-t)(1+t)}$

113. $\dfrac{1}{x\ln 5}$ **115.** $\dfrac{x}{x^2+5}$ **117.** $x^n\ln x$ **119.** Definition
of logarithm; Product Rule for exponents; definition of logarithm;
substitution   **121.** Definition of logarithm; if $u=v$, then
$u^c=v^c$; Power Rule for exponents; definition of logarithm; substitution and the commutative law for multiplication   **123.** 1
125. e^π **127.** Minimum: $-e^{-1}\approx-0.368$
129. (a)

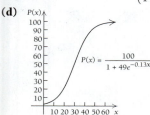

(b) $-\frac{1}{3},-\frac{1}{2},-1$
(c) They are the opposites
of those slopes. **(d)** ✏️

Technology Connection, p. 326

1. 312 billion   **2.** 179 trillion   **3.** 1.97 quadrillion
4. 7.91×10^{22} **5.** Left to the student
6. $y=12{,}767.750(1.0320889)^x$ **7.** \$18,651.75; \$19,250.27;
\$21,163.53

Exercise Set 3.3, p. 331

1. $f(x)=ce^{4x}$ **3.** $A(t)=ce^{-9t}$ **5.** $Q(t)=ce^{kt}$
7. (a) $N(t)=483{,}000e^{0.058t}$; **(b)** $N(11)=914{,}170$; **(c)** 12 yr
9. (a) $P(t)=P_0e^{0.059t}$; **(b)** \$1060.78, \$1125.24; **(c)** 11.7 yr
11. (a) $G(t)=4.7e^{0.093t}$; **(b)** 48.07 billion gallons; **(c)** 7.5 yr
13. 4.62%   **15.** 6.9 yr after 2012   **17.** 11.2 yr; \$102,256.88
19. \$7,500; 8.3 yr   **21. (a)** $k=0.151$, or 15.1%,
$V(t)=30{,}000e^{0.151t}$; **(b)** \$549,188,702; **(c)** 4.6 yr; **(d)** 69 yr
23. (a) $F(t)=2.3e^{0.092969t}$; **(b)** \$3.34 trillion; **(c)** after about
15.8 yr, or in 2026   **25. (a)** $y=1.60680697(1.60414481)^x$,
$y=1.60680697e^{0.4725068483x}$; **(b)** 47.3%; **(c)** 43.9 EB;
(d) about 7.3 yr; **(e)** about 1.47 yr   **27.** Approximately \$8.6 billion
29. (a) $S(t)=4e^{0.0482t}$; **(b)** 4.82%/yr; **(c)** 54¢, 59¢, 65¢;
(d) \$11,000; **(e)** ✏️   **31. (a)** 2%; **(b)** 3.8%, 7%, 21.6%, 50.2%,
93.1%, 98%; **(c)** $P'(x)=\dfrac{637e^{-0.13x}}{(1+49e^{-0.13x})^2}$

(d)

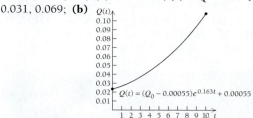

$P(x)=\dfrac{100}{1+49e^{-0.13x}}$

33. (a) $V(t)=0.10e^{0.227t}$;
(b) \$12,132,700; **(c)** 3.05 yr;
(d) after about 86 yr, or in 2024
35. 2019   **37.** 1%/yr
39. 4%/yr   **41. (a)** 2.23%;
(b) ✏️

43. (a) 20, 46, 119, 146; **(b)** $P'(t)=\dfrac{83{,}460e^{-0.214t}}{(20+130e^{-0.214t})^2}$
(c) 3.71 tortoises/yr, 6.86 tortoises/yr, 5.28 tortoises/yr,
0.93 tortoise/yr; **(d)** 150   **45.** $N(t)=48{,}869e^{0.0367t}$, where
$t_0=1930$; exponential growth rate $=3.67\%$ **47. (a)** 11.8,
221.4, 547.2; **(b)** 4.4 new cases per week; **(c)** ✏️
49. (a) $N(t)=\dfrac{29.47232081}{1+79.56767122e^{-0.809743969t}}$; **(b)** 29 students;
(c) $N(t)=\dfrac{29.47232081}{1+79.56767122e^{-0.809743969t}}$

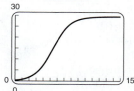

(d) $N'(t)=\dfrac{1898.885181e^{-0.809743969t}}{(1+79.56767122e^{-0.809743969t})^2}$; **(e)** ✏️
51–59. ✏️   **61.** $\ln 4=kT_4$ **63.** 2 yr   **65.** ✏️   **67.** 10 yr
69. 9.9 yr, 19.8 yr   **71. (a)** $R(0)=\$2$ million; this repre-
sents the initial revenue of the corporation at its inception.
(b) $\lim\limits_{t\to\infty}R(t)=\$4000$ million $=R_{\max}$; this represents the
upper limit of the revenue of the company over all time. It is
never actually attained.   **(c)** $t=24$

Exercise Set 3.4, p. 343

1. 5.78 yr   **3.** 86.64 months   **5.** 23.1 yr   **7.** 36.48 wk
9. (a) $N(t)=N_0e^{-0.096t}$; **(b)** 341 g; **(c)** 7.2 days
11. (a) $A(t)=A_0e^{-kt}$; **(b)** 11 hr   **13.** 23.1%/min   **15.** 22 yr
17. 42.9 g   **19.** 4223 yr   **21.** 25 days   **23.** 3965 yr
25. \$13,858.23   **27.** \$6,788,463   **29.** \$42,863.76
31. (a) \$40,000; **(b)** \$5413.41; **(c)** ✏️   **33. (a)** 0.022,
0.031, 0.069; **(b)**

Q(t) graph with axis values 0.01–0.10; $Q(t)=(Q_0-0.00055)e^{0.163t}+0.00055$; t axis 1–10

35. (a) $N(t)=5{,}650{,}000e^{-0.015t}$; **(b)** 2,099,000; 1,977,000
(c) about 2065   **37. (a)** $B(t)=64.6e^{-0.0068t}$; **(b)** 58.3 lb;
(c) 2172   **39. (a)** $k=-0.00855$, $P(t)=51.9e^{-0.00855t}$;
(b) 42.63 million; **(c)** 2025   **41. (a)** 27; **(b)** 0.05878;
(c) 83°; **(d)** 28.7 min; **(e)** ✏️   **43.** The murder was committed
at 7 P.M.   **45. (a)** 145 lb; **(b)** -1.2 lb/day   **47. (a)** 11.2 W;
(b) 173 days; **(c)** 402 days; **(d)** 50 W; **(e)** ✏️   **49.** (c)
51. (e)   **53.** (f)   **55.** (d)   **57.** (a)   **59. (a)** 12.05 wk;
(b) about 24.1 wk   **61. (a)** $x=\$166.16$, $q=292$ printers;
(b) 0.3   **63.** and **65.** ✏️

Exercise Set 3.5, p. 352

1. $(\ln 6)6^x$ **3.** $(\ln 8)8^x$ **5.** $x^5\cdot\ln 3.7\cdot(3.7)^x+5x^4(3.7)^x$
7. $(\ln 7)\cdot 7^{x^4+2}\cdot 4x^3$ **9.** $2xe^{x^2}$ **11.** $(\ln 3)\cdot 3^{x^4+1}\cdot(4x^3)$
13. $\dfrac{1}{x\cdot\ln 8}$ **15.** $\dfrac{1}{x\cdot\ln 17}$ **17.** $\dfrac{9}{(9x-2)\ln 32}$
19. $\dfrac{6}{(6x-7)\ln 10}$ **21.** $\dfrac{4x^3-1}{(x^4-x)\ln 9}$ **23.** $\dfrac{2}{(x-2\sqrt{x})\ln 7}$

25. $\dfrac{5^x}{x\cdot\ln 2} + 5^x\cdot\ln 5\cdot\log_2 x$ **27.** $5(\log_{12} x)^4\left(\dfrac{1}{x\ln 12}\right)$

29. $\dfrac{(5x-1)6^x\cdot\ln 6 - 5\cdot 6^x}{(5x-1)^2}$

31. $\dfrac{6\cdot 5^{2x^3-1}}{(6x+5)(\ln 10)} + (\ln 5)5^{2x^3-1}\cdot 6x^2\cdot\log(6x+5)$

33. $6\cdot\log_9 x\cdot 4^{6x}\cdot\ln 4 + \dfrac{4^{6x}}{\ln 9\cdot x}$

35. $5(3x^5+x)^4(15x^4+1)\cdot(\log_3 x) + \dfrac{(3x^5+x)^5}{\ln 3\cdot x}$

37. (a) $V'(t) = 5200(\ln 0.80)(0.80)^t$; **(b)** ✏️
39. (a) $N'(t) = 400{,}000(\ln 0.341)(0.341)^t$; **(b)** ✏️
41. (a) 22.4 million; **(b)** 187,000/yr; **(c)** ✏️
43. (a) $A(t) = 78{,}947.36842[(1.012666667)^{3t} - 1]$
(b) \$45,255.66;   **(c)** \$4690.07/yr   **45.** 7.0
47. (a) 63 times more intense; **(b)** 398 times more intense
49. (a) $I = I_0 10^7$; **(b)** $I = I_0 10^8$; **(c)** $I_0 10^8$ is 10 times
greater than $I_0 10^7$; **(d)** $dI/dR = I_0 10^R(\ln 10)$
51. (a) $\dfrac{dR}{dl} = \dfrac{1}{(\ln 10)I}$; **(b)** ✏️   **53. (a)** $\dfrac{dy}{dx} = \dfrac{m}{(\ln 10)x}$;
(b) ✏️   **55.** $\ln 5 \approx 1.6094$ **57.** $\frac12$ **59.** $(\ln 2)2^{x^4}\cdot 4x^3$
61. $\dfrac{1}{\ln 3\cdot\log x\cdot\ln 10\cdot x}$ **63.** $\ln a\cdot a^{f(x)}\cdot f'(x)$
65. $\left(\dfrac{g(x)\cdot f'(x)}{f(x)} + g'(x)\cdot\ln(f(x))\right)\cdot[f(x)]^{g(x)}$ **67.** ✏️

Technology Connection, p. 358

1. 5 months, about \$243.27   **2.** 14 months, about \$644.14

Exercise Set 3.6, p. 359

1. \$135.33   **3.** \$593.88   **5.** \$42.60   **7.** \$4936.26
9. \$8762.22   **11. (a)** \$355.13; **(b)** \$21,307.80; **(c)** \$3157.80
13. (a) \$803.07; **(b)** \$289,105.20; **(c)** \$142,855.20
15. (a) \$10.59; **(b)** \$1270.80; **(c)** \$770.80
17.

Balance	Payment	Portion to interest	Portion to principal	New balance
\$18,150.00	\$355.13	\$98.31	\$256.82	\$17,893.18
\$17,893.18	\$355.13	\$96.92	\$258.21	\$17,634.97

19.

Balance	Payment	Portion to interest	Portion to principal	New balance
\$146,250.00	\$803.07	\$633.75	\$169.32	\$146,080.68
\$146,080.68	\$803.07	\$633.02	\$170.05	\$145,910.63

21.

Balance	Payment	Portion to interest	Portion to principal	New balance
\$500.00	\$10.59	\$9.48	\$1.11	\$498.89
\$498.89	\$10.59	\$9.46	\$1.13	\$497.76

23. \$18,205.38   **25.** \$370,291.65   **27. (a)** \$227.56, \$193.26;
(b) \$13,653.60, \$13,914.72; **(c)** option 1, \$261.12
29. (a) \$1073.64; **(b)** \$1199.10; **(c)** \$45,165.60
31. (a) \$89,955.16; **(b)** \$88.39   **33.** \$151,212.94
35. (a) \$1073.64; **(b)** \$186,510.40; **(c)** about 22.5 yr;
(d) about \$133,366; **(e)** about \$53,144

37. Start with $P\left(1 + \dfrac{r}{n}\right)^{nt} = \dfrac{p\left[\left(1 + \dfrac{r}{n}\right)^{nt} - 1\right]}{\dfrac{r}{n}}$. Multiply both

sides by $\dfrac{r}{n}$, distribute p to clear the parentheses, and rearrange
terms so that the two terms containing $\left(1 + \dfrac{r}{n}\right)^{nt}$ are on one
side of the equation. Isolate the expression $\left(1 + \dfrac{r}{n}\right)^{nt}$, and take
the natural logarithm of both sides of the equation.

39.

Balance	Payment	Portion to interest	Portion to principal	New balance
\$18,150.00	\$355.13	\$98.31	\$256.82	\$17,893.18
\$17,893.18	\$355.13	\$96.92	\$258.21	\$17,634.97
\$17,634.97	\$355.13	\$95.52	\$259.61	\$17,375.37
\$17,375.37	\$355.13	\$94.12	\$261.01	\$17,114.35
\$17,114.35	\$355.13	\$92.70	\$262.43	\$16,851.93
\$16,851.93	\$355.13	\$91.28	\$263.85	\$16,588.08
\$16,588.08	\$355.13	\$89.85	\$265.28	\$16,322.80
\$16,322.80	\$355.13	\$88.42	\$266.71	\$16,056.08
\$16,056.08	\$355.13	\$86.97	\$268.16	\$15,787.92
\$15,787.92	\$355.13	\$85.52	\$269.61	\$15,518.31
\$15,518.31	\$355.13	\$84.06	\$271.07	\$15,247.24
\$15,247.24	\$355.13	\$82.59	\$272.54	\$14,974.70

41.

Balance	Payment	Portion to interest	Portion to principal	New balance
\$146,250.00	\$803.07	\$633.75	\$169.32	\$146,080.68
\$146,080.68	\$803.07	\$633.02	\$170.05	\$145,910.63
\$145,910.63	\$803.07	\$632.28	\$170.79	\$145,739.84
\$145,739.84	\$803.07	\$631.54	\$171.53	\$145,568.30
\$145,568.30	\$803.07	\$630.80	\$172.27	\$145,396.03
\$145,396.03	\$803.07	\$630.05	\$173.02	\$145,223.01
\$145,223.01	\$803.07	\$629.30	\$173.77	\$145,049.24
\$145,049.24	\$803.07	\$628.55	\$174.52	\$144,874.72
\$144,874.72	\$803.07	\$627.79	\$175.28	\$144,699.44
\$144,699.44	\$803.07	\$627.03	\$176.04	\$144,523.40
\$144,523.40	\$803.07	\$626.27	\$176.80	\$144,346.60
\$144,346.60	\$803.07	\$625.50	\$177.57	\$144,169.03

43.

Balance	Payment	Portion to interest	Portion to principal	New balance
$500.00	$10.59	$9.48	$1.11	$498.89
$498.89	$10.59	$9.46	$1.13	$497.76
$497.76	$10.59	$9.44	$1.15	$496.60
$496.60	$10.59	$9.41	$1.18	$495.43
$495.43	$10.59	$9.39	$1.20	$494.23
$494.23	$10.59	$9.37	$1.22	$493.01
$493.01	$10.59	$9.35	$1.24	$491.77
$491.77	$10.59	$9.32	$1.27	$490.50
$490.50	$10.59	$9.30	$1.29	$489.21
$489.21	$10.59	$9.27	$1.32	$487.89
$487.89	$10.59	$9.25	$1.34	$486.55
$486.55	$10.59	$9.22	$1.37	$485.19

Chapter 3 Review Exercises, p. 366

1. (b)   **2.** (e)   **3.** (f)   **4.** (c)   **5.** (a)   **6.** (d)   **7.** False
8. True   **9.** True   **10.** False   **11.** True   **12.** False
13. True   **14.** False   **15.** True   **16.** (a) -3; (b) 81;
(c) 1024 **17.** $\dfrac{1}{x}$ **18.** e^x **19.** $\dfrac{4x^3}{x^4+5}$ **20.** $\dfrac{e^{2\sqrt{x}}}{\sqrt{x}}$
21. $\dfrac{1}{2x}$ **22.** $3x^4e^{3x}+4x^3e^{3x}$ **23.** $\dfrac{1-3\ln x}{x^4}$
24. $2xe^{x^2}(\ln 4x)+\dfrac{e^{x^2}}{x}$ **25.** $4e^{4x}-\dfrac{1}{x}$ **26.** $\dfrac{1-x}{e^x}$
27. $(\ln 9)9^x$ **28.** $\dfrac{1}{(\ln 2)x}$
29. $3^x(\ln 3)(\log_4(2x+1))+\dfrac{(3^x)2}{(2x+1)(\ln 4)}$

30.
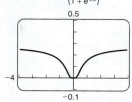

31.

32. 6.93 **33.** -3.2698 **34.** 8.7601 **35.** 3.2698
36. 2.54995 **37.** -3.6602 **38.** $Q(t)=25e^{7t}$ **39.** 4.3%
40. 18.24 yr   **41.** (a) $C(t)=15.81e^{0.023t}$; (b) $30.27, $34.19
42. (a) $C(t)=2.69e^{0.0282t}$; (b) $6.26, $8.31
43. (a) $N(t)=60e^{0.12t}$; (b) 97 franchises; (c) 13.4 yr after 2014
44. (a) $N(t)=24e^{0.07t}$; (b) about 39; (c) in about 9.9 yr
45. 5.3 yr   **46.** 18.2%   **47.** (a) $A(t)=800e^{-0.07t}$;
(b) 197 g; **(c)** 9.9 days   **48.** (a) 0.50, 0.75, 0.97, 0.999, 0.9999;
(b) $p'(t)=0.7e^{-0.7t}$; **(c)** ✏️ **(d)**

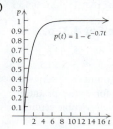

49. $186,373.98   **50.** (a) $5813.33; (b) $871.52/yr
51. (a) $190.77; (b) $1696.20   **52.** $508,192.17
53. (a) $23.79; (b) $20.75 in interest; $3.04 to principal
54. $\dfrac{-8}{(e^{2x}-e^{-2x})^2}$ **55.** $-\dfrac{1}{1024e}\approx 0$
56.

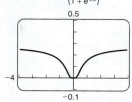
$$f(x)=\dfrac{e^{1/x}}{(1+e^{1/x})^2}$$

57. 0   **58.** (a) $y=169.3145859(1.151556523^x)$,
$y=169.3145859e^{0.1411145255x}$, 14.1%; **(b)** $395 billion,
$694 billion; **(c)** 12.6 yr after 2010; **(d)** 4.9 yr

Chapter 3 Test, p. 368

1. [3.1] $6e^{3x}$ **2.** [3.2] $\dfrac{4(\ln x)^3}{x}$ **3.** [3.1] $-2xe^{-x^2}$
4. [3.2] $\dfrac{1}{x}$ **5.** [3.1] e^x-15x^2 **6.** [3.1, 3.2] $\dfrac{3e^x}{x}+3e^x\cdot\ln x$
7. [3.5] $(\ln 7)7^x+(\ln 3)3^x$ **8.** [3.5] $\dfrac{1}{(\ln 14)x}$ **9.** [3.2] (a) 5;
(b) 2   **10.** [3.2] 1.0674   **11.** [3.2] 0.5554   **12.** [3.2] 0.4057
13. [3.3] $M(t)=2e^{6t}$ **14.** [3.3] 23.1%/hr   **15.** [3.3] 10.0 yr
16. [3.3] (a) $C(t)=3.22e^{0.012t}$; (b) $3.63, $3.81
17. [3.4] (a) $A(t)=3e^{-0.1t}$; (b) 1.1 cc; (c) 6.9 hr
18. [3.4] About 16.47 centuries, or 1647 yr
19. [3.4] 4.0773%/sec   **20.** [3.3] (a) 4%; (b) 5.2%, 14.5%,
40.7%, 73.5%, 91.8%, 99.5%, 99.9%;
(c) $P'(t)=\dfrac{672e^{-0.28t}}{(1+24e^{-0.28t})^2}$; **(d)** ✏️;
(e)

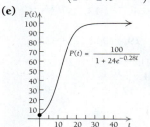
$$P(t)=\dfrac{100}{1+24e^{-0.28t}}$$

21. [3.5] (a) $1057.01; (b) $1093.81/yr   **22.** [3.6] (a) $1667.22;
(b) $600,199.20; **(c)** 240,199.20   **23.** [3.6] $20,815.15
24. [3.2] $(\ln x)^2$ **25.** [3.1] Maximum is $\dfrac{256}{e^4}\approx 4.689$;
minimum is 0.   **26.** [3.1]

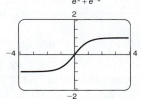
$$f(x)=\dfrac{e^x-e^{-x}}{e^x+e^{-x}}$$

27. [3.1] 0   **28.** [3.3] (a) $y=726522.8277(1.076351493)^x$,
$y=726522.8277e^{0.073577t}$; **(b)** $5.3 million, $6.6 million;
(c) 98 yr; **(d)** 9.4 yr; **(e)** ✏️

Extended Technology Application, p. 370

Graphs left to the student.
1. Linear: $G(t) = -14.745t + 98.91$; quadratic:
$G(t) = 3.694t^2 - 47.994t + 154.324$; cubic:
$G(t) = -0.544t^3 + 11.039t^2 - 76.012t + 181.254$;
exponential: $G(t) = 240.282(0.527)^t$. The dramatic decrease in revenue and its "flattening" suggests that an exponential model is most appropriate. Note that the linear model will eventually give negative revenue values, as will the quadratic and cubic models. **2.** $0.75, $0.40, $0.21, $0.11, $0.06 (millions)
3. $261.21, $261.61, $261.82, $261.93, $261.99 (millions)
4. $R(t) = 260.36/(1 + 3.187e^{-0.982t})$; about $260 to $262 million

5. $R'(t) = \dfrac{814.832e^{-0.982t}}{(1 + 3.187e^{-0.982t})^2}$; $\lim\limits_{t\to\infty} R'(t) = 0$

6. The release date may have been delayed to coincide with the holiday shopping season. The date is close to Halloween, a natural "tie-in" with the movie's theme.
7. Linear: $G(t) = 0.053t + 1.693$;
quadratic: $G(t) = -0.24t^2 + 2.216t - 1.912$;
cubic: $G(t) = 0.007t^3 - 0.332t^2 + 2.565t - 2.248$;
exponential: $G(t) = 1.022(1.088)^t$. None are satisfactory. The limited early release resulted in small revenue figures for weeks 1–3.
8. Linear: $G(t) = -0.827t + 7.314$;
quadratic: $G(t) = 0.009t^2 - 0.938t + 7.63$;
cubic: $G(t) = -0.009t^3 - 0.174t^2 - 1.897t + 9.423$;
exponential: $G(t) = 25.302(0.655)^t$. The exponential model fits the data best.
9. About $17 million

CHAPTER 4

Exercise Set 4.1, p. 378

1. $\dfrac{x^7}{7} + C$ **3.** $2x + C$ **5.** $\frac{4}{5}x^{5/4} + C$ **7.** $\frac{1}{3}x^3 + \frac{1}{2}x^2 - x + C$

9. $\frac{2}{3}t^3 + \frac{5}{2}t^2 - 3t + C$ **11.** $-\dfrac{x^{-2}}{2} + C$ **13.** $\frac{3}{4}x^{4/3} + C$

15. $\frac{2}{7}x^{7/2} + C$ **17.** $-\dfrac{x^{-3}}{3} + C$ **19.** $10 \ln |x| + C$

21. $3 \ln |x| - 5x^{-1} + C$ **23.** $-21x^{1/3} + C$ **25.** $e^{2x} + C$
27. $\frac{1}{3}e^{3x} + C$ **29.** $\frac{1}{7}e^{7x} + C$ **31.** $\frac{5}{3}e^{3x} + C$
33. $\frac{3}{4}e^{8x} + C$ **35.** $-\frac{2}{27}e^{-9x} + C$ **37.** $\frac{5}{3}x^3 - \frac{2}{7}e^{7x} + C$
39. $\dfrac{x^3}{3} - x^{3/2} - 3x^{-1/3} + C$ **41.** $3x^3 + 6x^2 + 4x + C$

43. $3 \ln x - \frac{5}{2}e^{2x} + \frac{2}{9}x^{9/2} + C$ **45.** $14x^{1/2} - \frac{2}{15}e^{5x} - 8 \ln x + C$
47. $f(x) = \frac{1}{2}x^2 - 3x + 13$ **49.** $f(x) = \frac{1}{3}x^3 - 4x + 7$
51. $f(x) = \frac{5}{3}x^3 + \frac{3}{2}x^2 - 7x + 9$ **53.** $f(x) = x^3 - \frac{5}{2}x^2 + x + 4$
55. $f(x) = \frac{5}{2}e^{2x} - 2$ **57.** $f(x) = 8x^{1/2} - 13$
59. $D(t) = 16.714t^2 + 71.143t + 2555.229$

61. $C(x) = \dfrac{x^4}{4} - x^2 + 7000$ **63. (a)** $R(x) = \dfrac{x^3}{3} - 3x$; **(b)** ✎

65. $D(x) = \dfrac{4000}{x} + 3$ **67. (a)** $E(t) = 32 + 30t - 5t^2$;
(b) $E(3) = 77\%, E(5) = 57\%$ **69. (a)** $R(t) = 59e^{-0.796t} + 66$;
(b) $R(4) = 68.4$ beats/min; **(c)** $R'(4) = -1.95$ beats/min/min;
(d) ✎ **71. (a)** $h(t) = -16t^2 + 75t + 30$; **(b)** $h(2) = 116$ ft,

$h'(2) = 11$ ft/sec; **(c)** $t = \dfrac{75}{32} \approx 2.344$ sec;

(d) $h(2.344) \approx 117.89$ ft; **(e)** 5.06 sec;
(f) $h'(5.06) = -86.92$ ft/sec **73. (a)** Alphaville:
$P(t) = 45t + 5000$, Betaburgh: $Q(t) = 3500e^{0.03t}$;
(b) Alphaville: $P(10) = 5450$, Betaburgh: $Q(10) = 4725$;
(c)

Populations are the same (~ 5700) in 2016 ($t \approx 16.5$).

75. $f(t) = \frac{2}{3}t^{3/2} + 2t^{1/2} - \frac{28}{3}$ **77.** $\frac{25}{7}t^7 + \frac{20}{3}t^6 + \frac{16}{5}t^5 + C$

79. $\frac{2}{5}t^{5/2} + 4t^{3/2} + 18t^{1/2} + C$ **81.** $\dfrac{t^4}{4} + t^3 + \frac{3}{2}t^2 + t + C$

83. $3x^4 - \frac{8}{3}x^3 - \frac{17}{2}x^2 - 5x + C$ **85.** $\dfrac{x^2}{2} - x + C$ **87.** ✎

Exercise Set 4.2, p. 387

1. $45,000 **3.** $1460 **5.** $2260 **7.** $8400 **9.** $2800
11. $2^1 + 2^2 + 2^3 + 2^4$, or 30 **13.** $0^2 + 1^2 + 2^2 + 3^2 + 4^2$
$+ 5^2 + 6^2 + 7^2 + 8^2 + 9^2 + 10^2 = 385$
15. $f(x_1) + f(x_2) + f(x_3) + f(x_4) + f(x_5)$ **17. (a)** 1.4914;
(b) 1.1418 **19.** 3,166,250¢, or $31,662.50 **21.** 247.68
23. 124 **25.** 23,302.4¢, or $233.02 **27.** $471.96 **29.** 4
31. 12 **33.** $\frac{9}{2}$ **35.** 25

37. $\displaystyle\sum_{i=1}^{4} kf(x_i) = kf(x_1) + kf(x_2) + kf(x_3) + kf(x_4)$

$$= k[f(x_1) + f(x_2) + f(x_3) + f(x_4)]$$

$$= k\sum_{i=1}^{4} f(x_i);$$

$$\sum_{i=1}^{n} kf(x_i) = kf(x_1) + kf(x_2) + kf(x_3) + \cdots + kf(x_n)$$

$$= k[f(x_1) + f(x_2) + f(x_3) + \cdots + f(x_n)]$$

$$= k\sum_{i=1}^{n} f(x_i)$$

39. (a) $\frac{1}{4}$; **(b)** 2; **(c)** 4; **(d)** $2\sqrt{3}$ **41.** 47.5 **43.** 5.641

45. ✎ **47.** 37.96 (Exact area is $\frac{1}{2} \cdot 25\pi$.)

Technology Connection, p. 394

1. (a) Left to the student; **(b)** 400; **(c)** the area is the square of x; **(d)** $A(x) = x^2$; **(e)** $A(x)$ is the antiderivative of $f(x)$.
2. (a) Left to the student; **(b)** 60; **(c)** the area is 3 times x;
(d) $A(x) = 3x$; **(e)** $A(x)$ is the antiderivative of $f(x)$.
3. (a) Left to the student; **(b)** $20^3 = 8000$; **(c)** the area is the cube of x; **(d)** $A(x) = x^3$; **(e)** $A(x)$ is the antiderivative of $f(x)$.

Technology Connection, p. 398

1. 0 **2.** 13.75 **3.** 0.535 **4.** 27.972 **5.** -260

Exercise Set 4.3, p. 400

1. 8 **3.** 8 **5.** $41\frac{2}{3}$ **7.** $\frac{1}{4}$ **9.** $10\frac{2}{3}$ **11.** $e^3 - 1 \approx 19.086$
13. $-3 \ln 6 \approx -5.375$ **15.** Total cost, in dollars, for t days
17. Total number of kilowatts used in t hours **19.** Total revenue, in dollars, for x units produced **21.** Total amount of the drug, in milligrams, in v cubic centimeters of blood **23.** Total number of

words memorized in t minutes   **25.** 4   **27.** $9\frac{5}{6}$ **29.** 12
31. $e^5 - e^{-1}$, or approximately 148.045   **33.** 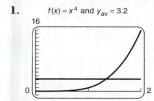   **35.** 0;
the area above the x-axis is the same as the area below the x-axis.
37. 0; the area above the x-axis is the same as the area below it.
39–41. Left to the student   **43.** 40   **45.** $\frac{5}{3}$ **47.** $\frac{637}{6}$
49. $e^2 - e^{-5}$, or approximately 7.382   **51.** $\dfrac{b^3 - a^3}{6}$

53. $\dfrac{e^{2b} - e^{2a}}{2}$ **55.** $\dfrac{e^2 + 1}{2}$, or approximately 4.195   **57.** $\frac{8}{3}$

59. \$1330.63   **61. (a)** \$5763.42; **(b)** \$1603.42
63. (a) \$1474.13; **(b)** \$1456.95   **65.** \$363.86 billion
67. 18.69 hr; 20.12 hr   **69.** 7 words   **71.** About 5 words
73. $s(t) = t^3 + 4$ **75.** $v(t) = 2t^2 + 20$

77. $s(t) = -\dfrac{t^3}{3} + 3t^2 + 6t + 10$ **79. (a)** 104.17 m;

(b) 229.17 m   **81.** $\frac{1}{8}$ mi   **83. (a)** 16.67 km/hr;
(b) 0.1875 km   **85.** $s(t) = -16t^2 + v_0 t + s_0$ **87.** 1320 ft,
or $\frac{1}{4}$ mi   **89.** 148 mi   **91. (a)** 114,688 lb; **(b)** after 11 months
93. 3.5   **95.** $359\frac{7}{15}$ **97.** 30   **99.** 8   **101.** ✏
103. 4068.789   **105.** 885.333   **107.** ✏

Technology Connection, p. 408

1. $\frac{4}{3}$ **2.** Left to the student

Technology Connection, p. 411

1.

$f(x) = x^4$ and $y_{av} = 3.2$ Over $[0, 2]$, the areas under
$f(x) = x^4$ and $y_{av} = 3.2$ are equal.

Exercise Set 4.4, p. 412

1. 22   **3.** $18\frac{1}{6}$ **5.** $89\frac{11}{12}$ **7.** 5   **9.** $\frac{7}{2}$ **11.** $\frac{3}{2}$ **13.** 62.5
15. $\frac{1}{4}$ **17.** $4\frac{1}{2}$ **19.** $20\frac{5}{6}$ **21.** $4\frac{1}{2}$ **23.** $\frac{3}{10}$ **25.** $41\frac{1}{3}$
27. $10\frac{2}{3}$ **29.** 3   **31.** $20\frac{5}{6}$ **33.** $85\frac{1}{3}$ **35.** $\frac{8}{3}$
37. $-e^{-1} + 1$, or approximately 0.632   **39.** $\frac{16}{3}$ **41.** $2a + 5$

43. $\dfrac{2^{n+1} - 1}{n + 1}$ **45. (a)** \$2,201,556.58; **(b)** \$220,155.66

47. \$26,534.37   **49.** \$32,781.35   **51. (a)** 2 more words;
(b) 0.7 word per minute; **(c)** 0.9 word per minute   **53. (a)** 90
words per minute; **(b)** 96 words per minute, at $t = 1$ min;
(c) 70 words per minute   **55. (a)** 42.03 μg/mL;
(b) 22.44 μg/mL   **57. (a)** 31.7°C; **(b)** −10°C; **(c)** 46.25°C
59. $40\frac{8}{15}$ **61.** 16   **63.** 6   **65.** 4   **67.** 4   **69.** 5.886
71. 0.237

Technology Connection, p. 420

1. 1.177

Exercise Set 4.5, p. 421

1. $\frac{1}{6}(8 + x^3)^6 + C$ **3.** $\frac{1}{16}(x^2 - 6)^8 + C$ **5.** $\frac{1}{24}(3t^4 + 2)^2 + C$
7. $\ln |2x + 1| + C$ **9.** $\frac{1}{4}(\ln x)^4 + C$ **11.** $\frac{1}{3}e^{3x} + C$
13. $3e^{x/3} + C$ **15.** $\frac{1}{5}e^{x^5} + C$ **17.** $-\frac{1}{2}e^{-t^2} + C$
19. $\frac{1}{2}\ln(5 + 2x) + C$ **21.** $\frac{1}{3}\ln(12 + 3x) + C$
23. $-\ln(1 - x) + C$ **25.** $\frac{1}{12}(t^2 - 1)^6 + C$

27. $\frac{1}{8}(x^4 + x^3 + x^2)^8 + C$ **29.** $\ln(4 + e^x) + C$
31. $(\ln |x|)^2 + C$ **33.** $\ln(\ln x) + C$ **35.** $\dfrac{1}{3a}(ax^2 + b)^{3/2} + C$

37. $\dfrac{P_0}{k}e^{kt} + C$ **39.** $\dfrac{1}{24(2 - x^4)^6} + C$ **41.** $\frac{5}{6}(1 + 6x^2)^{6/5} + C$

43. $e - 1$ **45.** $\frac{21}{4}$ **47.** $\ln 5$ **49.** $\ln 19$ **51.** $1 - e^{-b}$
53. $1 - e^{-mb}$ **55.** $\frac{208}{3}$ **57.** $\frac{1640}{6561}$ **59.** $\frac{315}{8}$
61–77. Left to the student   **79.** $x + 5\ln |x - 5| + C$
81. $-\frac{1}{4}x - \frac{1}{16}\ln |1 - 4x| + C$ **83.** $\frac{2}{3}x + \frac{13}{9}\ln |3x - 2| + C$
85. $\frac{1}{11}(x + 2)^{11} - \frac{3}{5}(x + 2)^{10} + \frac{4}{3}(x + 2)^9 - (x + 2)^8 + C$
87. $D(x) = 2000\sqrt{25 - x^2} + 5000$ **89. (a)** About 40 per
1000 people; **(b)** about 75 per 1000 people   **91.** $5\frac{1}{3}$

93. $\dfrac{1}{a}\ln |ax + b| + C$ **95.** $2e^{\sqrt{t}} + C$ **97.** $-e^{1/t} + C$

99. $-\frac{1}{3}(\ln x)^{-3} + C$ **101.** $\frac{2}{9}(x^3 + 1)^{3/2} + C$
103. $\frac{1}{3}(\ln x)^3 + \frac{3}{2}(\ln x)^2 + 4\ln x + C$ **105.** $\frac{1}{8}[\ln(t^4 + 8)]^2 + C$
107. $x + \dfrac{9}{x + 3} + C$ **109.** $t - 4 - \ln |t - 4| + C$, or
$t - \ln |t - 4| + K$, where $K = -4 + C$
111. $-\ln(1 + e)^{-x} + C$ **113.** $\dfrac{(\ln x)^{n+1}}{n + 1} + C$

115. $\dfrac{9}{14(n + 1)}(7x^2 + 9)^{n+1} + C$ **117.** ✏

Technology Connection, p. 427

1. 1.941

Exercise Set 4.6, p. 429

1. $xe^{4x} - \frac{1}{4}e^{4x} + C$ **3.** $\dfrac{x^6}{2} + C$ **5.** $\frac{1}{5}xe^{5x} - \frac{1}{25}e^{5x} + C$

7. $-\frac{1}{2}xe^{-2x} - \frac{1}{4}e^{-2x} + C$ **9.** $\dfrac{x^3 \ln x}{3} - \dfrac{x^3}{9} + C$

11. $\frac{1}{4}x^2 \ln x - \frac{1}{8}x^2 + C$ **13.** $(x + 5)\ln(x + 5) - x + C$
15. $\left(\dfrac{x^2}{2} + 2x\right)\ln x - \dfrac{x^2}{4} - 2x + C$
17. $\left(\dfrac{x^2}{2} - x\right)\ln x - \dfrac{x^2}{4} + x + C$
19. $\frac{2}{3}x(x + 2)^{3/2} - \frac{4}{15}(x + 2)^{5/2} + C$
21. $\dfrac{x^4}{4}\ln(2x) - \dfrac{x^4}{16} + C$, or $\dfrac{x^4 \ln 2}{4} - \dfrac{x^4 \ln x}{4} + \dfrac{x^4}{16} + C$
23. $x^2 e^x - 2xe^x + 2e^x + C$ **25.** $\frac{1}{2}x^2 e^{2x} - \frac{1}{2}xe^{2x} + \frac{1}{4}e^{2x} + C$
27. $-\frac{1}{2}x^3 e^{-2x} - \frac{3}{4}x^2 e^{-2x} - \frac{3}{4}xe^{-2x} - \frac{3}{8}e^{-2x} + C$
29. $\frac{1}{3}(x^4 + 4)e^{3x} - \frac{4}{9}x^3 e^{3x} + \frac{4}{9}x^2 e^{3x} - \frac{8}{27}xe^{3x} + \frac{8}{81}e^{3x} + C$
31. $\frac{8}{3}\ln 2 - \frac{7}{9}$ **33.** $14\ln 14 - 10\ln 10 - 4$ **35.** 1
37. $\frac{1192}{15}$ **39.** $C(x) = \frac{8}{3}x(x + 3)^{3/2} - \frac{16}{15}(x + 3)^{5/2}$

41. (a) $-e^{-kt}\left(\dfrac{T}{k} + \dfrac{1}{k^2}\right) + \dfrac{1}{k^2}$; **(b)** about 14.850 mg

43. $\frac{2}{125}(5x + 1)^{5/2} - \frac{2}{75}(5x + 1)^{3/2} + C$; they are the same.
45. $2\sqrt{x}e^{\sqrt{x}} - 2e^{\sqrt{x}} + C$ **47.** $\frac{2}{3}x^{3/2}\ln x - \frac{4}{9}x^{3/2} + C$
49. $2\sqrt{x}(\ln |x|) - 4\sqrt{x} + C$
51. $\frac{2}{7}(27x^3 + 83x - 2)(3x + 8)^{7/6}$
$- \frac{4}{91}(81x^2 + 83)(3x + 8)^{13/6}$
$+ \frac{1296}{1729}x(3x + 8)^{19/6} - \frac{2592}{43,225}(3x + 8)^{25/6} + C$
53. $\dfrac{x^{n+1}}{n + 1}(\ln |x|)^2 - \dfrac{2x^{n+1}}{(n + 1)^2}\ln |x| + \dfrac{2x^{n+1}}{(n + 1)^3} + C$

55. Let $u = x^n$ and $dv = e^x\, dx$. Then $du = nx^{n-1}\, dx$ and $v = e^x$.
Next, use integration by parts. **57.** ✎ **59.** $\dfrac{3^x e^x}{1 + \ln 3} + C$

61. $\dfrac{10^x e^{3x}}{3 + \ln 10} + C$ **63.** About 355,986

Exercise Set 4.7, p. 435

1. $-\frac{1}{9}e^{-3x}(3x + 1) + C$ **3.** $\dfrac{6^x}{\ln 6} + C$ **5.** $\frac{1}{10}\ln\left|\dfrac{5 + x}{5 - x}\right| + C$

7. $3 - x - 3\ln|3 - x| + C$ **9.** $\dfrac{1}{8(8 - x)} + \frac{1}{64}\ln\left|\dfrac{x}{8 - x}\right| + C$

11. $(\ln 3)\,x + x\ln x - x + C$ **13.** $\dfrac{x^5}{5}(\ln x) - \dfrac{x^5}{25} + C$

15. $\dfrac{x^4}{4}(\ln x) - \dfrac{x^4}{16} + C$ **17.** $\ln|x + \sqrt{x^2 + 7}| + C$

19. $\dfrac{2}{5 - 7x} + \frac{2}{5}\ln\left|\dfrac{x}{5 - 7x}\right| + C$ **21.** $-\frac{5}{4}\ln\left|\dfrac{x - 1/2}{x + 1/2}\right| + C$

23. $m\sqrt{m^2 + 4} + 4\ln|m + \sqrt{m^2 + 4}| + C$

25. $\dfrac{5}{2x^2}(\ln x) + \dfrac{5}{4x^2} + C$ **27.** $x^3 e^x - 3x^2 e^x + 6xe^x - 6e^x + C$

29. $\frac{1}{15}(3x - 1)(1 + 2x)^{3/2} + C$

31. $S(x) = 100\left[\dfrac{20}{20 - x} + \ln(20 - x)\right]$

33. $-4\ln\left|\dfrac{x}{3x - 2}\right| + C$ **35.** $\dfrac{-1}{2(x - 2)} + \frac{1}{4}\ln\left|\dfrac{x}{x - 2}\right| + C$

37. $\dfrac{-3}{e^{-x} - 3} + \ln|e^{-x} - 3| + C$ **39.** ✎

Chapter 4 Review Exercises, p. 443

1. True **2.** False **3.** True **4.** False **5.** False **6.** (e)
7. (d) **8.** (a) **9.** (f) **10.** (b) **11.** (c) **12.** $102,500
13. $4x^5 + C$ **14.** $\frac{3}{4}e^{4x} + 2x + C$ **15.** $6t^3 + 3t^2 + \frac{1}{2}\ln t + C$
16. 15 **17.** 39 **18.** Total number of words keyboarded in t
minutes **19.** Total sales in t days **20.** $\dfrac{b^6 - a^6}{6}$ **21.** $-\frac{57}{20}$

22. $\frac{1}{2}e^2$ **23.** $2\ln 4$, or $4\ln 2$ **24.** $\frac{22}{3}$ **25.** $\frac{9}{2}$ **26.** Zero
27. Negative **28.** Positive **29.** $\frac{1}{6}$ **30.** $\frac{1}{4}e^{x^4} + C$

31. $\ln(4t^6 + 3) + C$ **32.** $\frac{1}{4}(\ln 4x)^2 + C$
33. $\frac{1}{2}\ln(e^{2x} + 2) + C$ **34.** $\frac{3}{4}xe^{4x} - \frac{3}{16}e^{4x} + C$
35. $-\dfrac{2x}{3} + x\ln x^{2/3} + C$ **36.** $\frac{5}{3}x^3\ln|x| - \frac{5}{9}x^3 + C$
37. $e^{3x}\left(\frac{1}{3}x^4 - \frac{4}{9}x^3 + \frac{4}{9}x^2 - \frac{8}{27}x + \frac{8}{81}\right) + C$
38. $\dfrac{1}{14}\ln\left|\dfrac{7 + x}{7 - x}\right| + C$ **39.** $\frac{2}{735}(21x - 4)(2 + 7x)^{3/2} + C$
40. $\dfrac{1}{49} + \dfrac{x}{7} - \dfrac{1}{49}\ln|7x + 1| + C$ **41.** $\ln|x + \sqrt{x^2 - 36}| + C$
42. $-\dfrac{4}{\sqrt{10}}\ln\left|\dfrac{\sqrt{10} + \sqrt{10 - x^2}}{x}\right| + C$ **43.** About $96,000
44. $\frac{1}{2}(1 - 3e^{-2})$, or approximately 0.297 **45.** 80 mi
46. About $162,753.79 **47.** $\ln|4t^3 + 7| + C$
48. $\frac{2}{75}(5x - 8)\sqrt{4 + 5x} + C$ **49.** $e^{x^5} + C$ **50.** $\ln|x + 9| + C$
51. $\frac{1}{96}(t^8 + 3)^{12} + C$ **52.** $x\ln|7x| - x + C$
53. $\dfrac{x^2}{2}\ln|8x| - \dfrac{x^2}{4} + C$ **54.** $\frac{1}{10}\left[\ln|t^5 + 3|\right]^2 + C$
55. $-\frac{1}{2}\ln|1 + 2e^{-x}| + C$ **56.** $(\ln\sqrt{x})^2 + C$, or $\frac{1}{4}(\ln x)^2 + C$

57. $x^{92}\left(\dfrac{\ln|x|}{92} - \dfrac{1}{8464}\right) + C$
58. $(x - 3)\ln|x - 3| - (x - 4)\ln|x - 4| + C$
59. $\dfrac{1}{3(\ln|x|)^3} + C$ **60.** $\frac{3}{7}(x + 3)^{7/3} - \frac{9}{4}(x + 3)^{4/3} + C$
61. $\frac{1}{16}(2x + 1)^2 - \frac{1}{4}(2x + 1) + \frac{1}{8}\ln|2x + 1| + C$ **62.** 1.343

Chapter 4 Test, p. 444

1. [4.2] 95 **2.** [4.1] $\dfrac{2\sqrt{3}}{3}x^{3/2} + C$ **3.** [4.1] $50x^6 + C$
4. [4.1] $\frac{1}{5}e^{5x} + \ln x + \frac{8}{11}x^{11/8} + C$ **5.** [4.3] $\frac{1}{6}$
6. [4.3] $4\ln 3$ **7.** [4.3] Total miles run in t hours **8.** [4.3] 21
9. [4.3] $\dfrac{1 - e^{-2}}{2}$ **10.** [4.3] 2 **11.** [4.4] $\frac{61}{6}$ **12.** [4.2] $\frac{7}{2}$
13. [4.3] Positive **14.** [4.5] $\frac{1}{3}\ln(x + 4) + C$
15. [4.5] $-2e^{-0.5x} + C$ **16.** [4.5] $\dfrac{(t^4 + 3)^8}{32} + C$
17. [4.6] $\frac{1}{5}xe^{5x} - \frac{1}{25}e^{5x} + C$
18. [4.6] $\dfrac{x^4}{4}\ln x^4 - \dfrac{x^4}{4} + C$, or $x^4\ln|x| - \dfrac{x^4}{4} + C$
19. [4.7] $\dfrac{1}{18}\ln\left|\dfrac{x - 9}{x + 9}\right| + C$ **20.** [4.7] $\dfrac{1}{7}\ln\left|\dfrac{x}{7 - x}\right| + C$
21. [4.4] 6 **22.** [4.4] $\frac{5}{14}$ **23.** [4.4] $118,750
24. [4.3] 94 words **25.** [4.3] 7.47 km
26. [4.5] $\frac{6}{7}\ln(5 + 7x) + C$ **27.** [4.6] $x^5 e^x - 5x^4 e^x +$
$20x^3 e^x - 60x^2 e^x + 120xe^x - 120e^x + C$ **28.** [4.5] $\frac{1}{6}e^{x^6} + C$
29. [4.6, 4.7] $\frac{2}{3}x^{3/2}(\ln x) - \frac{4}{9}x^{3/2} + C$
30. [4.7] $\frac{2}{9}(1 + e^{3x})^{3/2} + C$ **31.** [4.6, 4.7] $-10x^4 e^{-0.1x}$
$- 400x^3 e^{-0.1x} - 12,000x^2 e^{-0.1x} - 240,000xe^{-0.1x}$
$- 2,400,000e^{-0.1x} + C$ **32.** [4.6] $\dfrac{x^2}{2}\ln(13x) - \dfrac{x^2}{4} + C$
33. [4.6] $\frac{1}{15}(3x^2 - 8)(x^2 + 4)^{3/2} + C$
34. [4.5] $\dfrac{(\ln x)^4}{4} - \dfrac{4}{3}(\ln x)^3 + 5\ln x + C$
35. [4.6] $(x + 3)\ln|x + 3| - (x + 5)\ln|x + 5| + C$
36. [4.6, 4.7] $\frac{3}{10}(8x^3 + 10)(5x - 4)^{2/3}$
$- \frac{108}{125}x^2(5x - 4)^{5/3} + \frac{81}{625}x(5x - 4)^{8/3}$
$- \frac{243}{34,375}(5x - 4)^{11/3} + C$
37. [4.6] $\frac{2}{27}(3x - 2)^{3/2} + \frac{4}{9}(3x - 2)^{1/2} + C$
38. [4.6] $x + 8\ln|x| - \dfrac{16}{x} + C$
39. [4.5] $\dfrac{1}{\ln 5}e^{(\ln 5)x} + C$, or $\dfrac{5^x}{\ln 5} + C$ **40.** [4.4] 16

Extended Technology Application, p. 446

1. (a) 36%; **(b)** 33.3 **2. (a)** 16.7%; **(b)** 55.5
3. (a) $f(x) = x^{1.75}$, where $0 \le x \le 1$; **(b)** 0.272, 27.2;
(c) about 59% **4. (a)** $f(x) = x^{2.34}$, where $0 \le x \le 1$;
(b) 0.2, 20; **(c)** about 15.4%; **(d)** about 21.9% **5.** Left to the
student **6. (a)** $f(x) = x^{2.77}$; **(b)** about 19.1%
7. (a) $f(x) = x^{1.86}$; **(b)** about 32.9% **8. (a)** 34.6%;
(b) $y = 0.0001428182959(5362.9137820952)^x$; **(c)** 0.411;
(d) 0.822, 82.2; **(e)** 0.1878%; **(f)** 78.9%

CHAPTER 5

Technology Connection, p. 450

1.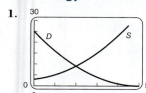
The point of intersection is $(2, 9)$; this is the equilibrium point.

Exercise Set 5.1, p. 454

1. (a) $(1, \$4)$; (b) $\$1.50$; (c) $\$1$ **3.** (a) $(1, \$4)$; (b) $\$2.33$;
(c) $\$1.67$ **5.** (a) $(4, \$16)$; (b) $\$85.33$; (c) $\$42.67$
7. (a) $(40, \$7600)$; (b) $\$24,000$; (c) $\$12,000$ **9.** (a) $(3, \$4)$;
(b) $\$4.50$; (c) $\$2.67$ **11.** (a) $(899, \$60)$; (b) $\$50,460$;
(c) $\$17,941.33$ **13.** (a) $(8, \$5)$; (b) $\$32$; (c) $\$3.40$
15. (a) $\$81$; (b) $\$30$; (c) equilibrium point: $(3, 140)$, consumer
surplus: $\$137.25$, producer surplus: $\$67.50$; (d) 🖊
17. (a) $(5, \$0.61)$; (b) $\$86.36$; (c) $\$2.45$; **19.** 🖊
21. (a) $(6, \$2)$; (b)
(c) $\$7.62$;
(d) $\$7.20$

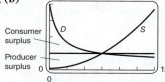

23. (a) Linear; (b) $y = -25x + 225$; (c) $\$450$; (d) $\$242$

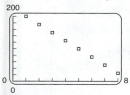

Exercise Set 5.2, p. 464

1. $\$119,721.74$ **3.** $\$235,955.31$ **5.** $\$83,527.02$
7. $\$223,130.16$ **9.** $\$2,004,166.02$ **11.** $\$12,255,409.28$
13. $\$3,207,798.40$ **15.** $\$12,824,916.68$ **17.** $\$585,877.18$
19. $\$386,076.11$ **21.** $\$100,918.91$ **23.** (a) $\$1,476,052.70$;
(b) $\$6,615,209.23$ **25.** $\$99,486.60$ **27.** A: $\$656,292.19$,
B: $\$647,939.34$; B is the better buy.   **29.** (a) Crunchers:
$\$2,565,959.80$, Radars: $\$2,623,269.10$; (b) the difference
between the accumulated present values of the two offers, or
$\$57,309.30$ **31.** (a) $\$62,144.41$; (b) $\$4796.74$
33. $\$326,606.91$ **35.** (a) $\$3,676,620$; (b) $\$1,652,010$
37. (a) $\$1,603,300$; (b) $\$665,019.42$; (c) 3%: $\$751,980.61$, 4%:
$\$688,338.79$, 5%: $\$632,120.56$; (d) 🖊   **39.** $\$7990.75$/yr
41. 1279.1 trillion cubic feet   **43.** In 42.7 yr, or in 2055
45. (a) Approximately 36.99 billion barrels;
(b) approximately 36.8 yr after 2013   **47.** 16.031 lb
49. $\$535,847$ **51.** $\$732,121$ **53.** 🖊

Exercise Set 5.3, p. 470

1. Convergent; $\frac{1}{5}$ **3.** Divergent   **5.** Convergent; 1
7. Convergent; $\frac{1}{2}$ **9.** Divergent   **11.** Convergent; 2
13. Divergent   **15.** Divergent   **17.** Divergent
19. Convergent; 1   **21.** Covergent; $\dfrac{1000}{\pi^{0.001}}$ **23.** Divergent

25. 2   **27.** 1   **29.** $\$6,250,000$ **31.** $\$4,500,000$
33. $\$72,000$ **35.** $\$135,135.14$ **37.** $\$900,000$
39. $33,333\frac{1}{3}$ lb   **41.** (a) 4.20963; (b) 0.702858 rem;
(c) 2.37551 rems   **43.** Divergent   **45.** Convergent; 2
47. Convergent; $\frac{1}{2}$ **49.** $\dfrac{1}{k^2}$ is the total dose of the drug.

51. 🖊   **53.** 4   **55.** 🖊   **57.** About 1.252
59.

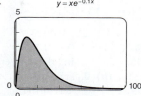

Exercise Set 5.4, p. 480

1. $\displaystyle\int_0^1 2x\,dx = \left[2 \cdot \frac{x^2}{2}\right]_0^1 = [x^2]_0^1 = 1^2 - 0^2 = 1$

3. $\displaystyle\int_0^{1/3} 3\,dx = [3x]_0^{1/3} = 3\left(\frac{1}{3} - 0\right) = 1$

5. $\displaystyle\int_1^3 \frac{3}{26}x^2\,dx = \left[\frac{3}{26} \cdot \frac{x^3}{3}\right]_1^3 = \left[\frac{x^3}{26}\right]_1^3 = \frac{3^3}{26} - \frac{1^3}{26} = \frac{26}{26} = 1$

7. $\displaystyle\int_1^e \frac{1}{x}\,dx = [\ln x]_1^e = \ln e - \ln 1 = 1 - 0 = 1$

9. $\displaystyle\int_{-2}^1 \frac{1}{3}x^2\,dx = \left[\frac{1}{3} \cdot \frac{x^3}{3}\right]_{-2}^1 = \frac{1^3}{9} - \frac{(-2)^3}{9} = \frac{1}{9} - \frac{-8}{9} = 1$

11. $\displaystyle\int_0^\infty 3e^{-3x}\,dx = \lim_{b\to\infty}\int_0^b 3e^{-3x}\,dx = \lim_{b\to\infty}[-e^{-3x}]_0^b$

13. $\frac{2}{15}$; $f(x) = \frac{2}{15}x$ **15.** $\frac{3}{2}$; $f(x) = \frac{3}{2}x^2$ **17.** $\frac{1}{6}$; $f(x) = \frac{1}{6}$
19. $\frac{1}{2}$; $f(x) = \dfrac{2-x}{2}$ **21.** $\dfrac{1}{\ln 2}$; $f(x) = \dfrac{1}{x\ln 2}$
23. $\dfrac{1}{e^3 - 1}$; $f(x) = \dfrac{e^x}{e^3 - 1}$ **25.** $\frac{8}{25}$, or 0.32
27. $\frac{11}{16}$, or 0.6875   **29.** 0.3297   **31.** 0.999955
33. 0.3935   **35.** (a) $\displaystyle\int_0^{10} 0.23e^{-0.23t}\,dt = 0.9$; (b) 0.0286
37. 0.950213   **39.** 0.049787   **41.** $\frac{1}{2}$
43. (a) $k = 0.357$, $f(t) = 0.357e^{-0.357t}$, $[0, \infty]$ (b) 0.2427
45. (a) $c = \frac{1}{2}$; (b) $f(x) = \frac{1}{8}x - \frac{1}{8}$, $[1, 5]$; 0.1875, or $\frac{3}{16}$
47. $c = 4/(3e^4 - e^2) \approx 0.02557$
49–59. Check using the answers to Exercises 1–11.

Technology Connection, p. 490

1. (a) 0.6295, or 62.95%; (b) 0.0228, or 2.28%
2. (a) 0.621, or 62.1%; (b) 0.353, or 35.3%

Technology Connection, p. 492

1. $x = 0$ **2.** $x = -0.674$ **3.** $x = 9.935$ **4.** $x = 2.477$
5. $x = 103.201$

Exercise Set 5.5, p. 493

1. $\mu = E(x) = \dfrac{11}{2}$, $E(x^2) = \dfrac{97}{3}$, $\sigma^2 = \dfrac{25}{12}$, $\sigma = \dfrac{5}{2\sqrt{3}}$

3. $\mu = E(x) = \dfrac{8}{3}$, $E(x^2) = 8$, $\sigma^2 = \dfrac{8}{9}$, $\sigma = \dfrac{2\sqrt{2}}{3}$

5. $\mu = E(x) = \dfrac{14}{9}, E(x^2) = \dfrac{5}{2}, \sigma^2 = \dfrac{13}{162}, \sigma = \dfrac{1}{9}\sqrt{\dfrac{13}{2}}$

7. $\mu = E(x) = 0, E(x^2) = \dfrac{3}{5}, \sigma^2 = \dfrac{3}{5}, \sigma = \sqrt{\dfrac{3}{5}}$

9. $\mu = E(x) = \dfrac{2.4}{\ln 4}, E(x^2) = \dfrac{4.8}{\ln 4}, \sigma^2 = \dfrac{4.8 \ln 4 - 5.76}{(\ln 4)^2}$,

$\sigma = \dfrac{\sqrt{4.8 \ln 4 - 5.76}}{\ln 4}$ **11.** 0.1406   **13.** 0.4147

15. 0.9756   **17.** 0.0150   **19.** 0.1004   **21.** 0.0013
23. (a) 0.9544; **(b)** 95.44%   **25.** 0.3413   **27.** 0.5762
29. 0.4974   **31.** 0.6554   **33–53.** Check using the answers
to Exercises 11–31.   **55. (a)** −0.52; **(b)** 0; **(c)** 1.645
57. (a) −15.04; **(b)** −14.44   **59.** 0.62%   **61.** 0.7910
63. (a) 1364; **(b)** 1586; **(c)** 1987   **65. (a)** 84; **(b)** 71

67. 90.8th   **69.** $\mu = E(x) = \dfrac{b + a}{2}, E(x^2) = \dfrac{b^2 + ab + a^2}{3}$,

$\sigma^2 = \dfrac{(b - a)^2}{12}, \sigma = \dfrac{b - a}{2\sqrt{3}}$ **70.** $\mu = E(x) = \dfrac{3a}{2}$,

$E(x^2) = 3a^2, \sigma^2 = \dfrac{3a^2}{4}, \sigma = \dfrac{a\sqrt{3}}{2}$ **71.** $\sqrt{2}$ **73.** $\dfrac{\ln 2}{k}$

75. 7.801 oz   **77.** ✏

Exercise Set 5.6, p. 500

1. $\dfrac{\pi}{3}$, or about 1.05   **3.** $\dfrac{104\pi}{3}$, or about 108.91

5. $\dfrac{\pi}{2}(e^{10} - e^{-4})$, or about 34,599.06   **7.** $\dfrac{3\pi}{4}$, or about 2.36

9. $4\pi \ln \dfrac{9}{4}$, or about 10.19   **11.** 50π, or about 157.08

13. $\dfrac{32\pi}{5}$, or about 20.11   **15.** 6π, or about 18.85

17. $\dfrac{32\pi}{3}$, or about 33.51   **19.** 36π, or about 113.1

21. 2048π, or about 6434   **23.** 8π, or about 25.133

25. $\dfrac{33\pi}{2}$, or about 51.8   **27.** $\dfrac{844\pi}{5}$, or about 530.301

29. $(2\pi/3)(10^{3/2} - 1)$, or about 64.1   **31. (a)** $\dfrac{648\pi}{5}$, or

about 407.15 **(b)** $\dfrac{81\pi}{2}$, or about 127.23 **(c)** ✏

33. 1,703,703.7π ft^3 **35.** $\dfrac{8775\pi}{4}$ ft^3, or about 6891.9 ft^3

37. $V = \pi \displaystyle\int_{-r}^{r} [\sqrt{r^2 - x^2}]^2 \, dx$

$= \pi \displaystyle\int_{-r}^{r} (r^2 - x^2) \, dx$

$= \pi \left[r^2 x - \dfrac{1}{3}x^3 \right]_{-r}^{r}$

$= \pi \left[\left(r^3 - \dfrac{1}{3}r^3 \right) - \left(-r^3 + \dfrac{1}{3}r^3 \right) \right]$

$= \pi \left(\dfrac{2}{3}r^3 + \dfrac{2}{3}r^3 \right) = \dfrac{4}{3}\pi r^3$

39. $2\pi e^3$, or about 126.20   **41.** $\dfrac{108\pi}{5}$, or about 67.858

43. (a) $\dfrac{7\pi}{3}, \dfrac{5\pi}{3}$; the volume around the x-axis is larger.

(b) $\dfrac{26\pi}{3}, \dfrac{28\pi}{3}$; the volume around the y-axis is larger.

(c) $a = \sqrt{3}$ **45.** π

Exercise Set 5.7, p. 509

1. $y = \dfrac{10}{3}x^3 + C; y = \dfrac{10}{3}x^3 + 1, y = \dfrac{10}{3}x^3 + 5, y = \dfrac{10}{3}x^3 - \dfrac{1}{2}$
(answers may vary)

3. $y = -2e^{-x} + \dfrac{x^2}{2} + C; \; y = -2e^{-x} + \dfrac{x^2}{2} + 1$,

$y = -2e^{-x} + \dfrac{x^2}{2} - 5, \; y = -2e^{-x} + \dfrac{x^2}{2} + 47$
(answers may vary)

5. $y = 4 \ln |x| + \dfrac{1}{x}; y = 4 \ln |x| + \dfrac{1}{x} + 1$,

$y = 4 \ln |x| + \dfrac{1}{x} - 2, y = 4 \ln |x| + \dfrac{1}{x} + 17$
(answers may vary)

7. $y' = 4 + \ln x$ and $y'' = \dfrac{1}{x}$, so $\left(\dfrac{1}{x}\right) - \dfrac{1}{x} = 0$

9. $y' = 3xe^x + 4e^x$ and $y'' = 3xe^x + 7e^x$, so $(3xe^x + 7e^x) - 2(3xe^x + 4e^x) + (3xe^x + e^x) = (3xe^x - 6xe^x + 3xe^x) + (7e^x - 8e^x + e^x) = xe^x(3 - 6 + 3) + e^x(7 - 8 + 1) = 0$

11. (a) $y' = -4e^{-4x}$, so $(-4e^{-4x}) + 4(e^{-4x}) = 0$

(b) $y' = -4Ce^{-4x}$, so $(-4Ce^{-4x}) + 4(Ce^{-4x}) = 0$

13. (a) $y' = 6e^{6x}, y'' = 36e^{6x}$, so $(36e^{6x}) - (6e^{6x}) - 30(e^{6x}) = e^{6x}(36 - 6 - 30) = 0$;

(b) $y' = -5e^{-5x}, y'' = 25e^{-5x}$, so $(25e^{-5x}) - (-5e^{-5x}) - 30(e^{-5x}) = e^{-5x}(25 + 5 - 30) = 0$;

(c) $y' = 6C_1e^{6x} - 5C_2e^{-5x}, y'' = 36C_1e^{6x} + 25C_2e^{-5x}$, so $(36C_1e^{6x} + 25C_2e^{-5x}) - (6C_1e^{6x} - 5C_2e^{-5x}) - 30(C_1e^{6x} + C_2e^{-5x}) = C_1e^{6x}(36 - 6 - 30) + C_2e^{-5x}(25 + 5 - 30) = 0$

15. (a) $M(t) = M_0 e^{0.05t}$; **(b)** $dM/dt = 0.05M_0e^{0.05t}$, so $0.05M_0e^{0.05t} = 0.05(M_0e^{0.05t})$ **17. (a)** $R(t) = R_0e^{0.35t}$;

(b) $dR/dt = 0.35R_0e^{0.35t}$, so $0.35R_0e^{0.35t} = 0.35(R_0e^{0.35t})$

19. (a) $G(t) = G_0e^{0.005t}$; **(b)** $dG/dt = 0.005G_0e^{0.005t}$, so $0.005G_0e^{0.005t} = 0.005(G_0e^{0.005t})$ **21. (a)** $R(t) = R_0e^t$;

(b) $dR/dt = R_0e^t$, so $R_0e^t = (R_0e^t)$

23. (a) $y = \dfrac{1}{3}x^3 + x^2 - 3x + 4$;

(b) $y' = \dfrac{1}{3}(3x^2) + 2x - 3 = x^2 + 2x - 3$

25. (a) $f(x) = \dfrac{3}{5}x^{5/3} - \dfrac{1}{2}x^2 - \dfrac{61}{10}$;

(b) $f'(x) = \dfrac{3}{5}(\dfrac{5}{3}x^{2/3}) - \dfrac{1}{2}(2x) = x^{2/3} - x$

27. (a) $B(t) = 500e^{0.03t}$; **(b)** $dB/dt = 15e^{0.03t}$, so $15e^{0.03t} = 0.03(500e^{0.03t})$ **29. (a)** $S(t) = 750e^{0.12t}$;

(b) $dS/dt = 90e^{0.12t}$, so $90e^{0.12t} = 0.12(750e^{0.12t})$

31. (a) $T(t) = 50e^{0.015t}$; **(b)** $dT/dt = 0.75e^{0.015t}$, so $0.75e^{0.015t} = 0.015(50e^{0.015t})$

33. (a) $M(t) = 6e^t$; **(b)** $dM/dt = 6e^t$, so $6e^t = (6e^t)$

35. $y = Ce^{x^4}$ **37.** $y = \sqrt[3]{4x^2 + C}$ **39.** $y = \pm\sqrt{2x^2 + C}$

41. $y = \pm\sqrt{12x + C}$ **43.** $y = 8e^{x^2/2} - 3$ **45.** $y = \sqrt[3]{15x - 3}$

47. (a) $dy/dx = y^2$, or $y' = y^2$; **(b)** $y = \dfrac{1}{C - x}$

49. (a) $dy/dx = 1/y^3$, or $y' = y^{-3}$; **(b)** $y = \pm\sqrt[4]{4x + C}$

51. (a) $dy/dx = xy, y(2) = 3$; **(b)** $y = 0.406e^{x^2/2}$

53. (a) $dA/dt = 0.0375A$; **(b)** $A(t) = 500e^{0.0375t}$;
(c) $A(5) = \$603.12; A'(5) = \$22.62/yr$;

(d) $\dfrac{22.62}{603.12} = 0.0375$, the continuous growth rate

55. (a) $I = Ce^{hkt}$; **(b)** $I = I_0e^{hkt}$ **57.** $V = -4.81e^{-kt} + 24.81$

59. $q = 2e^{(4/x)-1}$ **61.** $q = \dfrac{C}{x^2}$ **63.** (a) $dP/dt = 0.0175P$;
(b) $P(t) = 17{,}000e^{0.0175t}$; (c) $P(10) \approx 20{,}251$ people,
$P'(10) = 354.4$ people/yr; (d) $\dfrac{354.4}{20{,}251} = 0.0175$, the
continuous growth rate **65.** (a) $k = 0.296, P(t) = 24e^{0.296t}$;
(b) $P(41) = 4{,}475{,}165$ rabbits, $P'(41) = 1{,}324{,}649$ rabbits/yr
67. (a) $P = Ce^{kt}$; (b) $P = P_0e^{kt}$ **69.** (a) 4.18%/yr;
(b) $A(t) = 3200e^{0.0418t}$; (c) \$3200
71. $y = -\dfrac{4}{4x^5 + x^4 + C}$ **73.** 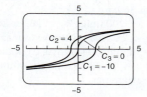 **75.** $y = 4250e^{0.05t}$;
(b) $dA/dt = 0.05A, A_0 = 4250$
77. $y = \sqrt[3]{6x + C}$

Chapter 5 Review Exercises, p. 519

1. (d) **2.** (e) **3.** (f) **4.** (b) **5.** (c) **6.** (a)
7. False **8.** False **9.** True **10.** True **11.** False
12. True **13.** (2, \$16) **14.** \$18.67 **15.** \$5.33
16. \$6255.36 **17.** \$8065.41 **18.** \$23,820.45
19. \$8787.35/yr **20.** \$626,142.24 **21.** 30.99 billion metric
tons **22.** In 25.11 yr **23.** Convergent; 1 **24.** Divergent
25. Convergent; $\frac{1}{2}$ **26.** $k = \frac{8}{3}; f(x) = \dfrac{8}{3x^3}, 1 \le x \le 2$ **27.** 0.6
28. $\frac{3}{10}$ **29.** $\frac{1}{2}$ **30.** $\frac{1}{2}$ **31.** $\frac{1}{20}$ **32.** $\dfrac{1}{2\sqrt{5}} \approx 0.224$
33. 0.4678 **34.** 0.8827 **35.** 0.1002 **36.** 0.5000
37. 0.3085 **38.** \$6199.06 **39.** $\dfrac{127\pi}{7} \approx 66.497$
40. $\dfrac{\pi}{6} \approx 0.524$ **41.** $40\pi \approx 125.664$ **42.** $\pi(e^9 - e^1) \approx 25{,}448.05$
43. $y = Ce^{x^{11}}$ **44.** $y = \pm\sqrt{4x + C}$
45. $y = 5e^{4x}$ **46.** $y = \sqrt[3]{15t + 19}$ **47.** $y = \sqrt{3x^2 + C}$
and $y = -\sqrt{3x^2 + C}$ **48.** $y = Ce^{-x^2/2} + 8$
49. $q = 100 - x$ **50.** $V = -6.37e^{-kt} + 36.37$
51. $\sqrt[3]{9} \approx 2.08$ **52.** Divergent **53.** Convergent; 3 **54.** 1
55. 3.142

Chapter 5 Test, p. 521

1. [5.1] (3, \$16) **2.** [5.1] \$45 **3.** [5.1] \$22.50
4. [5.2] \$17,459.90 **5.** [5.2] \$53,879.14 **6.** [5.2] 1813.86
million metric tons **7.** [5.2] About 33 yr after 2011
8. [5.2] \$4502.09/yr **9.** [5.2] \$1,903,296.31
10. [5.2] \$1,403,270.10 **11.** [5.3] Convergent; $\frac{1}{4}$
12. [5.3] Divergent **13.** [5.4] $\frac{1}{4}, f(x) = \frac{1}{4}x^3$ on $[0, 2]$
14. [5.4] 0.117 **15.** [5.5] $E(x) = \frac{13}{6}$ **16.** [5.5] $E(x^2) = 5$
17. [5.5] $\mu = \frac{13}{6}$ **18.** [5.5] $\sigma^2 = \frac{11}{36}$ **19.** [5.5] $\sigma = \dfrac{\sqrt{11}}{6}$
20. [5.5] 0.4032 **21.** [5.5] 0.1365 **22.** [5.5] 0.9071
23. [5.5] 0.3085 **24.** [5.5] \$14.59/lb **25.** [5.6] $\pi \ln 5$
26. [5.6] $\dfrac{5\pi}{2}$ **27.** [5.6] $\dfrac{339\pi}{2} \approx 532.5$
28. [5.6] $\dfrac{8000\pi}{3} \approx 8377.58 \text{ ft}^3$

29. [5.7] $y = Ce^{x^8}$ **30.** [5.7] $y = \sqrt{18x + C}$ and
$y = -\sqrt{18x + C}$ **31.** [5.7] $y = 11e^{6t}$
32. [5.7] $y = \pm Ce^{-x^3/3} + 5$ **33.** [5.7] $y = \pm\sqrt[4]{8t + C}$
34. [5.7] $y = Ce^{4x+x^2/2}$ **35.** [5.1] $q = \dfrac{C_1}{x^4}$, where $C_1 = e^{-C}$
36. [5.7] (a) $V(t) = 36(1 - e^{-kt})$; (b) $k = 0.12$;
(c) $V(t) = 36(1 - e^{-0.12t})$; (d) $V(12) = \$27.47$;
(e) $t \approx 14.9$ months **37.** [5.4] $\sqrt[4]{4}$, or $\sqrt{2}$
38. [5.3] Convergent; $-\frac{1}{4}$ **39.** [5.3] 3.628

Extended Technology Application, p. 523

1. $y = 0.0052548858x^3 - 0.2722052955x^2 + 3.486640971x + 4.056787769$ **2.** 12,348.28 cm^3 **3.** $y_1 = 0.00023310023x^4 - 0.0027292152x^3 - 0.0133741259x^2 + 1.074786325x + 1.13986014$
4. 34.703 in^3 **5.** 19.23 fl oz; yes
6. $y_3 = 0.0002310023x^4 - 0.0027292152x^3 - 0.0133741259x^2 + 0.1074786325x + 1.15486014$
7. 1.0317704 in^3

CHAPTER 6

Exercise Set 6.1, p. 531

1. $0; -14; 250$ **3.** $1; -\frac{125}{9}; 23$ **5.** $9; 66; 128$ **7.** $6; 12$
9. $\{(x, y) | -\infty < x < \infty, -\infty < y < \infty\}$ **11.** $\{(x, y) | y \ge 3x\}$
13. $\{(x, y) | y \ne -x^2\}$ **15.** 39.19 **17.** \$193,318.20
19. (a) \$282.60; (b) \$23,738.40; (c) 6.5%; (d) \$254.40
21. 0.025 **23.** 1.915 m^2 **25.** (a) 2.56; (b) 1984 min;
(c) $g \ge 0, m > 0$ **27.** 110.6 ft **29.** 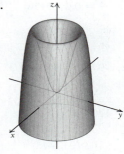 **31.** 0°F
33. -22°F **35.** Left to the student
37.
39.

41.
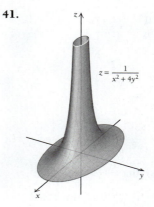

Exercise Set 6.2, p. 539

1. $7; -5; 7; -5$ **3.** $6x - 2y; -2x + 1; -6; 1$ **5.** $2 - 5y;$
$-5x; -18; -20$ **7.** $\dfrac{x}{\sqrt{x^2 + y^2}}; \dfrac{y}{\sqrt{x^2 + y^2}}; \dfrac{-2}{\sqrt{5}}; \dfrac{-2}{\sqrt{13}}$

9. $3e^{3x-2y}; -2e^{3x-2y}$ **11.** $ye^{xy}; xe^{xy}$

13. $\dfrac{x}{x - y} + \ln(x - y); \dfrac{-x}{x - y}$ **15.** $1 + \ln(xy); \dfrac{x}{y}$

17. $\dfrac{1}{y} + \dfrac{y}{3x^2}; -\dfrac{x}{y^2} - \dfrac{1}{3x}$ **19.** $24(3x + y - 8); 8(3x + y - 8)$

21. $4m^2 + 10m + 2b - 22; 3m^2 + 8mb + 10b + 26m - 56$
23. $5y - 2\lambda; 5x - \lambda; -(2x + y - 8)$
25. $2x - 10\lambda; 2y - 2\lambda; -(10x + 2y - 4)$
27. $f_{xx} = 0; f_{yy} = 0; f_{xy} = 2; f_{yx} = 2$
29. $f_{xx} = 0; f_{yy} = 14x; f_{xy} = 14y + 5; f_{yx} = 14y + 5$
31. $f_{xx} = 20x^3y^4 + 6xy^2; f_{yy} = 12x^5y^2 + 2x^3;$
$f_{xy} = 20x^4y^3 + 6x^2y; f_{yx} = 20x^4y^3 + 6x^2y$ **33.** $0; 0; 0; 0$
35. $4y^2e^{2xy}; 4xye^{2xy} + 2e^{2xy}; 4xye^{2xy} + 2e^{2xy}; 4x^2e^{2xy}$

37. $0; 0; 0; e^y$ **39.** $\dfrac{-y}{x^2}; \dfrac{1}{x}; \dfrac{1}{x}; 0$ **41. (a)** About 3,888,064 units

(b) $\dfrac{\partial p}{\partial x} = 1117.8\left(\dfrac{y}{x}\right)^{0.379}, \dfrac{\partial p}{\partial y} = 682.2\left(\dfrac{x}{y}\right)^{0.621};$

(c) $\left.\dfrac{\partial p}{\partial x}\right|_{(2500, 1700)} \approx 966, \left.\dfrac{\partial p}{\partial y}\right|_{(2500, 1700)} \approx 867;$ **(d)** ✎

43. (a) 1.274 million;

(b) $\dfrac{\partial P}{\partial w} = -0.005075w^{-1.638}r^{1.038}s^{0.873}t^{2.468},$

$\dfrac{\partial P}{\partial r} = 0.008257w^{-0.638}r^{0.038}s^{0.873}t^{2.468},$

$\dfrac{\partial P}{\partial s} = 0.006945w^{-0.638}r^{1.038}s^{-0.127}t^{2.468},$

$\dfrac{\partial P}{\partial t} = 0.019633w^{-0.638}r^{1.038}s^{0.873}t^{1.468};$ **(c)** ✎

45. $99.6°F$ **47.** $121.3°F$ **49.** ✎ **51. (a)** $\dfrac{\sqrt{w}}{120\sqrt{h}};$

(b) $\dfrac{\sqrt{h}}{120\sqrt{w}};$ **(c)** -0.0243 m^2 **53.** 78.244 **55.** -0.846

57. $f_x = \dfrac{-4xt^2}{(x^2 - t^2)^2}; f_t = \dfrac{4x^2t}{(x^2 - t^2)^2}$

59. $f_x = \dfrac{1}{\sqrt{x}(1 + 2\sqrt{t})}; f_t = \dfrac{-1 - 2\sqrt{x}}{\sqrt{t}(1 + 2\sqrt{t})^2}$

61. $f_x = 4x^{-1/3} - 2x^{-3/4}t^{1/2} + 6x^{-3/2}t^{3/2};$
$f_t = -4x^{1/4}t^{-1/2} - 18x^{-1/2}t^{1/2}$

63. $f_{xx} = \dfrac{-6y}{x^4}; f_{xy} = \dfrac{-2}{y^3} + \dfrac{2}{x^3}; f_{yx} = \dfrac{-2}{y^3} + \dfrac{2}{x^3};$

$f_{yy} = \dfrac{6x}{y^4}$ **65.** ✎

67. $f_{xx} = \dfrac{-2x^2 + 2y^2}{(x^2 + y^2)^2}$ and $f_{yy} = \dfrac{2x^2 - 2y^2}{(x^2 + y^2)^2}$, so $f_{xx} + f_{yy} = 0$

69. (a) $\lim\limits_{h\to 0} \dfrac{y(h^2 - y^2)}{h^2 + y^2} = -y;$ **(b)** $\lim\limits_{h\to 0} \dfrac{x(x^2 - h^2)}{x^2 + h^2} = x;$

(c) $f_{yx}(0, 0) = 1$ and $f_{xy}(0, 0) = -1$; at $(0, 0)$, the mixed partial derivatives are *not* equal.

Exercise Set 6.3, p. 547

1. Relative minimum $= -33$ at $(6, -1)$
3. Relative maximum $= \frac{4}{27}$ at $\left(\frac{2}{3}, \frac{2}{3}\right)$
5. Relative minimum $= -8$ at $(2, 2)$
7. Relative minimum $= -10$ at $(2, -1)$
9. Relative minimum $= -41$ at $(-4, 5)$
11. No relative extrema
13. Relative minimum $= e$ at $(0, 0)$
15. 6 thousand of the \$17 sunglasses and 5 thousand of the
\$21 sunglasses **17.** Maximum value of $P = \$55$ million
when $a = \$10$ million and $p = \$3$ **19.** The bottom
measures 8 ft by 8 ft, and the height is 5 ft.
21. (a) $R = 64p_1 - 4p_1^2 - 4p_1p_2 + 56p_2 - 4p_2^2;$
(b) $p_1 = 6$, or \$60, $p_2 = 4$, or \$40;
(c) $q_1 = 32$, or 3200 units, $q_2 = 28$, or 2800 units;
(d) \$304,000 **23.** No relative extrema; saddle point at $(0, 0)$
25. Relative minimum $= 0$ at $(0, 0)$; saddle points at $(2, 1)$ and
$(-2, 1)$ **27.** ✎ **29.** Relative minimum $= -5$ at $(0, 0)$
31. No relative extrema

Technology Connection p. 551

1. $x = -2, y = 3$ **2.** $x = -10, y = 37$ **3.** $x = -13, y = 19$
4. $x = 1, y = 4, z = 3$

Exercise Set 6.4, p. 553

1. $y = 0.6x + 0.95$ **3.** $y = \frac{36}{35}x - \frac{29}{35}$ **5.** $y = 9.75(2.049)^x,$
or $y = 9.75e^{0.717x}$ **7.** $y = 9.93(0.76)^x$, or $y = 9.93e^{-0.27x}$
9. (a) $y = 0.26x + 4.36;$ **(b)** \$10.34, \$11.64
11. (a) $y = 0.108x + 78.667;$ **(b)** 81.9, 82.4
13. (a) $y = 1.07x - 1.24;$ **(b)** 85% **15. (a)** $y = 1.469e^{-0.0173x};$
(b) 0.62 million, 0.57 million **17.** ✎
19. (a) $y = -0.0059379586x + 15.57191398;$ **(b)** 3:36.4,
3:34.6; **(c)** the value predicted is 3:42.1, about a second less
than the actual record.

Exercise Set 6.5, p. 564

1. Maximum $= \frac{25}{3}$ at $\left(\frac{5}{3}, 5\right)$ **3.** Minimum $= 20$ at $(4, 2)$
5. Maximum $= -16$ at $(2, 4)$ **7.** Minimum $= -96$ at $(8, -12)$
9. Minimum $= \frac{3}{2}$ at $\left(1, \frac{1}{2}, -\frac{1}{2}\right)$ **11.** 25 and 25 **13.** 3 and -3
15. $\left(\frac{3}{2}, 2, \frac{5}{2}\right)$ **17.** $x = \frac{9}{5}, y = \frac{3}{5}$ **19.** $9\frac{3}{4}$ in. by
$9\frac{3}{4}$ in.; $95\frac{1}{16}$ in²; no **21.** $r = \sqrt[3]{\dfrac{27}{2\pi}} \approx 1.6$ ft, $h = 2r \approx 3.2$ ft;
about 48.3 ft² **23.** Maximum value of S is 1012.5 at
$L = 22.5, M = 67.5$ **25. (a)** $C(x, y, z) = 7xy + 6yz + 6xz;$
(b) $x = 60$ ft, $y = 60$ ft, $z = 70$ ft; minimum cost is \$75,600
27. 10,000 units on A, 100 units on B **29.** About 0.707
mi, \$5657 **31.** Absolute maxima at $(-1, 2, 9)$ and $(1, 2, 9)$;
absolute minimum at $(0, 0, 0)$ **33.** Absolute maximum at
$(2, 2, 8)$; absolute minimum at $(0, 6, -12)$ **35.** \$3400 when
$x = 50$ and $y = 40$ **37.** Minimum $= -\frac{155}{128}$ at $\left(-\frac{7}{16}, -\frac{3}{4}\right)$
39. Maximum $= \frac{8}{27}$ at $\left(\pm\sqrt{\frac{2}{3}}, \pm\sqrt{\frac{2}{3}}, \pm\sqrt{\frac{2}{3}}\right)$
41. Maximum $= 2$ at $\left(\frac{1}{2}, \frac{1}{2}, \frac{1}{2}, \frac{1}{2}\right)$ **43.** $\lambda = \dfrac{p_x}{c_1} = \dfrac{p_y}{c_2}$ **45.** ✎
47. $x \approx 1.389, y \approx -0.293$ **49–55.** Left to the student

Exercise Set 6.6, p. 571

1. 54 **3.** 14 **5.** -63 **7.** $\frac{1}{8}$ **9.** $\frac{3}{20}$ **11.** $\frac{1}{2}$ **13.** 4
15. $\frac{4}{15}$ **17.** $-\frac{1}{2}$ **19.** $\frac{108}{5}$ **21. (a)** 18,000 fireflies;
(b) 30 fireflies/yd² **23.** 39 **25.** $\frac{13}{240}$ **27.** ✎

29. (a) $\displaystyle\int_{3}^{5}\int_{1}^{2}\left(x^2 - 3x + \tfrac{1}{3}xy - \tfrac{1}{3}y + 2\right)dx\,dy$

$\displaystyle= \int_{3}^{5}\left[\tfrac{1}{3}x^3 - \tfrac{3}{2}x^2 + \tfrac{1}{6}x^2y - \tfrac{1}{3}xy + 2x\right]_{1}^{2}dy$

$\displaystyle= \int_{3}^{5}\left[\left(\tfrac{8}{3} - 6 + \tfrac{2}{3}y - \tfrac{2}{3}y + 4\right) - \left(\tfrac{1}{3} - \tfrac{3}{2} + \tfrac{1}{6}y - \tfrac{1}{3}y + 2\right)\right]dy$

$\displaystyle= \int_{3}^{5}\left(\tfrac{1}{6}y - \tfrac{1}{6}\right)dy$

$\displaystyle= \left[\tfrac{1}{12}y^2 - \tfrac{1}{6}y\right]_{3}^{5} = \left[\left(\tfrac{25}{12} - \tfrac{5}{6}\right) - \left(\tfrac{3}{4} - \tfrac{1}{2}\right)\right] = 1;$

(b) $\tfrac{5}{12}$; **(c)** $\tfrac{7}{12}$; **(d)** $\tfrac{7}{24}$ **31.** Left to the student

Chapter 6 Review Exercises, p. 577

1. (e) **2.** (h) **3.** (f) **4.** (g) **5.** (b) **6.** (c) **7.** (d)
8. (a) **9.** 1 **10.** $3y^3$ **11.** $e^y + 9xy^2 + 2$
12. $9y^2$ **13.** $9y^2$ **14.** 0 **15.** $e^y + 18xy$
16. $D = \{(x,y)\,|\,x \neq 1, y \geq 2\}$ **17.** $6x^2 \ln y + y^2$
18. $\dfrac{2x^3}{y} + 2xy$ **19.** $\dfrac{6x^2}{y} + 2y$ **20.** $\dfrac{6x^2}{y} + 2y$ **21.** $12x \ln y$
22. $\dfrac{-2x^3}{y^2} + 2x$ **23.** Relative minimum $= -\tfrac{549}{20}$ at $\left(5, \tfrac{27}{2}\right)$
24. Relative minimum $= -4$ at $(0, -2)$ **25.** Relative maximum $= \tfrac{45}{4}$ at $\left(\tfrac{3}{2}, -3\right)$ **26.** Relative minimum $= 29$ at $(-1, 2)$
27. (a) $y = \tfrac{3}{5}x + \tfrac{20}{3}$; **(b)** 9.1 million
28. (a) $y = 19.584x + 121.975$; **(b)** \$533;
(c) $y = 131.834e^{0.083x}$; **(d)** \$757   **29.** Minimum $= \tfrac{125}{4}$ at $\left(-\tfrac{3}{2}, -3\right)$ **30.** Maximum $= 300$ at $(5, 10)$ **31.** Absolute maximum at $(3, 0, 9)$; absolute minimum at $(0, 2, -4)$ **32.** $\tfrac{5}{4}$
33. $\tfrac{1}{60}$ **34. (a)** About 1533 students; **(b)** About 767 students/mi^2 **35.** 0   **36.** The cylindrical container, with 4.074 in^2 less surface area   **37.**
38.

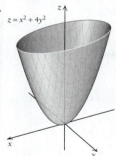

$z = x^2 + 4y^2$

Chapter 6 Test, p. 578

1. [6.1] $e^{-1} - 2$ **2.** [6.2] $e^x + 6x^2y$ **3.** [6.2] $2x^3 + 1$
4. [6.2] $e^x + 12xy$ **5.** [6.2] $6x^2$ **6.** [6.2] $6x^2$ **7.** [6.2] 0
8. [6.3] Minimum $= -\tfrac{7}{16}$ at $\left(\tfrac{3}{4}, \tfrac{1}{2}\right)$ **9.** [6.3]
Maximum $= 18$ at $(-1, -2)$, minimum $= -18$ at $(1, 2)$
10. [6.4] **(a)** $y = \tfrac{9}{5}x + \tfrac{17}{3}$; **(b)** \$24 million;
(c) $y = 7.462e^{0.321x}$; **(d)** \$26.9 million
11. [6.5] Maximum $= -19$ at $(4, 5)$ **12.** [6.6] 720
13. [6.6] $\tfrac{45}{8}$ **14.** [6.5] \$400,000 for labor, \$200,000 for capital
15. [6.2] $f_x = \dfrac{-x^4 + 4xt + 6x^2t}{(x^3 + 2t)^2}; f_t = \dfrac{-2x^3 - 2x^2}{(x^3 + 2t)^2}$

16. [6.1]

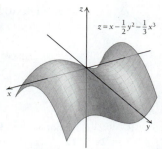

$z = x - \tfrac{1}{2}y^2 - \tfrac{1}{3}x^3$

Extended Technology Application, p. 580

1.

Case	Bldg.	n	k	A	h	$t(h, k)$
1	B1	2	40	3200	15	21.5
	B2	3	32.66	3200	30	19.4
2	B1	2	60	7200	15	31.5
	B2	3	48.99	7200	30	27.5
3	B1	4	40	6400	45	24.5
	B2	5	35.777	6400	60	23.9
4	B1	5	60	18000	60	36
	B2	10	42.426	18000	135	34.7
5	B1	5	150	112500	60	81
	B2	10	106.066	112500	135	66.5
6	B1	10	40	16000	135	33.5
	B2	17	30.679	16000	240	39.3
7	B1	10	80	64000	135	53.5
	B2	17	61.357	64000	240	54.7
8	B1	17	40	27200	240	44
	B2	26	32.344	27200	375	53.7
9	B1	17	50	42500	240	49
	B2	26	40.43	42500	375	57.7
10	B1	26	77	154154	375	76
	B2	50	55.525	154152	735	101.3

2. Yes **3.** $t(h, k) = \dfrac{h}{10} + \dfrac{k}{2}$ **4.** $A = \left(1 + \dfrac{h}{12}\right)k^2 = 40{,}000$
5. About 57.7 ft by 57.7 ft by 132.2 ft (11 floors)
6. Left to the student

CUMULATIVE REVIEW, p. 581

1. [R.4] $y = -4x - 27$ **2.** [R.4] $m = \tfrac{1}{3}$, y-intercept $= \left(0, -\tfrac{3}{2}\right)$
3. [R.2] $x^2 + 2xh + h^2 - 5$
4. [1.2] **(a)**   **(b)** 3; **(c)** -3; **(d)** no

$f(x)$

5. [1.2] (a)

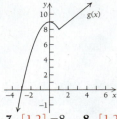

(b) 8; (c) 8; (d) yes

6. [1.2] 3 **7.** [1.2] −8 **8.** [1.2] Does not exist **9.** [2.3] 4
10. [2.3] 0 **11.** [1.4] $f'(x) = 2x$ **12.** [1.1] a_6
13. [1.2] a_2, a_6 **14.** [1.4] a_1, a_4, a_6 **15.** [1.5] −9
16. [1.5] $2x − 7$ **17.** [1.5] $\frac{1}{4}x^{-3/4}$ **18.** [1.5] $-6x^{-7}$
19. [1.7] $\frac{10}{3}x^4(2x^5 − 8)^{-2/3}$ **20.** [1.6] $\dfrac{20x^3 − 15x^2 − 8}{(2x − 1)^2}$

21. [3.2] $\dfrac{2x}{x^2 + 5}$ **22.** [3.2] 1 **23.** [3.1] $3e^{3x} + 2x$

24. [3.1] $\dfrac{e^{\sqrt{x-3}}}{2\sqrt{x − 3}}$ **25.** [3.2] $\dfrac{e^x}{e^x − 4}$ **26.** [1.8] $2 − 4x^{-3}$

27. [2.5] −$0.04/pair **28.** [2.7] $3xy^2 + \dfrac{y}{x}$

29. [3.1] $y = x − 2$ **30.** [1.5] $x = 1, x = \dfrac{1}{3}$

31. [2.2] Relative maximum
at (−1, 3), relative minimum at
(1, −1); point of inflection at
(0, 1); increasing on $(-\infty, -1)$
and $(1, \infty)$, decreasing on
$(−1, 1)$; concave down on
$(-\infty, 0)$, concave up on $(0, \infty)$

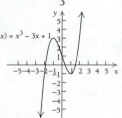

$f(x) = x^3 − 3x + 1$

32. [2.2] Relative maxima at
(−1, −2) and (1, −2), relative
minimum at (0, −3); points of
inflection at $\left(-\dfrac{1}{\sqrt{3}}, -\dfrac{22}{9}\right)$ and
$\left(\dfrac{1}{\sqrt{3}}, -\dfrac{22}{9}\right)$; increasing on
$(-\infty, -1)$ and $(0, 1)$, decreasing
on $(−1, 0)$ and $(1, \infty)$; concave down on $\left(-\infty, -\dfrac{1}{\sqrt{3}}\right)$ and
$\left(\dfrac{1}{\sqrt{3}}, \infty\right)$, concave up on $\left(-\dfrac{1}{\sqrt{3}}, \dfrac{1}{\sqrt{3}}\right)$

$f(x) = 2x^2 − x^4 − 3$

33. [2.2, 2.3] Relative maximum at
(1, 4); relative minimum at (−1, −4);
points of inflection at $(-\sqrt{3}, -2\sqrt{3})$
and (0, 0) and $(\sqrt{3}, 2\sqrt{3})$; decreasing
on $(-\infty, -1)$ and $(1, \infty)$, increasing on $(−1, 1)$; concave down on
$(-\infty, -\sqrt{3})$ and $(0, \sqrt{3})$, concave up
on $(-\sqrt{3}, 0)$ and $(\sqrt{3}, \infty)$; horizontal
asymptote at $y = 0$

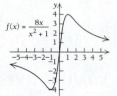

$f(x) = \dfrac{8x}{x^2 + 1}$

34. [2.3] Relative maximum at
(0, −2); no points of inflection; vertical asymptotes at $x = −2$ and $x = 2$;
horizontal asymptote at $y = 0$;
increasing on $(-\infty, -2)$ and $(−2, 0)$,
decreasing on $(0, 2)$ and $(2, \infty)$; concave up on $(-\infty, -2)$ and $(2, \infty)$,
concave down on $(−2, 2)$

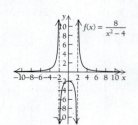

$f(x) = \dfrac{8}{x^2 − 4}$

35. [2.4] Absolute minimum $= −7$ at $x = 1$ **36.** [2.4]
No absolute extrema **37.** [2.4] Absolute maximum $= 6\frac{2}{3}$ at
$x = −1$; absolute minimum $= 4\frac{1}{3}$ at $x = −2$
38. [2.5] 15 sweatshirts **39.** [2.5] 30 times; lot size of 15
40. [2.6] $\Delta y \approx 0.5$; $f'(x)\Delta x = 3$ **41.** [3.3]
(a) $N(t) = 8000e^{0.1t}$; (b) 17,804; (c) 6.9 yr **42.** [3.6]
(a) $E(x) = \dfrac{x}{12 − x}$; (b) $E(2) = \frac{1}{5}$, inelastic; (c) $E(9) = 3$,
elastic; (d) increase; (e) $x = \$6$ **43.** [2.6] (a) $\Delta A \approx \pm 50$ ft²;
(b) 6 squares, $48 **44.** [3.6] (a) $A(t) = 89{,}552.23881$
$(1.034019181)^t − 89{,}552.23881$; (b) $9453.92; (c) $3312.09/yr
45. [3.6] (a) $382.01/month (b) $25,504.72; (c) $3954.72
(d)

Balance	Payment	Portion of Payment Applied to Interest	Portion of Payment Applied to Principal	New Balance
$23,550	$382.01	$103.03	$278.98	$23,271.02
$23,271.02	$382.01	$101.81	$280.20	$22,990.82

46. [4.2] $\frac{1}{2}x^6 + C$ **47.** [4.3] $3 − \dfrac{2}{e}$
48. [4.7] $\dfrac{7}{9(7 − 3x)} + \dfrac{1}{9}\ln|7 − 3x| + C$ **49.** [4.5] $\frac{1}{4}e^{x^4} + C$
50. [4.6] $\left(\dfrac{x^2}{2} + 3x\right)\ln x − \dfrac{x^2}{4} − 3x + C$
51. [4.2] $75\ln|x| + C$ **52.** [4.3] 2 **53.** [4.3] $\frac{232}{3}$, or $77\frac{1}{3}$
54. [5.2] $41,324.72 **55.** [5.2] (a) $28,992.09;
(b) $21,007.91 **56.** [5.2] $711,632.48
57. [5.3] Convergent, $\dfrac{1}{4374}$ **58.** [5.1] (7, $169); $751.33
59. [5.6] $-\dfrac{\pi}{2}\left(\dfrac{1}{e^{10}} − 1\right)$, or approximately 1.571
60. $124\pi/3$, or approximately 129.85
61. [6.4] (a) $y = 2x + \frac{5}{3}$; (b) $216,666.67;
(c) $y = 3.05e^{0.275x}$; (d) $477,100.27 **62.** [6.2] $8xy^3 + 3$
63. [6.2] $e^y + 24x^2y$ **64.** [6.5] Minimum $= −10$ at (3, −2)
65. [6.5] Maximum $= 4$ at (3, 3) **66.** [6.6] $3(e^3 − 1)$
67. [6.6] 1467 shoppers **68.** [5.7] $y = Ce^{x^2/2}$
69. [5.7] $y = \frac{9}{4}e^{2-2x^2} + \frac{3}{4}$ **70.** [5.7] (a) $\dfrac{dy}{dx} = \sqrt{y}$;
(b) $y = \left(\dfrac{x}{2} + C\right)^2$ **71.** [5.4] (a) $\frac{12}{47}$; (b) $\frac{27}{47}$; (c) $\frac{35}{47}$; (d) 0
72. [5.4] 0.12 **73.** [5.4] (a) 0.593; (b) 0.105
74. [5.5] (a) $\frac{3}{2}$; (b) $\mu = \frac{3}{2}\ln 3, \sigma = 0.5333$; (c) 0.4
75. [5.5] (a) 0.644; (b) 99.4th; (c) $87,322.49

APPENDIX A

Exercise Set A, p. 595

1. $5 \cdot 5 \cdot 5$, or 125 **2.** $7 \cdot 7$, or 49 **3.** $(−7)(−7)$, or 49
4. $(−5)(−5)(−5)$, or −125 **5.** 1.0201 **6.** 1.030301
7. $\frac{1}{16}$ **8.** $\frac{1}{64}$ **9.** 1 **10.** $6x$ **11.** $−t$ **12.** $−1$ **13.** 1
14. $\frac{1}{3}$ **15.** $\dfrac{1}{3^2}$, or $\frac{1}{9}$ **16.** $\dfrac{1}{4^2}$, or $\frac{1}{16}$ **17.** 8 **18.** $\frac{4}{49}$
19. 9 **20.** 4 **21.** 0.1 **22.** 0.0001 **23.** $\dfrac{1}{e^b}$ **24.** $\dfrac{1}{t^k}$
25. $\dfrac{1}{b}$ **26.** $\dfrac{1}{h}$ **27.** x^5 **28.** t^7 **29.** x^{-6}, or $\dfrac{1}{x^6}$ **30.** x^6

31. $35x^5$ **32.** $8t^7$ **33.** x^4 **34.** x **35.** 1   **36.** 1

37. x^3 **38.** x^4 **39.** x^{-3}, or $\dfrac{1}{x^3}$ **40.** x^{-4}, or $\dfrac{1}{x^4}$ **41.** 1

42. 1   **43.** e^{t-4} **44.** e^{k-3} **45.** t^{14} **46.** t^{12} **47.** t^2

48. t^{-4}, or $\dfrac{1}{t^4}$ **49.** a^6b^5 **50.** x^7y^7 **51.** t^{-6}, or $\dfrac{1}{t^6}$

52. t^{-12}, or $\dfrac{1}{t^{12}}$ **53.** e^{4x} **54.** e^{5x} **55.** $8x^6y^{12}$

56. $32x^{10}y^{20}$ **57.** $\dfrac{1}{81}x^8y^{20}z^{-16}$, or $\dfrac{x^8y^{20}}{81z^{16}}$

58. $\dfrac{1}{125}x^{-9}y^{21}z^{15}$, or $\dfrac{y^{21}z^{15}}{125x^9}$ **59.** $9x^{-16}y^{14}z^4$, or $\dfrac{9y^{14}z^4}{x^{16}}$

60. $625x^{16}y^{-20}z^{-12}$, or $\dfrac{625x^{16}}{y^{20}z^{12}}$ **61.** $\dfrac{c^4d^{12}}{16q^8}$ **62.** $\dfrac{64x^6y^3}{a^9b^9}$

63. $5x - 35$ **64.** $x + xt$ **65.** $x^2 - 7x + 10$
66. $x^2 - 7x + 12$ **67.** $a^3 - b^3$ **68.** $x^3 + y^3$
69. $2x^2 + 3x - 5$ **70.** $3x^2 + x - 4$ **71.** $a^2 - 4$ **72.** $9x^2 - 1$
73. $25x^2 - 4$ **74.** $t^2 - 1$ **75.** $a^2 - 2ah + h^2$
76. $a^2 + 2ah + h^2$ **77.** $25x^2 + 10xt + t^2$
78. $49a^2 - 14ac + c^2$ **79.** $5x^5 + 30x^3 + 45x$
80. $-3x^6 + 48x^2$ **81.** $a^3 + 3a^2b + 3ab^2 + b^3$
82. $a^3 - 3a^2b + 3ab^2 - b^3$ **83.** $x^3 - 15x^2 + 75x - 125$
84. $8x^3 + 36x^2 + 54x + 27$ **85.** $x(1 - t)$ **86.** $x(1 + h)$
87. $(x + 3y)^2$ **88.** $(x - 5y)^2$ **89.** $(x - 5)(x + 3)$
90. $(x + 5)(x + 3)$ **91.** $(x - 5)(x + 4)$ **92.** $(x - 10)(x + 1)$
93. $(7x - t)(7x + t)$ **94.** $(3x - b)(3x + b)$
95. $4(3t - 2m)(3t + 2m)$ **96.** $(5y - 3z)(5y + 3z)$
97. $ab(a + 4b)(a - 4b)$ **98.** $2(x^2 + 4)(x + 2)(x - 2)$
99. $(a^4 + b^4)(a^2 + b^2)(a + b)(a - b)$ **100.** $(6y - 5)(6y + 7)$
101. $10x(a + 2b)(a - 2b)$ **102.** $xy(x + 5y)(x - 5y)$
103. $2(1 + 4x^2)(1 + 2x)(1 - 2x)$ **104.** $2x(y + 5)(y - 5)$
105. $(9x - 1)(x + 2)$ **106.** $(3x - 4)(2x - 5)$
107. $(x + 2)(x^2 - 2x + 4)$ **108.** $(a - 3)(a^2 + 3a + 9)$
109. $(y - 4t)(y^2 + 4yt + 16t^2)$ **110.** $(m + 10p)$
$(m^2 - 10mp + 100p^2)$ **111.** $(3x^2 - 1)(x - 2)$
112. $(y^2 - 2)(5y + 2)$ **113.** $(x - 3)(x + 3)(x - 5)$
114. $(t - 5)(t + 5)(t + 3)$ **115.** $\frac{7}{4}$ **116.** $\frac{79}{12}$ **117.** -8
118. $\frac{4}{5}$ **119.** 120   **120.** 140   **121.** 200   **122.** 200
123. $0, -3, \frac{4}{5}$ **124.** $0, 2, -\frac{3}{2}$ **125.** $0, 2$ **126.** $3, -3$
127. $0, 3$ **128.** $0, 5$ **129.** $0, 7$ **130.** $0, 3$ **131.** $0, \frac{1}{3}, -\frac{1}{3}$
132. $0, \frac{1}{4}, -\frac{1}{4}$ **133.** 1   **134.** 2   **135.** No solution
136. No solution   **137.** -23 **138.** -145 **139.** 5
140. $-\sqrt{7}, \sqrt{7}$ **141.** ± 8 **142.** ± 12 **143.** $\pm\sqrt{30}$

144. $\pm\sqrt{97}$ **145.** $\pm\frac{7}{2}$ **146.** $\pm\frac{13}{10}$ **147.** $\pm\dfrac{\sqrt{20}}{3}$

148. $\pm\dfrac{\sqrt{3}}{5}$ **149.** $\pm\dfrac{2}{\sqrt{5}}$, or $\pm\dfrac{2\sqrt{5}}{5}$ **150.** $\pm\dfrac{12}{\sqrt{3}}$, or $\pm 4\sqrt{3}$

151. $1, \frac{1}{3}$ **152.** $\frac{5}{2}, -\frac{25}{6}$ **153.** $\dfrac{2 + \sqrt{10}}{5}, \dfrac{2 - \sqrt{10}}{5}$

154. $\dfrac{1 + \sqrt{2}}{4}, \dfrac{1 - \sqrt{2}}{4}$ **155.** $x \geq -\frac{4}{5}$ **156.** $x \geq 3$

157. $x > -\frac{1}{12}$ **158.** $x > -\frac{5}{13}$ **159.** $x > -\frac{4}{7}$ **160.** $x \leq -\frac{6}{5}$
161. $x \leq -3$ **162.** $x \geq -2$ **163.** $x > \frac{2}{3}$ **164.** $x \leq \frac{5}{3}$
165. $x < -\frac{2}{5}$ **166.** $x \leq \frac{5}{4}$ **167.** $2 < x < 4$ **168.** $2 \leq x \leq 4$
169. $\frac{3}{2} \leq x \leq \frac{11}{2}$ **170.** $\frac{6}{5} \leq x < \frac{16}{5}$ **171.** $-1 \leq x \leq \frac{14}{5}$
172. $-5 \leq x < -2$ **173.** $650   **174.** $800   **175.** More
than 7000 units   **176.** More than 4200 units   **177.** 480 lb

178. 340 lb   **179.** 810,000   **180.** 5200
181. $60\% \leq x < 100\%$ **182.** $50\% \leq x < 90\%$
183. 18 people   **184.** 75 tiles

APPENDIX B

Exercise Set B, p. 601

1. $y = 3.45x + 0.6$ **2.** $y = 2.5283x + 13.34$
3. $y = 0.0732x^2 + 2.3x + 11.813$

DIAGNOSTIC TEST, p. xv

Part A

The red bracketed references indicate where worked-out solutions can be found in Appendix A: Review of Basic Algebra. For example, [Ex.1] means that the problem is worked out in Example 1 of the appendix.

1. 64 [Ex. 1]   **2.** -32 [Ex. 1]   **3.** $\frac{1}{8}$ [Ex. 1]   **4.** $-2x$ [Ex. 2]

5. 1 [Ex. 2]   **6.** $\dfrac{1}{x^5}$ [Ex. 3]   **7.** 16 [Ex. 3]

8. $\dfrac{1}{t}$ [Ex. 3]   **9.** x^{11} [Ex. 4]   **10.** x [Ex. 4]   **11.** $\dfrac{10}{x^7}$ [Ex. 4]

12. a [Ex. 5]   **13.** e^7 [Ex. 5]   **14.** $\dfrac{1}{x^6}$ [Ex. 6]

15. $\dfrac{y^{15}}{8x^{12}z^9}$ [Ex. 6]   **16.** $3x - 15$ [Ex. 7]

17. $x^2 - 2x - 15$ [Ex. 7]   **18.** $a^2 + 2ab + b^2$ [Ex. 7]
19. $4x^2 - 4xt + t^2$ [Ex. 8]   **20.** $9c^2 - d^2$ [Ex. 8]
21. $h(2x + h)$ [Ex. 9]   **22.** $(x - 3y)^2$ [Ex. 9]
23. $(x + 2)(x - 7)$ [Ex. 9]   **24.** $(2x - 1)(3x + 5)$ [Ex. 9]
25. $(x - 7)(x + 2)(x - 2)$ [Ex. 10]   **26.** $x = 6$ [Ex. 11]
27. $x = 0, 2, -\frac{4}{5}$ [Ex. 12]   **28.** $x = 0, \frac{1}{2}, -\frac{1}{2}$ [Ex. 13]
29. No solution [Ex. 15]   **30.** $x \leq \frac{21}{13}$ or $\left(-\infty, \frac{21}{13}\right]$ [Ex. 17]
31. 660 lb [Ex. 18]   **32.** 351 suits [Ex. 19]

Part B

The red bracketed references indicate where worked-out solutions can be found in Chapter R. For example, [Ex. R.2.5] means that the problem is worked out in Example 5 of Section R.2.

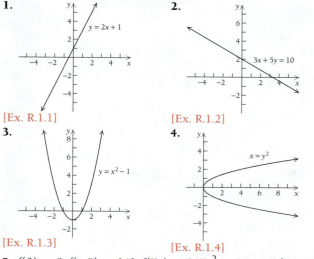

1.
$y = 2x + 1$
[Ex. R.1.1]

2.
$3x + 5y = 10$
[Ex. R.1.2]

3.
$y = x^2 - 1$
[Ex. R.1.3]

4.
$x = y^2$
[Ex. R.1.4]

5. $f(0) = 8; f(-5) = 143; f(7a) = 147a^2 - 14a + 8$ [Ex. R.2.4]

6. $\dfrac{(f(x + h) - f(x))}{h} = 1 - 2x - h$ [Ex. R.2.5]

7.   [Ex. R.2.9]

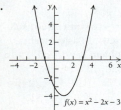

$$f(x) = \begin{cases} 4, & \text{for } x \geq 0, \\ 3 - x^2, & \text{for } 0 < x \leq 2, \\ 2x - 6, & \text{for } x > 2. \end{cases}$$

8. $(-4, 5)$ [Ex. R.3.1a]   **9.** $\{x \mid x$ is any real number and $x \neq \frac{5}{2}\}$ or $\{x \mid (-\infty, \frac{5}{2}) \cup (\frac{5}{2}, \infty)\}$ [Ex. R.3.4]

10. Slope $m = \frac{1}{2}$; y-intercept: $(0, -\frac{7}{4})$ [Ex. R.4.4]

11. $y = 3x - 2$ [Ex. R.4.5]   **12.** $m = -\frac{3}{2}$ [Ex. R.4.7]

13.

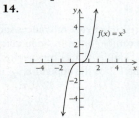

$f(x) = x^2 - 2x - 3$

[Ex. R.5.1]

14.

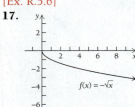

$f(x) = x^3$

[Ex. R.5.4]

15.

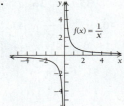

$f(x) = \frac{1}{x}$

[Ex. R.5.6]

16.

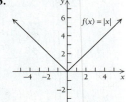

$f(x) = |x|$

[Ex. R.5.8]

17.

$f(x) = -\sqrt{x}$

[Ex. R.5.9]

18. $1,166.40 [Ex. R.1.6]

Index of Applications

Index

Summary of Important Formulas for Differentiation

1. Power Rule. For any real number k, $\dfrac{d}{dx} x^k = kx^{k-1}$.

2. Derivative of a Constant Function. If $F(x) = c$, then $F'(x) = 0$.

3. Derivative of a Constant Times a Function. If $F(x) = cf(x)$, then $F'(x) = cf'(x)$.

4. Derivative of a Sum. If $F(x) = f(x) + g(x)$, then $F'(x) = f'(x) + g'(x)$.

5. Derivative of a Difference. If $F(x) = f(x) - g(x)$, then $F'(x) = f'(x) - g'(x)$.

6. Derivative of a Product. If $F(x) = f(x)g(x)$, then $F'(x) = f(x)g'(x) + g(x)f'(x)$.

7. Derivative of a Quotient. If $F(x) = \dfrac{f(x)}{g(x)}$, then $F'(x) = \dfrac{g(x)f'(x) - f(x)g'(x)}{[g(x)]^2}$.

8. Extended Power Rule. If $F(x) = [g(x)]^k$, then $F'(x) = k[g(x)]^{k-1}g'(x)$.

9. Chain Rule. If $F(x) = f[g(x)]$, then $F'(x) = f'[g(x)]g'(x)$.

Or, if $y = f(u)$ and $u = g(x)$, then $\dfrac{dy}{dx} = \dfrac{dy}{du} \cdot \dfrac{du}{dx}$.

10. $\dfrac{d}{dx} e^x = e^x$

11. $\dfrac{d}{dx} e^{f(x)} = e^{f(x)} \cdot f'(x)$

12. $\dfrac{d}{dx} \ln x = \dfrac{1}{x}, \quad x > 0$

13. $\dfrac{d}{dx} \ln f(x) = \dfrac{f'(x)}{f(x)}, \quad f(x) > 0$

14. $\dfrac{d}{dx} \ln |x| = \dfrac{1}{x}, \quad x \neq 0$

15. $\dfrac{d}{dx} \ln |f(x)| = \dfrac{f'(x)}{f(x)}, \quad f(x) \neq 0$

16. $\dfrac{d}{dx} a^x = (\ln a)a^x, \quad a > 0$

17. $\dfrac{d}{dx} \log_a |x| = \dfrac{1}{\ln a} \cdot \dfrac{1}{x}, \quad a > 0 \text{ and } x \neq 0$